AF519469

Advances in Plant Protection Sciences

Advances in Plant Protection Sciences

EDITORS
D. PRASAD
AMERIKA SINGH

Foreword by

Dr. Mangla Rai
Secretary, D.A.R.E. & D.G., I.C.A.R.

AKANSHA PUBLISHING HOUSE
NEW DELHI

AKANSHA PUBLISHING HOUSE
R 37-B, Vani Vihar, Uttam Nagar
New Delhi-110059
Email: ektabooks@yahoo.com
Ph. : 9212065579 / 9811582579

Showroom
4649/21B, Ansari Road
Darya Ganj
New Delhi - 110 002

ADVANCES IN PLANT PROTECTION SCIENCES

First Edition 2004
ISBN 81–87606–53-3

PRINTED IN INDIA

Published by M.P. Misra for Akansha Publishing House, New Delhi and Printed at Tarun Offset Press, Delhi.

Foreword

Modern agriculture is input intensive and highly dependent on chemical pesticides for the management of various pests. However, indiscriminate and excessive use of pesticides in agriculture has played havoc with the agro-ecosystem by polluting air, water and food chains. Effective and timely management of a vast array of insect pests and diseases of crops by the use of bio-agents and bio-pesticides in the **Integrated Pest Management (IPM)** recognizes the importance of friendly insects to be judiciously conserved promoters. This in consort with plant based pesticides forms the basic tenet of IPM approach which is knowledge based, has greater flexibility and contributes to the preservation of eco-system and sustainable agriculture.

The editors of the book **Advances in Plant Protection Sciences**, Drs. D. Prasad and Amerika Singh have made a concerted effort to compile the work done at various institutions under the National Agricultural Research System (NARS) into 33 thought provoking chapters. The authors have put forth new impetus on the plant protection subject as a whole. This book would not only serve as a reference source but also as a guide for shaping future research in the area of plant protection to achieve the most cherished goal of ecologically and economically sustainable agriculture.

I congratulate the authors and editors for this timely publication.

Dated the 8th June, 2004

MANGALA RAI
Secretary,
Deptt. of Agricultural Research & Education
Director General,
Indian Council of Agricultural Research,
Ministry of Agriculture
Government of India
Krishi Bhawan
New Delhi-110 001

Preface

Protection of crops is an integral part of crop production. The purpose of pest control is to improve yield and quality of crops. In a broader sense pests (insects, fungi, bacteria, viruses, nematodes, rodents and weeds) control aims to maximize the efficient and effective use of arable land to meet the increasing demand for food and fibre for the ever growing population in India. Hence, their control measures are prime importance. Most of these methods rely on the use of toxic chemicals which although give a quick and convincing control that also involve immediate and/or cumulative environmental hazards. Concern for the environment is not limited to the agriculturists and in particular plant protectionists. A large number of eco-friendly methods are also known, some of which are the results of long human experience which other have been formulated after scientific studies on the ecological relationships between the crop-pest-pathogen environment complex. These methods individually may give a less spectacular and limited degree of control but combinations of more than one of them, can provide practically feasible, economical and eco-friendly methods of integrated multiple pest management. During the recent years, bollworms and leaf curl virus in cotton; stem borers, blast and bacterial leaf blight in rice; rusts and leaf blight in wheat; white grub and bud necrosis virus in groundnut; mosaic virus and caterpillar in tobacco; fruit borers and leaf diseases in vegetables; pod borers and wilt in pulses; fruit flies and powdery mildew in mango; mites and wilt virus in coconut, to name a few, have caused enormous productivity losses. Beside insect pests and diseases, weeds compete with crop plants for moisture, nutrients and light. Some weed also release certain chemicals in the soil that are antagonistic to the soil environment and cause productivity losses.

Integrated Pest Management (IPM) promoted by GOI since 1985, is an eco-friendly approach which aims at keeping insect pests and diseases at economic threshold levels by employing all available alternate pest control methods and techniques such as cultural, mechanical and biological control with greater emphasis on use of bio-pesticides and pesticides of plant-origin like Neem formulations. The use of chemical pesticides is advised as a last resort when pest crosses economic threshold level (ETL). It is an economically viable and sustainable system of crop protection that leads to maximum productivity with the least possible adverse impact on the total environment. India is a signatory to the Agenda 1 of United Nations Conference on Environment Development (UNCED) 1992, which has also approved the acceptance of IPM to reduce the use of pesticides in agriculture. IPM is a priority area for research and development efforts to provide safe and effective technologies for plant protection. The need of hour is to comprehensively view the effects of various management practices on pests and pathogens. The present book, **Advances in Plant Protection Sciences** is an attempt to bring together the available literature in a book form. The experts in the relevant field, have given a critical appraisal of the various aspects of plant protection in general and extended useful suggestions in particular.

In all, 33 various chapters have been included in this book such as conservation of natural enemies, bio-agents for IPM, *Bt* transgenic, use of botanicals, eco-friendly pest management, systemic acquired resistance, fungal bio-agents for nematode management, microbes for weed management, rodent pest management, system analysis and application in IPM, grub management, impact of climatic change of pests and diseases and IPR issues in Indian Agriculture.

The book **Advances in Plant Protection Sciences** thus, attempts to consolidate knowledge on management strategies adopted under Indian conditions. The book may not only provide reference but also serve as guide and inspiration for future research and onslaughts of modern plant protection tactics. The scientists, teachers, scholars and administrators dealing with bio-agents, plant protection and environmental protection are expected to find this book useful.

We are grateful to all the authors for their kind co-operation and valuable contributions and to Shri M.P. Misra, Akansha Publishing House, New Delhi and Shri S.K. Gupta for their technical help and guidance in the finalisation of the book.

DELHI **D. PRSAD**
12/7/2004 **AMERIKA SINGH**

Contents

Conservation of Natural Enemies of Crop Pests in Different Agro-ecosystems

R. K. TANWAR,[1] O. M. BAMBAWALE[1] AND AMERIKA SINGH[1]

In agriculture, many of the tactics currently used for pest management, although resulting in high productivity, may be very perilous for the environment, consumers and farmers. Widespread reliance on chemicals alone has brought national and international crisis due to their toxic effect on non-target beneficial and endangered species, to wild life and to humans. Excessive use of pesticides has also led to develop resistance in pest and created new pest complexes. Many chemical pesticides are no longer available due to registration revocations. To restore and maintain the ecological integrity of the environment alternate approaches are urgently needed. In recent years there has been increasing interest in approaches that promote sustainable agroecosystem. Biological control, and in particular conservation biological control, can be an important alternative to wide spread use of pesticide as well as significant component of sustainable agriculture. De Bach (1964) considered conservation to mean environmental modification to protect and enhance natural enemies, a definition that is consistent with the major theme. The process of conservation involves the use of tactics and approaches that involve the manipulation of the environment (i.e. habitat) to favour natural enemies either by removing or mitigating adverse factors or by enhancing or induction of favourable factors which are lacking in the habitat, or present at inadequate levels.

CURRENT STATUS

The use of conventional pesticides in the USA dramatically doubled during 1964 to 1978 (Report of U.S. Congressional Office of Technology Assessment, 1995) and became stabilized since 1980. Similar pattern are evident in global pesticide use (Waage, 1997). About the conservation biological control OTA reported that the approach is rarely used as a major and deliberate component of pest management. In Indonesia, excessive use of pesticides eliminated natural enemies of brown plant hopper in rice ecosystem during 1976-77 and 1986. Realising this fact, an exemplary bold political decision was taken by the Government of Indonesia through Presidential order on November 6, 1986 banning 57 pesticides used in rice and permitting use of only four pesticides under the strict supervision of the trained entomologists. In India the use of pesticides which had increased steadily to a level of about 75,000 Tons Tech. Gr. in 1990-91 has declined to the level of about 46,195 Tons Tech. Gr. during 1999-2000. A wide gape still exists between research and implementation in conservation biological control.

The control of *Pyrilla perpusilla* Walker in sugarcane by conservation of nymphal and adult parasitoid, *Epiricania melanoleuca* Fletcher is the most successful example. In UP and other adjoining states it was a common practice to spray insecticides in the month of April or May against *Pyrilla,* at the stage when the parasitoid population had just started building. After *Pyrilla* epidemic in 1985, a team constituted by

---

1. National Centre for Integrated Pest Management, L.B.S. Building, IARI Campus, New Delhi 110012

the Govt. of India had recommended that if sufficient number of *E. melanoleuca* is noticed in *Pyrilla* infested area, then no chemical control is warranted. It has brought very encouraging results and since then no *Pyrilla* epidemic has been observed. During July- October, 1999 Pyrilla outbreak was noticed in sugarcane and fodder sorghum crops in North India, especially in Uttar Pradesh . The survey team visited areas and observed the presence of *E. melanoleuca* and egg parasitoids. Based on their observation the aerial spraying was not recommended. By November 1999 the *Pyrilla* was completely checked. This has resulted into net saving of more than Rs. 17.56 crores besides social, environmental and ecological benefits (Pawar *et al.*, 2002).

ECOLOGICAL APPROACHES IN CONSERVATION

The most important step to conserve natural enemies in agricultural system is to address the root cause of parasitoid failure i.e. disturbance and lack of population persistence in different levels. Ecologists define disturbance as "any relatively discrete event in time that disrupt ecosystem community or population structure and changes resources, substrates availability, or the physical environment" (Pickett and White, 1985). The most common abiotic factors include fires, windstorm, floods, landslides and other physical forces such as insect or disease outbreaks. Frequently disturbed terresterial ecosystem may experience only one disturbance event every several years (fire in forests) or multiple disturbance in each growing season. In addition, many of these events that are originated by human activities, are severe and occur uniformly over large areas. The implications of these disturbances can be examined at three different levels:within the crop level, at farm level and the landscape level.

Crop level disturbance

Agricultural production in a season involves different operations such as removing vegetation canopy, mixing of upper layer of soil profile, secondary tillage events followed by various nutrient and pesticide application, cultivation and harvest. Each of these operations is a significant disturbance element to some portion in the field for microbial, plant and animal community. Failure of parasitoids in these system is often a result of the direct and indirect effect of pesticides, tillage cultivation, lack of adequate food resources, scarcity of alternate hosts and lack of shelter (Powell, 1986; Dutcher, 1993).

Farm level disturbance

Production technologies within a farm such as mechanization, presence of hedgerows, and field size and shape set up a particular disturbance scenario. Production practices also determine the spatial distribution of crops and refuge habitats for natural enemies (Drinkwater *et al.*, 1995). Lack of nectar and honey sources in large maize fields, combined with high temperature before canopy closure forced the Ichneumonid larval parasitoid, *Erioborus terebrans* (Gravenhorst) to migrate to wooden habitats to seek food and shelter (Landis and Menalled, 1998). Therefore, presence or absence of wooded hedgerows plays an important role.

Landscape level disturbance

The landscape may be simple or complex. The complex landscape contains small crop fields with varied rotation and considerable crop diversity. Whereas the simple landscape consists of large crop fields with no uncultivated habitats, simplified crop rotation and low crop diversity. Ryszkowski *et al.*, (1993) in Eastern Europe found that mean parasitoid biomass was always greater in the perennial crop alfalfa compared to the annual crop of wheat, barley, maize and sugar beet. It is necessary to maintain an effective metapopulationa (a series of genetically interconnected local populations) at the landscape level to use parasitoids to control insect pests in higher disturbed agricultural system. Hanski *et al.*, (1996) defined the minimum viable metapopulation (MVM) size as the minimum number of interacting local population necessary for long-term persistence of a metapopulation. This concept is directly linked to the Minimum Amount of Suitable Habitats (MASH).

Influence of Plants on Parasites and Predators

Plant diversity may be a critical feature for determining the survival and abundance of parasitoids and predators. Individual plants may provide essential food in the form of nectar (floral or exta floral) and pollen (Jervis *et al.*, 1993; Stapel *et al.*, 1997) to adult parasitoids and predators including coccinellids and mites (Huffaker *et al.*,1970, Hagen,1976). Food can also be provided indirectly by the herbivores that feed on honey dew produced by plant or other sugar secretions used by parasitoids (Powell, 1986; Jervis and Kidd, 1996). The availability of food can increased the fecundity, longevity, survival and level of parasitism for many parasitoid species (Olson and Nechols,1995). Reduction in predator density had been observed in nectarless variety of crops such as cotton (Schuster *et al.*, 1976; Adjei-Maafo and Wilson, 1983; Scott *et al.*, 1988).

Morphology and structure of plants can also affect parasitoids and predators in several ways. Factors such as surface area, foliar pubescence, glandular trichomes, waxy leaf surfaces, and leaf toughness can impede or facilitate parasitoid search (Vinson, 1976, Obrycki, 1986). A negative influence of plant pubescence on levels of parasitism has been shown in number of crop plants such as tobacco (Elsey and Chaplin, 1978), potato (Obrycki, *et al.*, 1983), cotton (Treacy *et al.*, 1986), and peanuts and soyabeans (Mc Auslane *et al.*, 1995). In cotton for example there is an inverse relationship between trichome density and parasitism by *Trichogramma pretiosusm* (Treacy *et al.*, 1986). First instar *Chrysopa rufilabris* destroy eight and four times more eggs on the smooth leaf of a glabrous variety than on hirsute (moderately hairy) and pilose (hairy) variety of cotton, respectively (Treacy *et al.*, 1985). Studies conducted in Tamil Nadu in rice indicated that variety TRY 1 harboured more web–spinning spiders such as *Agrippa* and *Tetragnatha* than variety CO_3 and variety CO_3 harboured more non-webbing spiders such as *Lycosa* (Jalaluddin, 2000).

Nevertheless, some studies have shown that parasitoids respond to colour and other visual cues (Vinson, 1976, Wackers and Lewis, 1994). The allelochemicals produced by the plants may serve as cues to aid in finding prey or may have detrimental consequences. Tomatine, an allelochemical present in tomato plants fed upon by the noctuid, *Helicoverpa zea* (Boddie), has detrimental effects on the time of development, survival and adult size of the ichneumonid, *Hyposoter exiguae.* (Duffey *et al.*, 1986)

TACTICS FOR CONSERVATION

Altering Pesticide Use

The only acceptable approach to conserve natural enemies is to employ selective pesticides, at the least injurious time and in the least disruptive formulation. Even then they should be used when it is found to be absolutely necessary, based on scientifically established economic injury thresholds. Diamond back moth (*Plutella xylostella*), a serious pest of cabbage and cauliflower throughout India, has not attained pest status in middle Gujarat for last 20 years because fo the fact that the pest is heavily parasitised by its larval and pupal parasitoids, *Apanteles plutellae* and *Tetrstichus sdkoloyosky*, respectively (Yadav *et al.*, 1973). Whereas the pest is causing severe damage in cabbage and cauliflower at north and south Gujarat where the chemical pesticide use is very heavy and the parasitoids are totally killed by chemicals (Yadav, 2001). There is an inverse relationship between the importance of natural enemies and that of insecticides (Fig. 1). When the natural enemy complex is ineffective, pesticides provide one of a few means of controlling the pest. On the other hand of the spectrum where the natural enemy complex is highly effective, pesticides are unnecessary. The intermediate region of this continuum is where IPM can be effectively employed. There are several means of integrating natural enemies with pesticides in pest management programmes. In certain cases changing the active ingradients, rates, formulations, timing and or location of pesticide application can allow natural enemies to remain effective.

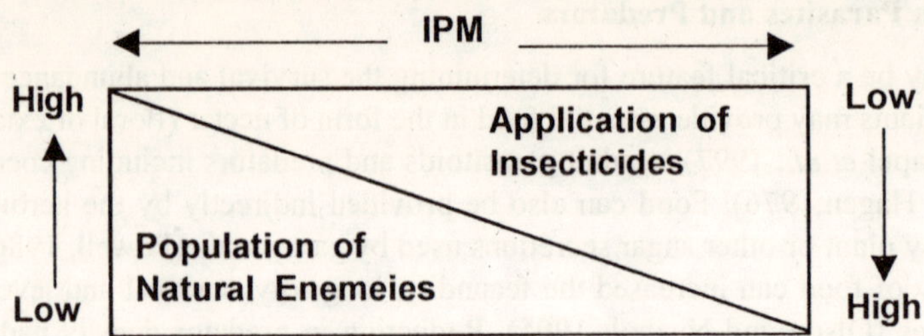

Figure 1. Inverse relationship between pesticides application and population of natural enemies in agriculture. Pesticides are unnecessary when natural enemies are effective (left). On the other hand pesticides dominate when natural enemies are ineffective (right). Pests falling into the intermediate zone are candidates for IPM practices integrating natural enemies and pesticides (Ruberson *et al.*, 1998).

Pesticide selectivity

Use of selective pesticides is the most powerful means of integrating pesticide use and natural enemies (Hull and Beers, 1985). Selective pesticides can be particularly valuable in crop systems with multiple pests, especially where natural enemies can wholly suppress potential pests. With the advent of pyrethroids, more selective chemical have entered the market. Recently, several new classes of chemical pesticides have been developed that exhibit high specificity for target pest but little activity against natural enemies. There are, however, economic constraints limiting the degree to which specificity will be persued by the pesticide industry. Pesticides those are highly specific, by definition, narrow target ranges. Thus, the expense of using several pesticides versus a single broad-spectrum pesticide can be prohibited to growers. Selective pesticides can be particularly valuable in crop system with multiple pests, where natural enemies can wholly suppress potential pests.

Temporal separation of natural enemies and pesticides

It can be accomplished in various ways, but is rarely practiced. For example the pesticides can be applied when key natural enemies are absent or when they are present in more tolerant life stages to conserve natural enemies. If the natural enemies are to be released the releases may be conducted after residual toxicity has declined sufficiently to allow natural enemies to survive. Using pesticides with short residual effect can also reduce temporal synchrony of pesticides and natural enemies.

Spatial separation of natural enemies and pesticides

It has been practiced in various ways for many years. For example, spot treatment has long been advocated as a means of targeting pest and conserving natural enemies. Other practices that minimize spatial concurrence of pesticides and natural enemies include treatments of alternate rows, use of bait formulations that are attractive only to target pests, and the use of trap crops that attract pest populations away from the crop where they can be treated without disrupting the natural enemy complex within the crop. Surulivelu and Kumaraswami (1989) showed that skip row coverage at Coimbatore during 1981-82 with monocrotophos was as effective as full coverage with monocrotohos or endosulfan against the cicadellid, *Amrasca biguttula* and *Aphis gossypi*. Its sprays also spared coccinellids, *Coccinella sexmaculata* compared to full coverage. Surulivelu and Kumaraswami (1990) while studying skip row coverage against bollworm damage also recorded more parasitoids *Apanteles angaleti* emerged with carbaryl skip coverage than with full coverage.

Improving knowledge of ETL of non-target pest and natural enemies

Conservation of natural enemies is limited by inadequate information on compatibility of pesticides with parasitoids and predators. It is often possible to reduce pesticide use by 50-80%, with no reduction in crop yield or quality. We can also reduce pesticide application and rates with improved knowledge of economic injury levels, particularly in case of indirect pests where the threshold for control could be much higher. Currently registered and experimental pesticides should be evaluated to determine whether changes in application and dosage could be made. A natural database of the impact of pesticide on target and non target pest species (including natural enemies) should be developed so that duplication and deficiencies in research efforts could be avoided and information made more readily available.

Use of insect growth regulators (IGR)

Use of insect growth regulators (IGR) may become an important component of IPM for suppression of insect pests, as they have been found safer to natural enemies. IGR served a critical role in overcoming the whitefly resistance crisis in 1995 in Arizona, USA. As the findings from Israel and Netherlands have demonstrated rapid development of whitefly resistance to buprofezin (Applaud R) and pyriproxfen (Knack R), therefore, it was recommended that Arozona cotton growers continue to limit their use to once per season, while strongly emphasizing delaying the uses of pyrethroids, for conservation of natural enemies (Simmons *et al.*, 1997). Large scale, multi-institutional experiments conducted in USA in 1996 for management of *Bemisia tabacii* indicated that parasitoid abundance and rates of parasitism were depressed in plots receiving rotation of conventional insecticides in comparison to those receiving growth regulators, buprofezin (Applaud R) and pyriproxfen (Knack R) (Naranjo and Hagler, 1997).

Use of Neem

Neem has emerged as a single important alternative to synthetic insecticides. Neem has been identified a source of environmentally 'soft' natural pesticide with diverse biological effects on insects such as antifeedent, oviposition deterrent, repellent, vigour, fitness, flight, longevity, IGR, toxic, mating disruptor etc. A recent compilation (Schmutterer and Singh, 1995) indicated that 417 species of insect pests, one species of ostracod, several species of mites, and nematodes, and even noxious snails and fungi, including aflatoxin- producing *Aspergillus* spp. (Saxena, 1989) are sensitive to neem. However, all these species cannot be very successfully managed with neem, therefore, the integration becomes necessary. There are various tactics (cultural, mechanical, host plant resistance, biocontrol agents, botanical insecticides, Bt and NPV etc.) which can be integrated. Among various biological effects, insect growth regulator property of neem is the most important effect and almost all the insects ultimately succumb to this effect. However, adults generally remain unaffected as its effect is limited to developmental stages. Field and laboratory evaluation have shown that neem extracts are generally safe to parasitoids and predators and non–insect predators (Singh and Singh, 1996). Neem has been found safe to foraging bees also.

Use of pesticide-resistant natural enemies

Pesticide resistant strains can be developed by selection in the field or by genetic improvement project in the laboratory (Hoy, 1985). This area is currently limited to only a few natural enemy species, but there are opportunities for expansion in the future (Narang *et al.*, 1994). Endogram, the endosulfan tolerant strain of *T. chilonis* was developed at PDBC, Bangalore by treating the parasitoids with endosulfan initially at 0.004%, gradually increasing the dose to 0.008%, 0.017%, 0.026%, 0.035%, 0.043%, 0.052%, 0.061% and 0.071%. Using this process, it took more than 325 generations, and about 8 years to develop endosulfan tolerant strain. Although use of resistant natural enemies holds promise, grower acceptance likely will be limited until releasing resistant natural enemies is demonstrated to be consistently cost effective. Widespread use of this technique may have some detrimental effects on the use of highly specific pesticides as it may encourage the growers to continued use of broad-spectrum pesticide.

HABITAT ENHANCEMENT

Natural habitats in the vicinity of crop lands can be quite important in providing alternative foods and refugia for natural enemy populations. Natural enemies require hosts or prey, sources of nectar and/

or free water and refugia that are available in habitats with greater vegetational diversity, temporal stability and sheltering capacities. However, large scale monoculture resulting from agricultural expansion and commodification of agricultural enterprises (that is the institutional and socioeconomic changes associated with a shift from subsistence production to local and global markets) are characterized by homogenous vegetation, frequent disruptions due to harvest and fallow cycles, and a lack of shelter from biocides harmful to natural enemies and their alternative hosts. Thus in the same way the protected habitats preserve viable populations of threatened and endangered species. Non-crop vegetation such as hedgerows or even unsprayed crop habitats can be crucial for maintaining and fostering population growth of predators and parasitoids in agricultural landscape.

Natural areas between agroecosystem, sometimes referred to as ecological compensation areas consist of separate or interconnected fields forming a network (Thomas *et al.*, 1991; Raskin *et al.*, 1992). This buffer zone acts as refuge area and /or dispersal center and offers many species adequate niches; compensating, at least partly, for the negative efforts of agriculture.

Sown weed strip is one type of artificial ecological compensation area having many advantages and offer good opportunities to combine agricultural use and natural conservation. Sown weed strips are highly attractive for many species and increase biodiversity.

Creation of Sown Weed Strips

The minimum required ecological compensation areas (i.e. totally unmanaged and unused or only partly managed areas) is considered to be 5 to 10% of an agricultural landscape. In many parts of India such ideal situation is already available especially in Central India. However, in certain states intensively used landscapes dọn't provide such an ideal situation. Therefore, additional ecological compensation areas are necessary in such landscapes. There are two optons:separating selected areas and allowing normal plant succession to proceed or sowing a given seed mixture. In extensively used large scale landscape the first possibility may be feasible. In intensively managed small-scale areas, however, establishment of semi natural compensation areas may be preferred. In "semi natural" compensation areas plant development is faster and problematic weeds can be avoided more easily.

In Switzerland sown weed strips usually have a width of 3 to 8 m (Heitzmann-Hofmann, 1995) which are generally sown at the border of a field or large fields are divided in small parts so that the distance between strips does not exceed 50 to 100 m. The length of strip depends on the field length and may be in the range of several hundered metres. Weed strip should start or end at other ecological compensation areas such as road sides, field margins, hedges, forest remnants, dry slopes, and so on. Thus the total system of sown strip forms a network of ecological compensation areas that connects several types of natural, semi-natural and artificial habitats.

A seed mixture is sown after careful soil preparation. The composition of mixture is based on the results of a long screening process in which specific characters of about 100 species of wild flowers and cultivated plants are analyzed. Important plant properties were investigated in monoculture plots of $10m^2$ (3 to 5 replications) sown in large wheat field (Heitzmann- Hofmann, 1995). These properties included insect diversity and abundance of beneficial species in patches of the test plant species (Frei and Manhart, 1992), length of flowering period, survival capacity in agricultural soil, length in the complex plant mixture as well as the tendency of dispersal into adjacent fields (Heitzmann- Hofmann,1995). Currently recommended seed mixture avoids *Brassica* species (but still contains *Sipapis alba*) because of the potential of the former to act as reservoir for crop pathogens. In Switzerland seed mixture consist of 29 species of wild flowers and cultivated plants and is available from several commercial distributors. Although the experience with the actual complex seed mixture for weed strip is restricted to only a few years but it is assumed that these ecological compensation areas can stay for many year. This may be accomplished by some kind of regeneration techniques such as minimum soil treatment or by a complete ploughing followed by additional or new, sowing. In this way steady state between young and old sown weed strip would be achieved within a landscape (Nentwig *et al.*,1998) .

HABITAT MANIPULATION

Intercropping

Studies were conducted (Prasifka *et al.*, 1999) on movement of natural enemies between adjacent cotton and sorghum. Predators move between cotton and sorghum in both directions throughout cotton maturation, but cotton received more immigrant predators from sorghum than losses due to emigration. Further, the strength of predators from sorghum migration changed with sorghum phenology, with the greatest benefit to cotton coming during the soft dough stage of sorghum. Planting rows of sorghum around cotton fields is currently an accepted practice in some parts of Mexico (Debach and Rosen, 1991). The presence of maize and cowpea crops within the cotton system in Maharashtra has significantly enabled establishment of high population of beneficial insects in the cotton field on time, prior to the arrival of sucking pests and *Helicoverpa armigera* (Singh *et al.*, 2002). Cotton intercropping with cowpea (Table 1) resulted in a significant increase in activity of coccinellids (*Mencochilus sexmaculatus* Fabricius and *Coccinella* sp.) and higher parasitism (*Rogas aligarhensis* Qadri and *Agathis fabiae* Nixon) on *Earias* spp. (Natarajan and Seshadri, 1988).

Table 1. Abundance of parasites of spotted bollworm under different cropping systems (after Natarajan and Seshadri,1988)

Cropping system	1985		1986	
	Parasitism on Earias spp.	*Coccinellids*	*Parasitism on Earias spp.*	*Coccinellids*
Cotton +Cowpea	24.8ª	4.0 ª	35.2 ª	6.4 ª
Cotton +Soybean	40.6ª	2.0 ᵇ	32.9 ᵇ	2.0 ᵇ
Cotton +Onion	26.9ᵇ	1.5 ᵇ	13.9 ᵇ	2.7 ᵇ
Cotton alone	25.0ᵇ	2.0 ᵇ	18.2 ᵇ	1.7 ᵇ

Means followed by the same letters are not significantly different (P=0.05) by LSD

Planting a green cover

In China a green manure crop (*Ageratum conyzoides*) planted in citrus groves not only provided pollen as food for the phytoseiid predators of the citrus red mite (*Panonychus citri*), but also lowered the air temperature on groves by an average of 5 degrees and raises humidity. This proved very important, as the predators are rendered ineffective by high temperature and low humidity.

Use of trap crops

Planting castor around tobacco attracts *Spodoptera litura* to deposit egg masses that are either collected manually or egg parasioid, *Trichogramm remus* is released. Similarly planting of marigold with tomato attracts *Helicoverpa armigera*. A large number of eggs are deposited on marigold plant. Use of castor in cotton as a trap crop for management of *Spodoptera litura* in Andhra Pradesh and Tamil Nadu is now an established practice followed by many farmers. Marigold is also being used as a trap crop for *H. armigera* by Gujarat farmers.

Weed conservation

Weeds are important source of nutrients to the parasitoids and provide alternate host to many insect pest and their parasitoids. Weed conservation has resulted in more diversified arthropod fauna, higher natural enemy population and reduced damage by insect pests. Weeds if kept below economically competitive densities, may be ecologically and financially sound assets to sugarcane pest management tactics. Weeds in Louisiana sugarcane field increased the abundance of arthropod prey and predators and caused at least 25% less injury from *Diatraea saccharalis* (Showler and Reagan, 1991).

Crop sanitation

This practice has been successful against a number of insect pests. Destruction of crop residues or plant refuse may prevent damage to future crops as it may cause direct mortality to vulnerable stages of insects or remove breeding or hibernating sites. Substantial reduction of cotton boll weevil populations was observed in Texas by early-autumn destruction of cotton plants, soon after harvesting but before first killing frost. This method resulted in only 27 per cent punctured squares by late July of the next season as compared to 63 per cent punctured squares where the plant were not destroyed. For the pink bollworm control, area-wise stalk shredding following harvest, along with ploughing of the refuse is looked upon as the best technique (De Bach and Rosen, 1991). Overwintering sugarcane borers may be killed in large numbers by shaving the stubbles with stubble shaver or by ploughing the stubbles left in the fields after harvest. Regular removal of fallen fruits may be carried out to increase the effectiveness of egg parasitoid due to overall lowering of population of the pests. The fallen fruits can be stored in steel drums covered with fine wire mesh so that the insect larvae may not escape whereas the emerged parasitoids can go out through the wire mesh.

Retention of crop stubbles

Increasing the height of stubbles (10 to 15 cm high) in rice fields in southern Karnataka during summer and *kharif* seasons resulted an increase in the spider population (Venkateshalu and Viraktamath, 1998).

Providing supplementary diet

Dusting pollen on crop plants may serve to increase the population of certain predatory insects and mites, either directly as a supplementary food source for natural enemies themselves or indirectly, as food for their alternative prey. Similar spraying artificial honeydew prepartions, consisting mainly of yeast hydrolysates, sugar and water, with some tryptophan added as an attractant, may attract adults green lacwings to a given field and increase their fecundity. By using such treatments, a significant increase in predation of alfalfa, cotton, bell peppers and other crops has been demonstrated (Hagen, 1986).

PROVIDING ARTIFICIAL STRUCTURES

Rice straw bundles

The use of bundles of rice straw as habitats for arthropod predator and prey communities during crop free periods in irrigated rice was evaluated in Philippines in 1989. The bundles were prepared by arranging rice straw into cone shaped tents about 30 cm in diameter at the bottom. One tent was placed in the center of each of 24 plots (10×7 m) in a newly harvested field. Significantly more arthropods were collected 10 days after placing the tents in the field. Spiders were the most abundant group, followed by Orthoptera and Coleoptera (Shepard *et al.*, 1989). Straw bundles are used in China to provide shelter for spiders in rice fields, and this practice alone has reportedly reduced the use of insecticides by 50 to 60 per cent. (Sparks *et al.*, 1982).

Bamboo cage cum percher

In sugarcane collection and destruction of egg masses of top borer (*Scirpophaga excerptalis* Walker), gurdaspur borer (*Acigona steniellus* Hmpsn.) and plassey borer (*Chilo tumidicostalis* Hmpsn.) are common practices for managing the bore infestation. But these practices lead to mortality of natural parasitoids parasitising the egg masses. These parasitoids can be conserved by simple techniques. The technique involves the collection of egg masses and putting them in a jar covered by a brass mesh. This jar is kept in a water trough containing pesticide (0.07% endosulfan or any other). The egg parasitoids, just after emergence, escaped through the brass mesh and fly back to fields whereas the larvae hatched from the unparasitised eggs get killed while crossing the pesticide solution in the trough. The parasitoids can also be conserved by bamboo cage cum bird perches. This structure permits the escape of egg parasitoids and allows the predatory birds to perch, and trap and kill the hatching larvae. This perches has been found effective under field conditions.

Nesting boxes for wasps and predatory birds

Availability of nesting sites for predatory *Polistes* wasps, as well as for insect-eating birds, improves their efficiency. There are major projects in Spain and other European countries to control forest pests by providing nesting sites and drinking water for insectivorous birds (De Bach and Rosen, 1991).

CULTURAL PRACTICES

Cultural practices used in cropping as well as cultural pest control techniques may be detrimental from overall viewpoint. For example, regular disc harrowing, which produces a lot of dust, may lead to substantial increase in mites and certain insect populations. The reason is that dust kills small predators and parasites, while having little effect on the pests. By washing the dust from the citrus trees at weekly intervals, population of California red scales was reduced by about one- half due to higher efficiency of parasites (De Bach and Rosen, 1991).

Conservation tillage

A negative correlation between the intensity of tillage and the number of soil and litter inhabiting arthropod is documented (Stinner and House, 1990). Robertson *et al.* (1994) reported higher incidence of ants in no tillage fields of grains in Australia. The most obvious impact of plough on the soil inhabiting ants is the destruction of their nests. This destruction causes some mortality but many colonies simply move their nests to the edge of the field (Perfecto, 1991).

Avoidance of trash burning

Trash burning is a common practice after crop harvest in sugarcane, rice and wheat to kill the larvae and to clean the field. However, this practice destroys only the parasitized eggs and larvae present on leaves and stubbles whereas the pupating population entering the soil escapes. By simple manipulation of burning i.e. sparing a patch from burning and spreading the trash in sugarcane can lead to the promotion of the activity of parasitoids and predators (Joshi and Sharma, 1989). Mohyuddin (1991) while investigating the conservation of the egg parasitoids, *Parachrysocahris javensis* (Girault) and the encyrtid *Ooencyrtus papilionis* Ashmead attacking *Parallax perpusilla* (Walker) in sugarcane reported that these parasitoids overwinter in crop residues where 100% of host eggs may be parasitised. The practice of burning crop residue after harvest destroys such overwintering sources and delays colonization of fields during the spring. By delaying burning at the field edges until the following spring, parasitisation was increased to nearly 80% versus less than 5% in fields where the residue was completely burned in the previous season (Mohyuddin, 1991).

Strip harvesting

In Alfalfa it has been observed (De Bach and Rosen, 1991) that harvesting large acreage all at one time greatly decimated the exposed natural enemies, whereas if the fields were harvested in alternate strips on a chronological basis, predators and parasitoids would be conserved and natural biological control greatly improved.

Use of organic manure

Use of organic manure instead of chemical fertilizers helps in conservation of predaceous arthropods such as carabid beetles and improves efficiency of plant disease antagonists. Increasing organic matter in rice fields resulted an increase in the populations of detritivores and plankton feeders, and in the abundance of generalist predators (Settle *et al.*, 1996).

CONSERVATION OF BIOLOGICAL CONTROL IN GENETICALLY ENGINEERED CROPS

Pest resistance Bt expressing cultivars developing through genetic engineering have been one of the first commercial products. Insect toxins can affect biological control agents when they feed on toxic plants or intoxicated prey. Direct effect can also accrue on biological control agents from exposure to the resistant mechanism expressed on the plant surface such as plant trichome exudates. The insect natural enemies are also affected through complex modification of the resistant mechanism. Pests that are targeted by the resistance mechanism are likely to reduced in number. A reduction in pesticide use may result in increased number and biodiversity of other arthropods in the crop that could have either a positive or negative affect on biological control. Conservation of biological control agents and their potential importance in avoiding the evolution of resistant pests will be a useful consideration in developing new genetically engineered cultivars.

Plant resistance traits may affect natural enemies directly or their affect may be indirect mediated through effects of resistance trait on insect hosts/prey of natural enemies.

Direct effects on biological control agents

Many insect parasitoids and predators have an intimate association with the plants upon which their insect hosts or prey feed (Barbosa and Letourneau, 1988). Plants provide a habitat as well as behavioural cues that mediate searching behaviour and a source of water or nutrition for some species which feed in floral or extra floral nectar, pollen or pollen sap (Hagen, 1986). The primary toxins responsible for resistance in the transgenic plants produced to date are lepidopteran- and Coleopteran specific Bt. Predators and parasitoids may be affected by these toxins expressed in the nectar, pollen or sap on which they feed. For example, predaceous hempteran, *Podisus maculiventris* (Say) that fed on plant sap as well as insects experienced reduced growth and prolonged development on resistant soybean foliage (Orr and Boethel, 1986).

Indirect effects on biological control agents

Indirect effects of resistance trait on natural enemies are those mediated through the phytophagous insects that serve as hosts or prey. Indirect effects are more likely to be of consequence in situations in which herbivores experience sub lethal effect on the transgenic plants than in situations in which they are killed quickly, leaving little opportunity for interactions with natural enemies. In small scale field scale trial transgenic tobacco lines expressing high level of Bt provided excellent control of tobacco hornworm, *Manduca Sexta* L. and tobacco bollworm *Heliothis virescens* (L.), whereas lines expressing lower level of toxin reduced larval growth and delayed development. Larval parasitization of *H. virescens* tended to be higher (56%) on plant lines that expressed low level of Bt than on lines, which did not express the toxin (43%) (Warren *et al.*, 1992).

CHALLENGE FOR THE FUTURE

At the beginning of the new millenium, crop protection speciliststs throughout the world have been challenged to reduce pesticide usage through IPM approaches. Most of the developed countries have already set their goal to cut down the use of pesticides by 50 per cent or more (Matteson, 1995). The developing countries like India the economy in future would be closely linked with the export of agricultural commodities, and therefore, important steps have been initiated to radically reduce the use of pesticides in export commodities to capture the world market. Under these and related initiatives, biological control is expected to play a key role in achieving the stated goals. In principle, conservation biological control appears to be well suited for the challenge at hand, however, its future remains uncertain, as it is likely to face challenges from an array of competing management tactics.

The biggest challenge for conservation biological control will be to shed its image of being not implementable. This will require integration of conservation biological control into IPM programme. For implementation of both conservation biological control and IPM, long-term field research is necessary. Training of cadre of specialists and broad training at the graduate level is also required that encourage a holistic approach to agro ecosystem management. Currently IPM practitioners focus on monitoring pest population densities and then recommending control tactics and /or cultural practices to be used. IPM practitioners will have to work with growers to create a diversified nursery of resource plants outside of the crop or selectively use of agrochemicals within the agricultural landscape to favour resource plants. This would be an additional cost to the growers and would require a high level of education for the growers and practitioners.

The other challenge which will be faced by those seeking to implement habitat manipulation in a agricultural system is the clinging tendency of those who employ high doses of pesticides. Though habitat manipulation may provide refugia for natural enemies from pesticide treated areas, the toxic residue left within the crop may repel or disrupt subsequent activity. Under most circumstances, therefore, use of conventional insecticides will be antagonistic to aim of conservation biological tools in control. Newer chemicals such as IGR will be important tools in habitat manipulation in a variety of agro ecosystems.

Present crop support system is focused on a particular commodity that encourages only individual and the regional specialization rather than diversified agriculture. There is a need for policy change to encourage agricultural diversification by providing incentives for multiple land use options that in turn would improve the conservation of natural enemies.

REFERENCES

Adjei-Maafo, I. K. , and I. T .Wilson 1983. Factors affecting the relative abundance of arthropods on nectaried and nectariless cotton. *Environ. Entomol.* 12:349-352

Barbosa, P. and D. K Letourneau 1988. *Novel Aspects of Insect- plant Interactions*. John Wiley and Sons. New York, NY.

De Bach, P. 1964. The scope of biological control. *In* "Biological control of Insects pests and Weeds." (P. DeBach. ed.), pp. 3-20. Chapman and Hall. London, U.K.

De Bach, P. and D. Rosen 1991. "Biological Control of Natural Enemies."2nd ed. Cambridge University Press. New York.NY.

Drinkwater, L. E., D. K. Letourneau, F. Workneh, A.H.C Brugan and C, Shennan 1995. Fundamental differences between conventional and organic tomato agroecosystems in California. *Ecol. Appl.* 5:1098-1112.

Duffey, S. S., K. A .Bloem and B. C.Campbell 1986. Consequence of sequestration of plant natural products in plant- insect- parasitoid interactions. *In: Interaction of Plant Resistance and Parasitoids and Predators of Insects* (D. J. Boethel and R. D. Eikenbary, eds.), pp. 31-60. John Wiley and Sons. New York. NY.

Dutcher, J. D. 1993. Recent examples of conservation of arthropod natural enemies in agriculture. *In: Pest Management:Biological Based Technology*." (R. D. Lumsden, and J. L. Vaughn, eds), pp. 101-108 Amer. Chem. Soc. Washington, DC.

Elsey, K. D. and J-F.Chaplin 1978. Resistance of tobacco introduction 1112 to the tobacco budworm and green peach aphid, *J. Econ. Entomol.* 71: 726-725.

Frei, G. and C. Manhart 1992. Nutzlinge und Schadlinge an kunstlich angelegten.Ackerkrautstreifen in Getreidefelden. *Agrarokologie* 4:1-14.

Goeden, R. D. and L. T. Kok 1986. Comments on proposed 'new' approach for selecting agent for biological control of weeds. *Canad. Entomol.* 118:51-8

Hagen, D. S. 1976. Role of nutrition in insect management. *Proc. Tall Timbers Conf. Ecol. Anim. Contr. By Habitat manag.* 6:261-262.

Hagen, K. S. 1986. Ecosystem analysis, plant cultivars (HPR), entomophagous species and food supplements. *In: Interactions of Plant Resistance and Parasitoids and Predators of Insects*. (D. J. Boethel and R. D. Eikenbary, eds.), p.151-187. John Wiley and Sons. New York. NY.

Hanski, I. 1989. Metapopulation dynamics:does it help to have more of the same ? *Trends Ecol. Evol.* 4:113-114.

Hanski, I., A. Moilanen and M. Gyllenberg 1996. Minimum viable metapoopulation size. *Amer. Nat.* 147:527-541.

Heitzmann-Hofmann, A.1995. Angesate Ackerkraustreifen–Veranderungen des Pflanzenbestandes wahrend der naturlichen Sukzession. *Agrarokologies* 13:1-152.

Hoy, M.A. 1985. Recent advances in genetic and genetic improvement of the phytoseiidae. *Annu. Rev. Entomol.* 30:345-370.

Hoy, M.A. 1990. Genetic improvement of arthropod natural enemies:becoming a conventional tactic? *In* "New Direction in Biological Control:Alternative for suppressing Agricultural Pests and Diseases." (R. Baker and P. Dunn, eds.), pp. 405-417. Alan R. Liss. New York, NY.

Huffaker, C. B., M. van de Vrie and J. A. McMurtry 1970. Tetranychid populations and their possible control by predators:an evaluation . *Hilgardia* 40:391-458.

Hull. L. A. and E. H. Beers 1985. Ecological Selectivity:Modifying chemical practices to preserve natural enemies. *In: Biological Control of Agricultural IPM Systems*. (M.A. Hoy and D. C. Herzog, eds.), pp. 103-122. Academic Press. Orlando, FL.

Jalaluddin, S. M. 2000. Conservation of predatory spiders in rice varieties. *Insect Environment*. 6:57.

Jervis, M. A. and N.A. C. Kidd 1996. Phytophagy. *In: Insect Natural Enemies Practical Approaches in Their Study and Evaluation.* (M. Jervis and M .Kidd, eds.), Chapmann and Hall, London, U.K., pp. 375- 394.

Jervis, M. A. ,N.A. C. Kidd , M. G, Fitton, T. Huddleston and H. A. Dawah 1993. Flower visiting by hymenopteran parasitoids. *J. Nat. Hist.* 27:67-105.

Joshi, R. K., and S. K. Sharma 1989. Augmentation and conservation of *Epiricania melanoleuca* Fletcher, for the population management of sugarcane leafhopper, *Pyrilla perpusilla* Walker. *Indian Sug*. 39:625-628.

Landis, D. A. and F. D. Menalled 1998. Ecological considerations in the conservation of effective parasitoid communities in agricultural systems. *In: Conservation Biological Control* (Pedro Barbosa ed.) pp. 101-121, Academic Press, New York.

McAuslane, H. J., F. A. Johnson, D. L. Colvin and B. Sojack 1995. Influence of foliar pubescence on abundance and parasitism of *Bemisia argentifolii* (Homoptera:Aleyrodidae) on soyabean an peanut. *Environ. Entomol.* 92:641-652.

Mohyuddin, A. I. 1991. Utilization of natural enemies for the control of insect pests of sugarcane. *Insect Sci. Applic.* 12:19-26.

Naranjo, S. E. and Hagler, J. R, 1997 *Bemisia* growth regulators:conservation of natural enemies? *Proc. Beltwide Cotton Conf.* Vol. 2, 932-934, National Cotton Council; Memphis, USA.

Narang, S. K., A. C. Barlett and R. M. Faust 1994. "Application of Genetics of Arthropods of Biological Control Significance." CRC Press. Boca Raton, FL.

Natarajan, K. and V. Seshadri 1988. Abundance of natural enemies of cotton insects under intercropping system. *J. Biol. Control* 2:3-5.

Nentwig, W., T. Frank and C. Lethmayer 1998. Sown weed strips:artificial ecological compensation areas as an important tool in conservation biological control. *In*:"Conservation Biological Control (Pedro Barbosa ed.) pp.133-154, Academic Press, New York.

Obrycki, J. J. 1986. The influence of foliar pubescence on entomophagous species. *In* "Interactions of Plant Resistance and Parasitoids and Predators of Insects." (D. J. Boethel and R. D. Eikenbary, eds.), p.61-83. John Wiley and Sons. New York. NY.

Obrycki, J. J., M. J. Tauberand and W. M. Tingey 1983. Predators and parasitoid interaction with aphid- resistance potatoes to reduce aphid densities:a two –year old field study. *J. Econ. Entomol.* 76:456-462.

Olson, D. J. and J.R. Nechols 1995. Effects of squash leaf trichome exudates and honey on adult feeding, survival, and fecundity of the squash bug (Heteroptera:Coreidae) egg-parasitoid *Gryon pennsylvanicum* (Hymenopterae:Scelionidae). *Environ. Entomol*. 24:454-458.

Orr, D. B., and D. J. Boethel 1986. Influence of plant antibiosis through four trophic levels. *Oecologia* 70:242-249.

Pawar, A. D., M. P. Misra and J. Singh 2002. Role of *Epiricania melanoleuca* (Fletcher) in biological control of sugarcane pyrilla, *Pyrilla perpusilla* (Wlk.) outbreak in 1999 in Western Uttar Pradesh. *J. Bio. Control*, 16:71-76.

Perfacto, I. 1991. Dynamics of *Solenopsis geminata* in a tropical fallow after ploughing. *Oikos.* 62:139-144

Pickett, S.T.A. and P.S. White 1985. "The ecology of Natural Disturbance and Patch Dynamics." Academic Press. San Diego, CA.

Powell, W. 1986. Enhancing parasitoid activity on crops. *In* "Insect Parasitoids." (J. Waage ,and D. Greathead, eds.), pp.319-340, Academic Press. London, UK.

Prasifka, J. R., K. M. Heinz , P. C. Krauter, C. G. Sansong, R. R. Minzenmayer and P. Dugger , D. Richter1999. Natural enemy movement between adjacent sorghum and cotton fields. *Proc. Beltwide Cotton Conf.*, Orleando, Florida, USA, Jan.,3-7 ,1999:Vol. 2. 1999, 1112-1114. National Cotton Council, Memphis, USA.

Raskin, R., E. Gluck and W. Pflug 1992. Floren-und Faunenentwicklung auf hebizidfrei gehaltenen Agrarflachen des ackerrandstreifenprogramms. *Natur und Landschaft* 67:7- 14.

Robertson, L. N., B.A. Kettle. and G. B. Simpson 1994.The influence of tillage practices on soil macrofauna in semi-arid agroecosystems in northeastern Australia. Agric. *Ecosys. Environ.* 48:149-156.

Ruberson, J. R. , H. Nemoto and Y. Hirose 1998. Pesticides and conservation of natural enemies in pest management. In *In* "Conservation Biological Control" (Pedro Barbosa ed.) pp207-220, Academic Press, New York.

Ryszkowski, l., J. Karg, G. Margalit, M.G. Paoletti and R. Zlotin 1993. Above ground insect biomass in agricultural landscapes in Europe. In" Landscape Ecology and Agriecosystems." (R.G.H. Bunce, L. Ryszkowski, and M.G. Paoletti. Eds.), pp.71-82, Lewis Publishers. Boca Raton, FL.

Saxena, R. C. 1989. Insecticide from neem. *In* "insecticides of Plant Origin" (J. T. Arnason., B. J. R. Philogene, and P. Morand eds.) ACS Symp.Ser. 387. pp.110-135, American Chemical Society, Washington, DC.

Schmutterer, H., and R. P. Singh 1995. List of insect pests susceptible to neem products. *In* "The Neem Tree, *Azadirachta indica* A. Jusss., and other Meliaceous Plants:Source of Unique Natural Products for Integrated Pest management, Medicine, Industry and Other Purposes. pp. 326-365. VCH, Weinheim.

Schuster, M. F. , M. J .Lukefahr and F. G. Maxwell 1976. Impact of nectariless cotton on plant bug and natural enemies. *J. Econ. Entomol.* 69:400-402.

Settle, W. H., H. Ariawan , E. T. Astuti, W. Cahyana, A. L. Hakim, D. Hindayana, A. S. Lestari and Pajarningsih 1996. Managing tropical rice pets through coservation of generalist natural enemies and alternative prey. *Saranto Ecology.* 77:1975-1988.

Shepard, B. M., H. R. Rapusas and D. B. Estano 1989. Using rice straw bundles to conserve beneficial arthropod communities in rice fields. *Rice Res, Newsletter* 14:30-31.

Showler, A. T. and T. E. Reagan 1991. Effect of Sugarcane borer, weed and nematode control strategies in Louisiana sugarcane. *Entomology* 20:358-370.

Showler, A. T., T. E. Reagan and R. M. Knaus 1990. Sugar-cane weed community interactions with arthropods and pathogens. *Insect Sci. Applic.* 11:1-11.

Simmons, A.L., Williams, L. III, Dennehy, T. J., Antilla, L., Jech, L. E., and Husman, S. 1997. Investigations of two insect growth regulators against Arizona whitefly populations. Proc. Beltwide Cotton Conf., New Orleans, LA. USA, Jan.,6-10,1997:Vol. 2. 1248-1252. National Cotton Council, Memphis, USA.

Singh, Amerika, O. P. Sharma, R. C. Lavaekaar, O. M. Bambawale. K. S. Murthy and A. Dhandapani 2002. IPM Technology for Rainfed Cotton . *Tech. Bull.* 11:1-36.

Singh, R. P., and S. Singh 1996. Neem in agriculture-advantages and disadvantages. *In* "Recent Advances in Indian Entomology" (O.P. Lal ed.) pp. 67-82, APC Press, New Delhi.

Sprks, A. N., J.R.Ables and R. L. Jones 1982. Notes on Bioklogical control of stem borer in corn, sugarcane, and rice in the People's Republic of China. In:Biological Control of Pests in China, pp.193- 215. Washington, D C:US Department of Agriculture.

Stapel. J. O., A. M. Cortesero, C.M. De Moraes, J. H. Tumlinson and W. J. Lewis 1977. Effects of extrafloral nectar, honey dew, and sucrose on searching behaviour and efficiency of *Microplitis croceipes* (Hymenopterae:Braconidae) in cotton . *Environ. Entomol.* 26:617-623.

Stinner, B. R. and G. J. House 1990. Arthropods and other invertebrates in conservation tillage agriculture . *Annu. Rev. Entomol.* 35:299-318.

Surulivelu,T., and T. Kumaraswami 1989. Effect of 'skip row coverage' of insecticide application on some sucking pests and their predators in cotton. *J. Biol. Cont.* 3:17-19.

Surulivelu, T. and T. Kumaraswami 1990. Studies on 'skip row coverage' against bollworm damage and parasite emergence in cotton. *Entomon.* 15:69-73.

Thomas, M. B., S. D. Wratten and N. W. Sotherton 1991. Creation of 'island' habitats in farmland to manipulate populations of beneficial arthropods:predator densities and emigration. *J. Appl. Ecol.* 28:906-917.

Treacy, M. F., G.R. Zummo and J. H. Benedict 1985. Interactions of host-plant resistance in cotton with predators and parsites. *Agric. Ecosys. Environ.* 13:151-157.

Treacy, M. F., J. H. Benedict, J. C. Segers, R. K. Morrison. and J. D. Lopez 1986. Role of cotton trichome density in bollworm (Lepidoptera:Noctuidae) egg parasitism. *Environ. Entomol.* 15:365-368.

Venkateshalu, Gubbaiah and C. A. Viraktamath 1998 Conservation of spiders on rice ecosystem. *Entomon.*23, 147-149.

Vinson, S. B. 1976. Host selection by insect parasitoids. *Annu. Rev. Entomol.* 21:109-133.

Waage, J. 1997. "Yes, but does it work in the field?" The challenge of transferring biological control technology. *Entomophaga* 41:315-332.

Wackers, F. L., and Lewis, W. J. 1994. Olfactory and visual learning and their combined influence on host site location by *Microplitis croceipes*. *BioCont.* 4:105-112.

Warren, G. W., N. B., Carozzi, N. Desai and M. G. Koziel 1992. Field evaluation of transgenic tobacco containing a *Bacillus thuringiensis* insecticidal protein gene. *J. Econ. Entomol.* 85:1651-1659.

Yadav, D. N. , R. C. Patel and T. M Manjunath 1973. Seasonal activity of *Apanteles plutellae* (Kurdj.) a laral parasite of *Plutella xylostella* (L.) a Anand (Gujarat). *Indian. J. Pl. Prot.*, 3(2), pp 111-115.

Yadav, D. N.2001. Biocontrol based Integrated Pest Management in agricultural crops. *IPM Mitr* 11:76-84.

2

Recent Trends in Utilization of Sex Pheromones in Insect Pest Management

A.J. TAMHANKAR[1]

With the increasing awareness about the safety of our environment and the damage caused to it by insecticides, increasing attention is being paid to environmentally safe pest management strategies. The problem of resistance development to insecticides in ever-increasing number of insects is also of major concern and this has focused the attention of researchers to developing newer methods of pest control. Amongst the array of such methods available to us, pheromones are a major component of the now evolving ecologically safe integrated pest management strategy. Control of insect populations through species-specific manipulation of sexual communication, without adversely affecting other beneficial organisms, has been the guiding principle for research on insect pheromones. The female-produced sex pheromones of lepidopteran insects are the most widely used chemical stimuli for pest management, mainly a) by monitoring adult insect populations in agricultural crops for taking up control interventions and b) by directly controlling them by the mating disruption technique. Associated with these two are the evolving techniques of mass trapping, lure and kill and lure and infect. While research in these techniques gets major attention by research workers, the primary use of pheromones in detecting mere presence of an insect species in an area remains ever useful. For example, Kindi *et al.* (2002) used pheromone monitoring very successfully to detect *Cameraria ohridella* (Lepidoptera, Gracillariidae) in earlier un-infested areas in Europe, while Berger *et al.* (2002) using sex pheromone traps showed that several parts of Europe were free of the destructive pest *Diabrotica virgifera virgifera* LeConte.

The first identification of sex pheromone was from a lepidopteran insect and even now discoveries of sex pheromones are largely from the same order. Though identification of pheromones from other orders of insects is also forthcoming, the trend still shows dominance by lepidopteran insects. For example, sex pheromones from more than 1500 moth species were identified till 1986 (Arn *et al.*, 1986) and record of pheromones from other orders was negligible. The general trend still continues to be the same. In the last 4 years, the ratio of reports on pheromones from lepidopterans: non-lepidopterans was about 5:1. It is not surprising therefore that the greatest use of pheromones is against lepidopteran pests. However use of pheromones of other insects is now getting increased attention and to indicate this trend, many examples from other orders have been included in the discussion presented here.

MONITORING

Sex pheromone traps have been widely used for decision support, usually with thresholds for intervention for using insecticides. Many of the issues associated with this type of application are getting resolved, but trapping systems need to ensure that each component of trapping system-lure integrity and

1 Nuclear Agriculture and Biotechnology Division, Bhabha Atomic Research Centre, Mumbai-400085.

longevity, trap efficiency, and other factors such as correct establishment and interpretation of thresholds-are addressed.

PHEROMONE BLEND VARIATION

After the initial identification of pheromones and the commencement of their use under field conditions, in many -a species it was noticed from time to time that the number of insects trapped-employing the known pheromone component blend-was not representative of the population density in the field. Reinvestigations in this context in the recent years have revealed that this was due to geographical/ host specific variation in the female sex pheromone of a species. The Sex pheromone components of the rice leaf folder moth, *Cnaphalocrocis medinalis* (Guenee) (Lepidoptera: Pyralidae) of either Philippine or Indian origin were shown to be (Z)-11 hexadecenyl acetate and (Z)-13-octadecenyl acetate in a ratio of 98:2 in the Philippine and 1:10 in the Indian species. But in Japan, neither the Philippine blend nor the Indian blend showed any attraction in the field. Investigations have revealed the sex pheromone components of the Japanese insects to be (Z)-11-octadecenal, (Z)-13-octadecenal, (Z)-11-octadecen-l-ol and (Z)-l 3-octadecen-l ol in a ratio of 11:100:24:36 (Kawazu *et al.* 2000). Field bioassays in Japan showed that the two aldehydes were essential for male attraction and the alcohols may have a synergistic effect on the aidehydes. Further, field bioassays using the three sex pheromone blends (Indian, Philippine and Japanese) at I I sites in Japan and Hangzhou in China showed that only the Japanese pheromone blend attracted a significant number of male moths in china, while the Indian and Philippine pheromone blends showed no marked activity (Kawazu *et al.* 2002). In Japan however, both the Philippine and the Indian blends attracted males of a different species, C. *pilosa,* which is closely related to *C. medinalis* (Kawazu *et al.* 2001). These results suggest that there are remarkable geographical variations in the sex pheromone composition of this species or there are several distinct races using different sex pheromone blends. Needless to say, that for obtaining trapping that is representative of the population-density of the target area, a location specific blend must be used. The pine sawfly, *Neodiprion sertifer* (Geoffroy) (Hymenoptera: Diprionidae), is a widespread and economically important forest insect. The sex pheromone communication system of this species has been previously investigated in North America, Japan and Europe, with the acetate or propionate of the alcohol (2S,3S,7S)-3,7-dimethyl-2-pentadecanol (diprionol) shown to be the main pheromone component. In some locations, male attraction either increased or decreased by the addition of the (2S,3R,7R)-diprionyl acetate isomer. However, these studies were made with different batches of synthetic pheromone, with different types of traps and according to different procedures, so the observed differences might not have reflected true geographic variation. Studies by Anderbrant *et al.* (2000) show the importance of using unified techniques in this context. These investigations conducted to establish the geographic pattern of male sawfly pheromone response were conducted by using identical chemicals, traps and experimental procedures at eight field sites ranging from Japan in the east to Canada in the west and an increased inhibitory effect of the (2S,3R,7R)-isomer from Japan and Siberia to Europe was noticed. At the eastern sites, increasing amounts of the (2S,3R,7R)-isomer up to and equal to the amount of the (2S,3S,7S)-isomer, did not influence the trap catch, whereas at sites in Europe, as little as 1% of the (2S,3R,7R)-isomer almost completely inhibited the attraction. The response of the North American population was intermediate. The only site in which the (2S,3R,7R)-isomer was essential for the attraction of males was in Siberia. A similar pattern was found for the (2S,3R,7S)-isomer. Both the acetate and the propionate form of the (2S,3S,7S)isomer were attractive by themselves in Japan, Europe and North America, and neither the (2S,3R,7S)-isomer nor the (2S,3R,7R)-isomer alone were attractive, in the acetate or propionate form. Such investigations result in the development of an appropriate location specific pheromone blend and therefore more efficient monitoring/ trapping schemes. For *Helicoverpa armigera* Hubner (Hardwick) (Lepidoptera: Noctuidae) also it was experienced that the number of insects trapped-employing the known pheromone component blend- did not represent the population density in the field. Using identical chemicals, traps and experimental procedures Tamhankar *et al* (2003) discovered a significant variation in the pheromone component ratios attractive to males at different locations in India and use of this information for trapping resulted in capture of significantly increased number of mates.

KAIROMONES, HOST PLANT CUES, CO-ATTRACTANTS

Insects and plants have co-evolved and it is likely that the plant volatiles may be involved in the sexual communication of insects in several ways. In the pheromone-producing sex, these volatiles may affect pheromone production or release and, in the receptive sex, plant volatiles may have a synergistic effect on the attraction to sex pheromone. In case of *Earias vittella* (Fabricius) (Lepidoptera: Noctuidae) the females release the sex pheromone for a longer period when in the presence of the host plant than in its absence (Tamhankar, 1994). Experiments by Groot and Visser (2001) indicated that leaves on which *Lygocoriv pabulinus* females were earlier sitting, attracted more males, indicating that attractive components from females were deposited and adsorbed to the substrate. Ethyl (2E, 4Z)-2,4-decadienoate, a pear-derived kairomone was found to be a species-specific, durable, and highly potent attractant for the males and females of the codling moth, *Cydia pomonella* L. (Tortricidae: Olethreutinae) (Light *et al.*2001). This kairomone was highly attractive in both walnut and apple orchards, but showed limited effectiveness in a pear orchard. This particular attribute of attracting females, renders this kairomone as a novel tool for monitoring flight and mating-ovipositional status of females, and potentially is a major new weapon for directly controlling populations. The role of host plant volatiles is also evident from the results of Krishnaiah *et al.* (1998) who found that addition of both antioxidant Vitamin E and Pusa Basmati Plant extract could improve the attractancy of the pheromone of yellow stem borer *Scirpophaga incertulas* Walker (Lepidoptera: Pyralidae) by two to three times. Recent experiments suggest that volatiles from some non-host plants may also have a role to play in pheromone communication. Virgin females of *H. armigera* responded to the volatiles from wilted leaves of a non-host plant, Chinese wing-nut tree (VWLCT), *Pterocarya stenoptera* but mated females and males did not (Xiao *et al.*, 2002). The virgin females responded at a time, which coincided with their calling behavior. Males were strongly attracted to a combination of a virgin female and VWLCT volatiles.

TRAPS/DISPENSERS

Beside the lure, appropriate pheromone trap and pheromone dispenser are crucial to obtain optimal trapping performance by a pheromone trapping system. Presently use of dry, liquid and sticky traps is in vogue and continuing efforts are always needed to improve trapping efficacy. Tamhankar *et al.* (2001) developed a trap (BARC trap) more efficient than delta sticky trap and liquid trap for *Pectinophora gossypiella* (Saunders) (Lepidoptera: Gelechiidae). The performance of Delta trap in catching *Leucinodes orbonalis* (Lepidoptera: Pyralidae) was significantly improved when 'windows' were cut in the side pane s of delta sticky traps (Cork *et al.*, 2003). Khoo et at. (2000) observed that a syringe driver, being a controlled release device, was more satisfactory than rubber septum, for dispensing sex pheromone for attraction of female fruit fly *Bactrocera papayas* (Diptera: Tephritidae).

Monitoring for control intervention

Pheromone based monitoring is spreading all over the world and is finding useful application in guiding timing of pest management tactics for varying insect pests and in varying host environments. From time to time there appear reports that pheromone trap catches failed to reflect the true population levels in the field, however a very large body of reports speak to the contrary and they are supported by a significant correlation with the visual sampling methods, and also with economic damage and this is true not only for Lepidoptera but also for other orders (Ebenebe *et al.* 2000, Visalakshmi *et al.*, 2000, Kindi et at, 2002, Millar *et al.* 2002). Pheromone trap based monitoring has also been useful in establishing migratory pattern of insects (Zhou and Applebaum, 2000). The reasons for improper performance of traps could be many and some of the following examples illustrate how currently researchers are resolving them.

For pheromone trap catch based control intervention, proper establishment and interpretation of thresholds is very important. The points that need to be paid greater attention to achieve this are i) establishment of a proper dose and composition for lures, ii) taking into account variation in trap catch thresholds in different host environments, iii) discriminating between tactics to be employed when populations are at outbreak and non-outbreak levels etc. Grafton *et al.* (2000) studied the effects of dose and field longevity of sex pheromone lures for monitoring flight activity of male yellow scale, *Aonidiella citrina* (Coquillett) (Homoptera: Diaspididae). Pheromone doses of 10-200 microgram per lure attracted large numbers of males. Lower doses (1-5 microgram per lure) generally attracted fewer males, but these trap-counts unequivocally showed the beginnings and peaks in male flight activity at both low and high population densities. Furthermore, the low-dose pheromone traps that collected fewer individuals were easier and less time-consuming to assess. The lower doses were therefore recommended for monitoring phenology and population densities. Similarly, in case of *P. gossypiella,* while the Delta sticky trap and the Omni directional liquid trap could not, the BARC trap proved capable of demonstrating locational differences in population levels and it maintained this sensitivity at pheromone loadings of 0.5 and 1.0 mg levels of pheromone and not when pheromone dose was increased beyond these levels (Tamhankar *et al.* 2001). An important issue was addressed by these workers, because in many a cases the traps are needlessly cluttered with large number of insects which make it difficult to establish a linear relationship between immature field stages and trap catch. Similarly attention needs to be paid to establishing trap-catch thresholds for different crops. Reddy and Guerrero (2001) conducted Field trials in cabbage *(Brassica oleracea var capitata), cauliflower (B. oleracea var botrytis)* and knol khol *(B. oleracea gongylodes)* crops to optimize the timing of insecticide applications to control the diamondback moth, *Plutella xylostella* Linneus (Lepidoptera: Yponomeutiidae) and found that pheromone trap catch levels varied in these crops at 8, 12 and 16 males per trap per night respectively and using these varying thresholds for control intervention resulted in optimum use of insecticides.

An important aspect of monitoring is discriminating between populations at outbreak and non-outbreak levels. Schmidt *et al.* (2003) evaluated sex pheromone baited traps for of forest tent caterpillar moths to trap adults under these two conditions and found that lures dosed with 11 microgram or less of pheromone failed to capture moths at non-outbreak population levels, while lures dosed at 390 microgram gave the highest mean trap catch and the lowest rate of zerocaptures. Tamhankar *et al.* (2003) conducted similar experiments to test most attractive pheromone component ratios in case of *H. armigera* and discovered that at outbreak levels the males responded to wide ranging pheromone component ratios than at non-outbreak levels.

Monitoring for dispersal ability

Another use of sex pheromone traps is to assess the dispersal ability of insects. An eradication program is being implemented for sweet potato weevil, *Cylas formicarius* (Coleoptera: Curculionidae) in the Ryukyu Islands of Japan using the sterile insect technique (SIT). In this programme to assess the dispersal ability of the weevils, mark-recapture studies using sex pheromone of *C. formicarius* were conducted which indicated that a distance of at least 2 km is needed as a buffer zone to separate the SIT target area from other areas (Moriya and Miyatake, 2001).

In the case of stored product insects, their presence in warehouses and flourmilis needs to be monitored. Doud and Phillips (2000) conducted investigations to determine the distribution of *male Plodia interpunctella* (Hubner) (Lepidoptera: Pyralidae) at different locations in and around the mills throughout the season, and to monitor moth activity before and after the mill was fumigated with methyl bromide to assess efficacy of treatment. The highest numbers of *P. interpunctella* were caught outside the facility and at ground floor locations near outside openings. There was a negative correlation between moth capture and distance from the facility, which suggested that moth activity was concentrated at or near the flour mill. This study also documents high levels of *P. interpunctella* outdoors, relative to those recorded inside a food processing facility. The study concludes that potential for immigration of *P. interpunctella* into flourmilis and other stored product facilities from other sources may be greater than previously recognized.

ESTIMATING/FORECASTING

The usefulness of pheromone-trap based monitoring to predict population densities of the pests is another facet, which is getting attention by researchers to improve the system's capabilities. It is now being realized that along with pheromone trap captures, temperature accumulations need to be taken into account to increase the reliability of the forecasts. The relationship between temperature accumulations and pheromone trap captures of the males of *Lobesia botrana (Den.* and Schiff.) (Lepidoptera: Tortricidae) in the vineyards of Spain presented a statistically acceptable linear behaviour. The reliability of the forecasts based on the resulting logprobit lines was highly acceptable as a tool to improve the effectiveness of integrated pest management in the area (Del *et al.*, 2001). Studies on *P. gossypiella* in Turkey also revealed that effective total thermal summation could be employed together with catches in sex pheromone traps to predict first adult emergence of the over-wintering generation and also the seasonal population dynamics of the pest.

Multi-species trapping

Another interesting aspect in the area of monitoring is multi-species trapping (trapping of more than one species in the same trap). This is possible when more than one species respond to the same chemical(s) or a trap/dispenser can be designed to capture more than one species. Johansson (2002) explored multispecies sex pheromone trapping for the pine shoot moth *Rhyacionia buoliana,* the pine moth, *Dendrolimus pini,* the nun moth, *Lymantria monacha, the* pine beauty moth, *Panolis flammea* and the pine sawflies *Diprion pini and Neodiprion sertifer.* The catch from traps baited with the pheromone of a single species was compared with the catch from traps baited with pheromones for several species. The catch in the multispecies trap was significantly reduced in comparison with the single species traps for *D. pini, L. monacha and N. sertifer. D. pini* pheromone most likely inhibited *N. sertifer.* A follow-up study of *D. pini and* L. *monacha* showed no pheromonal interference between them. However, for practical applications a slight decrease in catch, due to pheromone] interference, could probably be tolerated as long as the catch reflects total population density. Males of *Callosobruchus maculatus* (Fabricius) (Coleoptera: Bruchidae) respond to sex pheromone of *C. subinnotatus,* but males of *C. subinnotatus* do not respond to that of *C. maculatus.* The two sex pheromone components of *C. subinnotatus* are also constituents of *C. maculatus* sex pheromone. These two components may be potentially useful in monitoring the populations of both the species in stored beans (Mbata *et al.* 2000). Further studies are needed to evaluate the feasibility of multispecies monitoring in these and other insects to refine this novel approach, as it has the potential to bring down the cost of pheromone use.

Sex pheromones in insect biocontrol

Pheromone trapping for insect biological control agents is a valuable tool that can help to determine and improve parasitoid success. Further, affractants to beneficial insects such as sex pheromones of wasp parasitoids can offer additional benefits, including detecting the level of abundance and demonstrating synchrony of the biological control agent with the target pest. For example, it was discovered that sex pheromone and larval frass volatiles from the diamondback moth, as well as volatile compounds from host (cabbage), are used by the natural enemies -the parasitoids *Trichogramma chilonis* (Hymenoptera: Trichogrammatidae) and *Cotesia plutellae* (Hymenoptera: Braconidae)- and the predator *Chrysoperla carnea* (Neuroptera: Chrysopidae), to locate the host. (Reddy *et al.* 2002). Suckling *et al.* (2002) used synthetic female sex pheromone to monitor the phenology of male *Ascogaster quadridentata* (Wesmael) (Hymenoptera: Braconidae), an egg-larval parasitoid of codling moth, *C. pomonella,* in conjunction with pheromone trapping of the moth and banding of the trees to determine larval parasitism. Catch of this parasitoid was typically approximately I 0 times less than the moth-catch, and larval parasitism varied from 7 to 12%. The emergence of the parasitoid occurred slightly later than the female moth emergence, by both outdoor emergence cages and pheromone trapping. The seasonal activity of male parasitoids was otherwise synchronous with male codling moth flight.

Direct control applications

Pheromones have enjoyed considerable success in direct control of pests of economic importance, mainly through mating disruption. There has been a recent upsurge of interest in lure and kill tactic internationally and also in mass trapping and lure and infect.

MATING DISRUPTION

Pheromone synthesis, pheromone dispensers and dispenser application methodology dictate the cost of mating disruption and knowledge of behavioural mechanisms governs its successful use.

Major achievements have been made with large-scale industrial synthesis of insect pheromones (e.g. Yamamoto and Ogawa 1989) and significant reduction in the cost of production has been achieved. It is important to realize that commercial use of codling moth mating disruption became possible only after economic synthesis of codlemone became available. The price for this compound has reduced by a factor of 50 to 100 since the eighties. As the quantities of pheromone required for mating disruption are large and costs are prohibitive, if a similar effect can be obtained with an alternate and cheap compound, cost of mating disruption can be brought down. In this direction, efforts are underway to determine if an analogue of a pheromone can be used instead of a pheromone. For example, investigations by Renou *et al.* (2002) on the effects of (Z)-11-hexadecenyl trifluoromethyl ketone (Z11-16: TFMK) a fluorinated pheromone analogue, on the responses to sex pheromone of the male cabbage armyworm, *Mamestra brassicae, using* an actograph and electroantennograph (EAG) provide new pointers to the possible use of Z 11-16:TFMK in pest-control strategies. Tamhankar et al (patent pending) have reported a process for obtaining an insect communication modifier from a cheap and renewable vegetable source and have shown that this material can cause a significant disruption in *P. gossypiella.*

A major flaw in the currently used commercial pheromone dispensers is that pheromone release depends on ambient temperature. For example, take a typical case of codling moth. Ninety per cent of pheromone applied is released outside its diel flight period, mainly during daytime at peak ambient temperatures (Witzgall *et al.* 1999). In addition, dispensers must be applied early in season when population densities are still low, while their release rates decrease during the season, as population densities start to increase (Ogawa, 1997). These problems could be circumvented by using active pheromone "puffers", releasing large amounts of pheromone when the insects are active (Shorey *et al.*, 1996 a and b, Baker *et al.*, 1997).

Hand-application of dispensers does not pose a problem in labour-intensive crops, but precludes the use of mating disruption in most field crops. The formulation of sprayable pheromone formulations has been attempted in the past (Hall *et al.* 1982, Weatherston 1990), and is also a subject of recent efforts for improvement (Polavarapu *et al.,* 2001).

In general, mating disruption efforts are directed towards a single species, however, the crop environment is of a multispecies nature. A multispecies mating disruption would be more economical and practical. It is obviously more economical to use one formulation including a single compound rather than impregnating complex blends into dispenser formulations. Towards this aim, Gronning *et al.* (2000) studied relative abilities of several sex attractant blends to reduce captures of 4 leaf roller species in pheromone traps, with encouraging results in reducing orientation of males. This is a requisite first step in the development of multispecies mating disruption blend. Downham *et al.* (2001) conducted a trial for the control of sweet potato weevils, *Cylas puncticollis and C. brunneus,* by mating-disruption using the synthetic sex pheromone of *C. brunneus* formulated in PVC resin dispensers. Almost no *C. brunneus males* were caught in pheromone traps in the treatment plot, whereas catches in the control plot ranged up to 30 weevils per trap over each two-night monitoring period. Suppression of *C. puncticollis* catches was weaker. Mating of *C. puncticollis* was significantly suppressed on one sample date, 18 weeks after planting, but not on two subsequent occasions, at 19 and 33 weeks. Yields of roots in the two plots were similar, but root infestation was lower in the plot treated with pheromone on four out of five sample harvest dates, from 17 to 36 weeks after planting.

Whether sex pheromones can be used for mating disruption of soil dwelling insects? A positive answer to this was obtained due to the experimentation of Polavarapu *et al.* (2002). They evaluated the feasibility of disrupting sexual communication in oriental beetle, *Exomala orientalis* (Waterhouse) with a microencapsulated sprayable formulation of (Z)-and (E)-7-tetradecen-2-one, in blueberries and ornamental nurseries and found significantly fewer tethered virgin female oriental beetles in copula/ contact with males in treated fields.

Although atmospheric treatment with synthetic pheromone is used commercially to control several lepidopteran pests, how mate-finding behavior is altered by this procedure is not very certain. To resolve this, mechanisms of disruption of the mate-finding behavior of *Choristoneura rosaceana* Harris (Lepidoptera: Tortricidae) were examined in wind tunnel bioassays by Evenden et al (2000). Male moths were exposed to synthetic sex pheromone or pheromone components prior to or during upwind flight to a calling virgin female. The proportion of males successfully contacting a calling female was reduced only when the synthetic pheromone treatment was presented simultaneously. A synthetic source of the attractive fourcomponent pheromone placed upwind of a calling female was more effective than a less attractive two-component blend in achieving disorientation. Habituation of the central nervous system did not appear to be a significant factor in disruption of mate-finding behavior of male *C. rosaceana,* as exposure to pheromone prior to the bioassay did not alter the proportion of males that achieved subsequent upwind flight. Disruption of mate-finding behavior of *C. rosaceana* is probably the result of a combination of mechanisms including adaptation of antennal receptors, camouflage of the female-produced plume, and false-trail following, which contributed, to the additive disruption effect observed with the most attractive four-component pheromone.

A standard tool to determine effectiveness of a disruption treatment is the use of pheromone traps. However, the placement of the traps is almost on an adhock basis and no serious experimentation has been involved in determining the placement of traps. Knight *et al.* (1999) evaluated relative placements of mating disruption dispensers and pheromone traps for codling moth and recommended that traps should be placed higher than the dispensers in the canopy.

In the same context a new approach of Passive Intervention (PI) traps was introduced to study the behaviour of the target species in pheromone treated and untreated fields (Weisling and knight, 1994). Using PI traps in *C. pomonella* it was found that moth catch and fruit injury were significantly higher in the edge versus the center plots. The numbers of total and female moths caught with PI-traps were significantly correlated with fruit injury for each generation (Knight, 2000).

Svensson *et al.* (2001) attempted a novel approach of using harmonic radar for monitoring mating disruption in *Agrotis segetum* (Denis and Schiffermuller) (Lepidoptera: Noctuidae). Tagging indivi dual male moths with radar transponders they monitored the longrange, pheromone-mediated, flight behaviour of male moths under natural and mating disruption conditions. In addition, male attraction to pheromone-baited traps and mating of calling females in treated and untreated plots were investigated. High doses of a four-component pheromone blend reduced trap catch by 79% and mating of females by 62% when compared with control plots in pre-radar experiments. Surprisingly, this effect was not associated with any pronounced differences in the flight behaviour of males between a treatment and a control plot as revealed by harmonic radar recordings. Moths could be followed for up to 77 min, corresponding to a track length of 7350 in. Mean ground speed ranged from 0.7 in s-1 to 5.4 in s-1. There was a strong trend for a greater number of male orientations to traps from downwind in the control field compared to the treatment field. Although, many flight tracks were fragmented due to radar shadow, further investigations may be taken up for using harmonic radar to study the pheromonemediated flight behaviour of nocturnal moths.

Mating disruption is generally being recommended and is considered potentially successful for contiguous large crop areas. However results of Zhang et al. (2001), who employed mating disruption treatment for the Chinese tortrix *Cydia trasias* (Meyrick) (Lepidoptera: Olethreutidae) infesting shade-trees planted along streets, reveal that mating disruption is also promising as an effective control agent against pest infestations occurring in discontinuous areas.

There have been instances where inconsistent results have been obtained for mating disruption. Large-scale field tests have been performed on mating disruption of *P. xylostella and* the results have been variable and often ambiguous. Schroeder et al (2000), after conducting field trials for 2-years, found that although significant differences in the numbers of male *P. xylostella* caught in pheromone-baited traps occurred between pheromone-treated and untreated fields, they did not represent the true levels of field infestations. Such differences in trap catches are often thought of as indications of mating disruption, and Schroeder et al caution that this may not be always the case. They have opined that in *P. xylostella,* mating disruption with the present technology does not appear to work and improvements need to be made. This warns us that the technical and biological issues are more challenging in direct control applications as there remain major knowledge gaps in the amount of pheromone required to achieve disruption, as well as the plume structure and behavioural mechanisms that operate with different types of formulations, blends and delivery systems.

Insecticides do not work as well as often believed, and the recurrent use of insecticides year after year is a clear indication that a long-term population decrease has never been achieved, not in any species. In comparison, a well-implemented mating disruption programme may clearly be more efficient than insecticides in some species. An observation shared by many working with mating disruption is that its long-term use results in continuously decreasing populations of the target species. One of the contributing factors for this may be a recovery of the beneficial insect fauna and an overall increasing stability of the ecosystem.

More often than not, population biology rarely receives sufficient importance in mating disruption applications. Models on insect population dynamics have been elaborated for many insects and should always be given due importance before applying control by mating disruption . As long as population fluctuations are ignored or are poorly understood, it is probably impossible to answer the question, whether a full, partial or off-blend of pheromone chemicals is appropriate for mating disruption. It goes without saying that a great many factors contribute to insect attack, many of which are not directly related to the efficacy of the mating disruption method.

Mass trapping

Whether mass trapping by sex pheromone traps can alone become an exclusive method of control will always remain in doubt as it involves removal of only one sex from the target area and that too partially, as pheromone traps may not ever become so efficient as to remove all the males. However, as a component of IPM and particularly at low population densities mass trapping can have a very good potential. Cork *et al.* (2003) conducted IPM trials for *L. orbonalis*, using a combination of pheromone traps placed at a density of I 00 per ha and weekly removal of infested shoots. Pheromone trap catches were reduced significantly from 2.0 to 0.4 moths per trap per night respectively in check and IPM plots in a young crop and 1.1 to 0.3 moths per trap per night in check and IPM plots respectively in a mature crop. Fruit damage was significantly reduced from an average of 41.8% and 51.2% in check plots of young and mature crops respectively to 22% and 26.4 respectively in the associated IPM plots. An estimated increase of 50% in marketable fruit was obtained by the combination of control techniques compared to insecticide treatment alone. The shade trees planted along streets are patchy in distribution, however, experiments by Zhang et al (2002) on mass trapping of the Chinese tortrix *C. trasias by* removal of males using sex pheromone traps deployed in the trapping plots at a density of one trap per two trees showed that male removal using sex pheromone traps can be a promising means of effectively controlling *C. trasias* on street-planted Chinese scholar-trees. Mating rates of caged virgin females were reduced by 70.65% and 66.48% when compared with the control plots during the over-wintering- and second-generation moth flight-period, respectively. Population densities of hibernating larvae on tree trunks during the third generation were significantly lower than those of the over-wintered generation. Tunneled petioles caused by first-and second-generation larvae were reduced by 47-79% and 62-81 %, respectively. Bored seedpods caused by third-generation larvae were reduced by 87-96%.

Lure and kill/Attract and kill

The attract and kill strategy, combines the selective attraction of the sex pheromone with the insecticide treatment, as a means of controlling a species. A viscous formulation based on castor oil containing the insecticide cyfluthrin and the main component of the codling moth sex pheromone, was developed by Loesel *et al.* (2000). The formulation, when applied to the bark of apple trees *(Malus domestics)* gave a good level of control, comparable with a spray treatment, of an insect growth regulator. Results of field trials showed that efficacy of the attracticide depended on population density of the pest and 4500 sources per ha was required to attain satisfactory levels of codling moth control (Loesel *et al.* 2002). Similar studies conducted by Krupke et al (2002) using Last Call CM- a newly registered attracticide product- showed that individual attracticide droplets were significantly more attractive to male moths than calling females. In plots with 500 droplets/ha, male moth catch was suppressed significantly compared with catches in control plots, and the number of females attracting at least one male was reduced significantly as well. Experiments investigating sub lethal physiological effects of attracticide exposure upon mating competency of male codling moths demonstrated male leg autotomy at 1, 24, 48, and 72 h after exposure. Male codling moth at 1, 24, 48, and 72 h after exposure placed near calling virgin females exhibited significant behavioral differences from sham-treated mates in courtship and mating. These results clarify some of the possible mechanisms, and strengths and weaknesses of the attract-and-kill management strategy. Mitchell (2002) evaluated a promising new technology-a novel, new matrix system for delivering sex pheromone- to suppress sexual communication in diamondback moth. The slow-release formulation contained a combination of diamondback moth pheromone, and the insecticide permethrin. Field trapping experiments showed that the lure-toxicant combination was highly attractive to male moths for at least four weeks and the insecticide had no adverse influence on response of the moths. The results indicated that the slow release formulation effectively suppressed this pest in cabbage and other crucifers, although as many as three applications probably would be required for suppression over an entire growing season.

Will attract-and-kill technology be potent against insecticide-resistant insects? Pouliot *et al.* (2001) tested this hypothesis on insecticide resistant codling moths using a commercial formulation. The resistant strains were significantly less affected than the susceptible strain by contact with the attracticide. No sub lethal effects in terms of pheromone response, mating or fecundity occurred in moths surviving contact with the attracticide. The resistance background of the target pests will influence choice of insecticides in attracticide formulations. Principles of insecticide resistance management may also be applied to attract-and-kill technology by alternating with other insecticides or control methods. Insecticide resistance may, however, limit the effectiveness and even the attractiveness of such formulations where resistance pleiotropic effects influence pheromone perception.

Lure and infect - Auto-infection system

Instead of applying a pathogen artificially in an ecosystem if the insects can themselves be made the carrier, it would save a lot of efforts and the inoculum would get applied automatically. The advantages of such a system are: 1) Attracted insects are efficiently autoinfected with the pathogen 2) only the target species is infected by this system 3) the cost of this system is low as compared with conventional field sprays 4) once attracted and infected, insects can be expected to return to their habitat, which accelerates dispersal of the fungus. Hartfield *et al.* (2001) tested two types of pheromone traps, modified as delivery systems for entomopathogens, against aphids in plum orchards using an isolate of the pathogen *Verticillium lecanii.* Out of a sample of 16 live aphids removed from a trap dispensing V *lecanii,* 9 died from infection by the pathogen and 5 of the 16 initiated colonies of the fungus after they walked on a sterile agar plate for five minutes. Yasuda (1999) developed an autoinfection system consisting of a modified sex pheromone trap and a bottle with exit holes containing conidia of the fungus *Beauveria bassiana* (9.3 × 1010/l g medium) and tested it to control the sweet potato weevil, C. *formicarius.* Male weevils were attracted to the system by the pheromone and exited the bottle after infection. In the sweet potato field where the systems were placed, the highest mortality of males by the disease was 96.2% on 21 days after treatment and that of females was 33.0% on 49 days. Rates of fungus carrier were 57.9% in males and 31.6% in females. In the treated field, the sex ratio of the SPW population was greatly skewed to females. Females seemed to be infected through mating with fungus carrier males.

Genetically engineered crops - Bt cotton

During the last few years genetically engineered crops have made entry into the agroecosystem and Bt cotton is the most prominent amongst them. How the entry of Bt cotton has impacted pheromone use can be seen from the following example-An early and most successful development of use of pheromones for mating disruption was achieved with pink bollworm but the availability of Bt-transgenic cotton has led to a dramatic decrease in the use of pheromones against it (Jenkins 2002). In contrast, the area treated by mating disruption against codling motha pest having no genetically engineered host- increased from ca. 1000 ha in 1991 to 9000 ha in 1996 in the United States. What looked like a rather linear increase in 1996 then turned into an almost exponential increase - the area under mating disruption expanded to 45000 ha in 2000 (Doerr and Brunner 2001, Brunner *et al.* 2002). The utility of pheromone traps in monitoring however, remains unchallenged even in Bt cotton. Micinski (2001) demonstrated unequivocally the importance of bollworm control in Bt cotton during periods of high bollworm presence by using pheromone traps.

Pheromone communication modification - insecticide resistance

Whether there can be a modification of the sex-pheromone communication system in insecticide resistance strains? This was examined by El-Sayed *et al.* (2001) in the oblique banded leaf roller. The pheromone glands and pheromone effluvium of azinphosmethyl-resistant females was approximately one-half as much as that of susceptible females. The rate of capture of marked and released susceptible males was greater than that of resistant males in traps baited with susceptible females, but not in traps baited with resistant females or in traps baited with synthetic pheromone. The reduced ability to locate virgin females suggests that the presence of resistant males in an apple orchard may result in a reduction in the capture of moths in pheromone-baited traps. The differences observed in the pheromone communication systems of susceptible and resistant *C. rosaceana* are likely pleiotropic effects associated with the selection for insecticide resistance.

Resistance to mating disruption

Whether insects will develop resistance to mating disruption or will the females exhibit altered behaviour due to expose to pheromone, is a question, which has no unambiguous answer as yet. Rahalkar *et al.* (1985) showed that when *Trogoderma granarium* Everts (Coleoptera: Dermestidae) males were subjected to a selection pressure of pheromone for several generations, the threshold concentration of pheromone required for attraction increased as the generations progressed. Shani and Clearwater (2001) studied the effect of high dosages of pheromone applied for mating disruption on sex pheromone production of Ephestia cautelia. A higher pheromone titer was observed in the disrupted population relative to that in the control population and this may represent an evasion of the selection pressure caused by mating disruption. Mochizuki *et al.* (2002) have reported that in tea fields where a mating disruptant had been used for controlling *Adoxophyes honmai* (Yasuda) (Lepidoptera: Tortricidae) annually since 1983, 15 years after the treatment was started, the disruption percentage of pheromone trap catches became less than 50%. When the disruptant was applied in other tea fields where it had not been previously used, the disruption percentage was over 99%. These results strongly suggest the possibility that a selection pressure caused by continuous use of the disruptant might have caused the low disruption percentage in the field continuously treated with a mating disruptant.

Future outlook

There is data to support both- positive and negative projections.

For practical utility, pheromones will have to be competitive in cost and efficacy to other methods of control. Failures of mating disruption at high population densities, high costs, and the occurrence of other insects, which can only be controlled with insecticides, are presently serious handicaps.

However it is also apparent that there is great potential in pheromone-based methods, once the methodology has been developed properly, which constitutes the knowledge of the pheromone chemistry and biology of a given species, economic synthesis, appropriate dispensers and the knowledge and experience of how to use them in the field. One of the successful examples of the use of mating disruption technique in Europe is against codling moth on >10.000 ha and the grape berry moth on >30.000 ha (Arn and Louis 1996, Waidner 1997, Kast 2001, Zingg 2001) and the figures are encouraging. One of the most important factors in this success is the determination of the growers' association in Northern Italy, to reduce the use of insecticides for obvious reasons and to replace it with mating disruption (Waldner, 1997). The most heartening aspect of this example is that- 'next to insecticides the farmers believe that pheromones will control insects'.

The current work on attracticides is also promising (Charmillot and Hofer 1997; Suckling and Brockerhoff, 1999). Attracticide lures can perhaps also be used for egg-laying females. A volatile compound from pear has recently been shown to attract codling moth males and females (Light *et al.* 2001). Semiochemical-based methods directed at females are expected to become a most important complement to mating disruption. Further, more emphasis should be placed on the integration of different biological control methods like microbial pesticides and resistant plants and the available biological tools should be developed as components of an integrated crop management strategy, rather than as sole agents. Pheromone will have a bright immediate future under such an IPM strategy.

REFERENCES

Anderbrant, O., Lofqvist, J., Hogberg, H.E., Hedenstrom, E., Baldassari, N., Baronio, P., Kolmakova, G., Lyons, B., Naito, T., Odinokov, V., Simandi. J., Supatashviii, A., Tai, A. and Tourianov. 2000 Geographic variation in the field response of male European pine sawflies, *Neodiprion sertifer,* to different pheromone stereoisomers and esters. *Entomologia Experimentali,v et Applicata* 95: 229-239.

Arn, H. and Louis, F. 1996 Mating disruption in European vineyards, in: Pheromone research new directions, Card6, R.T. and Minks, A.K. (eds.), pp. 377-382. Chapman and Hall, New York.

Arn, H., Toth, M. and Priesner, E. 1986 List of sex pheromone of Lepidoptera and related attractants. *IOBC WPRS.*

Baker, T.C., Mafra, N.A., Dittl, T. and Rice, M.E. 1997 A novel controlled release device for disrupting sex pheromone communication in moths. *IOBC wprs Bulletin* 20: 141-149.

Berger, H.K., Baufeld, P., Pajmon, A., Reynauld, P., Sivicek, P., Ulvee, C.C. and Urek, G. 2002 1998-1999 detection survey results of *Diabrotica virgifera virgifera* LeConte in non-infested European countries. *Acta Phytopathologica et Entomologica* 37: 183-192.

Brunner, J., Welter, S., Calkins, C., Hilton, R., Beers, E., Dunley, J., Unruh, T., Knight, A., Van, S. R. and Van, B. P. 2002 Mating disruption of codling moth: a perspective from the western united states. *IOBC wprs Bulletin* 25, (in press).

Charmillot, P.J. and Hofer, D. 1997 Control of codling moth, *Cydia pomonella* L., by an attract and kill formulation. *IOBC wprs Bulletin* 20: 139-140.

Cork, A., Alam, S. N., Rouf, F. M.A. and Talekar, N.S. 2003 Female sex pheromone of brinj'al fruit and shoot borer, *Leucinodes orbonalis* (Lepidoptera: Pyralidae): Trap optimization and application in IPM trials. *Bulletin of Entomological Research* 93: 107-113.

Del, T.R., Martinez, J.L., Ocete, R. and Ocete. M.E. 2001 Study of the relationship between sex pheromone trap catches of *Lobesia botrana (Den.* and Schiff.) (Lep., Tortricidae) and the accumulation of degree-days in Sherry vineyards (SW of Spain). *J. of Applied Ent.* 125: 9-14.

Doerr, M. and Brunner, J. 2001 Update of codling moth mating disruption. Washington State University, Tree Fruit Research and Extension Centre. http:/entomology.tfrec.wsu.edu/jfbhome/growerarticles/ CMMD200I.pdf

Doud, C.W. and Phillips, T.W. 2000 Activity of *Plodia interpunctella* (Lepidoptera: Pyralidae) in and around flour mills. *J. of Economic Ent.* 93: 1842-1847.

Downham, M.C.A., Smit, N.E.J.M., Laboke, P.O., Hall, D.R. and Odongo, B. 2001 Reduction of pre-harvest infestations of African sweet potato weevils *Cylas brunneus and C. puncticollis* (Coleoptera: Apionidae) using a pheromone mating-disruption *technique. Crop-Protection* 20: 163-166.

Ebenebe, A.A., van den Berg. J., van der Linde and T.C. de K. 2000 Seasonal flight activity of the maize stalk borer, *Busseolafusca* (Fuller) (Lepidoptera: Noctuidae), in Lesotho. *African Ent.* 8: 63-68.

El-Sayed, A.M., Fraser, H.M. and Trimble, R.M. 2001 Modification of the sex-pheromone communication system associated with organophosphorus-insecticide resistance in the obliquebanded leafroller (Lepidoptera: Tortricidae). *Canadian Ent.* 133: 867-881.

Evenden, M.L., Judd, G.J.R. and Borden, J.H. 2000 Investigations of mechanisms of pheromone communication disruption of *Choristoneura rosaceana* (Harris) (Lepidoptera: Tortricidae) in a wind tunnel. *J. Insect Behavior* 13: 499-510.

Grafton, C.E.E., Millar, J.G., Connell, N.V. and Hanks, L.M. 2000 Sex pheromone of yellow *scale, Aonidiella citrina* (Homoptera: Diaspididae): Evaluation as an IPM tactic. *J. Agricultural and Urban Ent.* 17: 75-88.

Gronning, E.K., Borchert,D.M., Pfeiffer,D.G., Felland, C.M., Walgenbach, J.F., Hull, L.A. and Killian J.C. 2000 Effect of specific and generic sex attractant blends on pheromone trap captures of four leafroller species in mid-Atlantic apple orchards., *J. Eco. Ent.* 93: 157-164.

Groot, A.T. and Visser, J.H. 2001 Influence of host plants on sexual communication in the herbivorous bug *Lygocoris pabulinus. Chemoecology* 11: 161-166.

Guerin, P. M., Arn, H., Busar, H. R. and Charmillot, P. J. 1986 Sex pheromone of Adoxophyes orana: additional components and variability in ratio of Z-9 and Z-1 I tetradecenyl acetate. *J. Chcmical Ecology* 12: 763-772.

Hall, D.R., Nesbitt B.F., Marrs, G.J., Green, A.S.J., Campion, D.G. and Critchley, B.R. 1982 Development of microencapsulated pheromone formulations, In: *Insect pheromone technology: chemistry and applications*. ACS Symp. Ser. 190, Washington DC., pp. 131-143.

Hartfield, C.M., Campbell, C.A.M., Hardie, J., Pickett, J.A. and Wadhams, L.J. 2001 Pheromone traps for the dissemination of an entomopathogen by the damson-hop aphid *Phorodon humuli. Biocontrol Sci. & Tech.* 11: 401-41 0.

Jenkins, J.W. 2002 Use of mating disruption in cotton in North and South America. *IOBC wprs Bull.*, 25 (in press).

Johansson, B.G., Anderbrant, O. and Sierpinski, A. 2002 Multispecies trapping of six pests of scots pine in Sweden and Poland. *J. Applied Ent.* 126: 212-216.

Kast, W.K. 2001 Twelve years of practical experience using mating disruption against *Eupoecilia ambiguella and Lobesia botrana* in vineyards of the Wuerttemberg region, Germany. IOBC *wprs Bulletin* 24:71-73.

Kawazu, K., Hasegawa, J., Honda, H., Ishikawa, Y., Wakamura, S., Sugie, H., Kamiwada, H., Kamimuro, T., Yoshiyasu, Y. and Tatsuki, S. 2000 Identification of pheromone components in Japan. *Entomologia Experimentalis el Applicata* 96: 103-109.

Kawazu, K., Nagata, K., Zhang, Z., Sugie, H. and Tatsuki, S. 2002 Comparison of attractiveness in Japan and China of three synthetic pheromone blends based on geographic variations in the rice leaffolder, *Cnaphalocrocis medinalis* (Lepidoptera: Pyralidae). *Bull. of Ent. Res.* 92: 295-299.

Kawazu, K., Setokuchi, O., Kohno, K., Takahashi, K., Yoshiyasu, Y. and Tatsuki, S. 2001 Sex pheromone of the rice leaffolder moth, *Cnaphalocrocis medinalis* (Lepidoptera: Crambidae): Synthetic Indian and Philippine blends are not attractive to male *C. medinalis,* but are attractive to *C. pilosa* in the South-Western islands in Japan. *Applied Ent. & Zool.* 36: 471-474.

Khoo. C.C.H., Yuen, K.H. and Tan, K.H. 2000 Attraction of female *Bactrocera papayae* to sex pheromone components with two different release devices. *J. of Chemical Eco.* 26: 2487-2496.

Kindl. J., Kalinova. B., Freise, J., Heitland, W., Augustin, S., Guichard, S., Avtzis, N. and Svatos, A. 2002 Monitoring the population dynamics of the horse chestnut leafminer *Cameraria ohridella* with a synthetic pheromone in Europe. *Plant Prot. Sci.* 38: 131-138.

Knight, A.L. 2000 Monitoring codling moth (Lepidoptera: Tortricidae with passive interception traps in sex pheromone-treated apple orchards. *J. Econ. Ent.*, 93: 1744-1751.

Knight, A.L., Croft, B.A. and Bloem, K. 1999 Effect of mating disruption dispenser placement on trap performance for monitoring codling moth (Lepidoptera: Tortricidae). *J. the Entomol.Soc. of British Columbia* 96: 95-102.

Koch, U.T., Farbert , P., Termer, A., Sauer, A., Milli, R., Karg, G. and de Kramer, J.J. 1992 Pheromone density measurements by EAG method in vineyards. *IOBC wprs Bulletin,* 15, 145.

Koch, U.T., LUder, W., Clemenz, S. and Cichon, L.I. 1997 Pheromone measurements by field EAG in apple orchards. *IOBC wprs Bulletin* 20: 181-190.

Krishnaiah, K., Varma, N.R.G., Pasalu, I.C. and Zainulabeuddin, S. 1998 Improvement in the efficacy of sex pheromone of yellow stem borer *Scirpophaga incertulas* Walker. with coattractants and plant extracts. *Indian J. Plant Protection* 26: 131-137.

Krupke, C.H., Roitberg, B.D. and Judd, G.J.R. 2002 Field and laboratory responses of male codling moth (Lepidoptera: Tortricidae to a pheromone-based attract-and-kill strategy. *Environmental Entom.* 31: 189-197.

Light, D.M., Knight, A.L., Henrick, C.A., Rajapaska, D., Lingren, B., Dickens, J.C., Reynolds, K.M., Buttery, R.G., Merrill, G., Roitman, J. and Campbell, B.C. 2001 A pear-derived kairomone with pheromonal potency that attracts male and female codling moth, *Cydia pomonella (L.). Naturwissenschafien* 88, 333-338.

Loesel, P.M., Potting, R.P.J., Ebbinghaus, D. and Scherkenbeck, J.2002 Factors affecting the field performance of an attracticide against the codling moth *Cydia pomonella. Pest Management Sci.* 58: 1029-1037.

Loesel, P.M., Penners, G., Potting. R.P.J., Ebbinghaus, D., Elbert, A. and Scherkenbeck, J. 2000 Laboratory and field experiments towards the development of an attract and kill strategy for the control of the codling moth, *Cydia pomonella. J. Entomologia Experimentalis et Applicata,* 95: 39-46.

Mbata, G.N., Shu, S., and Ramaswamy, S.B. 2000 Sex pheromones of *Callosobruchus subinnotatus and C. maculatus* (Coleoptera: Bruchidae): Congeneric responses and role of air movement. *Bull. Ento. Res.* 90: 147-154.

Micinski, S. 2001 Relationship between bollworm (Lepidoptera: Noctuidae pheromone trap catches and yield differences in sprayed and nonsprayed *Bt* cotton. *Southwestern Entomologi,vt,* 26: 137-142.

Millar, J.G., Daane, K.M., McElfresh, J.S., Moreira, J.A., Malakar, K.R. and Guillen, M. and Bentley, W. 2002 Development and optimization of methods for using sex pheromone for monitoring the mealybug *Planococcus ficus* (Homoptera: Pseudococcidae) in California vineyards. *J. Eco. Ent.* 95: 706-714.

Mitchell, E.R. 2002 Promising new technology for managing diamondback moth (Lepidoptera: Plutellidae) in cabbage with pheromone. *J. Environmental Science and Health. Part B. Pesticides Food Contaminants and Agricultural Wastes* B37: 277-290.

Mochizuki, F., Fukumoto, T., Noguchi, H., Sugie, H., Morimoto, T. and Ohtani, K. 2002 Resistance to a mating disruptant composed of (Z)- I I -tetradecenyl acetate in the smaller tea tortrix, *Adoxophyes honmai* (Yasuda) (Lepidoptera: Tortricidae). *Applied Ent. and Zoo.* 37: 299-304.

Moriya, S. and Miyatake T. 2001 Eradication programs of two sweet potato pests, *Cylas formicarius and Euscepes postfasciatus,* in Japan with special reference to their dispersal ability. *JARQ* 35: 227-234.

Ogawa, K. 1997 The key to success in mating disruption. *IOBC wprs Bulletin* 20: 1-9.

Polavarapu, S., Lonergan, G., Peng, H. and Neilsen, K. 2001 Potential for mating disruption of *Spar-ganothis sulfureana* (Lepidoptera: Tortricidae) in cranberries. *J. Eco. Ent.* 94: 658-665.

Polavarapu, S., Wicki M., Vogel, K., Lonergan, G. and Nielsen, K. 2002 Disruption of sexual communication of oriental beetles (Coleoptera: Scarabaeldae) with a microencapsulated formulation of sex pheromone components in blueberries and ornamental nurseries. *Environmental Ent.* 31: 1268-1275.

Pouliot, D., Beslay, D., Bouvier J.C. and Sauphanor, B. 2001 Is attract-and-kill technology potent against insecticide-resistant Lepidoptera? *Pest Management Sci.* 57: 729-736.

Rahalkar G.W., Tamhankar A.J. and Gothi K.K. 1985 Selective breeding for reduced male response to female sex pheromone in *Trogoderma granarium Everts. J. Stored Product Res. 21*:123-126.

Reddy, G.V.P. and Guerrero, A. 2001 Optimum timing of insecticide applications against diamondback moth *Plutella xylostella* in cole crops using threshold catches in sex pheromone traps. *Pest Management Sci.* 57: 90-94.

Reddy, G.V.P., Holopainen, J. K. and Guerrero, A. 2002 Olfactory responses of *Plutella xylostella* natural enemies to host pheromone, larval frass, and green leaf cabbage volatiles. *J. Chemical Eco.* 28: 131-143.

Renou, M., Berthier, A. and Guerrero, A. 2002 Disruption of responses to pheromone by (Z)I 1-hexadecenyl trifluoromethyl ketone, an analogue of the pheromone, in the cabbage armyworm *Mamestra brassicae. Pest Management Sci.* 58: 839-844.

Schmidt, B.C., Roland, J., Wakarchuk, D. 2003 Evaluation of synthetic pheromones for monitoring forest tent caterpillar (Lepidoptera: Lasiocampidae) populations. *Environmental Ent.* 32: 214-219.

Schroeder, P.C., Shelton, A.M., Ferguson, C. S., Hoffmann, M. P. and Petzoldt, C.H. 2000 Application of synthetic sex pheromone for management of diamondback moth, *Plutella xylostella,* in cabbage. *Entomologia Experimentalis et Applicata* 94: 243-248.

Shani, A. and Clearwater, J. 2001 Evasion of mating disruption in *Ephestia cautella* (Walker) by increased pheromone production relative to that of undisrupted populations. *J. Stored Products Res.* 37: 237-252.

Shorey, H.H. and Gerber, R.G. 1996a Use of puffers for disruption of sex pheromone communication of codling moths (Lepidoptera: Tortricidae) in walnut orchards. *Environmental Ent.* 25: 1398-1400.

Shorey, H.H., Sisk, C.B. and Gerber, R.G. 1996b Widely separated pheromone release sites for disruption of pheromone communication in two species of Lepidoptera. *Environmental Ent.* 25: 446-45 1.

Suckling, D.M. and Brockerhoff, E.G. 1999 Control of light brown apple moth (Lepidoptera: Tortricidae) using an attracticide. *J. Economic Ent.* 92: 367-372.

Suckling, D.M., Gibb, A.R., Burnip, G.M. and Delury, N.C. 2002 Can parasitoid sex pheromones help in insect biocontrol? A case study of codling moth (Lepidoptera: Tortricidae) and its parasitoid *Ascogaster quadridentata* (Hymenoptera: *Braconidae). Environmental Ent.* 31: 947-952.

Svensson, G.P., Valeur, P.G., Reynolds, D.R., Smith, A.D., Riley, J.R., Baker T.C., Poppy, G.M. and Lofstedt, C. 2001 Mating disruption in *Agrotis segetum* monitored by harmonic radar. *Entomologia Experimentalis et Applicata* 101: 111-1 2 1.

Tamhankar, A J., Rajendran, T P. and Mamdapur, V.R. 2001 Evaluation of a pheromone trap for Cotton Pink bollworm *Pectinophora gossypiella Saunders. International J. Pest Management* 47: 79-80

Tamhankar, A. J. 1994 Influence of host-presence, age and mating on the pheromone release behaviour of female *Earias vittella* (Fabricius) (Lepidoptera: *Noctuidae). Insect Science and its Application* 15: 275-280.

Tamhankar. A. J., Rajendran, T.P., Hariprasad Rao, N., Lavekar, R.C. Jeyakumar, P., Monga, D. and Bambawale, O.M. 2003 Variability in response of *Helicoverpa armigera* males from different locations in India to varying blends of female sex pheromone suggests male sex pheromone response polymorphism. *Current Sci.* 84: 448-450

Varner, M., Lucin, R., Mattedi, L. and Forno, F. 2001 Experience with mating disruption technique to control grape berry moth, *Lobesia botrana,* in Trentino. *JOBC wprs Bull.* 24: 81-88.

Visalakshmi, V., Rao, P.A. and Krishnayya, P.V. 2000 Utility of sex pheromone for monitoring *Heliothi.5 armigera* (Hub.) infesting sunflower. *J. Ent. Res.* 24: 255-258.

Waidner, W. 1997 Three years of large-scale control of codling moth by mating disruption in the South Tyrol, Italy. *IOBC wprs Bull.* 20: 35-44.

Weatherston, 1. 1990 Principles of design of controlled-release formulations, pp.93-112 in Ridgway, R.L., Silverstein, R.M., Inscoe, M.N. (eds.) Behavior-modifying chemicals for insect management: applications of pheromones and other attractants. *Marcel Dekker,* New York. pp 55.

Weissling, T. J. and Knight, A.L. 1994 Passive trap for monitoring codling moth (Lepidoptera: Tortricidae) flight activity. *J. Eco. Ent.* 87: 103-107.

Witzgall, P., Bengtsson, M., El Sayed, A., Biickman, A.C., Rauscher, S., Borg, K.A.K., Uneli-us, C. R. and Lifqvist, J. 1999 Chemical communication in codling moth: towards environmentally safe control methods. *IOBC wprs Bulletin* 22: 57-65.

Xiao, C., Gregg, P.C., Hu, W., Yang, Z. and Zhang, Z. 2002 Attraction of the cotton bollworm, *Helicoverpa armigera* (Hubner) (Lepidoptera: Noctuidae), to volatiles from wilted leaves of a non-host plant, *Pterocarya stenoptera. Applied Ent. and Zoo.* 37: 1-6.

Yamamoto, A. and Ogawa, K. 1989 Chemistry and commercial production of pheromones and other behaviour modifying chemicals, in: *Insect pheromones in plant protection,* Jutsum, A.R. and Gordon, R.F.S. (eds.), John Wiley, London, pp. 123-148.

Yasuda, K. 1999 Auto-infection system for the sweet potato weevil, *Cylas formicarius* (Fabricius) (Coleoptera: Curculionidae) with entomopathogenic fungi, *Beauveria bassiana* using a modified sex pheromone trap in the field. *Applied Ent. and Zoology* 34: 501-505.

Zhang, G. F., Yan, X. H, Sheng, C.F. and Meng, X.Z. 2001 Mating disruption of the Chinese tortrix *Cydia trasias* (Lepidoptera: Olethreutidae) by atmospheric permeation with synthetic sex pheromone. *Environmental Ent.*, 30: 7-1 1.

Zhang, G.F., Meng, X.Z., Han, Y. and Sheng, C.F. 2002 Chinese tortrix *Cydia trasias* (Lepidoptera: Olethreutidae): Suppression on street planting trees by mass trapping with sex pheromone traps. *Environmental Ent.* 31: 602-607.

Zhou, X. and Applebaum, C. M. 2000 Overwintering and spring migration in the bollworm *Helicoverpa armigera* (Lepidoptera: Noctuidae) in Israel. *Environmental Ent.* 29: 1289-1294.

Zingg, D. 2001 Mating disruption in Switzerland. *IOBC wprs Bull.* 24: 65-69.

3

Status, Scope and Issues in Effective Utilization of *Bt* Transgenic Cotton in India

AMERIKA SINGH, O.M. BAMBAWALE, O.P. SHARMA AND R.K. TANWAR

The insecticidal bacterium *Bacillus thuringiensis* (*Bt*) is the most commercially successful biological control agent of insect pests (Glare and O'Callaghan, 2000). Tools of biotechnology have overcome the limitations associated with the use of conventional *Bt* insecticides. In the past few years more than 30 plant species have been transformed by using a range of *Bt* genes (Schuler *et al.*, 1998; de Maagd *et al.*, 1999). Currently three *Bt* crops (cotton, corn and potato) are under cultivation in countries such as USA, China, South Africa, Australia, Argentina, Mexico, Indonesia and India. Approximately 12 million hectares of insect protected transgenic crops incorporating *Bt* insecticidal crystal proteins (ICPs) are now planted annually worldwide (American Academy of Microbiology, 2002) with an expected increase of 10% or more since 2000. *Bt* cotton is genetically enhanced to resist major caterpillar pests, including the three major bollworms, *Helicoverpa armigera*, *Earias insulána* / *E. vitella*, and *Pectinophora gossypiella.*

On a worldwide basis, cotton is grown on >32 million ha with approximately 71 % of the production in developing countries (Food Agric. Organ. UN (FAO) 2000). Conventional cotton receives the most insecticide use of any crop worldwide. Of the $ US 8.1 billion spent annually on all insecticides worldwide, it has been estimated that nearly $ 2.7 billion could be substituted with *Bt* biotechnology products (Krattiger, 1997). Insects targeted for control by *Bt* plants are primarily Lepidoptera. Reduction in the use of broad-spectrum insecticides on *Bt* cotton would likely result in conservation of natural enemies and non-target organisms, decreased potential of soil and water contamination, and benefits to farm workers and others likely to come into contact with these insecticides.

While transgenic *Bt* cotton is useful in several ways and is likely to be adopted at a high pace, the technology remains controversial due to misconceptions among individuals/opposing organizations and inadequate information on various facets of *Bt* technology when in use in the farmers' fields. There have been set backs to the *Bt* technology in India in the first year (2002) of commercialization. The 2002 cotton season was marked by scanty rains/uneven distribution and under these condition the *Bt* cotton did not perform to the expectation of the farmers as per media reports. Debate has arisen on implications and utility. Concerns are mainly about the proneness of the technology to other biotic and abiotic stresses, likely harm and side effects to natural enemies and non-target organisms including human health, development of resistance in populations of the target insects and possible ecological consequences of gene flow to non-engineered crops and wild relatives. In countries where *Bt* cotton is at present widely grown, published data has been reviewed (Shelton *et al.*, 2002; American Academy of Microbiology, 2002), which shows that the adoption of GM technology can lead to a marked reduction in pesticide use and improved control of target pests with minimal risks to the ecosystem. Many of the short and long-term implications have been predicted based on past and ongoing experiences, but much remains to be learned in the Indian

National Centre for Integrated Pest Management, LBS Bldg, IARI campus, Pusa, New Delhi – 110 012.

context where *Bt* cotton has been recently introduced and the diversity of cotton cultivation poses an uphill task for proper and effective utilization of *Bt* transgenic technology.

Bt AS A BIOPESTICIDE

Bacillus thuringiensis (*Bt*) is an ubiquitous gram +ve soil bacterium isolated from soil, stored grain, insect cadavers and the phylloplane (Martin and Travers, 1989). Thus, three prevailing niches of *Bt* can be envisaged viz., as an entomopathogen, as a phylloplane inhabitant and as a soil microorganism. *Bt* is recognized by its parasporal body (known as the crystal) that is proteinaceous in nature and possesses insecticidal properties. Recent information about *Bt* proteins/genes is available at http://www.biols.susx.ac.uk/Home /Neil Crickmore/*Bt*/. Considerable amount of information with respect to various aspects of *Bt* such as fermentation (Bryant, 1994), biology and genetics (Aronson *et al.*, 1986), molecular biology (Hofte and Whiteley, 1989; Schnepf *et al.*, 1998), mechanism of action (Knowles, 1994), application as biopesticide (Federicii, 1999), and insecticidal proteins and their application in agriculture (Kumar and Bambawale, 2002) is available.

Over 60,000 isolates of *Bt* are available in culture collections worldwide with 67 H-serotypes and eight are non-flagellated (Zeigler, 1999). Known *Bt* toxins kill subsets of insects among the Lepidoptera, Coleoptera, Diptera (Hofte and Whiteley, 1989), and nematodes (Feitelson *et al.*, 1992). By virtue of the lack of toxicity toward other species of animals, human beings, and plants, there is tremendous potential for exploiting *Bt* as a biological control agent (Federici, 1999).

Bt is the most popular biological control agent with nearly 200 registered *Bt* products (Schnepf *et al.*, 1998). The commercial *Bt* formulations mostly consist of the viable spores and parasporal crystals produced during sporulation process. The parasporal crystals when ingested by the insect larvae feeding on plant surfaces sprayed with *Bt* formulation dissolve and insecticidal crystal proteins (ICPs) are activated by proteases in the juices of the midgut, which typically are alkaline (pH 8-10.5). The active ICP then binds to specific receptors on the midgut epithelium, forming pores and leading to loss of the transmembrane potential, cell lysis, leakage of the midgut contents, paralysis, and death of the insect. Insects that develop resistance to *Bt* most commonly exhibit decreased or altered receptor binding, although altered proteolytic activation also has been reported. The use of conventional *Bt* insecticides was found to have limitations like narrow specificity, short shelf life, low potency, lack of systemic activity, low persistence and uniformity of coverage and the presence of viable spores (Lambert and Peferoen, 1992). As a result of these limitations the *Bt* biopesticide sales have not picked up and comprise only 1-2 per cent of the global insecticide spray market.

Bt IN THE TRANSGENIC CROPS

The limitations of *Bt* as a biopesticide are now overcome by various approaches that utilize the tools of molecular biology and genetic engineering expressing *Bt* gene in the plants. The first *Bt*-transgenic plants were made in 1987 (Barton *et al.*, 1987). More than 30 plant species have now been transformed by using a range of modified *Bt* genes (Schuler *et al.*, 1998; de Maagd *et al.*, 1999). In *Bt* crops, by virtue of the presence of *Cry* gene, the plant tissues produce specific ICPs in a soluble form. *Bt* crops currently are engineered to produce a single insecticidal crystal protein (ICP) continuously throughout all parts of the plant. Considerable amount of information with respect to *Bt* transgenic plants (Schuler *et al.*, 1998; de Maagd *et al.*, 1999) is available.

The most widely deployed *Bt* transgenic crops are corn, cotton and potato. Many additional *Bt* crops are close to release, including rice, canola, and various vegetables and fruits. Monsanto markets *Cry1Ac* cotton, Event 531, under different trade names in several countries. Since 2002 there are nine countries where *Bt* cotton has been commercialized: Argentina, Australia, China, Colombia, India, Indonesia, Mexico, South Africa and USA). An improved *Bt* cotton with two genes (Bollgard II®) event 15985 was developed (Rahn *et al.*, 2001) that encode protein from two genes: *Cry2Ab* and *Cry 1Ac*. The dual gene cultivars are expected to provide broader control over a wide variety of insects including *Spodoptera* spp. Bollgard II was approved for use in Australia in Sept 2002 covering an area of about 5000 ha. In India field trials of Bollgard II are underway. In 2002 Dow AgroSciences announced development of new *Bt* cotton, which contained the dual genes *Cry 1Ac* and *Cry1F*. Syngenta also plans to release cotton with a novel VIP (Vegetative Insecticidal Protein) gene in the US in 2004.

Bt biotechnological research and developmental work was initiated in the country in the 1990's. The National Agricultural Research System (NARS) in India has developed three varieties of *Bt* cotton. These are in the process of rigorous lab and pothouse testing and would be available for commercial cultivation in another three years (Anon., 2002). Pental (personal communication with P.A. Kumar) has synthesized new *cry1Ac* gene, which would be utilized in the future Public Sector GE programmes of the country. Bambawale (1995) developed a collection of > 100 native *Bt* strains. Out of these, five strains, found to be novel, were characterized at molecular level and the location of the *CryI* and *Cry II* genes effective in *H. armigera* control were identified (Bambawale, 1995, Bambawale *et al.*, 1997, Das *et al.* 1998). The isolate of *Bt* strain AAG1 (Autoagglutinating strain of *Bt*, Bambawale, 1995) produced cryIIa crystal protein, which is known to be toxic to many lepidopterans and dipterans. The *cryIIa* gene was isolated from the new isolate, AAG1 and the 2.269 kb fragment containing 1.9 kb coding region of the *CryIIa* was cloned as a BamH1 fragment in pUC18 which produced the 65 kDa toxin. The cloned gene has differences of 4 nucleotides with the published data of the gene sequence and these differences resulted in alteration of 3 amino acids in the deduced peptide chain (Das, 1999). Such genes can be directly utilized without any modification in insertions involving chloroplast transformation procedures.

BIOSAFETY OF BT

Safety and benefits of *Bt* as a biopesticide

Bt has a long history of use in spray formulations and is generally considered safe insecticide because it is highly specific. Glare and O'Callaghan (2000) exhaustively reviewed the literature on the biosafety of *Bt*, with a view to assess the environmental and health impacts possible from broadcast application of various strains in formulated products. Exceptionally few problems associated with human health have been reported and proven cases of *Bt* causing clinical disease in humans remain extremely rare, despite large scale use of *Bt* for over 60 years. Unlike some insecticides that accumulate in biological systems, there are no data to indicate that *Bt* proteins do so. Studies examining the impact of *Bt* formulations in aquatic environments have failed to show adverse effects.

Toxicity tests on a standard set of organisms, including avian and aquatic species, beneficial insects, soil invertebrates, and mammals are required before pesticide registration, providing a basis for assessing potential toxicity to non-target species. Fact sheet summarizing results from the testing of commercial *Bt* crops can be found at: www.epa.gov/pesticides/biopesticides /factsheets.

In a number of studies, *Bt* was shown to be least toxic to non-target organisms among a range of insecticides. An IOBC/WPRS working group conducted a series of bioassays of 40 pesticides, including *Bt* (Dipel), on beneficial arthropods (Hassan *et al.,* 1983). Dipel rated "harmless" to 8 of 9 test insects, which made it the least harmful of the 20 insecticides tested. Further tests (Hassan , 1992) also rated *Bt* as a safer biopesticide.

Remarkably, *Bt* is a pathogen almost entirely of herbivores, with virtually no carnivorous hosts, resulting in little effect on predator species, which contribute to natural insect control. Biocontrol through insect parasitoids or fungal pathogens, mating disruption, crop rotation, adjustment in date of planting, polyculture, and conventional chemical sprays, are all compatible with *Bt* use. In particular, indigenous insect populations and introduced biocontrol agents that are harmed by broad spectrum chemical pesticides are not generally harmed by *Bt*. *Bt* is thus well suited to IPM programs that preserve large segments of indigenous invertebrate predator and parasitoid populations. While competition for hosts and reduced host food quality can indirectly retard development or cause premature death of mature stages of predators and parasitoids, *Bt* has been shown to be useful in IPM systems in most situations. Biotic community may sometimes be affected indirectly because of perturbations in insect populations.

Safety and benefits of *Bt* transgenic crops

Food safety issues involve the direct potential toxicity and allergenicity of proteins, changes in the nutritional composition of plants, and the safety of the antibiotic resistance marker genes used during the process of genetic engineering. A testament to its safety is that *Bt* is the only insecticide for which there are no mandated tolerance limits for residue in food. *Bt* proteins in all *Bt* plants break down rapidly, do not resemble any known food allergen or protein toxin and have no oral toxicity. In a survey of 283 cotton farmers in China, Pray *et al.* (2001) reported that "farmers using *Bt* cotton reported less pesticide poisonings than those using conventional cotton." Farmers in China currently growing *Bt* cotton applied 80% less insecticides than those growing conventional cotton varieties. The farmers growing *Bt* cotton also reported more than four-fold fewer instances of symptoms, such as headache, nausea, skin pain, or digestive problems from applying pesticides. Small farmers in developing countries are at particular risk of exposure to pesticides because they often apply pesticides by hand, without safety precautions, such as respirators, goggles and gloves. Such farmers would therefore benefit most by use of transgenic cotton. It also has been shown that farm workers do not develop respiratory, coetaneous, or eye diseases from exposure to large amounts of *Bt* in sprays. They do develop antibodies to ICPs , but in no case has the presence of these antibodies been linked to acute or chronic disease.

In some cases, *Bt* crops can improve food safety. For example, approximately 59 % of corn grown globally is contaminated with fumonisins, fungal toxins produced by the genus *Fusarium,* that can cause liver and kidney damage in many species, and are probable human carcinogens. These fungal pathogens enter the plant through wounds caused by boring insects that are among the most important insect pests of corn worldwide. *Bt* corn is protected against damage from corn borers and consistently has 90% less fumonisin than conventional plants (Munkvold *et al.*, 1999). Thus, protection against insect damage and subsequent fungal infection may have important health implications for consumers and farm animals exposed to fumonisins in their diet.

ENVIRONMENTAL IMPACT OF *Bt* COTTON

In recent years significant progress has been made in the area of *Bt* deployment in transgenic crops and their cultivation, which has aroused concerns about environmental impacts. The major benefits of this system are economic, environmental, and qualitative. In addition to the reduced input costs to the farmer, the transgenic plants provide season-long protection independent of weather conditions, effective control of burrowing insects difficult to reach with sprays, and control at all of the stages of insect development.

Field studies of transgenic crops have failed to demonstrate significant detrimental impacts of *Bt* on related non-target organisms or their predators. Because chemical insecticides are generally used less frequently on *Bt* crops, it is likely that such crops will benefit non-target populations, especially those of parasitoids and predators that control insect pests. Studies in the U.S.(Head *et al.*, 2001), China (Pray *et al.*, 2001) and elsewhere have documented larger populations of predatory bugs, spiders, and ants, and enhanced biodiversity of beneficial insects in *Bt* cotton fields as compared with conventional fields treated with chemical insecticides.

A more complete assessment of the environmental impact of *Bt* plants would include the fate of *Bt* proteins in the environment, their direct and indirect impact on natural enemies and non-target organisms, the frequency with which pollen from *Bt* plants fertilizes other plants, and horizontal gene transfer in which the plant genes may move into other organisms. EPA, as the agency charged with regulating the use of *Bt* plants in the US, has served as the clearinghouse for studies examining these various aspects. Additional synthesis studies have been published by Snow and Palma (1997), Traynor and Westwood (1999),

Wolfenbarger and Phifer (2000) and WHO (2000). The National Research Council's Standing Committee on Biotechnology, Food and Fiber Production and the Environment sponsored a workshop in July 2000 on ecological monitoring of GE plants to discuss what is known and what need to be done (Natl. Acad. Sci., 2000).

Work on some of the important and relevant aspects of environmental impact of the *Bt* transgenic technology is reviewed here.

Outcrossing and geneflow between closely related plant species

Gene transfer between closely related species of plants occurs via pollen under natural conditions all the time. Hybrids of conventional crops and wild relatives form in 12 of the 13 most common crops worldwide in some part of their agricultural range, and this process has been implicated in enhanced weediness about half of the time. Gene flow of *Bt* ICP genes from *Bt* crops to wild relatives could occur where sexually compatible wild relatives are within the pollination range of the crop. Ecological consequences could result if the gene confers resistance to a pest that is a significant challenge to the wild relative.

EPA reviewed the potential for gene capture and expression of *Bt* endotoxins by wild or weedy relatives of corn, cotton and potatoes and concluded that there is "not a reasonable possibility" of passing their traits to wild relatives because of differences in chromosome number, phenology, and habitat (US EPA, 2000). The only exception is cotton in Florida and Hawaii, where feral populations exist of related *Gossypium* species, and EPA has prohibited or restricted the use of cotton in these areas.

Outcrossing and gene transfer to soil organisms

Because *Cry* genes originate from soil bacteria, it is important to evaluate the potential for gene transfer from GE plants to other organisms, primarily bacteria, as well as soil organisms. Crop residue, pollen, or root exudates may be potential sources of DNA in the soil. These numbers vary considerably based on soil characteristics. DNA from crop plants can remain in the soil from several months to several years (Gebhard and Smalla, 1999) if protected from soil nucleases, but the amounts are small (US EPA, 2000). Additionally, even with much higher concentrations of DNA, "transformation of bacteria with plant transgenes has only been accomplished at low frequencies and under optimum conditions," and therefore "DNA transfer occurs rarely if at all from plants to bacteria" (US EPA, 2000).

Although it is theoretically possible for microorganisms to incorporate DNA from transgenic plants, this has only been accomplished under specifically optimized laboratory conditions and at a very low frequency. The probability that such transfers will occur in nature, and involve functional genes is even lower. There is no evidence that genes from non-engineered crops have been transferred to microorganisms and there is no basis to suggest that transgenes would transfer more readily than any other genes. Microbes are so much more likely to obtain genes, such as those for antibiotic resistance from other microbes in the environment that the probability for such genes to be acquired from genetically engineered plant DNA is insignificant.

Impact on soil organisms

Bt is a natural inhabitant of soil and therefore, soil organisms have a long evolutionary association with *Bt*. Laboratory studies have shown that actively growing *Bt* plants can increase the levels of ICPs in soil, and that soil can bind active *Bt* proteins for extended periods of time. Available data indicate that half-life of cry proteins incorporated into the soil in corn plants, the most extensively studied, is 1.6-22 days but can be as long as 46 days, although the amounts will be small (US EPA, 2000).

Exposure to soil microorganisms and invertebrates can result from crop residues as well as from the roots themselves during and after the growing season. Whereas most of the concern of *Bt* plants has focused on crop residues, actively growing plants can increase the level of Cry1Ab protein in the soil. Saxena *et al.* (1999) found Cry1Ab protein in the exudates of 13 *Bt* corn hybrids, representing three transformation events. They note that the toxin could accumulate in the soil during plant growth as well as crop residue stage. To assess the effects of Cry1Ab toxin released in the root and from biomass on soil organisms, they introduced earthworms into soil grown to *Bt* and non-*Bt* corn or amended with biomass of *Bt* or non-*Bt* corn. Although the protein was present in the casts and guts of worms in the *Bt* treatments, there were no significant differences in mortality or weight of the earthworms, nor in the "total numbers of nematodes and culturable protozoa, bacteria (including actinomytcetes), and fungi between rhizosphere soil of *Bt* and non-*Bt* corn or between soils amended with *Bt* or non-*Bt* biomass" (Saxena and Stotzsky, 2001). In another report (Escher *et al.*, 2000) , foliage of Cry1Ab corn was compared with foliage of the corresponding nontransgenic maize variety in laboratory feeding experiments to study the effects of the *Bt* protein on the decomposer *Porcellio scaber* and on leaf litter-colonizing microorganisms. The studies could not detect significant differences in populations of this single decomposer.

EPA (US EPA, 2000) concluded the findings to indicate that *Bt* soils "show no effect of total biomass, bacteria, actinomyces, fungi, protozoa or nematodes.. (and) the C/N ratio is not changed." The EPA report concluded by noting that "sufficient data exist to suggest that adverse effects of currently commercialized *Bt* Cry1Ab and Cry1Ac proteins in the soil are not likely, although the levels of expression in the root, where not currently available, should be determined to assure that unexpectedly high levels of root expression are not found." Although the data to date do not indicate striking problems with *Bt* proteins in the soil, they point out the difficulty in working in a complex soil system. Studies often focus on single organisms under specific environmental conditions and over an often-short period of time. Under these conditions the power to test for differences is relatively low, and longer-term and more complex studies are needed to ensure the integrity of important soil organisms.

Impact on other nontarget organisms

Prior to the registration of the first *Bt* crop in 1995 (Event 176 corn), EPA evaluated potential effects of *Bt* endotoxins on a series of nontarget organisms including birds, fish, honey bees, ladybugs, parasitic wasps, lacewings, springtails, aquatic invertebrates, and earthworms (US EPA, 1995). Organisms were chosen as indicators of potential adverse effects when these crops are used in the field. These studies consisted of published reports as well as company reports. An extensive collection of reports can be seen electronically (Ammann, 1999). From this review of existing data, EPA concluded that there were "no unreasonable adverse effects to humans, nontarget organisms, or to the environment…" (US EPA, 1995).

Since registration of *Bt* transgenic crops, there have been two major debates related to their likely impact on nontarget organisms. The first report, indicating that consumption of corn pollen affects the development and mortality of lacewing larvae (Hilbeck *et al.*, 1998a; Hilbeck *et al.*, 1998b; Hilbeck *et al.*, 1999) has galvanized discussion focusing on the compatibility of *Bt* plants and biological control. The report indicated increased mortality and prolonged development when lacewing larvae were reared on either ECB, *Ostrinia nubilalis* or *Spodoptera littoralis* that had ingested corn leaves expressing Cry1Ab. As noted by EPA (US EPA, 2000), the "experimental design did not permit a distinction between a direct effect due to the *Bt* protein on the predator versus an indirect effect of consuming a sub-optimal diet consisting of sick or dying prey that had succumbed to the *Bt* protein." Hilbeck *et al.* (1998a) noted that ECB will be unlikely hosts for lacewing larvae in *Bt* fields because ECB "will almost completely be eradicated" by the *Bt* plants. Although Hilbeck *et al.* (1998a) state that "no conclusion can be drawn at this point as to how results from our laboratory trials might translate in the field," they recommend that such tritrophic effects be studied but concluded that *Bt* transgenic plants "are still more environmentally friendly than most if not all chemical insecticides." Interactions in the laboratory, although dramatic, may not be realistic in the field. Similarly, testing only a single factor in isolation in the laboratory may not produce the subtle effects that may arise in the field.

The second report is on likely impact of ICP on monarch butterfly, *Danus plexippus*. Losey *et al.*

(1999) reported that *Bt* corn pollen may become hazardous to monarch butterfly larvae. The study consisted of depositing an unspecified amount of corn pollen from N4640 (a *Bt* hybrid producing Cry1Ab protein) onto milkweed leaves and placing three-day-old larvae on the plants, recording leaf consumption, larval survival , and final larval weight over a four-day period. The authors found lower survival of larvae feeding on leaves dusted with *Bt* pollen compared with leaves dusted with untransformed pollen or on control leaves with no pollen. Based on these laboratory data, the authors hypothesized that there could be "potentially profound implications for the conservation of monarch butterflies" with widespread use of *Bt* corn. The report was criticized for its inappropriate design, methodology, and interpretation (Hodgson, 1999), but had worldwide repercussions (Shelton and Roush, 1999) primarily with regard to biosafety issues.

INSECT RESISTANCE TO *Bt*

Past experience has shown that insects have developed resistance to many organic insecticides and it can be assumed that resistance to *Bt* in transgenic crops will also develop eventually. Numerous *Bt*-resistant populations were developed in the laboratory (Tabashnik, 1998). However, only one insect species (Diamondback moth; *P. xylostella*) developed resistance under natural conditons following sprays of *Bt* products (Shelton *et al.*, 2000). Many possible mechanisms can be envisaged to explain resistance development in insects (Frutos *et al.*, 1999). The first is the activation of the protoxins by gut proteinases. The proteolytic processing of protoxins into toxins is necessary for the insecticidal activity of *Bt* formulations. It is also required by some transgenic plants expressing combinations of *Bt* toxins as fusion proteins or toxins extending beyond the protease activation site. The most frequently observed mechanism of resistance among insect pests is the modification of the receptor site (Tang *et al.*, 1996). This mechanism was studied extensively on a large range of insect species and toxins, and most of these studies pointed out a change in the level of affinity of the receptor for the toxin or to a decreased number of receptor sites (Frutos *et al.*, 1999). The third mechanism of resistance is not related to reduced binding or altered protein activation and corresponds to a still unknown mechanism or perhaps to several unknown mechanisms, and was reported for the first time in *H. virescens* (Gould *et al.*, 1992). A detailed account of stability and genetics of resistance was given by Frutos *et al.* (1999).

Bt Resistance Management strategies

Many strategies to manage resistance to *Bt* toxins were proposed and reviewed (Gould, 1998; Frutos *et al.*, 1999). Field data and experimental evidence are now available to help designing resistance management strategies (Roush, 1997; Shelton *et al.*, 2000). These strategies fall into several categories:

Tissue- or time-specific expression of toxins
Gene stacking/pyramiding
Mixtures, rotation or mosaics of transgenic plants
Combination of toxins with different modes of action
Refuge strategy
"Trap plants" strategy

Each *Bt*-based product (sprays or transgenic plants) has the potential to trigger evolution of resistance and it is essential that the end users, authorities, and industry will be important players in the successful implementation of resistance management and therefore sustainability of transgenic products (Kennedy and Whalon, 1995). There is also a need for training and information and extension workers will have to play a major role in the successful implementation of resistance management. To attempt to delay the development of *Bt* resistance, federal regulations in the US allow growers to choose one of the following three *Bt*-growing plans (Hardee *et al.*, 2001):

- Plant a maximum of 80 per cent *Bt* cotton along with 20 per cent non-*Bt*, with the option of spraying the latter as needed with anything but foliar *Bt* insecticides. (currently followed in India).

- Plant a maximum of 95 per cent *Bt* cotton along with 5 per cent non-*Bt*, but the latter cannot be sprayed with any insecticides.
- Plant a maximum of 95 per cent *Bt* cotton along with 5 per cent non-*Bt*, with the latter planted in the same field with *Bt* and sprayed only if the 95 per cent is sprayed with non-*Bt* foliar insecticides.

Such plans would have varying impacts with regard to the population densities of pests, natural enemies and the soil microflora and fauna.

CONSERVATION BIOLOGY

Natural habitats in the vicinity of croplands can be quite important in providing alternative foods and refugia for natural enemy populations. However, trends in agriculture over the past several decades include decreasing landscape heterogeneity, increasing agrochemical inputs, increasing mechanization, and decreasing genetic diversity (Letourbeau, 1998). In conservation biology jargon, the landscape mosaic has been transformed over time from one that featured "disturbance patches" (i.e., agricultural fields in a sea of natural habitats) to the one that features "remnanat patches" (i.e., remnants of natural habitats in a sea of agricultural enterprises). For the Central zone cotton growing region of India, the habitat is still more natural with plenty of diversity available compared to low diversity and high levels of mechanization in the North Zone cotton conditions. The habitat would have to be basically considered while studying the impact of the *Bt* technology.

In particular, natural enemies require hosts or prey, sources of nectar and/or free water, and refugia. Thus, to the extent that noncrop habitats act as insular refugia for natural enemies, which then colonize or visit agricultural fields and inflict mortality on target pests, the size, design, abundance, and location of noncrop habitats are relevant to conservation biological control.

Conservation biologists have acknowledged that all species in a community are not equal in their contributions to community structure and processes, and not all interactions are vital to the integrity of the ecosystem. Some species are particularly important because they interact with and affect many other species, process materials disproportionately to their numbers, or because their functions cannot be substituted or compensated for by other species in the community (Letourbeau, 1998).

Direct and indirect effects of *Bt* technology on biocontrol agents and impact on food web interactions

Insect toxins can affect biological control agents when they feed on toxic plants or on intoxicated prey. More likely are effects on insect natural enemies through complex modification of the food web. Pests that are targeted by the resistance mechanism are typically reduced in number. A concomitant reduction in pesticide use, however, may result in increased number and diversity of other arthropods in the crop. These changes in availability of various arthropods could have either positive or negative impacts on biological control. Many insect parasitoids and predators have an intimate association with the plants upon which their insect hosts or prey feed (Barbosa and Letourneau, 1988). These plants provide not only a habitat but also behavioral cues that mediate searching behavior and a source of water or nutrition for some species which feed on floral or extrafloral nectar, pollen, or plant sap (Hagen, 1986). Plant resistance factors can potentially affect natural enemies directly through any of these avenues. Plant chemicals produced constitutively or in response to herbivore feeding play an important role in host searching behavior by many entomophagous arthropods (Vet and Dicke, 1992). There is no evidence to date that any plant traits altered through genetic engineering to increase pest resistance have directly affected plant attractiveness to natural enemies.

Indirect effects of resistance traits on natural enemies are those mediated through the phytophagous insects, which serve as hosts or prey. These effects can result from changes in vulnerability of phytophagous insects to natural enemies resulting from altered behavior or development rates on resistant plants or through changes in their suitability as hosts/prey resulting from intoxication or the sequestration of plant-produced toxins.

Resistant cultivars negatively affect natural enemies through reduced prey population density but affect them positively when pesticide use is reduced. The impact of genetically engineered cultivars on conservation biological control will likely vary according to the arthropod community in a particular cropping system and the changes in pesticide use permitted by the resistance trait (Hoy *et al.*, 1998). As the acreage grown to *Bt* transgenic crops increases, and population densities of targeted pests are reduced, other sources of hosts for natural enemies (particularly lepidopterans) will become more critical for maintaining natural enemy populations. In addition to weeds and other crops, any acreage of the target crop that is not planted with a toxin-producing cultivar and is not treated with broad-spectrum insecticides could boost natural enemy population densities.

A cornerstone of resistance management for transgenic crops is a refuge, generally thought of as acreage of a nontransgenic cultivar (or noncrop) that is grown so that there will be a source of toxin-susceptible genotypes of the pest insect to mate with resistant individuals. Because the objective is to produce susceptible pests, these refuges should be treated minimally or not at all with insecticides and so may also be a source of natural enemies that feed on the toxin susceptible pests. Nontransgenic refuges established near transgenic crops for resistance management will, therefore, contribute to the conservation and abundance of insect natural enemies. The area devoted to such refuges and the extent of their contribution to conservation biological control will depend on their contribution to suppressing resistance development.

DIVERSITY OF COTTON CULTIVATION AND ITS PEST PROBLEMS IN INDIA

Cotton is an important commercial crop in India occupying around 9 million hectares and supports 60 million people. The country's share of the cotton area is 27 percent of the 33 million ha in the world, topping the list of 80 countries. However, we compare poorly; in productivity we are one of the lowest with 293 Kg lint/ha compared to 609 Kg lint/ha world average.

In India cotton is grown under diverse agroecological conditions. All the four cultivated species and intra and inter specific hybrids are cultivated with no apparent systematic control over their deployment.

The Northern irrigated Zone comprises of Punjab, Haryana and parts of Rajasthan and Uttar Pradesh. Though insect pest and disease problems were low till the last decade, these are now threatening the cotton cultivation and there is a downward trend in terms of productivity. The major problems are American bollworm, spotted bollworm and Cotton Leaf Curl Virus.

The Central Zone primarily comprises rainfed areas of Maharashtra, Gujrat and Madhya Pradesh. These areas are repeatedly exposed to the vagaries and uncertainties of Monsoon pattern in terms of total precipitation and the distribution. Farmers have little control over executing sowing, interculture and crop protection (spraying) operations within the ideal time. The insect pest and disease scenario depends on the kind of weather conditions prevalent in a particular year and therefore, the zone is most vulnerable. The problems are compounded as the farmers in this zone are resource poor and are not free to execute the proper crop management practices. Cotton productivity is just around 100 kg lint/ha.

The Southern Zone consists of Andhra Pradesh, Tamil Nadu and Karnataka. Orissa is also having sizable area under cotton and has a potential for future increase in area under the crop. In this zone all the four cultivated species of *Gossypium* are grown. It is possible to grow cotton round the year due to diverse climatic conditions available. The major pest problem is American bollworm and some diseases like wilts and blights.

Around 125 varieties/hybrids have been released apart from cultivars promoted by > 50 pvt. seed companies to cater to the needs of such diverse climatic zones. The area under hybrids is estimated to be around 45 per cent in 2002. The multiplicity of varieties and hybrids due to their unregulated and unrestricted cultivation led to the occurrence of early, medium and late maturing groups contiguously, favoring serious bollworm outbreaks and resurgence of whitefly in many areas. This disturbs the whole plant protection planning. To cover sizable area under *Bt* cotton the *Bt* hybrids would have to be evolved for each of the agroecological zone and inbuilt host resistance for sucking pests and diseases as per the requirements of the zones would be an added advantage for the *Bt* hybrids.

The annual losses due to pests are estimated to the tune of $ 300 million and 21,500 metric tons active ingredient of insecticides with a total cost of $ 340 million was utilized on cotton in 2001 in the country.

The insect pest scenario on cotton changed over the years since independence, owing mainly to the kind of genetic material cultivated during those periods, diversity of cultivation, agronomic practices and plant protection interventions The American bollworm remains the number one pest on cotton in India. Successive cyclic national episodes of *H armigera* epizootics on cotton (1987, 1992, 1997 and recently in 2001) have shown that mitigation of this pest remains a complex and central issue to the overall sustainability of the cotton-based cropping system. The pest has built up widespread resistance against the available chemical insecticides (Armes *et al*; 1996). To overcome this resistance problem, various State Agri. Universities, ICAR institutes and State Agri. Depts advocate the principles of IPM. *Bt* cotton would form an important component of IPM.

COMMERCIALIZATION OF *Bt* COTTON IN INDIA AND IMMEDIATE MAJOR CONCERNS

On March 26, 2002, the Indian government allowed commercial use of genetically engineered crops by licensing *Bt* cotton and as a result 40,000 ha area has been planted in 2002-03 in central and southern zones under three *Bt* hybrids viz., MECH-12, MECH-162 and MECH-184. These hybrids have been developed by Mahyco (Maharashtra Hybrid Seed Company). MMBiotech Ltd., a joint venture of Mahyco and Monsanto has also sublicensed this technology to other seed companies such as Raasi Seeds, Krishidhan, Ankur Seeds, Ajeet Seeds and Emergent Genetics who will incorporate the gene in their own hybrids. Thus, a few more *Bt* transgenic cotton hybrids developed by the private sector, such as RCH 134, RCH 138, MRC-6301, MRC-6322, MRC-6201, MRC-6703, MRC-6304, MRC-915, Ankur 651 etc. are now under various stages of development and commercialization. In 2003 crop season supply of *Bt* cotton seed for covering 2,85,000 ha has been made. By 2005 around 25 per cent of the total area is likely to be covered by *Bt* cotton.

Initial setback to *Bt* Cotton

India is the only country where the technology has been introduced in hybrid cotton. The extent of uniformity of *Bt* toxin expression in the F1 population (hybrid *Bt* cotton) would depend on the purity of the parents and would have implications on various efficacy and ecological aspects consequences.

Background of *Bt* hybrids with regard to the susceptibility to pests and diseases other than bollworms and vulnerability to abiotic stress is important to know their behavior in different agro climatic conditions. Such background would influence the need for management of these problems and their influence on the whole gamut of cotton protection as well as environmental impact. In the very first year of its commercialization, MECH-162 *Bt* was found to be prone to "Para wilt", a genetically controlled physiological disorder which expresses when there is prolonged drought followed by downpour rain, and sizable area was reported to be affected by the malady in the central zone. MECH-12 is known to be highly sensitive to jassids whereas MECH-915 (now MRC-915) is susceptible to leaf curl. Some reports in the Indian press (Ghadyalpatil, 2002) have even indicated failure of the *Bt* crop (particularly, MECH-162) to protect from ABW. These aspects have raised additional doubts about the utility of the technology in the Indian context. Other *Bt* hybrids in the pipeline would also have to be critically evaluated.

Illegal *Bt* cotton

The greatest threat to the *Bt* cotton technology is the illegal variant of *Bt* cotton itself. The same *Cry1Ac* gene was clandestinely introduced in Navbharat-151, an established commercial hybrid developed by Navbharat Seeds of Gujrat. The seed of this illegal *Bt* cotton variant was sold for commercial cultivation in 2001 season in Gujrat. This marked a spectacular success for this hybrid in the wake of extreme American bollworm pestilence in that year. An account of the origin, methods of production and spread of illegal *Bt* seeds is available from a number of newspaper reports and business magazines (Latha Jishnu, 2003, Indian Express, May 29, 2003, Mumbai – Source: http://www.indianexpress.com/full story.php?content id = 24785, Financial express, May 30, 2003, online – Source: http://www.financialexpress.com/fe full story.php?content id = 35200, Indian Express, May 28, 2003, Online, Sonu Jain – Source: http://www.Indianexpress.com/full_story.php?content_id=24759). As per these reports the suspect illegal *Bt* brands available in the Gujrat markets are Maxi 151, BesT, *Bt* Cotton Trial Farm, Rakshak, Maharakshak, Krushi and Suraksha; and in Andhra Pradesh four illegal *Bt* brands – AUM –3, BB5, MI-20 and Thunderball – 4 are being sold in the rural market. The origin of all these brands is the Navbharat-151 *Bt* cotton. The illegal *Bt* is further spreading to other states such as Maharashtra, MP, Rajasthan, Haryana and Punjab. It is estimated that nearly half of Gujrat's area under cotton is covered by illegal *Bt* hybrids in 2003 season. Government on its part has been unable to contain the scourge. Farmers prefer illegal *Bt* mainly because the seed of these costs substantially less compared to approved *Bt* cotton.

The spread of illegal *Bt* is a damper for companies seriously going through stringent procedures laid down by the Genetic Engineering Approval Committee (GEAC). Legal safeguards may take their course and time for containing the illegal *Bt* but it may be too late to prevent the spread of too many variants of *Bt* and there are indications that in Gujrat nearly 90 percent area of the nearly 1.47 million ha cotton may come under the illegal *Bt* cotton in 2004 season. *Bt* cotton cultivation in India needs to be regulated through legal interventions and policy modification. Otherwise the situation may go out of hand very soon.

The consequences of growing illegal *Bt* would soon be visible. Though the first illegal *Bt* has been quite successful, it is not guaranteed that all other *Bt* hybrids would also be so. Whereas Navbharat-151 has been an established hybrid, the other *Bt* hybrids have been developed hurriedly without proper evaluations. Moreover, F1, F2 and even F3 seeds are all being foisted as *Bt* cotton as there is no control over sale of seeds of such cotton. Yield performance and *Bt* ICP expression of these hybrids would therefore always be in doubt.

A comparison is drawn between legal *Bt* (such as the first three *Bt* hybrids, MECH-12, MECH-162 and MECH-184) and the illegal *Bt* (Navbharat-151) while giving reasons for spread of illegal *Bt* like a wildfire. Navbharat-151 had performed well in 2001 (prior to commercial approval by GEAC) when the American bollworm infestation was high in Gujrat. In the first year of commercialization (2002), as per the press reports, the legal *Bt* cotton had performed poorly in some parts of Maharashtra and Gujrat owing mainly to the drought conditions. However, the On-Farm field trial evaluation of *Bt* cotton done by NCIPM in 2002 season (unpublished) in Nanded distt. has shown that MECH-162 could perform well when used as one of the critical components of IPM. In 2003 season also evaluation of the *Bt* hybrids MECH-184 and MECH-162 is under progress and it is seen from the first pickings so far made that MECH-184 is quite suitable for the central zone and the farmers of Nanded distt. (Villages Hotala, Naigaon, Palasgaon, Khairgaon and Navandi) are much convinced about the utility of MECH-184 (unpublished). Farmers in these villages are expecting 18-20 q/ha seed cotton yield from MECH-184, which is a record for the yields obtained by the farmers under rainfed situations particularly when there were two episodes of high incidence of *H. armigera* in the region. Quite a large area is likely to be brought under *Bt* cotton MECH-184 in 2004 season. There are a good number of legal *Bt* cotton hybrids in the pipeline, which have shown good performance at several locations in the country under field evaluation tests and highly productive hybrids would soon be available to the farmers once cleared for commercialization. We expect that farmers by and large would soon realize the utility of adopting the well-evaluated *Bt* hybrids and the dangers of uncertainty of using illegal *Bt* cotton.

Refugia requirements, pest related issues and economic considerations

Structured refugia may not be required in the immediate future in India, particularly in the central and south zones where sufficient crop diversity exists and where more than required refugia is available in the form of companion *H. armigera* host crops, weed hosts and conventional cotton. The current mandated 20 per refugia requirements are causing additional burden on the farmers for taking special plant protection care of the small area or else lose the yield substantially.

Bt cotton is mainly targeted to manage ABW and other bollworms. ABW appears in cyclic form every 4-5 years. In the intervening years, when the pest is in moderate to low intensity, the *Bt* cotton has no distinct advantage. We have to critically analyze the economics of growing *Bt* cotton in non-pest/ low pest years. To reap best harvest in the epidemic year we may have to grow *Bt* cotton for 4 more years. We don't have forecasting system for American Bollworm in place and so we cannot decide in which year to go for *Bt* cotton. There are several endemic and hot spot areas where ABW causes sizable losses every year. *Bt* cotton would be useful in such areas but we need to observe added precaution for the resistance management of the pest.

The future of *Bt* cotton

Bt cotton provides protection only from the lepidopteran pests and therefore it still needs to be managed for other insect pests and diseases. Farmers were seen spraying even *Bt* cotton for bollworm control in the past two years owing to their compulsive clinging to the old habit. Thus, two aspects of immediate concern are education of the farmers for scouting and monitoring and keeping a close watch on other sucking pests, diseases, abiotic stress, etc. We have noticed that *Bt* cotton in 2003 season had high levels of infestation of thrips in central zone. Also *Bt* cultivars showed high intensity of grey mildew disease and certain leaf spots. Some of the *Bt* cotton cultivars also showed water stress / stress of high retention of boll load, which might have led to premature drying of the leaves and wilt like appearance of the plants. Rainfed situation is particularly sensitive for getting best performance of *Bt* cotton and water management strategies also need to be evolved and incorporated into our ongoing efforts. Availability of best agronomic practices for *Bt* cotton is also lacking. Nutritional requirements of *Bt* cotton have also not been properly understood. We need to develop practical strategies through directed research on these aspects, identifying suitable *Bt* hybrids for different ecoregions.

Bt cotton provides a good foundation for IPM as it takes care of bollworms and it would be comparatively easier to manage the pest problems of cotton, provided the farmers follow the IPM principles.

Cropping system-based community approach to rainfed cotton IPM in India was successfully validated from 1998-2001 at village Ashta in Nanded district of Maharashtra (Singh *et al.*, 2002) and was found to be sustainable even in the face of *Helicoverpa* epidemic situation of 2001. The effort is an important landmark in the history of cotton IPM in the country and the concept is relevant for utilization with *Bt* cotton. An effort has already been initiated at NCIPM to develop IPM technology for *Bt* cotton, based on the Ashta model and the results indicate high performance of *Bt* cotton in an IPM mode during 2002 and 2003 crop seasons.

Bt cotton appears to hold a high promise under the Indian scenario in the management of bollworms and in improving seed cotton yield at competitive cost without causing any major detriment to the natural enemies.

REFERENCES

American Academy of Microbiology 2002. Report on a colloquium, " 100 years of Bacillus thuringiensis, a Paradigm for Producing Trasgenic Organisms: A Critical Scientific Assessment," sponsored by the American Academy of Microbiology and held November 16-18, 2001, in Ithaca, New York. www.asmusa.org

Ammann, K. 1999. Debate 991120b. Non-target organisms, literature survey Novartis. Klaus.ammann@ips.unibe.ch

Anonymous 2002. Indian scientists develop 'superior' *Bt* cotton seed. *Financial Express*, April 9, 2002.

Armes, N.J., D.R. Jadhav and K.R. Desouza 1996. A survey of insecticide resistance in *Helicoverpa armigera* in the Indian Subcontinent. *Bulletin of Entomological Research* 86: 499-514.

Aronson, A.I., W. Beckman and P. Dunn 1986. *Bacillus thuringiensis* and related insect pathogens. *Microbiol. Rev.* 50: 1-24.

Bambawale, O.M. 1995. Molecular characterization of Indian *Bacillus thuringiensis* strains for their insectidal proteins. Report on the SERC Visiting Fellowship of DST, Govt. of India from May-August, 1995 at Bose Institute, Calcutta. pp. 32.

Bambawale, O. M., D. Das, S.K. Sen, T.P. Rajendran, Sheo Raj and D. Basu 1997. Exploration of diversity in indigenous *Bacillus thuringiensis* strains and their molecular characterization. *Indian Phytopathological Society Golden Jubilee International Conference*, 1997, New Delhi, India

Barbosa, P. and D.K. Letourneau 1988. Novel aspects of insect-plant interactions. John Wiley and Sons, New York, NY.

Barton, K. A., H.R. Whiteley and N.S. Yang 1987. *Bacillus thuringiensis* d-endotoxin expressed in transgenic *Nicotiana tabacum* provides resistance to lepidopteran insects. *Plant Physiol.* 85: 1103-1109.

Bryant, J.E. 1994. Commercial production and formulation of *Bacillus thuringiensis*. *Agric. Eco. Environ.* 49: 31-35.

Das, D. 1999. Cloning and strategic transfer of a *Bacillus thuringiensis* ICP gene in plant to minimise chances for development of resistance by the target insect against ICP sensitivity. Ph.D thesis,1999, University of Kalyani, India.

Das, D., O.M. Bambawale, D. Basu, T.P. Rajendran, and S.K. Sen 1998. Isolation and characterization of *Bacillus thuringiensis* strains from soil samples. National Symposium on Microbial Biotechnology, Organized by IIT-BREF-Biotek, IIT, Kharagpur, India, Feb 16-17, 1998.

De Maagd, R. A., D. Bosch and W. Stiekema 1999. *Bacillus thuringensis* toxin-mediated insect resistance in plants. *Trends Plant Sci.* 4: 9-13.

Escher, N., B. Kach, and W. Nentwig 2000. Decomposition of transgebucBacillus thuringiensis maize by microorganisms and woodlice Porcellio scaber (Crustacea:Isopoda). *Basic Appl. Ecol.* 1: 161-69.

Federici, B.A. 1999. *Bacillus thuringiensis* in biological control, In: Handbook of Biological Control, Academic Press, New York, pp.575-593.

Feitelson, J. S., J. Payne and L. Kim 1992. *Bacillus thuringiensis*: insects and beyond. *BioTechnol.* 10:271-275.

Food Agric. Organ. U.N. (FAO) 2000. Agriculture data. http://apps.fao.org/page/ collections?subset=agriculture.

Frutos, R., C. Rang and M. Royer 1999. Managing Insect resistance to plants producing *Bacillus thuringiensis* toxins. *Crit. Rev. Biotech.* 19: 227-276.

Gebhard, F. and K. Smalla 1999. Monitoring field releases of genetically modified sugar beets for persistence of transgenic plant DNA and horizontal gene transfer. *FEMS Microbiol*. 28: 261-72.

Ghadyalpatil, A. 2002. Bollworm eats into *Bt* cotton's pride. *The Hitvada*, Nagpur,Oct. 10, 2002.

Glare, T. R., and M. O'Callaghan 2000. *Bacillus thuringiensis:* Biology, Ecology and Safety. John Wiley and Sons, Chichester, 350p.

Gould, F. 1998. Sustainability of transgenic insecticidal cultivars : integrating pest genetics and ecology. *Annu. Rev. Entomol*. 43: 701-726.

Gould, F., A. Martinez-Ramirez A. Anderson, J. Ferre, F.J. Silva and W.J. Moar 1992. Broad spectrum resistance to *Bacillus thuringiensis* toxins in *Heliothis virescens. Proc. Natl. Acad. Sci. USA.* 89: 7986-7990.

Hagen, K.S. 1986. Ecosystem analysis, plant cultivars (HPR), entompophagous species and food supplements. In: *Interaction of host plant resistance and parasitoids and predators of insects*. Eds. D.J. Boethel and R.D. Eikenbary, pp. 151-187, Ellis Horwood Ltd. Chichester, U.K.

Hardee, D.D., D.V. Sumerford, J.J. Jr. Adamczyk and L.C. Adams 2001. Protecting farmers' investment in *Bt* cotton. Agricultural Research Magazine. www.ars.usda.gov/is/ AR/archive/feb01/farm0201.htm

Hassan, S.A. 1992. Testing methodology and the concept of the IOBC/WPRS working group. Pesticides and Non-target invertebrates. Ed. Jepson. P, C., Intercept, Wimborne, Dorset, pp 1-18.

Hassan, S.A., F. Bigler, H. Bogenschutz, J.U. Brown, S.I. Firth, P. Huang, M.S. Ledieu, E. Naton, P.A. Oomen, W.P.J. Overmeer, W. Rieckmann, W.S. Peterson, G. Viggiani and A.Q. van Zon 1983. Results of the joint pesticide testing programme by the IOBC/WPRS-Working group "Pesticides and Beneficial Arthropods. *Z. Angew. Entomol*. 95: 151-158.

Head, G. B., W. Freeman, J. Moar, F. Ruberson and S. Turnipseed 2001. Natural enemy abundance in commercial bollgard and conventional cotton fields. In: *Proc. of the Beltwide Cotton Cconference*. National Cotton Council, Memphis. TN.

Hilbeck, A., M. Baumgartner, P.M. Freid, and F. Bigler 1998a. Effects of *Bacillus thuringiensis* corn-fed prey on mortality and development time of immature *Chrysoperla carnae. Environ. Entomol*. 27: 480-87.

Hilbeck, A., W.J. Moar, M. Pusztai-Carey, A.M. Filippini and F. Bigler 1998b. Toxicity of *Bacillus thuringiensis* Cry1Ab toxin to the predator *Chrysoperla carnae. Environ. Entomol*. 27: 1255-63.

Hilbeck, A., W.J. Moar, M. Pusztai-Carey, A.M. Filippini and F. Bigler 1999. Prey-mediated effects of Cry1Ab toxin and protoxin and Cry2A protoxin on the predator *Chrysoperla carnae. Entomol. Exp. Appl*. 91: 305-16.

Hodgson, J. 1999. Monarch *Bt* corn paper questioned. *Nat. Biotechnol*. 17: 627.

Hofte, H., and H.R. Whiteley 1989. Insecticidal crystal proteins of *Bacillus thuringiensis. Microbiol. Rev.* 53:242-255.

Hoy, C.W., F. Feldman, G.G. Gould, G. Kennedy, G. Reed and J.A. Wyman 1998. Naturally occurring biological controls in genetically engineered crops. pp.185-205. In: *Conservation biological control*, Ed. Barbosa, P. Academic Press, New York.

International Cotton Advisory Committee (ICAC) 2000. Report on expert panel on biotechnology in cotton. 1-17. (http://www.icac.org/icac/meetings/plenary/59cairns/biotech.pdf).

Krattiger, A.F. 1997. Insect resistance in crops: a case study of *Bacillus thuringiensis* (*Bt*) and its transfer to developing countries. ISAAA Briefs No. 2, p. 42. Ithaca, NY: ASAAA.

Kennedy, G.G., and M.E. Whalon 1995. Managing pest resistance to *Bacillus thuringiensis* endotoxins: constraints and incentives to implementation, *J. Econ. Entomol*. 88: 454-460.

Knowles, B.H. 1994. Mechanism of action of *Bacillus thuringiensis* insecticidal d-endotoxin. *Adv. Insect Physiol*. 24: 275-308.

Kumar, P.A. and O.M. Bambawale 2002. Insecticidal proteins of *Bacillus thuringiensis* and their application in agriculture. In : Advances in microbial toxin research and its biotechnological exploitation. Ed. R.K. Upadhyay. pp. 259-280. Kluwer Academic Publishers, USA.

Lambert, B. and M. Peferoen 1992. Insecticidal promise of *Bacillus thuringiensis*. Facts and mysteries about a successful biopesticide. *BioSci*. . 42: 112-122.

Latha Jishnu 2003. The *Bt* cotton wildfire. *Businessworld.* 8 Sept. 2003. 38-42.

Letourbeau, D.K. 1998. Conservtion biology: Lessons for conserving natural enemies. In Conservation biological control. Ed. P.D. Barbosa, pp.9-38. Academic Press.

Losey, J.E., L.S. Raynor and M.E. Carter 1999. Transgenic pollen harms monarch larvae. *Nature* 399: 214.

Martin, P. A. W., and R.S. Travers 1989. Worldwide abundance and distribution of *Bacillus thuringiensis* isolates. *Appl. Environ. Microbiol*. 55: 2437-2442.

Munkvold, G.P., R.L. Hellmich and L.G. Rice 1999. Comparison of fumonisin concentrations in kernels of transgenic *Bt* maize hybrids and non-transgenic hybrids. *Plant Dis*. 83: 130-138.

Natl. Acad. Sci. 2000. Ecological Monitoring of Genetically Modified Crops: A Workshop Summary, ed. R. Pool, J. Esnayra. Washington, DC; Board Agric. Nat. Resourc. 60 pp. http://books.nap.edu/nap-cgi/

newsrch.cgi?term=ecological+monitoring.

Pray, C.E., J.D. Ma Huang and F. Qiao 2001. Impact of *Bt* cotton in China. *World Dev.* 29: 813-825.

Rahn, P.R., L. Ruschke and Z.W. Shappley 2001. Efficacy and agronomic performance of Bollgard® II. Proceedings of the Beltwide Cotton Conference. Memphis, TN, USA. *National Cotton Council*. 2 : 832.

Roush, R. T. 1997. Managing risk of resistance in pests to insect-tolerant transgenic crops. In: Commercialization of transgenic crops: Risks, Benefits and Trade considerations. Eds. Waterhouse, P.M., Evans, G. and Gibbs, M.J. Cooperative Research Center for Plant science and Bureau of Statistics, Canberra, Australia, pp.259-271.

Saxena, D., S. Flores and G. Stotzsky 1999. Insecticidal toxin in root exudates from *Bt* corn. *Nature* 402: 480.

Saxena, D. and G. Stotzsky 2001. *Bacillus thuringiensis* (*Bt*) toxin released from root exudates and biomass of *Bt* corn has no apparent effect on earthworms, nematodes,protozoa, bacteria, and fungi in soil. *Soil Biol. Biochem.* 33: 1225-1230.

Schnepf, E., N. Crickmore, J. Van Rie, D. Lereclus, J. Baum, J. Feitelson, D.R. Zeigler and D.H. Dean 1998. *Bacillus thuringiensis* and its pesticidal crystal protein. *Microbiol. Mol. Biol. Rev.* 62: 775-806.

Schuler, T.H, G.M. Poppy and I. Denholm 1998. Insect-resistant transgenic plants. *Trends Biotech.* 16:168-175.

Shelton. A.M., D. Juliet, J. D. Tang, R.T. Roush, T.D. Metz and E.D. Earle 2000. Field tests on managing resistance to *Bt*-engineered plants. *Nature Biotech.* 18: 339-342.

Shelton, A.M. and R.T. Roush 1999. False reports and the ears of men. *Nat. Biotechnol.* 17: 832.

Shelton, A.M., J.Z. Zhao and R. T. Rousch 2002. Economic, ecological, food safety, and social consequences of the deployment of *Bt* transgenic plants. *Annu. Rev. Entomol.* 47: 845-881.

Singh, A., O.P. Sharma, R.C. Lavekar, O.M. Bambawale, K.S. Murthy and A. Dhandapani 2002. IPM technology for rainfed cotton. Technical Bulletin No. 11, NCIPM, New Delhi. pp. 36.

Snow, A.A. and P.M. Palma S. 1997. Commercialization of transgenic plants: potential ecological risks. *BioScience* 47: 86-96.

Tabashnik, B. E. 1998. Transgenic crops for the pacific basin: prospects and problems. In: Proceedings of the Australian Applied Entomology Research Conference Vol. 1. University of Queensland, Australia, pp. 161-161.

Tang, J. D., A.M. Shelton, J. van Rie, S. de Roeck, W.J. Moar, R.T. Roush and M. Peferoen 1996. Toxicity of *Bacillus thuringiensis* spore and crystal protein to resistant diamondback moth (*Plutella xylostella*). *Appl. Environ. Microbiol.* 62: 564-569.

Traynor, P.L. and J. H. Westwood Eds. 1999. Proc. Workshop Ecol. Effects Pest Resistance Genes Managed Ecosystems, Jan.31-Feb 3. Bethesda, MD: Blacksburg, VA: Inf. Syst. Biotechnol. 133 pp. http://www.isb.vt.edu.

US EPA 1995. Pesticide fact sheet for Bacillus thuringiensis subsp. Kurstaki CryI(A)b delta-endotoxin and the genetic material necessary for the production (plasmid vector pCIB4431) in corn. EPA Publ. EPA731-F-95-004.

US EPA 2000. Off. Pestic. Programs. Biopesticides, and Pollut. Prev. Div. Biopesticides registration document; preliminary risks and benefits sections: *Bacillus thuringiensis* plant pesticides. U.S. Environmental Protection agency, Washington, DC.

Vet. L.E.M. and M. Dicke 1992. Ecology of infochemical use by natural enemies in a tritrophic context. *Annu. Rev. Entomol.* 17: 141-172.

Wolfenbarger, L.L. and P.R. Phifer 2000. The ecological risks and benefits of genetically engineered plants. *Science* 290: 2088-93.

World Health Organization 2000. Safety aspects of genetically modified foods of plant origin. Report joint FAO/WHO expert consultation on foods derived from biotechnology, Geneva, Switzerland, 29 May- 2 June 2000. http://www.who.int/fsf/GMfood/FAO-WHO_cosultation_report_2000.pdf

Zeigler, D.R. 1999. *Bacillus thuringiensis* and *Bacillus cereus* Catalog of Strains, Bacillus Genetic Stock Center, Columbus, p. 56.

4

Current Trends in Use of Molecular and Novel Approaches in Plant Disease Management

P. K. CHAKRABARTY[1] AND O.M. BAMBAWALE[2]

Plants are constantly under attack by a variety of pathogens including bacteria, viruses, fungi, insects and nematodes. The percentage of crop losses caused by plant pathogens, insect pests and weeds worldwide has been estimated to 42%, accounting for $ 500 billion worth of damage (Oerke *et al.,* 1994). Out of this, global losses caused by pathogens alone are estimated to be 12 per cent of the potential crop production (FAO, 1993). Presently most of the control measures rely on the use of conventional methods of control including quarantine, soil sanitation, seed treatment, chemical pesticides and resistant varieties. The most economical and effective control is the host plant resistance. Breeding a new variety with improved resistance however, requires considerable time and efforts, particularly when the source of resistance is a poorly adapted wild species and requires a lengthy backcrossing / breeding programme to eliminate deleterious genes. Often, pathogens can develop virulence even before a new variety is released, leading to boom and bust cycles whereby varieties possessing effective resistance when grown under a vast area (boom) selects for the evolution of matching virulence which spreads within the pathogen population (bust) overcoming the resistance (Pink and Puddephat, 1999).

Recent survey has shown that the world population has almost doubled in past 50 years, while the food production has increased meagerly by 25% per head. To meet the food requirement for 1.5 billion populations by 2020, the production has to be raised by 30% (Anonymous, 2000). One possibility to achieve this demand is by striving to prevent 42% damage caused due to pests and diseases. The biotic losses of this magnitude occur despite the use of conventional management practices including resistant varieties and chemical control measures. Plant transformations through development of transgenic offers efficient means to introduce specific genes, determining disease resistance into crop species without the problem of introducing deleterious genes and the linkage drag that occurs in resistance breeding. Plant transformation thus has become a tool for crop improvement and lines expressing genes conferring resistance against viruses, insects, herbicides have all found commercial applications (Birch, 1997). Several exciting approaches and candidates are now available for developing designer plants capable of providing effective resistance against diseases. This paper presents the current perspectives of characterization of resistance genes or the effectors of resistance, their deployment for transgenic resistance, and future options for application of molecular and novel approaches for plant disease management.

Engineering universal resistance with genes from plant kingdom

First group of genes that have been exploited in agriculture from time immemorial through conventional breeding are the plants' own defence genes. Advances in molecular biology has made it possible to isolate

1. Central Institute for Cotton Research, Shankar Nagar, Nagpur-440 010.
2. National Centre for Integrated Pest Management, IARI, New Delhi- 110 012.

and clone these *R* genes and use them for engineering resistance against different groups of pathogens, for several exciting reasons mentioned in subsequent sections.

Cloning and Characterization of R genes

More than 30 genes have been cloned and characterized from different plant species both monocots and dicots, that provide resistance against a variety of different pathogens (Hulbert *et al.*, 2001). Structures of most of these genes are consistent with a role in pathogen recognition and defence response signaling because they encode protein with motifs found in receptor and signal transduction proteins. Such genes are abundant in plant genomes and most belong to tightly linked gene families. No function other than disease resistance has yet been ascribed to them. Various *R* genes cloned so far from different plant species have many structural and functional features like, NBS domain and C-terminal LRR (Penuela *et al.*, 2002) in common, regardless of the plants from which they were isolated or the pathogens against which they are effective viz., bacteria, fungi, viruses or nematodes, implying a common related mechanism of resistance (Table 1). Genetic analyses of plant pathogen interactions have revealed that resistance reactions in most of these cases are controlled by two dominant loci: an avirulence (*Avr*) gene in the pathogen (Chakrabarty and Raj 2002) and the corresponding resistance (*R*) gene (Hulbert *et al.*, 2001) in the plant. The interactions have been genetically defined by the gene-for-gene theory (Flor, 1971). It is envisaged that *R* gene interacts with specific *Avr* gene of the pathogen. *R* genes are either involved in detoxifying or neutralizing virulence factors produced by the pathogens or they recognize pathogen legends (Jones, 1996; Dangl 1995). Recognition of specific legends (possibly the products of *Avr* genes of pathogen) triggers a cascade of signalling, involving multiple signal transduction pathways that eventually lead to manifestation of resistance, characterized by hypersensitive response (HR) or by other means. HR involves the death of first cell or cells infected and the local accumulation of antimicrobial compounds. Many *R* genes cloned are components of signal transduction pathways. Isolation and characterization of *R* genes from various sources have indicated that these genes have remarkable structural similarities, indicating that plants evolve and use similar signal transduction pathways to trigger resistance against a large number of unrelated pathogens. The absence of either member of the gene pair results in the development of disease. The NBS and LRR motifs are known for protein-protein and protein-nucleic acid interaction between the host and pathogens.

R genes are the potential candidates for cloning in plants through genetic engineering. Most of them are cloned by map-based, positional cloning or T-DNA tagging. Using degenerate PCR primers directed towards conserved motifs of the *R* genes it is now possible to isolate resistance gene analogues (RGAs) and then test them for their association with *R* genes (Meyers *et al.*, 1999). Employing this approach RGAs have been amplified and cloned from a number of diverse crop species including corn (Collins *et al.*, 1998), flax (Dodds *et al.*, 2001), soybean (Graham *et al.*, 2000), citrus (Deng *et al.*, 2000), wheat and barley (Penuela *et al.*, 2002), lettuce (Shen *et al.*, 1998) and several other plants including cotton. RGAs amplified from plants species can be used as DNA probes to clone *R* genes from the BAC and cDNA libraries of plants. Using similar approach, RGAs were cloned from cotton possessing different bacterial blight resistance genes (Chakrabarty, unpublished). Twelve isogenic lines of cotton each having an individual blight resistance gene, viz., *B1, B2, B3, B4, B5, B6, B7, Bin3* were analysed for polymorphism of *R* genes. Four sets of degenerate primers based on conserved nocleotide binding motifs of *R* genes were designed to amplify resistance gene analogues from the isogenic lines including the bacterial blight immune IM216 parent line. Southern hybridisation of the isogenic lines with four RGA probes was done. Polymorphism was detected in case of *Bin3* line with only one of the four probes tested. Cloning and characterization of *R* genes having broad-spectrum resistance against bacterial blight of cotton is contemplated using a BAC library prepared from the resistant line.

Table 1. List of R genes characterized to govern pathogen resistance

Pathogen group	Genes	Host	Effective against	Protein motifs
Fungi	*L*	Linseed	*Melampsora lini*	TIR-NBS-LRR
	M	Linseed	*Melampsora lini*	TIR-NBS-LRR
	P	Linseed	*Melampsora lini*	TIR-NBS-LRR
	RPP1	Arabidopsis	*Peronospora parasitica*	TIR-NBS-LRR
	RPP5	Arabidopsis	*Peronospora parasitica*	TIR-NBS-LRR
	Dm3	Lettuce	*Bremia lactucae*	NBS-LRR
	I2	Tomato	*Fusarium*	NBS-LRR
	Mla	Barley	*Blumeria*	NBS-LRR
	Pib	Rice	*Magnaporthe grisea*	NBS-LRR
	Pi-ta	Rice	*Magnaporthe grisea*	NBS-LRR
	Rp1	Maize	*Puccinia*	NBS-LRR
	RPP8	Arabidopsis	*Peronospora parasitica*	NBS-LRR
	RPP13	Arabidopsis	*Peronospora parasitica*	NBS-LRR
	Cf-2/5	Tomato	*Cladosporium fulvum*	LRR-TM
	Cf-4/9	Tomato	*Cladosporium fulvum*	LRR-TM
	Rpw8	Arabidopsis	*Erysiphe*	Unique
	mlo	Barley	*Blumeria*	Membrane protein
	Hm1	Maize	*Cochliobolus*	HC-Toxin reductase
		Bean	*Colletotichum lindemuthianum*	NBS-LRR
Bacteria	*RPS4*	Arabidopsis	*Pseudomonas*	TIR-NBS-LRR
	Bs2	Pepper	*Xanthomonas*	NBS-LRR
	Prf	Tomato	*Pseudomonas*	NBS-LRR
	RPM1	Arabidopsis	*Pseudomonas*	NBS-LRR
	RPS5	Arabidopsis	*Pseudomonas*	NBS-LRR
	Xa1	Rice	*Xanthomonas*	NBS-LRR
	Pto	Tomato	*Pseudomonas*	Protein Kinase
	Xa21	Rice	*Xanthomonas*	LRR-TM-Kinase
Virus	*N*	Tobacco	*TMV*	TIR-NBS-LRR
	Rx1	Potato	*PVX*	NBS-LRR
	Rx2	Potato	*PVX*	NBS-LRR
	Sw-5	Tomato	*Tospovirus*	NBS-LRR
Nematode	*Gpa2*	Potato	*Globodera rostrochinesis*	NBS-LRR
	Mi	Tomato	*Meloidogyne*	NBS-LRR
	HS1	Beet	*Heterodera*	Unique

Deployment of cloned Resistance (R) genes through genetic engineering

Molecular characterization of several resistant genes from different plant species led to identification of several mechanistic classes and provides opportunity for transgenic disease control strategies for generating resistant plants having different specificities. Transgenic resistance thus can be deployed through inter-specific transfer of *R* genes in various plant species (Hulbert, 2001).

The first plant resistance gene that was cloned was the maize *Hm1* gene (Table 1) that conferred resistance against the leaf spot fungus *Cochliobolus carbonum*. Unlike most of the characterized *R* genes, which are implicated in signal transduction or defence signaling pathway, *Hm1* codes for an enzyme HC-toxin reductase that detoxifies the toxin made by race 1 of *C. carbonum* (Johal and Briggs 1992). This gene could be useful for engineering resistance in other crops affected by this pathogen or against other related toxin producing pathogens (Mitra, 2001). *Pto* is a serine-threonine kinase gene of tomato that confers resistance against *Pseudomonas syringae* pv. *syringae* strains containing *avrPto*, in gene-for-gene manner (Martin *et al.*, 1993). The transgenic tobacco containing *Pto* showed race-specific resistance against *P. syringae* pv. *tabaci* harbouring the avirulence gene *avrPto* (Rommens *et al.*, 1995). A barley *R* gene *mlo*, confers durable field resistance to powdery mildew fungus *Erysiphe graminis* fsp. *hordei* (Jorgenson 1992). This is an unusual gene where the allele conferring resistance is functionally recessive. The gene probably works through activation of host's inherent defence system. Most other *R* genes cloned recently possess NBS-LRR sequences, which are known for protein-protein and protein-DNA interactions and are the components of defense signaling cascades.

N gene governs resistance against TMV and possesses NBS-LRR motifs (Whitham *et al.*, 1994). In addition to these two conserved motifs, which it shares with most of the cloned *R* genes, *N* also possesses a TIR domain, homologous to the *Drosophila Toll* and mammalian *interleukin-1* receptor at N terminus. Several species of tobacco that possess *N* gene are resistant against TMV. Transformation of this gene in susceptible tobacco as well as tomato rendered the plants, resistant against TMV (Whitham *et al.*, 1996). Several other genes (Table 1) governing resistance against fungal, viral, bacterial and nematode pathogens share the same conserved motifs characteristics of *R* genes (Hulbert *et al.*, 2001; Tolin 2001; Creusot 1999).

Bacteria belonging to genus *Xanthomonas*, is a highly destructive pathogen as it can infect atleast 124 monocot and 268 dicot plant species, causing variety of diseases including leaf spots and streaks, wilts and blights, rots and cankers (Starr, 1981; Hayward, 1993). *Xa21* gene is the first *Xanthomonas R* gene that was characterized at the molecular level (Song *et al.*, 1995). Transgenic plants expressing the cloned *Xa21* gene conferred resistance to 31 strains of *X. oryzae* pv. *oryzae* collected from all over the world (Wang *et al.*, 1996). Several research groups in different countries have tried and introduced the cloned *Xa21* gene into elite rice varieties and evaluated its field performance (Zhang *et al.*, 1998; Tu *et al.*, 1998). *Xa21* transgenic rice proved excellent in terms of their yield and broad-spectrum resistance against different strains of *X. o.* pv. *oryzae*. Other cloned *R* genes against *Xanthomonas* include the rice gene *Xa1* conferring resistance to Japanese race 1 of *X. o.* pv. *oryzae* (Yoshimura *et al.*, 1998) and the pepper gene *Bs2* controlling resistance to strains of *X. campestris* pv. *vesicatoria* containing *avrBs2* (Tai *et al.*, 1999). *Bs2* functions both in pepper and tomato. These *R* genes of rice and pepper that provide resistance against several pathovars of Xanthomonas compestsis are potential candidates for cloning into citrus, cotton, beans and other crops which too are invaded by Xanthomonads possessing the highly homologous members of *Xanthomonas avr/pth* gene family (Chakrabarty and Raj 1992; Chakrabarty *et al.*, 1997, 2002; Gabriel 1999). Inter-specific transfer of *R* genes is expected to be effective because of several reasons.

I) Many plants *R* genes are known to interact with avirulence genes from different pathogens, inciting HR responses.

II) Several *R* genes have been demonstrated to function after transfer as transgenes to different but related plant species (Hulbert *et al.*, 2001). *Bs2* gene cloned from pepper (Tai *et al.*, 1999) can interact with several members of *Avr* gene family widespread amongst pathovars of *X. campestris* including pathovar *malvacearum* that infects cotton (Chakrabarty *et al.*, 1997).

III) *R* genes cloned from a number of plant species and effective against diverse array of pathogens including viruses, bacteria, fungi, nematodes (Baker *et al.*, 1997) show significant similarities in conservation of their structural motifs.

Merits of deployment of R genes

It is possible to pyramid resistance genes with each gene effective against different strains of a single pathogen or different pathogens. Transgenic approaches overcome two objections to pyramiding resistance genes by classical methods.

I. Unlike sequential introduction of multiple genes through traditional back crossing, several *R* genes can be introduced at one time through molecular approach. Such introduction of transgenes will allow different types of resistance or mechanistic classes of resistance genes to be combined.

II. Transfer of *R* genes by transformation is an efficient substitute to back crossing and will overcome the possible linkage drag of deleterious genes.

Despite significant theoretical feasibility, it remains to be seen if *R* gene with a known function in one species will have a role in different species, if the recipient does not share the same pathogen.

Deployment of novel genes other than R genes

R genes are known to be the initial players in signal transduction pathway of disease resistance. Their function is believed to intercept the pathogens possessing corresponding *Avr* genes and transcend signal to trigger plant's inherent defence system. In the absence of cognate *Avr* genes, pathogens go unrecognised leaving the plants unchallenged and defence responses untriggered and resulting in manifestation of disease. Since specific '*R*' genes in plants confer resistance to pathogen containing specific '*Avr*' genes, the resistance of the host plant can also be compromised if the pathogen sheds or mutates the cognate *Avr* gene (Hirt, 2002). Therefore, besides *R* genes, which act mostly in gene-for-gene specific manner, there are host of other defence genes and non-host *R* genes (Chisholm, 2000), which can confer general resistance that is independent of interaction with pathogen *Avr* genes. Transformation of plants with novel genes that act downstream *R* gene signal transduction pathway appear to be a potential strategy for developing disease resistant transgenic plants. Asai *et al.,* (2002) described flagellin, a mitogen activated protein kinase (MAPK) signaling cascade as a potential tool for engineering crops with broad-spectrum resistance against bacterial and fungal pathogens. Although the defence genes in plants and animals differ considerably, the signal transduction mechanism connecting the activation of host receptors to induction of defence responses shares similarities between the kingdoms. In both animal and plants, kinases of MAPK class are activated by pathogen factors. The MAPKs are the ideal intracellular mediators of information between cytoplasm and nucleus where they act upon their targets- the transcriptional factors that in turn trigger expression of defense genes/ responses. Transient over-expression of MAPKs and other related kinases conferred resistance to *Aradiopsis thaliana* leaves upon infection by the bacterial and fungal pathogens (Asai *et al.,* 2002). Expression of these novel proteins like MAPKs and other receptor like kinases in plants thus offers a challenging strategy for disease control.

Engineering broad-spectrum durable resistance in plants by over-expression of signalling components and defence response genes that act downstream of *R* genes (Rommens and Kishore, 2000) in signal transduction pathway can provide effective resistance. Plant intrinsic responses that can be engineered to attain a wider, more durable resistance include systemic acquired resistance (SAR) response. The potential candidate in this category is the *NPR1* gene, which manifests resistance in plants by regulation of SAR. *NPR1* is the first SAR-regulating gene cloned from *A. thaliana* and is also known as *NIM1* and *SA1* gene (Cao *et al.,* 1994; Delaney *et al.,* 1995; Glazebrook *et al.,* 1996; Shah *et al.,* 1997; Cao *et al.,* 1997; Ryals *et al.,* 1997). Mutations in this *Arabidopsis* gene resulted in plants that were no longer responsive to SAR induction. PR gene expression and disease resistance were both abolished in the *npr1* mutants. Over-expression of this gene in *Arabidopsis* increased resistance to virulent isolates of *P.syringae* and *Peronospora parasitica* in a dose-dependent manner (Cao *et al.,* 1998). Interestingly, over expression of NPR1 did not lead to constitutive PR gene expression prior to pathogen challenge, indicating that the NPR1 protein requires activation to be functional. This active-only-upon-induction characteristic of NPR1 makes it favourable target for engineering broad-spectrum disease resistance. Also no obvious detrimental effects to the plants like constitutive PR gene expression or spontaneous cell death occurs. The gene has been over-expressed successfully in wheat and rice and as such is an exciting candidate for development of disease resistant crop plants. Although a tangible approach to boost plant's resistance through deployment of various components of resistance in holistic manner is yet to be achieved, individual components of signal transduction pathways for resistance are being tried with different degree of success.

Engineering resistance against fungal pathogens using antimicrobial proteins

Genetic engineering of plants with traits like over-expression of anti-microbial proteins (e.g. chitinases, glucanases, etc.) or induction of key plant defence pathways through signalling molecules (e.g. salicyclic acid, jasmonic acid or ethylene) are other novel strategies for manipulation of resistance. Genes encoding hydrolytic enzymes such as chitinases, chitosanase and glucanase, which can degrade fungal cell wall components, have been specifically used for fungal disease resistant transgenic plants (Mitra, 2001). Broglie *et al.*, (1991) demonstrated for the first time that transgenic tobacco and canola plants expressing a chitinase gene were highly resistant to a root pathogen *Rhizoctonia solani* in green house tests. Chitin is a major component of fungal cell walls and degradation of chitin has been suggested to have protective effects against chitin containing, plant pathogenic fungi. Chitinases hydrolyze the *ß*-1, 4 linkage of the N-acetyl –D-glucosamine polymer chitin. Several chitinase genes have been cloned from insects and rice plants (Lin *et al.*, 1995). Transgenic rice plants constitutively expressing these chitinase genes showed a high degree of resistance to the sheath blight fungus, *R. solani*. Co-expression of chitinase and glucanase in transgenic crops heightened resistance to a number of fungal pathogens including species of *Alternaria* and *Cercospora* (Zhu *et al.*, 1994). This strategy will be more useful in case of the diseases caused by polyphagous pathogens (like species of *Rhizoctonia* and *Fusarium* as well as other pathogens that cause root rot and seedling diseases) against which the inherent resistance is scanty. Gene encoding a strong antifungal endochitinase from *Trichoderma harzianum* provided broad-spectrum fungal resistance in transformed tobacco and potato plants. Transgenic plants were highly resistant to *Alternaria alternata, A. solani, Botrytis cinerea* and *R. solani* (Lorito *et al.*, 1998). Apple scab caused by the fungus *Venturia inaequalis* is one of the greatest limitations to successful cultivation of this fruit in India as well as in the whole world. The pathogen drastically affects the productivity and quality of the affected fruits. Growers almost exclusively rely upon multiple applications of fungicides to control this disease during the growing season. Transgenic 'McIntosh' apple trees expressing either the endo- or exochitinase gene or both genes have increased resistance to apple scab (Bolar *et al.*, 1999, Bolar, 1999). These results suggest the potential broad usage of chitinase transgenes to control fungal diseases of plants.

Osmotin, a thaumatin-like protein isolated from plants demonstrated antifungal activity against *Phytophthora infestans*, a destructive pathogen of potato and several other solanaceous plants. (Liu *et al.*, 1994). Nakajima *et al.*, (1997) expressed a human lysozyme in tobacco to obtain resistance against Powdery mildew causing fungus *Erysiphe cichoracearum*.

The production of active oxygen species like superoxide anions, hydroxy radicals and hydrogen peroxide (H_2O_2), have been observed in many plant-pathogen interactions and are known to play an important role in plant defence. Plants have been engineered to continuously produce active oxygen species. Transgenic potatoes constitutively expressing the glucose oxidase gene, responsible for generating H_2O_2 and cloned from *Aspergillus niger* showed apoplastic accumulation of peroxide ions. The production of the free oxygen species in transgenic plants resulted in enhanced resistance against diseases such as late blight *(Phytophtora infestans)*, wilt (*Verticillium dahliae)*, and early blight (*Alternaria solani)* of potato (Wu *et al.*, 1997). Transgenic tobacco plants expressing a ribosome inactivating protein (RIP) from barley were shown to be resistant against *R. solani*. RIPs inhibit eukaryotic protein synthesis by modification of 28s RNA of ribosomes. These proteins show specific characteristics in that they are inactive in the species from which these are isolated but can inactivate non-self ribosomes (Longemann *et al.*, 1992). Jach *et al.*, (1992) expressed three anti-fungal genes in tobacco individually or in tandem. Expression of a barley type II chitinase, a ß-glucanase and a type I, RIP resulted in synergistic disease resistance in transgenic tobacco inoculated with *R. solani*. Transgenic tomato plants expressing yeast –9 desaturase gene have shown to have enhanced resistance against *Erysiphe polygoni*, the powdery mildew pathogen of tomato. Resistance was attributed to the increased levels of fatty acids in transgenic tomato (Wang *et al.*, 1998).

Expression of genes encoding hydrolytic enzymes from infection inducible promoters in transgenic plants would provide ingenious and effective method of resistance. Production of hydrolytic enzymes, anti-nutrients, enzymes for phytoalexin synthesis and viral components have been shown to protect against pathogens that have been typically difficult to control (Michelmore, 1995), thereby building durable resistance.

Since most cultivated plants are either consumed fresh or processed for human or animal consumption, it is essential that the protein expressed by non-plant transgenes in engineered crops should be non-allergenic and non-toxic to non-target organisms as well as to humans and animals who consume them.

Engineering resistance against bacterial pathogens

Application of genetic engineering to combat bacterial diseases is in the stage of infancy. Most phytopathogenic bacteria thrive and multiply in the intercellular spaces in plants. Therefore signal peptide fusions have been used to direct transgene products to intercellular spaces. A variety of strategies have been tried to engineer bacterial resistance in transgenic plants, namely expression of non-plant antibacterial proteins, detoxification or inhibition of bacterial virulence factors, expression of antibacterial proteins of plant origin and over expression of cloned resistance genes in transgenic plants.

Expression of non-plant antibacterial proteins

Several natural and synthetic polypeptides capable of disrupting bacterial membranes and eventually killing the bacterium have been expressed in transgenic plants. Lytic peptides are one of then and are small proteins with an amphipathic a-helical structure, which makes pores in membranes resulting in the lysis of the bacterial cell membrane (Boman, 1991). Cecropins are antibacterial lytic peptides produced in the hemolymph of the giant silk moth, *Hyalophora cecropia.* Transgenic tobacco plants expressing cecropins conferred increased resistance to *P. syringae* pv. *tabaci*, the causal agent of the devastating wild fire disease (Huang *et al.*, 1994; Huang and McBeath, 1997). Similarly, the bacterial blackleg of potato caused by *Erwinia carotovora* subsp. *atroseptica* is a potential menace to potato production worldwide. The disease can cause up to 30% loss in field as well as in storage despite the use of chemicals and resistant varieties. Synthetic lytic peptide analogs, Shiva-1 and SB-37, produced from transgenes in potato plants reduce bacterial infection caused by *E. carotovora* subsp. *atroseptica* in transgenic potato plants (Wilson and Conner, 1995; Arce *et al.*, 1999). Similarly, fire blight of apple caused by *E. amylovora* is a difficult disease to control by conventional management strategies. Limited success of antibiotics, development of antibiotic-resistant strains and lack of commercially acceptable fire blight resistant varieties makes the conventional management strategies less effective. Transgenic apple expressing the SB-37 lytic peptide analog showed increased resistance to *E. amylovora* in field tests (Norelli *et al.*, 1998).

Attacins are another group of antibacterial proteins produced by *H. cecropia* pupae (Hultmark *et al.*, 1983). Proteins of this group act by inhibiting the synthesis of outer membrane protein of Gram-negative bacteria (Carlsson *et al.*, 1998). Transgenic potato expressing Attacin exhibited enhanced resistance to bacterial infection by *E. carotovora* subsp. *atrospetica* (Arce *et al.*, 1999). Attacin gene expressed in transgenic pear and apple also lead to significant enhancement of resistance against *E. amylovora* (Norelli *et al.*, 1994; Norelli *et al.*, 1999; Reynoird *et al.*, 1999; Ko *et al.*, 2000). Targetting attacin to intercellular spaces- a site of multiplication of *E. amylovora* in transgenic apple significantly reduced the incidence of fire-blight in apple (Ko *et al.*, 2000).

Lactoferrin, another antibacterial protein, is an iron binding glycoprotein, highly effective against both Gram- positive as well as Gram-negative bacteria. Expression of a mammalian lactoferrin protein in transgenic tobacco plants resulted in delayed symptom expression in plants inoculated with *R. solancearum*. Quantification of lactoferrin in transgenic plants indicated a positive correlation between lactoferrin accummulationa and level of disease resistance (Zhang *et al.,* 1998). Lactoferrin is effective at a very low concentration and affects bacterial cells by altering the permeability of cell membranes resulting in disruption of bacterial cells (Montville and Bruno, 1993). Thus this antimicrobial metabolite qualifies to be a novel and potential gene for engineering resistance against bacterial diseases. Tachyplesin is another antimicrobial peptide produced by Asian horseshoe crabs. Expression of Tachyplesin gene in transgenic potato resulted in resistance against soft rot disease caused by *Erwinia* spp. Lysozyme is a common antibacterial protein that causes lysis of cell walls through cleavage of the murein layer of bacterial peptidoglycan. Lysozyme genes from diverse sources, from bacteriophages (Jaynes *et al.,* 1987), plants (Broekaert *et al.,* 1997), animals (Vunnam *et al.,* 1997), and humans (Mitra and Zhang, 1994, Nakajima *et al.,* 1997) have been cloned and can serve as the potential transgene for cloning into plants. Expression of T4L lysozyme from T4 bacteriophage in transgenic potato plants demonstrated significant reduction in tuber tissue maceration by soft rot pathogen *E. caotovora* (During *et al.,* 1993). Similarly, transgenic apple with T4L lysozyme showed significant resistance to fire blight disease caused by *E. amylovora* (Ko 1999). Lysozyme genes have also been used to confer disease resistance against plant pathogenic bacteria in transgenic tobacco plants (Trudel *et al.,* 1995, Kato *et al.,* 1998). Human lysozyme transgenes have conferred disease resistance in tobacco through inhibition of fungal and bacterial growth, suggesting the possible use of the human lysozyme gene for controlling plant disease (Nakajima *et al.,* 1997).

Another effective gene that seems to hold promise in development of transgenic plants not only resistant to fungal disease as described earlier but also to bacterial diseases, is the glucose oxidase gene cloned from *Aspergillus niger*. The gene when expressed in transgenic potato imparted resistance against the bacterial soft rot pathogen *E. carotovora*. Expression of the transgene led to enhancement in the levels of hydrogen peroxide, which in turn activated the cellular defence genes and triggered a programmed cell death (Wu *et al.,* 1995).

Expression of antibacterial plant proteins

Plants are the sources for many antimicrobial proteins, which are usually small, cationic, amphipathic and have open chain forms. The amphipathic structure with an a-helix and an anti-parallel b-sheet is highly conserved among plants, insects and animals. Most of these antimicrobial peptides contain cysteine residues, which are likely to form disulfide bonds leading to a rigid and compact structure. Antimicrobial peptides from plants possess 3-4 disulfide bonds such as thionin from barley (Reimann-Philipp *et al.,* 1989) and plant defensin from radish (Thomma *et al.,* 2002). Various kinds of antimicrobial peptides called defensins have been identified in plants including thionins (Bohlmann 1994), maize zeamatin (Malehorn *et al.,* 1994), coffee circulin (Tim *et al.,* 1999) and wheat puroindoline (Krishnamurthy *et al.,* 2001). Radish defensins (AFP1, 2,3,4) are small, 45-54 amino acid cysteine-rich peptides (Broekaert *et al.,* 1995) conserved in several plant species including members of Brassicaceae and inhibit growth of a broad range of microbes. Transgenic rice expressing AFP1 homologs of radish defensin cloned from *Brassica* spp. conferred effective resistance against paddy blast and blight caused by *Magnaporthe grisea* and *X. oryzae*, respectively (Kawata *et al.,* 2003). Simlarly expression of barley thionin gene resulted in significant reduction in growth of *P. syringae pv. syringae* and *P. syringae pv. tabaci.* Extracts of the transgenic plants when bioassayed against *Clavibacter michiganensis* pv. *sepedonicus*, also exhibited high antibacterial activity (Carmona *et al.,* 1993).

Expression of Bacterial genes

Toxins are virulence factors produced by many plant pathogens and are responsible for the damage or disease caused by these pathogens (Bender *et al.,* 2002; Chakrabarty *et al.,* 2003). Because these toxins are deleterious to the pathogens in addition to their host plants, the bacteria produce compounds that detoxify these toxins or produce toxin resistant target enzymes. Phaseolotoxin produced by the phyto-pathogenic bacteria *P. syringae* pv. *phaseolicola* is rendered inactive by a peptidase, octase, both of plant and bacterial origin. Expression of octase of *P. syringae* pv. *phaseolicola* in transgenic tobacco confers resistance against the same pathogen (Fuente-Martinez, 1992). Tabtoxin is a dipeptide toxin produced by *P. syringae* pv. *tabaci,* the causal agent of Wild fire disease of tobacco. The toxin acts by inhibiting the activity of gutamine synthetase in plants. Expression of a tabtoxin-inactivating enzyme, tabtoxin acetyl transferase, from *P syringae* pv. *tabaci* enhanced resistance against the same bacterium in transgenic tobacco (Anazi *et al.,* 1989).

Engineering resistance against viral pathogens

Pathogen derived resistance

Transformation of plants with pathogen derived resistance (PDR) factors has become a potential tool for engineering resistance against diseases especially those that are caused by viruses. The term used for transformation of plants with portion of pathogen genome, frequently gives rise to lines of plants that are resistant to pathogens from which the sequence was derived. The nature of resistance obtained is variable and can either be RNA or protein mediated. RNA mediated resistance protects against very high levels of inoculum but is highly specific, whereas protein mediated resistance in general offers low levels but broader spectrum of resistance (Pang *et al.,* 1993, Lomonosoff, 1995, Baulcombe, 1996a, Dawson, 1996).

The number of plant species amenable to transformation and regeneration has increased greatly in recent years and it will likely be possible to produce fertile transformation of most if not all, significant crops in near future. Expression of literally any part of a plant virus gene can confer PDR (Lomonosoff, 1995). Most common are the transgenic plants expressing viral coat protein. Coat protein-mediated protection has been reported for TMV (Nelson *et al.,* 1988), tomato mosaic virus (Sanders *et al.,* 1992), CMV (Namba *et al.,* 1991, Quemada *et al.,* 1991), alfalfa mosaic virus (Loeshc-Fries *et al.,* 1987, Tumer *et al.,* 1987), potato virus X (Hemenway *et al.,*1988), potato virus Y (Perlak *et al.,* 1994), and potato leaf roll virus, PLRV (Kaniewski *et al.,* 1993). Resistance has also been generated by transforming plants by expressing viral replicases, proteases, movement proteins, satellite virus RNA, viral antisense RNA and untranslated regions of viral genes (Gosal and Gosal 2000; Chawla 2000).

Examples of successful deployment of virus resistant transgenic plants that have been released in commercial scale include two cultivars of papaya ('Sunrise' and 'Rainbow') that are resistant to Papaya ring spot virus (PRSV) and a cultivar of squash (Freedom II), resistant to Zucchini yellow mosaic and watermelon mosaic viruses. PRSV is a highly destructive viral pathogen of papaya in several countries of the world (Lius *et al.,* 1997) and had been the biggest threat to the cultivation of papaya in Hawaii for several decades (Gonsalves, 1998 a and b). The papaya industry was revived during 1990s by release of PRSV transgenic cultivars developed through coat protein mediated resistance. Freedom II, a virus resistant squash, is another transgenic cultivar that has been developed through coat-protein mediated PDR approach and has been commercially released at large scale (Fuchs and Gonsalves 1995). Similarly, transgenic tomato engineered to resist cucumber mosaic virus (CMV), a virus causing severe stunting and yield reductions throughout the world, showed high levels of resistance to CMV under field conditions (Fuchs *et al.,* 1996). Research is underway to generate transgenic tomato plants that are resistant to different strains of tomato spotted wilt virus and different tospoviruses, a group of viruses that seriously limit tomato production in the glasshouse and field, by combining virus transgenes with the native plant resistance gene Sw-5 (Gubba, 2000).

RNA based methods for symptom attenuation

Double stranded RNA interference mediated silencing of the viral gene has recently become very potential mechanism of development of disease resistant transgenic plant. The method is based on cloning a part of the viral gene in sense and antisense orientation in such a way that a double stranded RNA interference (dsRNAi) construct is formed. Transformation of plants with dsRNAi construct makes the plant resistant to subsequent infection by the virus having sequence complementary to the construct. This strategy is being exploited widely to engineer resistance in plants against RNA viruses (Hannon, 2002; Tenllado and Diaz-Riaz, 2001). Most recently this phenomenon has been demonstrated successfully to engineer resistance against the bipartite DNA containing *Vigna mungo* yellow mosaic gemini virus from mung bean (Pooggin *et al.*, 2003).

Genetic engineering is proving to be highly effective for controlling virus diseases in a wide range of crops grown worldwide (Wilson, 1993; Gonsalves and Slightom, 1992). Compared to conventional breeding for virus resistance, genetic engineering provides a quicker and more precise technology to obtain plants that are resistant to viruses, although most transgenic virus-resistant plants are still under laboratory development. There have been concerns regarding a potential risk of recombination between the invading virus and the transgene leading to development of virus strains with altered specificities like wide host range and higher virulence.

Plantibodies: a novel strategy to defend plants against pathogens

Production of antibodies in response to invasion of an antigen (proteins or antigenic components of pathogen origin) was considered a domain of animal immune system. The remarkable potential of the recombinant DNA technology has made it possible for the plants to express the antibodies against pathogen proteins, which in turn enable them to defend against the target pathogen. The expression of pathogen specific antibody in plants, termed plantibody (Smith 1996; Gibbs, 1997) is a promising approach for defending plants against potential pathogens (Schots *et al.*, 1992). The strategy for this unique approach includes fusion of the antibody molecules with carrier protein followed by its cloning and expression in transgenic plants. The plantibodies produced in the cell cytosol are expected to interact with their targets, rendering them inactive (Zhang and Wu, 1998). Most encouraging results so far have come from the cases of plantibodies generated against human pathogens, such as oral vaccines (Tacket and Mason, 1999).

Drawing a clue of the potential of plantibodies in combating human disease, plant scientists are now geared to extend this remarkable technology for plant disease control. The technology has been contemplated to control Tomato Spotted Wilt Virus, TSWV (Franconi *et al.*, 1999) and root knot caused by *Meloidogyne* spp. (Baum *et al.*, 1996). TSWV is a destructive pathogen of tomato and is one of the limitations to its successful cultivation. Transgenic plants expressing TSWV monoclonal antibody producing mammalian genes are expected to inactivate the biological activities of series of viral proteins (Franconi *et al.*, 1999). In the pathogenesis of root knot nematodes *Meloidogyne incognita* and other species of *Meloidogyne* cellulases are known to play important role. Anti-cellulase plantibodies expressed from transgenic plants are expected to interfere with pathogenesis of these nematodes by inactivating cellulases produced by them, thereby affecting their migration inside the plant roots (Baum *et al.*, 1996). In addition, antibodies against cell cycle proteins are being generated to block initiation of giant cell formation through which these nematodes feed (Vrain, 1999). Stabilising motifs have been found to ensure significant expression levels of single chain antibody fragments in the cytosol of plant cells. Although research on this novel approach of plant defence mechanism has been initiated, their practical application and commercialisation is yet to be realised.

Induced resistance in plant health

Introduction of the first reliable chemical ASM (acibenzolar-s-methyl), an activator of broad-spectrum disease resistance in crops provides an additional tool of induced resistance in crop health management (Tally *et al.*, 1999). This chemical induction of plant inherent resistance strategy can be integrated with available genetic, cultural and biological methods judiciously to optimize their benefits and devise a sustainable disease control strategy (Staub, 2001). The strategy is based on the principle that of many natural defence mechanisms plants have evolved to survive in nature, only few are inducible and can be triggered by biological or chemical agents without any deleterious side effects and in a sufficiently controlled way for practical use. The best inducible defence mechanism is the systemic acquired resistance (SAR) which is triggered by localized infection by pathogen and non-pathogen including viruses, bacteria or fungi. The activation of SAR results in manifestation of broad-spectrum resistance against a wide array of pathogens. ASM is the first commercially available product that activates SAR in plants like other biological inducers with same spectrum of protection against pathogens (Ruess *et al.*, 1996; Staub *et al.*, 2001; Tally *et al.*, 1999). The spectrum and duration of SAR activation are crop-specific. The role of ASM in induction of defence in crop plants against various pathogens needs further experimentation and evaluation.

Future prospects

Newly acquired expertise in plant transformation, continued discovery of plant *R* genes and novel genes as well as improvement of gene isolation techniques have intensified genetic engineering of plants for resistance against devastating diseases. Engineering both *R* genes and downstream defence responses will further enhance resistance. In addition to the progress being made on the plant components contributing to resistance, a thorough understanding of the pathogen side of the equation that contributes to its pathogenicity as well as the critical genes involved in pathogenicity are expected to open new horizon in plant protection. In future an integrated approach based on combined knowledge of the defence systems of the plant and assault systems of the pathogen should lead to development of designer genes for effective resistance against diseases. Plant biotechnology has already made significant strides towards development of disease resistant transgenic plants with several varieties being used commercially.

However, transgenic crops are not the panacea for all problems and will not single-handedly end world hunger. Like any other new technology it may have its own unforeseen and unrealised limitations. As more and more products of biotechnology go to field or come to the market, environmental impact, risk analysis and public concerns should be explicitly addressed. Adoption of such techniques should qualify political, social and ethical issues. In a developing country like India where more than 30 million people live below the poverty line, we are not left with much choice while adopting the products of biotechnology. Researchers have to weigh benefit-to-risk ratio (Marvier *et al.,* 2001). Being the first generation of a new and largely untested area of scientific intervention, perceived benefits of biotechnology is apparently clearer than the perceived risk.

REFERENCES

Anazi, H., K. Yoneyama, and I. Yamaguchi 1989. *Mol. Gen. Genet*. 219: 492-494.

Anonymous 2000. Agriculture and Technology: Growing pains". *The Economist*, March 25, 2000, pages 1-16, available at: (www.economist.com).

Arce, P., M. Moreno, M. Gutierrez, M. Gebauer, P. Dell'Orto, H. Torres, Acuna, I., P. Oliger, A. Venegas, X. Jordana, J. Kalazich, and L. Holuigue 1999. Enhanced resistance to bacterial infection by *Erwinia carotovora* subsp. *atroseptica* in transgenic potato plants expressing the attacin or the cecropin SB-37 genes. *American Journal of Potato Research* 76: 169-177.

Asai, T, G. Tena, J. Plotnikova, M.R. Willmann, W.L. Chiu, L.G. Gomez T. Boller, F.M. Ausubel, and J. Sheen 2002. MAP kinase signalling cascade in A*rabidopsis* innate immunity. *Natúre* 425: 977-983.

Baker, B., B. Zambryski, R. Staskawicz, and S.P. Dinesh-Kumar 1997. Signaling in plant-microbe interactions. *Science* 276: 726-733.

Baulcombe, D. C. 1996a. Mechanisms of pathogen-derived resistance to viruses in transgenic plants. *Plant Cell* 8: 1833-1844.

Baulcombe, D. C. 1996b. RNA as a target and an initiator of post-transcriptional gene silencing in transgenic plants. *Plant Mol. Biol*. 32: 79-88.

Baum, T.J., A. Hiatt, W.A. Parrott L.H. Pratt and R.S. Hussey 1996. Expression in tobacco of a functional monoclonal antibody specific to stylet secretions of the root-knot nematode. *Mol. Plant Microbe. Interact.* 9:382-387.

Bender, C.L., F. Alarcon-Chaidez and D.C. Gross 1999. *Pseudomonas syringae* phytotoxins: Mode of action, regulation and biosynthesis by peptde and polyketide synthetases.Microbiol. *Mol. Biol. Rev.* 63: 266-292.

Birch, R.G. 1997. Plant transformation: Problems and strategies for practical application. *Annu. Rev. Plant Physiol. Plant Mol.* 48: 297-326.

Bohlmann, H. 1994. The role of thionins in plant protection. *Crit. Rev. Plant Sci.* 13: 1-16.

Bolar, J.B. 1999. Genetic engineering of 'McIntosh' apple for resistance to scab (*Venturia inaequalis).* Ph.D Dissertation. Cornell University, Ithaca, NY. 131 pp.

Bolar, J.B., J.L. Norelli, K.W. Wong, C.K. Hayes, G.E. Harman and H.S.Aldwinckle, H.S. 1999. Expression of endochitinase from *Trichoderma harzianum* in transgenic apple increases resistance to apple scab and stunts growth. *Phytopathology* 90:72-77.

Boman, H.G. 1991. Antibacterial peptides: Key components needed in immunity. *Cell* 65:205-207.

Broglie, K, I. Chet, M. Holliday, R. Cressman, P. Biddle, S. Knowlton, C. Mauvais, and R. Broglie 1991. Transgenic Plants with Enhanced Resistance to the Fungal Pathogen Rhizoctonia solani. *Science* 254: 1194-1197.

Broekaert, W.F., B.P.A. Cammue, M.F.C. De Bolle, K. Thevissen, G.W. De Samblanx and R.W. Osborn 1997. Antimicrobial peptides from plants. *Crit. Rev. Plant Sci.* 16: 297-323.

Cao, H., S.A. Bowling, A.S. Gordan and X. Dong 1994. Characterization of an Arabidopsis mutant that is nonresponsive to inducers of systemic acquired resistance. *Pl. Cell* 6: 1583-1592.

Cao,H., J. Glazebrook, J. Clarke, S. Volko and X. Dong 1997. The Arabidopsis *npr1* gene that controls systemic acquired resistance encodes a novel protein containing ankyrin repeats. *Cell* 88: 57-63.

Cao,H., X. Li and X. Dong 1998. Generation of broad-spectrum disease resistance by overexpression of an essential regulatory gene in systemic acquired resistance. *Proc. Natl. Acad. Sci.USA* 95: 6531-6536.

Carlsson A., T. Nyström, H. de Cock and H. Bennich 1998. Attacin-an insect immune protein-binds LPS and triggers the specific inhibition of bacterial outer-membrane protein synthesis. *Microbiology* 144: 2179-2189.

Carmona, M.J., A. Molina, J.A. Fernandez, J.J. Lopez-Fando and F. Garcia-Olmedo 1993. Expression of the alpha-thionin gene from barley in tobacco confers enhanced resistance to bacterial pathogens. Plant J. *3 : 457-462.*

Chakrabarty, P.K., and S. Raj 2002. Molecular analysis of avirulence gene with special reference to Plant Pathogenic Xanthomonads. *In* Molecular Approaches in Plant Disease Management (Eds. G.S. Shekhawat, B.P. Singh and R.A. Singh). Malhotra Publishing House, New Delhi, pp. 63-87.

Chakrabarty, P.K., A. Chatterjee, W. Gassmann and A.K. Chatterjee 2003. Bacterial products important for plant disease- Toxins and Growth factors. *In* Encyclopedia of Plant and Crop Science (R.M. Goodman, Ed.). Marcel Dekker Inc., New York (in press).

Chakrabarty, P.K., Y.P. Duan and D.W. Gabriel 1997. Cloning and characterization of a member of the *Xanthomonas avr/pth* gene family member that defeats all commercially available *R* genes in cotton in USA. *Phytopathology* 87: 1160-1167.

Chakrabarty, P.K., D.W. Gabriel, V. Sravan Kumar and S. Raj 2001. Molecular basis of pathogenicity of *Xanthomonas axonopodis* pv. *malvacearum* and use of cloned pathogenicity genes as diagnostic tools. In: *Role of Resistance in Intensive Agriculture* (eds: S. Nagarajan and D.P. Singh), Kalyani Publishers, New Delhi, India. pp.115-135.

Chawla, H.S. 2000. Biotechnology in the production of disease resistance plants. In: *Plant Biotechnology-Recent Advances* (P.C. Trivedi, Ed.). Panima Publishing Corporation, New Delhi, pp. 407-434.

Chisholm, S.T., S.K. Mahajan, S.A. Whitham, M.L. Yamamoto and J.C. Carrington 2000. Cloning of the Arabidopsis *RTM1* gene, which controls restriction of long-distance movement of tobacco etch virus. *Proc. Natl. Acad. Sci. U S A.* 97: 489-94.

Collins, N.C., C.A. Webb, S. Seah, J.G. Ellis, S.H. Hulbert and A. Pryor 1998. The isolation and mapping of disease resistance analogs in maize. *Mol. Plant-Microbe Interact.* 11: *968-978.*

Dangl, J.L. 1995. Piece de Resistance: Novel Classes of Plant Disease Resistance Genes. *Plant Cell* 80, 363-366.

Dawson, W.D. 1996. Gene silencing and virus resistance: A common mechanism. *Trends Pl. Sci.* 1:107-108.

Delaney, T.P., L. Friedrich and J.A. Ryals 1995. Arabidopiss signal transduction mutant defective in chemically and biologically induced disease resistance. *Proc. Natl. Acad. Sci. USA* 92: 6602-6606.

Deng, Z., S. Huang, P. Ling, C. Chen, C. Yu, C.A. Weber, G.A. Moore and F.G. Gmitter 2000. Cloning and characterization of NBS-LRR class resistance gene candidate sequences in citrus. *Theo. Appl. genet.* 101: 814-822.

Dodds, P.N. G.J. Lawrence and J.G. Ellis 2001. Contrasting modes of evolution acting on the complex N locus for rust resistance in flax. The Plant J. 27: 439-453.

Düring, K., P. Porsch, M. Fladung and H. Lörz 1993. Transgenic potato plants resistant to the phytopathogenic bacterium *Erwinia carotovora. Plant J.* 3:587-598.

Flor, H.H. 1971. Current status of the gene for gene concept. *Annu. Rev. Phytopath.* 9: 275-296.

Flores, A., I. Chet and A. Herrera Estrella 1997. Improved biocontrol activity of *Trichoderma harzianum* by over-expression of the proteinase-encoding gene *prb1. Curr. Genet.* 31: 30-37.

Food and Agriculture Organization (FAO) 1993. "Production year book" FAO, Rome.

Franconi, A., P. Roggero, P. Pirazzi, F.J. Arias, A. Desiderio, O. Bitti, D. Pashkoulov, B. Mattei, L. Bracci, V. Masenga, R.G. Milne and E. Benvenuto 1999. Functional expression in bacteria and plants of an scFv antibody fragment against tospoviruses. *Immunotechnology* 4:189-201.

Fuchs, M, R. Provvidenti J.L. Slightom and D. Gonsalves 1996. Evaluation of transgenic tomato plants expressing the coat protein gene of cucumber mosaic virus WL under field conditions. *Plant Dis.* 80: 270-275.

Fuchs, M. and D. Gonsalves 1995. Resistance of transgenic hybrid squash ZW-20 expressing the coat protein genes of zucchini yellow mosaic virus and watermelon mosaic virus 2 to mixed infections by both potyviruses. *Bio/ Technol.* 13: 1466-1473.

Fuente-Martinez, J., G. Mosqueda-Cano, A. Alvarez-Morales and Herrera-Estrella 1992. Expression of a bacterial phaseolotoxin-resistant ornithil transcarbamylase in transgenic tobacco confers resistance to Psedumonas syringae pv. phaseolicola. *Bio/ Technol.* 10: 905-907.

Gabriel, D.W. 1999. The *Xanthomonas avr/pth* gene family. In: *Plant-Microbe Interactions* (G. Stacey and N.T. Keen, Eds.), Vol. 4, APS Press, St. Paul, pp. 39-55.

Gibbs, W.W. 1997. Biotechnology – Plantibodies. *Sci. Am.* 277: 44.

Glazebrook, J., E.E. Rogers and F.M. Ausubel 1996. Isolation of Arabidopsis mutants with enhanced disease susceptibility by direct screening. *Genetics* 143:973-982.

Gonsalves, D. 1998a. Transgenic virus resistant papaya: new hope for controlling papaya ringspot virus in Hawaii. APSnet Feature article, September 1 through September 30, 1998.

Gonsalves, D. 1998b. Resistance to papaya ringspot virus. *Annu. Rev. Phytopath.* 36:415-437.

Gonsalves, D. and J.L. Slightom 1993. Coat-protein mediated protection: analysis of transgenic plants résistance in a variety of crops. *Sem. Virol.* 4: 397-406.

Gosal, S.S. and S.K. Gosal 2000. Genetic transformation and production of transgenic plants. *In* Plant Biotechnology-Recent Advances (P.C.Trivedi, Ed.). Panima Publishing Corporation, New Delhi, pp. 29-40.

Graham, M.A., L.F. Marek, D. Lohnes, P. Cregan and R.C. Shoemaker 2000. Expression and genome organization of resistance gene analogs in soybean. *Genome* 43: 86-93.

Gubba, A. 2000. Transgenic and natural resistance: Towards development of tomato and pepper plants with broad resistance to virus infection. PhD Dissertation, Cornell University, 196 pp.

Hannon G. J. 2002. RNA interference. Nature 418 : 244-251.

Hayward, A.C. 1993. The hosts of Xanthomonas. In: *Xanthomonas* (J.G.Swing and E.L.Civerolo Ed.). Chapman & Hall, London, p. 1.

Hemenway, C., R.X. Fang, W. Kaniewski, N.H. Chua and N.Tumer 1988. Analysis of the mechanism of protection in transgenic plants expressing the potato virus X coat protein or its antisense RNA. *EMBO J.* 7: 1273-1280.

Hirt, H. 2002. A new blue print for plant pathogen resistance. *Nature biotech.* 20: 450-451.

Huang, W., X. Cui, Y. Tian, M. Lin and X. Peng 1994. Cloning of T7 lysozyme gene and construction of the vector for transgenic plants resistant to bacterial infection. *Weishengwu Xuebao* 34: 261-265.

Huang, Y., and J.H. McBeath 1997. Plant promoters controlled expression of cecropin B gene in transgenic potatoes renders disease resistance to soft rot bacteria. *Phytopathology* 87:S45. (Abstr.)

Hulbert, H. S; Craig, A. W; Shavannor, M. S and Sun, Q. 2001. Resistance gene complexes: Evolution and utilization. *Annu. Rev. Phytopathol*. 39: 285-312.

Hultmark, D., A. Engstrom, K. Andersson, H. Steiner, H. Bennich and H.G. Boman 1983. Insect immunity. Attacins, a family of antibacterial proteins from *Hyalophora cecropia. EMBO J.* 2:571-576.

Jach, G., S. Longemann, G. Wolf, A. Oppenheim, I. Chet, J. Schell and J. Longemann 1992. Expression of a bacterial chitinase leads to improved resistance of transgenic tomato against fungal infection. *J.Biopractice* 1: 33-40.

Jaynes, J. M., K.G. Xanthopoulos, L. Destefano-Beltrán and J.H. Dodds 1987. Increasing bacterial disease resistance in plants utilizing antibacterial genes from insects. *Bioassays* 6:263-270.

Jorgensen J.H. 1992. Discovery, characterization and exploitation of *Mlo* powdery mildew resistance in barley. *Euphytica* 63, 141-152.

Johal, G.S. and S.P. Briggs 1992. Reductase activity encoded by the Hm1 disease resistance gene in maize. *Science* 258: 985-987.

Jones, J.D.G. 1996. Plant disease resistance genes: structure, function and evolution. *Curr. Opin. Biotech*. 7: 155-160.

Kaniewski, W., C. Lawson and P.Thomas 1993. Agronomically useful resistance in Russet Burbank potato containing a PLRV CP gene. Abstract. IX International Congress of Virology, Glasgow, Scotland.

Kato, A., S.Nakamura, H. Ibrahim, T. Matsumi, C. Tsumiyama and M. Kato 1998. Production of genetically modified lysozymes having extreme heat stability and antimicrobial activity against Gram-negative bacteria in yeast and in plants. *Nahrung* 42: 128-130.

Kawata, N., Y. Nakajima, T. Yamamoto, K. Mori, T. Oikawa, F. Fukumoto and S. Kuroda 2003. Genetic Engineering for disease resistance in rice (*Oryza sativa*) using antimicrobial peptides. *JARQ* 37: 71-76.

Ko, K. 1999. Attacin and T4 lysozyme transgenic 'Galaxy' apple: Regulation of transgene expression and plant resistance to fire blight (*Erwinia amylovora*) Ph.D dissertation, Cornell University, Ithaca, NY. 194 pp.

Ko, K., J.L. Norelli, J.P. Reynoird, E. Boresjza-Wysocka, S.K. Brown and H.S. Aldwinckle 2000. Effect of untranslated leader sequence of AMV RNA 4 and signal peptide of pathogenesis-related protein 1b on attacin gene expression, and resistance to fire blight in transgenic apple. *Biotech. Lett*. 22:373-381.

Krishnamurthy, K., C. Balconi, J.E. Sherwood and M.J. Giroux 2001. Wheat puroindolines enhance fungal disease resistance in transgenic rice. *Mol. Plant Microbe Interact.*14:1255-1260.

Lin, W., C.S. Anuratha, K. Datta, I. Potrykus, S. Muthukrishnan and S.K. Datta 1995. Genetic engineering of rice for resistance to sheath blight. *Bio/Technol*. 13: 686-691.

Liu, D., K.G. Gaghothama, P.M. Hasegawa and R.A. Bressan 1994. *Proc. Natl. Acad. Sci. USA* 91: 1888-1892.

Lius, S., R.M. Manshardt, M.M.M. Fitch, J.L. Slightom, J. C. Sanford and D. Gonsalves 1997. Pathogen-derived resistance provides papaya with effective protection against papaya ringspot virus. *Mol. Breed*. 3: 161-168.

Loesch-Fries, L. S., D. Merlo, T. Sinnen, L. Burhop, K. Hill, K. Krahn, N. Jarvis, S. Nelson and E. Halk 1987. Expression of alfalfa mosaic virus RNA 4 in transgenic plants confers virus resistance. *EMBO J* 6: 1845-1852.

Lomonossoff, G. P. 1995. Pathogen-derived resistance to plant viruses. Annual Review of *Phytopathology* 33: 323-343.

Longemann, J., G. Jach, M. Tommerup, J. Mundy, J. Schell 1992. Expression of a barley ribosome inactivating proteinleads to increased fungal protection in transgenic tobacco plants. *Bio/Technol*. 10: 305-308.

Lorito M., S.L. Woo, I.G. Fernandez, G. Colucci, G.E. Harman, J.A. Pintor-Toro, E. Filipponem, S. Muccifora, C.B. Lawrence, A. Zoina, S. Tuzun and F. Scala 1998. Genes from mycoparasitic fungi as a source for improving plant resistance to fungal pathogens. *Proc. Natl. Acad. Sci. USA* 95:7860-7865.

Malehorn, D. E. 1994. Characterization and expression of an antifungal Zeamatin like protein (Zlp) gene from Zea mays. *Plant Physiol*. 106: 1471-1481.

Martin, G.B., S.H. Brommonschenkel, J. Chuwongse, A. Frary, M.W. Ganal, R. Spivey, T. Wu, E.D. Earle and S.D. tanksley 1993. Map based cloning of a protein kinase gene conferring disease resistance in tomato. *Science* 262: 1432-1436.

Marvier, M. 2001. Can risk analysis "colorize" the black and white transgenic crops? On line. Plant Health Progress doi: 10.1094/PHP-2001-0831-01-RV.

Meyers, B.C., A.W. Dickerman, R.W. Michelmore, S. Sivaramakrishnan, B.W. Sobral, and N.D. Young 1999. Plant disease resistance genes encode members of an ancient and diverse protein family within the nucleotide-binding superfamily. *Plant J.* 20: 317-332.

Michelmore, R. 1995. Molecular approaches to manipulation of disease resistance genes. *Annu. Rev. Phytopathol.* 33: 393-427.

Mitra, A. 2001. Genetic Engineering. In: *Encyclopedia of Plant Pathology* (O.C. Maloy and Murray, D. Eds.), VolI, pp. 507-512, John Wiley and Sons Inc., New York.

Mitra, A. and Z. Zhang 1994. Expression of a human lactoferrin cDNA in tobacco cells produces antibacterial protein(s). *Plant Physiol.* 106: 977-981.

Montville, T.J. and M.E.C. Bruno 1994. Evidence that dissipation of proton motive force is a common mechanism of action for bacteriocins and other antimicrobial proteins. *Int. J. Food Microbiol.* 24:53-74.

Nakajima H., T. Muranaka, F. Ishige, K. Akutsu, K. Oeda 1997. Fungal and bacterial disease resistance in transgenic plants expressing human lysozyme. *Plant Cell Rep.* 16: 674-679.

Namba, S., K. Ling, C. Gonsalves, D. Gonslaves and J.L. Slightom 1991. Expression of the gene encoding the coat protein of cucumber mosaic virus (CMV) strain WL appears to provide protection to tobacco plants against infection by several different CMV strains. *Gene* 107: 181-188.

Nelson, R. S., S.M. McCormick, X. Delanney, P. Dube, J. Layton, E.J. Anderson, M. Kaniewska, R.K. Prosksch. R.B. Horsch, S.G. Rogers, R.T. Fraley and R.N. Beachy 1988. Virus tolerance, plant growth, and field performance of transgenic tomato plants expressing coat protein from tobacco mosaic virus. *Bio/Technol.* 6: 403-409.

Norelli, J.L., H.S. Aldwinckle, L. Destéfano-Beltrán and J.M. Jaynes 1994. Transgenic 'Malling 26' apple expressing the attacin E gene has increased resistance to *Erwinia amylovora*. *Euphytica* 77:123-128.

Norelli, J.L., E. Borejsza-Wysocka, M.T. Momol, J.Z. Mills, A. Grethel, H.S. Alkwinckle, K. Ko, S.K. Brown, D.W. Bauer, S.V. Beer, A.M. Abdul-Kader, and V. Hanke 1999. Genetic transformation for fire blight resistance in apple. *Acta Hort.* 489:295-296.

Norelli, J.L., J.Z. Mills, M.T. Momol and H.S. Aldwinkle 1998. Effect of cercropin-type transgenes on fire blight resistanc of apple. *Acta Hort.* 489:273-278.

Oerke, E. -C., H. –W. Dehne, F. Schonbeck and A. Weber 1994. "Crop production and crop protection: estimated losses in major food and cash crops." Elsevier, Amsterdam.

Pang, S-Z., J.L. Slightom and D. Gonsalves 1993. Different mechanisms protect transgenic tobacco against tomato spotted wilt and Impatiens necrotic spot tospoviruses. *Bio/Technol.* 11: 819-824.

Peneula, S., d. Danesh and N.D. Young 2002. Targeted isolation, sequence analysis and physical mapping of non TIR NBS_LRR genes in soybean. *Theo. Appl. Genet.* 104: 261-272.

Perlak, F., W. Kaniewski, C. Lawson, M. Vincent and J. Feldman 1994. Genetically improved potatoes: their potential role in integrated pest management. In: *Proceed. 3rd EFPP Conference* (M. Manka, Ed. *J Phytopath.*, pp 451-454.

Pink, D. and I. Puddephat 1999. Deployment of disease resistance genes by plant transformation-a "mix and match" approach. *Tr. Plant. Sci.* 48: 71-75.

Pooggin, M., P.V. Shivprasad and K. Veluthambi 2003. RNA targeting of DNA virus in plants. *Nature Biotech.* 21: 1312-132.

Quemada, H. D., D. Gonsalves and J.L. Slightom 1991. Expression of coat protein gene from cucumber mosaic virus strain C in tobacco: Protection against infections by CMV strains transmitted mechanically or by aphids. *Phytopathology* 81: 794-802.

Reimann-Philipp, U., S. Behnke, A. Batschauer, E. Schafer and K. Apel 1989. The effect of light on the biosynthesis of leaf-specific thionins in barley, *Hordeum vulgare*. *Eur. J. Biochem.* 182: 283-289.

Reynoird, J.P., F. Mourgues, J.L. Norelli, H.S. Aldwinckle, M.N. Brisset and E. Chevreau, 1999. First evidence for differences in fire blight resistance among transgenic pear clones expressing attacin gene. *Plant Science* 149:23-31.

Rommens, C.M., J.M. Salmeron, G.E. Oldroyd and B.J. Staskawicz 1995. Intergenic transfer and functional expression of the tomato disease resistance gene *Pto*. *Pl. Cell.* 7: 1537-1544.

Rommens, C. M. and M.G. Kishore 2000. Exploiting the full potential of disease-resistance genes for agriculture use. *Curr.Opin. Biotech.* 11: 120-125.

Ruess,W., K. K.G. Mueller, Knauf-Beiter and T. Staub 1996. Plant activator CGA-245704: An innovative approach for disease control in cereals and tobacco. In: *Proceedings of the 1998 Brighton Conference: Pests and diseases* (British Crop Protection Council Ed.), Brighton,UK, pp. 53-60.

Ryals, J., K. Weymann, K. Lawaton, L. Friedrich, D. Ellis, H.Y. Steiner, J. Johnson, T.P. Delaney, T. Jesse, P. Vos and S. Uknes 1997. The Arabidopsis NIM1 protein shows homology to the mammalian transcription factor inhibitor IkB. *Pl. Cell* 9:425-439.

Sanders, P. R., B. Sammons, W. Kaniewski, L. Haley, J. Layton, B.J. Lavallee, X. Delannay, and N.E. Tumer 1992. Field resistance of transgenic tomatoes expressing the tobacco mosaic virus or tomato mosaic virus coat protein genes. *Phytopathology* 82: 683-690.

Schots, A., J. deBoer, A. Schouten, J. Roosien, J.F. Zilverentant, H. Pomp, L. Bouwmansmits, H. Overmars, F.J. Gommers, B. Visser, W.J. Stiekema and J. Bakker 1992. Plantibodies- a flexible approach to design resistance against pathogens. *Neth. J. Plant Pathol.* 9:39-46.

Shah, J., F. Tsui and D.F. Klessig 1997. Characterization of a salicyclic acid insensitive mutant (*sai1*) of *Arabidopsis thaliana*, identified in a selective screen utilizing the SA-inducible expression of the *tms2* gene. *Mol. Plant-Microbe Interact.* 10: 69-78.

Shen,K.A., B.C. Meyers, M.N. Islam-Faridi, D.B. Chin, D.M. Stelly and R.W. Michelmore 1998. Resistance gene candidates identified by PCR with degenerate oligonucleotide primers map to clusters of resistance genes in Lettuce. *Mol. Plant-Microbe Interact.* 11: 815-823.

Smith, M.D. 1996. Antibody production in plants. *Biotechnol. Adv.* 14:267-281.

Song, W.Y., G. Wang, L. Chen, H. Kim, L. Pi, J. Gardner, B. Wang, T. Holsten, W. Zhai, L. Zhu, C. Fuaguent and P.C. Ronlad 1995. A Receptor Kinase like protein encoded by the rice disease resistance gene *Xa21*. Science 270: 661-667.

Starr, M.P. 1981. The genus Xanthomonas. In: *Phytopathogenic bacteria* (M.P. Starr, Ed.), Springer-Verlag, NewYork, pp. 742-746.

Staub, T. 2001. Induced disease resistance in crop health management. Online. *Plant health progress*, doi: 10.1094/PHP-2001-0913-01-PS.

Staub, T., W. Kunz and M. Oostendorp 2001. Chemical activators of disease resistance in crop Protection. In: *Encyclopedia of agrochemicals*, John Wiley & Sons, New York.

Tacket C.O., and H.S. Mason 1999. A review of oral vaccination with transgenic vegetables; *Microbes and Infection*.1: 777-83.

Tai, T.H., D. Dahlbeck, E.T. Clark, P. Gajiwala and R. Pasion 1999. Expression of the Bs2 pepper gene confers resistance to bacterial spot disease in tomato. *Proc. Natl. Acad. Sci. USA* 96: 4153-58.

Tally, A., M. Oostendorp, K. Lawton, T. Staub and B.Bassi 1999. Commercial development of elicitors of induced resistance to pathogens. In: *Induced plant defenses against pathogens and herbivores.* Agrawal, A.A. Tuzun, S. and Bent,E. (eds.) Ameriacn Phytopathological Society, St. Paul, Mn., pp. 357-369.

Tam, J. P. 1999. An unusual structure motif of antimicrobial peptides containing end to end macrocycle and cystine knot disulfides. *Proc. Natl. Acad. Sci. USA* 96, 8913-8918.

Tenllado, F. and J.R. Diaz-Ruiz 2001. Double-stranded RNA mediated interference with plant virus infection. *J. Virol.* 75: 12288-12297.

Thomma, B.P., B.P. Cammue and K. Thevissen 2002. Plant defensins. Planta 216: 193-202.

Tolin S.A. 2001. Disease Resistance Genes. In: *Encyclopedia of Plant Pathology* (O.C. Maloy and Murray, D. Eds.), John Wiley and Sons Inc., New York, Vol I, pp. 364-366.

Trudel. J., C. Potvin and A. Asselin 1995. Secreted hen lysozyme in transgenic tobacco: Recovery of bound enzyme and in vitro growth inhibition of plant pathogens. Plant Sci. 106: 55-62.

Tu, J., I. Ona, Q. Zhang, G. Zhang, G.S. Khush and S.K. Datta 1998. Transgenic Rice variety "IR72" with *Xa21* is resistant to bacterial blight. *Theor. Appl. Genet.* 97:31-36.

Tumer, N. E., C. Hemenway, K. O 'Connell, M. Cuozzo, R.X. Fang, W. Kaniewski, W. and N.H. Chua 1987. Expression of coat protein genes in transgenic plants confers protection against alfalfa mosaic virus, cucumber mosaic virus and potato virus X. In: *Plant Molecular Biology* (D. von Wettstein and N-H. Chua, Eds. Plenum Publishing Corporation. Pp. 351-356.

Vrain, T.C. 1999. Engineering natural and synthetic resistance for nematode management. *J. Nematol*. 31: 424-436.

Vunnam, S., P. Juvvadi and R.B. Merrifield 1997. Synthesis and antibacterial action of cecropin and proline-arginine-rich peptides from pig intestine. *J. Peptide Res*. 49:59-66.

Wang, C., C.K. Chin and A. Chen 1998. Expression of the yeast ?-9 desaturase gene in tomato enhances its resistance to powdery mildew. *Physiol. Mol. Plant Pathol*. 52: 371-383.

Wang, G., W.Y. Song, D. Ruan, S. Sideris and P.C. Ronald 1996. The cloned gene *Xa21*, confers resistance to multiple *Xanthomonas oryzae* pv. *oryzae* isolates in transgenic plants. *Mol. Plant -Microbe Interact*. 9: 850-855.

Whitham, S., S. McCormick and B. Baker 1996. The N gene of tobacco confers resistance to tobacco mosaic virus in transgenic tomato. *Proc. Natl. Acad. Sci. USA* 93: 8776-81.

Whitham, S., S.P. Dinesh-Kumar, D. Choi R. Hehl, C. Corr and B. Baker 1994. The product of the tobacco mosaic virus resistance gene *N*: similarity to Toll and the interlukin-1 receptor. *Cell* 78: 1011-1115.

Wilson, C.R. and A.J. Conner 1995. Activity of antimicrobial peptides against the causal agents of common scab, black leg and tuber soft rot diseases of potato. *New Zealand Natural Sciences* 22: 43-50.

Wilson, T.M.A. 1993. Strategies to protect crop plants against viruses: pathogen-derived resistance blossoms. *Proc. Nat. Acad. Sci. USA* 90: 3134-3141.

Wu, G., B. J. Shortt, E.B. Lawrence, E.B. Levine, K.C. Fitzsimmons and D.M. Shah 1995. Disease resistance conferred by expression of a gene encoding H_2O_2- generating glucose oxidase in transgenic potato plants. *Plant Cell* 7: 1357-1368.

Wu, G.S., B.J. Shortt, E.B. Lawrence, J. Leon, K.C. Fitzsimmons, E.B. Levine, I. Raskin, and D.M. Shah 1997. Activation of host defense mechanisms by elevated production of H_2O_2 in transgenic plants. *Plant Physiol*. 115:427-435.

Yoshimura, S., U. Yamanouchi, Y. Katayose, S. Toki, Z. X. Wang, I. Kono, N. Kurata, M. Yano, N. Iwata and T. Sasaki 1998. Expression of *Xa1*, a bacterial blight-resistance gene in rice, is induced by bacterial inoculation. *Proc. Natl. Acad. Sci. USA* 95:1663-1668.

Zhang, S. P., W.Y. Song, L.L. Chen, D.L. Ruan, P.C. Ronald, R. Beachy and C. Fauquet 1998. Genetic engineering of agronomical important Indica rice varieties for resistance to bacterial blight disease. *Mol. Breed*. 4:551-558.

Zhang, Z., D. Coyne, A. Vidaver and A. Mitra 1998. Expression of human lactoferrin cDNA confers resistance to *Ralstonia solanacearum* in transgenic tobacco plants. *Phytopathology* 88: 730-734

Zhang, Z.H. and L.P. Wu 1998. Research and development of expressing antibodies in plants. *Prog. Biochem. Biophys*., 25:136-139.

Zhu, Q., E.A. Maher, S. Masoud, R.A. Dixon and C.J. Lamb 1994. Enhanced protection against fungal attack by co-expression of chitinases and glucanase genes in transgenic tobacco. 1994. *Bio/Technol*. 12: 807-812.

Systemic Acquired Resistance: A Defense Mechanism in Plants against Biotic Stress

PANKAJ[1] AND RAKESH KUMAR[2]

Plants have evolved complex and varied defense mechanisms for protection against biotic and abiotic stresses. Plant defense to biotic stress results from a complex combination of structural plant characteristics and induced biochemical reactions. The structural characteristics, including the externals cuticle and the cell wall, constitute a preformed plant defense that acts as a physical barrier to prevent the entry and movement of pathogens within the plant. In addition to this constitutive defense, plants may perceive directly or indirectly the presence of a pathogen and subsequently induce plant defense responses.

Inducible defenses are often localized at the site of attack, plant defense mechanisms may be activated systemically throughout the plant following a localized infection or attack (Kessman *et al.*, 1994). One of the first published reports of systemic resistance in plants was by Chester (1933), who used the term "acquired physiological immunity". Later, Ross (1961a,b) reported that tobacco plants exhibited "systemic acquired resistance" following local infection with tobacco mosaic virus. Other terms that have been used to describe systemic resistance in plants include "translocated resistance" (Hubert and Helton 1967), "plant immunization" (Kuc, 1987), and "induced systemic resistance" (Hammerschmidt and Baker, 1997). ISR, which can be defined as the process of active (systemic) protection of a plant dependent on the host plant's physical or chemical barriers activated by an inducing agent when applied to a single part of the plant (Kloepper *et al.*, 1992). Various "classical" inducers of ISR have been reported, including pathogens, attenuated pathogens, synthetic chemicals, and metabolic products of hosts or infectious agents, however reports of induced resistance in the field by classical inducing agents have been infrequent.

These inducible plant defense responses include synthesis of signals such as salicylic acid, ethylene and jasmonates that regulate gene expression, and the production of defense molecules such as reactive oxygen species, phenylpropanoids, phytoalexins and pathogenesis-related proteins. All these inducible biochemical reactions tend to create protective physiological conditions to limit pathogen growth and invasion in the host tissues. The final result of the host-pathogen interaction, plant disease (compatibility), or plant resistance (incompatibility), depends on the combination of a number of different variables. These variables include the genetic characteristics and physiological state of both the plant and the pathogen, and several environmental conditions including light, temperature, and humidity, among others.

HOST-PARASITE INTERACTION

In some plant-pathogen interactions, the ability of the host plant to recognize the pathogen and activate these responses is regulated in a gene-for gene-specific manner by the direct or indirect interaction

[1]Division of Nematology, Indian Agricultural Research Institute, New Delhi – 110 012.
[2]Department of Botany, J.J. College, Jhumari Tellayia, Jharkhand – 825 409.

between the products of a plant resistance (*R*) gene and a pathogen avirulence (*Avr*) gene. When either the plant or the pathogen lacks its cognate gene, activation of the plant's defense responses either fails to occur or is delayed sufficiently so that pathogen colonization ensues. In contrast to this race-cultivar-specific form of resistance, which is relatively rare, many plant species exhibit non-host resistance. Non-host resistance is characterized by the activation of many of the same defense responses as are associated with race cultivar-specific resistance; however, it occurs in the absence of any known *RyAvr* gene combination.

One of the earliest responses activated after host plant recognition of an *Avr* protein or a non-host specific elicitor is the oxidative burst, in which levels of reactive oxygen species (ROS) rapidly increase. Other rapid responses include the cross linking of cell wall proteins, the activation of protein kinases, and the increased expression of various plant protectant and defense genes. Some of these genes encode peroxidases, glutathione *S*-transferases, proteinase inhibitors, and various biosynthetic enzymes, such as phenylalanine ammonia lyase (PAL). PAL is the first enzyme in the phenylpropanoid pathway, which is involved in the synthesis of low molecular weight, antimicrobial compounds known as phytoalexins. At somewhat later times, resistant plants often develop a hypersensitive response (HR), in which necrotic lesions form at the site(s) of pathogen or elicitor entry. The localized cell death associated with the HR resembles animal programmed cell death, and it may help prevent the pathogen from spreading to uninfected tissues. Just before or concomitant with the appearance of a HR is the increased synthesis of several families of pathogenesis-related (PR) proteins in the inoculated leaves. Many of these proteins have been shown to exhibit antimicrobial activity either *in vitro* or *in vivo*. PR-proteins are subsequently expressed in the uninoculated portions of the plant, concurrent with the development of a long lasting, broad-based resistance known as systemic acquired resistance (SAR). Because of this correlation, increased *PR* gene expression is frequently used as a marker for SAR.

TERMINOLOGY

The term induced resistance emphasizes the fact that a triggering factor ('inducing agent') is needed to achieve this enhanced defensive capacity. The term 'acquired resistance', advocated by Ross (1966), points to a change in the physiology of the plant resulting from an added property. Meanwhile, the term 'systemic acquired resistance' (SAR) has found widespread acceptance in describing the state of enhanced defensive responsiveness throughout a plant resulting from local infection with a pathogen inducing necrotic lesions, such as in a HR. The term is appropriate and emphasizes the historic link with the pioneering work of Ross (1966). Whereas, Kloepper *et al.* (1992) suggested the use of the term 'induced systemic resistance' as an all compassing designation, this has not been generally adopted. However, it can be appropriately used in a broader sense for phenotypically similar phenomena resulting from different types of agents or treatments. Another way to define induced resistance would be to allude to the induced state as the 'enhanced defensive capacity' (EDC) of the plant. This term avoids any confusion that may arise about the meaning of 'induced' or 'resistance', while stressing that it is a 'capacity' that is utilized only upon challenge inoculation.

MECHANISM OF ACTION OF SAR

Inducing materials are released at the site of injury due to persistent irritation that leads to necrosis. The inducer is transported via the vascular system to cells through the plant where it interacts with receptors on the plasma membrane. This message is conveyed to the nucleus and new mRNA is synthesized and translated to produce new proteins. Proteins are released into the vacuole and exported to the cell wall where they are bound ionically and act against further attack.

TIME REQUIRED FOR THE INDUCTION OF SAR

The induction of SAR depends upon the plant or the host and the type of inducing organism. Very rapid induction has been noticed in cucumber where SAR was found to set in 7 hours after inoculation with *Pseudomonas syringae.* Late induction was observed in tobacco plants where SAR was observed in tobacco plants where SAR was found to set in only 2-3 weeks after primary inoculation with the blue mould pathogen *Perenospora parasitica* pv. *tabacii.* The factors that are responsible for early or late induction are unknown. But a common feature in most inducting treatments appear to be the local development of necrosis. Type of necrosis and its time of development are critical.

1. Triggered by viral, bacterial and fungal pathogens, nematodes and to a limited extent by insects.
2. Results in systemic changes in physiology and gene expression
3. Involves salicylic acid accumulation.
4. Resistance is effective against diverse pathogens. SAR is non specific in terms of the organisms that induce the response and of the range of pathogens against which the plant is protected.
5. SAR is time dependent i.e. resistance is established only after certain metabolic changes occur in the host during a specific interval following the inducting inoculation.
6. SAR is temperature depended and does not work above 32ºC.
7. SAR is affected by light to different degrees but is not essential in all cases of systemic resistance.
8. Resistance is persistant once the machinery for induction of the plant response has been turned on.
9. Rapidity and intensity of induced resistant response depended on the concentration of inducers and the number of host cells that are affected. Higher the concentration of inducers, faster and more intense reaction.
10. SAR is an energy demanding process and it places certain demands on the plant that cannot be satisfied if the plant is growing under nutritional conditions that are already limiting. For this reason supplemental nutrients must be applied to plants during the induction period to maintain yields at normal levels.

ASSOCIATION OF PATHOGENESIS-RELATED PROTEINS WITH INDUCED RESISTANCE

The notion that the enhanced resistance apparent upon challenge inoculation depends on the same defense mechanisms as expressed after primary infection led to the identification of common metabolic alterations induced systemically in response to local infection. Whereas induction of phytoalexins and cell wall rigidification are local reactions, accumulation of pathogenesis-related proteins (PRs) extends into non-inoculated plant parts that, upon challenge, exhibit acquired resistance.

Induction of PRs has since been found to be invariably linked to necrotizing infections giving rise to SAR, and has been taken as a marker of the induced state. This notion has been reinforced by the characterization in *Arabidopsis* of mutants that either are compromised in both the production of PRs and the induction of SAR (*npr1*; Cao *et al.*, 1994, *nim1*; Delaney *et al.,* 1995), or are constitutive expressors of PR genes as well as SAR (*cpr1*; Bowling *et al.,* 1994). PRs have been defined as plant proteins that are induced in pathological or related situations. Although they are implicated in plant defense, they have not been identified because of their anti pathogenic action, but solely because of their accumulation in infected plants. Eleven families of PRs have now been officially recognized (Van Loon *et al.*, 1994), but additional pathogen-induced proteins with potential anti-pathogenic action keep being described. PRs have been identified in at least nine plant families, with those in tobacco and tomato characterized best. It is now known that they comprise four families of chitinases (PR-3, -4, -8 and -11), one of _-1,3-glucanases (PR 2), one of proteinase inhibitors (PR-6), and one specific peroxidase (PR-9), as well as the PR-1 family with unknown biochemical properties, the thaumatin-like PR-5 family, and the birch allergen *Betv1*-related PR-10 family.

Together the PRs form a set of pathogen-induced proteins that may be considered as stress proteins. In the past decades, it has become evident that plants, when exposed to various environmental stresses, respond by synthesizing sets of specific proteins. Well-known are the heat-shock proteins, that appear to be common to all living organisms, and are transiently induced when ambient temperature exceeds some critical limit. The plant hormone abscisic acid (ABA) induces a partly similar set of proteins and increases resistance to freezing stress, indicating that acclimation is hormone-controlled. Similar proteins are induced by ABA during the acquisition of desiccation tolerance in developing seeds and upon drought stress of leaves. PRs may be considered as stress proteins produced in response to, particularly necrotizing, infections by viruses, viroids, fungi and bacteria, and thought to function in the acquired resistance against further infection. Some of the PRs possess potential anti-pathogenic activities (Van Loon *et al.*, 1994). Chitinases, together with glucanases, could be directed against fungal cell walls and, perhaps, insects. Insects are likely to be affected by proteinase inhibitors. Bacteria may be inhibited by the PR-8 family of chitinases, which also possesses lysozyme activity. Proteinase inhibitors may also act to restrict infection by some nematodes. Production of these inhibitors is highly regulated by a signal transduction pathway that is initiated by predation and transduced as a wound response (Koiwa *et al.*, 1997).

GENE-FOR-GENE INTERACTIONS

The gene-for-gene type of plant-pathogen interactions involve the specific plant recognition of a protein encoded by a dominant avirulence gene (*avr*) from the pathogen, via a complementary interacting protein encoded by a dominant plant resistance gene (*R*). When one of the *avr* or the complementary *R* genes is not expressed, plant recognition does not occur, the plant-pathogen interaction is compatible and disease may be the result (Flor, 1971). When both interacting members, Avr and the complementary R proteins, are present, recognition occurs resulting in an incompatible interaction where plant resistance is the result. The hypersensitive response (HR) is a typical plant defense response associated with resistance in these incompatible plant-pathogen interactions, and is characterized by the local death of plant cells around the site of infection, which is thought to restrain pathogen growth and spreading.

Several plant *R* genes and the corresponding *avr* genes from the pathogen have been identified from many plant-pathogen interactions. Based on the structural domains that different R protein sequences exhibit, they have been classified into different groups. Most of the R genes encode proteins containing leucine rich repeats (LRR) that are thought to mediate protein-protein interactions (Ellis *et al.*, 2000). In addition to the LRR regions, some R proteins contain nucleotide-binding sites (NBS) and they are classified as NBS-LRR. The NBS-LRR proteins are further subdivided according to the characteristics of their amino-terminal sequences, featuring either a coiled coil (CC) or a Toll- Interleukin receptor (TIR) homology domain.

PLANT DEFENSE RESPONSES

Early events of plant defense

A series of cytological changes and biochemical responses have been identified in plant cells following pathogen or elicitor recognition. The cytological changes include papilla formation, increased cytoplasmic streaming and nuclear migration, which are associated with depolymerization of microtubules and microfilaments (Kombrink and Schmelzer, 2001). The biochemical responses include changes in the H+, K+, Cl-, and Ca^{2+} fluxes across the plasma membrane, and the formation of reactive oxygen species (ROS) that occur within 2-5 minutes after elicitor treatment. Protein phosphorylation/ dephosphorylation is another early event that follows pathogen recognition and is involved in signal transduction cascades that trigger plant defense responses (Dixon *et al.*, 1994). The identification of several MAP kinases and kinase receptors associated with defense responses highlight the relevance of the phosphorylation/dephosphorylation processes during plant defense signaling. Recently, all the components of the kinase-signaling cascade associated with the recognition of flagellin, a peptide of bacterial flagella, have been identified (Asai *et al.*, 2002).

Local and systemic defense responses

Local defense responses to pathogens or elicitors involve the regulation of several genes, which contribute to generate protective physiological conditions against the invading pathogens. At the site of infection these responses include the generation of ROS, the synthesis of proteins involved in the production of signals such as salicylic acid (SA), jasmonates, and ethylene (Creelman and Mullet, 1995) as well as enzymes related to the phenylpropanoid metabolism (Dixon and Paiva, 1995) and the biosynthesis of phytoalexins, and pathogenesis-related (PR) proteins (Linthorst, 1991). Eventually, cells at the site of infection could undergo programmed cell death that often becomes visible as a hypersensitive response (HR) (Dangl *et al.*, 1996). A local infection often leads to the induction of similar defense responses in uninfected systemic plant tissues that result in broad spectrum disease resistance to subsequent infections. Additionally, systemic resistance can also be induced by wounding and non pathogenic rhizobacteria (Van Loon *et al.*, 1998). Furthermore, the application of an extensive amount of natural and synthetic compounds including SA, jasmonates, ethylene, 2,6-dichloroisonicotinic acid (INA) induce similar defense responses leading to a subsequent plant resistance (Kuæ, 2001; Oostendorp *et al.*, 2001). The different systemic defense responses associated with pathogen infections include the induction of several PR genes, accumulation of phytoalexins, induction of ROS and micro-HR (Alvarez *et al.*, 1998). Interestingly, the systemic resistance to virulent pathogens generated by previous treatment of a plant with biotrophic bacteria (*Pseudomonas fluorescens*) is not associated with induction of PR genes in systemic tissues. Different signal molecules originated at the local sites of infection are responsible for the systemic responses. The systemic responses induced depend on the pathogen and signal pathways triggered by this pathogen. For example, the systemic acquired resistance (SAR) has been associated with the signal molecule SA, which in turn is required for the induction of SAR genes associated with resistance. However, other systemic responses resulting in resistance are SA-independent. In addition, an extensive amount of emerging evidence suggest that the induction of plant disease resistance results from a complex signaling network involving cross-talk between different pathways (Feys and Parker, 2000)

Wound response

Wound stress and pathogen attack are closely related. Pathogens may gain entrance into plant tissues through wounds produced by mechanical stress and pest injury, e.g. wind, rain, feeding insects, etc. Some of the plant defense responses induced by pathogens are similar to those produced by wounding. For example, defense responses to wound damage include generation of ROS, and induction of signals such as ethylene, and JA and derivatives such as methyl jasmonate (MeJA) (Bowles, 1991, 1998). These similarities may reflect the fact that common elicitors such as plant oligogalacturonides are generated during tissue damage produced either by wound or pathogens. Recently, an overlap in pathogen-specific resistance and wound response gene expression profiles has recently been shown (Durrant *et al.*, 2000).

The best characterized event during the wound response is the accumulation of proteinase inhibitor (PIN) in both local and systemic leaves. The induction of wound responses has been associated with several molecules including, oligogalacturonides (OGAs), systemin, abscisic acid, jasmonates, and electrical pulses (León *et al.*, 2001). Oligogalacturonides have been shown to induce local and systemic expression of PIN and they are thought to be formed during leaf wounding. However, although oligogalacturonides may be responsible for the induction of PIN at the local level, it is unlikely that they could be the signal traveling from local to systemic leaves (Bowles, 1991). In contrast, there is evidence that systemin, a wound inducible 18-amino-acid oligopeptide inducing PIN locally and systemically, moves from local wounded leaves to systemic leaves (Ryan and Pearce, 1998). Electric pulses have also been suggested to be a mobile signal responsible for the systemic induction of wound responses (Wildom *et al.*, 1992). However, it may be possible that both local and systemic wound responses result from a complex regulatory network that involves several components (León *et al.*, 2001).

Reactive oxygen species

The production of reactive oxygen species (ROS) including hydrogen peroxide (H_2O_2), superoxide (O_2 -), hydroxyl radical (OH-) and singlet oxygen ($1O_2$), normally occurs in the metabolism of plant cells. They are usually generated by the electron transport activities of chloroplasts and mitochondria, and by enzymes in other cell compartments and the apoplast, involved in reduction-oxidation processes of the plant cell. The hydroxyl radical is believed to be the most reactive species among the ROS mentioned, especially for its ability to initiate radical chain reactions responsible for the irreversible modifications of macromolecules and damage to cell organelles. At the other end, hydrogen peroxide is a relatively stable ROS and it is able to diffuse across cell membranes and reach cell locations remote from the site of its generation. ROS exhibit a dual function in the normal metabolism of plant cells. On one hand, they are toxic compounds that damage plant cells; on the other hand, they may be beneficial as for example, in the lignin formation in cell walls (Foyer *et al.*, 1994).

In plant cells under normal physiological conditions these systems provide sufficient protection against ROS. However, the increased generation of ROS induced by external stimuli, may overcome these systems and produce oxidative stress. In host-pathogen interactions, accumulation of ROS has been associated with local defense responses (Wojtaszek, 1997). In some type of plant-pathogen interactions two phases of oxidative burst have been observed. The first phase occurs within the first hour after elicitor treatment of the plant cells and the second phase starts approximately after 4h of treatment and continues for several hours. In addition, accumulation of ROS has also been reported in systemic tissues of plants responding to pathogen infection (Alvarez *et al.*, 1998). The most studied sources of ROS formation during plant pathogen-interactions are the NADPH oxidase complex in the plasma membrane and the generation of hydrogen peroxide by cell-wall peroxidases (Grant and Loake, 2000). The production of ROS during plant defense responses to pathogens is associated with both local and systemic HR (micro-HR) (Alvarez *et al.*, 1998; Kombrink and Schmelzer, 2001). Furthermore, the oxidative burst has been proposed as a prime candidate for triggering HR (Lamb and Dixon, 1997; Kombrink and Schmelzer, 2001). Other plant defense responses associated with ROS include a direct antimicrobial activity, the crosslinking of plant cell-wall proteins, and the induction of defense related genes (Baker *et al.*, 1997; Lamb and Dixon, 1997).

Defense signals

Salicylic acid

Salicylic acid (SA) is a phenolic compound that participates in several physiological processes of plant cells including defense responses to pathogens. The biosynthesis of SA is initially derived from the shikimate pathway. A number of studies carried out in several plants including tobacco, potato and *Arabidopsis* indicate that SA is synthesized from phenylalanine via cinnamic acid and benzoic acid, and that SA forms different conjugated products (Shulaev *et al.*, 1997; Coquoz *et al.*, 1998). However, recent studies in *Arabidopsis* have shown that SA is synthesized from chorismate via isochorismate and that SA made by this pathway is required for local and systemic defense responses against pathogens (Wildermuth *et al.*, 2001).

These proposed cascades together with the question marks that they still have and the complexity of the phenylpropanoid pathway, probably reflect the fact that SA may be generated from alternative pathways. Several pieces of data indicate that SA plays critical roles in plant defense responses. During plant defense responses to some pathogens an increase in SA levels has been observed in local and systemic tissues. For example, in tobacco plants treated with tobacco mosaic virus (TMV), SA increases 20 fold in locally treated leaves and 5-10 fold in systemic tissues (Malamy *et al.*, 1990; Yalpani *et al.*, 1991). The increase of SA acid during plant defense responses is correlated with the induction of PR proteins and resistance, and the exogenous application of SA or functional analogs of SA to plants induce the synthesis of PR proteins and resistance (Ward *et al.*, 1991). These lines of evidence suggest that SA plays an important role plant defense response (Fig 1). SA was originally proposed to be the mobile signal for systemic acquired resistance (SAR) (Malamy *et al.,* 1990). Several studies supported this hypothesis. In addition, SA has been shown to be transported to uninfected leaves in tobacco and cucumber (Mölders *et al.,* 1996; Metraux *et al.*, 1990). However, grafting experiments using tobacco plants expressing nahG or reduced levels of phenylalanine ammonia-lyase (PAL), and wild-type plants, suggested that other primary systemic signals are involved in the onset of SAR (Vernooij *et al.*, 1994; Pallas *et al.*, 1996). Scions of chimeric tobacco plants exhibited SAR after TMV treatment of nahG rootstock leaves (Vernooij *et al.*, 1994). However, several studies suggest that SA action during plant defense responses may involve the inhibition of catalase and ascorbate peroxidase, which link SA action with the action of ROS. In addition, other SA-binding proteins have been identified although the role of these proteins remains elusive. Furthermore, protein phosphorylation/dephosphorylation has also been shown to play a role in SA signal transduction. The identification of *Arabidopsis* mutants affected in SA-mediated responses is contributing in clarifying the mechanism of action of SA during plant defense responses (Dempsey *et al.*, 1999; Cameron, 2000).

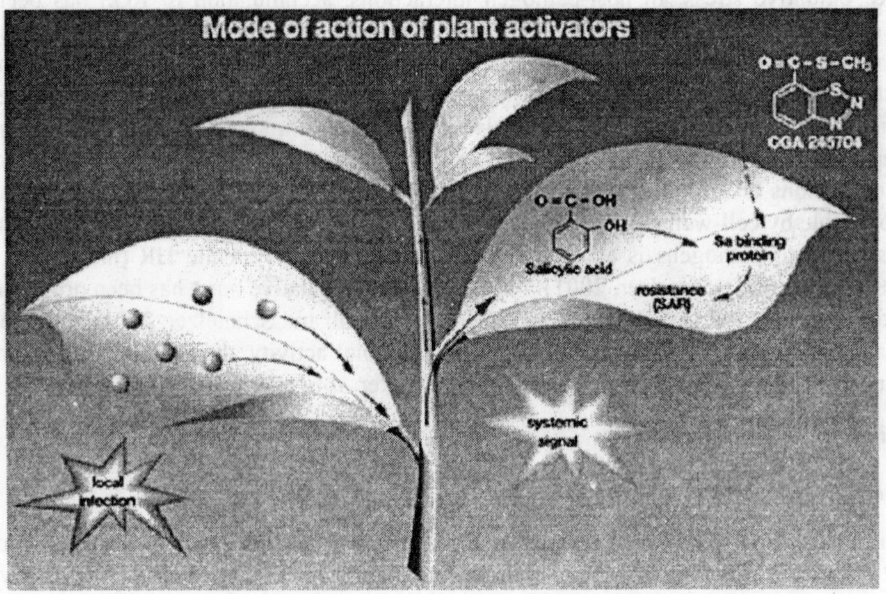

Fig 1. Role of plant activators in plant defense

Exogenous foliar application of SA on tomato @ 50ug/ml affected the development of *M. incognita* J2 and increase in the male population indicating sex reversal (Ganguly *et al.*, 1999). The galling on the roots also reduced with the foliar spray of SA in tomato as well as cowpea (Ganguly *et al.*, 1999; Nandi *et al.*, 2000; Sirohi and Pankaj, 2001; Pankaj and Sirohi, 2003). Since SA is known to induce production of PR-proteins, a similar kind of response in tomato/cowpea may affect the nematode feeding and development of nematode. SA applied as root drench in white clover against *Heterodera trifolii* reduces fecundity of the nematode increase the proportion of distorted females and of females with fewer eggs (Kemspter *et al.*, 2001).

Jamonates and C-6 volatiles

Jasmonic acid (JA) and other structural derivates including methyl jasmonate (MeJA), function as plant signals involved in plant defense to wounding and pathogens and developmental processes such as tuberization and pollen growth (Creelman and Mullet 1995). Jasmonates are synthesized from linolenic acid that is oxidized by lipoxygenase (LOX) to form 13 hydroperoxylinolenic acid. Jasmonic acid is produced from 13-hydroperoxylinolenic acid via a series of enzymatic reactions including allene oxide synthase (AOS), allene oxide cyclase (AOC) and other enzymes (Creelman and Mullet 1997; Mueller, 1997). Increased levels of jasmonates have been found in several plant species after wound and pathogen as well as elicitor treatment (Creelman and Mullet 1995). In addition, treatment of plants with JA increases resistance to pathogens/nematodes and improves growth of the plants (Cohen *et al.,* 1993; Anon, 2002). JA and MeJA regulate the expression of several genes including proteinase inhibitors, PR genes, and genes involved in phytoalexin biosynthesis (Wasternack and Parthier, 1997). Furthermore, octadecanoid precursors of JA and MeJA also activate the synthesis of proteinase inhibitors (Farmer and Ryan, 1992). On the other hand, photosynthesis related genes such as the small and large subunits of rubisco are downregulated by jasmonates (Creelman and Mullet, 1997). The JA signal transduction pathway is mainly unknown. However, several *Arabidopsis* mutants have been identified, which exhibit reduced or increased sensitivity to jasmonates (Creelman and Mullet, 1997; Vijayan *et al.*, 1998).

Current evidence suggests that the production of inhibitors occurs via the octadecanoid pathway, which catalyses the breakdown of linolenic acid and the formation of Jasmonates to induce proteinase inhibitor gene expression (Koiwa *et al.*, 1997). Furather, accumulation of lectins in barley and wheat roots and of the wheat germ agglutinin (WGA) was observed after the exposure to jasmonic acid (Oka *et al.*, 1997).

Another set of signal molecules, the C6-volatiles, is related to the jasmonate pathway. C6- volatiles are present in several plants and have been associated with defense responses upon tissue damage. C6-volatiles derive from 13- hydroperoxy linolenic or linoleic acids, which are formed by LOX from linolenic or linoleic acids respectively. These C6- volatile aldehydes are substrates for an alcohol dehydrogenase and an isomerization factor that produce other C6-volatiles such as *trans*-3-hexenal, *trans*-2-hexenal, *cis*-3-hexenol, *trans*- 3-hexenol, *trans*-2-hexenol and hexanol (Hatanaka, 1993). The increased production of some of these volatiles including *cis*-3-hexenol and *trans*-2-hexenal has been shown during plant defense responses to pathogens. Some of the C6-volatiles have been shown to induce expression of genes and the production of phytoalexins. Furthermore, C6- volatiles have been found to exhibit antimicrobial activity and reduce insect feeding rates (Hildebrand *et al.,* 1993).

Ethylene

Ethylene is a gaseous signal molecule that regulates numerous processes in plants including normal growth and development, and defense responses to biotic and abiotic stresses. Ethylene is synthesized from methionine via Sadenosyl- L-methionine (AdoMet) and 1-aminocyclopropane-1-carboxylic acid (ACC). The conversion of AdoMet to ACC is catalyzed by ACC synthase, and ACC oxidase catalyzes the conversion of ACC to ethylene. Ethylene exhibit the capacity to stimulate its own synthesis via a positive feed back loop, and for many biological responses it is effective at nano-molar concentrations. ACC synthase is the key enzyme responsible for the regulation of ethylene production and it is encoded by a multigene family. The expression of ACC genes is differentially regulated by diverse stimuli. Extensive data indicate that ethylene plays important roles in plant defense responses. The levels of ethylene have been shown to increase upon wounding, pathogen infection or treatment with elicitors of defense responses (O'Donnell *et al.*, 1996). In addition, ethylene regulates the expression of several genes including those participating in plant defense responses to biotic stresses. For example, PR proteins such as glucanase and chitinase are induced by ethylene.

Microbes

Besides SAR induced by pre-infection with pathogens forms of induced systemic resistance were observed that are triggered by plant growth promoting rhizobacteria (PGPRs) known from biocontrol studies.

In addition, PGPRs can also induce systemic defences in the plants against foliar pathogens. Kloepper and Schroth (1978) reported that root-colonizing bacteria (PGPRs) could promote radish growth in greenhouse and field trials. Results of early studies with PGPR also demonstrated control of soil borne pathogens (Défago *et al.*, 1990). PGPR act as antagonists against soil pathogens through competition or production of bacterial metabolites deleterious to the pathogen e.g., siderophores, HCN, antibiotics. More recently, three laboratories independently demonstrated that certain PGPR strains protected plants through mechanisms associated with ISR against pathogens that cause foliar disease symptoms. Subsequent research at Auburn University has demonstrated PGPR-ISR against various fungal and bacterial pathogens of cucumber. Following these studies, we were interested in testing some of the same PGPR strains for ISR against three intractable insect-transmitted diseases; bacterial wilt of cucurbits caused by *Erwinia tracheiphila*, and cucumber mosaic virus and tomato mottle Gemini virus on tomato.

Systemic resistance induced by PGPR has also been considered as a mechanism for biocontrol of plant parasitic nematodes. The two rhizosphere bacteria, *Bacillus sphaericus* strain, B 43 and *Agrobacterium radiobacter* (G12) significantly reduce *G. pallida* infection on potato. Induced resistance was triggered by vital and heat-killed bacterial cells of both the strains (Hasky, *et al.*, 1998). Heat stable bacterial surface carbohydrates like exopolysaccharides (EPS) and lipopolysaccharides (LPS) are essential for recognition in symbiotic interaction between Rhizobium and legumes, their role in *R. etli*-potato interaction. Soil treatment with LPS reduce the *G. pallida* infection of potato roots (Reitz *et al.*, 2000). *Pseudomonas fluorescens* (Pf_1 isolate) reduced nematode (*R. reniformis*) penetration in tomato roots ranging from 42 – 46.5% (Niknam and Dhawan, 2002).

Plants inoculated with necrotizing pathogens can express resistance against the same or different pathogens in parts of the plant physically distinct from the first inoculation. Induction of SAR is dependent on salicylic acid, Mutants of *Arabidopsis thaliana* have confirmed the involvement of a salicylic acid - dependent pathway in SAR. A second type of induced resistance, termed induced systemic resistance (ISR) was found to be induced in leaves by root colonizing fluorescent Pseudomonads. Unlike SAR, ISR is dependent on ethylene and jasmonate. During ISR, pathogenesis-related proteins (PRs) are not induced but, interestingly, it was shown that SAR and ISR pathways are not totally independent. For instance, Arabidopsis mutants defective in the NPR1 (non inducer of PR1) protein show neither induction of SAR nor ISR. The presence of yet another signal transduction pathway leading to resistance was demonstrated. It has been demonstrated that b-aminobutyric acid (BABA) has a action independent of the known signals involved in induced resistance. It was found that BABA protected Arabidopsis against the oomycete pathogen *Peronospora parasitica* or *Botrytis cinerea* through activation of natural defense mechanisms of the plant such as callose deposition, the hypersensitive response (HR), and the formation of trailing necrosis. BABA was still fully protective against *P. parasitica* in transgenic plants or mutants impaired in the salicylic acid, jasmonic acid and ethylene signaling pathways. Against *Botrytis*, BABA was also active in jasmonic acid and ethylene-insensitive mutants but was inactive in plants impaired in the SAR pathway. Treatment with BABA did not induce the accumulation of mRNA of the systemic acquired resistance-associated PR-1 and the ethylene- and jasmonic acid-dependent PDF1.2 genes. However, BABA potentiated the accumulation of PR-1 mRNA after attack by virulent pathogenic bacteria. As a result, BABA-treated Arabidopsis plants were less diseased compared to the untreated control. In the case of bacteria, BABA protected mutants insensitive to jasmonic acid and ethylene, but was not active in plants impaired in the SAR transduction pathway. Thus, BABA protects Arabidopsis against different virulent pathogens by potentiating pathogen-specific plant resistance mechanisms. In addition, we have provided evidence that BABA-mediated papilla formation after *P. parasitica* infection is independent of the classical SAR signaling pathway. Mutants insensitive to the action of BABA were also isolated and are now available to investigate the proteins involved in the mode of action of BABA.

Foliar spray and soil drenching with BABA reduced rot-galling 7 days after inoculation, as well as number of eggs 30 days after inoculation. Resistance phenomenon in seedlings lasted 5 days after spraying with BABA (Oka *et al.*, 1999).

CONCLUSION

Systemic acquired resistance is a general rather than elegant response developed by plants against various biotic stresses. Studies on SAR have made progress against number of plant invaders, and can be a useful tool in developing new strategies for crop protection. But very little is known against plant parasitic nematodes. New chemical inducers of resistance have already been developed commercially with potential application in the market. The use of genes or combination of genes in transgenic plants might be another approach to improve their agricultural value. Genes closely tied to SAR could also be useful for marker-assisted breeding of varieties with improved potential of SAR.

REFERENCES

Alvarez, M.E., R.I. Pennel, P.J. Meijer, A. Ishikawa, R.A. Dixon and C. Lamb 1998. Reactive oxygen intermediates mediate a systemic signal network in the establishment of plant immunity. *Cell,* 92 : 773-784.

Anonymous 2001. Annual Scientific Report of Division of Nematology, Indian Agricultural Research Institute, New Delhi (2000-2001) pp 45.

Asai, T., G. Tena, J. Plotnikova, M.R. Willman, W.L. Chiu, L. Gomez-Gomez, T. Boller, F.M. Ausubel and J. Sheen 2002. MAP kinase signalling cascade in *Arabidopsis* innate immunity. *Nature,* 415 : 977-983.

Baker, B., P. Zambryski, B. Staskawicz and S.P. Dinesh-Kumar 1997. Signaling in plant-microbe interactions. *Science* 276 : 726-733.

Bowles, D. 1991. The wound response of plants. *Curr. Biol.* 1: 45-48.

Bowles D 1998. Signal transduction in the wound response of tomato plants. Phil. Trans. R. Soc. Lond. UK, 353:1495-1510.

Bowling S.A., A. Gua, H. Cao, A.S. Gordon and D.F. Klessig 1994. A mutation in *Arabidopsis* that leads to consecutive expression of systemic acquired resistance . *Plant Cell.* 6: 1845-1857.

Cameron, R.K. 2000. Salicylic acid and its role in plant defense responses: what do we really know? *Physiol. Molec. Plant Pathol.* 56 : 91-93.

Cao, H., S.A. Bowling, A.S. Gordon and X. Dong 1994. Characterization of an Arabidopsis mutant that is non responsive to inducers of systemic acquired resistance. *Plant Cell* 6 : 1583-1592.

Chester, K. 1933. The problem of acquired physiological immunity in plants. *Quart. Rev. Biol.* 8: 129-151.

Coquoz J.L., A. Buchala and J.P. Métraux 1998. The biosynthesis of salicylic acid in potato plants. *Plant Physiol.* 117:1095-1101.

Creelman, R.A. and J.E. Mullet 1997. Biosynthesis and action of jasmonates in plants. *Annu. Rev. Plant Physiol. Plant Mol. Biol.* 48:355-381

Creelman, R.A. and J.E. Mullet 1995. Jasmonic acid distribution and action in plants: Regulation during development and response to biotic and abiotic stress. *Proc Natl Acad Sci USA* 92: 4114-4119.

Dangl, J.L., R.A. Dietrich and M.H. Richberg 1996. Death don't have no mercy: cell death programs in plant-microbe interactions. *Plant Cell* 8:793-807.

Defago, G., C.H. Berling, V.Burger, D. Haas, G. Kahr, C. Keel, C. Voisard, P.H. Wirthner, B. Wuthrich 1990. Suppression of black root rot of tobacco by a Pseudomonas strain: Potential applications and mechanisms. In: *Biological Control of Soil borne Plant* Pp. 93-108, In D. Hornby, R.J. Cook and Y. Henis, eds.

Delaney, T.P., S. Uknes and B. Vernooij 1994. A central role of salicylic acid in plant resistnace. *Science* 266:1247-1250.

Dempsey, D.A., J. Shah, D.F. Klessig 1999. Salicylic acid and disease resistance in plants. *Crit. Rev. Plant Sci.*18:547-575

Dixon, R.A. and N.L. Paiva 1995. Stress-induced phenylpropanoid metabolism. *Plant Cell* 7:1085-1097

Dixon, R.A., M.J. Harrison and C.J. Lamb 1994. Early events in the activation of plant defense responses. *Ann. Rev. Phytopathol.* 32:479-501.

Durrant, W.E., O. Rowland, P. Piedras, K.E. Hammond-Kosack and J.D. Johnes 2000. cDNA-AFLP reveals a striking overlap in race-specific resistnace and wound esponse gene expression profiles. *Plant Cell* 12:963-977.

Ellis, J., P. Dodds and T. Pryor 2000. Structure, function and evolution of plant disease resistance genes. *Curr. Op. Plant Biol.* 3:278-284

Farmer, E.E. and C.A. Ryan 1992. Octadecanoid precursors of jasmonic acid activate the synthesis of wound inducible proteinase inhibitors. *Plant Cell* 4:129-134

Feys, B.J. and J.E. Parker 2000. Interplay of signaling pathways in plant disease resistance. *TIG* 16:449 -455

Flor, H.H. 1971. Current status of the gene-for-gene concept. *Ann. Rev. Phytopathol.* 9:275-296

Foyer, C.H., M. Lelandais and K.J. Kunert 1994. Photoxidative stress in plants. *Physiol. Plant.* 92:696-717.

Ganguly, A.K., A. Sirohi, Pankaj and V. Singh 1999. Salicylic acid induced resistance in tomato against *Melidogyne incognita. Indian J. Nematol.* 29: 182-184.

Grant, J.J. and G.J. Loake 2000. Role of reactive oxygen intermediates and cognate redox signaling in disease resistance. *Plant Physiol.* 124:21-29.

Hammerschmidt, R. and J.S. Becker 1997. Acquired resistance to disease in plants. *Horticultural Rev.* 18: 247-289.

Hasky, G.K., H.S. Hoffmann and R.A. Sikora 1998. Resistance against potato cyst nematode, *Globodera pallida* systemically induced by rhizobacteria, *Agrobacterium radiobacter* (G12) and *Bacillus sphaerius* (B43). *Fundum. Appl. Nematol.* 21 : 511-517.

Hatanaka, A, 1993. The biogeneration of green odour by green leaves. *Phytochem.* 34:1201-1218

Hildebrand, D.F., G.C. Brown, D.M. Jackson, T.R. Hamilton-Kemp 1993. Effects of some leaf-emitted volatile compounds on aphid population increase. *J. Chem. Ecol.* 19:1875-1887.

Kempster, V.N., K.A. Davies and E.S. Scott 2001. Chemical and biological induction of resistance to the clover cyst nematode (*Heterodera trifolii*) in white clover. *Nematology* 3 : 35-43.

Kessman, H., T. Staub, C. Hofmann, T. Maetzke, G. Herzog, E. Ward, S., Uknes, and J. Ryals 1994. Induction of systemic acquired disease resistance in plants by chemicals. *Annu. Rev. Phytopathol.* 439-460.

Kombrink, E. and E. Schmelzer 2001. The hypersensitive response and its role in local and systemic disease resistance. *Eur. J. Plant Pathol.* 107:69-78

Kloepper, J.W. and M.N. Schroth 1978. Plant growth-promoting rhizobacteria in radish. pp 879-882. *Proc. 4th Int'l. Conf. Plant Pathogenic Bac*teria Gilbert-Clarey, Tours, France.

Kloepper, J.W., S. Tuzun, and J. Kuc 1992. Proposed definitions related to induced disease resistance. *Biocontrol Sci. Technol.* 2: 349-351.

Koiwa, H., R.A. Bressen and P.M. Hasegawa 1997. Regulation of protease inhibitor and plant defense. *Trends Pl. Sci.* 2 : 379-384.

Kuæ, J. 2001. Concepts and direction of induced systemic resistance in plants and its application. *Eur. J. Plant Pathol.* 107:7-12

Kuc, J. 1987. Plant immunization and its applicability for disease control. In: Innovative approaches to plant diseases (Ed. Chet, I.) John Wiley and Sons. Pp- 255-274.

Lamb, C. and R.A. Dixon 1997. The oxidative burst in plant disease resistance. *Annu. Rev. Plant Physiol. Plant Mol. Biol.* 48:251-275.

León, J., E. Rojo and J.J. Sánchez-Serrano 2001. Wound signaling in plants. *J. Exp. Bot.* 52:1-9.

Linthorst, H.J.M. 1991. Pathogenesis related proteins of plants. *Crit Rev. Plant Sci* 10:123-150.

Malamy J., J.P. Carr, D.F. Klessig, I. Raskin 1990. Salicylic acid: a likely endogenous signal in the resistance response of tobacco to tobacco mosaic virus. *Science* 250:1002-1004.

Mehdy, M.C. 1994. Active oxygen species in plant defense against pathogens. *Plant Physiol.* 105:467-472

Métraux J-P, H. Signer, J. Ryals, E. Ward, M. Wyss-Benz, J. Gaudin, K. Raschdorf, E. Schmid, W. Blum and B. Inverardi 1990. Increase in salicylic acid at the onset of systemic acquired resistance in cucumber. *Science* 250:1004-1006.

Mölders W., A. Buchala and J.P. Métraux 1996. Transport of salicylic acid in tobacco necrosis virus-infected cucumber plants. P*lant Physiol.* 112:787-792

Mueller, M.J. 1997. Enzymes involved in jasmonic acid biosynthesis. *Physiol. Plantarum* 100 : 653-663.

Nandi, B. N.C. Sukul and S.P.S. Babu 2000. Exogenous salicylic acid reduces *Meloidogyne incognita* infestation of tomato. *Allelopathy J,* 7 : 285-288.

Niknam G. R. and S. C. Dhawan 2002. Systemic resistance induced by *Pseudomonas fluorescens* isolate Pf_1 in tomato against *Rotylenchulus reniformis. Int. J. Nematol.*-12 :203-208.

O'Donnell, P.J., C. Calvert, R. Atzorn, C. Wasternack, H.M.O. Leyser and D.J. Bowles 1996. Ethylene as a signal mediating the wound response of tomato plants. *Science* 274:1914-1917.

Oka, Y., I. Chet and Y. Spiegel 1997. Accumulation of lectins in cereal roots invaded by the cereal cyst nematode, *Heterodera avenae. Physiol. Mol. Pl. Pathol.* 51 : 333-345.

Oka, Y., Y. Cohen and Y. Spiegel 1999. Local and systemic induced resistance to the root-knot nematode in tomato by DL-beta-amino-n-butyric acid. *Phytopathology* 89 : 1138-1143.

Oostendorp, M., W. Kunz, B. Dietrich and T. Staub 2001. Induced resistance in plants by chemicals. *Eur. J. Plant Pathol.* 107:19-28

Pallas, J.A., N.L. Paiva, C. Lamb and R.A Dixon 1996. Tobacco plants epigenetically suppressed in phenylalanine ammonia-lyase expression do not develop systemic acquired resistance in response to infection by tobacco mosaic virus. *Plant J.* 10:281-293.

Pankaj and Anil Sirohi 2003. Role of Salicylic Acid, Giberrelic Acid and Rose Bengal in inducing resistance in cowpea against root-knot nematode. In *Proceedings of National Symposium on Biodiversity and management of nematodes in cropping systems for sustainable agriculture.* ARS Durgapura, Jaipur (11-13 Nov.), pp 62-65.

Reitz, M., K. Rudolph, I. Schroder, H.S. Hoffmann, J. Hallman and R.A. Sikora 2000. Liposaccharides of *Rhizobium etli* (G12) act in potato roots as an inducing agent of systemic resistance to infection by the cyst nematode, *Globodera pallida. Appl Envtl. Microbiol.* 66 : 3515-3518.

Ross, A. F. 1961a. Localized acquired resistance to plant virus infection in hypersensitive hosts. *Virology* 14: 329-339.

Ross, A. F. 1961b. Systemic acquired resistance induced by localized virus infections in plants. *Virology* 14: 340-358.

Ross, A.F. 1966. Systemic effects of local lesion formation. In: Viruses of Plants (eds A.B.R. Beemster and J. Dijkstra) pp. 127-150, Amsterdam, North Holland.

Ryals, J., U. Neuenschwander, M. Willitis, A. Molina, H.Y. Steiner 1996. Systemic acquired resistance. *Plant Cell* 8: 1899-1919.

Ryan, C.A. and G. Pearce 1998. Systemin: A polypeptide signal for plant defensive genes. *Ann. Rev. Cell Dev. Biol.* 14:1-17.

Shulaev V, P. Silverman and I. Raskin 1997. Airborne signaling by methyl salicylate in plant pathogen resistance. *Nature* 385:718-721

Shulaev, V., J. León and I. Raskin 1995. Is salicylic acid a translocated signal of systemic acquired resistance in tobacco? *Plant Cell* 7:1691-1701.

Sirohi, A. and Pankaj 2001. Salicylic acid enhances PAL and PO activity in tomato and reduces root-knot nematode infection. Abstract :National Congress on Nematology : Appraisal and future plans, IARI, New Delhi (5-7 Dec, 2001).

Van Loon, L.C., P.A.H.M. Bakker and M.J. Pieterse 1998. Systemic resistance induced by rhizosphere bacteria. *Ann. Rev. Phytopatol.* 36 :453-483

Van Loon L.C., W.S. Pierpoint, T. Boller and V. Conejero 1994. Recommendations for naming plant pathogenesis related proteins. *Plant Mol. Biol. Rep.* 12: 245-264.

Vernooij, B., L. Friedrich, A. Morse, R. Reist, R. Kolditz-Jawhar, E. Ward, S. Uknes, H. Kessmann and J. Ryals 1994. Salicylic acid is not the translocated signal responsible for inducing systemic acquired resistance but is required in signal transduction. *Plant Cell* 6:959-965.

Vidal, S., I. Ponce de León, J. Denecke and E.T. Palva 1997. Salicylic acid and the plant pathogen *Erwinia carotovora* induce defense genes via antagonistic pathways. *Plant J.* 11:115-123

Vijayan, P., J. Shockey, C.,A. Lévesque, R.J. Cook, J. Browse 1998. A role for jasmonate in pathogen defense of *Arabidopsis. Proc. Natl. Acad. Sci.* USA 95:7209-7214 .

Ward, E.R. , S.J. Uknes, S.C. Williams, S.S. Dincher, D.L. Wiederhold, D.C. Alexander, P. Ahl-Goy, J.P. Métraux and J.A. Ryals 1991. Coordinated gene activity in response to agents that induce systemic acquired resistance. *Plant Cell* 3:1085-1094.

Wasternack, C. and B. Parthier 1997. Jasmonate-signal plant gene expression. *Trends Plant Sci.* 2:302-307

Wildermuth, M.C., J. Dewdney, G. Wu and F.M. Ausubel 2001. Ischorismate synthase is required to synthesize salicylic acid for plant defence. *Nature* 414:562-565

Wildom, D.C., J.F. Thain, P.E.H. Michin, I. Gubb, A. Reilly, Y. Skipper, H. Doherty, P. O'Donnel, D.J. Bowles 1992. Electrical signaling and systemic proteinase inhibitor induction in the wounded plant. *Nature* 360:62-65.

Wojtaszek, P. 1997. Oxidative burst: an early plant response to pathogen infection. *Biochem. J.* 322:681-692

Yalpani, N., P. Silverman, T.M.A. Wilson, D.A. Kleier and I. Raskin 1991. Salicylic acid is a systemic signal and an inducer of pathogenesis-related proteins in virus-infected tobacco. *Plant Cell* 3:809-818.

6

Induced Resistance vs Biological Control against Plant Pathogens

PRATIBHA SHARMA[1]

With the upcoming new biotechnological and molecular biological approaches, the plant disease managers are trying to reshape or rather say develops disease management strategies based on new technologies. Current concern about the environment indicates a need to limit application of chemicals for plant disease control and therefore, In this scientific renaissance, very importantly are coming two encouraging concepts: Biological control and Induced resistance. These are the most encouraging techniques for plant protection. There are a variety of terms used to describe the phenomenon of induced plant defenses against pathogens, phenomenon of induced plant defenses against pathogens, including immunization, induced systemic resistance (ISR) and systemic acquired resistance (SAR). Traditionally, the term IR had been used by entomologists (e.g. Karban and Baldwin 1997) and ISR and SAR by phytopathologists, each having the same meaning. Advancements in the biological control of plant diseases using microbial agents contribute directly to the development of biological control products for improved plant health, but also indirectly through the identification of promising physiological or molecularstrategies. A researcher working on biological control trespasses induced resistance, as one of the mechanism, hence these two phenomenon can be understood as two sides of a coin and are interrelated. This review is an attempt to understand these phenomenon and their interrelationships.

PRINCIPLES OF INDUCED RESISTANCE AND BIOLOGICAL CONTROL

The induced responses were highly ignored by the general scientific community as a tool to manipulate plant resistance to control diseases. A systematic definition is not yet understood, as the phenomenon. Though, induced resistance is only one of several types of resistance to microbial disease that flowering plants process. The other major categories are basic resistance, organic specific, age related and parasite specific resistance. Induced resistance differs from other types of resistance as its expression relies on some previous treatment of the plant that "sensitizes" it so that what should be susceptible tissue resists invasion by parasites that previously were successful pathogens. A single induction treatment can elicit resistance to a wide variety of parasites and in this lack of specificity, induced resistance resembles basic resistance more than parasite-specific resistance. Indeed, induced, age-related and organ-specific resistance do not require the presence of known parasite-specific genes for resistance against the parasite in question, although some forms of parasite-specific resistance may also be age-related in that are only expressed at a certain stage of plant development (Bhattacharyya and Ward 1986).

Induced resistance is distinguished from conventional chemical as well as biological procedures in plant protection by the lack of toxicity of the inducing agents towards the pathogens. The protection of

1. Division of Plant Pathology, Indian Agricultural Research Institute, New Delhi - 110 012.

the plants is not based on the elimination of the pathogens but rather on the activation of plant defense mechanisms or on the enhancement of their activity. The basic idea behind induced resistance is that genes for resistance or defense reactions exist in all plants. These genes are not expressed until after a resistance inducing treatment activates or enhances their expression, or that changes in the plant metabolism modify the activity of such genes. Induced resistance is considered to be a biological plant protection procedure in which the plant is the target of the procedure, not the pathogens.

Many definitions have been proposed for biological control.The term biological control was coined by the late Harry Smith of the University of California, who defined it as "the suppression of insect populations by the actions of their native or introduced enemies". There have been debates regarding the scope and definition of biological control ever since, mainly to accommodate the technological advances in the tools available for pest management. The definition presented by Van Drieshce and Bellows [1998], "the use of parasitoid, predator, pathogen, antagonist, or competitor populations to suppress a pest population making it less abundant and thus, less damaging than it would otherwise be," appears to convey the current thinking on biological control by those researchers involved in insect research.

In their earlier definition of biological control, Baker and Cook [1974] described it as "the reduction of inoculum density or disease-producing activities of a pathogen or a parasite in its active or dormant state, by one or more organisms, accomplished naturally or through manipulation of the environment, host or antagonist, or by mass introduction of one or more antagonists." This concept is closely related to the SAR concept where host is the target. The manipulation of environment reflects the use of various cultural practices or using tolerant cultivars or in other words it does not allow the bioagent to be used solo. of This definition was subsequently modified to read "the reduction in the amount of inoculum or disease-producing activity of a pathogen accomplished by or through one of more organisms other than man" [Cook and Baker 1983]. This definition clearly widens the scope of biological control by allowing the researchers to use microbials, plant products cultural practices into a biological control system. These are perhaps the most widely quoted and accepted definitions of biological control. Some time later, the U.S. National Academy of Sciences introduced some modifications to the definition, referring to biological control as "the use of natural or modified organisms, genes or gene products to reduce the effects of undesirable organisms and to favour desirable organisms such as crops, beneficial insects and microorganisms", Obviously, due consideration has been given to the advances by the advent of molecular biology to plant pathology and to research in biological control.

There have been other recent definitions, many of which still reflect the basic idea presented in the classical definitions of biological control. According to Shurtleff and Averree [1997], biological management or control refers to "disease or pest control through counter balance of microorganisms and other natural components of the environment. It involves the control of pests (bacteria, fungi, insects, mites, nematodes, rodents, weeds, etc.,) by means of living predators, parasites, disease-producing organisms, competitive microorganisms, and decomposing plant material, which reduce the population of the pathogen. This definition has widened the scope of the use of bioagents in IPM systems where we need to apply and protect the crop as a whole and therefore a single application or approach does not help the researcher in using the bioagents. Agrios [1997] defined biological control as the total or partial destruction of pathogen populations by other organisms.All these definitions revolve around the suppression of plant pathogens and management of host.The modifications introduced to the basic definition of biological control by different workers indicate its changing scope and perspective. This is largely attributed to the insights obtained by the use of various molecular and biochemical tools for the study of this hitherto poorly understood phenomenon involving interactions within a multicomponent system. Therefore, biological control of plant pathogens has now emerged as a broad concept, and encompasses several mechanisms. How best we define the biological control but basically researches conducted till today are based on the disease control and growth promotion. Amongst the most usable and practical approach of biocontrol mechanism is IR .The other mechanisms of biological control are antibiosis, mycoparasitism and competition which are still referred as possible mechanisms. Therefore, we need to understand the mechanisms underlying biologically induced resistance and biological control

PREVIEW OF BIOLOGICALLY INDUCED SAR OR BIOLOGICAL CONTROL

Phenotypically, biocontrol agent induced systemic resistance and pathogen induced SAR are similar n in that both mechanisms develop as a result of the interaction of plant with a microorganism and the resulting enhanced defensive capacity is expressed both locally and systemically against a broad spectrum of attacking organisms.To determine whether biocontrol agent induced resistance is not only phenotypically but also mechanistically similar to pathogen induced SAR , induced plants can be analyzed for the presence of specific PRs and induction of resistance can be compared in untransformed and Nah- G transformed plants. Several ways exist for inducing resistance in plants. Among them are: inoculation with pathogen, avirulent races or nonpathogenic strains, inactivated pathogens, nonpathogens , mirobial metabolites , i.e. biogenic elicitors , chemical elicitors signaling resistance. All these factors are being worked out by various workers in one form or other and being explained as induced resistance in the category of general mechanism of biological disease control. Below are explained such examples for inducing resistance against phytopathogens, a means of biological control.

Pathogenic Organisms: Development of resistance of diseases in plants following infection has been known for over 60 year (Chester, 1933). In order to achieve successful protection, plants must recover from initial infection or must be infected at a site that does not normally become diseased. For example, immunization against blue mold in tobacco can be achieved by stem injection with live sporangiospores of the causal organism of blue mold, *Peronospora tabacina* (Tuzun and Kuc 1985a). The site of injection appears to be important since only delivery of spores into the tissue external to the xylem induces systemic resistance without stunting. In a similar way, protection can be achieved in tobacco varieties with the N gene for resistance to tobacco mosaic virus (TMV), in which the virus does not become systemic, by inoculation of two or three bottom leaves with TMV (Ye *et al.* 1989). More interestingly, both of these pathogens induce resistance against either pathogen equally well, which may indicate activation of multiple mechanisms once the state of resistance is achieved.

The systemic protection of tobacco against blue mold is somehow unique since it was first reported on systemically stem-infected and stunted field-grown tobacco in Australia. Extensive field tests conducted over a 3 year period in Kentucky and Puerto Rico utilizing a modified technique (which employs injection of a sporangiospore suspensions of *P. tabacina* into stem tissue external to the xylem) further indicated that tobacco plants can be protected can be protected against blue mold under various field conditions. These tests indicated that immunized plants had reduced lesion number and size compared to the nontreated controls, and the level of protection obtained by stem injections with *P. tabacina* was comparable to the obtained by treatment with metalaxyl. Furthermore, even in the absence of disease, immunized plants grew more vigorously and yields were increased up to 20%, compared to controls. Further field experiments were conducted in Mexico for three growing season to test the effectiveness of immunization against-metalaxyl-tolerant strains of *P. tabacina* (Tuzun *et al.* 1992). Highly significant reductions in the number and size of blue mold lesions were observed on plants injected with *P. tabacina,* compared to noninjected controls, regardless of metalaxyl applications. Some vigorously growing plants with necrotic stem lesions but markedly reduced blue mold on the foliage were observed in heavily infected commercial fields indicating the natural occurrence of immunization against blue mold in the gulf coast of Mexico (Tuzun *et al.* 1992). Tolerance to immunization was not evident, although conditions for its development were very favorable. In field trials, plants derived via tissue culture from immunized parents were also protected against blue mold compared to plants derived from control parents. Cucurbits have been utilized as model system to establish many aspects of plants immunization, including applicability of the phenomenon to the real world. Cucumbers can be protected in the field by application of biological as well as chemical inducers (Doubrava *et al.* 1989; Mucharromah and Kuc 1989).

These studies provide evidence that immunization of plants with biotic or abiotic inducers effectively controls disease in the field. Utilization of pathogenic organisms in the field, however, may create problems if handled carelessly. Culture filtrates of pathogenic organisms may provide a safer alternative, and several studies have indicated induction of resistance in many crop species by microbial metabolites (Trivedi and Sinha 1976; Doke *et al.* 1987; Kopp *et al.* 1989; Maiss 1987; Ozeretskovskaya *et al.* 1988). Avirulent, hypovirulent and/or attenuated strains of pathogenic organisms (Kroon *et al.* 1991; Martyn *et al.* 1991; Sneh *et al.* 1989;) as well as L-forms of pathogenic bacteria (Amijee *et al.* 1992) may provide additional alternatives. Cabbage plants were successfully protected against the black rot pathogen, *Xanthomonas campestris pv.campestris* (XCC), under field conditions by prior application of attenuated stains of XCC or an isolate of *Xanthomonas campestris pv. malvacearum,* nonpathogenic on cabbage . In these experiments, conducted at several locations over two years, inducing bacteria were introduced into natural openings of leaves utilizing an organosilicone surfactant (Silwet) prior to transplanting and three weeks prior to pathogen challenge. Utilization of silwet to introduce organisms into leaf tissue may have a large-scale practical application by eliminating the need for injection.

Nonpathogenic Organisms: Non-pathogenic organisms that have been utilized by several research groups to induce resistance include suspensions and/or culture filtrates of saprophytic bacteria (Schmidt 1988; Schonbeck *et al.* 1980) such as *Bacillus subtilis* (Mais 1987), *B. thuringensis* (Roveratti *et al.* 1989), B. *pumulis* and Erwinia herbicola (Reiss *et al.* 1988), and saprophytic fungi (Gregersen and Smedegaard –Peterson 1989). There are also several reports that suggest induction of resistance by mycorrhiza against fungi (Rosendahl 1985), bacteria (Garcia-Garrido and Ocampo 1988) and nematodes (Carling *et al.* 1989). Although most of these studies are conducted in controlled envirorments, spraying plants with *B. subtilis* effectively controlled powdery mildew on barley under field conditions (Oerke *et al.* 1989).

Applications of inducing agents to leaf surfaces makes the inducing agents more susceptible to stress conditions such as UV irradiation, rain washing, and temperature fluctuations. Application of good root colonizers to seed or soil may protect inducing agents from some of these negative effects. Recently, rhizosphere bacteria, and fungi have effectively induced resistance in plants while providing other beneficial effects to plants (Kloepper *et al.* 1993, Meera *et al.* 1993; Tuzun and Kloepper 1994). Free-living root and soil bacteria have been studied for the past century as possible inoculants for enhancing crop productivity. Utilization of natural microbial strains as inducers of plants defensive responses may increase the chance of their applicability and offer a practical way to deliver plant immunization.

Plant Growth-Promoting Rhizobacteria (PGPR): Beneficial effects of PGPR that have been documented by many research groups can be summarized as: (1) direct plant growth promotion (Kapulnik 1991; Schroth and Becker,1990), (2) biological disease control (Bakker *et al.*, 1991; Kloepper 1991, 1993; Schippers et al, 1991; Schroth and Becker, 1990), and (3) inducing host resistance (Kloepper *et al.* 1993; Tuzun and Kloepper, 1994). The reports summarized in these reviews include the effects of many different bacterial groups on many host plants. Plants growth-promoting fungi (PGPF), however, have been described more recently from the rhizosphere of turf grass and such fungi also appear to induce systemic resistance against antrachnose disease caused by *Collectotrichum orbiculare* in cucumber plants. Although it was suggested previously (Scheffer 1983; Voisard *et al.* 1989), direct pathological evidence for the induction of systemic resistance by PGPR was first published for three plant-pathogen systems in 1991. In a carnation system applications of *Pseudomonas* sp. strain WCS417 to rockwool cubes resulted in protection from wilt caused by *Fusarium oxysporum f. sp. dianthi* (van peer al. 1991). The pathogen was spatially separated from the PGPR strain by inoculating into stems one week after PGPR application, and separation was confirmed by the failure to isolate WCS417 from stems. In a bean system, seed treatment with a *P. Fluorescens* PGPR strain led to reductions in the numbers of foliar lesions caused by subsequent inoculations of P. *syringae pv. Phaseolicola* (Alstrom 1991). In another report with cucumber, 94 known PGPR strains were examined for ability to control C. *orbiculare* (Wei *et al.* 1991). Six PGPR strains applied as seed treatments consistently resulted in significant reductions in anthracnose lesion diameters and lesion numbers when the pathogen was applied 21 days after planting. In a subsequent study with the 6 inducing PGPR, none of the inducing strains was recovered from leaf petioles, confirming the spatial separation of pathogen and PGPR. These reports with carnation, bean and cucumber demonstrate that saprophytic root-associated bacteria may act as agents of induced systemic resistance, and hence, they serve to expand the potential mechanisms by which PGPR may exert biological disease control.

Protection achieved with two of the inducing PGPR strains used by Wei *et al.* (1991) on cucumber. , applied as seed treatments, significantly reduced mean diameter of lesions induced by foliar-applied C. *orbiculare*. PGPR-mediated protection against cucumber mosaic virus (CMV) is affected at the viral inoculation site. One PGPR strain significantly protected against CMV that was mechanically inoculated onto cotyledons (Liu *et al.* 1992). Protection was evident as prevention of symptom development on PGPR induced plants, which differs from protection reported against CMV for classical induced systemic resistance (Bergstrom *et al.* 1982). With classical induced resistance, CMV symptoms were delayed but not prevented. With the PGPR system, both strains delayed, but did not stop, symptom development when CMV was inoculated onto the first, second, or third leaves, which is equivalent to results with classical induced resistance. The same two PGPR strains induced protection of cucumber against *Pseudomonas syringae pv. Lachrymans* as seen by a significant reduction in mean lesion number and lesion size compared to noninduced controls (Liu *et al.* 1993a).

Preliminary studies with Fusarium wilt demonstrate that one PGPR strain reduced the rate of symptom development and plant death (Liu *et al.* 1993b). With Fusarium, a split root system was used to ensure spatial separation of PGPR and the pathogen. PGPR were applied to one-half of the roots after splitting, and *Fusarium* was incorporated into the soil in which the other half of the roots were growing. While this system demonstrates the potential to use PGPR-mediated induced resistance for protection from a soilborne vascular wilt pathogen, it is fundamentally different from the other PGPR systems which use seed treatments. PGPR applied as seed treatments may induce systemic resistance earlier in the plant's life, and hence, more work is needed to compare the biochemical responses of plants induced by seed and root treatments with PGPR.

Several studies have indicated that specific PGPR may stimulate the production of biochemical compounds associated with host defense. Van Peer and his colleagues (1991) observed increased accumulation of phytoalexins in carnation plants treated with PGPR (isolate WCS417) following pathogen inoculation. In a bean system, Hynes & Lazarovits (1989) found that levels of a PR-protein increased in leaves following seed treatment with PGPR strains. Plants root colonization by PGPR was associated with increased peroxidase activity and enhanced lignification of stems or leaves in bean (Anderson and Guerra, 1985) and wheat (Frommel *et al.* 1991). Inoculation of bean root with a *P. putida* PGPR strain led to an increased abundance of mRNA encoding PRIa protein in leaves. These reports clearly demonstrate that particular bacteria inoculated onto seeds or roots may elicit systemic physiological changes in plants.

Early investigations with PGPR-mediated induced resistance in cucumbers suggested that the biochemical response of the plant may depend on the inducing PGPR strain. Some inducing PGPR were associated with enhanced peroxidase activity similar to that observed with classical induced controls. Some, but not all, inducing PGPR strains were associated with enhanced mRNA encoding acidic chitinases. It will be necessary to conduct further biochemical investigations of how PGPR-mediated and classical induced resistance affect host defense-related compounds.

Biocontrol Agents

Disease suppression: The known bioagents, defined by the researchers are mainly fungal and bacterial in origin. The key feature to demonstrate the existence of SAR or ISR is that the subsequent challenge with a pathogen is either spatially separated form the inducing inoculation or, if the challenge is at the same position as the inducing inoculation that a period elapses where novel genes are switched on and characteristic resistance moieties are produced. Unfortunately, most of the work on induced resistance has concerned foliar pathogens and only relatively recently has the potential of this mechanism been recognized for biocontrol of soil-borne pathogens. Consequently, in contrast to the vast literature covering induced resistance in relation to the control of foliar pathogens, unequivocal examples of SAR or ISR involving soil-borne pathogens are rare. Using split root systems, both bacteria and fungi have been shown to induce resistance in several plants when applied to roots. For example, *Pseudomonas putida* 89B-27 and *Serratia marcescens* 90-166 induced systemic resistance to *Fusarium oxysporum f. sp cucumerinum* in cucumber (Liu *et al.*, 1995a,b);' *Pseudomonas corrugata and Pseudomonas fluorescens* induced systemic resistance to *Pythium aphanidermatum* in cucumber (Zhou and Paulitz, 1994): and *P. fluorescens* induced systemic resistance to *F. oxysporum f.sp. raphanii* in radish (Leeman *et al.*, 1995a,b), with lipopolysaccharides of the bacterial wall strongly implicated as being involved in the induction process. Similarly, non-pathogenic *Fusarium oxysporum* isolates induced resistance to *F. oxysporum f.sp. cucumerinum* in cucumber (Mandeel and Baker, 1991), *F. oxysporum f. sp. dianthi* induced resistance to *F. oxysporium f.sp. lycopersici* in tomato (Kroon *et al.*, 1991). In an alternative treatment, inoculation of the first true leaves of cucumber with *Colletotrichum orbiculare* or tobacco necrosis virus resulted in induced resistance to *F. oxysporum f. sp. cucumerinum* on the roots.

The phenomenon of treating seeds, roots or cuttings with inducing bacteria or fungi and achieving induced resistance to subsequent stem or foliage challenge by a range of viral, bacterial and fungal pathogens is also known (Alstrom 1991, Wei *et al.*, 1991; Liu *et al.*, 1992, 1995a,b; Maurhofer *et al.*, 1994b; Meera *et al.*, 1994, 1995), suggesting an important role for induced resistance in bio control in general. There are few data on the molecular and biochemical changes occurring in roots as a result of induced resistance, although phytoalexins and peroxidases are thought to be involved in the response of carnation root to inoculation with *Pscudomonas* (Van Peer *et al.*, 1991; van Peer and Schippers, 1992) and pathogenesis-related proteins were synthesized in roots and leaves of cotton plants following inoculation with arbuscular mycorrhizal fungi (Liu *et al.*, 1995d). Based on studies from other SAR systems, responses may involve accumulation of antimicrobial low molecular weight chemicals such as phytoalexins, formation of protective layers via accumulation of polymers such as lignin, callose and hydroxyproline-rich glycoproteins, increases in activators of enzymes leading to the production of such materials, and increases in the amount of chitinases, B-1,3-glucanases, peroxidases and other pathogenesis-related proteins (Gawande and Sharma 2003).

In a study conducted in an intact host-pathogen interaction among cauliflower, tomato and *Trichoderma* species at Division of Plant Pathology IARI by the author the two *Trichoderma* species namely *T. harzianum* and *T. viride* were used against *Sclerotinia sclerotiorum.* Development of *S. sclerotiorum* infection was consistently reduced after a prior *T. harzianum* and *T. viride* treatment on plant parts spatially separated from the site of *S. sclerotiorum* inoculation. *T. harzianum* and *T. viride* seedling treatment reduced *S. sclerotiorum* stem and leaf infection in cauliflower and tomato. Though, the suppression in disease index was slightly greater in *T. harzianum* than that of *T. viride* treatment and cauliflower than that of tomato . Because in the these experiments *Trichoderma* species could not directly interact with *S. sclerotiorum* or affect the environment for *S. sclerotiorum* development for reasons of spatial separation, induction of systemic resistance is, by exclusion of alternatives, the most likely explanation of the *S. sclerotiorum* control. Moreover, the observed phenomenon shares other characteristics with ISR.First of all, *Trichoderma* spp not only effective against *S. sclerotiorum* but also worked, as like ISR and SAR (Kloepper *et al.;* 1997 and Ryals *et al.,* 1997) against several pathogen. *T. harzianum* seedling treatment reduced anthracnose symptoms of *Colletotrichum lindemuthianum* in bean, gray mould of *Botrytis cinerea* in tomato, lettuces, pepper, bean and tobacco and also reduced white mould in lettuce. Finally the effect of *Trichoderma* spp seedling treatment needed, just like ISR, some times to develop in the host plant. The *S. sclerotiorum* was only efficiently controlled when inoculated one or more days after *Trichoderm*a species treatment) compared with pathogen-induced systemic resistance this lag period of one day is rather short but nonetheless possible, since Smith *et al.*, (1991) detected systemic resistance one day after inoculation with *Pseudomonas syringae* pv. *syrirgae.* However, the real lag period for induction of systemic resistance by *Trichoderma* species could be more than one day because *S. sclerotiorum* infections might still be affected by induced resistance during this slow initial development. In this cause the induced resistance that developed after the *S. sclerotiorum* inoculation probably came too late to suppress spreading blackening formulation, but managed to delay the infection. On the other hand *Trichoderma* species treated leaves the modes of action like competition for nutrients and suppression of *S. sclerotiorum* pathogenicity enzymes, can be involved in *S. sclerotiorum* control.

Growth promotion: Over the last 20 years, there have been an increasing number of reports of promotion of plant growth following treatment of seeds, roots, cuttings, soil or artificial growing media with bacteria and fungi, particularly species of *Pseudomonas* and *Trichoderma* . Depending on the plant studied, growth promotion has been expressed in a variety of ways but most commonly as increases in germination, emergence fresh or dry weight of roots or shoots, root length, yield and flowering. Many aspects of this phenomenon have been reviewed in detail several times (Weller, 1988; Baker, 1989; Kloepper *et al.*,1989, 1991; Inbar *et al.*, 1994; Ryder *et al.*, 1994) and numerous large screening studies have been carried out specifically to search for such plant growth-promoting microorganisms (Ousley *et al.*, 1994a; Tang, 1994; Glick *et al.*, 1995). Indeed, the term plant growth-promoting rhizobacteria (PGPR) has been coined specifically to describe bacteria which colonize roots and have the ability to stimulate plant growth and this has led to a new area of work (Ryder *et al.*, 1994).

Frequently, growth promotion has involved application of known biocontrol agents, but even so, the modes of action involved in the plant growth promotion observed have not always been clear. In soils containing major pathogens such as *Gaeumannomyces graminis var. tritici, Pythium spp. and Rhizoctonia spp,* the growth promotion effect may well reflect biocontrol acting through mechanisms such as competition, antibiosis, parasitism and induced resistance, as described earlier. However, growth promotion can still occur in soils or environments lacking such major pathogens (Sharma 2003). It is then thought to be due either to control of minor pathogens such as deleterious rhizobacteria (Schippers *et al.*, 1987; Baker, 1989) in the same way as major pathogens or to a direct effect on the plant.

Direct effects are commonly thought to be mediated by production of plant hormones such as auxins, cytokinins or gibberellins (Arshad and Frankenberger, 1991). It is difficult to obtain unequivocal evidence for their production in non-sterile soil, although several studies carried out in sterile conditions have implied their involvement. For example, results from a gnotobiotic growth pouch assay showed that *Pseudomonas putida* GR12-2 consistently induced a significant elongation of roots of canola *(Brassicanapus)* compared with controls and was associated with enhanced shoot height and phosphorus uptake by roots and subsequent transfer to shoots (Lifshitz *et al.*, 1987). *Pseudomonas putida* GR 12-2 produced indole-1-acetic acid (IAA) and fixed nitrogen (although at rates 20-fold slower than *Azotobacter vinelandii* based on the acetylene reduction assay, (Hong *et al.*,1991), which may have influenced plant growth directly. However, in addition it also degraded 1-aminocyclopropane-1-carboxylate (ACC), the immediate precursor of ethylene, thereby potentially relieving the normal ethylene-induced inhibition of root development (Glick *et al.*, 1994, 1995). Similarly, work with corn (*Zea mays*), tomato (*Lycopersicon esculentum*) and tobacco (*Nicotiana tabacum)* cultivars suggested that a growth regulator was produced by *Trichoderma* spp. and that this was directly involved with the increased rate of seed germination and dry weight of shoots and stems following inoculation (Windham *et al.*, 1986 , Sharma 2003).

Associative nitrogen fixation may also occur with *Azospirillum* spp., *Azotobacter spp., Bacillus spp.* and possibly some *Pseudomonas* spp. (Kloepper *et al.*, 1989; Hong *et al.*, 1991) to increase plant growth directly. Production of vitamins, conversion of non-utilizable material into a form that can be used by the plant, and improved availability and uptake of some minerals may also contribute to the growth promotion phenomenon (Baker, 1989; Kleifeld and Chet, 1992). Significantly, especially for microbes that are to be used commercially in the future, it would seem important to understand the mode of action involved in growth promotion to ensure reproducibility of effect.

FEATURES OF INDUCED RESISTANCE

The above examples very well highlights the induced resistance concept due to the factors explained and therefore it is clear that till today the defined biological control mechanism in field works through IR than other mechanisms. Induced resistance can be elicited in otherwise susceptible plants by the application of living organisms also known as b c a s (Kuc 1982) or by abiotic treatments (e.g. Ye *et al.* 1989; Stevens *et al.* 1990), and may be expressed only in the treated tissue or considerably beyond the region of treatment. Localized resistance elicited by prior inoculation with incompatible pathogens can most easily be explained when these pathogens trigger defenses whose effects linger sufficiently to inhibit subsequent infection by a compatible pathogen. The more intriguing examples of induced resistance are those in which the inducing treatment does not directly elicit any of the defenses, such as cell death, phytoalexin accumulation, or detectable cell wall modifications, that usually are thought to account for resistance in genotypically resistant plants. These treatments must elicit other localized for systemic changes that either directly interfere with any subsequent invasion by the parasite, or alter the parasite-plant interaction in favour of the plantThat these changes may effect subsequent fungal development is indicated by the fact that, once inside the tissue, fungal growth often appears impaired in comparison with that in unprotected plants (Kovats *et al.* 1991; Ye *et al.* 1992); however, it has yet to be shown that these changes are more important in preventing fungal growth than antimicrobial compounds elicited during fungal infection. In addition, induced resistance also is commonly characterized by a decrease in the frequency of initial penetration of epidermal cells in association with an increased frequency of localized, modifications of the plant cell wall elicited after fungal inoculation (Conti *et al.*, 1990' Kovats *et al.* 1991). Such rapid and localized wall modifications are typical of the expression of basic resistance and it would seem that protected plants may respond more as nonhosts than hosts towards some of their normally successful pathogens. Significantly, compared to normal cucumber plants, protected ones have higher latent activity of membrane-bound *B*-1,3-glucan synthase (Schmele and Kauss 1990), which may account for the faster appearance of callose-containing papillae after fungal challenge of protected plants (Stumm and Gessler 1986; Kovats *et al.* 1991). Moreover, protected plants lignify faster than unprotected plants in response to wounding, probably because of the systemic increase in at least two enzymes, including peroxidase, essential for lignin formation (Dean and Kuc 1987). It seems likely, therefore, that some of the systemic changes induced in protected plants ensure that fungal-induced damage results in rapid wall-associated modifications that may help prevent further fungal ingress. The probable involvement in induced resistance of several pre-existing and fungal-induced defenses indicates that this form of resistance, like basic resistance (Heath 1991), is multicomponent (Ye *et al.* 1990; Kovats *et al.* 1991).

Interestingly, some of the systemic changes that occur in a plant after treatments that induce resistance also occur as a plant ages (Fluhr *et al.* 1991); Wyatt *et al.* 1991). This raises the possibility that the mechanisms of induced resistance and some forms of age-related resistance may be similar, a hypothesis supported in bean (*Phaseolus vulgaris*) by the similar susceptibility of these forms of resistance of heat treatment, and the differing effect of heat on some examples of parasite-specific or nonhost resistance. Studies based on following features are required to understand biologically induced SAR/biological control/IR

- Absence of toxic effects of the inducing agents on the pathogens
- Suppression of the induced resistance by a previous application of specific inhibitors, such as actinomycin D, which affect gene expression of the plant.
- Necessity of a time interval between application of the inducer and the onset of protection in the plant
- Absence of a typical dose-response correlation known for toxic compounds.
- Nonspecifity of protection
- Local as well as systemic protection

- Dependence on the plants genotype causing significant differences in level and type of protection in different cultivars.

CONCLUSION

Problems of the variability of effectiveness of induced systemic resistance under varying climatic and agronomic conditions, encountered with biological control in general, must be addressed. Induced systemic resistance depends upon a physiological response by the plant and therefore can, under certain conditions, be less consistent than chemical pesticides, which kill the infectious agent. Induced systemic resistance utilizes the plant's own defense mechanisms to restrict development of pathogens. It has the advantage of controlling a broad spectrum of plant diseases including those caused by viruses and bacteria for which effective and economically-sound chemical pesticides are not available. Continued research directed at understanding the induction and regulation biologically induced systemic resistance and the defense compounds would help in maximizing reliability. While all the above mentioned biological agents microbial in nature have proven to be useful in specific applications, it is believed the application of inducing chemicals (or compounds purified from micro organisms) will provide more consistent results and ultimately prove to be of greater use in large scale agriculture, and needs applied research in this direction. Study of the mechanisms involved in microbial mediated induction of plant defenses also called as biological control may also elucidate how plant immunization and growth promotion may occur simultaneously, and how these phenomena are affected by environmental factors. Further study of the mechanisms by which microbials affect plants may allow for the discovery of new elicitors, or means of delivering these to plants, in addition to examining the role soil microbes may play in the composition of plant communities.

REFRENCES

Agrios G.N., Plant Pathology, 4th ed. San Diego, CA: Academic Press, 1997.

Alstrom, S. 1991. *Journal of General and Applied Microbiology* 37, 495-501.

Amijee, F., Allan, E.J., Waterhouse, R.H., Glover, L.A., and Paton, A.M. 1992., Biocontrol Science and Technology 2, 203-214.

Anderson, A.J., Guerra, D. 1985. *Phytopathology*, 75, 992-995.

Arribas, A., Revilla, G., Zarra, I., and Lorences, E.P. 1991. *Journal of Experimental Botany*, 42, 1181-1187.

Arshad , M. and Frankenberger Jr., W.T. 1991. (1) L.Keister and P.B. Cregan., eds), pp. 327-334. Kluwer Academic Publishers, Dordrecht, The Netherlands.

Baker, R. 1989. *Trends in biotechnology* 7, 34-38.

Bakker, P.A.H.M., van peer, R., Schippers, B. 1991. in A.B.R. Beemster, G.J. Bollen, M. Gerlagh, M.A. Ruissen, B. Schippers, and A. Tempel, (eds.), *Biotic Interactions and soil-Borne Diseases*, Elsevier, Amsterdam, The Netherlands, pp. 217-223.

Bergstrom, G.C., Johnson, M.C. and Kuc, J. 1982., *Phytopathology*, 72, 922-925.

Bhattacharyya, M.K. and Ward, E.W.B. 1986. 'Expression of gene-specific and age-related resistance and the accumulation of glyceollin in soybean leaves infected with *Phytophthora megasperma f. sp. Glycinea'*, *Physiological and Molecular Plant Pathology* 29, 105-113.

Broglie, K., Broglie, R., Benhamou, N. and Chet, I. 1993." (I. Chet, ed.), pp. 139-156. Wilcy-Liss, Inc., New York.

Broglie, K., Chet, I., Holliday, M., Cressman, R., Biddle, P., Knowlton, S., Mauvais, C.J. and Broglie, R. 1991. *Science*, 254, 1194-1197.

Carling, D.E., Roncadori, R.W., and Hussey, R.S. 1989., *Plant Disease* 73, 730-733.

Carsolio, C., Gutierrez, A., Jimenez, B., van Montagu, M. and Herrera-Estrella, A. 1994. *Proceedings of the National Academy of Sciences,* USA 91, 10903-10907.

Chet, I., ed. 1993. Wiley-Liss, Inc., New York.

Conti, G.C., Bassi, M., Carminucci, D ., Gatti, L., and Bocci, A.M. 1990. *Journal of Phytopathology* 128, 191-202.

Cook, R.J. and Baker, K.R. 1983., *American Phytopathological Society*, St. Paul. MN.

De la Cruz, J., Hidalgo-Gallego, J., Lora, J.M., Benitez, T., Pintor-Toro, J.A. and Llobell, A. 1992. *European Journal of Biochemistry* 206, 859-867.

De la Cruz, J., Pintor-Toro, J.A., Benftez, T., Liobell, A. and Romero, L.C. 1995. *Journal of Bacteriology* 177, 6937-6945.

Destefano-Beltran, L., Nagpala, P.G., Centiner, S.M., Denny, T. and Jaynes, J.M. 1993. *In: Biotechnology in Plant Disease Control* (I. Chet, ed.), pp. 175-189. Wiley-Liss, inc., New York.

Doke, N., Ramirez, A.V., and Tomiyama, K. 1987. *Journal of Phytopathology* 119, 232-239.

Doubrava, N.S., Dean, R.A. and Kuc, J. 1989., *Physiological and Molecular Plant Pathology* 33, 69-79

Dowling, D.N. and O'Gara, F. 1994. *Trends in Biotechnology* 12, 133-141.

Fenton, A.M.. Stephens, P.M., Crowley, J., O'Callaghan, M. and O'Gara, F. 1992. *Applied and Environmental Microbiology* 58, 3873-3878.

Fluhr, R., Sessa, G., Sharon, A., Ori, N., and Lotan, T. 1991., in H.Hennecke and D.P.S. Verma (eds.), *Advances in Molecular Genetics of Plant-Microbe Interactions* Vol. I, Kluwer Academic Publishers, The Netherlands, pp. 387-394.

Frommel, M.I., Nowak, J., and Lazarovits, G. 1991., *Plant Physiology* 96, 928-936.

Gaffney, T.D., Lam, S.T., Ligon, J., Gates, K., Frazelle, A., di Maio, J., Hill, S., Godwin, S., Torkewitz, N., Allshouse, A.M., Kempf, H.J. and Becker, J.O. 1994. *Molecular Plant-Microbe Interactions* 7, 455-463.

Garcia-Garrido, J.M., and Ocampo, J.A. 1988., *New Phytologist* 110, 551-555.

Gawande, S. and Sharma, P 2003. *Annals. Agril. Res.*

Geremia, R.A., Goldman, G.H., Jacobs, D., Ardiles, W., Vila, S.B., Montagu, M.V. and Herrera-Estrella, A. 1993., *Molecular Microbiology* 8, 603-613.

Glick, B.R. , Karaturovic, D.M. and Newell, P.C. 1995. *Canadian Journal of Microbiology* 41, 533-536.

Glick, B.R., Jacobson, C.B., Schwarze, M.M.K. and Pasternak, J.J. 1994. *In* "Improving Plant Productivity with Rhizosphere Bacteria" (M.H. Ryder, P.M. Stephens and G.D. Bowen, (eds), CSIRO, Division of Soils, Adelaide, Australia, pp. 150-152.

Goldman, G.H., Vasseur, V., Contreras, R. and van Montagu, M. 1994b. *Gene* 144, 113-117.

Gregersen, P.L., and Smedegaard, V. 1989. *Journal of Phytopathology* 124, 128-136.

Guedes, M.E., Richmond, S. and Kuc, J. 1980., *Physiological Plant Pathology* 17, 229-233.

Gutterson, N. 1990. *Critical Reviews in Biotechnology* 10, 69-91.

Gutterson, N.I., Layton, T.J., Ziegle, J.S. and Warren, G.J. 1986. *Journal of Bacteriology* 165, 696-703.

Haran, S., Schickler. H., Pe'er, S., Logemann, S., Oppenheim, A. and Chet I. 1993. *Biological Control* 3, 101-108.

Hayes, C.K., Klemsdal, S., Lorito, M., Di Pietro, A., Peterbauer, C., Nakas, P., Tronsmo, A. and Harman, G.E. 1994. *Gene* 138, 143-148.

Heath, M.C. 1991. 'Evolution of resistance of fungal parasitism in natural ecosystems', (Tansley Review No. 33), *New Phytologist* 119, 331-343.

Hill, D. S., Stein, J.I., Torkewitz, N.R., Morse, A.M., Howell, C.R., Pachlatlko, J.P., Becker, J.O. and Ligon, J.M. 1994. *Applied and Enviornmental Micronb.* 60, 78-85.

Hong, Y., Pasternak, J.J. and Glick, B.R. 1991. *Canadian Journal of Microbiology* 37, 796-799.

Howie, W. J. and Suslow, T.V. 1991. *Molecular Plant-Microbe Interractions* 4, 393-399.

Hynes, R.K., and Lazarovits, G. 1989. *Canadian J. of Plant Pathology* 11, 191.

Inbar, J. and Chet, I. 1994. *Microbiology* 140, 651-657.

Jach, G., Logemann, S., Wolf, G., Oppenheim, A., Chet, I. , Schell, J. and Logemann. J. 1992. *Biopractice* 1, 33, 41.

Jenns, A. and Kuc, J. 1979. *Phytopathology* 69, 753-756.

Kapulnik, Y. 1991., in Y. Waisel, A. Eshel, and U. Kafkafi, (eds.), *Plant Roots: The Hidden Half*, Marcel Dekker, Inc., New York, NY, pp. 717-729.

Karban, R. 1987., *Oecologia* 73, 414-419.

Karban, R., and I.T. Baldwin. 1997. University of Chicago Press, Chicago.

Keel, C. and Defago, G. 1997. Interactions between beneficial soil bacteria and root pathogens: mechanisms and ecological impact. *In: Multitrophic Interactions in Terrestrial Systems* (A.C. Gange and V.K. Brown, eds) pp. 27-47. Blackwell Science, Oxford.

Kleifeld, O. and Chet, I. 1992. *Plant and Soil* 144, 267-272.

Kloepper, J.W. 1991. ' J. Bay-Petersen, (ed.), Biological Control of Plant Diseases. FFTC Book Series No. 42. Food and Fertilizer Technology Center, Taipei, Taiwan pp. 142-156.

Kloepper, J.W., Lifshitz. R. and Zàblotowicz, R.M. 1989. *Trends in Biotechnology* 7, 39-44.

Kopp, M., Rouster, J., Fritig, B., Darvill, A. and Albersheim, P. 1989. *Plant Physiology* 90, 208-216.

Kovats, K., Binder, A., and Hohl, H.R. 1991. *Planta* 183, 484-490.

Kraus, J. and Loper, J.E. 1995. *Applied and Environmental Microbiology* 61, 849-854.

Kroon, B.A.M., Scheffer, R.J. and Elgersma, D.M. 1991. *Netherlands Journal of Plant Pathology* 97, 401-408.

Kuc, J. 1982. Induced immunity to plant disease. *Bioscience* 32, 854-860.

Leeman, M., van Pelt J.A., den Ouden, F.M., Heinsbrock, M., Bakker, P.A.H.M.and Schippers, B. 1995a. Induction of systemic resistance by *Pseudomonas fluorescens* in radish cultivars differing in susceptibility to Fusarium wilt, using a novel bioassay. *European Journal of Plant Pathology* 101, 655-664.

Leeman, M., van Pelt, J.A., Hendrickx, M.J., Scheffer, R.J., Bakker, P.A.H.M. and Schippers, B. 1995b. Biocontrol of Fusarium wilt of radish in commercial greenhouse trials by seed treatment with *Pseudomonas fluorescens* WCS374. *Phytopathology* 85, 1301-1305.

Lifshitz, R., Kloepper, J.W., Kozlowski, M., Simonson, C., Carlson, J., Tipping, E.M. and Zaleska, I. 1987. *Canadian Journal of Microbiology* 33, 390-395.

Liu, L., Kloepper, J.W. and Tuzun, S. 1992. *Phytopathology* 82, 1108.

Liu, L., Kloepper, J.W. and Tuzun, S. 1993a, *Phytopathology* 83, 1340.

Liu, L., Kloepper, J.W. and Tuzun, S. 1993b. Proceedings of the 6th International Congress of Plant Pathology, p. 214.

Liu, L., Kloepper, J. W. and Tuzun, S. 1995a. *Phytopathology* 85, 843-847.

Liu, L., Kloepper, J.W. and Tuzun, S. 1995b. *Phytopathology* 85, 1064-1068.

Maiss, E. 1987. *Archiv fur Phytopathologie und Pflanzenschutz* 4, 275-283.

Mandeel, Q. and Baker, R. 1991. *Phytopathology* 81, 462-469.

Martyn, R.D., Biles, C.L. and Dillard, E.A. 1991. *Plant Disease* 75, 874-877.

Maurhofer, M., Hase, C., Meuwly, Ph., Metraux, J.P. and Defago, G. 1994b. *Phytopathology* 84, 139-146.

Maurhofer, M., Keel, C., Haas, D. and Defago, G. 1995. *Plant Pathology* 44, 40-50.

Meera, M.S., Shivanna, M.B., Kageyama, K. and Hyakumachi, M. 1994. *Phytopathology* 84, 1399-1406.

Meera, M.S., Shivanna, M.B., Kageyama, K. and Hyakumachi, M. 1995. *Crop Protection* 14, 123-130.

Meyer, G. De. Bigirmana , J. Elad, Y. and Hofte M, 1998. *Eur. J. Plant Pathology* 104 : 279-286

Mucharromah, E. and Kuc, J. 1989. *Phytopathology* 79, 1144.

Ousley, M.A., L:ynch, J.M. and Whipps. J.M. 1994a. *Scientia Horticulturae* 59, 147-155.

Ozeretskovskaya, O.L., Chaleva, L.I., Avduishko, S.A., Chalenko, G.I. and Karavaeva, K.A. 1988., *Soviet Plant Physilolgy* 35, 141-148.

Peter. S., Barak, Z., Yarden, O. and Chet, I. 1991. *Soil Biology and Biochemistry* 23, 1043-1046.

Ryals, J.K. Weymann, K.Lawton, L.Friedrich,D. Ellis, H-Y. Steiner,J.Johnson,T. Delaeney, J. Taco, P.Voc, and S. Uknes.1997. *The Plant Cell* 9: 425-439.

Reiss, E., Schubert, J. and Naumann, K. 1988., *Archiv fur Phytopathologie und Pflanzenschutz* 24, 335-340.

Research briefing. Washington, DC: National Academy Press pp. 191.

Rosendahl. S. 1985. *Phytopathologische Zeitschrift* 114, 31-40.

Roveratti, D.S., Teixeira, A.R.R. and Morai, W.B.C. 1989. *Journal of Phytopathology* 126, 149-159.

Ryder, M.H., Stephens, P.M. and Bowen, G.D., eds 1994. CSIRO, Division of Soils, Adelaide, Australia.

Scheffer, R.J. 1983., *Annals of Applied Biology* 103, 21-30.

Schinder. U., Keel, C., Defago, G. and Haas, D. 1995b. *Applied and Environmental Microbiology* 61, 3856-3864.

Schmele, I., and Kauss, H. 1990., *Physiological and Molecular Plant Pathology* 37, 221-228.

Schmidt, D. 1988., *Journal of Phytopathology* 122, 253-260.

Schonbeck, F., Dehne, H.W. and Beicht, W. 1980. 87, 654-666.

Schroth, M.N., and Becker, J.O. 1990., In D. Hornby, (ed.), *Biological Control of Soil-Borne Plant Pathogens*, CAB International, Wallingford, Oxon, U.K., pp. 389-414.

Shapira, R., Ordentlich, A., Chet, I. and Oppenheim, A. B. 1989. *Phytopathology* 79, 1246-1249.

Sharma, Pratibha 2002. *Annals plant Prot. Science* 10(1): 99-103

Sharma, Pratibha 2003. In Proceedings of International Seminar on Downsizing Technology for Rural Development at Bhubaneswer Vol. I pp. 193-202.

Shurtleff, M.C., Averre, C.W., III. *Glossary of Plant Pathological Terms*. APS Press, St. Paul, MN, 1997.

Smith, J. A., D. W. Fulbright and R. Hammerschmidt. 1991. *Physiological and Molecular Plant Pathology*. 38:223-235.

Sneh, B., Ichielevich-Auster, M. and Plaut, Z. 1989. *Canadian Journal of Botany* 67, 2135-2141.

Steiner, U., Oerke, E.C., and Schonbeck, F. 1988. *Journal of Plant Diseases and Protection* 95, 506-517.

Stevens, C., Khan, V.A., Tang, A.Y., and Lu, J.Y. 1990. *Journal of Food Protection* 53, 223-226.

Stumm, D., and Gessler, C. 1986., *Physiological and Molecular Plant Pathology* 29, 405-410.

Sundheim, L., Poplawsky, A.R. and Ellingboe, A.H. 1988. *Physiological and Molecular Plant Pathology* 33, 483-491.

Tang, W., Pasternak, J.J. and Glick, B.R. 1995. *Canadian Journal of Microbiology* 41, 445-451.

Tang, W.H. 1994. *In* "Improving Plant Productivity with Rhizosphere Bacteria" (M.H. Ryder, P.M. Stephens and G.D. Bowen, eds.), pp. 267-273, CSIRO, Division of Soils, Adelaide, Australia.

Thomashow, I. S. and Weller, D.M. 1988. *Journal of Bacteriology* 170, 3499-3508.

Thompson, C., Dunwell, J. M. Johnstone. C. E., Lay, V., Ray, J., Watson, H. and Nisbet. G., 1995. (J.M. Whipps and M. Gerlagh, eds), pp. 34-38. *IOBC/WPRS Bulletin* 18 (3).

Trivedi, N. and Sinha, A.K. 1976. *J. of Phytopathology* 86, 335-344.

Tuzun, S. 2001. *European J. Plant Pathol.* 107: 85-93

Tuzun, S. and Kuc, J. 1985a. 'A modified technique for inducing systemic resistance to blue mold and increasing growth of tobacco', *Phytopathology* 75, 1127-1129.

Tuzun, S. and Kuc, J. 1989. in W.E. McKeen (ed.), Blue Mold of Tobacco, *American Phytopathological Society Press*, pp. 177-200.

Tuzun, S. and Kuc, J. 1991 in J.-Bay-Petersen (ed.) Biological Control of Plant Diseases, *FFTC Book Series No.* 42, Taiwan, pp. 30-40.

Tuzun, S., Juarez, J., Nesmith, W.C. and Kuc, J. 1992. *Phytopathology* 82, 425-429.

Van Drieshce, R.G., 1996. Jr. Pest origins, pesticides and the history of biological control. In: RG Van Drieshce, TS Bellows, Jr., (eds.), Biological Control. New York: Chapman and Hall, pp 3-18.

Van Peer, R. and Schippers, B. 1992. *Netherlands Journal of Plant Pathology* 98, 129-139.

Van Peer, R., Niemann, G.J. and Schippers, B. 1991. *Phytopathology* 81, 728-734.

Vincent, M. N., Harrison, L. A. Brackin, J. M., Kovacevich, P. A., Mukerji, P., Weller, D. M. and Pierson, E. A. 1991. *Applied and Environmental Microbiology* 57, 2928-2934.

Voisard, C., Keel, C., Haas, D. and Defago, G. 1989. *European Molecular Biology Organisation J.* 8, 351-358.

Wei. G., Kloepper. J.W. and Tuzun, S. 1991. *Phytopatholgy* 81, 1508-1512.

Weller, D. M. 1988. *Annual Review of Phytopathology* 26, 379-407.

Windham, M. T., Elad, Y. and Baker, R. 1986. *Phytopathology* 76, 518-521.

Wyatt, S.E., Pan, S.Q., and Kuc, J. 1991. *Physiological and Molecular Plant Pathology* 39, 433-440.

Ye, X.S., Pan, S.Q, and Kuc, J. 1989. 'Pathogenesis-related proteins and systemic resistance to blue mould and tobacco mosaic virus induced by tobacco mosaic virus, *Peronospora tabacina* and aspirin', *Physiological and Molecular Plant Pathology* 35, 161-175.

Ye, X.S., Pan S.Q., and Kuc, J. 1990. '*Physiological and Molecular Plant Pathology* 36, 523-531.

Ye, X.S., U. Jarlfors, S. Tauzun, S. Q. Pan and J. Kuc. 1992. *Canadian J. of Botany* 70:49-57.

Yoneyama, K. and Anzai, H. 1993. Transgenic plants resistant to diseases by the detoxification of toxins. *In* "Biotechnology in Plant Disease Control" (I. Chet, ed.), Wiley-Liss, Inc., New York pp. 115-137.

Zhou, T. and Paulitz, T. C. 1994. *J. of Phytopathology* 142, 51-63.

Zhu, Q., Mahur, E. A. Masoud, S., Dixon, R.A. and Lamb, C. J. 1994. *Bio/Technology* 12, 807-812.

7

Breeding for Powdery Mildew Resistance in Pea

R.R. SHARMA[1] AND V. R. ANIL KUMAR[1]

The garden pea (*Pisum sativum* L.) is an important commercial vegetable grown in both hills and plans of India. It occupies an area of 0.2 million hectares with an annual production of 2 million tonnes making India first in the world in both area and production (FAO, 2000).

The importance of peas as a pulse and as a vegetable crop in human diet needs no emphasis. Peas are highly nutritive and contain a high percentage of digestible protein, along with carbohydrates and vitamins (Choudhury, 1967). Besides they contain a higher proportion of mineral matter. It is an excellent food for human consumption take either as a vegetable or in soup, canned, frozen or dehydrated. On account of its high nutritive value it is an important component for balanced nutrition. It has good marked potential in dehydrated and canned from (Nath *et al.*, 1976).

The garden pea is prone to attack of various diseases. Powdery mildew disease poses a serious limitation for the production of mid, late and seed crop of garden pea in north India plains (Tewatia and Kalloo, 1992). It is an important cause for reduction and instability in pea production (Narsinghani and Tiwari, 1993). The disease is cuased by *Erysiphe polygoni* DC. (Syn. *Erysiphe pisi* Syd.) It occurs late in the season reaching its maximum intensity at the time of pod formation (Nath *et al.*, 1976). It has been found to occur in almost all pea-growing regions of India (Narginghani, 1969). Wide occurate of powdery mildew disease of pea has been reported from Australia, Canada, France, Peru, South Africa, the UK, and the USA and in Zimbabwe, the disease has caused the cessation of pea production (Dixon, 1981).

POWDERY MILDEW

Powdery mildew of peas caused by *Erysiphe polygoni* has been reported as early as 1918 by E. J. Butler (Butler, 1918). Uppal *et al.*, (1935) reported pea powdery mildew in Bombay. According to Paul and Kapoor, 1983, the various species described under the name *Erysiphe polyoni* on different hosts in India were found to comprise 8 species, *E. polygoni*, *E. pisi*, *E. berberidis*, *E. betae*, *E. heraclei*, *E. martii*, *E. ranunculi* and *E. salviae*. This disease is more serious than downy mildew due to the fact that it occurs more frequently and covers a larger area of the host surface than does the downy mildew of the same corp. It usually develops late in the season reaching its maximum intensity when the pods are forming. It is worst in dry weather, unlike downy mildew which flourishes in humid weather.

Physiological races of the fungus attack beans, urd bean, lucerne, coriander, turnip, cabbage and many other plans of different families. The dieases is of worldwide occurrence. Early varieties are less damaged. Varieties maturing in January usually escape the maximum intensity of the disease (Singh, 1985). The loss is proportional to the disease intensity and varies considerably depending on the stage at which the disease occurs. In a hundred per cent infected crop, loss in yield in estimated to be about 21 to 31 per

1. Division of Vegetable Crops, I.A.R.I., New Delhi-110 012.

cent in terms of pod number and 26 to 47 per cent in terms of pod weight (Munjal *et al.*, 1963). The distribution, sympton, pathogen, disease cycle, host range, losses and control of powdery mildew of pea have been reviewed by Singh and Singh (1988).

Symptoms of the disease

The disease first appears on the upper surface of the lower most leaves as white almost circular colonies. White powdery patches appear on either side of leaves, on tendrils, pods and stems. These patches originate as minute discoloured specks from which a powdery mass radiates on all sides. When the attack has advanced, large areas on the aerial parts of the host may be covered with these white floury patches. The superficial mass consists of the mycelium and spores of the fungus. Histological symptoms include collapse of the necrotic epidermal cells in case of sub infection, collapse of tissues below the penetrated epidermal cells, and movement of the host nuclei towards the haustoria of the fungus. Physiology of the host is also affected. There is increased transpiration, especially during night. Respiration is also increased and photosynthesis is decreased. Plants become stunted if infected at young stage. On stems, the basal part is infected first. Immature pods are more prone to infection. Infected pods gradually shrivel and finally dry. The heavily infected plants exhibit apparently talcum powder like appearances from out side. Leaves become small, colour changes from green to pale yellow and become stunted (Singh, 2000). The vines dwarf, the tissue dries up and if infection is severe, the plant may die (Chupp, 1998). Yield reduction occurs due to reduction in pod number and number of seeds per sod (Singh *et al.*, 1978).

Disease cycle

White mold on the plant surface is made of short stalks on which conidia are borne. Black globose perithecia contain few to many four spored asci. Ascospores in the perithecia can over winter and serve as inoculum for primary infection. Conidia can germinate and give rise to secondary infection (Parry, 1990). The pathogen is also considered to be seed borne. When pods are affected by the mildew, the fungus may grow through the pod into the seeds so that the seeds adhere to the pods. Crawford (1927), Chupp and Sherf (1960) and Dixon (1981) documented seed borne nature of infection, but Jhooty and Munshi (1990) ruled out this possibility.

Predisposing factors

To have a suitable screening procedure against the disease, it becomes necessary to study the various environmental factors that affect the incidence and development of disease. Hagedorn (1973) reported that powdery mildew epidemics are favoured by prolonged warm days and nights cool enough for few formation.

Temperature

Disease development as a response to temperature should be the test criterion of temperature relations of a disease caused by a powdery mildew or any other pathogen (Corneli, 1934). Uppal *et al.*, (1935) reported 22°C while Yarwood *et al.* (1954) reported 25°C as optimum temperature for development of powdery mildew in peas.

Moisture

Powdery mildew spores do not require free water for germination unlike most other fungi. High humidity at leaf surface can facilitate spore germination (Westcott, 1971). In general the growth and spread of powdery mildew pathogen will be slow during high moisture conditions like rainfall.

Relative humidity

Powdery mildew is known to thrive under a wide range or relative humidity (Schnathorst, 1965). Conidia of *Erysiphe pisi* can germinate under a wide range of 20 to 100 per cent relative humidity (Paul and Kaushal, 1985). Sivapalan (1993) found negative relation of powdery mildew with relative humidity, which is associated with rainfall. Rainfall increase relative humidity and washers down as well as inhibits development of powdery mildew.

BREADING PASS FOR RESISTANCE TO POWDERY MILDEW

Von Hmammerlund (1925) first observed resistance to pea powdery mildew. Breeding for resistance requires a suitable and efficient screening and scoring techniques to identify sources of resistance. Knowledge of mode of inheritance of these resistance genes is also an important prerequisite. Various works like Harland (1948); Pierce (1948); Saxena *et al.* (1975); Narsinghani (1997) etc. have conducted studies so as to know the mode of inheritance of powdery mildew resistance.

Screening for resistance

Screening for resistance adopted should be quick and efficient. It should be able to identify critically the resistant genotypes in the field from disease escaping ones and other susceptible types. Screening can be done by different methods like:

Late planting in the field

In this method the breeding material is planted late in the season when the temperature starts rising. The environmental conditions will be congenial for the development of the disease. Susceptible border rows are also planted to maximize the spread of the infection.

Detached leaf culture technique

Here the leaflets from 30 to 45 days old plants were detached and inoculated by dusting inoculums on to then using tap water in glass petridishes at 22°C

Spraying powdery mildew suspension

Spraying the spore suspension prepared from previously infected plants can also create the epidemic. This suspension after testing for the contents of optimum conidia can be sprayed on to the standing breeding material to incite the infection by the powdery mildew fungus.

Border planting or spreader rows of highly susceptible varieties.

The border row planting of highly or know susceptible varieties will help to develop the epidemic in a large scale. It will then spread to the breeding materials in the field and hence can be used to screen the breeding material.

For scoring the disease, different scales have been used by different workers. Munjal *et al.* (1963) developed an easily recognizable five-grade scale (A to E) based on disease development as follows:

A. Plant completely healthy with no trace of infection or any part.

B. 25% infection, slight infection, fine coating of powdery growth on the upper surface of leaves, stems free from infection and plant size normal.

C. 50% infection, which is moderate, the upper leaves more severaly infected, stems free from infection and plant size normal.

D. Nearly 75% of the foliage infected, the whole plant appears to be covered with white coating, stem also severely infected and plant slightly stunted.

E. 100 per cent infection, all the leaves of the plant as well as stems heavily coated with the fungal growth, leaves turn pale green to yellow and start drying up and the plant conspicuous because of stunted growth.

Note: Yield of grade A (healthy plant) is taken as reference grade to estimate the loss in yield due to other grade of infection.

Singh and Singh (1977) used a modified form of scale suggested by Mains and Deitz (1930). Singh (1991) developed a zero to nine point scale for the evaluation of resistance to pea powdery mildew. It is suggested that genotypes with 0 to 3 ratings may be used as donors for effecting crosses in resistance breeding, 4 to 6 if agronomically superior may be used as cultivars and those with 7 to 9 rating should be restricted to germsplasm pool only.

Some other scales used to score powdery mildew disease in pea were 0 – 4 scale (Narsinghani, 1978) and 0 – 5 scale (Sokhi *et al.*, 1979, Ram, 1992) etc.

Sources of resistance

Table 1: Cultivars/lines which were reported to have resistance against powdery mildew

Sl. No.	Cultivars/lines	Reference
1.	EC 24959, Pefr. 326, IC 4604	Sohi *et al.* (1974)
2.	EC 33866, T 163, GC 3222, 6115	Datar and Bhalerao (1977)
3.	T 10, P-185, P 388, 6588, 6587, 404 B	Narsinghani (1978)
4.	P 185, 6587, PC/P6578B	Sharma *et al.*, (1980)
5.	P 388	Saharan & Gupta (1983)
6.	A 474-288, A474-289, S-143	Singh *et al.* (1983)
7.	Sr. No. 32, JP 501, VP 7906, PH 1-1, HUVP-1	Kallo and Singh (1988)
8.	P6588, Rachna, P6587, P383, P388, DPS3, DPS 6, HUP 2, DMR 9, P 185, P6583	Kamlesh *et al.*, (1991)
9.	EC 21857, EC269429, HFP-4, HFP-8712, HFP-8718	Dang *et al.*, (1994)
10.	HPPC-71, HPPC, 95, New Era, Plant P-9, Dark Skin Perfection, HPPC-42, HPPC-61, HPPC-63	Kapoor (1994)
11.	DPH 19, JP 4, JP 71, JP 83, PM 5, Pant P-5, PMR-5, PMR-3, PMR-4, Sugar Giant Tall, PLP 555, DPH 46, HFP-4	Jindal *et al.* (1995)
12.	Sugar Giant, Pant P-8, DDP-54, PMR-3, JP-71, DPP-26, HPPC-95	Thakur *et al.*, (1996)
13.	VG-1, HPPC-16, HPCC-17, HPCC-95, Rachna, Pant P-9, PH-1, JP 514-A, VG-9, Sugar Giant	Malhotra *et al.*, (2000)

Genetic basis of resistance

Plant disease results from interplay of three factors viz., host plant, pathogen and environment. Development of a resistant variety aims at genetic manipulation of the host component of disease triangle. Mode of inheritance of resistance gene plays an important role in choosing the breeding method to be followed. Studies on inheritance of resistance to powdery mildew in pea have been carried out under field conditions except for some of the experiments of Heringa *et al.*, (1969) which were carried out in a heated house where the air temperature could be maintained at about 20°C.

Harland (1948) was the first to study the mode of inheritance of powdery mildew resistance in pea. He observed that a single recessive gene governs resistance and there is tight linkage between the gene for susceptibility and the anthocyanin factor, 'A.' Single receive nature of powdery mildew resistance was also reported by Pierce (1948); Saxena *et al.* (1975); Narsinghani (1979); Mishra & Shukla (1984); Singh *et al.* (1986); Kalia and Sharma (1988); Narsinghani and Bedi (1990); Singh and Singh (1991); Sharma *et al.*

(1996) and Vaid and Tyagi (1997) while Heringa *et al*. (1969); Sokhi *et al*. (1979); Kumar and Singh (1981) and Ram (1992) attributed resistance due to duplicate recessive genes. Hence there is controversy regarding the mode of resistance gene(s) inheritance (Table 2).

Table 2: Controversies regarding Inheritance

Resistant gene involved	Mode of inheritance	Reference
er	Monogenic recessive	Harland (1948)
er	Monogenic recessive	Saxena *et al*., (1975)
er	Monogenic recessive	Singh *et al*., (1983)
er	Monogenic recessive	Kalia and Sharma (1988)
er	Monogenic recessive	Vaid and Tyagi (1997)
er	Monogenic recessive	Janila *et al*. (2001)
er-1, *er*-2	Double recessive	Heringa *et al*. (1969)
er-1, *er*-2	Double recessive	Sokhi *et al*. (1969)
er-1, *er*-2	Double recessive	Kumar and Singh (1981)
er-1, *er*-2	Double recessive	Ram (1992)
er-1, *er*-2	Double recessive	Tiwari *et al*. (1977)

There is lack of diversity for resistance to *E. polygoni* at global level since the same recessive gene confers resistance in material collected from different geographical areas. The pattern of inheritance of resistance gene was also similar with five isolates of pathogen. Hence, it is suggested that the same recessive gene offers resistance to all the five isolates of *E. polygoni* (Vaid and Tyagi, 1997).

Biochemical basis of resistance

Accumulation of phenolic compounds in host parasite reaction is a general phenomenon of disease resistance and the degree of resistance is determined by the rate of accumulation and breakdown of these compounds (Frakas and Kiraly, 1962).

There is higher levels of total phenols and orthodihydroxy phenols (Sindhan and Parashar, 1984; Kalia and Sharma, 1988; Kalia, 1989; Guleria *et al*., 1998 and Kalia and Kalia 1998) as well as higher activity of phenol oxidizing enzymes like peroxidase and polyphenol oxidase in resistant cultivars compared to susceptible cultivars (Kirik *et al*., 1974; Kalia and Sharma, 1988; Kalia, 1989; Guleria *et al*., 1998 and Kalia and Kalia, 1988). Production of pistain (phytoalexin) in response to infection by *E. polygoni* on pea have also been reported (Oku *et al*., 1975).

Powdery mildew infection induces the synthesis of pathogenesis related proteins like beta-1,3 gluconase and the synthesis is more conspicous and drastic in resistant as compared to powdery mildew susceptible genotypes (Sujay *et al*., 2000a). A critical limite of beta-1, 3 gluconase for resistance response exist between 1.64 and 2.05 micro mol glucose eq. per min. per mg proteins (Sujay *et al*., 2000b).

Stability of resistance

Breaking down of resistance conferred by *er* gene was encountered by some investigators. Schroader and Provvidenti (1965) were the first to encounter this in completely immune lines. They attributed this due to yet unidentified reasons.

Narasinghani and Bedi (1990) also observed breaking down of resistance in advanced generation lines of garden pea, which also could not be explained.

Durability of genes is important since resistance mediated by single gene is limited by the appearance of new pathogen races that are able to overcome host resistance (Johnson, 1984).

BREEDING METHODOLOGIES AND ACHIEVEMENTS

Powdery mildew resistance gene in pea being monogenic, back cross breeding can be use to transfer the gene. Apart from this pedigree method can also be used (Ram *et al.,* 1981). Biparental approach can supplement the classical breeding methods since the biparental progenies are superior in resistance to powdery mildew by the accumulation of favourable low frequency genes (Kalia and Sharma, 2000). Some of the resistance varieties developed are listed in Table 3.

Table 3: Resistant varieties developed

Variety	Method	Remarks
JP 83	Double cross	Mid season, vegetable type
JP 71	Double cross	Mid season, vegetable type
PRS4	Hybridization	Batri, vegetable type
JP 179	Hybridization	Batri for upland culvation
JP 169-1	Hybridization	Duel purpose Batri
JP 501	Hybridization	High resistant, vegetable type
JP 293	Hybridization	Batri for utea cultivation (Substitute for *Lathyrus*)
JP 54	Double cross	Early variety, vegetable type
JM 5	Hybridization	Late and vegetable type
Taichung 12	Hybridization	Vegetable type
Pusa pragati	Introduciton from Germany	Vegetable type
Arka Ajit	Hybridization	Vegetable type
Arka Karthik	Hybridization	Vegetable type
K.S. 245	Hybridization	Mid season, Vegetable type
Arka Sampoorma	Hybridization	Edible podded

USE OF MARKERS IN RESISTANCE BREEDING

Apart from conventional breeding techniques biotechnological tools were also utilized in powdery mildew resistance breeding. Morphological as well as molecular markers are found associated with powdery mildew resistance gene(*er*).

The obligate parasitic nature of *E. polygoni* makes it impossible to maintain the pathogen in culture and difficult to apply in screening progenies in controlled growth conditions. The occurrence of the disease under field conditions is dependent upon the occurrence of appropriate environmental conditions. However, selection of pea genotypes containing both *er*-1 and *er*-2 on the basis of visual scoring is difficult, since *er*-1 alone provides a high level of resistance. Therefore, the development of reliable and user friendly specific primers closely linked to er-1 provides useful tools for breeders for gene pyramiding. The pyramiding of er-1 and er-2 in cultivars would increase the durability of resistance (Tiwari *et al.*, 1997).

Hence, it is clear that identification of markers linked with resistance gene facilities in marker-assisted selection. Kumar and Singh (1981) observed three markers, namely afila (*af*, chromosone 1), stipule reduced (*st*, chromosome 3), and clvivula (*tl*, chromosome 7) in powdery mildew resistant culviar S-143. They also found monogenic recessive segregation of these three markers.

A Repulsion phase Random Amplified Polymorphic DNA (RAPD) marker for powdery mildew resistance gene have been identified by Thimmerman *et al.* (1994) while Tiwari *et al.* (1998) reported the development of three RAPD markers closely linked to er-1, of which one is in coupling phase and two are in repulsion phase. Linkage analysis by Sakr and Muehlbaur, 1996 also revelaed the location of *er* gene on chromosome 7.

CONCLUSION

Breeding for powdery mildew resistance is faced with the controversial reports regarding the mode of inheritance. It is quite likely that the difference may be due to difference in the material used by investigators.

- There is need for developing cultivars having quantitative resistance to powdery mildew in pea, as qualitative resistance may not be durable/stable.
- Breeding work for powdery mildew should be carried out in well-identified hot spots or under controlled condition to avoid the masking effect caused by other gene(s).
- Although most of the scientists till date believe that *er*-1 gene controls powdery mildew resistance in pea, there is lot of disparity regarding number of gene(s) governing the nature of inheritance and its expression and can be eluciadated by the use of modern molecular techniques.
- Construction of genetic and linkage maps with other morphological characters should be identified for wide applicability in conventional breeding.

REFERENCES

Butler, E.J. 1918. *Fungi and disease in plants*. Thacker, Spink and Co., Calcutta, pp. 457.

Choudhury, B. 1967 *Vegetables*. National Book. India. pp. 230.

Chupp, C. 1998. *Manual of vegetables plant diseases*. Discovery Publishing House. New Delhi. pp. 668.

Chupp, C. and A.F. Sherf 1960. *Vegetable diseases and their control*. Constable and Company Ltd. London. pp. 693.

Corneli, 1934. Temperature of germination of fungal spores in relation to infection of the hosts. Re*v. Appl. Mycol.*, 13: 719.

Crawford, R.F. 1927. Powdery mildew of peas. *Agr. Exp. Sta. Bull.*, 163. New Mexico.

Dang, J.K., M.S. Sangwan., R.K. Yadav and R.K. Chowdhury 1994. Sources of resistance against podery mildew in pea. *Legume Res.*, 17: 231-232.

Datar, V.V. and T.S. Bhalerao 1977. Verietal reaction of pea to powdery mildew in Maharashtra. *Indian Phytopath.*, 30: 283.

Dixon, G.R. 1981. *Vegetable crop diseases*. Macmillan publishers Ltd.P.404.

FAO. 2000. *Quarterly bulletin of statistics*. Food and Agriculture Organization, United Nations, Rome. 54

Farakas, G.L. and Z. Kiraly 1962. Role of phenolic compounds in physiology of plant disease and disease resistance. *Phytopathology*, 44: 105-150.

Guleria, S., B. Paul and K.L. Bajaj 1998. Levels of phenols and phenol metabolizing enzymes in powdery mildew (*Erysiphae polygone* D.C.) of pea. *Pl. Dis. Res.*, 13: 181-183.

Hagaedorn, D.J. 1973. Peas. In: *Breeding plants for disease resistance:concepts and applications*. R.R.Nelson (Ed.), State Univ. Press. Pennysylvania.

Harland, S.C. 1948. Inheritance of immunity to mildew in peruvian forms of *Pisum sativum, Heredity*, 2:263-269.

Heringa, R.J.A, Vannorel and M.F. Tazelaar 1969. Resistance to powdery mildew (*Erysiphe Polygoni* D.C.) in peas (*Pisum sativum* L.) *Euphytica,* 18: 163-169.

Janila, P., B. Sharma and S.K. Mishra (2001). Inheritance of powdery mildew resistance inpea (*Pisum sativum* L.) *Indian J. Genet.*, 61: 129-131.

Jhonson, R. (1984). A critical analysis of durable resistance. *Ann. Rev. Phytopath.*, 22: 309-330.

Jhooty, J.S. and G.D. Munshi (1990). Mode of perpetuation of powdery mildew of pea in Northwest India. *Indian Phytopath.*, 43: 455-456.

Jindal, K.K., S.K. Gupta, K.R. Shyam and K.B. Rastogi (1995). Evaluation of pea germplasm against bacterial blight and powedery mildew. *Pl. Dis. Res.*, 10: 35-38.

Kalia, P (1989). Genertics of powdery mildew resistance in pea. Egypt. *J. Genet. Cytol.*, 18: 17-25.

Kalia, P and A. Sharma (2000). Genetic analysis for quality traits and powdery mildew incidence in bipaental progenies of garden pea. *Acta Hori.*, 521: 111-117.

Kalia, P and P. Kalia (1998). Enzymic association of powdery mildew resistance in garden pea. *Veg. Sci.*, 25: 166-168.

Kalia, P and S.K. Sharma (1988). Biochemical genetics of powdery mildew resistance in pea. *Theor. Appl. Gen.*, 76: 795-799.

Kalloo, G and Singh, A (1988-89). Annual Report, Dept. *Vegetable crops*, HAU, Hisar.

Kamlesh, Mathur., K.S. Shekhawat., K.S. Bhatia, and L.K. Mathur, (1991) performance of pea varietires against powdery mildew. *Indian Phytopath.*, 44: 137.

Kapoor, A.S. (1994). Kitsno and G.F. Basyuk (1974). The role of redox enzymes in the resistance of pea to Erysiphe polygoni DC. *Nauk. Pr. Ukr. Sils-Kogospod Akad.*, 84: 198-201.

Kumar, H. and R.B. Singh 1981. Genetic anylysis of adult plant resistance to powdery mildew in pea (*pisum sativum L.*) *Euphytica*, 30: 229-239.

Mains, E.B. and S.M. Deitz 1930. Physiologic forms of barley mildew. Erysiphe graminis hordei marchal. *Phytopathology*, 20: 229-239

Malhotra, S.K. and A. Singh 2000. Diversity for resistance to powdery mildew in peas. *Legume Res.*, 23: 60-70.

Mishra, S.P. and Shukla 1984. Inheritance of powdery mildew in pea. *Z. pflanzenzucht.*, 93: 251-254.

Munjal, R.L., V.V. Chenulu, and T.S. Hora 1963. Assessment of losses due to powdery mildew (*Erysiphe polygoni* DC.) on pea. *Indian Phytopath.*, 16: 268-270.

Narsinghani, V.G. 1978. The reaction of diverse population of peas to powdery mildew desease. *Indian J Mycol. Pl. Pathol.*, 8: 199-200.

Narsinghani, V.G. 1979. Inheritance of powdery mildew in peas (*Pisum sarvum* L.) *Indian J. Hort.*, 36: 471-472.

Narsinghani, V.G. and A. Tiwari, 1993. Improvement of Garden pea in: Advances in Horticulture. Vol. 5. Vegetable crops: Ed. K.L. Chanda and G. Kalloo. Malhotra Publishing House, New Delhi, 217-233.

Narisinghani, V.G. and R. Bedi, 1990, Genetic inheritance of powdery mildew resistance and its breakage in peas. *Proc. Nat. Acad. Sci.*, (India)., 60: 73-75

Nath, P., S. Velayudhan, and D.P. Singh 1976. Vegetables for the tropical region. ICAR New Delhi, pp. 230.

Ohu, H., S. Ouchi., T. Shirashi and T. Baba 1975, Pistain production in powdery mildew pea seedlings. *Phytopathology*, 65: 1263-1267.

Parry, D. 1990, *Plant pathology in Agriculture*. Cambridge University Press, Cambridge.

Paul, Y.S. and R. P. Kaushal 1985, Effect of relative humidity on germination of powdery mildew conidia. *Indian Phytopath.*, 38: 757-758.

Peirce, W.H. 1948, Resistance of powdery mildew in pea. *Phytopathology*, 38: 21-23, pp. 385.

Ram, H.H., R.D. Singh and Y.V. Singh, 1981, Note on inheritance of resistance of powdery mildew and days to flowering in peas. *Curr. Sci.*, 50: 782-784.

Ram, R.S. 1992, Inheritnace of powdery mildew resistance in peas (*Pisum sativum* L.). *Experimental Genet.*, 8: 37-40.

Saharan, G.S. and V.P. Gupta 1983, Search for stable source of resistance in pea to powdery mildew. *Indian Phytopath*, 36: 144-146.

Sakr, B. and M. Muehlbaue 1996, Inheritance and linkage relationships of resistance to powdery mildew of peas. *Al-Awamita*, No. 94: 9-18

Saxena, J.K., R.M. Tripathi and R. Srivastava, 1995, Powdery mildew resistance in pea (*Pisum sativum* L.). *Curr. Sci*., 44: 746.

Schnathorst, W.C. 1965, *Ann. Rev. Phytopath*., 3: 353-366.

Schroeder, W.T. and R. Provvidenti, 1965. Breakdown of the *er er* resistance to powdery mildew in *Pisum sativum* L., *Phytopathology*, 55: 1075.

Sharma, A.P. Kalia and J. Kumar 1996, Inheritance of resistance of powdery mildew in garden pea, *Him. J. Agric. Res*., 22: 139-141

Sharma, L.C., A.K. Mathur and A. Kaur 1980, Performance of pea varieties against powdery mildew in Rajasthan, *Madras Agric. J*., 67: 483.

Sindhan, G.S. and R.D. Parashar 1984, A comparative study of pea varieties resistance and susceptible to powdery mildew diseases, *Prog. Hort*., 16: 137-139.

Singh, B.B. and D. P. Singh 1991, Inheritance of resistance of powdery mildew in peas. *Indian J. Pulses Res*., 4: 197-198.

Singh, D.V. and R.R. Singh 1991, Varietal resistance of pea of powdery mildew, *Indian Phytopath*, 30: 139-140.

Singh, H.B. and U.P. Singh, 1988, Powdery mildew of pea, *Int. J. Tropical Plant Dis*., 6: 1-18.

Singh, L., V.G. Narisinghani and A.S. Tiwari 1978, Analysis of yield losses by powdery mildew in different cultivars of pea. *Indian J. Agric. Sci*., 48: 86-88.

Singh, R.B. M.N. Singh, U.P. Singh and R.M. Singh 1983, Inheritance of resistance to powdery mildew in pea and its use in breeding. *Indian J. agric. Sci*., 53: 855-856.

Singh, R.N. 1991, A zero to nine-point rating scale for the evaluation of pea genotypes for resistance to powdery mildew. *Narendra Deva J. agric res*., 6: 339-340.

Singh, R.S. 1985, Disease of vegetable crops, Oxford and IBH pub. Co., New Delhi, pp. 406.

Singh, U.P., C.P. Shrivastava, M.N. Singh and R.N. Singh 1986, Allelic test and inheritance of resistance of powdery mildew in pea, *Veg. Sci*., 13: 131-136.

Sivapalan, A. 1993, Effects of impacting rain drops on the growth and development of powdery mildew fungi. Plant path., 42: 256-263.

Sohi, H.S., S.S. Sokhi and R.D. Rawal 1974, Varietal reaction of peas of powdery mildew (*Erysiphe polygoni*. DC.) and rust (*Uromuces fabae*) *Mysore J. Agric. Sci*., 8: 529-532.

Sokhi, S.S., J.S. Jhooty and S.S. Bains 1979, Resistance in pea against powdery mildew. *Indian Phytopath, 32:* 571-572.

Sujay Rakshit, S.K. Mishra, S.K. Dasgupta and B. Sharma, 2000a, Dynamics of beta-1,3 gluconase activity in powdery mildew resistant and susceptible lines of pea (*Pisum sativum* L.) *J. Plant Biochem. Biotech*., 9: 95-98.

Sujay Rakshit, S.K., Mishra, S.K. Dasgupta and B. Sharma 2000b, Differential induction of beta-1, 3 glucose in powdery mildew resistant and susceptible lines of pea (*Pisum sativum* L.) in response to powdery mildew infection. *India J. Plant Physiol*., 5: 209-213.

Tewatia, A.S., and G. Kalloo, 1992, Disease resistance in pea (*Pisum sativum* L.) A review. *Agric. rev*., 13: 81-94.

Thakur, B.R., A.S., Kapoor and R.S. Jamwal 1996, Varietal resistance of pea of powdery mildww in dry temperate zone of Himachal Pradesh, *Indian Phytopath*, 49: 92-93.

Timmerman, G.M., T.J. Frew, N.F. Weeden, A.F. Miller and D.S. Goulden 1994, Linkage analysis of er-1 a recessive *Pisum sativum* gene for resistnace of powdery mildew fungus (*Erysiphe Pisi* DC.) *Thero. Appl. Genet*., 88: 1050-1055.

Tiwari, K.R., G.A. Penner and T.D. Warkentin 1998, Identification of coupling and repulsion phase RAPD. markers for powdery mildew resistance gene er-1 in pea. *Genome*, 41: 440-444.

Tiwari, K.R., G.A. Penner, T.D. Warkentin and K.Y. Rashid 1997, Inheritance of powdery mildew resistance in pea. *Can. J. Plant Sci*., 77: 307-310.

Uppal, B.N., M.K. Patel and M.N. Kamat 1935, Pea powdery mildew in Bombay, *Bull. Dept. Agric*, Bombay, 177: 1-12.

Vaid, A and P.D. Tyagi 1997, Genetics of powdery mildew resistance *Euphytica*, 96: 203-206.

Von Hammerlund, C, 1995, ZurGenetik, Biologic and physiologic einiger Erysiphagen, *Herediatas*, 6: 11-26.

Westcoot, C., 1971, Plant disease hand book, 3rd ed. Van Nostrand Relanhold Company, pp. 400.

Yarwood, C.E., S. Sidky M. Cohen and V. Santilli 1954, Temperature relations of powdery mildews, *Hillgardi*, 22: 603-620.

The Impact of Climatic Change on Pests and Diseases and Strategies for their Management

H. PATHAK[1], A.K. GANGULY[2], S. CHANDER[1] AND P. SINHA[3]

All over the world there is a concern about the possible climatic change caused by the increase in the concentration of greenhouse gases such as carbon dioxide (CO_2), methane (CH_4) and nitrous oxide (N_2O) in the atmosphere. These gases contribute to warming of the earth by retaining more heat within the atmosphere. The current concentration of CO_2 in the atmosphere (356 ppm) is changing with an annual increment of 1.5 ppm reaching 397- 416 ppmv by 2010 and 605-755 by 2070 (Bachelet *et al.*, 1993). Concentration of methane has increased from 0.7 ppm during preindustrial times to 1.78 ppm at present. Similarly nitrous oxide with its current concentration of 310 ppb in the atmosphere is increasing at a rate of 0.22 ± 0.02% per year (Rodhe, 1990). Such build up of greenhouse gases will increase the global temperature (Fig. 1). It is a fact that earth's surface has warmed by 0.5+0.2°C during the past century. The expected rate of increase is now 0.1°C per decade (Kerr, 1998). The warming would increase evaporation rate, which would result in increase in global precipitation by 2±0.5% per 1°C warming. It is also being predicted that rainfall will be highly erratic. There will be less frequent rains but more intense leading to floods. The recurrence of climatic extremes such as droughts and floods has increased at certain places. Although the solar radiation received at the surface will be variable geographically but on an average it is expected to decrease by about 1%.

Climate change scenarios for India

Agriculture constitutes the single largest component of India's economy, contributing nearly 27% of the total Gross Domestic Product (CDP) in the year 1999-2000 (TERI, 2002). However, given that 62% of the cropped area is still dependent on rainfall (MOEF, 2002), Indian agriculture continues to be fundamentally dependent on the weather. Thus any change in the climate will have tremendous impact on Indian agriculture. For Indian subcontinent, it is predicted that there will be an increase of 1-4°C in the mean atmospheric temperature (Sinha and Swaminathan, 1991). However, the magnitude will be different in different seasons. The mean temperature is projected to increase by 0.1 to 0.3°C in kharif and 0.3 to 0.7°C in rabi by 2010 and to 0.4 to 2.0°C in kharif and to 1.1 to 4.5°C rabi by 2070 (IPCC, 1996). Another projection for Indian sub continent by middle of next century suggests an increase in annual mean surface air temperature of 1.0°C over land regions by 2040s with respect to the 1980s (Lal *et al.,* 1995). Mean rainfall is projected not to change by 2010 but increase by upto 10 per cent during kharif and by ± 10 per cent during rabi by 2070. At the same time there is an increased possibility of climatic extremes such as the timing of onset of monsoon, intensities and frequencies of droughts and floods. It is also assumed that there will an increase of 0.10°C in both minimum and maximum temperature for each degree increase of latitude. The

1Unit of Simulation and Informatics, 2 Division of Nematology, 3 Divison of Plant Pathology; Indian Agricultural Research Institute, New Delhi-110 012.

greater warning is thus expected at higher latitudes than in the tropics (Sutherst, 1991). Monsoon season warming will be less pronounced than in the winter months.

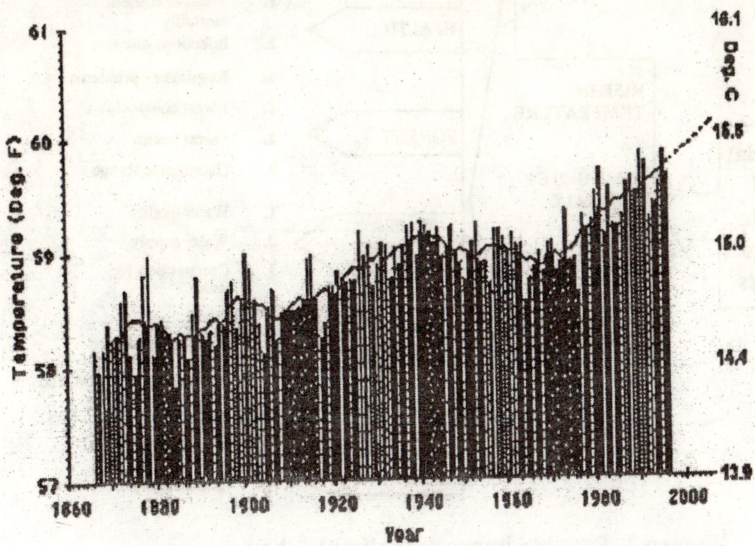

Figure 1. Rise of temprature of the earth with time
Source: IPCC (1996)

Potential impacts of climate change on agriculture

Agriculture is sensitive to variations in climate. The climatic elements, which affect the plant growth and development, hence the agriculture as a whole, are carbon dioxide concentration, temperature, radiation, precipitation and humidity. These are likely to be altered with the increased build up of greenhouse gases in the atmosphere and considerably affect the food supply and access through their direct and indirect effects on crops, soils, livestock, fisheries and pests (Fig. 2). Increase in atmospheric CO_2 promotes the growth and productivity of C3 plants. Increase in temperature, on the other hand, can reduce crop duration, increase crop respiration rates, affect the equilibrium between crops and pests, hasten nutrient mineralisation in soils, decrease fertiliser use efficiencies, and increase evapotranspiration. Indirectly, there may be considerable impact on agricultural land use due to snow melt, availability of irrigation, frequency and intensity of droughts and floods, soil organic matter transformations, soil erosion, decline in arable areas (due to submergence of coastal lands), and availability of energy. All these changes would have tremendous impact on agricultural production and hence food security of any region.

Increased in global temperature will result in increased global rainfall, although there will be regions that will receive less rainfall than before these changes took place. Annually, the number of days having intense rain or very high temperatures will both increase. Sea levels will rise by about 18 cm by 2040 and by 48 centimeters by 2100, effects due to mainly thermal expansion of water and melting of glacier (Smith and Dennis, 1989; IPCC, 1996). These will lead to the loss of productive land and contamination of freshwater sources. The much disused El-Nino effect, which invades world's oceans and continents every decade is suggested to get a big boost by the rising temperature of the atmosphere. All of these will have considerable effects on agriculture. As the detail discussion on the impact of climate change on crop productivity is beyond the scope of this chapter, it will be dealt with briefly while the impact on pest and diseases will be discussed rather in detail.

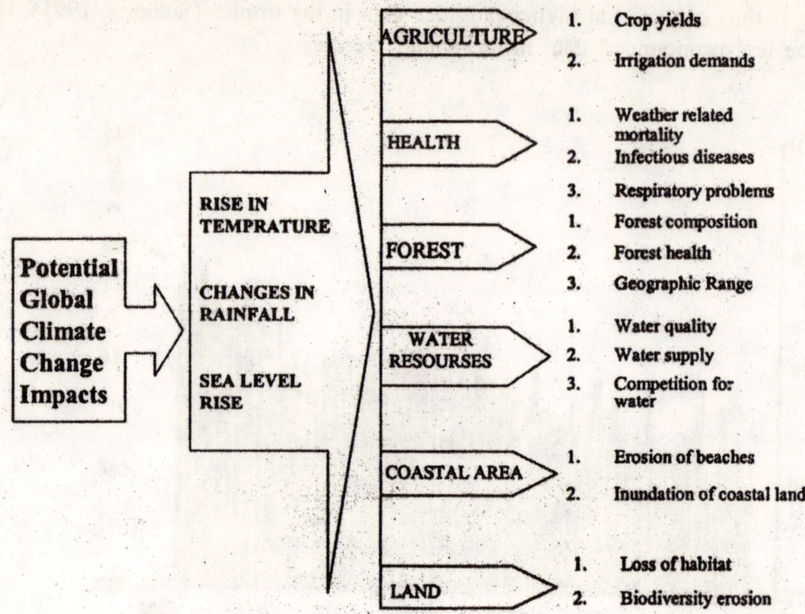

Figure 2. Possible impacts of climate change

Impacts on crop yield and quality

Effects of climate change may reduce yields of all C3 crops, both under rainfed and irrigated conditions. Moreover, climate change and climate variability will likely lead to decreased yield stability. However, some plants may increase photosynthesis because of higher temperature and CO_2, but all plants will not be benefited. Increase in temperature is likely to affect the crop calendars in low latitude regions, particularly when more than one crop is harvested each year. The duration of warm season will increase in many parts of the world, which may lead to increased production owing to the adaptation of agriculture through the introduction of some late ripening high yielding varieties. In general, the crop production may shift northward. Globally, agriculture, however, may remain stable at levels, which are sufficient for meeting human food requirements but the cost of the measures to maintain tnese levels are unclear. Serious negative impacts may occur in some regions, which are currently climate dependent. Many regions may expect an increase in precipitation, which will lead to changes in agricultural production and natural ecosystems.

Small changes in temperature, $C0_2$ and rainfall can have significant effect on quality of fruits, vegetables, tea, coffee, aromatic, and medicinal plants. There are reports that nutritional quality of cereals and pulses may also be moderately affected, particularly in areas where N fertilizer applications are already low. Since a large part of India has limited N fertilizer application, this will have consequences for our nutritional food security. Demand for irrigation water will increase in all regions, which in turn will increase competition for existing water supplies. Increase in temperature may also result in a higher amount of evapo-transpiration, leading to increased frequency of droughts. In some areas soil moisture may diminish, especially where moisture deficit in soil is already experienced, such as Africa.

Impacts of climatic change on pest and diseases

Pest and diseases are early indicators of climate change. This is because climate change will potentially affect the pest/weed - host relationship in one or more ways: by affecting the pest/weed population; by affecting the host population and by affecting the pest/weed - host interactions. Possible climate change due to increase in concentrations of greenhouse gases like CO_2, methane and nitrous oxide and resultant nse in temperature along with other atmospheric pollutants are thought to have direct or indirect consequences on pest and diseases (Chakraborty *et al.,* 1998; Coakley *et al.,* 1999).

Changes due to the effect of increase in CO_2 concentration

The elevated CO_2 will produce increased plant size and canopy density with high nutritional quality foliage and microclimates more conducive to developments of rusts, mildews, leaf spots and blights (Manning and Tiedemann, 1995). Necrotrophic pathogens, which will overwinter under wan-ner conditions, may increase in their severity as a result of enhanced inoculum survival on the greater amount of crop resi 'dues. There is increasing evidence that the most likely impact of elevated CO_2 on plant disease will be mediated through the changes in physiology and anatomy. In addition to effects on rubisco biochemistry and stomatal physiology, difference in anatomy, morphology and phenology have been found (Wolfe, 1995). Hibbered *et al.* 1996 related reduced rate of primary penetration by *Erysiphe graminis hordei* at 700 PPM CO_2 compared to ambient CO_2 to higher rates of net photosyntheiss allowing increased mobilization of resources into host resistance. For pathogens broad genetic diversity, increased population size and the number of generations in conducive microclimates may lead to the development and proliferation of well adapted and possibly more destructive sub-populations (Sutherst *et al.,* 1996).

Under higher CO_2, there is an increase in C/N ratio of litter and litter decomposes at slower rate (Ball, 1997). Higher plant biomass, slower litter decomposition and higher winter temperature might increase pathogen survival on overwintering crop residues and increase amount of initial inoculumn available to infect subsequent crops. Moreover, at elevated CO_2 there is increased partitioning of assimilates to roots in crops such as carrot, radish and sugar beet. Due to more carbon storage in roots, losses from soil home diseases and pests of root crop may be diminished under climate change (Coakley *et al.*, 1999).

Changes due to the effect of increase in temperature

Increases in temperature may alter host physiology and resistance. Evidences are ample of temperature effects on host as well as on pathogens. Elevated temperature may cause the breakdown of temperature-sensitive resistance. Conversely, in some forage species, there is increased lignification at higher temperatures (Wilson *et aL,* 1991) and this can enhance the level of host resistance to pathogens (Strange 1993). Impacts would, therefore, depend on the nature of the interactions in the pathosystem and mechanism of resistance. Both temperature and the length of exposure are important in deten-nining the effect of climate change on physiological process, such as photosynthesis (Wolfe, 1995). Even if the temperature change may be well within the limits of current climatic variability, a modest warming can cause a significant increase in heat sums above a critical temperature threshold to affect crop physiology and resistance to a disease. Temperature change might lead to appearance of different races of the pathogen hitherto not active and might cause sudden epidemic. Change in temperature will directly influence infection, reproduction, dispersal, survival between seasons and other critical stages in the life cycle of a pathogen (Coakley and Schenn, 1996). The impact will be different for different pathogen groups, such as root infecting vs. shoot -infecting pathogens and biotrophs vs. nectrotrophs. Increase in temperature with sufficient soil moisture may increase evapotranspiration resulting in humid microclimate in crop canopy may lead to incidence of diseases favoured under wann and humid conditions. Some of the soil home diseases may increase at the rise of soil temperature. Other major impacts on diseases will be through changes in host physiology and host-pathogen interactions. If climate change causes a gradual shift of cropping regions, pathogens will follow their host (Goudriaan and Zadoks, 1995).

Increase in food quality i.e., increase in nitrogen content in plants under stress condition due to high temperature can result in sudden resurgence of population of insect pests. Moreover under stress condition the plant defensive system is lowered and they become more susceptible to pest attack (EPA, 1989).

Changes due to increases in atmospheric pollutants (ozone and nitrous oxide)

Most air pollutant indirectly influences pest and diseases through their effect on host. Ozone induces reactions similar to those normally elicited by viral and other pathogens (Manning and Tiedemann, 1995). Pollutant concentrations which inhibit pathogen development also injure the host (Heagle, 1982). Similarly, infection by plant pathogens can alter ozone sensitivity of plants. Exposure to 5-10 pphm ozone for a few hours can cause visible injury to sensitive crops like barley, tomato, onion, potato, soybean, tobacco, and wheat. The general symptoms of ozone injury are appearance of white or pale lesions on the leaf surface. O_3 is likely to have adverse affects on plant growth. Necrotrophic pathogens may colonize plants weakened by O3 at an accelerated rate, while obligate biotroph infections may be lessened. O3 is unlikely to have direct adverse effects on fungal pathogens. O3 effects on plant diseases are host plant mediated.

Plants appear to be less sensitive to nitrous oxide higher concentrations can cause water-soaked lesions, which soon turn brown. Ozone and nitrous oxide injury on plants in turn may add new problem to pathologists in diagnosis. Atmospheric pollutants directly or indirectly affect the process of water and nutrient uptake from soil. It may not be possible to check the pollutants so easily. Therefore, breeding programs need to be oriented so that resistant or tolerant varieties of plants can be obtained. Adequate supply of organic manures can nullify any adverse effects on soil. Therefore, balanced nutrient application along with timely supply of water has to be ensured so that any adverse effects on plants are nullified to a greater extent.

Changes due to increases in ultraviolet B

Major effects of UV-B exposure on plants include stunting, increased branching, reduced leaf area (Runeckles and Krupa, 1994), and accelerated ripening and reproduction. UV-B radiation can alter the structure and composition of leaf waxes to modify wetness characteristics. There is stimulatory effect of near-UV light on reproduction (Vakalounakis and Christias, 1981) of many fungi, and spore production *in Leptosphaerulina trifoli* peaks at 287 nm (Leach and Anderson, 1982). Fungi differ in their sensitivity to UV-B. UV-B radiation can modify the relative composition of phylloplane organisms, such as pink and white yeast (Ayres *et al.,* 1996). Continued exposure to enhanced UV-B radiation lowers the level of antifungal compounds in foliar plant parts (Asthana *et al.,* 1993).

Effect of climate change on pests via host plants

Climate change will affect the pests through their host plants. Climate will effect the geographical distribution plant species and their growth patterns. Spatial distribution of crops and cropping system are also important in deciding insect pest distribution. Spatial changes of crops and cropping system for sustained productivity are imminent under climate change scenarios (Parry *et al.*, 1989; Carter *et al.*, 1991). Increased CO_2 levels would result in increased photosynthesis and water use efficiency (Goudriaan and Zadoks, 1995). This may result in higher production in most situations. However, now evidence is accumulating that crop production would be influenced depending on latitude (Rosenzweig and Hillel, 1998). Unfortunately almost all prevoious climate change- yield studies have been carried out excluding insects, pathogens and weeds (Coakley et al, 1999). Any direct gains caused by increased CO_2 or climate change could be neutralized by phytophagous insects, plant pathogens and weeds.

New diseases may arise while some diseases may become less important if wanning results in a northward shift of agro climatic zones and host plants migrate into new regions (Coakley *et al.*, 1999). Pathogens would follow the migrating hosts and may infect remnant vegetation of natural plant communities not exposed to the often more virulent strains from agricultural crops. Facultative parasites with wide host ranges would mostly fall in this category. However obligate parasites may also widen their host range to infect plants in their proximity (Eshed and Dinoor, 1981; Savile and Urban, 1982). Changes may occur in the type and relative importance of pathogens and affect the spectrum of diseases affecting a particular crop. Warming and other changes could also make plants more susceptible to damage from pathogens that are not currently important because of unsuitable climate.

The principal effects of increased UV-B on plant diseases would be via alterations in host plants. Increased flavonoids could lead to increased disease resistance. Reduced net photosynthesis and premature ripening and senescence could result in a decrease in diseases caused by biotrophs and an increase in those caused by necrotrophs. Microbial plant pathogens are less likely to be adversely affected by C02, 03 and UV-B than are their corresponding host plants. Changes in host plants may result in corresponding alterations of disease incidence, depending on host plant growth stages and type of pathogen.

Expansion in range of pests

Any increase in temperature is bound to influence the distribution of pests and diseases. It is predicted that 1°C in temperature would enable species to spread 200 km northwards or 140 km upwards in altitude (Parry *et al.*, 1989). Areas, which are not favourable at present due to relatively low temperature, may become favourable with rise in temperature. Minimum temperature plays important role in determining the global distribution of insect species rather than maximum temperature (Hill, 1987), hence any increase in temperature will result in a greater ability to over winter at higher latitudes ultimately causing pole wards extension of distribution i.e., there may be shift of pest intensity from south to north (Hill and Dymock, 1989). Mild winters could lead to higher incidence of virus diseases such as barley yellow dwarf, potato viruses and sugarbeet viruses following increased winter survival of their aphid vectors (Harrington *et al.*, 1995). On the other hand, areas with optimum temperature today may become less favourable with rise in temperature.

As the species richness of insects tends to increase with temperature (Turner *et al.*, 1987), it is presumed that with increase in temperature more species will be gained than lost. Over wintering survival and timing of commencement of spring is important in higher latitudes thereby population buildup of *Heliothis Zea,* endangering maize and soybean crops in Great Plains of North America are expected (EPA, 1989). It has been noted that insect population under increasing temperature move towards higher latitudes and elevation (Porter *et al.*, 1991 and Sutherst 1991). Presence of over wintering sites, soil types and moisture availability can become deciding factor for resurgence of insects under such newer areas (EPA, 1989). However, known pest species if introduced to new areas may encounter marginal conditions for their survival and suitable host plants may not be available. Besides natural enemies may also enter new areas thereby aiding in biological control of the pest. Among insects there is a distinction of specialists and generalists and response to changing climatic condition in both may be different. Generalist species have good dispersal capabilities and is best suited to track changing climate, and migrate with climatic shifts and thus feed on new part of the range. However this condition doesn't apply to specialist species, even if it is highly dispersive one because the plant species it feeds on may not be present in new region. Thus shifting will be under strong selection pressure, which may be different to those exerted when climate is stable (Travis, 2000).

Increased risk of invasion

Some important pests are long-range migrants and move into crop areas where they cannot overwinter successfully. However with increase in temperature, the unfavourable areas may become suitable for them thereby increasing their survival. Cabbage butterfly, *Pieris brassicae* migrates from hills to plains in winter and back to hills in summer (Atwal, 1986). The increase in temperature may affect this migration and extend distribution of the pest further North. However the activity of this pest may get reduced in plains.

Changes in over-wintering success

With increase in temperature hibernation may be suspended earlier than usual thereby initiating pest activity early. Rice stem borer, *Scirpophaga incertulas* becomes active with availability of rice nurseries in North India. With introduction of summer rice in Punjab and Haryana and with increase in temperature, we may expect higher population of this pest. Similarly with increase in temperature, the onset of hibernation of the pest may also be delayed. Therefore, period of activity of pests will be increased leading to more pest problems. On the other hand, the increase in temperature in summer months may reduce the activity of some pests.

Changes in interactions between species

Mustard crop is infested by two aphid species, *Lipaphis erysimi and Myzus persicae.* It has been observed that *Lipaphis erysimi* dominates during severe winters while *Myzus persicae* is relatively more abundant during mild winters (Chander and Phadke, 1994). With rise in temperature, we may witness higher incidence of *Myzus persicae.* Such faunal shifts may also occur in other crops. On the other hand with rise in temperature, activity of some pests in kharif may be reduced as temperature may cross favourable limit.

Effects on natural enemy-pest interaction

Bioclimatic studies on insect hosts and their natural enemies can provide insight into potential physiological limitations for their geographic spread. Temperature has been found to exert tremendous influence on growth and development of *Lydella jalisco,* a parasitoid of mexican rice borer, *Eoreuma Lofitini.* Its larval and pupal development was retarded with increasing temperature (Lauziere *et al.*, 2001).

With increase in temperature during winter months, the predators of aphids such as coccinellids and syrphids on mustard and wheat may be more active which at present remain inactive during winter. So these will help in reducing aphid incidence on the crops. However at the same time, aphid population will also be affected due to changes in temperature. Ultimate effect of climate change will depend upon pest-natural enemy interactions.

Changes in population growth rates

Insects are found active in a range of temperature, which is called range of favourable temperature. If with increase in temperature if temperature still remains favourable for the pest then population will grow quicker and we will experience higher pest populations. On the other hand if with increase in temperature, it crosses favourable limit then there may be reduction in pest population growth rate. Each species has to be analyzed separately in the light of its favourable range and quantum of temperature increase.

Change in crop-pest synchrony

Crop-pest synchrony may be altered due to changes in climate. During kharif we find heavy infestation of stem borer and shoot fly in maize and sorghum under North Indian conditions. However rabi maize and sorghum are not affected by these pests. In southern parts these pests are active during *rabi* season also. Therefore with increase in temperatures these pests may pose problem to these crops in *kharif* as well in North India.

Specific effects of climatic change on insects, diseases and weeds

Insects: As discussed earlier, climatic factors such as temperature, precipitation, humidity, wind speed and sunshine hours can directly influence insects by affecting their rate of development, reproduction, distribution, migration and adaptation and indirectly through host plants, natural enemies and inter-specific interactions with other insects. Owing to large body surface area relative to body volume, insects are extremely sensitive to desiccation. Light although not utilized directly by an insect as energy source but it controls daily behavioral rhythm. Movement of air masses in synoptic scale determines the destiny of insect migration as a more or less regular phenomenon. Climatic factors are thus important in determining the distribution, range and degree of infestation. Insect mortality due to climatic conditions that are not generally identifiable gives rise to the most frequently used concept of physiological death. The initial population is under the strong and direct influence of climatic factors and indirect influence of the seasonal change of habitat. Stilting (1993) concluded that reason for failure of 30% of the biological control programs is climate.

Among all the abiotic factors, temperature is the most important one affecting insect distribution and abundance in time and space since these are poikilothennic or cold-blooded animals. Insects cannot regulate their body temperature and it changes with the ambient temperature influencing their survival, growth, development and reproduction. Insects are found to be active in a range of temperature, which is called range of favourable temperature. On either side of the favourable range, there are zones of inactivity. Further there is lethal temperature for each species on minimum as well as maximum side. The organism dies once the temperature reaches lethal limits.

Implication of temperature for insect pest dynamics has always been found important as it determines the development of insect. It has been shown that insects remain active with in temperature range from 15 to 30 – 32°C (Phadke and Ghai, 1994). Within the range of favourable temperature, an increase in temperature increases the rate of development thereby decreasing the developmental period. In case of red cotton bug at constant temperatures of 20, 25 and 30°C, the average duration of life cycle was found to be 61.3, 38.3 and 37.6 days, respectively, while at 12.5 and 35°C the pest did not show any development (Bhatia and Kaul, 1966). Within the favourable limits the temperature increase also showed an increasing trend in egg laying and egg hatching. In the field conditions temperature and relative humidity were found to be positively correlated with incidence of infestation of sorghum shootfly, *Antherigona soccata* (Dubey and Yadav, 1980). On the other hand sorghum stem borer *(Chilo partellus),* incidence was positively related with temperature (Chelliah, 1988). Aphid populations on wheat thrive well in a temperature range of 5-28°C and further increase in temperature triggers their decline. Continuation of low temperature during March prolongs period of activity of aphids on wheat. Heavy rains in March considerably reduce population of aphids on winter crops. The effect of temperature, however, is liable to be modified by other factors.

The favourable range of temperature for rice stem borer, *Scirpophaga incertulas* varied from 15.1-38°C (Kalode, 1974). Perusal of temperature at Delhi during 1995-2000 showed that with 1-2°c rise in temperature, the minimum temperature during June will become more favourable for the pest while maximum temperature will become increasingly unfavourable. During months of July to October, the minimum temperature ranged from 17.3-28.4 while maximum temperature varied from 28.8-35.9°C. Both minimum and maximum temperature were thus well within favourable range of the pest. With presumed rise in temperature, these will still remain favourable. Therefore with rise in temperature the development rate of pest will increase and it may realize more number of generations. The incidence of stem borer thus may be expected to rise with rise in temperature. The expected reduction in rainfall may also increase incidence of the pest. Pasalu *et al.* (1999) opined that high temperature and low rainfall could cause severe stem borer infestation.

The favourable range of temperature for brown planthopper, *Nilaparvata lugens* lies between 15-36°C. During July-October, both minimum and maximum temperature are favourable for the pest. With rise in temperature the minimum temperature will prove more favourable while maximum temperature may cross favourable range. There may not be any effect of 1 or 2°c rise in temperature as beneficial effect of rise in minimum temperature may be nullified by adverse effect of rise in maximum temperature. Similarly, distribution and frequency of rainfall may also affect the incidence of pests through changes in humidity levels as well as directly. Small insects such as aphids, jassids and whiteflies are washed away by heavy rains thereby reducing their incidence of crops. Aphid population on wheat and other crops was adversely affected by rainfall and sprikler irrigation (Chander, 1998). Heavy incidence of annyworm in Delhi during rainy season of 1995 could be related to higher maximum and minimum temperatures and rainfall as compared to previous years (Chander and Sharma, 1997).

Diseases: Most agricultural diseases have greater potential to reach severe levels under warmer conditions. Fungal and bacterial pathogens are likely to increase in severity in areas where precipitation increases. Under warmer and more humid conditions cereals will be more prone to diseases like *Septoria.*

Increases in population levels of disease vectors may well lead to increased epidemics of diseases they carry. The potential impacts of climate change on major vector-bome tropical diseases are given in Table 1. Climate change could alter stages and rates of development of the pathogen, modify host resistance, and result in changes in the physiology of host-pathogen interactions. The most likely consequences are shifts in the geographical distribution of host and pathogen and altered crop losses, caused in part by changes in the efficacy of control strategies (Coakley, 1999).

Changes in atmospheric composition have also been shown to alter stages and/or rates of development in the life cycle and pathogenicity of causal organisms and to modify the physiology and resistance of the host plant. Evolution of pathogen populations may accelerate from enhanced UV-B radiation and/or increased fecundity in elevated CO_2 to rapidly overcome host resistances. The higher carbohydrate concentration within host tissue at high CO_2 promotes the development of some biotrophic pathogens such as rusts. This suggests that significant impacts of elevated CO_2 will be manifested through changes in host physiology. In addition, elevated CO_2 can increase the amount of usable host tissue due to increased biomass that influences the growth, sporulation, and spread of leaf infecting fungi. Increases in canopy size and density can modify microclimate and disease development. But realistic assessment of the impact of elevated CO_2 on plant diseases can not be made at present due to a paucity of knowledge on disease epidemiology (Chakrabarthy, 2002). Diseases are strongly dependent upon the humidity. Any increase in this, depending upon their base value, can significantly alter severity, which ultimately results in yield loss. Some livestock disease that at present are limited to tropical countries, may spread, causing serious economic loss. Predicting the impact of climate change, on the other hand, becomes speculative due to uncertainties in our ability to predict future climate change and a paucity of knowledge on the effect of climate change on individual plant diseases. As the gradual process of climate change interacts with other concomitant processes in the production system, it becomes difficult to isolate effects due to climate change alone. Only a handful of work has dealt with climate change impacts on plant diseases. Most probably, impacts will be felt in altered geographical distribution and increased or decreased crop loss, and efficacy of control measures due to changes in the physiology of host-pathogen interactions. Changes may occur in the complexes may arise while some diseases will cease to be economically important. Crops, especially perennials, may continue to be grown in marginal climates for economic reasons, and chronic stress would lead to progressive deterioration in plant health, increasing susceptibility. Disease management may be influenced due to altered efficacy of biological and chemical control options.

Weeds: All environmental factors that influence plant growth potentially can affect the ability of weeds and crops to exploit the environmental resources for which plants compete. Stressful levels of environmental factors such as temperature, light and water and nutrient availability influence weed/crop interactions directly and also may interfere with (or enhance) weed control.

Crop-weed competition will be affected depending upon their photosynthetic pathway. Since, most crops are C3 and most weeds are C4 plants, increased CO_2 would favour crops over weeds reducing needs for weed control. The accompanied temperature increase may further alter the competition depending upon the threshold ambient temperatures.

Weed population may change in the frequency and noxious weeds may become more common. Pest-Crop interactions will be directly affected by the rising levels through alteration of host plant attributes such as C/N ratios and secondary plant chemistry and through indirect modifications patterns of stomata opening, leaf water content and leaf temperature. Climate change is likely to cause a spread of tropical and subtropical species into temperate areas and to increase the numbers of many temperate species currently limited by the low temperature at high latitudes. An increase in temperature will lead to more opportunity for the population growth in areas already affected by the pest.

Nematodes: Climate change would cause spread and establishment of nematodes to new areas. For example, virus transmitted nematodes like *Xiphinema index and Longidorus macrosoma,* which were earlier found in northern Europe, have now spread to southern Europe due to increase in temperature (Boag *et al.*, 1991). Moreover, at higher temperature the nematodes will be able to complete more number of generations per year because of higher metabolic rate and increased fecundity especially in case of polypluagous nematodes. It is also suggested that evolution of new races of nematodes may take place by natural selection to tide over unfavourable conditions. Alteration in the expression of resistant genes in plants with changed temperature will also help nematodes in some cases. Mi gene, for example, becomes inactive above 28°C temperature. Survival of beneficial nematodes such as *Stenernema and Heterorhabditis sp.* the soil will be adversely affected as the soil temperature increases due to global wanning. However, there could be some benefits also. For example, temperature above 32°C decreases the population of *Meloidogyne incognita* and also reduces its penetration and fecundity. Moreover, elevation in soil temperature may lead to lowering of host compatibility reactions.

Accumulation of secondary metabolites due to high temperature stress may reduce the feeding by aerial nematodes because of thickening of leaves with waxy layer. Some secondary metabolites, e.g. terpenoides also act as repellent or deterrent in plants. Increased CO_2 concentration in the atmosphere will cause reduction in protein contents in the plants, which would force the nematodes to feed more quantity to meet their requirement. Thus the damage due to nematodes is likely to increase in the changed climate. Due to increase in temperature there would be more evaporation of nematicide, thereby reducing their effectiveness.

In summary the climate change may affect the insect pests by the following ways.

- Insect pests will generally become more abundant as temperature increases through a number of inter-related processes, including range extensions and phenological changes, as well as increased rates of population development, growth, migration and over wintering.
- Ranges and populations of agricultural pests may change. Temperature increases may extend the geographic range of some insect pest currently limited by temperature. As with crops these effects would be greatest at higher altitudes.
- The rise in atmospheric carbon dioxide will also affect pest species directly and indirectly. However, individual species responses to elevated carbon dioxide vary. Climate change is also considered from the perspective of changes in the distribution and abundance of species and communities. Migrant pests are expected to respond more quickly to climate change than plants, and may be able to colonize newly available crops/habitats.
- Climate change is likely to alter the balance between insect pests, their natural enemies and their hosts. One of the most important effects of climate change will be to alter the synchrony between host and insect pest development. The rise in temperature will favour insect development and winter survival.
- It is suggested that under a warmer climate, exotic pests could establish populations.
- Rising atmospheric carbon dioxide concentrations may lead to a decline in food quality for plant-feeding insects, as a result of reduced foliar nitrogen levels.
- Insect populations are strongly dependent upon the humidity. Any alteration in humidity, depending upon its base value, can significantly alter their population, which ultimately results in yield loss.
- The epidemiology of plant diseases will be altered. Prediction of disease outbreaks will be more difficult in periods of rapidly changing climate and unstable weather.
- Environmental instability and increased incidence of extreme weather may reduce the effectiveness of pesticides on targeted pests or result in more injury to nontarget organisms. Biological control may be affected either negatively or positively.

Approaches for analysis of climate change impacts on pests

The effect of climate change on pests would involve complex interactions of climatic and biological factors with technological and socioeconomic changes. Such complex phenomenon cannot be analyzed through routine experiments, which can probe effect of one or two factors in isolation without addressing multiple interactions among various factors. Climate change impact assessment is thus amenable to modeling approaches, which can help investigate multiple scenarios and interactions. Empirical models and simulation models have been in vogue albeit on limited scale for climate change impact assessment.

Empirical models

The effect of recent increase in mean and minimum temperature on four diseases of rice and wheat each in China was analyzed through regression analysis (Coakley *et al.*, 1999). It was found that rice blast and wheat scab increased sharply since 1970s, powdery mildew became more extensive while stripe rust decreased steadily. Jahn *et al.* (1996) utilized long-term plant disease monitoring records to develop empirical climate-disease models for 15 individual host-pathogen combinations, which were then used to predict possible changes in infestation levels under various climate change scenarios. There are a lot of empirical studies on insects where correlation has been established with weather parameters. However most of them are location based and area specific. These may be of great help in exposition of various facets pertaining to particular pest in relation to crop.

Simulation models

Simulation models have been used widely for assessing yield of various crops in different agro ecological zones under climate change scenarios (Goudriaan, 1996; Romm *et al.*, 1998). However biotic yield reducing factors such as insects, diseases and weeds have been ignored in these studies (Teng *et al.*, 1996). Therefore, development of coupled pest dynamics-crop growth simulation models is needed to enable prediction of crop yield under climate change scenarios in presence of pests. The effect of rise in temperature on pest population can be simulated through pest population dynamics models. Such models need to be developed for important pests based on degree-days requirement of different development stages. Likewise effect of increase in CO_2 and temperature on crop production can also be addressed through crop growth simulation models. Pest models can be coupled to crop growth models. This will allow us to determine effect of climate change on population, which will ultimately affect crop production.

Impact assessment for pest and diseases

Adverse effects of atmospheric pollutants may be nullified by improving crop production strategies. However, the impact of changes due to changes in temperature, radiation and moisture profiles may be more than what has not been imagined so far. Therefore, impact assessment on climate change may help to visualize the consequences. Analytical models are necessary to quantify impacts, examine a range of probable scenarios and evaluate prospective strategies. Two approaches have been used, the climate matching algorithms and linked disease crop models. The simple climate matching algorithms provide rapid answers to questions such as likely changes in the potential distribution and areas of greatest severity under climate change. They are most useful where detailed data are not available to build processed based, linked disease - crop models. CLIMEX model (Sutherst and Marywald, 1985) can be used for predicting climate change impacts. On the basis of meteorological data and the observed distribution are used to derive a species or disease profile of climatic preferences. Quantitative information generated through modeling approach can provide answer about occurrence of a disease and its distribution pattern.

Climate change in general, directly might alter stages and or rates of development in the life cycles and pathogenicity of causal organisms. There may be change in geographical distribution of the pathogens in time and space. Nature and magnitude of climate change will influence changes in geographical distribution of pathogens especially those exists in marginal conditions of environmental factors. Modification of physiology and resistance of host plants may lead to increase or decrease crop loss. Impact assessment would provide the about change and occurrence of pest and disease phenomena needed for the future mitigation strategies. However, impact assessment requires an extensive search of the scientific literature and data banks to obtain as much data and information as possible which is supplemented by expert Judgements. Therefore, theory and methodology for matching climatic requirements of organisms has received much attention as a fool for predictive ecology, in view of present concerns about global climate change. Similar concept has been in used for the impact assessment for pest and disease risk.

Strategies of pest management under climatic change

Pests are known to be agricultural production reducers. In an estimated loss of about 30% of crop production due to biotic interference, 10% loss each is attributed to insect pests, pathogens and weeds. Avoidance of such loss constitutes one of the main sources of sustainability in crop production. Less than 1% of insect species are pests and many of them occur sporadically in nature and population cycles are often regulated by weather. Incidence of pest and diseases is most severe in tropical regions due to favorable climate/weather conditions, multiple cropping and availability of alternate pests throughout the year. Accordingly in the south Asian region incidence of pests and disease deleteriously affecting the crop yields is prevalent. Climate change will add another dimension to the problem. Therefore, strategies should be developed to control pest and diseases. But before developing the strategies, it is necessary to address the following questions.

1. Will climate change significantly affect domestic agricultural production? If yes, in which region (special) and in which season (temporal).
2. What are the 'hot-spots' in Indian agro-ecosystems in relation to fan-ning/cropping systems under changing climate realm? What component of farming/cropping system is particularly vulnerable in these hot spots that may require a special attention.
3. What is the effect of climate change, acting through agriculture, altering (increasing or decreasing) the pace of on going natural resource degradation by other driving forces, if yes, demarcate 'hot-spots' of natural resource degradation?
4. How demand and supply of irrigation is changing under these conditions and specification of the probable 'hot-spots' in relation to demand-supply gaps or probable number of water stress days in crop calendar?
5. Will climate change increase food prices to consumers? For this, expected change in yields has to be converted to shift in supply curve and finally to supply elasticity. Is the agricultural research/ extension system capable of providing adaptation advice to farmers?

Based on the estimates for above questions, site specific and general framework for policies in climate change-food-resources for food can be attempted. Some of the potential strategy options are suggested below.

Host resistance

Cultivars resistance to pests may become more effective due to increased defence mechanisms from changes in physiology, nutritional status and water availability (Coakley *et al.*, 1999). However durability of resistance may be threatened due to increase in number of generations durin growing season because of increased growth rates under higher temperature regimes. This may lead to more rapid evolution of pest biotypes or pathogen races. This way many resistant cultivars may be lost. Increased pest population will also result in more crop losses.

Biological control

Climate change may have effects on tritrophic interactions of host, pest and biocontrol agents. Temperature increase from 17 to 22°C caused 10% increase in aphid, *Sitobion avenae* reproduction on wheat (Triltsch *et al.,* 1996). However activity of predators, *Coccinella septenipunctata* was increased by 250% simultaneously. Besides aphid damage was further reduced by earlier maturity of the crop due to higher temperature. The ultimate effect of climate change will depend upon interaction among its effects on crop, pest and biocontrol agent.

Chemical control

Climate change could affect efficacy of crop protection chemicals in following ways (Coakley *et al.,* 1999): (a) changes in temperature and rainfall may affect pesticide residues on crop foliage and (b) morphological and physiological changes in crop plants resulting from growth under elevated CO_2 could affect uptake, translocation and metabolism of pesticides. Globally climate change models predict an increase in frequency of intense rainfall events, which could result in increased pesticide wash off and reduced control. However for certain pesticides rainfall after pesticide application may result in enhanced pest control due to redistribution of active ingredient on the foliage (Schepers, 1996). Increased thickness of epicuticular wax layer under higher CO_2 could result in slower or reduced uptake by host while increased canopy size may hinder proper spray coverage and lead to a dilution of the active ingredient in the host tissue. Both factors would lead to less efficacy of pesticides at increased CO_2 level (Bowes, 1993). In contrast increased metabolic rate at higher temperature could result in faster uptake by plants and higher toxicity to pests. The rates of pesticide application thus have to be modified according to new situations.

Changes in cultural practices

As agronomic practices with slight modifications are used to suppress pests, the changed pest problems under climate change may necessitate suitable modifications in sowing dates, changed crop species or varieties, changing the cropping boundaries and other cultural practices.

Pest forecasting and use of simulation modeling

Climate factors are the causative agents in determining the population fluctuations. They influence plant disease establishment, progression and severity. In fact, a clear understanding of populations dynamic as influenced by abiotic and biotic parameters of environment will be of much help in pest forecasting and to fon-nulate control measures. However, climatic phenomenon, ecosystem process and human activities are interactive and interdependent, making prediction extremely tenuous. The macroclimatic changes influence the dispersal of the organisms over long distances dispersal (LDD). The genesis and growth of these systems can be monitored by observing cloud formation and features associated with the cloud systems in successive scans of weather satellite (Nagrajan and Sigh, 1975).

Systems analysis and simulation have contributed considerably to plant protection by providing the insight which has led to crop management policies and techniques. Modem crop protection is a complicated affair in which risk, cost-effectiveness and environment must be considered. System analysis is an intellectual tool that can be used to bring order to this bewildering complexity.

Prioritizing research

Research on agricultural productivity should be maintained. Agricultural research on vulnerabilities should be increased, including heat and drought tolerant crops, and germ plasm maintenance.

Farm policy evaluation

Farm policy should be evaluated for ability to adapt to global warming. Flexibility will be needed in shifting crop types and farm practices, and consideration is needed for issues such as soil erosion and water quality. International balance of trade issues need to be evaluated. A national drought policy needs to be established.

CONCLUSION

Climate change would affect crop, pests and their natural enemies and ultimate effect on crop production will depend upon interactive effect of all above components. Most of climate change studies regarding pests have been carried out with respect to effect of temperature and rainfall individually. However in nature these factors do not function in isolation, so these results should be viewed with caution. As climate change effect will depend upon interaction among climatic, biological, technological and sociological factors, so simulation models can play important role in analyzing climate change scenarios. It is felt that in annual crops the effect of climate change may be offset by adoption of proper management options. It should not be forgotten that crops, pests and beneficial organisms are capable of adapting to slow and steady changes in climate. However, that may take long time and before that the loss will be substantial. Therefore, for increasing and sustaining the crop productivity we have to develop the strategies, preventive as well as curative, to manage the pest and diseases in the face o global climatic change.

REFERENCES

Achanta, A. and Kanetkar R. 1996. Impact of climate change on forest productivity: a case study of Kerala, India. In *Paper presented at the Asian and Pacific Workshop on Climate Change Vulnerability and Adaptation Assessment.* Manila, Philippines, 15-19 January pp. 61.

Asthana, A., McCloud, E. S., Berenbaum, M. R. and Tuveson, R. W. (993) Phytotoxicity of *Citrusjambhiri* to fungi under enhanced UV-B radiation: role of furanocoumarins. *J. of Chemical Eco.* 19: 2813-2830.

Atwal, A. S. 1986. *Agricultural pets of India and South-East Asia.* Kalyani Publishers, Ludhiana pp. 529.

Ayres, P. G., Gunasekera, T. S., Raqsanayagam, M. S. and Paul, N. D. 1996. Effects of UV-B radiation (280-320 m-n) on foliar saprophytes and pathogens. In: *Fungi and Environmental Change* (Eds J. C. Frankland, N. Magan and G. M. Gadd), Cambridge University Press, Cambridge, pp. 3-50.

Bachelet, D., Van Sickle J. and Gay, C.A., 1993. The impacts of climate change on rice yields: evaluation of the efficiency of different modelling approaches. In: Penning de Vries, F.W.T. *et al.*, (eds.), Systems approaches for agricultural development. Kluwer Academic Publishers, Netherlands, pp. 145-174.

Battle, M., Bender, M., Sowers, T., Tans, P.P., Butler, J.H., Elkins, J.W., Ellis, J.T., Conway, T., Zhang, N., Lang, P. and Clarke, A.D.: 1996, 'Atmospheric gas concentrations over the past century measured in air from fin at the south pole', *Nature* 383, 231-235.

Boag, B., Cravowford and Neilson, R. 1996. The effect of potential climatic changes on the geographical distribution of the plant parasitic nematodes *Xinphinema and Longidorus* in Europe. *Nematologica* 37:312-323.

Bowes, G. 1993) _ Facing the inevitable: Plants and increasing atmospheric CO_2 *Ann. Rev. of Plant Physiology & Plant Molecular Biology* 44: 309-332.

Calvero, S. B. and Teng, P. S. 1992 - Validation of BLASTSIM. 2. Model in IRRI blast (BI) nursery and Cavinti, Lafuna, Philippines. *Int. Rice Res. Newsl.* 17: 5.

Cannon, R.J.C. 1998. The implication of predicted climate change for insect pests in the U.K., with emphasis on non-indigenous species. *Global-Change Biology* 47, 785-796.

Chakraborty, S., Murray, G.M., Magarey, P.A. Yonow, T., O'Brien, R.G., Croft, B.J., Barbetti, M.J., Sivasithamparam, K., Old, K.M., Dudzinski, M.J. Sutherst, R.W., Penrose, L.J., Archer, C. and Emmett, R.W. 1998)- Potential impact of climate change on plant diseases of economic significance to Australia. *Australasian Plant Path.* 27: 15-35.

Chakraborty, S., Ray, G. and White, N. 2002. Impact of climate change on important plant diseases, RIRDC Publication No. 02/10.

Chander, S. and Phadke, K. G. 1994. Incidence of mustard aphid, *Lipaphis erysimi and* potato aphid, *Myzus persicae* on rapeseed crop. *Ann. Agric. Res.* 13 (3):385-387.

Coakley, S.M. and Schenn, H. 1996 - Plant disease in a changing global environment. *Aspects of Applied Bio.* 45: 227-237.

Coakley, S.M.,, Scherm, H. and Chakraborty, S. 1999 - Climate change and plant disease management. *Ann. Rev. PhytopathoL* 37: 399-426.

Dizon, T.O., Ensebio, J.E. *et al.* 1996. Monitoring high temperature tolerance of rice orthopods. Presented on 27 anniversary and annual scientific meeting of the pest management council of Philippines, 7-10 May, 220.

Finckh, M.R., Chavez, A.Q., Dai, Q. and Teng, P.S. 1995 - Effects of enhanced UV-B radiation on the growth of rice and its susceptibility to tice blast under greenhouse conditions. *Agriculture, Ecosystems and Environment* 52: 223 -23 3. Frans, H., Straw, M., Watt, A. 2002. Climate change: implication for insect pests, Forestry Commission Bulletin No. 125, Edinburgh, U.K., pp. 251.

Goakley, S.M. Schenn, H., Sulunas Chakraborty, Chakraborty, S. 1999. Climate change and plant diseases management, *Ann. Rev. of Phytopath.*, 37, 399-426.

Goudriaan, J. and Zadoks, J. C. 1995 - Global climate chagne: modelling the potential response of agro-ecosystems with special reference to crop protection. *Environmental Pollution* 87: 215-224.

Hart, L., Yonow, T., Lupton, J., Chakraborty, S. and Sutherst, R. 1997 - Impact of climate change on the anthracnose pathogen and its host, *Stylosanthes scabra. Phytopathology* 87: S39 (Abstracts)

Heagle, A.S. 1982 - Interactions between air pollutants and parasitic plant diseases. In *Effects of Gaseous Air Pollution in Agriculture and Horticulture* (Eds M. H. Unsworth and D. P. Ormond. Butterworth, pp. 116.

Hibberd, J.M., Whitbread, R. and Farrar, J.F. 1996 - Effect of elevated concentrations of CO_2 on infection of barley by *Erysiphe graminis. Physiological & Molecular Plant Path.* 48: 37-53.

IPCC. 2001. Climate Change 2001: Impacts, Vulnerability and Adaptation (Contribution of Working Group II to the Third Assessment Report of the Intergovernmental Panel on Climate Change) Cambridge: Cambridge University Press

Krupa, S. 1997 - *Air Pollution, People and Plants.* American Phytopatholgical Society, Scientist (Sr.scale. Paul, MN.

Lal, M., Cubasch, U., Voss, R., Waszkewitz, J., 1995. Effect of transient increase in greenhouse gases and sulphate aerosols on monsoon climate. *Current Science* 69(9): 752-763

Leach, C.M. and Anderson, A. J. 1982 - Radiation quality and plant diseases. In: *Biometeorology in Integrated Pest Management* (eds J.L Zhatfield and I.J. Thomason), Academic Press, NY. pp. 267-306.

Lupton, J., Chakraborty, S., Dale, M. and Sutherst, R. 1995. Assessment of the enhanced greenhouse effect on plant diseases - a case study of *Stylosanthes* anthracnose. 10therefore, Biennial Australian Plant Pathology Conference, Lincoln Univeraity, New Zealand, 28-30 August.

Magarey, P.A., Wachtel, M.F., Weir, P.C. and Seem, R.C. 1991. A computer based simulator for rational management of grapevine downy *mildew(Plasmoapra viticola). Plant Protection Quarterly* 6:29-33.

Manning, W.J. and Teidemann, A.V. l995. Climate cahnge: potential impacvt effects of increased atmosperic Coarbon dioxide(CO_2), Ozone (O_2) , and Ultravilolet B (UVB) radiation on plant diseases. *Environmental Pollution* 88.- 219-245.

MoEF. 2002. Agenda 21: an assessment New Delhi: Ministry of Environment and Forests, Government of India.

Patterson, D.T., Westbrook, J.K., Joyce, R.J.V. *et aL,* 1999. Weeds, insects and diseases. 43: 4, 711-727.

Pettitt, T.R. and Parry, D.W. 1996. Effects of climate cahnge on Fusarium foot rot of winter whaet in the United Kingdom. In: *Fungi and Environmental Change(Eds J.C. Frankland, N. Magan and G.M. Gadd), Cambridge Unbiversity Press, Cambridge.* pp. 20-31.

Pfender, W.F. 2001. A temperature-based model for latent- period duration in stem rust of perennial ryegrass and tall Fescue. *Phytopathology* 91(l): 111-116.

Rabbinge, R., Ward, S.A. and Vanlaar, H.H. 1989. Simulations and systems management in crop protection. Perdoc., Wageningen. pp 420.

Rasayanagam, M.S. Paul, N.D., Royle, D.J. and Ayres, P.G. 1995. Variation in response to spores of *Septoria tritici and S. nodorum* to UV-B irradiation *in vitro. Mycological Res.,* 25, 315.

Ravindranath N H and Sukumar R. 1998. Climate change and tropical forests in India. *Climatic change* 39(2-3): 563-581

Rodhe, A.L. 1990, A comparison of the contribution of various gases to the greenhouse effect, *Science* 248, 1217-1219.

Runeckles, V., C. and Krupa, S.V. 1994. The impact of UV-B radiation and ozone on terrestrial vegetation. *Environmental Pollution* 83: 191-213.

Singh 1998. Effect of global warming on the streamflow of high altitude Spiti river In *Enhydrology of High Mountain Areas*, pp. 103-114, edited by S R Chalise, A Hermann, N R Khanal, H Lang, L Molnar and A P Pokhrel. Kathmandu: International Centre for Integrated Mountain Development.

Sinha, S.K., and Swaminathan, M.S., 1991. Deforestation, climate change and sustainable nutritional security: a case study of India. Climatic Change 19, 201209.

Smith, Joel B. and Dennis A. T. 1989. The Potential Effects of Global Climate Change on the United States. US-EPA Office of Policy, Planning and Evaluation. Strange, R.N. 1993 - *Plant Disease Control: Towards Environmentally Acceptable Methods.* Chapman and Hall, London.

Sutherst, R.W. and Maywald, G.F. 1985. A computerised system for matching climates in ecology. *Agri. Ecosystems and Environment* 13: 281-299.

TERI. 2002. TERI Energy Data Directory and Year book 2001/2002 New Delhi. Tata Energy Research Institute, New Delhi.

Vakalounakis, D.J. and Christias, C. 1981. Sporulation in *Alternaria cichorii is* controlled bya a blue and near ultraviolet reversible photoreaction. *Canadian Journal ofbotany 5*9: 626-628.

Volney, W.J.A., Fleming, R.A. 1983. Climate change and impact of boreal forest insects. *Agric Eco. Environ*, 82, 283-294.

Watson, R.T., Zinyowera, M.C., Moss, R.H. and Dokken, D.J.: 1996, 'Climate change 1995, impacts, adaptations and mitigation of climate change: Scientifictechnical analyses, Intergovernmental Panel on Climate Change, Cambridge University Press, USA, pp. 879.

Wilson, J.R., Deinum, B. and Engels, F.M. 1991 - Temperature effects on anatomy and digestibility of leaf and stem of tropical and temperate forage species. *Netherlands J. of Agri/. Sci.* 39: 31-48.

Wolfe, D.W. 1995 - Physiological and growth responses to atmospheric carbon dioxide concentration. In: *Handbook of Plant and Crop Physiology* (Ed) M. Pessarakli, Marcel Dekker Inc., New York, pp. 223-242.

Yang, X. B., Dowler, W. M. and Tschanz, A. T. 1991. A simulation model for assessing soybean rust epidemics. J *Phytopath.* 133: 187-200.

Systems Analysis and its Application in Pest Management

SUBHASH CHANDER[1], NAVEEN KALRA[1], H. K. RAI[1] AND A. K. SHARMA[1]

Integrated pest management implies use of available effective control methods in a compatible manner to reduce pest population below those causing economic damage and to ensure favourable ecological, economic and sociological consequences. Amount of damage, which justifies use of control measures against the pest is called economic damage and minimum population density which causes economic damage is known as economic injury level or damage threshold. Pest management is generally considered as useful concept for sustainable crop protection (Rossing, 1993). Use of resistant cultivars along with cultural practices form the basis of integrated pest management, which are reinforced by biological control methods. The pesticides are used only when other methods fail to contain pests below economic injury level. Pest management is a complex system involving many components such as crop, pests, natural enemies, beneficial organisms and non-target organisms subjected to man's production oriented interventions under a variable weather (Teng and Savary, 1992). These components are interlinked to each other. On the advent of synthetic organic pesticides a lopsided approach of pest control was adopted where single pest was targeted with broard spectrum pesticides without due regard to other components of agroecosystem. This approach has been responsible for creation of several pest control and environment related problems. As a cosequence single-factor and single-pest based pest control approach expanded to multiple-factor, multiple-pest studies and strategies.

As pest management involves multiple components with complex interactions, the systems approach is very relevant to it. Systems approach is a holistic view, which presumes that overall systems behaviour is influence by changes in any system component.

System

A system is a limited part of reality with inter-related components. It represents more than the mere addition of its components. The components of a system interact in such a way that changes in any component has repercussions for the whole system. The dimensions of a system depend upon the purpose of study. It may be a field, a farm, a village or a region. In crop protection, the following four main system types are distinguished. These are pathosystem, cropping system, farming system and agro ecosystem. Pathosystem includes the host and pest population. The effect of host plants on population dynamics of pests and decline in the yield of host plants due to presence of pests are analyzed here. Cropping system consists of sequence of crops grown on a farm during the year, such as rice-wheat, maize-wheat, maize-mustard-wheat cropping system. Pathosystem are parts of cropping system. Cropping system thus include not only the crop protection aspects but also crop agronomic practices. Farming system consists

Unit of Simulation and Informatics, Indian Agricultural Research Institute, New Delhi-110 012.

of cropping system and animal husbandry practices such as cattle rearing, poultry, and fisheries, followed by farmers. Agro ecosystem is an ecosystem, which is manipulated by man to serve his own needs i.e. for getting maximum production.

Historical background

The adoption of systems approach in pest management commenced with formulation of simulation models for insects and diseases in the late 1960s (Ruesink, 1976; Teng, 1985). Prior to this, statistical modeling of disease and insect life cycles and development of simulation techniques was emphasized. A number of pest population simulation models came into being in the 1970s and early 1980s. In late 1980s crop and socio-economic factors were incorporated in pest models and simultaneously research was undertaken to devise decision-aids for farmers and extensionists. As a result several decision support systems (DSS) have been developed since then. Decision support system can be defined as an interactive computer programme that uses analytical methods and models to help decision makers to formulate alternatives for a problem, analyze their impacts and then select appropriate solution for implementation. The DSS will essentially solve or give options for solving a problem. Decision support systems range from simple decision rules to complex multiple criteria optimization software. In its simplest form a decision support tool can be an economic threshold level calculated from empirical relations based on yield infestation relationship.

The development of an IPM scheme that fits into a sustainable farming system requires thorough analysis of the system first. An improved understanding of the system may be useful in identifying new control techniques by indicating intervention points and can be helpful in decision making in pest management (Rabbinge *et al.*, 1989; Boote *et al.*, 1993; Pinnschmidt *et al.* 1994). Pest management research is required to derive practical tools for developing tactics and strategies for pest management. Systems approach provides such tools in the form of simulation and decision models. The systems approach or systems analysis is equated to the development, testing and evaluation of a simulation model (Teng and Savary, 1992).

Model

A model is a simplified representation of a system. Model can be qualitative or quantitative.

Qualitative model

These models represent a conceptual relationship between different variables of the system where relationship between variables is not quantified. For example experiences regarding the association of cloudy weather in winters with high aphid population on mustard; reduction in aphid population on rabi crops by heavy rainfall in February-March or association of coccinellids with aphids on crops can be termed as qualitative models.

Quantitative models

These models represent mathematical relationship between various variables of a system. These can be further classified as empirical and simulation or mechanistic models.

Empirical models

These are the regression equations developed between various variables, for example, $y = a-bx$. This equation represents a simple regression between yield of the crop (y) and pest incidence (x). The a and b are intercept and regression coefficient respectively. The b signifies the change in crop yield with a unit change in pest incidence. These models do not explain the mechanism of yield loss due to pest. Such relationships are liable to vary spatially and temporally. These are thus site specific and cannot be applied elsewhere.

Simulation models

These are called mechanistic or explanatory models. They explain the behavior of the system through its components and their relationships as they are based on physiological processes. For example, crop growth simulation model is based on plant physiological processes such as photosynthesis, respiration, growth, partitioning, water uptake, nutrient uptake etc. Similarly pest population simulation model consists of various biological processes such as fecundity, sex ratio, migration, rate of egg, nymph or larvae, pupae and adult production and their conversion to next stage. It also considers various mortality factors during pest life stages. A crop growth model coupled with pest damage mechanisms is known as crop-pest model or a coupled model (Fig. 1). Various plant physiological processes affected by pests are called damage mechanisms. A crop-pest simulation model explains the mechanism or process of the crop loss due to pest.

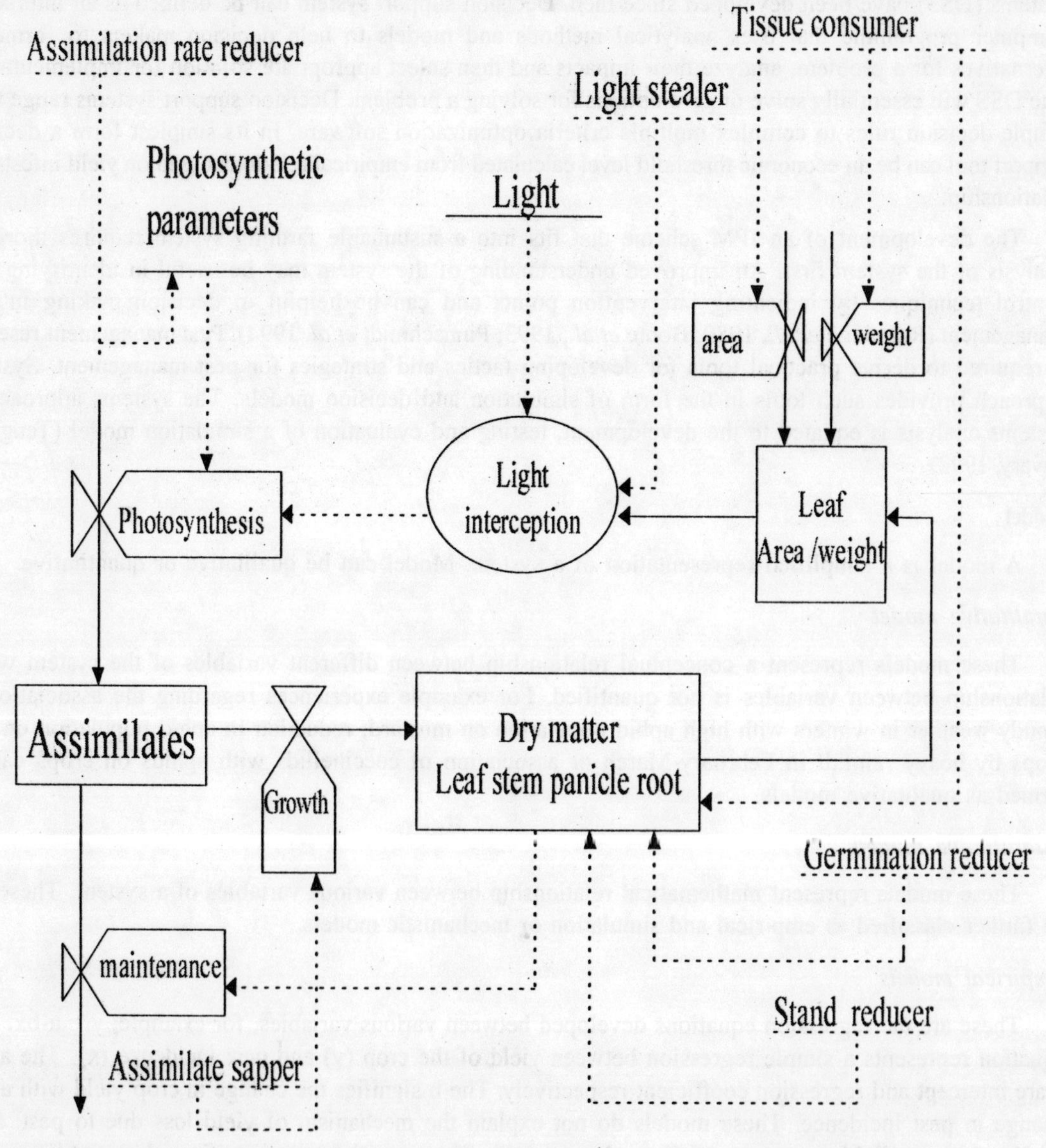

Fig 1 Coupling points of pest damage mechanisms in a crop growth model

The simulation models can be further classified as static and dynamic models. Static mechanistic models represent relations between variables, which do not involve time e.g. a model used to represent relation between respiration and growth, using knowledge of underlying biochemical processes. Dynamic mechanistic model describes the changes in system behaviour with time. These can be further categorized into continuous and discrete models. Dynamic continuous mechanistic models account for gradual changes in the system with time while dynamic discrete mechanistic models describe the changes in the system in discontinuities or in steps.

In agricultural systems all major elements such as leaf area, root length, weights of plant organs change only gradually with time in response to changes in weather, fertilizer or irrigation. Such systems are called dynamic continuous systems. The simulation models representing them are known as dynamic continuous mechanistic models. Crop growth simulation models thus are dynamic continuous mechanistic models. When such models are run with weather data along with various inputs these predict yield of crops at various intervals of time. These models are of two types: (a) models simulating gross photosynthesis and (b) models simulating net photosynthesis

First type of models simulate gross photosynthesis. Net photosynthesis is then calculated by subtracting respiratory requirement of plants from gross photosynthesis. These are very detailed models and require a large number of inputs. Examples are the SUCROS models, generic models that were parameterized for many crops like wheat, potato and maize (Spitters, 1989); MACROS models (Penning de Vries *et al.*, 1989); the ORYZA models that were derived from SUCROS and MACROS by Kroff *et al.* (1994); and the GRO models for crops like soybean, peanut and other leguminous crops.

Second type of models directly simulate net photosynthesis of the crop. These are relatively simple and require less number of inputs e.g. CERES (Crop estimation through resource and environment synthesis) models.

Pest population dynamics model

Population dynamics model is based on different biological processes of an insect written in computer programming language. Such a model when run with weather data predicts pest population at various time intervals

Simulation

The process of developing simulation models and studying systems through them is called simulation. Basic steps of systems simulation are as follows.

Define goals

Unit of Simulation and Informatics, Indian Agricultural Research Institute, New Delhi-110012 Define goals

Before any simulation work is attempted, the purpose of modeling should be clearly defined. All problems are not amenable to modeling. Similarly, many problems may not require a modeling solution.

Define system and its boundaries

Since a system is a limited part of the reality it should have a boundary which would isolate it from its surrounding. This boundary should be such that the environment influences the system but the system affects the environment as little as possible. In agricultural systems there are large number of climatic, soil, hydrologic, biotic, agronomic and socio-economic factors that influence crop growth and productivity. For understanding such complex systems different levels of crop production in the order of descending productivity are considered.

Crop growth can be simulated at three levels of production viz. potential, attainable and actual production. Potential production is determined by plant genotype and climatological factors and there are no constrains of nutrients, water and pests. Plant genotype and climatological factors are called yield-determining factors. For simulating potential production of the crop, crop genotype and climatological factors such as temperature, solar radiation is to be considered. Water, nutrients and pests are not taken into consideration.

Potential production gets reduced to attainable production level due to limitation of nutrients and water. Nutrients and water are called yield-limiting factors. There is not any pest constraint at this production level also. For simulating attainable production crop genotype, climatic factors, nutrients and water are to be considered. The pests are not taken into consideration here also.

When in addition to nutrient and water constraints, there is pest (insect, disease, weed, other harmful organism) attack, the yield gets further reduced from attainable level to actual yield level. Therefore for simulating actual yield, crop genotype, climatic factors, water, nutrients and pests are to be taken into consideration. The pests are known as yield reducing factors. Therefore we should clearly define the boundaries of the system i.e. which level of production we want to simulate and what factors are to be considered.

Define key variables in the system

After defining the system and its boundaries, it is important to define the key variables within the system. For crop growth simulation models, state and rate variable approach is used. In this approach it is assumed that the state of a system at any moment can be quantified and the changes in the system can be described by mathematical equations. The model variables can be distinguished as state, rate and driving variables. State variables describe the state or condition of the system. These are the quantities such as biomass, leaf area, amount of nitrogen, water in the soil etc. Similarly in a pest population dynamics model, number of eggs, larvae, pupae and adult are the state variables.

Driving variables are meteorological factors such as rainfall, solar radiation, temperature, relative humidity, wind speed etc. These are also called the forcing functions. These reveal the effect of the environment on the system.

Each state variable is related to a rate variable. The rate variable describes the rate of change of the state variable. For example rate of egg laying, rate of population change of larvae, pupae, adults etc. are rate variables. The egg population of a species will change based on the rate of egg laying which act as a rate variable. Models are simplified representation of reality and therefore these should be based on a limited number of the state variables.

Quantification of relationships

After defining the key variables of the system, the relationship between them is determined. Relational diagrams are used to indicate the relationships between variables. These diagrams show the relationship and feedback among the variables. Feedback is the change in rate variable based on state variable. Feedback can be negative of positive. If rate variable increases with an increase in state variable or vice-versa then feedback is termed as positive. On the other hand if rate variable decreases with an increase in state variable or vice-versa then feedback is called negative. The values of rate variables are calculated from field experiments and literature. These are then used to calculate state variables.

Calibration

After a crop simulation model has been developed, it needs to be calibrated. Calibration may be defined as the adaptation of crop growth simulation model to a particular experimental condition. Field experiments are conducted for the purpose of calibration. The data from field experiment such as sowing date, seed rate, amount of fertilizers with date, irrigation and other crop parameters are input into the model.

The model output (leaf area, biomass, yield) is then compared to experimental output. If these are not comparable then different parameters are adjusted such that model output becomes comparable to field output. The adjustment however be used with caution because it decreases the explanatory nature of the model.

Crop-pest models can also be used to study the effect of pest incidence on crop growth and yield. In such cases the calibration is first carried out for the healthy crop as discussed above. The model is then calibrated for pest damage mechanisms. For example, calibration of a crop-pest model for rice stem borer would be done as follows. The stem borer observed in a field would be input in the model. The model output is then compared with experimental output. If simulated and observed output do not compare well then damage mechanisms need to be modified.

Validation

After calibration, the model is validated for various pest infestation treatments. The pest infestation levels observed in the field are input in the model. If model output is comparable to field output then pest damage mechanisms are taken as validated. Validated model then can be used for various applications in pest management. If simulated and observed output do not match within certain limits then model needs improvement. Some of the probable reasons for non-agreement between these ouputs are that input data may not be accurately known and quantification of pest damage mechanisms may not be correct.

Sensitivity analysis

Sensitivity analysis is the procedure of increasing or decreasing the value of one parameter or a combination of parameters and simulating the results. If the change in output is large with change in the value of parameter them model is considered sensitive to the parameter. Sensitivity analysis can help in establishing the significance of different parameters.

Damage mechanisms

Major types of pest damage mechanisms (Rabbinge and Rijsdijk, 1981; Boote *et al.*, 1983) can be classified as follows.

Germination reduction,

Stand reduction,

Light stealing,

Assimilation rate reduction,

Assimilate sapping or sucking,

Tissue consumption and

Turgor reduction.

Accordingly the pests can be classified as germination reducers, stand reducers, assimilation rate reducers, assimilate sappers, tissue consumers and turgor reducers.

Germination reducers

These pests damage the seeds of the crop before their germination. Pests like mole cricket and seed maggots are included in this category. Their damage is simulated by reducing initial leaf weight of the crop. Their effect can be simulated by reducing initial leaf weight (WLVI). The WLVI is reduced by a reduction factor (1- RGMPST), where RGMPST is proportion of reduction in germination due to pest.

WLVI = SEEDRT*AFGEN(FCSDEP,SOWDEP)*FRLVWT*(1.-RGMPST)

In this equation SEEDRT, AFGEN(FCSDEP, SOWDEP) and FRLVWT represent seed rate, function for effect of sowing depth on germination and fraction of seed weight allocated to leaves respectively

Stand reducers

These are insects or diseases that kill whole plant or part of it. Pests like stem borer, cutworms, termites, damping off and wilts are classified as stand reducers. The severity of loss due to these pests depends upon compensation ability of remaining plants and development stage of the crop. Rice and wheat can compensate damage due to stand removers during the tillering stage of the crop but not after flowering. The crops like beets or maize compensate for pest damage less effectively. These pests reduce leaf area, leaf weight, stem weight, stem area, grain number and storage organ weight of the crop. Therefore their effect is modelled by reducing rates of leaf area, leaf weight, stem weight, stem area, grain number and storage organ weight of the crop in the model.

Light stealers

Light stealers intercept photosynthetically active radiation before it reaches photosynthesis site on the plant. Weeds and diseases such as rust, blight, mildew etc are classified as light stealers. Sooty mould developing on honeydew excreted by sucking pests also falls in this category of pests. Light interception by necrotic lesions on leaves of crop plant may also interfere with photosynthesis. Weeds besides light also compete for nutrients and water with crop plants and therefore are also sometimes termed as resource stealers. The light stealers reduce the amount of photosynthetically active radiation reaching by the crop. Their effect on crop growth is thus modeled by reducing availability of photosynthetically active radiation to the crop such that

PARCRP= 0.5*DTR*(1-EXP(-KDF*(EFFLAI-PSTPAR))

PSTPAR = EFFLAI * (HNYWT * 0.5/ (HNYWT + WLVG))

The PARCRP, DTR, KDF, EFFLAI, PSTPAR, HNYWT and WLVG refer to radiation interception by crop, daily terrestrial radiation, light extinction coefficient, effective leaf area, reduction in effective leaf area due to pests, weight of honey dew excreted by sucking pests and green leaf weight, respectively. To simulate effect of weeds besides radiation, uptake of nutrients and water is also taken into consideration.

Assimilation rate reducers

Many pathogens and insects affect the photosynthetic capacity of green tissue of the plant. They may affect the photosynthesis rate or the light use efficiency of the crop. Photosynthesis rate may be affected by i) reducing number of chloroplasts per unit leaf area ii) altering the chloroplast ultra structure and components of the electron transport chain and iii) causing malfunctioning of stomatal guard cells. Diseases such as powdery mildew, blight, rust and virus diseases belong to this category. Some insects also act as assimilation rate reducers by accelerating leaf senescence. The effect of assimilation rate reducer is modeled by reducing the radiation use efficiency of the crop such as

RUE = RUEMAX*AFGEN (RCFDS, DS)*RUEABI*RUEBIO

RUEBIO represents the reduction in the RUE due to biotic stresses.

RUEBIO= (1- BLIGHT/100)

Here RUE, RUEMAX, AFGEN (RCFDS, DS), RUEABI, RUEBIO, BLIGHT refer to radiation use efficiency, maximum radiation use efficiency on the crop, function for effect of development stage on radiation use efficiency, effect of abiotic factors on radiation use efficiency, effect of biotic factors on radiation use efficiency and severity of blight or other disease, respectively.

Assimilate sappers

Sucking pests like aphids, plant hoppers, whiteflies, mites, thrips, bugs etc. suck assimilates from plant and as a result less assimilates remain available for plant growth. These pests may suck sap from all plant parts or any of the plant parts. Their effect on plants is modeled by reducing the rate of leaf weight (RWLVG), stem reserve weight (RWIR) and panicle weight (RWSO) as follows.

RWLVG=GCROP*FSH*FLV- (DLV+SUCKLV)

RWIR = GCROP*FSH*FST*-SUCKST

RWSO= GCROP*FSH*FSO-SUCKSO

The GCROP, FSH,FLV,DLV,SUCKLV,FST,SUCKST,FSO,SUCKSO refer to daily rate of assimilates available for growth, assimilate fraction allocated to shoot, assimilate fraction allocated to leaves, leaf weight death rate, rate of assimilate sucking from leaves, assimilate fraction allocated to stem, rate of assimilate sucking from stem, assimilate fraction allocated to storage organs and rate of assimilate sucking from storage organs respectively.

For studying the effect of assimilate sappers on plant growth and yield, pest population per unit area and sucking rate per insect is required. These insects also drain plants of their nitrogen along with carbohydrates. The rate of nitrogen availability to plant is thus also reduced accordingly while simulating their effect on plant growth and yield.

Tissue consumers

These pests may feed on different plant parts like roots, leaves, stems and storage organs. However most evident among them are leaf consumers that reduce the capacity of the crop to capture light and to photosynthesize. Defoliating beetles, grasshoppers, leaf folders, leaf miners, node borers, grain feeders and foliar diseases such as leaf blight, blast and sheath blight can be grouped in this category. Their effect can be simulated by reducing growth rate of leaf area, leaf weight, stem area, stem weight or storage organ weight depending upon their nature of damage. The leaf area loss (LALOSS) and leaf weight loss (DLV) due to tissue consumers is calculated as follows.

LALOSS = LAI* RUST/100

DLV = WLVG * RUST/100

Here LAI, RUST and WLVG refer to leaf area index, rust or other pest severity and green leaf weight respectively.

Turgor reducers

Pests like nematodes and root pathogens affect the water balance of the plants. They also affect the crop nutrient balance by disrupting phloem transport to roots and reducing energy supply for active intake of nutrients. Effect of such pests is modeled by reducing root growth rate.

These categories have to be seen as a broad framework. Defining the category for a pest is difficult because some pests affect the crop in several ways like brown plant hopper in rice acts as assimilate sapper as well as light stealer. Likewise some of the foliar diseases act as light stealer, assimilation rate reducer as well as tissue consumers. For quantification purpose, all possible damage mechanisms of a pest are enlisted. Most important damage mechanisms are then coupled to crop growth model at appropriate plant growth processes. These are calibrated and validated through field experiments and if needed then are modified (Fig 2). After quantification and validation of damage mechanisms, crop-pest simulation models can be used for various applications in pest management.

Identification of pest problems

The systems approach in very handy in problem definition and prioritization in complex agricultura systems. The pest problems in an agro ecosystem can be identified and prioritized by two methods: (a semi-quantitative methods and (b) quantitative methods.

Semi-quantitative methods

Semi-quantitative methods used to address key issues in pest management were reviewed by Heor (1985). These include pest profiles, damage matrices and decision trees. A pest profile shows the peric of activity of pests in relation to various phenological stages of the crop. In any crop some pest damag the crop during initial stages while some harm it during reproductive stages only. However certain pes

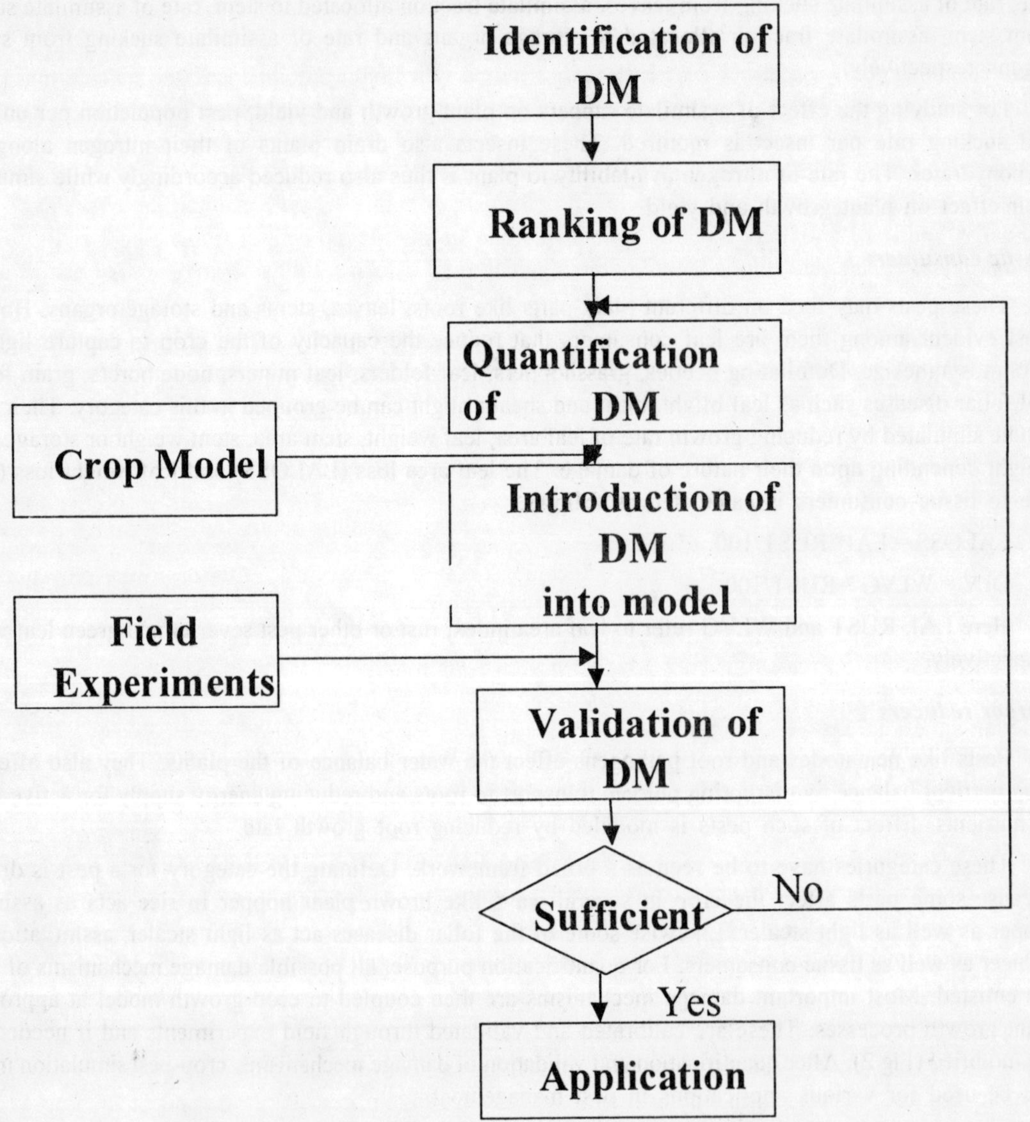

Fig 2. Schematic procedure for assessment of damage mechanisms (DM) on crop growth and production

amage the crop throughout its growth and development. A pest which remains active on the crop for nger period is liable to cause more damage than that which feed on the crop for a shorter period. Similarly pest which damages the crop during reproductive period would cause more loss than the one which amages during vegetative phase. The pest profile over time can help us to find relative importance of e pests associated with the crop.

A damage matrix reveals period of activity of the pest on the crop along with the plant parts damaged y the pest. It may allow identification of most important pests and help in prioritization of pests to be anaged and /or research action to be taken (Heong, 1985). Decision trees can be developed for comparing est control methods such as planting times or the damage thresholds for important pests and identifying uestions to be answered before executing these methods, (Norton, 1982; Mumford, 1981; Heong, 1983).

Adoption of systems approach in pest management furnishes a list of intervention points, which may elp to improve system behaviour. The intervention point may be in the form of research gaps needing tention, specific policy changes required to facilitate changes at a farm level or specific implementation eps (Teng and Savary, 1992).

uantitative techniques

With these techniques crop losses incurred by pests are quatified. These techniques can be classified empirical or regression approach and explanatory or mechanistic approach

rop loss assessment through empirical approach

Empirical approaches like regression analysis have been generally used for establishing yield-infestation lationship between pest incidence and crop damage. In this approach yield is regressed upon pest cidence (population or severity) observed at various crop growth stages. Pest is considered to inflict onomic loss only when regression coefficient of yield-infestation relation is significant. It is a two-step proach involving the pest incidence and crop yield. Pest incidence is related to crop yield without nsidering the effect of pests on plant physiological processes. These are liable to vary temporally and atially. The linear regression relationship between per cent folded leaves due to leaf folder and rice yield, r example, was found to be $y = 91.8182-1.0434x$ and $y = 100.1869-0.5036x$ during two years (Chander and ngh, 2000).

Several pests may infest the crop simultaneously or in sequence during the crop season. An early proach for handling several pests is the crop loss profile methodology using multiple regression (Pinstrup *al.*, 1976; Chiarappa, 1980). This approach quantified the contribution of each constraint to the total eld loss. With this approach several pest constraints, environmental variables and cropping practices ere considered simultaneously. This may also be considered an endeavour to include the concept of oduction situation. The production situation refers to crop management and environment variables that sult in to a given yield of the system (Penning de Vries, 1982). Statistical techniques such as principal mponents, canonical correlation and multiple regression analyses have been applied to analyze complex, ultivariate data generated by surveys (Stynes, 1980). In fact several statistical methods are available, nich can be exploited based on nature of data and objective of study (Madden, 1983).

sessment of crop loss through simulation model

Assessment of crop losses due to pests with mechanistic approach that is with simulation model is sed on damage

mechanisms of the pest. Damage mechanism of the pest can be defined as plant physiological process fected by the pest injury. Various plant physiological processes in crop simulation model are modified account for pest effects based upon damage mechanism of the pest. The assessment of crop losses th mechanistic approach is a three-step approach taking in to consideration, the pest incidence, damage echanism of the pest and the crop yield. The pest incidence has to be provided as input in the model.

However, pest dynamics model that could simulate pest population under given environmental conditions can be linked to crop-pest simulation model. The process of determining crop loss due to pests through simulation model is depicted in Figure 3. First the crop-pest model is run for healthy crop to obtain attainable yield. The model is then run with pest incidence level and the yield obtained in this case is called actual yield. The difference between attainable yield and actual yield is the yield loss due to pest.

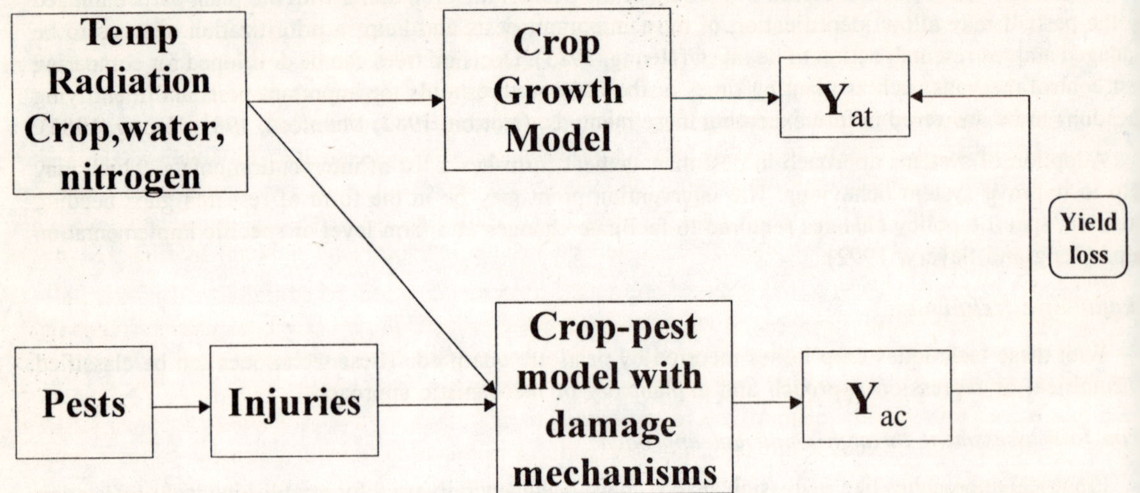

Fig 3. Assessment of crop loss due to pests with simulation model

Output of simulation models has been observed to be more reliable than empirical models. Simulation models are capable of addressing changes in weather, soil and management encountered at different locations. These can also simulate injury to the crop due to multiple pests, which is otherwise difficult to achieve in field experiments. Validated simulation models can be used to test numerous scenarios regarding effect of pests on crops. These models have great potential for applications in the field of pest management.

Decisions in pest management

There are mainly two types of decision in pest management (a) tactical and (b) strategic decisions. Tactical decisions are day to day decisions taken on farmers' level for a short time frame and smaller area such as decisions regarding insecticide application on the crop when pest population reaches economic threshold level. Strategic decisions are taken for a longer period and larger areas like decision regarding crop rotation for preventing build up of pests, breeding of resistant cultivars against the pests, introduction of natural enemies of the pests etc. In decision making in pest management the information on crop losses due to pests is of paramount importance in decision making as in problem prioritization.

Applications of simulation models in pest management

Crop-pest models can be used for rationalizing pesticide use, pest risk analysis, ecological pest zonation, assessing impact of climate change and pest forecasting.

Rationalizing pesticide use

Determination of economic injury levels

Economic injury level (EIL) is very useful concept is pest management. It helps to resort to need-based application of pesticides. Need based application of pesticides in turn avoids unnecessary expenditure and prevents environmental pollution. Coupled models have been used for simulating economic injury levels for bean leaf beetle infesting soybean (Nordh *et al.*, 1988). The coupled model was run from no pest to populations that resulted in no crop yield. Net return was determined by subtracting cost of production from gross return. The EIL was that pest density that produced a return similar to return from healthy crop after subtracting control expenditure. The EIL is a highly dynamic entity and varies with changes with yield-infestation relationship, cost of control and market value of the produce. Besides most of EILs are based on empirical relations, which are site specific. Such relations are thus required to be established for different locations through field experiments, which is impractical proposition. However with the help of a validated crop-pest model, economic injury levels can be easily worked out for a number of combinations of different factors. Simulation models can also be used for establishing economic injury level for multiple pests infesting the crop.

Formulation of iso loss curves

Crop-pest models may be used to generate iso-loss curves, which show various combinations of pest intensity and crop age that result in the same yield loss. These curves may then be used in surveillance programme for the pests. Daryaei (2003) used simulation model to devise iso loss curves for stem borer and bacterial leaf blight incidence in rice and stem borer and leaf rust incidence in wheat. These would help in need-base application of control measures and avoid unnecessary application of pesticides in the environment. Similarly simplified pest models or simlified decision rules from crop-pest models with economic values assigned to their outputs have been used for managing sugarbeet *Cercospora* leafspot (Shane *et al.*, 1985), sweet corn common rust (Teng, 1987), wheat diseases (Zadoks, 1984, 1989), rice blast (Surin *et al.*, 1991).

Pest risk analysis

Simulation models can also be used for analyzing the risk associated with the introduction of exotic pests into ecosystems in which they did not exist earlier. The foreign pests that are under quarantine cannot be brought deliberately into new countries to test them for their damage potential. Therefore models are very useful to evaluate situations where actual fieldwork on the pest is not possible (Teng, 1991). Yang *et al.* (1991) showed the possibility of the approach with a soybean rust model, SOYRUST, which when run with continental USA weather data predicted potential areas for epidemics. When model disease estimates were further linked to soybean crop model, the potential losses attributable to rust epidemics were determined.

Predictive pest zonation

Pest zonation is a concept that is particularly applicable for large area pest management, in which both tactics and strategies can be merged to achieve optimal management. Pest population dynamic model can be run with requisite weather data and probability of various levels of pest infestation for a site is determined (Yuen and Teng, 1990). Site predictions can then be extrapolated through geographic information system (GIS) to carve out the zones of equal epidemic potential for a pest. Knowledge of pest incidence in different zones would allow strategic decisions within each zone with respect to selection of crop cultivars. Probable life of resistant cultivars can also be determined. This would help extension workers in the selection of appropriate management options in different zones. Historical climate data from sites have been shown to be useful for classifying the conduciveness of a site for wheat diseases (Coakley, 1990).

Effect of climate change on pests and crops

There has been a rise in global temperatures and CO_2 since several previous decades (Rodhe, 1990; Mahlman, 1997). This is expected to continue and would affect climate of the earth. The change in temperature and CO_2 would influence development of crops and pests both. The impact of expected climate change during succeeding decades on crops can be assessed through crop growth simulation models.

Similarly effect of climate change on pests can be studied through population dynamics models. The ultimate effect on agricultural production would depend upon interaction between crops and pests.

Pest forecasting

Population of pests can be predicted through population dynamics models. These models represent different pest growth processes such as egg, larval, pupal and adult development in computer programming language. When these models are run with weather data of a place, they predict pest population at different time intervals. These models can thus play very important role in timely management of pests.

The development of an IPM scheme that fits into a sustainable farming system requires thorough analysis of the system first. An improved understanding of the system may be useful in identifying new control techniques by indicating intervention points and can be helpful in decision making in pest management. The systems approach is a holistic view, which presumes that overall systems behaviour is influence by changes in any system component. The adoption of systems approach in pest management commenced with formulation of simulation models for insects and diseases in the late 1960s. Pest management research is required to derive practical tools for developing tactics and strategies for pest management. Systems approach provides such tools in the form of simulation and decision models. The systems approach or systems analysis is equated to the development, testing and evaluation of a simulation model. The simulation models can be used for various application in pest management such as for rationalizing pesticide use, pest risk analysis, ecological pest zonation, assessing impact of climate change and pest forecasting. These can help to enhance efficiency of agricultural research greatly.

REFERENCES

Boote, K. J., J. W. Jones, J. W. Mishoe, and R. D. Berger 1983. Coupling pests to crop growth simulators to predict yield reductions. *Phytopathology*, 73, 1581-1587.

Chander, S. and V. S. Singh 2001. Distribution, economic injury level and sequential sampling of leffolder (Cnaphalocrosis medinalis) on rice. *Indian J. Agric. Sci.* 71 (12): 768-771.

Chiarappa, L. 1980. *Crop loss assessment methods. Supplement No. 3.* FAO /CAB, Farnham, England. pp. 123.

Coakley, S. M. 1990. Using historical weather and pest data for pest zoning. In *Crop Loss Assessment in Rice*(ed. Teng, P. S.).. IRRI, Manila , Philippines. pp. 303-312.

Daryaei, M. 2003. *Quantification of yield loss due to multiple pests in rice*. Ph. D. thesis, P. G. School, IARi, New Delhi.

Heong, K. L. 1983. Decision analysis in pest forecasting. *Mardi Res. Bull.* 11: 62-70.

Heong, K. L. 1985. Systems analysis in solving pest management problems. In *Integrated pst management in Malaysia* (eds. Lee, B. S., W. H. Loke and K. L. Heong). Malaysian Plant Protection Society, Kuala Lumpur, pp. 133-149.

Kropff, M. J., H. H. van Laar and R. B. Matthews 1994. ORYZA1 an ecophysiological model for irrigated rice production. SARP Research Proceedings. Wageningen Agricultural University, Wageningen, Netherlands and IRRI, Manila, Philippines. pp. 110.

Mahalman, J. D. 1997. Uncertainties in projections of humal-caused climate warming. Science 278: 1416-17.

Madeen, L. V. 1983. Measuring and modeling crop loss at field level. *Phytopthology.* 18: 67-83.

Mumford, J. D. 1981. Pest control and decision making: Sugar beet in England. *J. Agric. Econ.* 32: 31-41.

Nordh, M. B., L. R. Zavaleta and W. G. Ruesink 1988. Estimating multidimensional economic injury levels with simulation models. *Agric. Syst.* 26: 19-33.

Norton, G. A. 1982. A decision analysis approach to integrated pest control. *Crop Protection* 1: 147-164.

Penning de Vries, F. W. T. 1982. Systems analysis and models of crop growth. In *Simulation of plant growth and crop production* (eds. Penning de Vries, F. W. t. and H. H. van Laar). Pudoc, Wageningen. pp. 9-19.

Penning de Vries, F. W. T., D. M. Jansen, H. F. M. ten Berge and A. Bakema 1989. *Simulation of ecophysiological processes of growth in several annual crops.* Pudoc, Wageningen, Netherlands pp. 271.

Pinstrup, A. P., N. de Londono and M. Infante 1976. A suggested procedure for estimating yield and production losses in crops. *PANS* 22: 359-365.

Pinnschmimidt, H. O., W. D. Batchelor, and P. S. Teng1994. Simulation of multiple species pest damage on rice. *Agric. Syst.*, 48, 193-222.

Rabbinge, R. and F. H. Rijsdijk 1981. Disease and crop physiology: a modeler's poi of view. In Effects of disease on the physiology of the growing plants (ed. Ayres, P. G.).Cambridge University Press, Cambridge, U.K. pp. 201-220.

Rabbinge, R., S. A. Ward and H. H. van Laar 1989. *Simulation and systems management in crop protection.* Pudoc, Wageningen.

Rodhe, A. L. 1990. A comparison of the contribution of various gases to the greenhouse effect. *Science* 248: 1217-1219.

Rossing, W. A. H. 1993. Modeling effects of insects and pathogens on growth and yield of field crops. In *SARP Research Proceedings: Mechanisms of damage by stem bore, bacterial leaf blight and sheath blight, and their effects on rice yield* (eds. Rossing, W. A. H., Rubia, E. G., K. L. Heong, M. Heerati-Kasikorn and P. R. Reddy), DLO, Wageningen, IRRI, Los Banos. pp. 22-32.

Ruesink, W. G. 1976. Status of the systems approach to pest management. *Annual Rev. Entomol.* 21: 27-44.

Shane, W. W., P. S. Teng, A. Laney and B. Cattanach 1985. Management of *Cercospora* leafspot of sugarbeets: decision aids. *North Dakota Farm Res.* 43: 3-5.

Spitters, C. J. T. 1989. Weeds: population dynamics, germination and competition. In *Simulation and systems management in crop protection* (ed. Rabbinge, R., S. A. Ward and H. H. van Laar). Pudoc, Wnageningen, Netberlands. pp. 182-217.

Stynes, B. A. 1980. Synoptic methodologies for crop loss assessment. In *Crop loss assessment.* Misc. Publ. Univ. Minnessota. Agric. Exp. St. 7: 166-175.

Surin, A., P. Aruniyaanrt, W. Rojanahusdin, S. Munkong, R. Dhitikiattipong and S. Disthaporn 1991. Using empirical blast models to establish disease management recommendations in Thailand. In *Rice Blast Modelling and Forecasting*, IRRI, Los Banos, pp. 69-74.

Teng, P. S. 1985. A comparison of simulation approaches to epidemic modeling. *Annual Rev. Phytopathol.* 23: 251-279.

Teng, P. S. 1987. The systems approach to pest management. In *Crop loss assessment and pest management* (ed. Teng, P. S.). American Phytopathological Society Press, St. Paul, pp. 160-167.

Teng, P. S. 1991. Modeling and non modeling approaches to assessing the impact of field releases of bioengineered organisms. *Phytopathology* 80: 1137-1142.

Teng, P. S. and S. Savary 1992. Implementing the systems approach in pest management. *Agric. Syst.* 40: 237-264.

Yang, X. B., W. M. Dowler and M. H. Royer 1991. Assessing the risk and potential impact of an exotic plant disease. *Plant Disease* 75: 976-82.

Yuen, J. E. and P. S. Teng 1990. Assessing risk with stochastic dominance. *Phytopathology* 80: 997.

Zadoks, J. C. 1985. On the conceptual basis of crop loss assessment: the threshold theory. *Ann. Rev. Phytopathol.* 23: 455-73.

Zadoks, J. C. 1984. EPIPRE: a disease and pest management system for winter wheat developed in the Netherlands. *EPPO Bull.* 11: 365-369.

Mass Production of Insect Pathogens for Biological Control

O. M. BAMBAWALE[1], NASIM AHMAD[1], U.R. SANGLE[1] AND A.A. JAFRI[1]

Mass production of insect pathogens is essentially required in the current pest management scenario for catering to the growing need of the enlightened farmers. The efforts in the past two decades or so, of various public sector institutions, state and central agricultural departments, SAUs, NGOs and international agencies in working on various aspects of ecological approach to plant protection have helped farmers to learn about IPM and to implement crop protection decisions which are safe, productive and sustainable. There has also been a concomitant progress in the field of invertebrate pathology as a vocation because it was realized that entomopathogenic organisms have a future as important components of IPM of greenhouse and field crops, orchards, ornamentals and organic cultivation. Like all other living beings, insects are susceptible to a wide array of diseases caused by microbials. Many of these diseases are acute and fatal, and therefore, they can be used on a short and long term basis for the management of insect pests.

MICROBIALS FOR INSECT PEST MANAGEMENT

The microbials that cause disease in insects fall into four main groups. They are: viruses, bacteria, fungi and protozoa. Though hundreds of microbials are known to cause disease in insect pests, only about a dozen microbial agents (Table 1) have been commercially developed or are under development. Since they usually only infect specific groups of insects they often cause no disruption of naturally occurring biological control.

VIRUSES

Several virus isolates have been recorded causing disease in about 1100 species of insects and mites. These have been categorized into seven families and only baculoviruses have been given extensive consideration as microbial control agents. They share no obvious biochemical properties with viruses of vertebrates, plants or microorganisms. Being obligate intracellular pathogens, they can be produced only in larvae or insect cells in culture. Baculoviruses are classified into three subgroups. In two groups, granulosis (GV) and nuclear-polyhedrosis (NPV) viruses, virus particles are packaged within large proteinaceous occlusion bodies (OBs). These OBs are relatively stable and provide the means by which virus infectivity is preserved outside the host. Diseased larvae can produce $10^9 - 10^{11}$ OBs when they die and, under suitable conditions, these can promote the rapid spread of infection within susceptible insects. The OBs must be ingested by the larvae for contracting infection. Control is slow as death of an insect following virus infection is unlikely to occur before 3-4 days. Limited recycling of the pathogen once

National Centre for Integrated Pest Management, Indian Agricultural Research Institute campus, New Delhi 110 012.

applied in the field is possible. The effective dose of the NPV is generally 250 ml/ha (@ 2 x 10^9 OBs per ml). The NPV of *Helicoverpa armigera* (HaNPV) and *Spodoptera* spp. (SlNPV) are commercially mass produced. Nuclear Polyhedrosis Virus (NPV) is an important biological option available for *H. armigera* management (Jayaraj, *et al.*, 1994; Sathiah and Jayaraj, 1996) because of its well demonstrated proven efficacy on a number of crops all over the country as also its several attractive features such as specificity, ecological safety, recyclability, less chances of resistance development by the pest and long shelf life. The option remains grossly underutilized for want of a proper understanding of its weaknesses and strengths as regards ecological niches where it can be utilized and the lack of organized effort for its production and use.

Table 1. Microbials developed / considered for insect pest control*

Pathogen	Major targets	References
VIRUSES		
Nuclear Polyhedrosis Viruses	Caterpillars	Federici, 1998; Hunter-Fujita *et al.*, 1998; Treacy, 1998
Granulosis Viruses	Caterpillars	Tweeten *et al.*, 1981
Nonoccluded Viruses	Scarab larvae	Bedford, 1981, Zelazny *et al.*, 1992
BACTERIA		
Bacillus popilliae	Scarab larvae	Klein, 1997
Bacillus thuringiensis	CaterpillarsBeetle larvae	Baum *et al.*, 1998 Keller and Langenbruch, 1993
FUNGI		
Beauveria bassiana	Beetle larvae, caterpillars, grasshoppers, whiteflies	McCoy *et al.*, 1988; Wraight and Carruthers, 1998
Meterhizium anisopliae	Beetle larvae, caterpillars, leafhoppers	McCoy *et al.*, 1988
Verticillium lecani	Aphids, whiteflies	Hall, 1982; Hall and Paperiok, 1982
Nomuraea rileyi	Caterpillars	Vimla Devi and Prasad 1998
MICROSPORIDIA		
Nosema species	Grasshoppers, beetles	Brooks, 1988; Henry, 1990
Vairimorpha necatrix	Caterpillars	Brooks, 1988

* (Adapted from Federici 1999)

Whereas production of SlNPV is relatively easy as the larvae of *Spodoptera* can be reared gregariously, the production of HaNPV is far difficult as *Helicoverpa armigera*, though has high level of plasticity under natural conditions, is highly vulnerable when reared under lab conditions. The salient features of mass production of HaNPV are presented here.

Mass production of HaNPV

HaNPV can multiply only in the living host, as all viruses are obligate in nature. The virus produces resting structures in the infected larva. The resting structures look like a minute ball filled with rod shaped virus particles. Such resting balls of the virus are called Polyhedral Occlusion Bodies (POBs). The POBs are stable and provide the means by which the virus infectivity is preserved outside the host. It is the POBs that are applied as microbial biopesticide. Each POB measures about one micron in diameter and could be seen under a microscope at 400 X magnification. A diseased full-grown larvae releases on an average 200 crore of POBs of this virus (Pawar and Thombre , 1992) under Indian conditions of production.

Relevant information on mass production of NPV is available in the country suiting the local socio-economic needs.

Mass production of NPV can be done by two methods viz., by using the live larvae (*in vivo*) and by inoculating the cell line culture (*in vitro*) of the pest. The production of HaNPV on cell lines is not yet commercially viable (Hunter *et al.*, 1998). For the production in live larvae field collected larvae of *H. armigera* or the laboratory-reared larvae can be used using semi synthetic diet.

NPV production through lab-reared insects

A steady supply of healthy, disease free larvae of *H. armigera* is the first requirement of any programme to produce NPV. The establishment of such a culture facility is a long, input-intensive process. Several excellent review articles and production manuals (Shieh, 1989, Armes *et al.* 1992a, Sathiah and Jayaraj, 1998, Ramakrishnan and Paul, 1998, Gryzwacz *et al.* 1999) describe the steps involved in establishing such facilities. Even with such an apparently vigorous and successfully widespread insect pest such as *H. armigera*, it can prove difficult to establish a culture in the laboratory. While the major enemies of culture establishment are diseases, fungi and parasites, phenomena such as diapause and infertility can also be serious problems (Gryzwacz *et al.* 1999). Promotion of *H. armigera* mass rearing technology by the corporate sector has not taken off (Sathiah and Jayaraj, 1998) for various reasons. Investors are very few considering the risk factors involved in the project. Biocontrol Research Laboratories Ltd. Bangalore is the only private sector company, which has placed the HaNPV production from lab-reared larvae on a sound footing (Personal communication with Dr K.P. Jayanth). The BCRL Lab has a standardized protocol for rearing procedure of *H. armigera.* The host larvae are produced @ 10,000 per day and the unit can magnify the output as per demand. Since the lab reared larvae, free of other microbial contamination are used, the quality of the NVP produced is pure and exceptionally uniform. Whereas the unit was able to harvest only 2×10^9 POBs per LE during the beginning of the venture, now the productivity is around 3-4 $\times 10^9$ POB. Currently about > 1 crore ml (1 ml = 1×10^9 POBs) HaNPV is produced annually and sizable demand can be met if prior indent is available. This is one unit, which has shown the sustainability of the HaNPV production through lab reared larvae and is a role model for others to take up similar activities.

NPV production through field collected larvae

Since *H. armigera* is of polyhagous nature, the larvae of the pest are available for 4-6 months in overlapping generations on crops such as cotton, pigeon pea and chickpea and this natural resource can also be profitably harvested and put into use for NPV production.

NPV production through field-collected larvae is comparatively easy and requires fewer resources. However, such way of production is fraught with inherent problems: i) the activity is highly seasonal depending on the intensity of the pest on different crops, ii) there is usually a difficulty in getting the desired uniform stage of the larvae, iii) the larvae may be pre exposed to chemical pesticides, iv) the larvae may already be parasitised /infected with other pathogens or v) the product derived from such larvae may be contaminated with other microbes/viruses. As a consequence of all these factors, the mass producer has inadequate control on the quality and the quantity of the NPV produced.

NPV produced from the field-collected larvae is generally difficult to standardize on the basis of LE and counting of the POBs becomes inevitable. However, since the larvae are available in very large number in the field for at least 4-5 months in a year, "cottage industry for HaNPV production" would be more appropriate for entrepreneurs/ progressive farmers who have only limited resources.

Main steps in NPV production

The lab-reared or field-collected larvae are reared on a semi-synthetic diet consisting of chickpea grain flour, antibiotics, vitamins, live yeast and preservatives such as methyl paraben with agar-agar as a solidifying agent. Multicell cavity trays, which are now commercially available, are filled with chickpea flour based semi-synthetic diet @ 4-5 ml/cell. In place of cavity trays, many small-scale units use plastic vials of about 25—30 ml capacity, which cost around Rs 0.40 per piece. In place of artificial diet, chickpea grains (gram dal) soaked overnight in sterile water is also used as a cheaper source of food for the larvae. In this case the grains are inoculated with the virus and the grains are periodically added in the vial as per requirement. In case of artificial diet, after solidification the diet surface is inoculated with a standard dose of POBs with the help of a syringe (@ 20 μl per cell). Third to fourth instar larvae are released into the tray @ one per cell and lid is fixed.

Larvae get infected *per os*. Once the larvae get infected with the virus, the length of the disease depends on many factors. These include the host, the inoculum source and its concentration, the size of the larvae at the time of inoculation and temperature. At the advanced stage of infection as most of the polyhedra are formed inside the larvae, the host nuclei lyse and the larval body liquefies. Once liquefaction stage occurs, it is very difficult to harvest the entire amount of POBs formed. It is therefore, essential during commercial production system to harvest the diseased larvae just prior to terminal stage of the disease to avoid liquefaction. This also keeps the bacterial contaminants in the product to the minimum.

In case of the production through field collected larvae, the trays are observed for dead larvae after two days of inoculation and such dead larvae are promptly removed as these are not killed by the virus. After five days the trays are again watched and the dead larvae are collected in a container with distilled water in a ratio of 1 larvae : 1 ml. If there is need, the remaining larvae are transferred to other trays with fresh diet. Finally, the pupae and the larvae, dead or alive, are removed at the end of 7th or 8th day and kept in distilled water, which is stored for at least 4 - 5 days.

The collected dead larvae are macerated at the end of one week in a warring blender using distilled water @ 2 ml per diseased larva. Later the product is put into a centrifuge at 500 r.p.m. for 2 minutes and the settled debris is discarded and the product subjected again to centrifugation at 5000 rpm and the residue settled is collected in fresh distilled water. The stock solution of NPV is standardized for POB counts with the help of a haemocytometer and a light microscope. The whole process is over within 15 days - 8 days for incubation period, 5 days for processing and 2 days for standardization of the product, packing, etc. In India, liquid concentrate of the POBs in distilled water is developed without any additives. Alternatively wettable powders have also been developed (Ignoffo, 1973, Moscardii, 1999). Both such formulations can be used with commonly available spray equipment.

The issue of the use of Larval Equivalents (LE) as the unit of HaNPV activity has created much confusion for producers, customers and regulators. Kennedy *et al.* (1998) have thoroughly discussed this issue: 'A larval equivalent is the average amount of NPV produced in an infected larvae. In most cases, where LEs are quoted, they are not based upon actual counts of NPV but are merely a statement of how many infected insects were used to make up the product. However, the amount of NPV produced in larvae is highly dependent upon the quality of inocula, conditions of the larvae and production system. The use of larval equivalents as a standard measure of NPV activity is, therefore, very suspect and should be replaced by actual counts of NPV.' The Indian Standard prescribed for NPV is 1×10^9 POBs per ml of the product. Registration of the product is now mandatory.

Case study of a successful HaNPV producer (field collected larvae)

It is possible to produce HaNPV from the field-collected larvae on scientific lines on a sustainable basis as has been demonstrated from the successful continuation of such a unit for the past seven years. M/s J.R. Biocontrol Lab, Arni, Distt. Yavatmal, Maharashtra entered into an MoU with CICR, Nagpur in August, 1996 and started mass production of HaNPV through field collected larvae. The Unit was established on an area of 5000 sq. ft. Three tin shades were raised in an area of 1000, 600 and 600 sq.ft. The main shed housed the diet preparation, inoculation, sorting of larvae, incubation and harvesting. Another shed served the purpose of laboratory for centrifugation and counting of POBs, bottling, labeling, etc. The third shed is meant for the labor retiring room. *H. armigera* larvae were collected from an area of approximately

15 km radius from September to February on cotton, pigeon pea and chickpea each year.

The establishment cost of the laboratory included trays for holding plastic vials, racks for holding trays, microscope, haemocytometer and other miscellaneous related items, plastic vials, land cost and shed construction (Rs 4.15 Lakh). The staff included three regular employees. The daily paid labor ranged from 10-50 depending on the volume of work in the laboratory. Another 10-30 labor got employment by way of contract job for collecting larvae from crops in the fields. The total average employment generated was 25 persons for 6 months. The unit has been able to produce 40.888 Lakh LEs (2 x 10^9 POBs/LE) in six seasons from 1996 to 2001 with a net profit of 22.277 lakh rupees. (However, this does not include the salary of the entrepreneur and his family members involved in the project). Thus the unit appears to be sustainable and economically viable.

Constraints faced by the HaNPV producers

Constraints are faced at two levels. The technological constraints of lab production have been analysed (Sathiah and Jayaraj, 1998). These include sustainability of larval production and maintenance of targets of production (Armes *et al*, 1992a, Gryzwacz *et al*. 1999). These constraints have been mostly removed as is exemplified by the success of BCRL, Bangaloare, but their adaptation by a good number of entrepreneurs is now required. The constraints faced by the cottage industry producers are mostly related to uncertainties of production, registration and marketing and are as follows.

- High-risk enterprise
- Bank/institutional finance not available
- Scientific protocols are not followed
- Production procedure complicated
- Seasonal, uncertain activity
- Seasonal variation in POB productivity, input/output unpredictable
- Marketing of small volume of produce difficult, as own network not available
- Waiting period from production to sale is very long.
- Registration on individual basis for small entrepreneur expensive and absence of registration gives low credibility.

Registration requirement is now a major deterrent for the small-scale producers impending development of more units because registration involves not only time but also the cost component for data generation, which is conservatively estimated at Rs 4-5 lakh per microbial bioagent at the current situation. To overcome such problems, the bioagent producers of Maharashtra have taken a lead and formed Vidarbha-Marathwada Agro Biotech Association with > 20 members (personal communication with Dr R.K. Chintawar of JR Biocontrol Laboratories, Yavatmal). They have collectively obtained the relevant data necessary for registration and taken up processing of their individual cases in HaNPV and *Trichoderma harzianum* production. NCIPM, New Delhi is providing the necessary guidance in the matter.

Support system and quality control

One major problem after quality production is the marketing of the produce, particularly of the small-scale units, which are mostly managed individually without the availability of a substantial marketing network. Central Institute for Cotton Research, Nagpur realized the need for providing marketing support to the entrepreneurs in the initial stages of their establishment. Such system was developed and implemented from 1997 for a period of 4 years. It has been a small effort considering the requirements and needs to be magnified by developing a national grid for demand and supply. Public sector establishments can play a pivotal role in networking as they know various producers of HaNPV and the farmers who need the product.

BACTERIA

There are two main species of bacteria viz., *Bacillus thuringiensis* and *B. popilliae*, which are useful in insect pest management.

Bacillus thuringiensis

Bacillus thuringiensis (Bt) is a Gram Positive bacterium with unique insecticidal properties. It grows abundantly in the presence of oxygen (aerobic) and is a common soil inhabitant. Asiatic soils are particularly rich in Bt. In its life cycle Bt grows first as vegetative cells (rods of 2-5 u). These rods are motile through the presence of appendages called flagella. One division of a rod into two rods takes approximately 20 minutes only. After abundant exponential vegetative growth in a favorable medium (organic matter) for about 12-24 hours, the rods enter into sporulation phase. A sporangium is produced through transformation of the vegetative rod, which bulges in size and within this sporangium a spore on one side, and a crystal (mostly rhomboid) on the other side is formed. The sporulation stage lasts for 36-72 hrs. At the end of sporulation the walls of sporangium disintegrate and the spores and crystals get separated out in the substrate. Upon receiving favorable conditions the spores germinate again to form vegetative rods and the cycle continues. The significance of crystals in the life cycle of Bt is not very clear. However, these crystals, called delta-endotoxin, which are protenaceous in nature, happen to be toxic to a wide range of insect pests. Bt commercial product essentially contains a mix of spores and crystals.

Bt classification and host range

Classification of Bt is important to ascertain the identity and its spectrum of activity. The classification is attempted in three ways.

1. First, any unknown/wild strain is subjected to serovar classification. The swimming appendages (flagella) of vegetative rods have specific proteins on their surfaces which serve as antigens (called flagellar antigens). Using these antigens specific antibodies are produced in rabbits. Now with supportive biochemical properties and antigen-antibody reactions the existing Bt isolates have been differentiated into > 65 varieties. An unknown isolate is challenged with antibodies from known varieties of Bt. A precipitation reaction identifies the unknown strain as a particular variety. If there is no precipitation then such strains are called UNTYPIFIED (UTY) and subjected to development of a new set of antisera/biochemical tests and is then defined to be a new variety.
2. Bioassay: Bioassay of the protein toxin against different insect orders gives the host range of a particular isolate. A single isolate may exhibit activity against a single insect order, mainly Lepidoptera (L), Coleoptera (C) or Diptera (D), etc. or a group of orders such as L/D, L/C, etc.
3. Molecular characterisation: The delta-endotoxin can be classified into different *Cry* (crystal) genes on the basis of the sequence of the cry genes and the specificity of action of the gene product to different insect orders. Currently Bt toxins are classified into 22 classes. Molecular characterisation is an important step in getting the strain patented.

Isolation and identification

Since a large extent of the wild variability of Bt still remains to be explored or made public, there is a good scope for further improvement of Bt. This variability, if properly tapped, could be profitably utilized in the form of an increased potency, and/or wider spectrum of insecticidal activity. Strain search programs are important tools for improving chances of better insect-pest management tailored to local needs.

Bt is widely distributed in the environment and can be readily isolated from soil, insects and their habitats, stored grains and phylloplane. A commonly used procedure is the sodium acetate Bt selection procedure of Travers *et al.* 1990. A small soil sample (1 g) is placed in 25 ml of 0.25 M sodium acetate solution in L broth and shaken for two hours. At the end the sample is placed in a water bath at 80 °C for five minutes. In sodium acetate; all other spore-forming bacteria germinate whereas Bt doesn't. The heat shock kills the vegetative stages of other bacteria. A loopful of the suspension is then streaked on L-agar medium. The typical colonies of Bt are picked up and transferred to T3 slants. After three days the culture is observed for identification.

Preliminary identification of Bt isolates can be made on the basis of presence of parasporal crystals. For this the Bt culture is grown on a medium, the spore + crystal mixture harvested, spread over a microslide and subjected to differential staining.

***Bt* staining procedure**

Reagent A 1.5 g of Amidoschwartz stain (Naphthol Blue Black, Buffalo Black NBR, Amido Black are synonyms) is dissolved in 50 ml of 98 % methanol, 40 ml distilled water and 10 ml acetic acid. Best results are obtained when the reagent is stored for 3 days after preparation.

Reagent B It is Ziehl's Carbol Fuchsin.

1 g Basic Fuchsin is dissolved in 10 ml of 95 % ethanol and mixed with 5 g of phenol dissolved in 90 ml of distilled water. It is to be used at 30 % aqueous solution.

A thin layer of Reagent A is pippeted over the heat fixed Bt smear while the slide is stili warm. After 70 seconds or so the stain is washed off with cold tap water. While still wet, the smear is then covered with an even layer of Reagent B for 20 seconds, washed in cold tap water and dried on filter paper. Standard blue filter is used in light microscope. The slide is studied under immersion oil at 1000 X. Presence of crystals, which look violet to dark blue and spores, which look pink, confirm the isolate as Bt. The same staining procedure is used while quality control of the product.

Once identified as Bt, the isolate can be sent to one of the identification Centres for varietal identification. Alternatively, Bt cultures can be obtained from one of the available culture collections.

***Bt* culture collections:**

1. Bacillus Genetics Stock Centre, Ohio State University, Columbus, OH, USA.
2. Dr M. Ohba, Institute of Biological Control, Faculty of Agriculture, Kyushu University, Fukuoka, Japan
3. USDA, Brownsville, Texas, USA
4. USDA Forest Service, Hamden, Connecticut, USA
5. Chief de Laboratorie, Institut Pasteur, 25, Rue Du Docteur Roux, Paris, France

It is, however, important to have ones own isolates so that these can be commercialized without entering into MoU or IPR issues.

Storing of stock culture for use in mass production

Two methods are commonly used:

1. Pour autoclaved Liquid Paraffin over the 5-day old culture grown on TYG or T3 media slants so as to cover the top of the slants with 1 cm layer. Seal the cotton plug of the slants with contact film or aluminum foil and preserve at 4°C. such cultures stay easily for 3-5 years. Test viability periodically.
2. Prepare 15% glycerol solution in sterilized distilled water. Sterilize this solution in glass or polypropylene vials. Prepare broth cultures of Bt and pipette 0.5 ml of the growth (broth) in the vials. Store at -80 °C. Test viability periodically.

Lab production of Bt for bioassays

Bt strain is grown at 28°C on TYG medium (tryptone 5 g + yeast 2.5 g + glucose 1 g + agar 18 g in one L DH_2O) in sterile petri plates. At the end of 3 days of growth the spore + crystal mixture is harvested in DH_2O and washed thrice with 100 mM NaCl and 0.1 % triton-X. After each wash the mixture is spun at 9000 rpm at 4 °C. To remove the salt and triton the pellet is washed thrice in D DH_2O. The pellet is lyophilized to get powder of spore + crystal mixture and is used in bioassays.

To further improve bioassay, the crystal in the pellet is solublized in 100 mM Na_2CO_3, 0.1 % B-mercaptoethanol for 2 hrs at 37 °C. The spores and debris is pelleted out through centrifugation and the supernatant contains solublized toxin. The solublized protein quantity is estimated through Lowry's method or Bradford's method. In case of Bradford's Method, Biorad protein assay reagent (200 µl) is added to 800 µl of the sample solution and allowed to stand for 2-5 min. The absorbance is read at 595 µm. Protein concentration is deduced from the standard curve made with known amount of BSA. Solublized toxin protein is stored in vials under -20 °C for bioassay.

Standardized bioassay to calculate IUs: (Dulmage *et al.*, 1971)

Since standardization of preparations of Bt containing the delta-endotoxin cannot be accomplished by spore count or protein concentration alone, a bioassay is needed. A bioassay based on International Units (IUs) and using *Helicoverpa armigera* larvae can be taken up. In this assay the product is incorporated into a commonly used chickpea flour based semi-synthetic diet, and 10-15 mg larvae are allowed to feed on the diet for 5-6 days. At the end of the test period, activity is measured by determining LD50 of the test material and comparing it with that of an International Standard Bt (HD -1-S-1980 strain of *Bt kurstaki*). Potency is then expressed as IU/mg.

$$\text{IU/mg sample} = \frac{\text{LD50 Standard}}{\text{LD50 Sample}} \times \text{IU/mg Standard}$$

(IU/mg Standard, i.e., HD-1-S-1980 is 16,000 IUs/mg)

For each batch of commercially produced Bt formulation the IU/mg potency is to be found out and the information is to be used in the product standardizaton process.

Commercial production of *Bt*

Once a desirable isolate has been selected, the fermentation conditions for optimum production of insecticidal toxin can be investigated. While selection of a potent Bt isolate offers the potential for a highly potent end product, the fermentation conditions drastically alter the ability of the isolate to produce maximum potential potency. Thus, the consistency of the potency and spectrum of activity of the final product depends heavily on the quality and consistency of fermentation process.

In semi-solid fermentation procedure, semi-solid media nutrients are contained in a coarse porous matrix with a large surface area. Humidity has to be maintained within narrow limits to allow growth. In larger cultures, oxygen transfer, humidity and temperature are maintained by either placing the media on shallow trays, or keeping them in aerated drums.

Semi-solid medium of Mechalas, B.J., 1963 (US Patent 3,076,922) composed of 545 g wheat bran, 380 g expanded perlite, 62 g soybean meal, 36 g dextrose, 3.6 g lime, 0.9 g NaCl, 0.29 g $CaCl_2$ and 160 ml water. After partial sterilization with flowing steam for 60 min, the mixture weighs about 1 kg. This is inoculated with 400 ml of fermentor seed. At a time 250 kg of the bran medium is inoculated. The final product is the mixture of unutilized bran medium, spores, crystals and other byproducts of fermentation.

Semi-solid media are, however, difficult to sterilize and to keep sterile during production. Maintaining defined; uniform conditions during the production process as well as adjusting parameters like pH, is also difficult. Submerged culture is therefore, the preferred method for industrial mass production.

The submerged fermentation process involves a stepwise buildup of Bt cells. A liquid nutritive medium is used to suspend and propagate Bt biomass. Two such media compositions are given here.

Composition of media used in submerged fermentation (Dulmage *et al.* 1969).

Media A		Media B	
Ingredients	Weight (g)	Ingredients	Weight (g)
Tryptose	10.0	Proflo*	10.0
Dextrose	5.0	Peptone	2.0
Starch	5.0	Dextrose	15.0
Yeast extract	2.0	Yeast extract	2.0
K_2HPO_4	1.0	$MgSO_4.7H_2O$	0.3
KH_2PO	1.0	$FeSO_4.7H_2O$	0.02
DH_2O to	1000 ml	$ZnSO_4.7H_2O$	0.02
		$CaCO_3$	1.0
		DH_2O to	1000 ml

*partiallv defatted cooked cottonseed flour

Locally available organic sources of carbon and nitrogen can be utilized after standardization of protocols on pilot scale.

A typical Bt fermentation is started by inoculating a flask of culture medium with loopful of bacteria from an agar plate. The flask is then incubated at 28 °C on a shaker. At this stage spores germinate and the vegetative cells adopt to the culture medium. After 10-16 hrs, the cells are in mid-log phase. The culture is then used to inoculate a 20 ltr fermenter containing the same culture medium. Growth in the fermenter is monitored by counting cells microscopically. After 6-8 hrs, the culture is in the late log-phase and is transferred into the 200 ltr fermenter. Growth in the production fermenter is again microscopically monitored until the sporangia start to lyse.

During fermentation antifoaming agents need to be added and pH and oxygen supply monitored.

Recovery and formulation

The final contents of fermenter must first be stabilized and screened to remove large debris comprising of cells or growth medium. Initial screening may be done by ultrafiltration, screening or centrifugation. The product is then processed by centrifugation or evaporation to increase the potency of the broth and form an aqueous concentrate.

The aqueous concentrate is then formulated as i) liquid suspension concentrate , ii) wettable granules, iii) wettable powder or iv) oil flowable using a few additional steps.

Bacillus popillae

B. popilliae infects and kills coleopteran larvae, particularly of the family Scarabaeidae (Klein, 1997). As a result of infection milky disease develops in the scarabs. Opaque white color characterizes diseased larvae and results from the accumulation of sporulating bacteria in the haemolymph.

Suitable media for mass production *in vitro* are not available and have inhibited the commercial development of the bacterium.

To produce *B. popilliae*, scarab larvae are collected from field, injected with spores, and in 2 weeks the larvae get killed with the pathogen. Powdered formulations of the spores are then made. The application rate is one kg of formulation per ha.

FUNGI

Some 700 species of entomopathogenic fungi have been reported from over 100 genera, but only about 10 of these have been used or currently being developed for insect control. A broad range of biologies from obligate parasitism to opportunistic pathogens that can survive saprophytically in the absence of living hosts, have been documented. In most species, the access to the host is through the cuticle and may involve complex biochemical interactions between the host and the fungus before germination, penetration, growth, and reproduction of the fungus can occur. Of the entomopathogenic and disease antagonistic fungi, deuteromycete fungi have received most attention as they are among the easiest to produce in vitro and several have a broad host range. Due to their contact infection the fungal pathogens have potential for the control of sap-feeding arthropods, which do not ingest pathogens on the plant surface. High relative humidity is important for the germination of spores of fungi and this aspect becomes restricting for their wider use.

More than 30 trade-named mycoinsecticide products are now registered or are under development worldwide (Wraight *et al* 2001). In these commercial preparations the propagules of choice are conidia of various entomopathogenic fungi. This is because conidia are the means of dispersal and infection under natural conditions and also conidia of many species have strong hydrophobic walls, which confer environmental stability to the product. Since the fungal conidia are rarely able to cause infection singly even under favorable environmental conditions, there is a requirement of high doseses ($10^{13} - 10^{14}$ spores per ha) for pest control under field situations (Bartlett and Jaronski, 1988).

Important Deuteromycete fungi for pest management and their commercial level mass production status

S.No.	Species	Target pests	Status of commercialization in India
1.	*Beauveria bassiana*	Beetle larvae, caterpillars, grasshoppers, whiteflies	Commercially available
2.	*Metarhizium anisopliae*	Beetle larvae, caterpillars, leafhoppers	Production technology available; Commercially available through Public Sector/ SAUs
3.	*Verticillium lecanii*	Aphids, whiteflies	Limited availability through Public Sector/ SAUs
4.	*Nomuraea rileyi*	Caterpillars	Production technology available; Limited availability through Public Sector/ SAUs
5.	*Hirsutella thompsonii*	Mites	Production technology available with Public Sector Institute
6.	*Paecilomyces lilacinus*	Phytophagous nematodes	Commercially available

Nomuraea rileyi

The fungus is of cosmopolitan nature infecting mainly the lepidopteran larvae. In India, natural occurrence of the pathogen has been recorded on *Spodoptera* spp. (*exigua* and *litura*), *Helicoverpa armigera, Trichoplusia ni, Plusia orichalcea, Diacrisia obliqua, Achaea janata, Agrotis ipsilon*, etc. *N. rileyi* was noticed in epidemic proportions on *H. armigera* at the end of the epizootics of the pest on cotton and chickpea in 1987 and 1997 in the central and south regions of India and appear to be largely responsible for containing the carryover of the pest to the next season. The subsequent years of epizootics were invariably found to have very low incidence of the pest. The pathogen, therefore, deserves to be critically studied for ways of proper utilization. The pathogen has a wide host range in Lepidopteron pests, which can help it to get established in the fields much before the flare up of the main pets. Infection of the host is mainly through body wall, spiracles and mouthparts. Within 4-5 days the larval body is mummified and covered by a dense white mycelial mat from which conidia arise after another 2-4 days. The color of the dead larvae changes from milky white to green. Infection process is favored at 20-30C and RH >70 % and leads to the development of natural epizootics. The fungus is recommended for use at 2 x 10^{11} conidia/l spray solution for foliar application. The pathogen holds a high promise in utilization as a component of IPM for management of some of the most economically important and polyphagous noctuid pests.

The pathogen is readily amenable to mass production and needs to be promoted aggressively. *Nomurae rileyi* has a high market potential, and the production technology is now available for commercialization (Vimla Devi & Prasad 1997 and Vimla Devi *et al.* 2001).

Metarhizium anisopliae

Metarhizium anisopliae (Green muscardine fungus) is a widely distributed pathogen and the host range is very wide including Coleoptera, Lepidoptera, Hemiptera, Hymenoptera and Orthoptera. The important economic pests controlled by the pathogen are: the coconut rhinoceros beetle (*Oryctes rhinoceros*), spittle bug on sugarcane (*Mahanarva posticata*), *Nilaparvata lugens* on rice, scarabaeids (*Popillia japonica* and *Rhizotrogus majalis*), diamond back moth (*Plutella xylostella*), etc. Several secondary metabolites (cyclodepsipeptides, destruxins A, B, C, D and E and desmethyldestruxin B) are produced by the pathogen in culture which act as mycotoxins. They are also produced in the fungus-infected larvae and are important in development of symptoms. The pathogen can be used for both soil application and aerial spray. In case of *O. rhinoceros* all stages are susceptible to the pathogen. The manure pits and heaps require 3.4 x 10^{11} spores/tonne of FYM for management of the beetle and to achieve high mortality. The pathogen is readily amenable to mass production (Fargues *et al.* 2002, Patel, *et al.* 1990).

Beauveria bassiana

The pathogen is of cosmopolitan nature. The hosts include mainly Lepidoptera, coleoptera, Hemiptera, Diptera an Hymenoptera. The fungus has been recorded to cause infection of the lungs of wild rodents, nasal passages of humans, horses and giant tortoises and therefore, caution in use in necessary. Some of the pests of economic importance, which can be targeted for control through this pathogen, are: *H. armigera*, *Spodoptera* spp. *Plutella xylostella, Pieris brassicae, Odontotermes obeuss, Chilo infuscatellus, Idioscopus spp., Bemisia tabaci*, etc. *B. bassiana* also produces mycotoxins and the secondary metabolites prevent bacterial contamination and leads to fungal mummification of the infected insect. The pathogen is readily amenable to mass production (Mazundar *et al.* 1995, Puzari, *et al.* 1997) and is commercially available in India.

Hirsutella thompsonii

Several strains of *H. thompsonii* have been isolated from different species of mites. In artificial media the colonies of the fungus appear as a grey, loose, fluffy aerial growth slightly elevated above the medium. The pathogen sporulates only moderately and the conidia are spherical, ca. 2-4 um in diameter. Early symptoms of disease are difficult to see in small mites. Initially the infected mites show sluggish movement and later they change in color from a normal lemon yellow to a deep yellow to brown in citrus rust mite. Conidia usually germinate on contact with any part of the mite cuticle and invasion starts. Approximately 9 hrs are needed for the fungus to develop and sporulate outside an infected mite at 26 ^{0}C and 100 % RH. The technology of mass production is available with PDBC, Bangalore. Mass production protocols are available (Maimala *et al.* 1999).

Verticillium lecanii

The genus *Verticillium* is characterized by whorls of phialides formed at intervals along the conidiophore. The range of insect hosts of *V. lecanii* includes aphids, whiteflies, thrips and mites and it also hyperparasitizes plant pathogenic fungi such as the rusts. The fungus grows well in culture and a dose of 10^5 to 10^7 conidia/ml is effective in management of sucking pests. The fungus has been effectively used in greenhouses for the sucking pest control due to availability of high humidity. The pathogen can be mass produced (Fransen 1990) and is available from Public Sector Companies in limited quantities.

Mass production of fungal bioagents

Most of the deuteromycete fungi can be mass-produced on locally available organic substrates such as grains of millets (jowar, bajra), molasses and grain husks. These can be either produced through submerged fermentation procedure or on semi-solid substrates. In submerged fermentation, a diphasic procedure is followed. The mycelial mass is first produced in the broth cultures and then mycelial mass is harvested and placed under humid conditions on inert substrate and sufficient aeration for sporulation. In case of semi-solid fermentation procedure, the sporulation occurs sufficiently on the substrate itself and such substrate is pulverized with a wet grinder and after spore count the desired formulation is prepared using a carrier such as talk powder. A typical semi-solid fermentation production procedure followed for the mass production of *N. rileyi* is described hereunder.

N. rileyi can be readily multiplied on the food grains such as sorghum (Vimla Devi and Prasad 1997). Thirty grams of crushed sorghum is soaked overnight in a 250 ml conical flask in 27 ml of 0.5% yeast extract solution prepared in single DH_2O. The contents are then autoclaved. After cooling the clump of sorghum is broken aseptically and inoculated with 3 ml of spore suspension (10^6 conidia/ml). The flasks are agitated well for proper distribution of the spores and incubated at 25 °C for 8 to 10 days and then stored at 10 °C. The conidia remain viable even after one year at this temperature. The yield of conidia is 2-2.5 x 10^9 conidia/g of substrate.

N. rileyi can also be multiplied effectively on host insect. *Spodoptera litura* adult lay eggs in masses. The larvae are not cannibalistic and can, therefore, be easily reared in groups on castor leaves. Third instar larvae of the lab host are released on castor leaves sprayed with the pathogen (2 x 10^{10} conidia/l). The stalks of the sprayed castor leaves are kept in wet cotton so as to keep them fresh. After a contact period of 48 hrs, the larvae are transferred and fed on untreated castor leaves. Larval mortality occurs 6-7 days after exposure to treated leaves. The dead larvae become mummified. These are collected and placed on moist blotting paper in petri plates or plastic tubs. Mycelial growth followed by sporulation occurs in the next 24-72 hrs. Each larva yields about 2-4 x 10^{10} conidia.

Currently only a few pvt. entrepreneurs are known to produce fungal bioagents in India using submerged fermentation procedure. These include Agriland Biotech Ltd., Baroda and International Panacea Ltd., New Delhi and their range of bioagents include *Beauveria bassiana* and *Paecelomyces lilacinus.* Agriland Biotech ltd. has two fermenters of 1000 liter capacity each with computerized control for temperature and other inputs.

A thorough review (wraight *et al.* 2001) is available on the mass production of fungal biocontrol agents, product efficiency, quality, formulation and stabilization of the product.

PROTOZOA

Protozoa have not been used successfully as fast-acting microbial insecticides. Infection in the susceptible host is initiated by ingestion of spores. Microsporidia are obligate parasites that require living cells for their development, and the gut or fat body or both provide the main foci of infection. The disease is often not highly pathogenic, but significantly reduces the rate of development of the insect and lowers its fecundity. Natural spread and survival of the protozoan may be assured in many cases by vertical transmission within eggs of the host. Epizootics of microsporidia regularly occur in many insect species. The microsporidia, particularly, *Vairimorpha necatrix*, commonly occur in the late broods of *H. armigera* in the months of January and February. The pathogen weakens the insect, leading to reduced fecundity. The population of *H. armigera* crashes when the incidence of the protozoan *Vairimorpha* in the insect approaches significant levels. The pathogen has high potential in regulating the yearly cycle of the pest. Efforts are underway towards its multiplication and commercialization.

For microbial control, however, only *Nosema locustae* has received great attention. More than 60 species of grasshopper and crocket are susceptible to this protozoan. A commercial product (Noloc), based on *N. locustae*, is now manufactured on a small scale in the USA.

FUTURE THRUSTS

There is a growing awareness for the use of alternative means for insect pest management and microbials offer an attractive and broad range of options. Microbes have proven field efficacy, can be readily mass produced and have long shelf like. Since these are not detrimental to the natural enemies, their use helps in conservation and restoration of the balance of natural enemies. Technology of the mass production of most of the microbial biocontrol agents is simple and the activity can be taken up as a cottage industry in India. It is now mandatory for microbials to have registration and involves sizable expenditure. An organized effort is the logical solution for this situation where the small entrepreneurs can collectively generate information and share it for registration purposes. An effort has been made by the bioagent producers of Maharashtra who have formed Vidarbha-Marathwada Agro Biotech Association with > 20 members and taken up the process of registration collectively. More such joint efforts would be required in different regions for the spread of the mass production units. Excellent quality production of fungal microbes is now possible through biofermenters and in the near future more such units are likely to be established in the developing countries.

The high dose of microbials (10^{11}- 10^{13} propagules /ha) necessary to achieve effective control of pests in the field conditions warrants strict quality control and product stability. The cottage enterprise has high potential in generating rural employment in South Asia. Along with the establishment of production units, there is an urgent need for establishing quality control units powered by SAUs / public sector institutes so as to keep a check on supply of standard bioagents to the farmers.

REFERENCES

Ananthakrishnan, T.N., 1998. Technology in biological control. Oxford & IBH Publishing Co. Pvt. Ltd., New Delhi. pp.124.

Anon., 1997. Consultancy by CICR in mass production of Ha-NPV. CICR Newsletter XIII (1), Central Institute for Cotton Research, Nagpur. p 6.

Anon., 1999. Guidelines /data requirements for registration of bio-pesticides. Annexure VI Baculoviruses (NPV) and Granulosis viruses (GV). Letter No. 8-15/99 CIR dt. 02.08.1999. Directorate of Plant Protection, Quarantine and Storage, Govt. of India.

Armes, N.J. , Jadhav, D.R., Bond, G.S. and Kind, A.B.S., 1992b. Insecticide resistance in *Helicoverpa armigera* in south India. *Pesticide Sci.*, 34: 355-364.

Armes, N.J., Bond, G.S. and R.J. Cooter, 1992a. The laboratory culture and development of *Helicoverpa armigera.* Natural Resources Institute Bulletin 57, Chatham, United Kingdom. pp. 19.

Baum, J. A., Johnson, T. B. and Carlton, B. C., 1998. *Bacillus thuringiensis* : Natural and recombinant bioinsecticide products. In F. R. Hall and J. J. Menn (Eds.), *Biopesticides: Use and delivery*, Totowa: Humana Press, pp. 189-209.

Bedford, G.O., 1981. Control of the rhinoceros beetle by baculovirus. In H. D. Burges (Ed), *Microbial control of pests and plant diseases, 1970-1980,* London: Academic Press, pp. 409-426.

Bernhard, K. and Utz, R., 1993. Production of *Bacillus thuringiensis* insecticides for experimental and commercial uses. In *Bacillus thuringiensis*, An Environmental Biopesticide: theory and practice. Ed. P.F. Entwistle, J.S. Cory, M.J. Bailey and S. Higgs. John Wiley & Sov. Ltd. pp. 255-267.

Brooks, W. M., 1988. Entomogenous protozoa. In C. M. Ignoffo and N. B. Mandava (Eds.), *Handbook of natural pesticides*: Vol. 5. *Microbial Pesticides*, Part A, Entomogenous protozoa and fungi. Boca Raton, F L : CRC Press, pp. 1-149.

Bryant, J.E., 1991. Production and formulation of *Bacillus thuringiensis. Proceedings: Workshop on Bacillus thuringiensis*. 24-26 Sept., 1991. CSIRO, Divn. of Entomology, Canberra. pp. 22-24.

Bryant, J.E., 1994. Commercial production and formulation of *Bacillus thuringiensis*. Agric. Ecosystems Environ. 49: 31-35.

Currier, T.C. and Bruke, C., 1990. Commercial development of *Bacillus thuringiensis* bioinsecticide products. In: Biotechnology of Plant-Microbe Interactions (Eds. J.P. Nakas and C. Hagedorn). McGraw-Hill Publication Co. pp. 112-144.

Dulmage, H.T., Correa, J.A. and Martinez, A.J., 1970. Coprecipitation with lactose as a means of recovering the spore-crystal complex of *Bacillus thuringiensis*. J. invert. Path. 15:15-20.

Evans, H. and Shapiro, M., 1997. Virus. In: Lacy L.A. (ed.) A manual of techniques in insect pathology, Academic Press, San Diego.

Fargues, J., Smits, N., Vidal, C., Vey, A., Vega, F., Mercadier, G. and Quimby, P., 2002. Effect of liquid culture media on morphology, growth, propagule production and pathogenic activity of the Hyphomycete, *Metarhizium flavoviride*. Mycopathologia 154: 3, 127-138.

Federici, B.A., 1998. Naturally occurring baculoviruses for insect pest control. In F.R. Hall & J. J. Menn (Eds.), Biopesticides: use and delivery. Totowa: Humana Press, (pp. 301-320).

Federici, B.A., 1999. A perspective on pathogens as biological control agents for insect pests. In: Thomas, Bellows and Fisher (Eds), *Handbook of Biological Control*, Academic Press, New York, pp. 1046.

Fransen, J. J., 1990. Fungi on aphids, thrips and whitefly in the greenhouse environment. *Proceedings and abstracts. Vth International Colloquim on Invertebrate Pathology and Microbial Control*, Adelaide, August 1990, 376-380.

Gryzwacz, D. Brown, M, Cherry, A and Parnell, M., 1999. The *Helicoverpa armigera* NPV production manual. Natural Resources Institute, Chatham, UK.

Hall, R. A. and Paperiok, B., 1982. Fungi as biological control agents of arthropods of agricultural and medical importance. Parasitology, 84, 205-240.

Henry, J. E., 1990. Control of insect by protozoa. In R. Baker and P. Dunn (Eds.), New directions in biological control (pp. 161-176). New York: Alan R. Liss.

Hunter, F.R-Fujita, Entwistle, P.F., Evans, H.F. and Crook, N.E., 1998. Insect viruses and pest management . John Wiley & Sons, NY. Pp.518-541.

Hunter-Fujita, F.R. Entwistle, P. E., Evans, H. F. and Crook, N. E. (Eds.), 1998. Insect viruses and pest management. Chichester: John Wiley & Sons.

Ignacimuthu, S. and A. Sen (Eds.), 1999. Biopesticides in insect pest management. Phoenix Publishing House, Pvt. Ltd., New Delhi. pp.262.

Jayaraj, S., Sathiah, N. and Sundrababu, P.C., 1994. Biopesticides. In: B. Vasantharaj David (Ed.), *The pesticides industry*, Kothari's Desk Book Series, Madras. India, pp. 271-282.

Keller, B. and Lanlangenbruch, G. A., 1993. Control of Coleopteran pests by *Bacillus thuringiensis*. In: P. F. Entwistle, J. S. Cory, M. J. Bailey and S. Higgs (Eds.), *Bacillus thuringiensis, an environmental biopesticides: theory and practice*, London: J. Wiley and Sons, pp. 171-191.

Kennedy, J.S., Rabinddra, R.J., Sathiah, N. and Gryzwacz, D., 1998. The role of standardisation and quality control in the successful promotion of NPV insecticides. In: Ignacimuthu and Sen, A. (Eds.), *Biopesticides in insect pest management*, Phoenix Publishing House, New Delhi. pp. 170-174.

Klein, M. G., 1997. Bacteria of soil-inhibiting insects. In: L. A. Lacey (Ed.), *Mannual of techniques in insect pathology*, San Diego: Academic Press, pp. 101-116.

Lacey, L.A., Erutos, R., Kaya, H..K. and Valis, P., 2001. Insect pathogens as biocontrol agents. *Biological Control* 21: 230-248.

Lingappa, S. and Patil, R.K., 2002. *Nomuraea rileyi* - a potential mycoinsecticide. Department of Agricultural Entomology, University of Agricultural Sciences, Dharwad, Karnataka, India. pp.30.

Maimala, S., Chandrapatya, A., Chamswarng, C., McCoy, C.W., Saowanit Maimala, Angsumarn Chandrapatya, Chiradej Chamswarng, 1999. Preliminary testing of biomass production of *Hirsutella thompsonii* var. *synnematcsa. Thai Journal of Agricultural Science* 32(2), 229-239.

Mazumdar, D., Puzari, K.C., and Hazarika, L.K., 1995. Mass production of *Beauveria bassiana* and its potentiality on rice hispa. *Indian Psychopathology* 48(3), 275-278.

McCoy, C. W., Sampson, R. A. and Boucias, D. G., 1988. Entomogenous fungi. In: C. M. Ignoffo and N. B. Mandava (Eds.), *Handbook of natural pesticides*: Vol. 5. *Microbial Pesticides*, Part A, Entomogenous protozoa and fungi, Boca Raton, F L: CRC Press, pp. 151-236.

Narayanan, K., 1990. Training-cum group meeting on biological suppression of cotton pests. *Compendium of lectures*, PDBC, Bangalore. pp. 42-53.

Patel, K.C., Yadav, D. N., Dube, H.C. and Patel, R.J., 1990. Laboratory and mass production studies with *Metarhizium anisopliae*. *Annals of Biology* 6(2), 135-140.

Pawar, V.M. and Thombre, U.T., 1992. Role of microbial agents in the management of *Helicoverpa armigera* in pulses and cotton. In: Sachan, J.N. (Ed.), *Helicoverpa management : current status and future strategies*. Proceedings of first national workshop held at Directorate of pulses research, Kanpur, India, 30-31 August, 1990. pp. 143-149.

Puzari, K. C., Sarmah, D.K. and Hazarika, L.K., 1997. Medium for mass production of *Beauveria bassiana* (Balsamo) Vuillemin. *Journal of Biological Control* 11(1-2), 97-100.

Ramakrishnan, N. and paul, B. 1998. Technology of production of Baculoviruses on insect host. In:Ananthakrishnan, T.N. (Ed.), *Technology in biological control*. Oxford & IBH Publishing Co., New Delhi.

Sathiah, N. and Jayaraj, S., 1996. Technology for mass production of biopesticides. Silver Jubilee Seminar on "Employment opportunities for biologists", Guru Nanak College, Chennai.

Sathiah, N. and Jayaraj, S., 1998. Technology for mass production of *Helicoverpa armigera* and its NPV: a constraint analysis. In: *Technology in biological control*. Ed. pp.61-69.

Shieh, T.R., 1989. Industrial production of viral pesticides. *Advances in virus research* 36: 315-343.

Smirnoff, W.A., 1962. Staining procedure for *Bacillus thuringiensis*. *J. Invert. Path.* 4: 384-386.

Treacy, M. F., 1998. Recombinant baculoviruses. In: F.R. Hall and J. J. Menn (Eds.), *Biopesticides: use and delivery*, Totowa: Humana Press, pp. 321-340.

Tweeten, K. A., Bulla, L. A. and Consigli, R. A., 1981. Applied and molecular aspects of insect granulosis viruses. *Microbiology Review*, 45, 379-408.

Vimla Devi, P.S. and Prasad, Y.G., 1997. The entomofungal pathogen *Nomuraea rileyi*, In: *Information Bulletin*, Directorate of Oilseeds Research, Rajendranagar, Hyderabad, pp.5.

Vimla Devi, P.S. and Prasad, Y.G., 1998. Fungus to control bollworm in cotton, In: Science and Technology-Agriculture, *The Hindu*, dated 14th May.

Vimla Devi, P.S., Chowdary, A. and Prasad, Y.G., 2001. Cost effective multiplication of the entomopathogenic fungus *Nomuraea rileyi* (F) Samson, *Mycopathologia* 151: 1, 35-39.

Watkinson, I., 1991. *Bacillus thuringiensis*-product standardisation. Ibid. pp. 25-26.

Wraight, S. P. and Carruthers, R. I., 1998. Production, delivery and use of micoinsecticides for control of insect pests in field crops. In: F. R. Hall and J. J. Menn (Eds.), *Biopesticides: Use and delivery*. Totowa: Humana Press, pp. 233-269.

Wraight, S.P., Jackson, M.A., Kock, S.L., 2001. Production, stabilization and formulation of fungal biocontrol agents. In: Butt, T.M., Jackson, C.W. and Magan, N. (Eds.), *Fungi as Biocontrol Agents*.CABI Publishing, Wallingford, Oxon, U.K. pp. 385.

Zelazny, B., Lolong, A. and Pattang, B., 1992.*Oryctes rhinoceros* (Coleoptera: Scarabaeidae) population suppressed by a baculovirus. *Journal of Invertebrate Pathology*, 59, 61-68.

Technology for Commercial Production of Biological Control Agents

PRADYUMN KUMAR[1], R.K. SHARMA[1] AND PARMOD SIWACH[1]

Biological control agents (BCAs) infect or infest other organisms. Such organisms are termed as hosts. Populations of all the organisms are maintained by interactions among different species. Utilization of this phenomenon in pest management strategy call for the mass production of BCAs. These living entities are short lived, in other words their shelf life is very short; they may get perished during transportation. All these reasons suggest that they should be produced near the site of their use in small scale industries.

The present chapter discuss the technologies which can be imbibed by educated rural youth and implemented in raising mass production units of BCAs in rural areas. Such units will help increase the awareness among the farmers and hence their confidence in the use of BCAs. The production of BCAs near the site of their use can have accurate synchrony with their need in the fields which is very vital, both for pest management and the success of such industries. These industries will help generate rural employment and prevent urban agglomeration.

The production of BCAs is dependent on the production of their host organisms, therefore, mass production of healthy culture of host organisms is a prerequisite for the mass production of any biological control agent. There are large number of hosts and BCAs. Only a few important ones are being dealt here.

Mass production technology of *Trichogramma* sp

Trichogramma spp. are egg parsitoids of lepidopterous insects. They help managing lepidoptepterous pests in sugarcane, paddy, cotton, tomato, apple, and others. Different species of *Trichogramma* are recommended for different pests. The rearing technology for all species is by and large same. *Trichogramma chilonis* is effective against *Helicoverpa* in cotton, tomato, maize and for sugarcane borers. *T. japonicum* is used against rice stem borer, rice leaf folder, sugarcane top and shoot borers. *T. brasiliensis* for cotton bollworms. *T. embryophagum* is specific to the codling moth of apple. Commercial production of *Trichogramma* spp. is done on the eggs of *Corcyra cephalonica* Stainton. Mass production of this surrogate host, therefore, will be discussed first.

Mass Production of C. cephalonica

Rice meal moth, *C. cephalonica* is a stored grain pest on variety of food grains and processed foods. The eggs and larvae of this insect have been used as suitable surrogate host for many kind of parasitoids, predators and some entomopathogenic nematodes. *Corcyra* can be easily reared on any kind of coarse grain, but mass rearing requires selection of suitable rearing medium, congenial conditions, protection from natural enemies and efficient and safe handling of culture.

---

1 Directorate of Maize Research, IARI Campus, New Delhi-110 012, Email: pradyumn_kumar@yahoo.com

Food and Food Consumption

Sorghum, millets, maize, rice, wheat, groundnut, rice bran and wheat bran are generally used as rearing media. They are used in various combinations or otherwise. The food efficiency index was highest in case of sorghum followed by pearl millet, maize and rice (Kumar and Kumar, 2002). Yeast is used as an additive to improve the quality of the culture. The development was reported faster in ground grains rather than in whole grains

There have been several reports on quantitative food requirement for the larval development. The ratio varies from 1000 eggs per 1.5 kg of sorghum to 5000 eggs per kg rice. The most acceptable being 2000 eggs per kg food medium (Jalali and Singh, 1992). In less crowded condition, higher percentage of moth recovery is achieved. Development period increases in the crowded condition.

Environmental Conditions

The optimal condition for incubation was observed to be 25-30°C and 75-90% relative humidity. The incubation period reduces by increasing temperature but 40°C is lethal. The larval development is also greatly influenced by temperature and humidity. Larval development was indifferent to photoperiod, however, moths preferred to stay in dark.

Mating and Fecundity

Mating occurs during night, usually once, however, multiple mating are observed both in males and females. Males are reported to mate more often then females. Oviposition occurs 2 h after mating. The fecundity is directly correlated with the weight of females. Fecundity is also influenced by relative humidity. An average fecundity of 350 eggs per female is common with proper rearing method.

Rearing Methodology

Tribolium spp. is commonly present in the rearing medium. Larval food is, therefore, freed from it or any other stored grain pest by sterilizing the rearing medium in oven at 110°C for 4 h. It will be convenient and practical if freshly ground food is used for infesting the medium. The medium get sterilized by heat and sheering force during grinding. Four kilogram of this medium is uniformly mixed with 0.4 cc of *Corcyra* eggs. *Corcyra* larvae are known to emit some kairomone which attracts *Bracon* spp. To reduce the chances of Bracon invasion the special wooden box has been designed for *Corcyra* rearing (Kumar and Kumar, 2001). The lid of the box have two windows (10x10 cm). These windows are covered by double layer of fine copper wire-mesh (400 squares/inch^2). Open tray system of *Corcyra* rearing renders the culture vulnerable for *Bracon* infestation, which is a serious scourge to the culture. The boxes are periodically prepared as per the requirement of culture and stacked in the racks. Temperature of 28±1°C is maintained by using air conditioners or heat convectors, as the need may be. Both the cooling as well as heating appliances drastically reduce the humidity. Room cooler intercepted by humidistat is used very successfully to obtain 75±5% R.H. Under such conditions, the moth emergence starts after 30 days and continue for over 45 days. Soon after the emergence, the moths start mating and laying the eggs. The eggs laid before the collection of the moth in the cage hatch and develop on residual food which is insufficient, hence these eggs get wasted. It is, therefore, advisable to harvest the moth twice a day to prevent the loss of eggs. The cages should be discarded after 45 days.

The moths are collected by moth collection unit (MCU) developed by Kumar and Jalali (1993). The suction pressure of the vacuum cleaner (VC) is adjusted at 140 mm water column. With the help of this device it is possible to collect the moths at the rate of 1000 moths per hour directly into the oviposition cage (OC).

The eggs laid in the oviposition cage percolate down in the tray through the wire mesh. A thick mat of scales shed by moths covers the eggs. The eggs are cleaned and collected by another device patented by Kumar *et al.* (1999). The egg collection device offers easy collection of clean eggs and safe disposal of scales, which are health hazardous. The eggs and scales are sucked into the system where they are fluidized. The eggs are cleaned by the buff of air. The scales escape into the dust bag of the vacuum cleaner. Immaculate clean eggs are collected in the container.

Prevention from Natural Enemies

The *Corcyra* production is considerably affected by variety of natural enemies and competitors in the culture. More often than not, rearing medium is infested with *Tribolium*. This stored grain pest compete with the *Corcyra* culture. The larvae of *Tribolium* also predate upon *Corcyra* larvae. The use of freshly ground sorghum substantially reduce the *Tribolium* population. *Bracon* spp. are yet another serious pest of *Corcyra* culture. Apart from double-layered wire mesh in the windows of the cage, this parasitoid needs to be regularly monitored against the light. Further, the double door system in the culture room is also recommended to reduce the chances of contamination. A small glass window in the culture room smeared with poisoned sticky material helps monitoring as well as killing of the invading *Bracon* adults. Nearly ten species of spiders are reported feeding on *Corcyra* moths. A general cleanliness in the culture room solves spider problem. Rat is another potential pest, both for the culture and the food medium. It is, therefore, important to have the doors, windows, etc. rat proof. House lizards feed upon eggs and the moths. Oviposition cages are to be specially kept in the area free from the approach of lizards. Dead moths, if left for long, provide suitable condition for the development of mites in the culture. Mite problem is commonly encountered when the oviposition cages are not properly cleaned or the discarded cages languish in the laboratory for long. If the *Corcyra* culture gets infested with mites, it can be cured by treating it with sulphur powder.

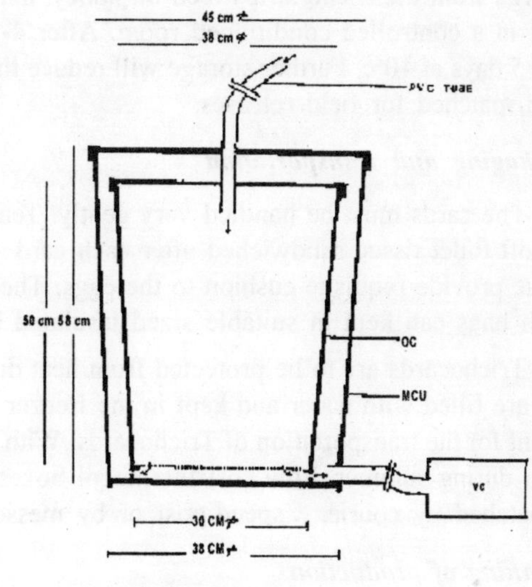

Moth Collection Unit

Mass Production of *Trichogramma* spp.

The Trichocards are generally post-card size cards, bearing host eggs parasitized by *Trichogramma* sp. One of the sides of the card is printed the name of the species, date of exposure and likely date of emergence of parasitoid. The other side remain blank. The cards are punched so that it can be divided into twelve equal pieces. Eggs are glued on blank side.

Gluing of eggs

A stencil of hard plastic sheet is made by cutting 12 slots in the fashion so that each slot fit on to the punched strips. The card is positioned under the stencil and the gum is pasted on the stencil with the help of paint brush. The pasted card is then kept in the tray. Clean *Cocyra* eggs are taken in a strainer and sprinkled over the card. Eggs get glued on the strips. Each card can hold one cubic centimeter (Approx. 20,000) eggs.

Sterilization of eggs

To prevent the embryonic development in *Corcyra* eggs, the cards are exposed to UV tube light of 30W at a distance of 35cm for a period of 10 minutes. During this period the glue also gets dried up. The sterilized eggs on the cards can be stored at 10°C for nearly fifteen days.

Parasitization

The date of exposure is marked on the card. Two cards are kept in a polythene bag (30 × 15cm) so as to expose their egg side. Two strips of tricho cards from which wasps are likely to emerge are also inserted in the bag. A thin strip of polythene sheet smeared with honey, folded and stapled in such a way so that smeared surface form the inner surface of the loop. One such loops is inserted in each bag. The bags are slightly bloated by making space with the help of hand and tied with rubber band. The wasps emerge from the tricho strips feed on honey, mate and parasitize the eggs glued on cards. Such bags are kept in a controlled conditioned room. After 4-5 days the eggs turn steel gray. They can be stored for 12-15 days at 10ºc. Further storage will reduce the efficacy of the parasitoid. During this period they must be dispatched for field releases.

Packaging and transportation

The cards must be handled very gently. Ten such cards can be kept one over the other with a layer of soft toilet tissue sandwiched after each card. This arrangement separate each card with other and the tissue provide requisite cushion to the eggs. The stack of ten cards are kept in one PVC bag (30x15cm). Such bags can kept in suitable sized insulated boxes.

Trichocards are to be protected from heat during transportation. For this purpose plastic bottles (500 ml) are filled with water and kept in the freezer overnight. One such bottle is kept in each insulated box meant for the transportation of Trichocards. With this arrangement the trichocards remain cool for 36 hours even during summer. Low cost thermocol boxes can be used for transportation and the material can be dispatched by courier / speed-post or by messenger.

Planning of production

Planning is first and foremost activity for any project. The host culture production has to be planned by keeping in view the quantity of parasitoid and the time at which they will be required.

Method of release

It is advisable to synchronize the release programme after monitoring the onset of pests by use of pheromone traps, light traps, etc. The trichocard strips are cut into pieces and stapled underside of the leaf. The release of parasitoid should be done preferably in the evening hours. During night, the parasitoids will emerge and search the host eggs. The parsitization may be completed before the next noon. The care should be taken that pesticide spray is not done immediately before or after release.

Expenditure estimate for setting up of mass production unit of *Trichogramma* (Trichocard)

One time expenditure (Assets)

Equipment

Item	Qty.	Rate	Cost
Cocyra cages	1,000	300	3,00,000
Racks	50	1,000	50,000
Oviposition cages	200	300	60,000
Racks (small)	40	800	32,000
Egg cleaning devices	2	1,000	2,000
Desert coolers	4	5,000	20,000
Heat convectors	6	1,000	6,000

Item	Qty.	Rate	Cost
Humidistat	2	1,000	2,000
Thermostat	2	1,500	3,000
Thermometer	2	1,000	2,000
Hygrometer	2	1,000	2,000
Vacuum Cleaner	4	7,000	28,000
Refrigerator	2	10,000	20,000
Egg cleaning device	1	5,000	5,000
UV hood	1	5,000	5,000
Trolley	2	2,000	4,000
Modification in Building			50,000
Misc. Petty Items			10,000
Training			20,000
Total			6,21,000

Recurring Expenditure for one month

A. Diet ingredients, etc. 12,000

B. Man Power

Manager	(B.Sc.)	One	12,000
Technical Officer	(Matric)	One	8,000
Moth collectors	(Labourer)	Five	15,000
Trichocard man		One	3,000
Cleaner		One	3,000
Total		Nine	41,000

C. Building rent 6,000

Rooms	3 (15'x20') 2 (12'x12')
Store	1
Washing place	1

D. Electric charges	6,000
E. Depreciation/breakage	3,000
F. Trichocards*	5,000

(*Tricho cards are highly perishable. The shelf life is only one week. However they can be stored for one month at 10°C but their efficacy get reduced.)

G. Contingency 5,000

Monthly expenditure

A+B+C+D+E+F+G = 78,000

Production 250 × 30 = 7,500 Trichocards per moth

N.B.

Mass rearing of *Chrysoperla*

Green lacewing or aphid lion, *Chrysoperla carnea* Stephens (Neuroptera: Chrysopidae) previously known by *Chrysopa scelestes* Banks is distributed in India, Romania, Morocco, Yugoslavia, Egypt and most part of Europe. Its larval stage is a general predator of small soft bodied insects and mites. It effectively decimates the population of aphids, whitefly, mealy bugs, mites, insect eggs and young insect larvae. In cotton, maize, pigeonpea and tomato, they are of great significance. Each larva has the potential to feed 12 aphids a day and 120 aphids during its development. Conservation of this predator pays rich dividend in term of prevention of crop loss. The release of these predatory larvae at the time of onset of pests can effectively suppress their population.

Mass rearing of *Chrysoperla* is carried out on eggs of *C. caphalonica.* It is, therefore, necessary to have a mass production unit of *Corcyra.* To initiate the *Chrysoperla* culture, adults can be collected from field or from light during night. Alternatively, eggs in fairly large number can be collected from crop. The eggs are green in colour and oblong. The characteristic feature of the egg is that they are stalked. The incubation period is 3-4 days. The eggs turn brown on third day. The young larvae are small white and have well developed legs. The mandibles are modified to pierce and suck the body fluid. The larvae are cannibalistic in nature so they cannot be reared in groups. After 10 days the full-grown larvae spin cocoon around them and pupate. The cocoon is spherical and of the size of a sorghum grain. After 7-8 days, adults emerge out from them. The adult is soft green about a centimetre in length. The wings have profuse venation. The antennae are filiform and longer then the body. Female adults have swollen abdomen.

Oviposition cage

The oviposition cage is a wooden box 50 cm × 30 cm × 25 cm. There are two circular windows one on each side. The top of the box has two lids one over the other. These lids are made of ply board and slide horizontally in the groove. The lower lid is covered with black cotton cloth. The other inner four sides and floor of the box have lining of plastic wire mesh. The females prefer to lay eggs on black cloth. Plastic wire mesh is unsuitable for egg laying. About 400 adults are kept in each cage. Honey solution-soaked cotton wade is kept in petri dish and the castor pollens in other. The edge of one side of the cage is fixed with a comb held vertically in such a way so that the teeth of the comb make a close contact with the black cotton cloth. When the lower lid is pulled out gently, the comb prevent escape of adults while the stalked eggs come out unscathed. The egg-laden cloth is removed and a new cloth is replaced. Handling of adults, cleaning of cage, removing of dead adults and change of adult food etc. is done through the windows provided in the sides of the cage. The windows are having sleeves to prevent escape of adults.

Handling of eggs

The eggs when laid are soft. After two days the egg-laden cloth is spread over the table with egg crop on upside. The eggs are removed by scraping the cloth with the help of dry sponge sheet. Some of the eggs get entangled in the sponge. These eggs can easily be extricated by shaking the sponge. This cloth is kept in a jar for 2 days.

Sterilization of* Corcyra *eggs

The egg content i.e. albumen and yolk of *Corcyra* eggs, act as food for *Chrysoperla* larvae. The embryo of the eggs, therefore should be killed at an early stage either by freezing them in the freezing chamber or by exposing the eggs to ultraviolet rays. The later method is better. *Corcyra* eggs are kept uniformly spread preferably single layer thick in a tray. The tray is kept under ultraviolet tube light (30 watt) at a distance of 50 cm for a period of 15 minutes. The eggs are then stored in the refrigerator at 5°C.

Rearing of early stage larvae

The *Chrysoperla* eggs mixed with *Corcyra* eggs are kept in circular trays. The content of the tray is uniformly spread and covered with the paper. The uniform spreading and paper covering reduce the chances of larvae meeting each other, therefore reducing the incidence of cannibalism. The circular trays are covered with black cloth and kept at 27°C for four days.

Rearing of Late-instar larvae

Late-instar larvae are reared singly as the cannibalism increases in the late-instars. Multi-cellular trays are used for this purpose. Each cell has 6-8 cc space. Sterilized *Corcyra* eggs are sprinkled uniformly over the multi-cellular tray so that each cell gets about 1000 eggs. One larva four day-old is transferred in each cell. The tray is then covered by the lid. A layer of paper is sandwiched between the tray and the lid. The paper provide tight ceiling to the cell top which prevent escape of the larvae. Also, the paper permit air diffusion. The trays are piled up in the rack in a room with controlled conditions. After about eight days the larvae complete their development and pupate in cocoons. The cocoons are collected with the help of fine camel hair brush and kept in a jar for adult emergence. If the sufficient number of trays are available, the tray are left as such for emergence of adults in the cell itself. By this method some labour is saved. Also, since the cocoon get stick to the substrate, some of the cocoons might get damaged while removing them from the cell. The adults from the trays are collected in the insect handling chamber to prevent the escape of the adults. Adults are then transferred to the oviposition cage.

Mass Production of HaNPV

Nuclear polyhedrosis virus (NPVs) infect commonly lepidopterous insects. They are also found to infect few hymenopterous insects. They are not known to infect non arthropod spp. Specificity is one attraction of NPVs which makes it a promising biopesticide. NPV of *Helicoverpa armigera* is a kind of baculovirus. Baculoviruses are DNA viruses occluded in silicaceous and proteinaceous bodies called polyhedra. Each polyhedron encloses rod shaped viruses, therefore, they are called polyhedral inclusion bodies (PIBs). The polyhedron is about one micron in size which can be seen by 400 × magnification. The PIBs get ingested by the larvae along with the food. In the gut, the silicaceous framework gets dissolved. The protein is digested by proteolytic enzymes and virus rods are released. They invade the epithelial cells of the gut; enter the nucleus and form extra cellular virus (ECV) or budded virus. These ECVs spread in haemolymph, fat cells and other cells of various visceras. The food intake of the larvae get reduced. The body become soft and it dies usually in 3-6 days. Since they are very host specific, thus can only be mass produced on their host larvae. For the mass production of NPV of *Helicoverpa,* mass production of larvae of *Helicoverpa* is an associated activity.

***Mass rearing of* H. armigera**

The *Helicoverpa* culture is highly susceptible to NPV infection. Mass rearing of *Helicoverpa* for NPV production requires a separate unit, which should be distantly located from the NPV production unit.

Oviposition

The pupae ready to emerge are kept in the oviposition cage. Pre-oviposition period is of three days. After three days of emergence, they start laying the eggs on the inside wall of the cloth cage. The female moths lay 1000-3000 eggs (Knipling and Stadelbacher, 1983 and Stinner *et al.*, 1974) in a span of 8-10 days. After 2 days of egg laying, the moths are transferred to another cage and the cloth of the old cage is removed. Most of the eggs are laid in first five days after the pre-oviposition period. The moths, therefore, should be discarded after this period. Temperature and humidity play very crucial role in realizing optimum egg-laying potential of the females. Under adverse conditions of temperature and relative humidity, the moths do not mate. Sometime females give unfertilized eggs, which results in serious reduction in the production. Kumar and Ballal (1990) developed an improved oviposition cage, which helps in maintaining the ambient conditions in summer, which is suitable for mating and oviposition. The entire cage gets sterilized after every use, which is very essential to keep the infection at bay. This cage was later modified by Kumar and Kumar (2001) for winter season. They could rear *Helicoverpa* for sixteen generations with average of 1289 eggs per female.

Handling of eggs and rearing of young larvae

Eggs are laid on the walls of the oviposition cage made of cloth. After two days, the moths are transferred to another cage and the eggs are extricated from the cloth component of the cage bearing eggs. For extricating the eggs, the cloth component of the cage is removed from the frame. The cloth is then soaked in 0.05 percent sodium hypochlorite solution in the wash tub of a simple washing machine and agitated for 5 minutes. Sodium hypochlorite dissolves the chorion of the egg shell thus help separating the eggs from the cloth. The water is drained through a big sieve at the end of the drainpipe. The eggs and moth debris are collected in the sieve. The contents in the sieve are then put in the bucket full of water. The debris float, while the eggs sink. Clean eggs are then collected and incubated in the folds of sterilized cloth. On third day, yellow band appear on the eggs.

Early-instar larval rearing:

Semi synthetic diet (Nagarkatti and Satyaprakash, 1974) is poured hot in sterilized petri dish and allowed to cool. On cooling, the diet gets solidified. It does not fall on inverting the petri dish. Approximately one hundred 2 day-old eggs are kept under the petri dish with diet on its ceiling. The petri dish is inverted on a tissue paper supported on a cloth lining and a glass plate. The arrangement makes the petri dish leak proof for the neonate larvae and also prevent moisture trapping inside the petri dish. Kumar and Ballal (1991) used this method of rearing early instar *Spodoptera litura* larvae. The young larvae are not cannibalistic, hence are group reared for five days in this manner. One tenth of these larvae are used for mass multiplication of host culture while ninety percent are sent in NPV production unit.

Later-instar larval rearing:

Five day-old larvae are individually transferred into culture tube (3" × 1") containing about ten gram semi-synthetic diet. The tubes are plugged with sterilized cotton wool and the larvae are reared until pupation. In some tubes, mould develop on excreta but that does not harm the pupating larvae. The pupae are collected, washed with water and then with 0.05 percent sodium hypochlorite solution for surface sterilization. They are again washed with tap water and kept at 27°C and 70% relative humidity for one week before they are transferred to oviposition cage.

NPV production unit

Semi-synthetic diet used for healthy culture is used for infected larvae too. Infected larvae can be reared in 30-35 ml plastic containers or even 20 ml used injection vials. The former are unbreakable, while the later are available at much lower cost.

Diet dispensing

The diet is prepared and poured hot @ approximately 7-8 ml in each container. They are allowed to cool and solidified. The containers are arranged in trays.

Inoculum dispensing and infection

Inoculum is prepared having 40 LE (8×10^{10} PIBs) NPV per litre. This solution is filled in a drip bottle having a discharge-regulating device. Two drops of this solution is added in each of the containers arranged in tray. Each container is then charged with 5-6 old larvae and covered with the help of lid. After four days these containers are checked everyday till the larvae die/pupated. If higher concentration of PIBs are used, the early death comes to the larvae. This result in the formation of less quantity of virus. Early death also occur if smaller larvae are used. It is, therefore, very essential to use suitable size of larvae and the proper dose of virus so as to get the optimum yield of virus. The larvae after their death usually hang with head down. This is the characteristic feature of NPV infection.

Processing the dead larvae

The dead larvae are collected in a plate, counted and weighed. They are then putrefied for 5 days in a china clay pot properly covered with a lid. The larvae are kept moist by sprinkling water every now and then. After 5 days the whole biomass is thoroughly macerated so as to get homogenous content. This will release polyhedra from the putrefied tissues. The macerated content is diluted in little water and sieved through fine cotton cloth. Polyhedra come in the filtrate. Little water is sprinkled over the putrefied tissue, 2-3 times to extract the residual polyhedra. For crude preparation the filtrate is diluted in water to make solution of 250 ml for every 100 dead larvae. This is called 100 larval equivalent (LE)

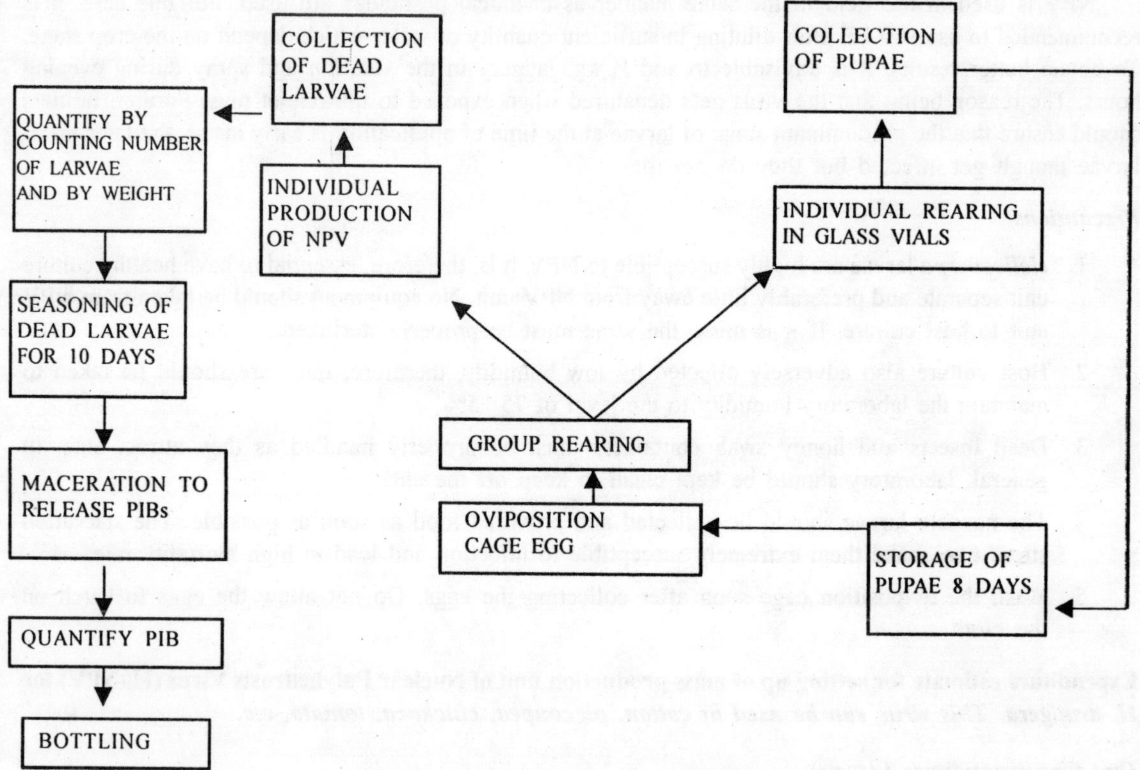

Schematic representation of NPV production

For purification, the filtrate is centrifuged at 400-500 rpm for ten minutes. The supernatant containing PIBs is decanted and the tissue precipitate discarded. The supernatant is again centrifuged at 6000 rpm for 30 minutes. PIBs settle in the form of white precipitate. This time the supernatant is discarded. Water is added to the precipitate to make 250 ml volume for each 100 larvae macerated.

Standardization of NPV

To determine the strength or active ingredient, the PIB count is necessary. One ml content is serially diluted in three steps to 10^{-3}. This process eliminates the error.

Now a drop of this solution is carefully kept on the haemocytometer. Cover slip is kept on it and number of PIBs are counted at least on 40 squares. The strength of the solution is determined by using formula:

$$\text{Number of PIBs} = \frac{D \times X}{N \times X}$$

D = Dilution factor

X = Number of PIB count

N = Number of squares

K =Volume of one square in cm^3

By counting number of PIBs per larvae, larval equivalent can be determined.

Method of field application

NPV is used in the field in the same manner as chemical pesticides are used. For one acre, it is recommended to use 100 LE after diluting in sufficient quantity of water which depend on the crop stage. To obtain better results, it is advisable to add ½ kg. jaggery in the solution and spray during evening hours. The reason being that the virus gets denatured when exposed to ultraviolet rays. Further, farmers should ensure that the predominant stage of larvae at the time of application is early instar, the later instar larvae though get infected but they do not die.

Precautions

1. *Helicoverpa* larvae are highly susceptible to NPV, it is, therefore, essential to have healthy culture unit separate and preferably little away from NPV unit. No equipment should be taken from NPV unit to host culture. If it is must, the same must be properly sterilized.
2. Host culture also adversely affected by low humidity, therefore, due care should be taken to maintain the laboratory humidity to the level of 75 ±5%.
3. Dead insects and honey swab containers must be properly handled as they attract ants. In general, laboratory should be kept clean to keep off the ants.
4. The neonate larvae should be collected and provided food as soon as possible. The starvation stress may make them extremely susceptible to infection and lead to high mortality rate.
5. Wash the oviposition cage soon after collecting the eggs. Do not allow the eggs to hatch on the cloth

Expenditure estimate for setting up of mass production unit of Nuclear Polyhedrosis Virus (HaNPV) for *H. armigera. This virus can be used in cotton, pigeonpea, chickpea, tomato, etc.*

One time expenditure (Assets)

Equipment

Item	Qty.	Rate	Cost
Diet blender	2	3,000	6,000
Heating appliance	1	2,000	2,000
Glass vials	2,000	1	2,000
Petri dishes	300	80	24,000
Plastic vials	70,000	1	70,000

Item	Qty.	Rate	Cost
Trays	500	50	25,000
Racks	20	1,000	20,000
Desert coolers	4	5,000	20,000
Heat convectors	4	1,000	4,000
Humidistat	2	1,000	2,000
Thermostat	2	1,500	3,000
Washing Machine	1	8,000	8,000
Oven	1	10,000	10,000
Thermometer	2	1,000	2,000
Hygrometer	2	1,000	2,000
Vacuum Cleaner	1	7,000	7,000
Refrigerator	1	10,000	10,000
Modification in Building			50,000
Misc. Petty Items			10,000
Training			20,000
Total			2,97,000

Recurring Expenditure for one month

*A. Chemicals**

(* The cost of chemicals for diet ingredients have been calculated based on the rates of laboratory supplies. For commercial production, the bulk supply rates of commercial grade chemicals will be substantially low.)

For rearing 5000 larvae per day for producing 4,000 LE per day 1,00,000

B. Man Power

Manager	(B.Sc.)	One	12,000
Technical Officer	(Matric)	One	8,000
(Handling of healthy culture)			
Labour (Diet preparation)		One	3,000
Labour (NPV Larval handling)		Six	18,000
Washer man/woman		One	3,000
Total		Ten	44,000

C. Building rent 4,000

Room	2 (15'x15')
	1 (15'x20')
	1 (12'x12')
Store	1
Kitchen	1
Washing place	2 (separately)

D. Electric charges	6,000
E. Depreciation/breakage	7,000
F. Bottling, packing, etc.	2,000
G. Contingency	5,000

Monthly expenditure

A+B+C+D+E+F+G = 1,68,000

Production 4,000x30 = 1,20,000 LE

N.B.

Mass Production of *Beauveria*:

More than 750 species of fungi have been reported pathogenic to insects but only a few have been considered seriously as biological control agents. The unique character of fungi as compared to bacteria and viruses is that they penetrate through insect cuticle or spiracles, therefore, need not to be consumed to cause infection. This characteristic make them virtually the only group of insect pathogens with potential against a whole range of insect pests such as aphids, termites, whiteflies, borers, thrips, mites, grubs, bugs attacking important agricultural and horticultural crops. The most important group of insect pathogenic fungi used in biological control are *Beauveria bassiana* and *B. brongniarti* belonging to Deuteromycotina and popularly known as white muscardine fungus. Their very obvious ability to destroy insect pests has triggered interest in increasing their incidence artificially. Therefore, development of economical mass production technique for *Beauveria* would be a step toward promotion of rational alternatives to chemical pesticides. The essential feature in mass production should be, use of low cost locally available materials. The various steps involved in mass multiplication of *Beauveria* are:

- Moistened and autoclaved grains of sorghum, rice, maize etc. can be used as substrate for mass production of fungus. Maize grains when used should be crushed to make small pieces and moistened and softened to facilitate the penetration of mycelia.
- Media is taken in 0.2 mm thick high-density polyethylene bags @200 g/bag. The open end of the bag is passed through a metal socket to make it a small rigid opening which is then plugged by cotton wool. The bag is autoclaved at 15 $lb/inch^2$ pressure for 35 minutes.
- After cooling, media is inoculated with pure culture by injecting the inoculum using surgical syringe and sealing the injection point with cellophane tape.
- Inoculated media is incubated at 22-30°C for about 10 days.
- Fungal mass is dried at 40°C and ground to get fine powder.
- Mix spore powder with appropriate carrier material like talc powder, sand or ash and then pack in airtight polythene bags for further use.

5. Mass Production of *Trichoderma* spp.

Trichoderma viridae and *T. harzianum* are two most important species which are used to suppress soil borne pathogenic fungi. They are often used for seed treatment or soil application to protect the rhizosphere from other invading fungal species. Their mass production is done by liquid fermentation method or on solid organic substrates like sorghum grains, wheat straw, wheat bran, spent tea leaf waste, coffee husk, wheat bran-saw dust, etc.

A. Liquid Fermentation Method

- Take molasses, yeast and water in the ratio of 30: 5: 1000 and autoclave the medium at 15 $lb/inch^2$ pressure for 20 minutes.
- After cooling, inoculate the media with pure culture of *Trichoderma* under aseptic condition.
- Incubate the culture at 30°C for 10 days and shake the culture periodically (three or four times daily) during this period. For Mass multiplication fomenters available in different capacity (50L, 100L, 200L)should be used depending upon the requirement.
- After 10 days, collect the fungal mats formed over the medium and macerate by using a blender.
- Mix this culture with carrier material like talc powder in 1:2 ratio and air dry it under shade.
- Carboxy methyl cellulose @ 5 kg of product may be added as sticker. Pack the material in airtight polythene bags. The product can be stored up to four months.

B. Mass Production on Solid Substrates

- Soak sorghum grains overnight in water. Decant water and boil these grains in 0.2% dextrose solution for 5 minutes.
- Transfer boiled grains in polypropylene bags (one kg. capacity) @ 200 g per bag and sterilize at 15 lb/ inch^2 pressure for 20 minutes in autoclave.
- After cooling, inoculate the medium with pure culture of *Trichoderma* and homogenize. The inoculation is done by injecting the inoculum using surgical syringe.
- Incubate the material at 28°C for three weeks.
- Dry the content at 40°C for 96 hours and grind to fine powder.
- Count the colony forming units (cfu) per g of powder with the help of haemocytometer.
- Mix spore powder with carrier material (talc) and carboxy methyl cellulose (sticker) in appropriate ratio so that the formulated product contain $>10^9$ cfu g^{-1}.
- Pack the final product in air tight polythene bags.

REFERENCES

Jalali, S. K. and Singh, S. P. 1992. Effect of infestation of sorghum grains by different dosages of *Corcyra cepalonica* adult emergence pattern. *Entomon,* 17 (1-2): 117-119.

Knipling, E. F. and Stadelbacher, E. A. 1983 The rationale of area-wide management of *Heliothis* (Lepidoptera: Noctuidae) populations, *Bull. Entomol. Soc. Am.* 29: 29-37.

Kumar Pradyumn and Chandish Ballal 1991. Improved technique for mass rearing of *Spodoptera litura* (Lepidoptera: Noctuidae). *Entomon.* 16 (1): 59-62

Kumar Pradyumn and Kumar Surender, 2001. Foolproof cage for rearing *Corcyra cephalonica Indian J. Ent.*, 63 (3): 211-214.

Kumar Pradyumn and Kumar Surender, 2002. Evaluation of different rearing media for *Corcyra caphalonica* (Stainton) *J. Biol. Cont.*, 16(2):117-120.

Kumar Pradyumn and Jalali, S.K. 1993. A hygienic and efficient method of *Corcyra* moth and egg collection. *Int. J. Pest Mangt.*, 39: 103-105.

Kumar, P. and Ballal, C. R. 1990. An improved oviposition cage and egg collection technique for the mass multiplication of *Heliothis armigera* Hubner, *Indian J. Plant Prot.* 18: 255-259.

Nagarkatti, S and Satyaprakash 1974. Rearing *Heliothis armigera* (Hb.) on artificial diet. *Tech. Bull. Commonwth. Int. Biol.Control.* 17: 169-73

Pradyumn Kumar, Surender Kumar and S.N. Puri 1999. An efficient device for collecting and immaculate cleaning of insect eggs mixed with loose scales. *Patent Application No.* 107/DEL/99 dt.20.1.99

Stinner, R. E., Rabb, R. L. and Bradley, J. R. 1974. Population dynamics of *Heliothis zea* (Boddie) and *H. virescens* (F.) in North Carolina: a simulation model, *Environ. Entomol.* 3: 163-168.

12

Helicoverpa (Heliothis) and Botanical Pesticides: An Overview

NUTAN KAUSHIK[1] AND VANIT KATHURIA[1]

A wide host range, which include many economically important crops such as cotton, legumes, cereals, oilseeds and vegetables, and versatility in its behaviour (Reed and Pawar, 1982; Fitt, 1989 and Cunningham *et al.*, 1999) have drawn special interest for the control of *Helicoverpa armigera*. Moreover, increasing concern for insecticide-resistance in *Helicoverpa* spp. to currently available groups of chemicals has necessitated to look for alternative strategies for its management. Many plant species belonging to different families and genera have been reported to possess pesticidal properties against various orders of insects (Singh, 2000). They produce an array of secondary metabolites contributing to pesticidal properties *viz.* alkaloids (Bentley, 1984), steroids (Bergamasco and Horn, 1983), terpenoids (Mordue and Blackwell, 1993), however, very few of them such as neem and pyrethrum have been exploited commercially (Mordue and Blackwell, 1993; Isman, 1994). Many other plant preparations are traditional in use (Ahmed *et al.*, 1984). In this endeavor, progress in production and use of biopesticides, both botanical and biocontrol agents, have played an important role in establishing their place as useful tools in Integrated Pest Management (IPM) and also as alternatives to synthetic chemicals (Menn, 1997). However, more research is needed to explore the potential species, which may prove better or at par with Neem for their utilization under sustainable agriculture. This chapter provides an overview of the different plant species tested against *H. armigera* along with their wide spectrum of activity.

JUVENILE HORMONE AND GROWTH REGULATORY ACTIVITY

In recent decades huvenile hormone and grwoth regulatory properties have found appropriate place in IPM. These substances initially affect the growth of the larvae at early stages and eventually result in larval mortality due to prolonged starvation. Life table studies with noctuids indicate that 80% of the larval mortality occurs before the third instar stage due to their small size and susceptibility to biotic and abiotic factors (Isman, 1994, Zalucki *et al.*, 2002). Dimlin and cutinase inhibitors are some of the examples of synthetic growth regulators.

Among the botanicals, azadirachtin from Neem is a good example, which is being utilized commercially. There is no dearth of literature on bioefficacy of Neem against *H. armigera*. It not only inhibite ecdysis but also affect growth of the larvae by more than 50 per cent at concentration around 1-ppm against *H. zea* (Boddie) and *H. virescens* (F.) (Kubo and Klocke, 1982). Aqueous methanolic extract of the fruits of *Melia azedarach* also effects the larval moulting of *H. zea* and *H. virescens* (Kubo and Klocke, 1984). Some recent literature shows efficacy of limonoids other than azadirachtin eg deacetylnimbin, 17-hydroxyazadiradione, gedunin, salannin and deacetylgedunin have shown to effect the growth, longevity

[2] TERI School of Advanced Studies, Habitat Place, Lodhi Road, New Delhi 110 003

and fecundity of adults when administered orally through cotton leaves (Jeyabalan and Murugan, 1997). However, many plant extracts, reported to possess growth inhibiting properties, have not found place under IPM program. Plumbagin [5-hydroxy-2-methyl-1, 4-napthalenedione] a compound isolated from the African medicinal plant *Plumbago capensis* is one such example. It has shown to have EC_{50} values of 150-350 ppm that inhibited ecdysis in 1st or 2nd instar larvae of *H. zea* (Boddie) and *H. virescens* (F.) (Kubo *et al*, 1983). When it was applied topically to final instar larvae at a concentration of 117.3 μg/g body weights, the concentration of both RNA and DNA were also significantly reduced (Krishnayya and Rao, 1995).

Similarly, the acetone extracts of *Catharanthus roseus* at 100, 250 and 500 ppm concentration significantly reduced the larval weight of 2nd, 3rd and 4th instars of *H. armigera*, although mortality was less. High mortality of 1st instar was observed at 500-ppm concentration (Deshpande *et al*, 1988).

Petuniasterone N, a novel steroid, isolated from the leaves of *Petunia hybrid* var. Royal Cascade, *P. parodii* and *P. integrifolia* in addition to larvicidal properties, is also reported to reduce the the larval growth of *Heliothis zea* by 50% at 120 ppm dietary concentration (Elliger *et al*, 1989). However, petuniasterone R was found less effective (Elliger *et al*, 1992). Effective concentration upto 500ppm for crude extract and upto 50ppm for pure compound have potential for commercial exploitation.

Leaves of *Liriope muscari* (Liliaceae) at 5% concentration (w/v) are reported to have significant effect on the overall development of *Heliothis zea* (Adeyeye and Blum, 1989). The methanolic leaf extracts of *Persea indica*, a species of the Canary Islands, is shown to have negative effects against *H. armigera* larvae (Gonzalez *et al*, 1992). On the other hand, Solsoloy and Rejesus (1993) observed the juvenile hormone effect from psychicnut, *Jatropha curcas* on *H. armigera* larvae in diet mediated bioassay. Although, *Helicoverpa* is not a pest of rice, however, an allelochemical identified as Pentadecenal from the *Chilo suppressalis* reistant rice variety found to disrupt the growth and development of *H. armigera* (Dhaliwal *et al*, 1990). Some plants like *Broussonetia papyrifera*, a flora local to Pakistan, are found to restrict the adult emergence of *H. armigera* (Anwar *et al*, 1992). Petroleum ether extracts of weed *Tribulus terrestris* is also reported to effect adult emergence beside to have juvenilizing effects, including larval mortality, in a dose-dependent manner against late instar larvae of *H. armigera* leading to only 41% adult emergence at a dose of 100μg/ insect (Gunasekaran *et al*, 1985). Experiments conducted by Chun and Co-workers (1994) suggested a different mode of action for the seeds of *Pentaclethra macroloba.* They found that seed not only contain factors that reduce larval growth, but also effect cell viability and inhibit the activity of proteinases, trypsin and chymotrypsin, from the larval midgut of *H. zea.* The crude ethanolic extracts of *Aglea elaeagnoidea* and *A. odorata* were active at 5 mg/g of dry diet weight, while *A. roxburghiana* was effective at half of the concentration for inhibiting the growth of *H. armigera* (Koul *et al*, 1997). Although literature suggest growth inhibitory and anti juvenile hormone activity of these plant species, however, their detailed mode of action is yet to be explained in order to get useful clues for future research.

ANTIFEEDANT ACTION

An antifeedant is a behaviour-modifying substance that acts directly on the chemosensilla, thereby results in the feeding deterrence (Isman, 1994). The concept of using feeding deterrents for pest management has drawn considerable interest ever since the striking antifeedant effects of azadirachtin was utilized successfully against desert locust, *Schistocerca gregaria* (Butterworth and Morgan, 1968). Although many other antifeedants from plant species have been the subject of several reviews (Frazier, 1986; Jermy, 1990) but very few attempts have been made to exploit their full potential in crop protection programmes (Murray *et al*, 1993; Isman, 1994).

There are numerous plant species that have shown to induce feeding inhibition in *Heliothis/ Helicoverpa*, a highly polyphagous noctuid. Limonoids from citrus seeds *viz.* nomilin and obacunone have shown significant antifeedant activity comparable to *azadirachtin* against larvae of *Heliothis zea* (Klocke and Kubo, 1982). These limonoids are availabe commercially in USA and have found a place as safe biopesticide in all models of sustainable agriculture. Following this, Kubo and Matsumoto (1984) isolated an antifeedant *abyssinin*, from the root bark of an African medicinal plant, *Bersama abyssinica*, having activity against cotton pest *H. zea*. In continuation to this, Kubo *et al* (1985) further claimed *abyssinin* (a 4-bufadienolide steroid) as the most potent antifeedant isolated from botanicals so far. However, its potential is yet to be exploited under field conditions.

Potential of drimane-related natural and synthetic compounds for their antifeedant action against the larvae of *H. armigera* and *H. virescens* have been investigated by Blaney and Co-workers (1987). Presence of decalin ring and the furofuran side chain in the clerodane diterpenoids was attributed for an important feeding inhibition effect against *Heliothis* spp. (Blaney *et al*, 1988). Subsequently, Simmonds *et al* (1989) explained the functional groups responsible for the biological action of clerodane diterpenoids, isolated from the labiatae plant *Teucrium*, against *H. armigera*. Antifeedant activity of *Teucrium haenseleri* has also been worked out for *H. armigera* (Gaspar *et al*, 1996). Studies have also been conducted to investigate the structural relationship between different polyhydroxy alkaloids on the feeding behaviour of *H. virescens* and *H. armigera* larvae (Simmonds *et al.*, 1991). However, contrary to the claim of Kubo in 1985 for *abyssinin*, Blaney and his associates (1990) proved azadirachtin and *dihydroazadirachtin* as the most potent feeding deterrents among the 40 tested compounds against *Heliothis* spp. A laboratory synthesized triterpenoid, 19 beta,28-epoxy-2-(beta-D-glucopyranosyloxy)-18alpha-olean-1-en-3-one, bearing the alpha beta –unsaturated A-ring functionality from *betulin*, isolated from bark of birch (*Betula* spp.), was evaluated for its potential against *H. zea* . In leaf-disc choice assay using maize foliage treated with 50ml of the test compound in acetone, this synthetic triterpenoid showed significant antifeedant activity against 4th instar larvae with an ED_{50} value of the 4 mg/leaf disc which is found to be better than limonin (ED_{50} = 70mg/ leaf disc) (Lugemwa *et al*, 1990). *Angulatin* A (sesquiterpene polyol ester), a compared isolated from the root bark of *Celastrus angulatus*, significantly affect the feeding of *H. armigera* and is also shown to have insecticidal properties (Wang *et al.*, 1991). Similarly, phenyl propanoids, elemicin and isoasarone, the major components of the essential oils from six insect resistance ecotypes (P_1 to P_6) of *Cymbopogon pendulus* are found responsible for feeding deterrence in *H. armigera* larvae (Koul *et al*, 1995). *Aglaia* (*Meliacea*) species revealed that the presence of complex limonoid fractions was found responsible for significant antifeedant effects against larvae of *H. armigera* (Koul *et al*, 1997) beside the growth inhibiting properties.

In view of significant antifeedant effect of aqueous extracts from the wild specie of pigeon pea (*Cajanus scarabaeoides*) on *H. armigera* larvae, Shanower *et al* (1997) have suggested the possibility of utilization of this species in breeding resistant varieties. Further the acetone extracts of seeds of *Trichilia havanensis* Jacq. (Meliaceae) gave the antifeedant activity at 5000-ppm against 5th instar *H. armigera* larvae. The mixture of havanensin-1,7-diacetate and havanensis,3,7-diacetate compounds fractionated by silica gel chromatography, had shown the maximum antifeedant activity at 1000-ppm under choice and no choice feeding assays (Lopez- Olguin *et al*, 1998). Scutecyprol B, the diterpenoid isolated from *Scutellaria rubicunda* subsp. *Linneana* showed potent antifeedant activity at 100-ppm against *H. armigera* (Bruno *et al*, 1999).

The ethyl acetate extract of *Delphinium* X *Cultorum* cv. Magic Fountains flowers displayed strong feeding inhibition action against corn earworm (*H. zea)* larvae under laboratory test. The magic fountain flowers upon chemical isolation gave the compound, 4-hydroxybenzoic acid and bis (4-hydroxyphenyl) methanol (Miles *et al,* 2000). Salannin isolated from the dichhloroethane (DCE) fractions of *Melia dubia* (Meliaceae) have also shown antifeedancy against *H. armigera* (Koul *et al*, 2000).

Apart from the extracts, essential oils owing to their volatile nature have also been evaluated extensively for antifeedancy. Some of these are registered in USA as plant protection agent, however, not against *H. armigera*. Essential oil of *Tagetes minuta* is reported to give highest antifeedant activity (86.3%) at 0.5 per cent concentration when 2nd instar *H. armigera* larvae were allowed to feed on cotton leaves dipped in the solution of essential oils (Rao *et al*, 2000). Significant antifeedant activity was also recorded with oils from *Cymbopogon flexous*, *C. martini*, *C. winterianus*, *Ocimum basilicum*, *O. gratissimum* and *O. sanctum*. However, essential oils need special formulations to prevent their loss due to volatilization.

Efforts have also been made to try out combinations of extracts with an aim to improve the efficacy. It is reported that a combination of that *B. thuringiensis* sub sp. Kurstaki and plant extracts greatly suppressed the feeding of *H. armigera* (Murugan et al., 1998).

Although antifeedants, do not kill the pests directly but lower their feeding potential and make them more vulnerable to other mortality factors. Unfortunately, the screening and evaluation of plant natural products based on their antifeedant effects alone may easily overlook associated properties physiological effects, a careful study of which could prove to be even more important and potentially useful. Recent studies have shown that many allelochemicals described as an antifeedants in the literature may have a broader range of physiological effects on insects (Perera *et al.*, 2000), thus needs to be studied in a more comprehensive program of insect bioassays for their further exploration for pest management programs.

LARVICIDAL ACTION

Most of the studies on biological activity of plants have shown killing potential against early instars, (Choi and Boo, 1989). This suggests the sensitivity of early instar larvae as target stage of pest for their possible application in pest management programmes. It has also been reported that late instar larvae are more tolerant to the insecticides as toxicity decreases with the age of insect due to the associated physiological changes (Anonymous, 1990), thus offering more resistance to the pressures induced due to external factors.

The most common symptoms consequent upon feeding of the larvae on toxicant are convulsions, copious diuresis, leading to dehydration, loss of body turgidity, and finally cause death. Elliger and Co-workers (1990) observed similar symptoms in *H. zea* larvae when fed on *Petunia parodii* leaves and presence of petuniasterones was attributed for the toxic action. They also isolated 3 new petuniolides. Sometimes the mixture of natural compounds shows more toxic action than the individual fractions. Thus, that could be the only reason that Berenbaum *et al.* (1991) observed the toxic effects of furanocoumarins compounds obtained from the fruits of parsnip (*Pastinaca sativa*; Apiaceae) against *H. zea.*

Plumbagin (2-methyl-5-hydroxyl-1,4-napthoquinone) from the medicinal plant *Plumbago* spp. showed to have dose dependent effect on the mortality of fifth instar *H. armigera* larvae using topical treatment besides the growth inhibitory properties as reported by Krishnayya and Rao in 1995 (Gujar, 1997). The methanolic extract of *Melia azedarach* stems at 7.5% concentration was observed to be larvicidal in action against *H. armigera* larvae, though the extract also showed ovicidal activity at 10% concentration (Kumar *et al.*, 1999). However, these concentrations are too high to make it a commercially viable option. The LC_{50} value of 1839 ppm of cinnamomin isolated from seeds of camphor tree *Cinnamomum camphora* was found against the larval stage of *H. armigera* in diet mediated bioassay and classified as type II ribosome inactivating protein may have some potential (Xin *et al.*, 2000). Thus very few compounds have been reported to have killing properties. However, LC_{50} values need to be brought down further in order to find some field application.

OVIPOSITIONAL DETERRENCY, OVICIDAL AND PUPICIDAL EFFECTS

The eggs and early instars are considered to be the most susceptible stages of the pest (Vekaria and Vyas, 1985). It may, therefore, be more useful to put equal emphasis on the search for bioactive compounds from plant materials having good ovicidal action so that pest can be controlled at the very early stage of growth. It has been reported that higher concentrations are required to achieve an antiovipositional effect than an ovicidal effectiveness. Jeyakumar and Gupta (1999) observed reduction in the oviposition of *H. armigera* at nearly at twice the concentration that required for ovicidal effects. Also, the anti-oviposition and ovicidal effects were observed to be higher in plant product extracts in alcohol as compared to other solvents. Besides growth and developments effects against *H. armigera*, Neem leaf extract observed to be ovicidal at 0.5-2.0% concentrations giving 20-74% egg mortality (Sharma *et al.*, 1997).

Larval-pupal intermediates and deformed wings of the adults are the most common features following toxicant treatment as ovicidal or antioviposition substances. However, very few attempts have been made to comprehend the effect against quiescent stage of the pest. Gupta and Co-workers (1998) treated the pupation site of *H. armigera* with 0.5 to 6.0% neem seed powder (w/w) and during eclosion, observed abnormal adults emergence and death during the prepupal stage with increasing NSP concentration. The calculated ED50 value was 2.902% of NSP in soil on w/w basis. Thus, application of neem seed cake, which is traditionally been advocated for soil application, will certainly help in managing the *H. armigera* population by preventing the formation of virulent adults.

CONCLUSION

Besides *Heliothine* group, many Lepidoptera are also pests of many economically important crops and a basic understanding of how they interact with their hosts is fundamental to development of sound sustainable management practices. The ever-increasing interest in biopesticide-based products in today's agriculture has slowly started changing the pest management practices from generic broad-spectrum chemicals to natural pesticides such as Neem and *Bt* based products. Much emphasis has been given on clean-technology where heavy premium is placed on protection of human health and the environment. The pace of prospecting of plant species both for biochemical and gene diversity at structural and functional level has increased for insect-pest management programs. The greatest impact of botanicals will be in organic food production systems in contexts of sustainable development. On the other hand, botanicals tend to be relatively expensive and face stiff competition from other products in the marketplace. It is evident from the literature that scientist are putting their efforts in exploring the plant diversity, however, these efforts need to be concretised for utilization in management of *H. armigera* population in field. The growth of the industry relating to future expectations and public acceptance will primarily depend on thrust given to research on new potential botanicals, focus on the long term benefits, fecilitation of registration, technology transfer process through a strong extension network, and the due encouragement by all concerned agencies will certainly help in recognizing these new molecules for their wider use.

Table 1: A brief overview of botanical pesticides tested against *Helicoverpa/Heliothis* over the years

Activity	Test material	Remarks	Reference
AT and GI	Limonoids (citrus seeds)	Limonin, nomilin and obacunone (the active limonoids)	Klocke and Kubo, 1982
EI	Azadirachtin (neem)	Conc. <1ppm inhibited growth rate by more than 50%	Kubo and Klocke, 1982
EI	Plumbagin (*Plumbago capensis*)	ED50= 150-350 ppm for 1st or 2nd instars in artificial diet bioassay	Kubo *et al.*, 1983
GI	Limonoid (*Melia azedarach*) and *Trichilia roka* (citrus spp.)	Inhibited moulting and reduced growth rate	Kubo and Klocke, 1984
AT	Abyssinin (*Bersama abyssinica*)	Bufadienolide-type steroid, the most potent antifeedant	Kubo and Matsumoto, 1984; Kubo *et al.*, 1985
JH	Petroleum ether extract (*Tribulus terrestris*)	Dose dependent response	Gunasekaran *et al*, 1985
AT	Drimane and clerodane related natural and synthetic compounds	Explained structure in relation to function	Blaney *et al.*, 1987; Blaney *et al.*, 1988

Activity	Test material	Remarks	Reference
L	Organo sulfur compounds (neem seeds)	Di-n-propyl disulfide proved effective	Balandrin *et al*.., 1988
GD	Acetone extracts (*Catharanthus roseus*)	Effective at 100 ppm	Deshpande *et al*., 1988
AT	Polyhydroxy alkaloids and clerodane diterpenoids (*Teucrium*)	Explained structural-functional groups	Simmonds *et al*., 1989 (a and b)
L, GI and OD	Wild plant species (about 7 species)	Activity with aqueous and ethanolic extracts	Choi and Boo, 1989
GI	Petuniasterone-N (*Petunia*)	50% growth reduction at 120 ppm	Elliger *et al*., 1989
GD	*Liriope muscari* extracts	Activity in the diet mediated bioassay	Adeyeye and Blum, 1989
AT	Azadirachtin and related limonoids	Aadirachtin and dihydro azadirachtin, the most potent	Blaney *et al*., 1990
AT	Isolated compounds	Triterpene betulin derived activity	Lugemwa *et al*., 1990
Tox	Ergostanoids (*Petunia parodii*)	Explained petuniasterone steroid	Elliger *et al*., 1990
Tox	*Pastinaca sativa*	Mixture of furanocoumarin	Berenbaum *et al*., 1991
AT	Angulatin-A (*Celastrus angulatus*)	Sesquiterpene polyol ester mixture showed activity	Wang *et al*., 1991
GI	Petuniasterone-R (*Petunia parodii*)	50% growth reduction at 400 ppm	Elliger *et al*., 1992
I	*Persea indica*	Diterpene attributed activity	Gonzalez *et al*., 1992
Tox	*Pentaclethra macroloba* seed extracts	Reduced larval viability and growth	Chun *et al*., 1994
Dev	Plumbagin (*Plumbago*)	ED_{50}=117.3 µg/ g body wt and reduced conc. of DNA and RNA	Krishnayya and Rao, 1995
AT	*Teucrium haenseleri*	Isolated 20-Deacetyleriocephalin	Gaspar *et al*., 1996
P	Aqueous extracts (*Syringa* leaves)	Effective at 10 g / 100 ml diet	Berg *et al*., 1997
GI	Azadirachtin	Decreased food utilization efficiency	Babu *et al*., 1996
Fd and Rep	Neem limonoids	Post-ingested toxic effects	Jeyabalan and Murugan, 1997

Activity	Test material	Remarks	Reference
GI and AT	Foliar and twig extracts (*Agalia*)	Activity due to limonoid fractions	Koul *et al.*, 1997
L	Azadirachtin and plumbagin	Larval mortality in Plumbagin	Gujar, 1997
GD	Neem leaf extracts	Toxic and Morphogenic effects	Sharma *et al.*, 1997
Qu	Neem seed powder	NSP ED_{50} = 2.902% in soil (w/ w)	Gupta *et al.*, 1998
GI and AT	Neem limonoids	Azadirachtin and other limonoids proved toxic	Murugan *et al.*, 1998
AT	Seed extracts (*Trichilia havanensis*)	Highest activity with acetonic extract against fifth instar	Lopez-Olguin *et al.*, 1998
G and Fd	*Bt* with Pongamia oil, *Vitex negundo* extracts and neem oil	Affected the gut homeostasis and suppressed the feeding	Murugan *et al.*, 1998
AT	Two diterpenoids from *Scutellaria rubicunda*	Scutecyprol-B at 100 ppm proved potent	Bruno *et al*, 1999
Ovi and L	Neem seed kernel extracts	Significant activity	Hassan, 1999
Ovi and OD	Neem seed kernel extract	Dose dependent action	Jeyakumar and Gupta, 1999
L & Ovi	*Melia azedarach* stem extracts	Methanolic extract showed activity	Kumar *et al.*, 1999
I	Kandhari chilli (*Capsicum frutescens*) extracts	Activity due to capsicin	Rethinaraja and Narayanasamy, 1999
AT	*Delphinium* X *cultorum* cv. Magic Fountain Flowers	Ethyl acetate extract displayed activity	Miles *et al.*, 2000
GI & AT	Extracts (*Melia dubia*)	Dichloroethane and methanol extracts showed activity	Koul *et al.*, 2000
AT & L	Aromatic oils (*Cymbopogon* spp., *Tagetes minuta* and *Ocimum* spp.)	*T. minuta* showed highest antifeedant activity and mortality	Rao *et al.*, 2000
Tox	Seeds of *Cinnamomum camphora*	Cinnamomin LC_{50} = 1839 ppm	Xin *et al.*, 2000
GI & Tox	Leaf extract (*Eucalyptus*)	Alcoholic extract proved better	Kaushik, 2001

AT: antifeedant, EI: ecdysis inhibitor, GI: growth inhibitor, JH: juvenile hormone, L: larvicidal, GD: growth and development, Ovi: ovicidal, Tox: toxicity, OD: ovipositional deterrency, I: insecticidal, Fd: feeding, Rep: reproduction, Qu: quiescent stage, Dev: development, P: pesticidal

Table 2: Plant species having biological activity against *Helicoverpa/ Heliothis* (Lepidoptera: Noctuidae)

Antifeedants	Growth Inhibitors	Larvicidal/ Insecticidal/	Ovicidal/ Pupicidal/ Toxicity Ovipositional Deterrency
Citrus seeds, *Bersama abyssinica*, Drimane and clerodane related natural and synthetic compounds, *Teucrium* spp., azadirachtin, azadirachtin-related and related limonoids, *Betula* spp., *Celastrus angulatus, Teucrium haenseleri Agalia* spp., *Trichilia havanensis ,Scutellaria rubicunda, Delphinium* X *Cultorum* cv. Magic Fountain Flowers, Melia dubia ,*Cymbopogon flexous, C. martini, C. winterianus, Tagetes minuta, Ocimum basilicum, O. gratissimum* and *O. sanctum*	citrus seeds, neem, *Plumbago capensis, Melia azedarach, Trichilia roka, Tribulus terrestris, Catharanthus roseus, Persicaria hydropiper, Forsythia koreana, Trifolium repens, Styrax japonica, Ginkgo biloba, Vitis amurensis, Petunia* species, *Liriope muscari, Petunia parodii, Plumbago* spp., Azadirachtin, Neem limonoids, *Agalia* spp., Pongamia oil, *Vitex negundo*, neem oil, *Melia dubia, Jatropha curcas, Cymbopogon pendulus*	Neem seeds, *Persicaria hydropiper, Forsythia koreana, Trifolium repens, Styrax japonica, Ginkgo biloba Vitis amurensis, Petunia parodii, Pastinaca sativa Persea indica, Pentaclethra macroloba, Syringa* sp., *Melia azedarach*, Kandhari chilli (*Capsicum frutescens*), *Tagetes minuta, Cinnamomum camphora, Tinospora crispa, lemon grass (Cymbopogon citrates), Alpinia galangal, Celastrus angulatus*	*Melia azedarach*, neem (leaves, seed powder, NSKE, limonoids, azadirachtin)

REFERENCES

Adeyeye, O.A. and Blum, M.S. 1989. Bioactivity of *Liriope muscari* (Liliaceae) extracts to corn earworm (Lepidoptera: Noctuidae). *J. Eco. Ent.* 82 (3): 942-947.

Ahmed, S., Grainage, M., Hylin, J.W., Mitchell, W.C. and Litsinger, J.A. 1984 Some promising plant species for use as pest control agents under traditional farming systems. *Proc. 2nd Int. Neem Conf.*, Rauischholzhausen, 1983, pp 565-580.

Anonymous 1990 Annual Progress Report, *Heliothis* Management Scheme (ICAR No. 1.39/88), A.P. Agricultural University RARS, Lam, Guntur: 32.

Anwar, T., Jabbar, A., Khalique, F., Tahir, S. and Shakeel, M.A. 1992 Plants with insecticidal activities against four major insect pests in Pakistan. *Trop. Pest Mgmt.* 38 (4): 431-437.

Babu, R., Muruggan, K. and VanithaKumaria, G. 1996 Interference of Azadirachtin on the food utilization efficiency and midgut enzymatic profiles of *Helicoverpa armigera* Hubner (Insecta: Lepidoptera: Noctuidae). *Indian J. Environ. Tox.* 6 (2): 81-84.

Balandrin, M.F., Lee, S.M., and Klocke, J.A. 1988 Biologically active volatile organosulfur compounds from seeds of the neem tree, *Azadirachta indica* (Meliaceae). *J. Agric. Food Chem.* 36 (5): 1048-1054.

Bentley, M.D., Leonard, D.E. and Bushway, R.J. 1984 *Solanum* alkaloids as larval feeding deterrents for spruce budworm, *Choristoneura fumiferana* (Lepidoptera: Tortricidae). *Ann. Entomol. Soc. Am.* 77: 401-403.

Berenbaum, M.R., Nitao, J.K., and Zangerl, A.R. 1991 Adaptive significance of furanocoumarin diversity in *Pastinaca sativa* (Apiaceae). *J. Chem. Eco.* 17(1): 207-215.

Berg, A.M. van-den, Schyf, A. van-der, Van- den Berg, A.M., Van-der Schyf, A., Kleeberg, H. (ed.) and Zebitz, C.P.W. 1997 Comparison between Neemazal-T/S and syringa extracts: laboratory trials. *Practice oriented results on use and production of neem-ingredients and pheromones. Proceedings 5th Workshop, Wetzlar, Germany, 22-25 Jan, 1996.* pp. 147-149.

Bergamasco, R. and Horn, D.H.S. 1983 In: *Endocrinology of Insects* p. 627-654 (Eds. G.H. Downer and H. Laufer). Aln R. Liss, New York.

Blaney, W.M., Simmonds, M.S.J., Ley, S.V. and Jones, P.S. 1988 Insect antifeedants: a behavioural and electrophysiological investigation of natural and synthetically derived clerodane diterpenoids. *Entomol. Expt. Applic.* 46(3): 267-274.

Blaney, W.M., Simmonds, M.S.J., Ley, S.V. and Katz, R.B. 1987 An electrophysiological and behavioural study of insect antifeedant properties of natural and synthetic drimane-related compounds. *Physiol. Ent.* 12(3): 281-291.

Blaney, W.M., Simmonds, M.S.J., Ley, S.V., Anderson, J.C., and Toogood, P.L. 1990. Antifeedant effects of azadirachtin and structurally related compounds on leipdopterous larvae. *Entomol. Expt. Applic.* 55(2): 149-160.

Bruno, M., Vassallo, N. and Simmonds, M.S.J. 1999 A diterpenoid with antifeedant activity from *Scutellaria rubicunda. Phytochemistry* 50(6): 973-976.

Butterworth, J.H. and Morgan, E.D. 1968 Isolation of a substance that suppresses feeding in locusts. *J. chem. Soc., Chem. Comm.* pp 23-24.

Choi, K.S. and Boo, K.S. 1989 Effects of extracts of some selected wild plants species on larval development and adult oviposition in *Heliothis assulta. Korean J. Appl. Ent.* 28(3): 113-119.

Chun, J.S., Goodman, C.L., Rice, W.C., McIntosh, A.H., Chippendale, G.M. and Schubert, K.R. 1994 *Pentaclethra macroloba* seed effect on larval growth, cell viability, and midgut enzyme activity of *Helicoverpa zea* (Lepidoptera: Noctuidae). *J. Eco. Ent.* 87(6): 1754-1760.

Cunningham, J.P., Jallow, M.F.A., Wright, D.J. and Zalucki, M.P. 1998a Learning in host selection in *Helicoverpa armigera* (Hübner) (Lepidoptera: Noctuidae). *Animal Behaviour* 55: 227–234.

Cunningham, J.P., Zalucki, M.P. and West, S.A. 1999 Learning in *Helicoverpa armigera* (Lepidoptera: Noctuidae): a new look at the behaviour and control of a polyphagous pest. *Bull. Ent. Res.* 89:201-207.

Deshpande, S.G., Joseph, M. and Sharma, R.N. 1988 Insect growth and development inhibition properties of *Catharanthus roseus. Int. J. Trop. Agri.* 6(3-4): 287-290.

Dhaliwal, G.S., Pathak, M.D. and Vega, C.R. 1990 Effect of a rice allelochemical on insect pest, predators and plant pathogens. *J. Insect Sci.* 3(2): 136-140.

Elliger, C.A., Haddon, W.F., Waiss, A.C.(Jr.) and Benson, M. 1989 Petuniasterone N, an unusual ergostanoid from Petunia species. *J. Nat. Products* 52(3): 576-580.

Elliger, C.A., Waiss, A.C.(Jr.) and Benson, M. 1992 Petuniasterone R, a new ergostanoid from *Petunia parodiii. J. Nat. Products* 55(1): 129-133.

Elliger, C.A., Waiss, A.C.(Jr.), Benson, M. and Wong, R.Y. 1990 Ergostanoids from *Petunia parodiii. Phytochemistry* 29(9): 2853-2863.

Fitt, G.P. 1989 The ecology of *Heliothis* in relation to agro-ecosystems. *Ann. Rev. Ent.* 34:17-52.

Frazier, J.L. 1986 The perception of plant allelochemicals that inhibit feeding. In: *Molecular aspects of Insect Plant Associations* (L.B. Bratsten and S. Ahmad Eds.) pp 1-42, Plenum Press, New York.

Gaspar, H., Palma, F.M.S.B., Torre, M.C. de-la., Rodriguez, B. and De, la-Torre, M.C. 1996 20-Deacetyleriocephalin: 1H- and 13C-NMR assignments. *Fitoterapia* 67(2): 188-189

Gonzalez, C.A., Cabrera, R., Castanera, P., Gutierrez, C. and Fraga, B.M. 1992 Insecticidal acitivity and diterpene content of *Persea indica. Phytochemistry* 31(5): 1549-1552.

Gujar, G.T. 1997 Biological effects of azadirachtin and plumbagin on *Helicoverpa armigera. Indian J. Ent.* 59(4): 415-422.

Gunasekaran, K., Chelliah, S., Regupathy, A. (ed.) and Jayaraj, S. (ed.) 1985 Juvenile hormone activity of *Tribulus terrestris* L. on *Spodoptera litura* F. and *Heliothis armigera* (Hb). *Behavioural and Physiological Approaches in Pest Management*, 146-149.

Gupta, G., Mahapatro, G.K., and Chandra, A. 1998 Neem seed powder: targeting the quiescent stages of *Helicoverpa armigera* Hubner. *Annals Plant Protec. Sci.* 6(2): 170-173.

Hassan, E. 1999 The insecticidal effects of neem seed kernel extract on eggs and larvae of *Helicoverpa armigera* (Hubner). *Zeitschrift fur Pflanzenkrankheiten und Plfanzenschutz* 106(5): 523-529.

Isman, M.B. 1994 Botanical insecticides and antifeedants: new sources and perspectives. *Pest. Res. J.* 6(1): 11-19.

Jermy, T. 1990 Prospects of antifeedant approach to pest control- a critical review. *J. Chem. Eco.* 16: 3151-3166.

Jevabalan, D. and Murugan, K. 1997 Effect of neem limonoids on feeding and reproduction of *Helicovarpa armigera* (Hubner) (Lepidoptera: Noctuidae). *Entomon* 22(1): 15-20.

Jeyakumar, P. and Gupta, G.P. 1999 Effect of neem seed kernel extract (NSKE) on *Helicoverpa armigera. Pestic. Res. J.* 11(1): 32-36.

Kaushik, N. 2001 Pesticidal activity of *Eucalyptus* leaf extracts against *Helicoverpa armigera* larvae. In: *3rd IUPAC Int. Conf. Biodiversity (ICOB-3)*, Nov 3-8, 2001, Antalya, Turkey

Klocke, J.A. and Kubo, I. 1982 Citrus limonoid by-products as insect control agents. *Entomol. Expt. Applic.* 32(3): 299-301.

Koul, O., Jain, M.P., Sharma, V.K. and Koul, O. 2000 Growth inhibitory and antifeedant activity of extracts from *Melia dubia* to *Spodoptera litura* and *Helicoverpa armigera* larvae. *Indian J. Expt. Biol.* 38 (1): 63-68.

Koul, O., Shankar, J.S., Mehta, N., Sharma, S.N., Singh, B. and Taneja, S.C. 1995 Resistance to insects in selective ecotypes of *Cymbopgon pendulus* (Nees Ex Steud) Wats. *Pest. Res. J.* 7(2): 132-139.

Koul, O., Shankar, J.S., Mehta, N., Taneja, S.C., Tripathi, A.K. and Dhar, K.L. 1997 Bioefficacy of crude extracts of *Aglaia* species (Meliaceae) and some active fractions against lepidopteran larve. *J. Appl. Ent.* 121(4): 245-48.

Krishnayya, P.V. and Rao, P.J. 1995 Plumbagin effects on *Helicoverpa armigera* Hubner (lepidoptera: Noctuidae) IV. Final instar haemolymph trehalose, cations and nucleic acids. *Entomon* 20 (3-4): 197-202.

Kubo, I. and Klocke, J.A. 1982 Azadirachtin, insect ecdysis inhibitor. *Agric. Biol. Chem.* 46(7): 1951-1953.

Kubo, I. and Klocke, J.A. 1984. Limonoids as insect control agents. *Les mediateurs chimiques agissant sur le comportement des insectes*. Symposium international. Versailles, 16-20 novembre 1981, 1982, recd. 1984, pp. 117-129

Kubo, I. and Matsumoto, T. 1984 Abyssinin, a potent insect antifeedant from an African medicinal plant, *Bersama abyssinica. Tetrahedron Lett.* 25(41): 4601-4604.

Kubo, I., Matsumoto, T., and Hedin, P.A. 1985 Potent insect antifeedants from the African medicinal plant *Bersama abyssinica. Bioregulators for Pest Control*, 183-200.

Kubo, I., Uchida, M. and Klocke, J.A. 1983 An insect ecdysis inhibitor from the African medicinal plant, *Plumbago capensis* (Plumbaginaceae); a naturally occurring chitin synthetase inhibitor. *Agric. Biol. Chem.* 47(4): 911-913.

Kumar, R., Singh, R., Rani, M, Suhag, P., Kalidhar, S.B., 1999 Biological efficacy of *Melia azedarach* stems against lepidopterous pests. *J. Medicinal Aromatic Plant Sci.* 21(4): 1074-1078.

Lopez, O.J.F., Torre, M.C. de-la., Vinuela, E., Castanera, P. and de-la, Torre, M.C. 1998 Activity of seed extracts of *Trichilia havanensis* Jacq. (Meliaceae) on larvae of *Helicoverpa armigera* (Hubner). *Boletin de Sanidad Vegetal, Plagas.* 24(3): 629-636.

Lugemwa, F.N., Huang, F.Y., Bentley, M.D., Mendel, M.J. and Alford, A.R. 1990 A *Heliothis zea* antifeedant from the abundant birchbark triterpene betulin. *J. Agric. Food Chem.* 38(2): 493-496.

Menn, J.J. 1997 Biopesticides-are they relevant? In: *Focus on Biopesticides*, The Royal Society of Chemistry, pp.1,2

Miles, J.E.C., Ramsewak, R.S. and Nair, M.G. 2000 Antifeedant and mosquitocidal compounds from *Delphinium Xcultorum* cv. Magic Fountains flowers. *J. Agric. Food Chem.* 48(2): 503-506.

Mordue (Luntz), A.J. and Blackwell, A. 1993 Azadirachtin: an Update. *J. Insect Physiol.* 39(11): 903-924.

Murray, K.D., Alford, A.R., Groden, E., Drummond, F.A., Storch, R.H., Bentley, M.D. and Sugathala, P.M. 1993 Interactive effects of an antiffedant used with *Bacillus Thuringiensis* var. San Diego delta endotoxin on Colorado potato beetle (Coleoptera: Chrysomelidae). *J. Eco. Ent.* 86: 1793-1801.

Murugan, K., Babu, R., Pandey, A. (ed.), Soccol, C.R. (ed.) and Josh, V.K. 1998 Impact of certain plant products and *Bacillus thuringiensis* Berliner subspecies *kurstaki* on the growth and feeding physiology of *Helicoverpa armigera* Hubner (Insecta: Lepidoptera: Noctuidae). *J. Sci. Industr. Res.* 57 (10-11): 757-765.

Murugan, K., Jeyabalan, D., Kumar, N.S., Babu, R., Sivaramakrishnan, S. and Nathan, S.S. 1998 Antifeedant and growth-inhibitory properties of neem limonoids against the cotton bollworm, *Helicoverpa armigera* (Hubner). *Insect Sci. Appli.* 18(2): 157-162.

Perera, D.R., Armstrong, G. and Senanayake, N. 2000 Effect of antifeedants on the diamondback moth (*Plutella xylostella*) and its parasitoid *Cotesia plutellae*. *Pest Mgmt.Sci.* 56:486-490.

Rao, M.S., Pratibha, G. and Korwar, G.R. 2000 Evaluation of aromatic oils against *Helicoverpa armigera*. *Annals Plant Protec. Sci.* 8 (2): 236-238.

Reed, W. and Pawar, C.S. 1982 *Heliothis*: a global problem. In: *Proceedings of the International Workshop on Heliothis Management*, ICRISAT, Patencheru, Andhra Pradesh, pp. 9-14.

Rethinaraja, R. and Narayanasamy, P. 1999 Kaandhari chilli *Capsicum frutescens* L.: a potential pesticidal plant. *Insect Environment* 5(3): 116-117.

Shanower, T.G., Yoshida, M. and Peter, A.J. 1997 Survival, growth, fecundity, and behavior of *Helicoverpa armigera* (Lepidoptera: Noctuidae) on pigeonpea and two wild *Cajanus* species. *J. Eco. Ent.* 90 (3): 837-841

Sharma, P.L., Chander, S. and Sheikher, C. 1997 Toxic and morphogenic effects of neem leaf extracts on *Helicoverpa armigera* (Hubner). *Pest Mgmt. Eco. Zoo.* 5(2): 95-100.

Simmonds, M.S.J., Blaney, W.M. and Fellows, L.E. 1991 Behavioral and electrophysiological study of antifeedant mechanisms associated with polyhydroxy alkaloids. *J. Chem. Eco.* 16 (11): 3167-3196.

Simmonds, M.S.J., Blaney, W.M., Ley, S.V., Savona, G., Bruno, M. and Rodriguez, B. 1989 The antifeedant activity of clerodane diterpenoids from Teucrium. *Phytochemistry* 28(4): 1069-1071.

Singh, D. 2000 Bioinsecticides from plants. *Curr. Sci.* 78 (1): 7-8.

Solsoloy, A.D. and Rejesus, B.M. 1993 Juvenile hormone effect of the insecticidal priciple from psychic nut, *Jatropha curcas* Linn. on cotton bollworm, *Helicoverpa armigera* (Hubn.). In: *24th Annual Scientific Meeting of the Pest Management Council of Philippines*, Inc. College, Laguna (Philippines), 1993. Pp 34.

Vekaria, M.V. and Vyas, H.N. 1985 Studies on ovicidal toxicity of certain insecticides against the eggs of *Heliothis armigera* Hubner. *Pestic.* 19(10): 43-45.

Wang, M.T., Qin, H.L., Kong M. and Ki, Y.Z. 1991 Insecticidal sesquiterpene polyol ester from *Celastrus angulatus*. *Phytochemistry* 30(12): 3931-3933.

Xin, Z., Li, X.Z., Yuan, J.Z., Tang, Z.H., Liu, W.Y., Xin, Z., Li, X.D., Yuan, J.Z., Tang, Z.H. and Liu, W.Y. 2000 Toxicity of cinnamomin - a new type II ribosome-inactivating protein to bollworm and mosquito. *Insect Biochem. Mol. Bio.* 30(3): 259-264.

Zalucki, M.P., Clarke, A.R. and S.B. Malcolm. 2002 Ecology and behaviour of first instar larval Lepidoptera. *Ann. Rev. Entomol.* 47: 361-393.

Entomopathogenic Nematodes in Integrated Pest Management

SUDERSHAN GANGULY, VISHAL S. SOMVANSHI AND K.S.RATHOUR

Nematodes are triploblastic, pseudocoelomates, bilaterally symmetrical, non-segmented vermiform organisms having well developed digestive, reproductive, nervous, excretory and muscular system, but lacking circulatory and respiratory system. Nematodes are commonly called as round worms due to their cylindrical body circular in cross-section. These are found in almost all the conceivable habitats, viz soil, fresh water, brackish water, sea, ocean beds and even in the polar glaciers. Their widespread distribution is mainly attributed to their adaptability for diverse mode of life. These may be fungal feeders, bacterial feeders, algal feeders, saprophagous, predaceous or even parasites of plants and animals including both vertebrates and invertebrates. The plant parasitic nematodes, being one of the limiting factors in agricultural production, are harmful to the mankind. Among the parasites of invertebrates, those associated with insects or other arthropods are called entomogenous (living in or on the insects) or entomophillic (having an affinity for insects). Presently, nematodes belonging to more than 30 families are known to be associated with insects. Although there are large number of nematodes associated with insects yet only a few of them belonging to nine families exhibit the biological control potential for insects. Nematode families in ascending order of biocontrol potential are Allantonematidae, Parasitylenchidae, Iotonchiidae, Tetradonematidae, Sphaerulariidae, Mermithidae, Phaenopsitylenchidae, Steinernematidae, and Heterorhabditidae.

Nematode-insect associations

The nematode-insect relationships may vary from fortuitous (accidental or by chance) to obligatory, and inquilinism (commensalism) to parasitism. Commensalism is the most common type of association, in which one species benefits from the other while the second is neither harmed nor benefitted. The commensal nematodes may have phoretic relationship with the insects wherein the insect acts as carrier of the dormant/dauer stage of nematode to the suitable environment. The fulfillment of Koch's postulates is an important prerequisite for establishing the parasitic relationship wherein the nematode is dependent on insect for its sustenance and cause harmful effects. The harmful effects due to parasitism by nematodes may vary from sterility, reduced fecundity, reduced longevity, reduced fight, delayed development, or other behavioural, morphological and physiological changes to mortality of the insect. The parasitism can be of two types – facultative parasitism and obligate parasitism. The nematodes of the families Steinernematidae and Heterorhabditidae, though highly pathogenic, are facultative parasites due to their capability to complete their life cycle on the symbiotic bacteria without the insect host (Gaugler and Kaya, !990; Gaugler, 2002; Kaya, 1993).

Entomopathogenic nematodes (EPN)

The nematodes belonging to Steinernematidae and Heterorhabditidae (Poinar, 1990) are commonly termed as entomopathogenic nematodes because of their capability to kill the insect host very quickly.

Division of Nematology, Indian Agricultural Research Institute, New Delhi-110 012.

These nematodes have very high potentialities as bioagents against the insect pests, due to their following attributes –i) Like parasitoids / predators, they have chemoreceptors and are motile; ii) like pathogens, they are highly virulent, killing their host within 24 to 48 h.; iii) can be cultured easily *in vitro*; iv) have a high reproductive potential; v) have broad host range; safe to the plant and animal health, and the environment; vi) can be easily applied using standard spray equipments; vii) have potential to recycle in environment; compatible with many chemical pesticides.

Till to date, there are two genera listed under Steinernematidae – *Steinernema* and *Neosteinernema,* with 35 species in the former and only one species in the latter. Under Heterorhabditidae, there is only one genus – *Heterorhabditis* with only 9 known species. The taxonomic aspects of EPN had been summarized by Adams and Nguyen (2002), Ganguly (2003).

The symbiotic bacteria

These nematodes have symbiotic relationships with the bacteria which is species specific (Table 1). The *Xenorhabdus* Thomas and Poinar, 1979 and *Photorhabdus* Boemare *et al.,* 1993 are Gram negative, anaerobic rods, classified under the family Enterobacteraceae. *Xenorhabdus* spp are associated with *Steinernema* while *Photorhabdus* spp are associated with *Heterorhabditis.*

Table 1. Some of the species of *Steinernema* and *Heterorhabditis* and their respective symbiotic bacteria species. (after Boemare, 2002)

Nematode species	Bacteria species
Steinernema affinis, S. kraussei,	*Xenorhabdus bovienii*
S. feltiae, S. intermedium,	
S. carpocapsae	*X. nematophila*
S. glaseri and S. cubanum	*X. poinarii*
S. kushidai	*X. japonica*
S. longicaudatum (?)	*X.beddingii*
Heterorhabditis bacteriophora subgroup Brecon	*Photorhabdus luminescens luminescence*
H. bacteriophora subgroup HP88	*P. luminescence laumondi*
H. dowensei	*Photorhabdus sp.*
H. indica	*P.luminescens akhurstii*
H. zealandica	*Photorhabdus temperata*
H. bacteriophora subgroup NC	*P. temperata*
H. megidis Nearctic group (Ohio, Wisconsin)	*P. temperata*
H. megidis Palaearctic group	*P. temperata temperata*

Nematode-bacterium complex in EPN

Excellent mutualism occurs between the nematode and the bacterium (Kaya, 1993). The nematode is dependent on the bacterium for: i) killing its host; ii) creating a suitable environment for nematode development and preventing the infection by secondary microorganisms; iii) serving as a food source; iv) breaking down the host tissues to serve as nutrient source for nematode. The bacterium needs the nematode for: i) protection from the external environment; ii) penetration into host's haemocoel; iii) inhibition of the host's antibacterial proteins.

Life cycle of steinernematids and heterorhabditids

The infective juveniles (IJs) carry bacteria in their digestive tract and penetrate the insect host through natural openings in case of steinernematids and also direct penetration through integument in case of heterorhabditids. Nematodes reach the haemocoel where the bacteria are released. The bacteria multiply rapidly and cause septicaemia thus killing the host within 24 – 48 h. The nematodes feed on the bacteria and host tissue, produce 2 to 3 generations , and emerge from the cadaver as IJs within 7 –10 days in case of steinernematids and 12 to 15 days in case of heterorhabditids. The emerged IJs are ready to infect the new hosts.

Symptoms of insects killed by EPN

The nematode-killed insects are flaccid and do not have a putrid odour because the mutualistic bacteria produce antibiotics which prevent the growth of secondary microorganisms. Insects killed by steinernematids turn yellow-brown, pale or black whereas those killed by heterorhabditids turn red, brick-red, purple, orange or sometimes green. In the dark, the heterorhabditid-infected insects will luminesce due to the presence of *Photorhabdus luminescens*, and the internal tissues are gummy or sticky.

MASS-PRODUCTION TECHNOLOGY

Various efforts are going on in various countries to standardize the mass production technology. Ehlers (2001) developed some mass production technology of *Steinernema* and *Heterorhabditis spp.* for the biological control of insect pests. Neves *et al* (2001) standardized effect of bioreactor conditions as airflow on the *Steinernema* liquid cultures. Chavarria and Torre (2001) studied other aspects of population growth kinetics of *Steinernema feltiae* in submerged monoxenic cultures. However most simple and widely adopted techniques for mass cultivation are following;

Culturing in vivo

Most steinernematids and heterorhabditids can easily be cultured on wax moth larvae (*Galleria mellonella*) or any other bait insect, as detailed below, for use in the laboratory and small field trials, but not for large scale production (Kaya and Stock, 1997).

a) Evenly distribute 200 IJs in 1 ml of nematode suspension on 9 cm filter paper placed on the lid of 100 × 15 mm petridish.

b) Add 10 *Galleria* larvae to the dish and cover the lid with inverted petridish.

c) Store the petridish in a plastic bag containing a wet paper towel to keep it moist, at 25°C.

d) After 5-7 days, place the cadavers on to the White Trap, which means place the petriplate containing cadavers in a bigger petriplate (150 × 20 mm) containing sterile water. Slowly the nematodes emerged from the cadaver move to the water in outer bigger petridish.

e) Harvest nematodes from White Trap (outer petridish), give 2- 3 washings with sterile water, concentrate and store in tissue culture flasks, keep horizontally at 15°C, till use.

Culturing in vitro:

The steinernematids and heterorhabditids (Fig. 1) can be cultured on a variety of substrates (Woodring and Kaya, 1988; Friedman, 1990), which can support the growth of their symbiotic bacteria. The ingredients for nematode culture include a source of nutrients for the symbiotic bacterium and a sterol source for the nematodes. *In vitro* nematode cultures can be obtained as per the following steps:

a) Place 1 g of sterile rabbit kidney on 10 ml of 1 % agar, or a small piece of autoclaved pig kidney placed aseptically on to Peptone- glucose- agar has also been used for steinernematids. Another artificial medium for steinernematid and heterorhabditid production uses dry dog food as the main ingredient (Hara *et al,* 1981; Ogura and Haraguchi, 1993).

b) After sterilisation and cooling the medium, streak the symbiotic bacteria (Phase I) from the species to be cultured.

c) Allow the bacterium to grow for 1-2 days, and then inoculate with sterilised IJs from another monoxenic culture.

d) After 2-3 weeks, harvest the IJs using White Trap.

Bedding (1981, 1984) developed a technique whereby large number of nematodes may be produced using chicken offal medium on a porous foam substrate such as polyether polyurethane. This substrate provides the largest surface to volume ratio while providing adequate interstitial space. Glass flakes are generally used. but autoclavable bags can also be used.

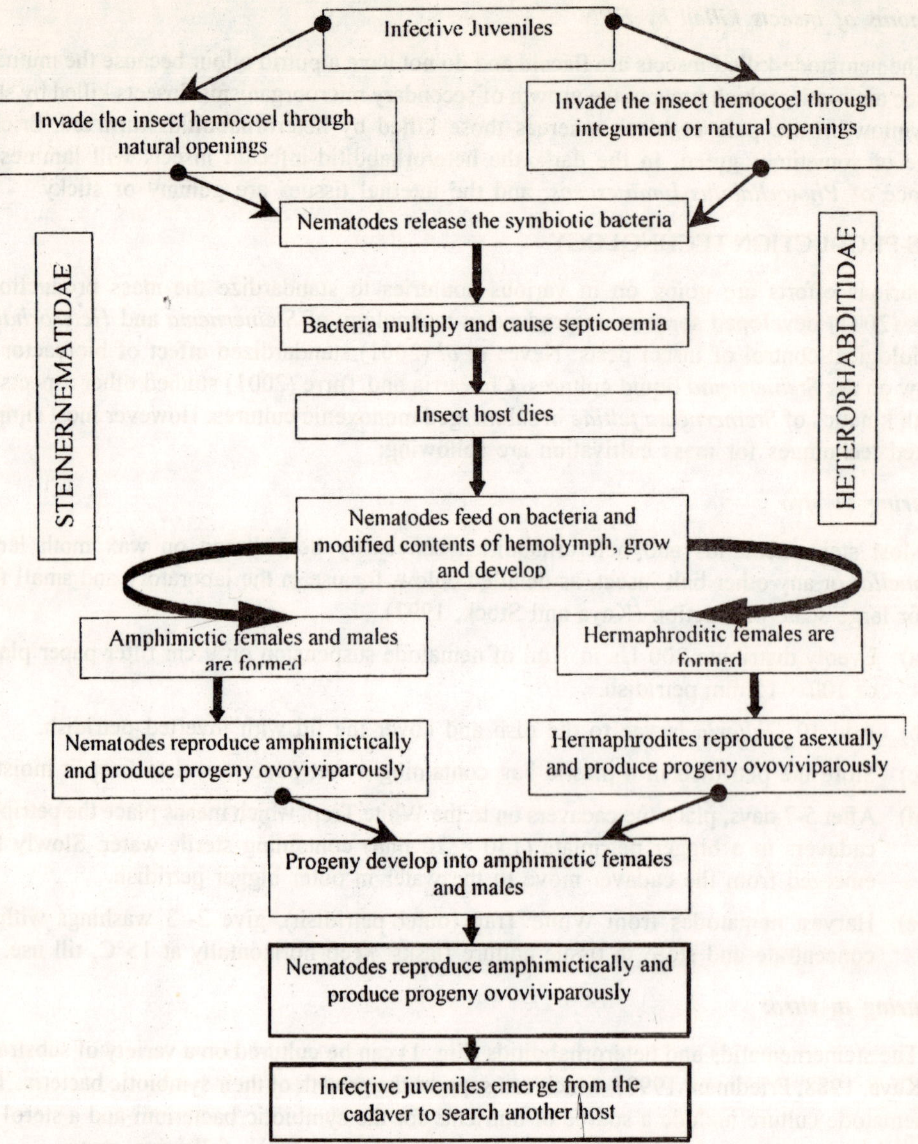

Fig.1: Life cycle of steinernematid and heterorhabditid nematodes

Surface sterilization of IJs

The IJs can be surface sterilized by using any one of the four methods- 0.1 % Hyamine 10X for 15 min; 0.1 % formaldehyde for 30 min; 0.1 % merthiolate solution (w/ v) plus streptomycin (5000 units per ml) for 2-3h; or 0.1 % sodium hypochlorite for 1h. after treating, the IJs should be given 2-3 washings with sterilized water.

Isolation of symbiotic bacteria

The symbiotic bacteria can be isolated either from IJs or from haemolymph of the infected host (Akhurst, 1980).

i) Isolation of bacteria from IJs

- Suspend the surface sterilized IJs in yeast salts or nutrient broth.
- Allow the bacteria to multiply in the medium kept on a shake for 24- 48 h

ii) Isolation of bacteria from haemolymph

- Infect 10 larvae of *G. mellonella* with 100 IJs.
- After 48 h, surface sterilize the dead larvae by 1 % sodium hypochlorite.
- Open the cadaver without rupturing the gut . Streak a drop of haemolymph on to NBTA or nutrient agar. Transfer the colony into YS broth and incubate for 1-3 days at 25 degree celcius. Fill a 2.2 ml cap with 0.3 ml sterile glycerol. Add 1.7 ml bacterial suspension.
- Store caps at -20 degree Celsius, till use.

Commercial scale production

Mass production of EPN has evolved from the first large scale *in vitro* solid-media production by Glaser *et al* (1940) to the *in vivo* production by Dutky et al (1964), to the 3 dimensional solid media *in vitro* process (Bedding, 1981; 1984). Production on commercial scale has been accompanied by using inexpensive protein and sterol source (Bedding 1981; 1984). This method is applicable for cottage industries or biotechnology companies. Various substances have been used for *in vitro* production such as pork kidney / peptone agar, pork kidney / beef fat, chicken heart and intestine, clotted cattle blood, wheat, corn fish several fly media and dog food (Bedding, 1976; Dutky *et al.*1964; House *et al.* 1965; Poinar, 1979). The EPN has also been produced monoxenically in large liquid fermenters using a combinations of plant and animal proteins (Ehlers *et al.,* 1998; Friedman, 1990). This method is highly efficient and quite suitable for companies with large capital investments because of economies of scale. Friedman *et al.* (1989) reported yields up to 110,000 IJs / ml in only 8 days using a medium containing soy flour, corn oil and egg yolk. In the scale up models, liquid monoxenic cultures provide the best economies of scale among all the production methods available. Besides a large array of literature on commercial production of EPN is available, which has been reviewed and discussed by Gaugler (2002a).

Formulation and storage

A major constraint on commercializing EPNs is their poor storage capability. It is extremely important that IJs and their mutualistic bacterium must possess a high and stable level of pathogenicity from production to application (Georgis, 1990). Active Ijs must be immobilised to prevent depletion of their lipids and glycogen reserves. Immobilisation has been accomplished by storing them in refrigerator (5 –15 °C) in oxygenated water where these can be stored up to 5 years (Friedman, 1990). But this is not commercially feasible. Diminishing the activity of nematodes through immobilisation and or partial desiccation in or on a substrate increase their viability. Therefore, nematodes formulated in various carriers such as alginate, clay, vermiculite and gel forming polyacrylamides can be stored for at least 6 months under refrigeration and at least 3 months at room temperature. Bait formulations generally consist of a mixture of carrier (e.g., corncob grits, wheat bran, and peanut hulls), a feeding stimulant (sucrose, malt extract, glucose, and molasses), and a toxicant. Gregoris et al (1995) have developed formulations for soil insects by immobilizing and/or partially desiccating infective stage. Nematodes and one granular formulation that allows nematode into anhydrobiotic state extending nematode survival and pathogenecity for upto six months at 4 to 25 degree Celcius, and upto 8 weaks at 30 degree celcius. Connick et al. 1993 patented a granular formulation product based on cohesive dough made of wheat flour, fillets and other additives, however lay emphasis on the storage temperature for the longevity of survival. Presently, there are about 75 companies in mostly in USA and other parts of the world (UK, Germany, Canada, Italy, Japan,Switzerland, Sweden, Denmark Australia) which are producing, formulating and marketing the EPN. Several commercial products of EPNs have been formulated by different companies, and are available in world pest control market (Table 2).

UTILIZATION OF ENTOMOPATHOGENIC NEMATODES IN BIOLOGICAL CONTROL OF INSECTS

Compatibility of EPNs with fertilizers and pesticides:

Infective juveniles of EPNs are tolerant of short exposures of 2-6 hours to most of the agricultural chemicals including herbicides, fungicides, acaricides and insecticides (Rovesti and Deseo, 1990; Ishibashi,

1993). EPNs are also compatible with most inorganic fertilizers but natural populations are negatively affected (Bednarek and Gaugler, 1997). However, heterorhabditids tend to be more sensitive to physical challenges including pesticides, than steinernematids. *H. bacteriophora* is quite sensitive to physical stress, but was found to be tolerant to 3 days of laboratory immersion in 58 out of 75 fungicides, herbicides, insecticides and nematicides assayed (Rovesti and Deseo, 1990). Das and Divakar (1987) found that *S. carpocapsae* was compatible with carbaryl, Dimethoate, malathion, Zineb at normal recommended doses. The compatibility of *S. glaseri* with urea, murate of potash and super-phosphate, and *S. feltiae* with atrazine, carbofuran, quinalphos and aldrin (Karunakar,1990). Margosan –o, a commercial neem based insecticide was toxic to *S. carpocapsae, S. feltiae* and *S. glaseri,* only at much higher concentrations than recommended field rate of 20 mg Azadirachtin/litre water (Stark, 1996).

Table 2. Some important EPN based commercial products

(Smart, 1994;Rahman et al,2000)

Product	Nematode species	Country of origin
ORTHO BioSafe (Alginate gel)	*Steinernema carpocapsae*	USA
BioVector (Polyacrylamide gel)	-do-	USA
Exhibit (Alginate gel)	-do-	USA
Sanoplant	-do-	USA
Boden Niitzlinge	-do-	Switzerland
Helix (Polyacrylamide gel)	-do-	Germany
Magnet	*S. feltiae*	USA
Nemasys (Clay)	-do-	USA
Stealth	-do-	UK
Proactant Ss	*S. carpocapsae*	USA
Otinem (Clay)	*Heterorhabditis bacteriophora*	USA
Nemasys (Clay)	*H. megidis*	UK
Vector MC	*S.riobrave*	USA
Green Commandos-Ecomax	*S. carpocapsae*	India
Soil Commandos-Ecomax	*H. bacteriophora*	India

Some chemical pesticides (Imidacloprid, tefluthrin) and biopesticides (*Paenibacillus popillae* and *Bacillus thuringiensis*) act synergistically with EPN and therefore improve nematode efficacy in inundative applications (Koppenhoefer and Kaya, 1997; 1998; Nishimatsu and Jackson, 1998; Thurston *et al.,* 1994). Combinations of Imidacloprid at concentrations ranging between 50-200 g (ai / ha) and nematodes had a strong synergistic effect on mortality of white grub (Scarabacidae: Coleoptera (Koppenhoefer and Kaya, 1998). Moreover, many chemicals noted to be toxic had only temporary effect only, as the nematodes recovered quickly when the exposure ended. Thus, simultaneous application of nematodes with pesticides and fertilizers should be practical when the exposures are short. Chemicals to be used with care or avoided include the insecticides bendiocarb, chloropyriphos, entoprop and isazophos, the fungicides anilazine, dimethyl benzyl ammonium chloride, fenarimol and mercurous chloride, the herbicides 2,4 D and trichlopyr and the nematicide fenamiphos.

B. Application Technology

Nematodes can be applied with common agrochemical equipment including hand guns, small pressurized sprayers, mist blowers, electrostatic sprayers, fan sprayers and helicopters. They stand pressure of 300 lb /sq. inch and can be delivered with all common nozzle type sprayers with openings as small as 50 microns diameter. Other methods of delivering nematodes to the target are injection, baits and alginate capsules. These strategies require fewer nematodes or lower spray volume for effective control because they provide better protection for nematodes from environmental extremes. Nematodes can also be applied by drip and sprinkler irrigation. This is attractive since it is labour saving, automatically provides sufficient moisture for the nematodes, provide flexibility of timing of application. Early morning and late evening applications are recommended. Applications should also be taken up when susceptible target stages are present in the environment. Similarly frequent applications or integration of the nematodes with other control measures would be needed for pests that are characterized by the presence of feeding stages throughout the year. Host behaviour may also greatly influence efficacy of nematodes. Field concentrations exceeding 2.5 billion nematodes / ha are usually applied to ensure that a sufficient nematode population will come in contact with the target insects / stages. The feeding behaviour and location of a particular insect pest may vary according to the crop, variety and the age. These factors exist among the species and strains of nematodes. Recent work is centered on the control of soil pests like white grubs, weevils, termites, fireants, mole cricket, cutworms, root maggots etc.,

Spraying: Like chemical insecticides, spraying of nematodes directly on to the soil surface is the most commonly used application method. A dosage of about 100, 000 Ijs / plant or 2.5 to 7.5 billion Ijs / ha is sufficient in the fields for pest control results. About 50 Ijs / flying insect are sufficient. However the application technology has to be standardized as per the crop to be sprayed. Baur et al. (1995) studied the factors like relative humidity, leaf type affecting entomophillic nematode infections of *Plutella xylostella* on a leaf surface.

Trickle irrigation: Button type of trickle irrigation outlet is used to deliver the inoculum to the area on one side of the plant .

Capsule: Capsule prepared from wheat bran (5 % w /w) with calcium alginate containing 1000-2000 nematodes / capsule are applied . Capsules are buried in soil or under dairy compost. About 70-80 capsules per plant could be used.

Liquid baits: Desiccated nematodes (*S. feltiae*) are mixed with 56 % sucrose solution (Georgis, 1990).

Pellet baits: Capinera *et al.* (1987) used wheat bait pellets from (1) wheat bran-wheat flour (50 % each); (2) locust bean gum (18mg / ml of water) and (3) corn oil.

Nylon pack cloth bands/terrycloth wrapping around trees: Nematodes can be applied to nylon pack cloth bands lined with fleece or terrycloth that is wrapped around tree trunk to control gypsy moth larvae (*Lymantria dispar*) (Kaya *et al*., 1984). This tactic may be useful when applied with other control measures.

Cardboard band: *S. carpocapsae* incorporated into cardboard band placed around the trunk of apple trees as an artificial bark substrate, infected 23-73 % of codling moth prepupae that moved under the barks (Begley, 1990).

Punch and syringe method: this method is used in case of forest trees. Inoculation hole is made by hammer and delivery of about 1 ml of nematode containing medium (aerated 12 % gelatin with 400 nematodes /ml) is done by syringe.

Guidelines for Field Application of EPN

Soil temperature at the time of application should range from 18-35°C. Nematodes should be applied to irrigated field and then should be again irrigated to maintain sufficient moisture. Application should be done preferably during evening to avoid UV radiation from the sun and high temperature. Sometimes application needs to be repeated for insects like white grubs. About 4week duration of trial is needed normally. For lepidopterous insects, 3 to 7 days is sufficient. Flood jet nozzle is fairly effective for spray, as it sprays more number of nematodes per unit area. Lello et al., 1996, worked on application technology of entomopathogenic nematode against foliar pests and assessed adjuvant toxicity, nozzle type, analyzed droplet spectra with *S. carpocapsae* infective juveniles against *Plutella xylostella.*

EPNs are being widely used in the developed countries for inundative biological control of insect pests (Bedding et al. 1993; Kaya and Gaugler, 1993), wherein the nematodes are applied in large numbers. Their use in augmentative / inoculative biological control have also been exploited, wherein the nematodes are added to the pest's environment in such a way that it enhances chances of their establishment. But it requires clear understanding about both, the parasites and environment (Parkman and Smart, 1996).

Classical cases of field efficacy of Entomopathogenic Nematodes (EPN):

Eidt and Dunphy (1991) reported 82% reduction of spruce bud moth *Zeiraphera haracteri* in white spruce populations by *S. carpocapsae* as measured by adult emergence in combination with *Bacillus thuringiensis* formulation and attributed the control to the nematode. Kim et al., 2002 reported 69-97% of field control of white grubs by *Heterodera* and other biocontrol agents. Other classical cases of the use of biocontrol agents are-

(a) Mole crickets, *Scapteriscus* spp. (Orthoptera: Gryllotalpidae): In field release trials in Florida, *Steinernema scapterisci* provided 74.8 % reduction in mole cricket damage at the application rate 2.5×10^9 infective juveniles (Ijs) per ha. The standard chemical , Acephate , at 3.40 a.i./ hectare caused a 68.4 % reduction . Late stage nymphs and adult mole crickets were found to be highly susceptible to the nematodes.

(b) Sugarcane root stalk borer, *Diaprepes abbreviatus* and blue-green weevil, *Pachnaeus litus* (Coleoptera: Curculionidae): Nematodes (*Steinernema carpocapsae, S. glaseri and H. bacteriophora*) are added to irrigation systems, resulting in a 72-98 % reduction of the emergence of *D.abbreviatus* and *P.litus* adults.

(c) In Hawaiian field trials, an average of 87.1 % larval mortality of the Mediterranean fruit fly, *Ceratitis capitata*, was observed following application *of Steinernema carpocapsae* Mexican strain at 5000 IJ / cm^2 of soil.

Presently formulations of entomopathogenic are being used against several insect pests of diverse crops, of which a few have been listed in Table 3.

INDIAN SCENARIO

In India, work on EPNs was first initiated by Rao and Manju Nath (1966) who demonstrated the use of DD-136 strain of *S. carpocapsae* in the control of insect pests of rice, sugarcane and apple. Since then, much applied work had been done on the exotic strains of EPN species (*S. carpocapsae, S.feltiae , S. glaseri* and *Heterorhabditis bacteriophora*) which has been reviewed by Rahman *et al.* 2000. But, the two decades of research on exotic or unidentified strains did not yield very encouraging results, due to the poor adaptability of exotic strains in the local soils. Search for indigenous strains which are adapted to local conditions, was felt extremely necessary. Sivakumar et al. 1989 isolated a strain from Tamil Nadu which was later described as *H. indica* by Poinar et al. 1992. Some more indigenous strains have been identified as *S. feltiae, S. riobrave, S. bicornutum, S. carpocapsae and H. indica* (Hussaini et al. 2000). Another indigenous strain (IARI-EPN1) was isolated from New Delhi which was described as *S. thermophilum* by Ganguly and Singh,2000, and it was the first new species of this genus from India. The species was so named due to its capability to infect and develop at a wide range of temperature (25-35°C). Due to this, it may prove to be a useful bioagent in tropical and subtropical parts of the world (Ganguly and Singh,2001). It has been found to have wide host range and can infect the host located at the soil surface as well as at a depth of 10 cms. Recently, experiments have been initiated to check the field efficacy of the *S. thermophilum* against diamond back moth and encouraging results were obtained. There is now urgent need to develop the mass culturing and formulation and application technology for this species in order to exploit its biocontrol potential by incorporation in the IPM schedules. There is also need to identify the other indigenous strains, identify them and characterize them ecologically before testing their field efficacies.

Table 3. Major target pests and commercial formulations of entomopathogenic nematodes (Modified after Grewal, 2002).

Crop	Insect pest	Nematode species	Formulation
Artichoke	Artichoke plume moth (*Sphenophorus parvulus*)	*Steinernema carpocapsae*	WG, WP,
Berries	Root weevil (*Diaprepes abbreviatus*)	*Heterorhabditis bacteriophora*	Sponge
Citrus	Root weevil (*Diaprepes abbreviatus*)	*S. riobrave*	WG and LC
Craneberry	Root weevil (*Diaprepes abbreviatus*)	*H. bacteriophora*	Sponge
	Craneberry girdler (*Chrysoteuchia topiaria*)	*S. carpocapsae*	WG, WP
Mushrooms	Sciarid flies (*Lycoriella* spp)	*S. feltiae*	Vermiculite
Ornamentals	Root weevils (*Diaprepes abbreviatus*)	*H. bacteriophora*	Sponge
		H. megidis	W P
	Wood borers (*Holcocercus insularis*)	*S. carpocapsae*	WG,WP
		H. bacteriophora	Sponge
	Fungus gnats (*Bradysia* spp.)	*S. feltiae*	Vermiculite
Turf grass	Scarabs (*Cyclocephala hirta*)	*H. bacteriophora*	Sponge
	Mole crickets (*Scapteriscus vicinus*)	*S. riobrave*	WG
		S. scapterisci	W P
		H. bactriophora	Sponge
	Billbugs (*Sphenophorus venatus*)	*S. carpocapsae*	WG,WP
	Army worm (*Spodoptera frugiperda*)	*S. carpocapsae*	WG,WP
	Cut worm (*Agrotis ipsilon*	*S. carpocapsae*	WG,WP
	Web worm (*Cephalcia abietis*)	*S. carpocapsae*	WG,WP

WG = Water dispersible granule; WP = Wettable powder; LC= Liquid concentrate

CONCLUSION

The steinernematids and heterorhabditids have been proved to be highly effective bioagents for managing the insect pests under field conditions. The future of nematode-based products for insect control is very promising. Their efficacy against the insect pests especially in soil and cryptic habitats is much better than the foliar habitats. Foliar habitat is hostile to the nematodes because of unfavourable environmental conditions including ultra-violet radiation, desiccation and high temperatures. The technology used currently for mass production, product formulation, storage and transportation of nematode products was developed hardly over the last 15 years, even though some of the technology is more than 60 years old, speaks of the volumes of lack of concerted efforts in this direction. Since the first commercial products were developed, vast technological improvements have been made. Future improvements may well make today's technology obsolete. More efficient methods of production, formulation etc will lower the cost of nematode products and make them more competitive economically. Even though a total of 45 species of the two genera of entomopathogenic nematode have been described, only six have been commercialized. *S.carpocapsae, S.feltiae, S. riobrave, S. scapterisci, H. bacteriophora* and *H. megidis*. These species, and perhaps some currently described, will add to the arsenal of nematode weapons aimed at pest insects. "Brighteners" which protect nematodes from harmful ultra-violet radiation (Nickle and Shapiro, 1994) and anti-desiccants may well be combined in the future to formulate nematode products to be used as sprays to control foliar insects. These techniques could well become fairly common in future. When one considers that the sale of commercial products, barring a few small producers for local markets, is no more than 15 years old, the potential market has not begun to be realized. While *S. carpocapsae* was the first nematode product marketed, *S. scapterisci* became available commercially only in 1993, and *S. riobrave* in 1994. Currently, the share of the pest control market captured by entomopathogenic nematode is very low. The share likely will remain relatively low for the foreseeable future, but will increase due to futuristic production methods and the demands of the public for safer, more environmentally acceptable products. As the user fully realizes that nematodes are biological organisms and must be handled as such to provide effective control, greater acceptance of nematode based products will occur. For control of certain insects, nematodes may replace or reduce the use of chemical pesticides and their scope in pest control market is definitely bright.

Keeping in view the diverse agro-climatic conditions and the large number of insect pests in this country, there is urgent need to explore, identify and characterize the indigenous entomopathogenic nematode biodiversity, before testing their field efficacy. Almost two and a half decades of EPN research (1966-1990) in India on either exotic or unidentified strains has gone futile as it did not yield very encouraging results in the field trials due to their poor adaptablity in local soils. The intensive and extensive basic and applied research on EPNs will definitely enable us to exploit their full biocontrol potential as one of the pest control strategies of integrated pest management schedules. Furthermore, identification of insect toxins and toxic genes in their symbiotic bacteria may prove to be of immense value in developing the biopesticides as well as insect resistant transgenic plants.

REFERENCES

Adams, B. J. and Nguyen, K.B. 2002. Taxonomy and Systematics. In: *Entomopathogenic Nematology* (Ed. R. Gaugler) CABI Publishing, Wallingford, Oxon, U.K., 1-34 pp.

Akhurst, R.J. 1980. Morphological and functional dimorphism in *Xenorhabdus* spp., bacteria symbiotically associated with the insect pathogenic nematodes *Neoaplectana* and *Heterorhabditis. Journal of General Microbiology* 121: 303-309.

Bauer, M.E., Kaya, H.K., Gaugler, R. and Tabashnik, B. 1997. Effects of adjuvants on entomopathogenic nematode persistence and efficacy against *Plutella xylostella. Biocontrol Science and Technology*. 7: 513-525.

Baur, M.E., Kaya, H.K. &Thurston, G.S. 1995. Factor effecting entomopathogenic nematode infection of *Plutella xylostella* on a leaf surface. *Entomologia Experimentalis et Applicata.* 77: 239-250.

Bedding , R.A., Akhurst, R and Kaya, H. 1993. "Nematodes for the biological control of insects". CSIRO, East Melbourne, Australia.

Bedding, R.A. 1976. New methods increase the feasibility of using *Neoaplectana* spp. (Nematoda) for the control of insect pests . *Proc. 1st. Int. Coll. Invertebr. Pathol., Kingston.* 250.

Bedding, R.A. 1984. Large scale production, storage and transport of the insect-parasitic nematodes, *Neoaplectana* spp. And *Heterorhabditis* spp. *Ann. Appl. Biol.* 104: 117 – 120.

Bedding, R.A. 1981. Low cost in vitro mass production of *Neoaplectana* and *Heterorhabditis* spp. (Nematode) for field control of insect pests. *Nematologica* 27:

109-114.

Begley, J.W. 1990. Efficacy against insects in habitats other than soil. In: *Entomopathogenic Nematodes in Biological Control* (Eds. R. Gaugler and H.K. Kaya) CRC Press, Boca Raton, Florida. Pp. 215-231.

Boemare, N. 2002. Biology, taxonomy and systematics of *Photorhabdus* and *Xenorhabdus.* In: Entomopathogenic Nematology (Ed. R. Gaugler). *CABI Publishing, Wallingford, Oxon, U.K.,* 35-56 pp.

Boemare, N.E., Akhurst, R.J. and Mourant, R.G. 1993. DNA relatedness between *Xenorhabdus* spp., (Enterobacteriaceae), symbiotic bacteria of entomopathogenic nematodes, and a proposal to transfer *Xenorhabdus luminescence* to a new genus, *Photorhabdus* gen.nov, *Int. J. Syst. Bacterial.* 43: 249-255.

Chavarria-Hernandez, N. and Torre, M, de-la. 2001. Population growth kinetics of the nematode, *Steinernema feltiae*, in submerged monoxenic cultures. *Biotechnol-lett.* 23 (4): 311-315.

Connick, W. J. Jr., Nickle, W. R. and Vinyard, B. T. 1993. "Pesta", a new granular formulation for *Steinernema carpocapsae. Journal of Nematology.* 25 20: 198-203.

Das, J.N. and Divakar, B.J. 1987. Compatibility of certain pesticides with DD-136 nematode. *Plant Protec. Bull.* 39: 20-22.

Dutky, S.R., Thompson, J.V. and Cantwell, G.E. 1964. A technique for the mass propagation of the DD-136 nematode. *Journal of Insect Pathology.* 6: 417-422.

Ehlers, R. U. 2001. Mass production of entomopathogenic nematodes for plant protection. *Applied Microbiology and Biotechnology.* 56 (5-6), 623-633.

Ehlers, R.U., Lunau, S., Krasomil-Osterfeld,K. and Osterfeld, K.H. 1998. Liquid culture of the entomopathogenic nematode-bacterium-complex *Heterorhabditis megidis / Photorhabdus luminescens*. *BioControl.* 43 1): 77 – 86.

Eidt, D.c. and Dunphy, G.B. 1991. Control of spruce bud moth, *Zeiraphera candensis* Mut. and free., in white spruce plantations with entomopathogenic nematodes, *Steinernema* spp. *Canadian Entomologist.* 123 2): 379-385.

Friedman, M.J. 1990. Commercial production and development. In: Entomopathogenic nematodes in biological control (Eds. R.Gaugler and H.K. Kaya) Boca. Raton, FL: CRC Press. Pp. 153-172

Friedman, M.J., Langston, S.L. and Pollitt, S. 1989. Mass production in liquid culture of insect-killing nematodes. *International Patent No. WO89/04602.*

Friedman, M.J., Langstron, S.E. and Pollitt, S. 1991. Mass production in liquid culture of insect killing nematodes. US patent No.5,023.183.

Ganguly, S. 2003. Taxonomy of Entomopathogenic nematodes and work done at IARI, New Delhi. In: *Current Status of Research on Entomopathogenic nematodes in India* (Eds. S.S. Hussaini, R.J. Rabindra and M. Nagesh) PDBC, Bangalore, ICAR publication, 69-108 pp.

Ganguly, S. and Singh, L.K. 2001. Optimum thermal requirements for infectivity and development of an indigenous entomopathogenic nematode, *Steinernema thermophilum* Ganguly and Singh, 2002. *Indian J. Nematol.* 31 2): 148-152.

Ganguly, S. and Singh, L.K. 2000. *Steinernema thermophilum* sp.n (Rhabditida: Steinernematidae)from India. *Int. J. Nematol* 10 2): pp. 183-191

Gaugler, R. and Kaya, H.K. 1990. Entomopathogenic nematodes in Biological Control. CRC Press, Boca Raton, Florida.

Gaugler, R. 2002. Entomopathogenic Nematology. *CABI Publishing, Wallingford, Oxon UK.* 388 pp.

Gaugler, R. and Han, R. 2002. Production technology. In: Entomopathogenic Nematology (Ed. R. Gaugler. *CABI Publishing, Wallingford, Oxon, UK.* 289-310 pp.

Georgis, R. 1990. Formulation and application technology. In: Entomopathogenic nematodes in biological control.(Eds. R. Gaugler and H.K Kaya), Boca Raton, Florida, CRC Press. Pp. 173-191.

Georgis, R. and Gaugler, R. 1990. Predictability in biological control using entomopathogenic nematodes. *J. Econ. Entomol.* 84: 713-720

Georgis, R., Dunlop, D.B. and Grewal, P.S. 1995. Formulation of entomopathogenicc nematodes. In: Hall, F.R. and Barry, J.W. (Eds). *Biorational Pest Control Agents: Formulation and Delivery.* American Chemical Society, Bethesda, Maryland. Pp. 197-205.

Glaser, R.W. 1940. Continued culture of a nematode parasitic in the Japanese beetle. *J. Exp. Zool.,* 84, 1.

Gregoris, R., Dunlop, D. B. and Grewal, P. S. 1995. Formulation of entomopathogenic nematodes in Biorational pest control agents; formulation and density. *American Chemical Society.* Pp. 197-205.

Grewal, P.S. 2002. Formulation and application technology. In: Entomopathogenic Nematology (Ed. R. Gaugler). *CABI Publishing, Wallingford, Oxon, U.K.,* 265-288 pp.

Hara, A.H., Lindegren, J.E. and Kaya, H.K. 1981. Monoxenic mass production of the entomopathogenic nematode, *Neoplectana carpocapsae* Weiser, on dog food / agar medium. USDA, Science and Education Administration, Advances in Agricultural Technology. ATT-W-16 / June,1981; page 1 – 8.

Hominick, W.M. and Briscoe, B.R. 1990. Occurrence of entomopathogenic nematodes (Rhabditida: Steinernematidae and Heterorhabditidae) in British soils. *Parasitology* 100: 295-302.

House, H.L., Welch, H.E. and Cleugh, T.R. 1965. Food medium of prepared dog biscuit for the mass-production of the nematode DD-136 (Nematoda: Steinernematidae). Nature 206: 847

Hussaini, S.S., Singh, S.P., Parthasarathy, R. and Shakeela, V. 2000. Virulence of native entomopathogenic nematodes against the black cut worms, *Agrotis ipsilon* (Hufnagel) and *Agrotis segetum* (Noctuidae: Lepidoptera). *Indian J. Nematol.* 30 1): 103-105.

Ishibashi, N. 1993. Integrated control of insect pests by *Steinernema carpocapsae.* In: Bedding, R., Akhurst, R. and Kaya, H.K. (eds). *Nematodes and Biological Control of Insects.* CSIRO, East Melbourne, Australia, pp. 105-113.

Kaya, H.K. and Gaugler, R. 1993. Entomopathogenic nematodes. *Ann. Rev. Entomol.* 38: 181 – 206.

Kaya, H.K. 1993. Entomogenous and entomopathogenic nematodes. In 'Plant parasitic nematodes in temperate agriculture' Eds. K. Evans, D.L.Trudgill and J.M. Webster, CAB Int. 648pp; 565 – 591.

Kaya, H.K. 1990. Soil ecology. In: *Entomopathogenic Nematodes in Biological Control* (*Eds.* Gaugler, R. and Kaya, H.K.). CRC Press, Boca Raton, Florida. Pp. 93-115.

Kaya, H.K. and Gaugler, R.1993. Entomopathogenic nematodes. *Annu.Rev. Entomol.* 38: 181-206

Kaya, H.K. and Stock, S.P. 1997. Techniques in insect nematology. In: Manual of techniques in insect pathology (ed. Lawrence Lacey) Biological Technique Series, Academic Press Ltd. Pp. 281-324.

Kim, H.H., Lee, S.M. and Choo, T.M. 2002. Entomopathogenic nematodes (*Steinernema* spp. And *Heterorhabditis bacteriophora*) and a fungus *Beauveria brongniartii* for biological control of white grubs, *Ectinohoplia rutipes* and *Exomala orientalis,* in Korean golf courses. *Biocontrol.* 47 2): 177-192.

Koppenhofer, A.M. and Kaya, H.K. 1997. Additive and synergistic interactions between entomopathogenic nematodes and *Bacillus thuriengiensis* for scarab grub control. *Biological Control.* 8: 131-137.

Koppenhofer, A.M. and Kaya, H.K. 1998. Synergism of imidacloprid and entomopathogenic nematodes: a novel approach to white grub (Coleoptera: Scarabaeidae) control in turf grass. *J. Econ. Entomol.* 91: 618-623.

Lello, G.R., Patel, M.N., Mathews, G.A. and Wright, D.J. 1996. Application technology for entomopathogenic nematodes against foliar pests. *Crop Protection* 15 (60): 567-574.

Neves, J.M., Teixeira, J.A., Simoes, N. and Mota, M. 2001. Effect of airflow rate on yields of *Steinernema carpocapsae* Az-20 in liquid culture in an external loop airlift bioreactor. *Biotechnol. Bioeng,* 72 (3): 369-373.

Nickle, W.R. and Shapiro, M. 1994. Effects of eight brighteners as solar radiation protectants for *Steinernema carpocapsae,* All strain. *Supplement to the Journal of Nematology.* 26: 786-794.

Nishimatsu, T. and Jackson, J.J. 1998. Interaction of insecticides, entomopathogenic nematodes, and larvae of the western corn rootworm (Coleoptera: Chrysomelidae). *Journal of Economic Entomology.* 91: 410-418.

Parkman, J.P. and Smart, G.C. Jr. 1996. Entomopathogenic nematodes, a case study: Introdcuction of *Steinernema scapterisci* in Florida. *Biocontr. Sci. and Technol.* 6: 413-419

Poinar, G.O. Jr. 1979. Nematodes for biological control of insects. CRC Press, Boca Raton, Florida.

Poinar, G.O. Jr., Karunakar, G.K. and David, H 1992. *Heterorhabditis indicus* n.sp. (Rhabditida:Nematoda)from India: separation of *Heterorhabditis* spp. by infective juveniles. *Fund. Appl. Nematol.* 15 (5): 467-472: 4 ref.

Poinar, J.O. Jr. 1990. Taxonomy and biology of Steinernematidae and Heterorhabditidae. In: "Entomopathogenic nematodes in biological control "(R.Gaugler and H.K. Kaya Eds), pp 23-61. CRC Press, Boca Raton, Florida.

Rahman, P.F., Sharma, S.B. and Wightman, J.A. 2000. A review of insect parasitic nematodes research in India: 1927-1997. *Int. J. Pest Managem.* 46: 19-28.

Rao, V.P. and Manju Nath, T.M. 1966. DD-136 nematode that can kill many insect pests. *Indian Farm.* 16: 43-44.

Ratansinghe, G. and Hauge, N.G.M. 1995. The susceptibility of diamond black moth, *Luella xylostylla* (Lepidoptera: Ponomeutidae) to entomopathogenic nematodes. *Afro Asian Journal of Nematology.* 5 1): 20-23.

Rovesti, L. and Deseo, K. V. 1990. Compatibility of chemical pesticides with the entomopathogenic nematodes, *Steinernema carpocapsae* Weiser and *S. feltiae* Filipjev (Nematoda: Steinernematidae). *Nematologica* 36: 237-245.

Schneider, D. and Grunder, J.M. 1993. Biological control of insects in forestry nurseries with entomopathogenic nematodes. In: Proceedings of European Meeting. *Microbial control of Pests, Zurich* (*Suisse*). Pp. 171.

Smart, G.C. Jr. 1994. Entomopathogenic nematodes for biological control of insects. *J. Nematol.* 28: suppl.

Stark, D. John 1996. Entomopathogenic nematodes (Rhabditida:Steinernematidae) Toxicity of Neem. *J. Econ. Entomol.* 89 1): 68-73.

Thomas, G.M. and Poinar, G.O.Jr. 1979. *Xenorhabdus* gen. nov., a genus of Entomopathogenic, nematophilic bacteria of the family Enterobacteriaceae. *Int. J. Syst. Bact.* 29: 352-360.

Thurston, G.S., Kaya, H.K. and Gaugler, R. 1994. Characterizing the enhanced susceptibility of milky disease-infected scarabaeid grubs to entomopathogenic nematodes. *Biological Control.* 4: 67-73.

Woodring, J.L. and Kaya, H.K. 1988. Steinernematid and heterorhabditid nematodes: a handbook of techniques. Southern Cooperative Series Bull. 331. Arkansas Agric. Exp. Stn. Fayetteville, A.K. 30pp.

Microbes in Weed Control

J.S. BEDI[1] AND BEANT B. SINGH[1]

Weeds are wild plants where they are not wanted. They are plants that are competitive, pernicious and persistent and have always been a menace to mankind. They grow in the agricultural fields at the expense of the cultivated plants and are, hence, costly pests for farmers as well as consumers, through out the world (McWhorter and Chandler, 1982). Therefore, attempts have always been made to control these weeds. Generally, three techniques are used to control weeds – a) Chemical control, b) Removal and disposal with mechanical devise and c) Biological control. Chemical herbicides are an effective immediate solution to most weed problems but they are not the only and necessarily the best solution. Like all other plants, weeds too have natural enemies, which can affect their growth and can be used as agents for their control. Thus biological control of weeds is the use of natural enemies to exert pressure on the population of their host plant to reduce it to below levels of economic importance. But now the question arises why we should look for alternatives or precisely the bioherbicides to chemical herbicides, which have been and are still used as effective means of weed control.

WHY BIOHERBICIDES

Bioherbicides offer similar advantages as herbicides do like selectivity, reduced erosion risk, easy application on the target, augmentation etc. for the last over 5 decades, chemicals have been in use and are still the biggest challenge to bioherbicide development. But many harmful effects accompany their use, which has necessitated research on bioherbicides. These include:

1. Pollution: one of the major problem with chemicals is the pollution of soil, air as well as ground water
2. Residue: Continuous application of chemicals over the years has percolated as residue through the plants to animals and human beings. This seriousness has now reached people's mind and they are becoming suspicious of the chemical use in agriculture.
3. Drift: it has been reported that most of the chemical applied does not reach the target. Only 5% is absorbed by the weeds. Drift is more in aerial spraying. In UK alone, it is estimated that 72 million liters of spray liquid go as long distance drift.
4. Recurrent weed problem: in developed as well as developing countries like India, a large area is sprayed with herbicides. But weds resurface every year. Continued use of chemicals has not diminished the problem. Farmers, too, look for a persistent alternative, which is sustainable.
5. Resistance: Chemical use has lead to resistance and cross resistance. Presently, more than 60 weed species have developed herbicide resistance.

1. Department of Plant Pathology, Punjab Agricultural University, Ludhiana 141 004, India.

6. Cost involved: the above factors also increase the cost involved in weed control, which the farmer has to bear year after year. He needs a solution which can last, at least, for a few years
7. Awareness: now people are becoming more and more aware to the problems associated with chemicals. Organic culture is spreading fast in western countries.
8. Public pressure: Up to 1980, no cases of herbicide poisoning of humans were known or reported. But now, the press is playing an important role in making people aware of the dangers of chemicals

Biological control has been used successfully throughout the world for about one century and is widely accepted as an economical, effective and environmentally safe method for controlling many weeds. Public interest in as alternatives to chemical weed control has steadily increased during recent years and as a consequence, the number of biological weed control projects initiated has increased during the last decade.

Table 1: Status of projects on weed control with micro-organisms

Approach	Total projects	No. of countries involved
Classical	32	8
Bioherbicidal	43	18

Unlike chemical compounds, biological control agents can be specific to the weed and do not cause any residue problems or environment pollution. The diversity of natural enemies may include vertebrates, antropods, herbivores, pathogens etc. Entomologists, who have introduced phytophagous insects from the native region of the weed to another geographical area where the plant had become a problem, have carried out most of the programs on biocontrol of weeds. However, some workers were aware of the role, diseases could play in weed control (King, 1966). Use of plant pathogens is a relatively new growing strategy in the management of weeds (McFadyan, 1998). A classical example is that of crofton weed (*Eupatorium adenophorum*) whose spread was restricted due to introduced gall forming lepidopteran fly (*Cactoblastis cactorum*), whose task was completed by a leaf spot fungus *Cercospora eupatorii,* which was accidentally introduced into Australia on the body the fly. Last two decades have witnessed and increasing interest in the deliberate use of pathogens as biocontrol agents. These can be viruses, bacteria or fungi. These pathogens are being used as sole agents or as part of a complex with other biocontrol organisms. They hold an important position among different control strategies because of their specificity, economics of their treatment and provide log and durable weed control. They are harmless to the non-target crops even if they are growing in close vicinity of the target weed. The idea of using plant pathogens in controlling weeds is not new and as previously occurred to scientists as well as farmers. Wilson (1969) cites a farmer who wrote to Dr. Halsted at the New Jesey Agricultural Experiment station in 1983 that his fields were attacked by Canadian Thistle *Cirsium* specie but in two years, were destroyed by a rust before reaching full bloom. Hardly any thistle was reported the following years.

There are two main situations where plant pathogens can be used as weed biocontrol agents. In the first case, where natural controls are absent e.g. when a plant is introduced from a different geographical region without its natural enemies, the weed can become so abundant that it becomes a serious problem The opportunity exists to redress the situation by introducing natural enemies for pathogens to restore the appropriate equilibrium, with a reduced plant population as the expected result. This is called the classical tactic. In this case, a search is made for effective plant pathogens in the native areas of the weed and if found safe, they are introduced in the region of weed introduction. Alternatively, if the herbivore/ pathogen is already present, or if introduced, but the resulting equilibrium still gives a higher plant population than desired, there is the possibility of displacing the equilibrium to the detriment of the plant e.g. by increasing the inoculums of the pathogen. Efforts are made to manipulate the infection build up to an extent that could cause marked damage to the weed. This is known as inundative or bioherbicide approach.

CLASSICAL STRATEGY

In the classical strategy, the ideal target is an aggressive, widespread, introduced perennial wed infesting large areas such a s rangeland and aquatic habitats. Traditionally, this strategy of weed control had been practiced by introducing exotic arthropods from the original habitat of the weed or elsewhere to control naturalized weeds commonly occupying wide areas of undisturbed land (Huffaker, 1970). In principle, an introduction on one occasion is adequate to allow the build up of a sufficiently large population to have a long term impact on the weed density. This strategy has more recently been applied to annual weeds (Andres *et al.*, 1976;) and extended mostly by the use of plant pathogens (Hasan, 1980, 1983, Hasan and Ayres, 1990; Hasan and Aracil, 1991; Hasan *et al.*, 1992).

BIOHERBICIDE STRATEGY

The mycoherbicide or inundative strategy has its primary application in cultivated crops. Now, the contemporary biological control approach is expanding and a new strategy has been developed using virulent strains of pathogens already occurring on the weed and enhancing their destructive action. This is called the bioherbicide tactic or inundative approach. A mycoherbicide is a preparation of living inoculum of fungal plant pathogens, often formulated, and applied in the same way as a "chemical" herbicide to control or suppress the growth of a weed species. Many of the host, pathogen, and environmental constraints on disease development are overcome by periodically dispersing an abundant supply of virulent inoculums uniformly onto the target weed population. Applications are timed to take advantage of favorable environmental conditions and the most susceptible stage of plant growth. Several mycoherbicides are being developed for use on weed seedlings with an aim to kill the weed or stunt it so that competition with infested crop plants is reduced or eliminated.

In this strategy, both endemic and exotic pathogens can be used as mycoherbicides though to date mostly indigenous fungi have been so employed (Hasan, 1983). Both alien and indigenous weeds may be controlled by mycoherbicides. Weed pathogens used as bioherbicides have the greatest potential for use on hand to control weeds in annual crops where specificity, immediacy and completeness of control are paramount. bioherbicides are also being developed for use against perennial weeds in orchard crops. They also have the potential for control of range and aquatic weeds. According to Templeton *et al.* (1979), the reasons for the lack of the effect of such endemic pathogens on indigenous weeds under normal conditions are multiple, including cultivation and the use of selective pesticides or management practices that disrupt or eliminate stages of the pathogenesis life cycle. In order to overcome these constraints, pathogens collected from undisturbed areas and multiplied in vitro. The inoculum containing spores and mycelium is applied so as to inundate the host plant with inoculum at the most appropriate susceptibility time for infection in order to eliminate the weed in its early stages of growth.

PROCEDURES

As the success of control depends on the ability of the pathogen to self perpetuate, disperse and reduce the host population, special care is taken to select the agents, which are strongly pathogenic to the plant and are ecoclimatically suitable for the target area. Biological control, more than any other method of weed control, has to consider the environment of the weed. Information on the climatic and geographical conditions existing in the area where a weed occurs and the crops in which it poses problems has to be gathered and borne in mind while searching for effective biocontrol agents. These agents may be indigenous or imported from the native area of the target weed.

Hasan described six steps in the development of a programme of biological wed control with insects. These are 1. Determine the suitability of a weed for biological control 2. Survey of plant's natural enemies

3. Study and evaluate ecology of several natural enemies 4. Determine their host specificity 5. Follow their introduction and establishment 6. Evaluate their effect. These steps were modified by Hasan (1980) to make their applicable to plant pathogens as biological control agents. Once, a weed has been identified as suitable for biological control, basic considerations in the selection of plant pathogens are:

i) Search for the pathogen:

All types of pathogens whether viruses, bacterial or fungi, which can cause reduction in plant growth, can be considered biological control agents. But so far, fungi have received maximum attention probably because they are easy to handle and identify than viruses or bacteria. Secondly their taxonomic position is well defined and in some cases, even the host range is known. Further, they do not depend on insects or other vertebrates for their dispersal or transmission.

Before making an extensive survey of the plant pathogens in the natural populations of a weed, a literature survey is necessary to collect all possible information on the occurrence, geographical distribution and diseases of the weed. It should be determined which organisms are already present and the extent to which they cause damage to the target weed in the area where weed control is desired. This information serves as a basis for selecting areas to be surveyed where it is an introduced species or within the same region where weed is a problem. Preliminary surveys should target all pathogens on the weed. Plant pathogens with wide spread occurrence and high virulence should be further evaluated for their potential as weed biocontrol agents. Both, obligate and facultative pathogens should be included in this screening. Some obligate pathogens like rusts are often highly damaging as well as host specific, one of the most important criteria for selecting a potential pathogen. On the other hand, facultative pathogens have the advantage that they can be cultured on artificial media and are ideal when mass culture of the organism is required for creating the epidemic.

ii) Ecological studies

The efficacy of a plant pathogen for the control of weed can be assessed in its original habitat by ecological observations to evaluate changes in the weed population due to infection. Weed density is measured at different intervals over the year to find out whether there is any relationship between infection levels and reduction in weed population. Hasan and Wapshere (1973) successfully demonstrated this phenomenon. Geographical factors and climate including that of acrosphere and rhizosphere affect the virulence, disease intensity, infectivity, sporulation and capacity of the pathogen to spread from one plant host to another. Observations should be made on the pathogenic behavior and virulence of the orgànism in the native areas and same behavior could be expected after introduction.. It has advantage that the weed population could be highly susceptible if it had never before been exposed to the pathogen.

iii) Determination of safety

Use of plant pathogens, whether exotic or indigenous, in the biological control of weeds is only possible when there is sufficient evidence that they will not be harmful to useful plants. This information can be generated from glass house tests but field observations can also confirm the infectivity range. The safety of a pathogen can be demonstrated by the following steps:

a) Taxonomic position

As soon as a pathogen has been found on weed, its taxonomic position must be determined. A literature scan will show on which plants, the pathogen has been reported. A higher degree of reliability can be assumed for the host specificity of a pathogen, which belongs to a group of fungi known for their host specialization e.g. Rust fungi. Further among these, monoecious rusts are more suitable than heteroecious rusts e.g. *Rumex* rust is a heteroecious rust on *Rumex crispus* but its alternate host *Ranunculus ficaria* could not be infected artificially (Inmon, 1991) whereas *Puccinia chondrillina*, a monoecious rust, did not present any such problems *vis a vis chondrillina*. Some other fungi, which are facultative pathogens also often, have specialized host-restricted forms like *Fusarium, Colletotrichum* and *Cercospora.*

b) Host specificity

This program of testing the pathogen against a selected group of wild and cultivated plants is usually undertaken in vitro. Most of the evidence can be obtained by glass house tests for host specificity but field observations can also provide supporting evidence. Firstly a suitable method for inoculation is developed and optimum conditions for disease development are determined. Through this technique a number of test host plants phyllogenetically related to the weed are exposed to the pathogen and plants scored for the extent of infection caused to them by the pathogen. However, it is impossible to test each and every plant species therefore methods have been developed to select representative species (Wapshere 1975). The list of plants should include (i) other forms of target species (2) other species of the same genus (3) other members of the tube (4) other members of the sub family (5) other members of the family and (6) other members of the order. This is also called centrifugal phyllogenetic method. However, Hasan (1980), suggested a three tier system which includes (1) Plants related to the weed species (2) Plant never exposed to the pathogen of disease but likely to be exposed. Once the pathogen has been introduced in its new environment (3) All crop plants in the native land or in areas where a pathogen is going to be introduced attacked by any other fungus species related to the pathogen i.e. of the same genus. Plants already present in the native region, exposed but not infected by the pathogen, are excluded.

iv) Release and establishment

It is important that once the pathogen has been released, care is taken that it establishes properly. Climate exerts definite impact on disease development; foci of introduction therefore have to be chosen in climatic areas favorable to the disease. Field inoculations should be done when the weather is suitable and at appropriates time of growth.

v) Assessment of Efficacy

a) Evaluation studies

After the pathogen release, the evaluation studies start and continue until it is clear whether or not the agent has been successful sometimes in the initial stages the infection is restricted and hence inoculum potential has to be supplemented by further releases of the pathogen.

b) Efficacy

Efficacy of the pathogen for weed control is defined as the ability to provide a satisfactory amount, speed and ease of weed control (Charudattan, 1988). Thus, the level or amount of weed control should be very high, control should be rapid circumscribed by the crop phenology and it should be easy to use and obtain satisfactory performance of the bioagent.

c) Parameters

The severity (amount per unit area of the plant), incidence and chronicity (severity and persistency) of the disease can be used to quantify its destructiveness and thereby derive an indirect measure of the efficacy of the pathogen. A number of weed control parameters can be chosen to assess the biocontrol agent efficacy (Charudattan, 1990). These include biomass, competitive ability, growth rate, specific leaf area, plant height, reproduction, survival and weed kill. Amongst these, assessments based on weed kill which can be quantified as percent kill or reduction in weed population, counts. Although assessing the efficacy of a pathogen capable of killing the weed is fairly straight forward, there are cases where complete kill is not achievable. In these pathosystems, disease may just reduce growth and/or reproduction of the weed, resulting in reductions in biomass competitiveness, flowering, seed set, seed bank etc. Studies of this type which measure the subtle non-lethal effects of diseases on natural plant population are likely to be of considerable importance in the future in suggesting innovative means of deploying pathogens against weeds (Charudattan 1988)

d) Scheme for rating efficacy

Although minor modifications and refinements may be needed for each pathosystem, a commonly acceptable scheme is needed to assure proper selection, evaluation and reporting of efficacy. Charudattan (1990) proposed a scheme for ranking candidates on the basis of efficacy (Table 1). The scheme when applied to different mycoherbicide agents provided different efficacy scores: Collego, 46; De Vine. 46; Aecass, 42; Cobecro, 42; Fuducro, 38; Pinigri, 31; and Cerodman, 30. He suggested that a score of 40-50 would represent a highly desirable level of efficacy in a candidate and or range of 30-39 useful if better candidates are not available.

Inundative v/s. classical Biological control

Choosing an approach for the control of a weed, depends on several factors like: whether the weed is native or introduced, the cropping system where it is to be applied, the amount of control desired, economics of the treatment or cost benefit analysis and who would be the beneficiaries (Tisdell, 1989). The classical biological control has the following disadvantages over inundative approach:

1. *Risk* – The biocontrol agent may attack non-target species as testing of all flora is not possible
2. *Low reversibility* – Once the pathogen is released and established, it is difficult to eradicate it if it negative consequences.
3. *Impact* – The impact of the biocontrol agent is beyond the control of the farmers.

The classical approach has some advantages too.

1. *Benefits* – Occur in perpetuity or at lease for long periods.
2. Extension or advisory costs are minimal. They are only in convincing the community or related bodies or concerned authorities.
3. Basically only an initial cost or outlay is required and no recurring expenses are involved.
4. *Automatic density adjustment* – The biological density tends to adjust automatically to the density of the weed host. An automechanism is involved which searches out the host.
5. *Effective at low densities* – A classical bioagent may extent control even at low densities of a weed where mechanical or inundative control would be uneconomical
6. *Benefits all* – All farmers, rich or poor are benefited.

Inundative biological control has the following disadvantages over classical approaches:

1. Limited effects; The biocontrol effects are usually limited to the areas in which it is applied and it does not search out its host on its own but requires human intervention.
2. No density adjustment – Inundative biocontrol agent does not adjust its density to the population of the host.
3. Limited effects – The effects are generally limited in time as well as in area. It thus requires continuous supplementation at an additional cost.
4. Uneconomical at low densities – They are not economical when the weed density is low or low value crop where weed control is required.
5. Higher extension and advertising costs are involved in bioherbicides.

Table 2: A scheme for rating efficacy* of mycoherbicide candidates.

Aspect of Efficacy	Assessment categories and sub categories	Importance factor	Rating scale
Amount	A.a. able to kill the weed	2 or 1	5
	A.a.1 Able to kill only seedlings	2	4
	A.a.1a Capable of >85% kill		3
	A.a.1.b. Capable of 50-85% kill		
	A.a.1.c Capable <50% kill		
	A.a.2 Able to kill older plants		5
	A.a.2.a Capable of >85% kill		4
	A.a.2.b Capable of 50-85% kill		3
	A.a.2.c Capable <50% kill		
			3-5
	A.b. Unable to kill the weed	1	
	A.b.1 Relative damage to the weed is mild to severe by the candidate causes measurable reductions in weed growth, reproduction and/or competitive ability.		1 or 2
	A.b.2 Effects are subliminal but results in stress to the weed or increase in crop yield		3-5
Speed		2 or 1	
	B.a. Able to kill quickly in 2-6 weeks post inoculation	2	3-5
	B.b. Unable to kill quickly; takes >6 weeks to kill	1	2
	B.b.1 Effects of disease fall in subcategories A.a.1 or A.a.2		1
	B.b.2 of disease fall in category A.b.1		
	B.b.3 of disease fall in category A.b.2		
Ease		2 or 1	3-5
	C.a. Candidate can be easily applied as mycoherbicide	2	1 or 2
	C.a.1 Can be applied as post emergence sprays		0 or 1
	C.a.2 Special formulation and application tools needed		
	C.b. Difficult to apply as mycoherbicide	1	5
	D.a. Candidate can be used over a range of environmental, host and crop conditions	2	3 or 4
	D.b. Candidate can be used only under specific but obtainable conditions	2	1 or 2
	D.c. Candidate needs highly specific conditions that are difficult to obtain	1	
			3-5
	E.a. Mycoherbicide can be easily integrated with on going pest and crop management practices	2	0-2
	E.b. Integration difficult in practice	1	

*Importance factor and Rating scale should be multiplied to derive the total score

The advantages of inundative biocontrol include:

1. Lower risk: This is because of the greater human control over the agent once it is released. It has unwanted side effects, it can be withdrawn or only applied in cases where side effects do not occur or are within limits. However, genetically engineered organisms need to be guarded against.
2. Less social conflict – It may cause less social conflicts where one group favours eradication of the weed and other favours its retention. Its application area is demarkated and whosoever wishes, can apply in his own territory.

CURRENT STATUS OF WEED BIOCONTROL WITH PLANT PATHOGENS

Currently several projects are underway on biocontrol of weeds with plant pathogens. The projects represent a diversity of plant species including algae, aquatic, semiaquatic, terrestrial, annual, perennial, herbaceous and woody plants and parasitic weeds. These weeds are found in different situations ranging from waterways, lakes, range land, pastures, cultivated crops and forests. Ellison and Evans (2001) have reported that till now, 83 projects have been undertaken which involved 71 species of fungi, 3 of bacteria, 3 of nematodes and 6 of viruses.

SOME SUCCESSFUL EXAMPLES

During the last three decades, many projects have been commenced in which plant pathogens alone or in addition to insects have been considered as biological control agents where natural enemies were absent in a weed population, exotic pathogens have been introduced from elsewhere. On the other hand, where potential control agents were already present, efforts have been made to increase their effectiveness by repeated applications of the virulent strains into the weed population. Julien and Griffith (1998) listed most of these pathogens, which have been introduced so far. This list has been updated in the present paper (Table 2). Some of the most successful examples are given here:

CLASSICAL APPROACH

Skeleton weed rust

One of the foremost examples of the use of exotic plant pathogens in the biocontrol of weeds is the successful introduction of *Chondrilla rust (Puccinia chondrillina)* into Australia for the control of skeleton weed, *Chondrilla juncea* the plant is a native of Mediterranean Europe and Middle East where it is not considered a serious weed due to the presence of its natural enemies. But it has now spread to North America and Australia.

After extensive investigations carried out in Europe, a rust, *Puccinia chandrillina,* was found to be most damaging (Hasan, 1983). It is a macrocyclic and autoecious rust, remains active throughout the year in the Mediterranean and infects all aerial parts of *Chandrilla* plants. Heavily infected plants are either killed or severely damaged and seed production and root reserves are reduced. Among the various strains of the rust collected, an Italian strain IT 32 was found to be highly pathogenic. Specificity of this strain was tested by inoculating all important crop pants grown in Australia as well as cultivated and wild composites closely related to *Chondrilla.* All these plants remained uninfected thus demonstrating host specificity of the rust. The rust was first released in the fields in fields in Australia in 1971 where it established itself successfully on narrow leaf form of skeleton weed. Spread of the disease was so rapid that in 12 generations, it was found as far as 320 km from the original site. Within a few months, it was wide spread and steadily reduced populations of this weed and now, in many areas, skeleton weed is not a problem.

Blackberry rust

The wild blackberries *Rubus constrictus* and *R. ulmifolius* are of European origin but have become serious weeds in many parts of the world, which are ecoclimatically similar to their areas of origin. A rust fungus, *Phragmidium violaceum* has been found very serious to the weeds in Europe. Under favorable conditions, it causes defoliation and severe reduction in the vigor of the canes.

P. Violaceum was released in Chile in 1973 for the control of these two blackberries. Within 2 years, the rust became wide spread and controlled the population of *R. constrictus*. No plants other than these two blackberries were found infected. Encouraged by these results, plants have been made to introduce strains of this rust into Australia.

Pamakani Leaf spot

Hamakua pamakani (*Ageratina* (*Eupatorium*) *riparia*), a composite of Maxican origin, was first observed in Hawaii in 1925 and since then, it has spread rapidly and become a serious problem. The Leaf spot fungus *Cercospora ageratina* was introduced to Hawaii from Jamaica and successfully controlled this weed. The fungus has been found to be host specific.

Table 3: Fungal pathogens introduced as classical biocontrol agents of exotic weeds

Fungal Pathogen	Exotic weed	Country of origin	Country of introduction	Control status
Maravalia cryptostegiae	*Crystogia grandiflora*	Madagasker	Australia (1994)	Spreading rapidity
Phaeoramularia eupatoriiodorati	*Ageratina adenophora*	Mexico	S.Africa (1987)	Limited impact
Entyloma ageratinae	*A. riparia*	Jamaica	Hawaii, USA (1975)	Substantial to complete control
			S.Africa (1989)	Established, spreading, severe
			New Zealand (1998)	Established
Puccinia corduorum	*Carduus nutans*	Turkey	USA (1987)	Decreases seed production
P. cardui-pycnocaphali	*C.pycnocephalus*	Italy	Australia (1993)	Impact unknown
Puccinia chondrillina	*Chondrilla juncea*	Italy/Turkey	Australia(1971,1996)	Extremely effective
			USA (1996)	Highly effective
			Argentina (1982)	Low impact
Puccinia abrupta var. parthenicola &	*Parthenium hysterophorus*	Mexico	Australia (1991)	Established, impact not high
P.melampodii			Australia (1999)	Still in quarantine
Puccinia xanthii	*Xanthium strumarium*	USA	Australia (1974)	Very effective in Queensland
Uromyces heliiotropaaa	*Heliotropium europaeum*	Turkey	Australia (1991)	Spreading, impact unknown
Uromyces galegae	*Galega officinalis*	France	Chile (1973)	Established, impact unknown
Colletotrichum gloeosporioides clidemiae	*Clidemia hirta*	Panama	Hawaii, USA (1986)	Established, die-back reported
Uromycladium tepperianum	*Acacia saligna*	Australia	S. Africa (1987)	Highly effective
Diabole cubensis	*Mimosa pigra*	Mexico	Austalia (1996)	Recent establishment
Sphaerulina mimosaepigrae (Phloeospora mimosaepigrae)		Mexico	Austalia (1994)	Established and spreading
Septoria pessiflorae	*Passiflora tripartita*	Colombia	Hawaii,USA (1996)	Established and spreading
Cercospora rodmanii	*Eichhornia crassipes*	USA	S.Africa (1991)	Established, being assessed
Phoma climatidina	*Climatis vitalba*	USA	New Zealand (1996)	Established, spreading rapidly
Phragmidium violaceum	*Rubus constrictus & R. ulmifolius*	Germany	Chile (1973)	Weed density decreasing
P. violaceum	*R. fructicosus*	France	Australia (1984,1981)	Causing population decline
Septoria sp	*Lantana camara*	Ecuador	Hawaii, USA (1997)	Status unknown
Puccinia evadens	*Baccharis halmifolia*	USA	Australia (1998)	Established
Septoria myricae	*Myrica faya*	USA	Hawaii, USA (1997)	Status unknown
C.gloeosporioides f.sp. miconiae	*Miconia calvescens*	Brazil	Hawaii, USA (1997)	Established, low impact
Sporisorium ophiuri	*Rottboellia cochinchinensis*	Madagascar	Costa Rica (2000 ?)	Status unknown
Maravalia cryptostegiae	*Cryptostegia grandiflora*		Australia (1995)	Established, weed spread checked
Ulex europaeus	*Uromyces pisi* f.sp.*europae*	U.K.	Hawaii, USA (1999)	Established, good impact
Asparagus asparagoides	*Puccinia myrsiphyllii*	S.Africa	Australia (2000)	Established and spreading
Mikania micrantha	*Puccinia spegazzinii*	Trinidad	India (2003)	Permission granted to import

Xanthium rust

Xanthium strumarium (Noogoora burr) and *X. spinosum* (Bathurst burr) are weeds of grasslands in Australia. Its fruits contaminate the fleece, thereby decreasing the value of the clip and increasing cost of wool processing. *Xanthium* rust *Puccinia xanthii*, of North American origin, now occurs widely in Europe.

A virulent strain collected from Po valley of Northern Italy has been found to infect Australian farms of noogoora and bathurst burr. It reduces the dry weight of the plant and also seed viability. It was unintentionally introduced into Australia but now it causes widespread and damaging infections on *X. strumarium* in Queensland.

Release of rust *Uromyces pisi* f.sp. *europali* of European origin in the U.S.A. for the control of gorse (*Ulex europaeus*) (Killgore 1999) and release of *Entyloma ageratinae* white smut in New Zealand in 1998 for the control of mistflower (Frohlich, 1999) have also yielded good results.

Classical biological control is the weed control method able to provide a long term, often permanent solution to a serious weed problem with limited or no further inputs after the agent has become established. Although preparation of benefit/cost ratios are complex tasks and open to criticism, a 200:1 benefit /cost ratio was calculated for the *Chondrilla juncea* project, a 10.5:1 to 48.1 for the *Echium* project and 56:1 ratio for the *Salvinia molesta* project. These values compare favorably to those of 4:1 and 5:1 commonly suggested for herbicides.

Annual weeds odf arable crops have been considered poor targets for classical biological control. However, it should be noted that some annual crop weeds are exotics and classical biological control has been effective against some annuals such as *Trbulus terrestris* and perennials such as *Chondrilla juncea.* The ecology of annual weeds has been noted as a constraint to the establishment and effectiveness of agents. Perhaps this is an area where our present, limited understanding of pathogen-weed dynamics is just sufficient to identify the problems, but where a better understanding may indicate the solution, in terms of what kind of agent we need and how these should be used.

INUNDATIVE OR BIOHERBICIDE APPROACH

Anthracnose of Northern Jointvetch:

Northern jointvetch, *Aeschynomene virginica* is an important leguminous weed in eastern Arkansas, USA, in rice fields, canals waste areas and to a lesser extent in soybean fields. *Colletotrichum gloeosporioides* f. sp. *aeschynomene,* which causes anthracnose and appears at all stages of growth from seedling to fully grown plant, but rarely kills the plant. Field observations indicated that the limited effect was due to low inoculum level in early stages of growth while the disease became epiphytotic in later stages. A water suspension of spores at a concentration of 2-15 x 10^6 spores/ml when sprayed on the weed at the rate of 94 to 374 litres/ha gave weed control in growth chambers, green house and in the field (Templeton *et al.*, 1976). Young seedlings were rapidly killed and 99% control was achieved in the field. Host specificity of the fungus was confirmed by exposing different crop plants and weeds to the inoculum. Now this fungus has been formulated ad commercialized under the trade name of "COLLEGO", as a mycoherbicide against northern jointvetch since 1982 (Templetan, 1984, Johnson, 1998).

Phytophthora rot of milkweed vine

A strain of stem and root rot fungus, *Phytophthora citrophthora* was isolated from drying milkweed vine *Morrenia odorata* a serious weed of citrus in Florida, U.S.A. The pathogen was evaluated as mycoherbicide. When applied to the infested fields as zoospores or chlamydospores, most of the milkweed vines were diseased and eventually

Table 4: Registered and unregistered bioherbicides

Biocontrol agent	Trade Name	Target weed	Company
Phytophthora palmivora	DeVine	Strangler vine	Abbot Labs, Chicago USA
Colletotrichum gloeosporioides f.sp aeschynomene	Collego	Northern jointvetch	Encore Technologies, USA
C.gloeosporioides f.sp. malvae	Biomal	Round leaved mallow	Philom Bios, Saskatoon, Canada
Puccinia canaliculata ATCC 40199	Dr.Biosedge	Yellow nutsedge	Tifton Innovation Corp., USA
C. gloeosporioides	LUBOA II	Dodder	Ningxia Region, China
	Hakahak	Silky hakea	Plant Protection Res Instt. S. Africa
C. coccodes	Velgo	Velvet leaf	–
C. truncatum	COLTRU	Hemp sesbania	–
Alternaria cassia	CASST	Sickle pod	Mycogen Corp. San Diego CA
Pseudomanas syringae pv tagetis	NA	Canada thistle	Encore Technologies
Xanthomonas compestris pv poaea	Camperico	Annual blue grass	Japan Tobacco, Kanagawa
Phomopsis amaranthicola	NA	Pigweed	University of Florida, USA
Chondrostereum purpureum	Biochon	Hardwood weeds	Koppert Biological Systems, The Netherlands
	Chontrol	Red alder and	Mycologic Inc. Canada
	Myco-Tech	Big leaf maple	Myco-Forestic Corp., Canada
Colletotrichum orbiculare	Conidial spray	Bathrust burr	Not registered but used in Australia
Exserhilum monoceras	Conidial spray	Swank (in rice)	To be available in Japan
Colletotrichum gloeosporioides f.sp. veronicae	Conidial spray	*Veronica perica* and *Galium aparine* var *tenerum*	Patent authorized Nanjing Agri. Univ. China
Valdensinia heterodoxa	Conidial spray	Salal	Patent applied, USA and Canada

Anthracnose of round-leaved mallow:

Another fungus *Colletotrichum gloeosporioides* f. sp. *malvae* was found pathogenic and highly virulent on round leaved mallow (*Malva pusilla*) in Canada. After conducting its host specificity and safety tests, it was registered in Canada in 1992 and has been commercialized under the trade name Bio Mal. In 1994, a patent to the use of this product was also given in the USA (Mortensen *et al.* 1994).

In addition, there are many other fungi and bacteria registered/unregistered, which are being used as bioherbicides against different weeds in different countries (Reeder, 1999; Sheng, 2003) (Table 3).

Among other pathogens, which have reached an advanced stage of development and are likely to be used as mycoherbicides are *Cercospora rodmanii* against water hyacinth, *Alternaria macrospora* against spurred anoda (*Anoda cristata*) (Hasan 1988), *Fusarium oxysporum* f. sp. *orthoceras* against sunflower broomrape (*Orobanche cumana*) (Bedi 1994) *Fusarium oxysporum* against *Striga hermonthica* (Abbasher and Sauerborn, 1994).

CONSTRAINTS

Numerous examples of successful use of biocontrol fungi have been earlier discussed where weed control has been achieved to significant level. However, a large number of factors affect the feasibility of any given biological control agent or product (Auld & Morin 1995). Even in the formulation and actively against weed species, a number of challenges and constraints are encountered. And they are more so in case of mycoherbicides.

BIOLOGICAL CONSTRAINTS

i) Host Variability

Weed species can be regarded as population of plant. Epidemic occurs due to plant-to-plant movement of the pathogen propagules. Existence of resistant biotypes anywhere in the host population can break the chain movement. However, direct application of the spores all over the target population overcomes this constraint.

ii) Host range

Host range of the bioherbicides is a complex issue, which impinges on its commercial aspects. Generally, the agents are chosen for their host specificity and ready wide exposure of flora to these pathogens has occurred. Non-target plant safety and possible latent colonization of crop plants by bioherbicides represents a risk that needs to be evaluated. Further, risk is also associated with the natural exchange of genetic material between bioherbicide strains and strains attacking distantly related crop plants.

iii) Resistant Mechanism

A number of barriers to infection on individual plant exist. They could be inherent or induced and include barriers like cuticle, has, antagonistic phyloplane microorganisms, or chemical barriers such as phytoalexins. In such cases, the bioherbicide efficacy gets reduced and to overcome it, higher dose of inoculum is applied. Ability of the host to tolerate low doses of inoculum or even age and phenological stage of plant may also influence the effect of bioherbicides (Makowski 1993).

iv) Interaction with other microorganisms

Interaction of the foliar bioherbicides with microorganisms inhabiting phyllosphere, their competition for nutrients and space, or direct antagonism as well as toxic leachates may reduce the efficacy of the bioherbicides.

ENVIRONMENTAL CONSTRAINTS

i) Aerial environment

The environmental conditions prevailing in the phyllosphere of plants are frequently hostile for the biocontrol agents. The presence of dew for longer duration, which is the requirement for most of the fungal agents, is usually not available (Andrews, 1992). Of course, the timing of application and formulation with water retaining materials has helped to solve this problem to some extent.

ii) Soil environment

Moisture, nutrient status and soil temperature can influence the physiology of the target weed as well as efficacy of the bioherbicidal pathogen particularly when the pathogen is applied to the soil.

TECHNOLOGICAL CONSTRAINTS

i) Mass production

The commercial feasibility to mass produce large number of viable, infective and genetically stable proagules of the bioherbicide can date pathogen, is a major requirement in the development of a bioherbicide. Proper selection of the growth medium is important because it can greatly influence the type, number, stability and pathogenicity of the pathogen propagules (Morin, 2002). The type of fermentation system selected also influences the formulation of this product.

ii) Formulation

Formulation is one of the major technological constraints to the effective biological herbicides. Its main aim is to keep the propagules viable and infective with a shelf life preferably 1-2 years. The most challenging aspect is to overcome the dew and free moisture requirements. Formulation with invert emulsion or encapsulation of propagules in organic polymers like Sodium alginate has somehow reduced the problem. The addition of exogenous nutrients to liquid or granular preparations of potential bioherbicides has shown promise in enhancing the activity of the pathogens.

Commercial Constraints

i) Market Size

Several efficient bioherbicides could not be developed commercially because of their market potential or their targets were quite restricted to only one weed commercially. It is possible to develop them only if the target weed is economically important. The small niche market of most potential bioherbicides has deterred industry involvement in their development due to limited opportunities of recovering cost involved and large scale production of the product (Heinz and Templeton 1993). Bioherbicides targeting small markets could be successfully developed by small companies which have low overhead costs or public sector could be associated for their commercialisation as competition from chemical herbicides represents a major hurdle in their commercialization.

ii) Plant Protection and Secrecy

It is an important requirement of potential bioherbicides to safeguard the recovery of development and registration costs. The aim of patent system is to promote science by recognizing and reading the inventors. This is achieved by giving inventors the exclusives rights to make sell or use the patented invention for a certain period. Public disclosure of the details of an invention further simulates the research and consequently guarantees the scientific and technological progress. However, in this system of patenting the cost involved are high and exchange of ideas and results amongst the scientists may be delayed.

iii) Cost of production

Cost of production is a major constraint in the development of bioherbicides. The type of fermentor, fermentation time, media and substrates used all determine the total cost (Charudattan 1982).

iv) Regulations

Bioherbicides are products that are generally applied similar to chemical herbicides. As such they have been treated as pesticides and subjected to similar regulations, which are often restricted and unrealistic (Auld and Morin, 1995). There are no guidelines in most of the cases especially when the pathogens are genetically improved agents. Concern also exists regarding the possible production of mycotoxins by fungi used as bioherbicides.

INTEGRATION OF BIO-CONTROL AGENTS

Irrigation of biological control with more conventional methods of weed control practice is getting attention recently. The use of plant pathogens has been combined with compatible herbicides and mechanical or cultural practices. The microbial herbicide, *Phytophthora palmivora* used for the control of *Morrenia odorata* in citrus groves of Florida, has been reported to be inhibited by treatment with some herbicides like diuron, glyphosate and paraquat (Smith, 1982). However, when the chlamydospore suspension was sprayed 3 weeks after spraying the herbicide, both herbicide and fungus together could control the weed. Similarly, the possibility of using a combination of the fungus *Cochliobolus lanatus* and a conventional herbicide against *Echinochloa crusgalli*, a major weed of maize in Europe, mainly because it has evolved hebicide-resistant strains has been suggested. The pathogen will kill seedlings with less than two leaves but not older plants. However, in experiments in controlled environments, the weed could be controlled by *C. lanatus* in the presence of low, sub-lethal doses of atrazine.

In trials in Mississippi, U.S.A, Imazaquin reduced the dry weight of sicklepod (*Cassia obtusifolia*) by 32% but did not kill plants, and *Alternaria cassiae* controlled 58% of sicklepod at the first true leaf stage. When the mycoherbicide and chemical were tank-mixed, the combination controlled 96% of the weed population.

The combination of plant pathogens and insects for the biological control of weeds is also a form of integrated control. The potential for success is demonstrated by the control of water hyacinth (*Eichhornia crassippes*). This weed of fresh waters in the sourthern U.S.A. can be effectively controlled by the host-specific fungus, *Cercospora rodmanii,* when concentrations of nutrients in the water are low but, in richer lakes and rivers, success is dependent upon the presence of arthropod parasites, *Neochetina eichhorniae* (mottled weevil), *N. bruchi* (chevroned weevil), *Someodes albiguttalis* (a moth), *Orthogalumna terebrantis* (a mite), and *Arzema densa* . Neither the pathogen nor the arthropods can kill the weed on their own; for example, 99% of plants died within 6 months after the fungus was applied to plants affected by the two weevils.

An integrated system of biological control thus holds a promising future for weed control. Selection andgenetic manipulation of the fungus pathogens may also give good results. Charudattan (1985) discussed selection of fungus strains for fungicide resistance. According to him, potentially all the fungi studied as biocontrol agents could be selected for more virulence, more host specificity better growth and sporulation, more environmentally optimal infection and disease development etc. The release of genetically altered organisms in the environment is not necessarily desirable, so that selection of naturally occurring strains is a desirable strategy. However, the potential of genetically altered organisms for biocontrol purposes should be further investigated.

CONCLUSION

The use of plant pathogens in the biocontrol of weeds has gained increasing importance in recent years. The foremost example of successful introduction of *P. chandrillina* into Australia to control *C. juncea* clearly demonstrated the feasibility of classical strategy of biocontrol, so far, used only by Entomologists, and opened the way for the development of several other projects. Classical biological control has the great advantage of providing continuing economic returns. After introduction of the pathogen, into an extensive weed population, it is allowed to perpetuate and become widely established on its own. This strategy is more suitable for exotic weeds, but in some cases, endemic weeds which have become cosmopolitan e.g. water hyacinth and field bind weed, may have well adapted biocontrol agents found elsewhere in their distribution. Also strains of endemic pathogens virulent to an endemic weed may be transferred from one part of the weed's distribution to another normally separated by natural barriers.

However, one important shortcoming of classical biological control is that once an agent is released in a country, it cannot be kept restricted to only those situations, where control is desired. This has, however, been overcome by the use of pathogens as bioherbicides or mycoherbicides. This technique has been developed mainly for endemic weeds, though it can also be used for exotic weeds if suitable pathogens are available. The pathogens have to undergo extensive studies to reach a level of host specificity and safety before they are finally developed. Since the pathogens are applied more than once, at particular intervals, it becomes much more of a commercial operation. Compared to chemical herbicides, bioherbicides offer one great attraction, that of not being harmful to the environment. The recent trend in the selection of non biotrophs as classical biocontrol agents particularly those with slime spores that have inefficient mode of dispersal, has resulted in the implementation of techniques usually reserved for bioherbicides. For example, coelomycete *Phloeospora mimosae-pigrae* was introduced in Australia in 1994 for the control of *Mimosa pigrae* and has been mass produced and sprayed on target to aid establishment and spread (Ellison and Evans, 2001).

In case of bioherbicides, most of the constraints are because of the desire to produce perfect analogues of chemical herbicides. This is unrealistic and in fact not the aim. Of late, there is an increasing trend in research to promote the development of pathogenic microorganisms that debilitate the weeds rather than killing tem as biocontrol agents. The agricultural community is also becoming more and more aware of the hazards of chemicals and is becoming more convinced in the use of other technologies, which do not harm mankind and environment.

In short, it appears that biological control agents hold a good promise in the present day world. Although they are not expected to replace the chemicals, which will dominate the market for many more years to come, their integration with chemicals should find an important place in weed management technology particularly in irrigated cropping systems where the environment can be manipulated.

REFERENCES

Abbasher, A.A. and Suerborn, J. 1992. *Fusarium nygamai* var. *redolens*, a potential bioherbicide for *Striga harmonthica* control in sorghum. *Biological Control.* 2: 291-296.

Andres. L.A., Davis, C.J. and Wapshere, A.J. 1976. Biological control of weeds. pp. 481-499 in *"Theory and Practice of Biological Control"*. C.B. Huffaker and P.S. Messenger, (eds.). Academic Press, New York.

Andrews, J.H. 1992. Biological control in the phyllosphere. *Ann. Rev. Phytopathology*. 30: 603-635

Auld, B.A. and Morin L. 1995. Constrains in the development of Bioherbicides. *Weed Technology* 9: 638-652.

Bedi. J.S. 1994. Further studies on the control of sunflower broomrape with *Fusarium oxysporum* f.sp. *orthoceras* – a potential mycoherbicide. In "*Biology and Management of Orobanche*" proceedings of the Third International Workshop on *Orobanche* and related *Striga*. A.H.Pieterse, J.A.C.Verklieg and S.J.ter Borg (eds.). Amsterdam, The Netherlands. pp 539-544.

Charudattan, R. 1982. Regulation of microbial weed control agents. In R. Charudattan and H.L. Walker eds. pp 175-188 "*Biological control of weeds with plant pathogens*". John Wiley and Sons Inc. New York.

Charudattan, R. 1985. The use of natural and genetically altered strains of pathogens for weed control. pp 347-372 in "*Biological Control in Agricultural IPM System"* M.A. Hoy and D.C. Herzog, (eds.) Academic Press, New York.

Charudattan, R. 1988. Integrated control of waterhyacinth (*Eichhhornia crassipes)* with a pathogen, insects and hrbicides. *Seed Sci.* 34 (1): 26-30. 1986

Charudattan, R. 1990. Assessment of efficacy of mycoherbicide candidates. Proc. "VII International symposium on biological control of weeds" 6-11 Mar. 1988. Delfosse, E.S. (ed) Rome, Italy. pp 93-100.

Christy, A.L., K.A. Herbst, S.J. Kostka, J.P. Mullen and P.S. Carlson. 1993. Synergizing weed biocontrol agents with chemical herbicides. P.87-100 in "Pest control with Enhanced Environmental safety". S. O.Duke, J. J. Menn, and J.R. Plimmer, (eds). ACS Symp. Ser. 534. American chemical society, Washington DC.

Dodd, A.P. 1961. Biological control of *Eupatorium adenophorum* in Queensland. *Aust. J.Sci.* 23:356

Ellison, C. and Evans, H.C. 2001. Biocontrol of weeds using pathogens: recent advances. Proc. "Symposium on biocontrol based pest management for quality crop protection in the current millenium" 18-19 July, 2001 Singh, D. *et al.* (eds). Ludhiana, India. pp 451-463.

Ellison, C. 2003. *Mikania micrantha. IBG. News* 12 (1): 12-13.

Evans, H.C. 1997. *Parthenium hysterophorus.* A review of its weed status and the possibilities for biological control. *Biocontrol News and Information.* 18(3): 89N-98N.

Frohlich, J. 1999. Bioherbicide Research – Status report, New Zealand. *International Bioherbicide Group News.* 8(1): 10-11.

Greaves, M.P., Auld, B.A. and Holloway, P.J. 1998. Formulation of microbial herbicides. In *"Formulation of Biopesticides, Beneficial Microorganisms, Nematodes and Seed Treatments"* (Ed. H.D. Burges), pp. 203-233. Kluwer Academic Publ. London and New York.

Hasan, S. 1980. Plant pathogens and biological control of weeds. *Rev. Plant Pathol.* 59: 349.

Hasan, S. 1983. Biological control of weeds with Plant pathogens-Status and Prospects. *"Proc. 10th Int. Cong. Plant Prot."*, Brighten, U.K., pp. 759.

Hasan, S. 1988. Biocontrol of weeds with microbes. In *"Biocontrol of Plant diseases"* vol. 1. K.G.Mukerji and K.L.Garg. (eds.). pp 129-152. CRC press, Boca Raton, Florida, USA.

Hasan, S. and P.G. Ayres 1990. The control of weeds through fungi: Principles and Prospects, *New Phytol.* 115: 201-222.

Hasan, S. and Cullen, J. M.1984. Studies in the Mediterranean region on biological control of Australian Weeds, *Proc. "European Weed Res. Soc. 3rd Symp. Weed Problems in the Mediterranean Area".* Oliras, Portugal, p. 365.

Hasan, S. and Aracil, E. 1991. Biology and effectiveness of *Uromyces heliotropii,* a potential biological control agent for *Heliotropium europaeum. New Phytol.* 118: 559-563.

Hasan, S., Delfosse, E.S., Aracil, E. and Lewis, R.C. 1992. Host specificity of *Uromyces heliotropii,* a fungal agent for the biological control of common heliotrope *(Heliotropium europaeum)* in Australia. *Ann. Appl. Biol.* 121: 697-705.

Hasan, S. and Wapshere, A.J. 1973. The biology of *Puccinia chondrillina*: a potential biological control agent of skeleton weed. *Ann. Appl. Biol.* 74: 325-332.

Heinz, D.K. and G.E. Templeten 1993. Economic comparisons of mycoherbicides to conventional berbicides. P. 395-408 in *"Pesticide Interaction in Crop Production: Beneficial and Deleterions".* J. Altman, (ed). Elsevier Science Publishers, Amsterdam.

Huffaker, C.B. 1970 Fudamentals of biological weed control, in *"Biological Control of Insect Pests and Weeds".* DeBach, P., (ed.). Chapman and Hall, London, p. 631.

Inman, R.E. 1971. A primary evaluation of *Rumex* rust as a biological control agent of curly dock, *Phytopathology.* 61: 102.

Johnson, D.R. 1998. Bioherbicides Research- Status report, Encore Technologies. *International Bioherbicide Group News* 7(2): 6.

Julien, M.H. and Griffiths, M.W. 1998. Biological control of weeds. A world catalogue of agents and their target weeds. Fourth edition, CABI Publishing: Wellingford, U.K. p 223.

Killgore, E. 1999. Hawaii Department of Agriculture Report, *International Bioherbicide Group News.* 8(1): 15

King, L.J. 1966. Weeds of the world, Biology and Control. Interscience, New York. USA. Pp 526-544

Makowski, R.M.D. 1993. Effect of inoculum concentration, temperature, dew period and plant growth stage on disease of round leaved mallow and valvet leaf by *Celletotrichum gloeosporioides* f. sp *malvae. Phytopathology* 83: 1229-1234.

McFadyen, R.E.C. 1998. Biological control of weeds. *Ann. Review of Entomology* 43: 369-393.

McWhorter, C.G. and Chandler, J.M. 1982. Conventional weed control technology. In: *Biological Control of Weeds with Plant Pathogens.* R.Charudattan and H.L.Walker (eds.). pp 5-27. J. Wiley and Sons, New York, USA.

Morin, L. 2002. Bridal creeper rust sets in the Australian environment. *International Bioherbicide Group News* 11(1): 6-7.

Mortensen, K. 1997. Biological control of weeds using microorganisms. In *"Plant –Microbe interaction and Biological Control"* G.J. and L.D. Kuykendall (eds.). pp 223-248. M.Dekker Inc. New York. USA.

Reader, R.H. 1999. Towards integrated control of itch grass weed. *International Bioherbicide Group News* 8(1):12.

Sheng, Q. 2003. On the potential of *Colletotrichum gloeosporioides* f.sp. *veronicae* as a mycoherbicides for *Venonica* sp and *Gallium aparine* var. *tenerum. International Bioherbicide Group News* 12 (1): 8-10.

Smith, R.J. Jr. 1982. Integration of microbial herbicides with existing pest management programs, in *"Biological Control of Weeds with Plant Pathogens"*. Charudattan, R. and Walker, H.L., (eds.) John Wiley & Sons. New York. 1982.

TeBeest, D.O., Yang, X.B. and Cisar, C.R. 1992. The status of biological control of weeds with fungal pathogens. *Annual Review of Phytopathology* 30:637-657.

Templeton, G.E. 1982. Status of weed control with plant pathogens in "*Biological control of weeds with Plant Pathogens"*. Charudattan. R and Walker, H.L. (eds. John Wiley & Sons, New York. 29.

Templeten, G.E., TeBeast, D.O., and Smith, R.J., Jr. 1976. Development of endemic fungal pathogen as jointvetch in rice. *Proc. 4th Int Symp. Biol. Control Weeds* Gainsville, Fla. USA. P. 214.

Templeten, G.E., TeBeast, D.O., and Smith, R.J., Jr. 1979. Biological weed control with mycoherbicides. *Ann. Rev. Phytopathology*. 17: 301.

Templeton, G.E., TeBeast, D.O. and Smith. R.J.Jr., 1984. Biological weed control in Rice with a strain of *Colletotrichum gloeosporioides* (Penz.) Sacc. used as a mycoherbicide. *Crop Prot.* 3:409.

Tisdell, C.A. and Auld, B. 1989. Evaluation of biological control projects. Proc. "*VII International symposium on biological control of weeds*" 6-11 Mar. 1988. Delfosse, E.S. (ed) Rome, Italy. pp 93-100.

Wapshere, A. J. 1975. Assessment of biological control potential of the organisms attacking *Chondrilla juncia L.* Proc. *Ist Int. Symp. Biol. Control Weeds* Delemont. Switzerland. p. 81.

Weaver, S.E. 1991. Size dependent economic thresholds for three broadleaf weed species in soybean. *Weed Technol.* 5:674-679.

Wilson, C.L. 1969. Use of plant pathogens in weed control. *Ann. Rev. Phytopathol.* 7: 411. 1969.

Woodhead, S. 1981. Field efficacy of *Phytophthora palmivora* for control of milkweed vine. *Phytopathology* 71:913.

Eco-friendly Approaches for Weed Management

N. T. YADURAJU[1] AND J. S. MISHRA[1]

Agriculture has gone through major changes during the last century. It has developed from a more or less extensive subsistence farming to a highly intensive, often mechanized, agricultural production. It relies heavily on the input of energy and other resources. Production has become dependent on chemical pesticides and commercial fertilizers. With the increased pressure on farmland worldwide, we are facing global problems such as soil erosion, leaching of minerals, salinisation, increasing pest problem and damage to humans and nature from chemical pesticides. A change from a high-input and chemically intensive agriculture to a more sustainable form of agriculture is not only desirable, but has become a necessity. Concern over environmental and human health impacts of conventional weed management practices, herbicide resistance in weeds and rising costs of crop production and protection have led agricultural producers and scientists to seek strategies for development of eco-friendly weed management practices. It is generally believed that certain herbicides can jeopardize environmental quality and human health. To minimize the potential for damage, effective weed management systems that are less reliant on herbicides are needed.

CROP LOSSES DUE TO WEEDS

Weeds are the major deterrent to the development of more sustainable agriculture systems. They cause enormous losses and sufferings to human beings by way of reduction in crop yield and quality, wastage of human energy and resources and increased expenditure to alleviate the problems caused by them. In general, impact of weeds on crop yields varies from high input to low input crop production systems. Unlike insect pests and pathogens, where the effect on crop is visible, the damage done by weeds is often noticed. In addition to direct effect on crop yield, weeds result in considerable reduction in the efficiency of inputs used. The precious and costly inputs such as fertilizers and irrigation water which are otherwise meant for realizing the potential yield will be usurped by the weeds. The yield losses due to weeds are presented crop-wise in Table 1.

In general, impact of weeds on crop yields varies from high input to low input crop production systems. It has been estimated that weeds caused 5% loss in agricultural production in most developed countries, while it caused 10 % loss in developed countries and 25% loss in least developing countries (Bhowmik, 1998). The world-wide loss of attainable production of eight principal food and cash crops (rice, wheat, barley, maize, potato, soybean, cotton and coffee) has been estimated to be 76.3 billion US dollar annually (Oerke *et al.* 1994). An analysis revealed that losses caused by weeds in India were to the tune of 9.28 million tonnes in cereals, 0.78 million tonnes in pulses, 0.57 million tonnes in oilseeds and 7.20 million tonnes in fibre and other commercial crops (Sahoo & Saraswat, 1988). The list of weeds acting as alternate host to insect pest, nematodes and diseases is given in Table 2.

1. National Research Centre for Weed Science, Maharajpur, Jabalpur – 482 004 India.

Table 1(a): Crop yield losses due to weeds in important field crops

Crop	No. of locations	No. of trials	Mean yield (kg/ha)		Mean loss (%)
			Weed free	Weed infested	
Rice (DS)	22	28	3365	1151	64.0
Rice (PS/PT)	20	30	4549	2620	43.9
Wheat	23	63	3241	2226	31.6
Maize	13	18	4067	2056	48.8
Groundnut	20	31	2486	934	63.6
Soybean	21	54	1986	874	53.0
Mustard	12	20	1433	889	37.7
Linseed	4	9	1110	399	62.7
Sesamum	5	6	616	288	54.2
Sunflower	4	5	1659	1117	32.2
Pigeon pea	16	21	1573	774	46.7
Mung bean	7	13	963	516	46.3
Urd bean	15	18	1001	474	55.4
Chick pea	10	13	1545	767	48.1
Lentil	6	6	1709	755	58.8
Field pea	9	13	2094	1054	47.1
Cotton	12	18	1234	824	60.4
Sugarcane	8	9	93.1*	64.4*	34.4
Potato	7	10	18.3*	12.8*	29.0

*t/ha, DS-Direct seeded, PS- Puddeled seeded, PT- Puddeled transplanted .

Table 1(b): Nutrient depletion (kg/ha) by weeds in different field crops

Crop	Nitrogen	Phosphorus	Potassium
Paddy (Direct seeded)	20-37	5-14	17-48
Paddy (Transplanted)	11	3	10
Maize	23-59	6-10	16-62
Sorghum	36-46	11-18	31-47
Wheat	20-90	2-13	28-54
Pigeon pea	28	24	14
Green gram	80-132	17-20	80-130
Cotton	67-75	20-38	30-109
Chick pea	29-55	3-8	15-72
Peas	61-72	7-14	21-105
Lentil	39	5	21
Soy bean	26-65	3-11	43-102
Groundnut	15-39	5-9	21-24
Rapeseed-mustard	22	3	12
Linseed	32	3	13
Sugarcane	35-162	22-44	135-242

Table 2: Weeds acting as alternate hosts of insects and diseases (Gupta, 1998)

Host weeds	Insect/disease organisms hosted	Crop affected
Weeds hosting Insects		
Aeschenomene sp.	Grasshoppers (Vector of *Colletotrichum*)	Rice
Achyranthus sp.	Leaf eating caterpillars	Lucerne and maize
Chenopodium album, Amaranthus and *Datura* spp.	Gram caterpillar	Redgram, cotton, okra, pea and brinjal
Crotolaria sp.	Hairy caterpillar	Castor
Echinochloa sp., *Panicum* sp.	Stem borer	Rice
Chenopodium album	Stalk borer	Tomato, maize
Brassica kaber	Aphids	Brassica crops and brinjal
Weeds hosting Diseases/Nematodes		
Agropyron and *Portulaca* spp.	Wilt	Tomato
Cynodon dactylon and *Digitaria spp.*	Sting nematode	Several vegetables
Commelina spp.	Certain viral diseases and parasitic nematodes	Banana
Saccharum spontaneum	Downy mildew	Maize
Solanum and *Physalis* spp.	Mosaic and cyst nematodes	Solanaceous vegetables, tobacco and cocoa

WEED COMPETITION

Competition is the rivalry between crop and weed for a common demand which becomes short in supply and eventually the demand and supply gap is reflected in the growth and productivity of weed and the crop. The competition occurs when weed and crop grow in a mixed stand, influence their growth mutually reduce growth of some plants in comparison to others. Competition is what weed control is all about. Weeds compete with the crop for nutrients, moisture, solar radiation and space when they are limited and reduce the crop yield and quality. Crop-weed competition is largely affected by nature and intensity of weeds, crop species and cultivar, light, moisture, nutrients and management practices.

CRITICAL PERIOD OF CROP-WEED COMPETITION

The time of weed removal has an important effect on growth and yield of the crop. Removing weeds at any time during growing season may not be beneficial. Stage of weed removal is as important as removal *per se*. Therefore, it is necessary to identify the critical period of crop-weed competition to render the weed control practices more effectively. Critical period of weed control is an important consideration in the development of alternative weed management strategies. It can be defined as "the shortest span of time in the ontogeny of crop growth when weeding will result in higher economic returns". Hall *et al.* (1992) described the critical period of weed control as representing the time interval between two separate components. The first component is the maximum length of time that weeds, which emerges with the crop, can remain before they reduce crop yield. The second component is the length of time a crop must be kept weed free after planting so that weeds emerging later do not reduce yield. The two components can be integrated to identify a single critical period of weed control, recognized as the duration over which neither component provides an acceptably low yield loss; weed control must be effective with in a period of overlap between the two components. A fundamental principle of plant competition is that early occupants on a soil tend to exclude the later ones. This principle finds application in practical weed control. Weeds are capable of accumulating dry matter faster than the crop plants. Thus, the duration of weed infestation and time of weed removal has a significant influence on crop growth and economic yield. Mishra (1997) reviewed the research work done under AICRP-Weed Control and else where since last 20 years and reported that the critical period varies from 15- 30 days after sowing (DAS) in greengram, blackgram, cowpea and okra; 15-45 DAS in direct seeded rice, maize, sorghum, sunflower, sesame and rapeseed-mustard; 30-45 DAS in transplanted rice, pearl millet, wheat, peas, jute, cauliflower, cabbage and tomato; 20-45 DAS in soybean and potato; 30-60 DAS in castor, lentil, chickpea, frenchbean, and groundnut; 15-60 DAS in pigeonpea and cotton; 30-75 DAS in onion and 30-120 DAS in sugarcane depending upon nature and intensity of weed flora, agro-climatic situations and management practices.

PRINCIPLES FOR WEED MANAGEMENT SYSTEMS

Every weed management program is only part of a total crop production system. Therefore, any combination of environmental manipulation and cultural techniques to reduce weed abundance must be compatible with other farm objectives. Thus, the interactions between soil, fertility, insects and diseases, and direct weed control procedures may be especially important. However, it is not possible to know everything about every (or any) weed species in order to develop effective weed management strategies. Zimdahl (1993) suggested following components in the development of weed management systems.

- incorporation of ecological principles
- use of plant interference and crop-weed competition
- incorporation of economic and damage thresholds
- integration of several weed control techniques including selective herbicides
- frequent supervision of weed management by a professional weed manager employed to develop a program for each crop-weed situation.

Altieri and Liebman (1988) suggested following principles in designing integrated weed management systems.

- Crop monocultures seldom use all the environmental resources available for plant growth. The resulting ecological niches, therefore, are susceptible to invasion by weeds and should be protected.
- Weed populations are either active, as photosynthesizing plants, or dormant, usually seeds. Thus, the seed bank as well as the above-ground vegetation should be considered when determining weed abundance.
- The reproductive capacity and seed survival of weeds determine the composition and abundance of the succeeding weed community in a cropping system. In intensive cropping systems, the type of weed community and its abundance are a direct product of the crop and how it is managed. In developing countries, where mixed cropping is traditionally practiced in contrast to more intensive forms of crop production, the complex cropping patterns appear to disfavor weeds. Additionally, in less intensive cropping systems than those presently practiced in many developed nations, weed management appears to have less relevance than overall cropping practices in determining the weed community composition.
- The cropping pattern can be a powerful agent in reducing weed densities by preemption of environmental resources to crop plants. Weed control tactics, while usually reducing the abundance of weeds, can shift the composition, density, and spatial distribution of weeds in fields. These shifts in the spatial or temporal dynamics of weed communities and populations ultimately may increase weed abundance and, therefore, crop competition over the long term. Although shifts, in species composition often appear inevitable, the practice of growing crops with divergent life cycles and using rotational sequences of crops and weed control tactics are effective strategies

to prevent any one weed species from becoming dominant. These practices reduce selection pressure on weed communities and populations.

- Single measurements of weed density are usually not adequate to determine weed impacts. Most crops have thresholds of weed tolerance, expressed as density, amount of biomass, or period of time before significant economic loss in crop yields results. However, these thresholds vary among cropping systems, weed species, and environmental constraints. The thresholds also may have to be adjusted when criteria other than crop loss are used to determine the economic, ecological, and social effects of weeds.
- Suppressed crop growth cannot always be explained by crop-weed competition. At times, weeds may simply indicate a deteriorated soil or resource base. Allelopathy also may be a mechanism through which weeds affect plant growth, and vice versa. In addition, crops and weeds may coexist without crop yields being reduced economically and beneficial effects are possible from some crop-weed associations.
- Weed populations and communities are regulated by a combination of factors (e.g., predation, parasitism, environmental stress, interference, and direct-control methods), and at several stages of the weed's life cycle. Thus, control of weeds can be achieved through direct or indirect means. During direct control the weed plant is physically or chemically removed, while indirect controls rely on biological functions of the crop, the weed, and their associates. Both strategies have been successful, although indirect regulation may be most appropriate for complex cropping systems.

INTEGRATED WEED MANAGEMENT SYSTEM

Better understanding of weeds will lead to better crop and weed management. A basic principle of weed control is that control measures should be directed at survival mechanisms of weeds in the soil. For annual weeds, the objective is to prevent seed production and to deplete seed reserves. In case of perennials, destruction of underground vegetative organs is sought. Most weed control practices and techniques exploit biological differences between the crop and the weed and are accomplished by manipulations of the shared habitat. Maintenance of the crop in a dominant or superior competitive position in the crop-weed association depend on a differential response of crop and weed to some factor of the habitat which can be modified or manipulated with predictable results. This basic concept underlies effective weed control practice. However, the selection of methods and techniques of agricultural weed control may vary with region, crop, nature of weed infestation and the level of agricultural mechanization. Integrated weed management systems are used as a part of pest management strategies and include:

- Use of multiple-pest-resistant strategies, high yielding, well adapted varieties that resist weed competition
- Pre-planting seed bed tillage
- Effective seed bed preparation and seeding methods that enhance crop growth and minimize weed growth
- Optimum plant populations including closer spacing in and between rows to optimize crop growth and minimize weed growth
- Use of crops that form a canopy
- Precision placement of fertilizer and timing the fertilizer application for maximum stimulation of crops and minimum stimulation of weeds
- Judicious use of irrigation
- Timely and appropriate cultivations
- Sound crop rotations

- Crop diversification
- Field sanitation
- Harvesting methods that do not spread weed seeds
- Use of biological agents such as insects and pathogens and effective chemical methods.

COMPONENTS OF INTEGRATED WEED MANAGEMENT

The weed management in field crops requires an integrated approach that utilizes effective cultural, mechanical, biological, ecological and chemical methods in a mutually supported manner into the crop production system with due consideration of economic, environmental and sociological consequences. Various methods of weed management such as cultural, mechanical, chemical etc. have been used with different degree of success in different agro-ecological zones and productions system, which are as follows.

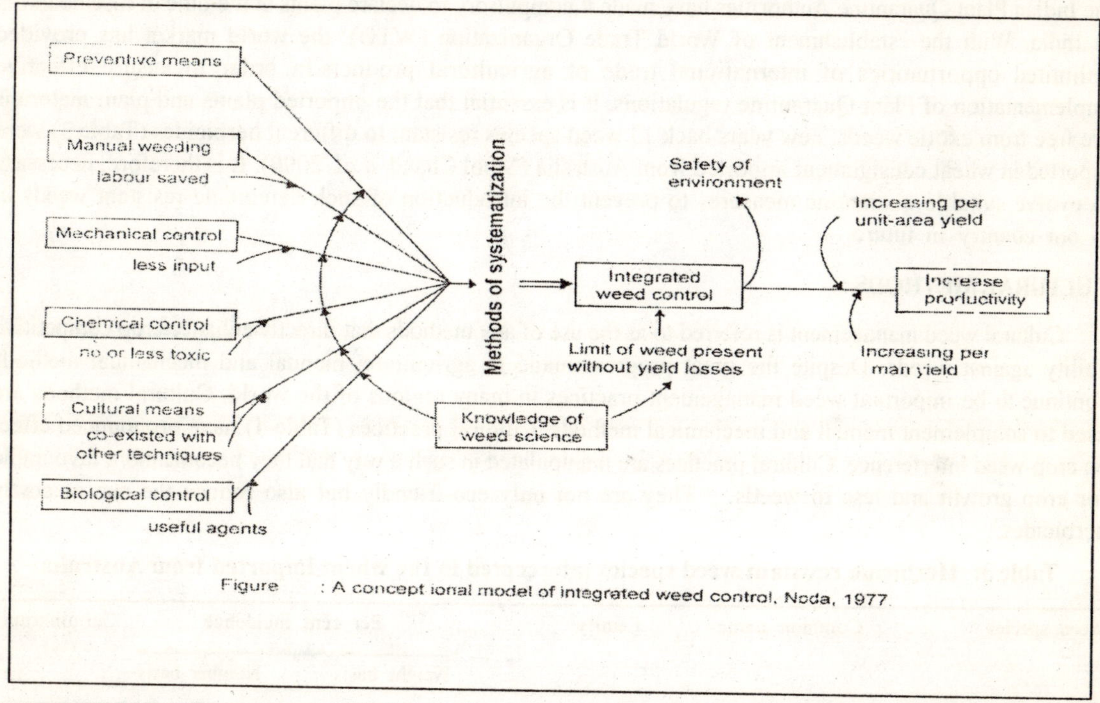

Figure : A concept ional model of integrated weed control, Noda, 1977

WEED PREVENTION

Preventive control encompasses all measures taken to forestall the introduction and spread of weeds. Success of a preventive programme varies with species and the amount of effort devoted to control. The prevention of a weed problem is usually easier and less costly than control or eradication. The following measures can be suggested to prevent the introduction of weeds into non-inhabited field.

- Use 'clean' (weed seed-free) crop seed for planting.
- Use organic manures only after thorough decomposition to kill weed seeds.
- Clean harvesters and tillage implements before moving to non-weed infested area.
- Avoid transportation or use of soil from weed infested area.
- Inspect nursery stock or transplant for seed and vegetative propagules of weeds.

- Remove weeds that are near irrigation ditches, fence rows, right-of-way and other non-crop land.
- Prevent reproduction of weeds.
- Use weed seed screens to filter irrigation water.
- Restrict live stock movement in to non-weed infested area.
- Other practices used to prevent and avoid potential weed problem at the state, regional or national level are weed laws, seed laws and quarantines.

LEGAL MEASURES

Legal measures are necessary to check inter-state and inter-country movement of noxious weeds if the cost of having to control additional alien weeds is to be saved. Unfortunately in most parts of the tropical and sub-tropical world, noxious weeds have not been subjected to strict quarantine laws. This has resulted in the introduction and spread of some of the costliest weeds of the world. Very recently the Indian Plant Quarantine Authorities have made it compulsory to declare plants at the time of disembarking in India. With the establishment of World Trade Organization (WTO), the world market has provided unlimited opportunities of international trade of agricultural products.In order to ensure effective implementation of Plant Quarantine regulations, it is essential that the imported plants and plant materials are free from exotic weeds. Few years back 13 weed species resistant to different herbicides (Table 3) were reported in wheat consignment imported from Australia (Mool Chand *et al.* 2000). It is therefore, necessary to evolve suitable quarantine measures to prevent the introduction of such herebicide resistant weeds in to our country in future.

CULTURAL METHODS

Cultural weed management is referred to as the use of any methods that directly enhance crop competitive ability against weeds. Despite the great progress made in agriculture, manual and mechanical methods continue to be important weed management practices in many regions of the world. Cultural methods are used to complement manual and mechanical methods. Cultural practices (Table 4) have pronounced effect on crop-weed interference. Cultural practices are manipulated in such a way that they become more favourable for crop growth and less to weeds. They are not only eco-friendly but also reduce the use of costly herbicides.

Table 3: Herbicide resistant weed species intercepted in the wheat imported from Australia

Weed species	Common name	Family	Per cent incidence		Germination (%)
			Weight basis	Number basis	
Avena fatua	Wild oat	Poaceae	0.024	0.038	60
A. sterilis	Sterile oat	Poaceae	0.0072	0.012	11
Brassica tournefortii	Wild turnip	Brassicaceae	0.0054	0.031	30
Echium plantagineum	Paterson's curse	Boraginaceae	0.023	0.012	60
Hordeum lepornium	Barley grass	Poaceae	0.0012	0.001	30
Lolium rigidum	Annual ryegrass	Poaceae	0.077	0.508	40
Phalaris paradoxa	Paradoxa grass	Poaceae	0.002	0.015	60
Polygonum convolvulus	Black bind weed	Polygonaceae	0.016	0.19	58
Raphanus raphanistrum	Wild radish	Brassicaceae	0.051	0.031	60
Rapistrum rugosum	Turnip weed	Brassicaceae	0.009	0.038	30
Sisymbrium officinale	Hedge mustard	Brassicaceae	0.002	0.192	10
Sonchus oleraceus	Common sawthistle	Asteraceae	0.002	0.384	15
Vulpia bromoides	Vulpia	Poaceae	0.009	0.004	33

Table 4: Classification of cultural practices potentially applicable in an integrated weed management system based on their prevailing effect (modified from Barberi, 2001)

Cultural practice	Impact	Example
Crop rotation	Reduction in weed emergence, Improving soil fertility	Alternation between winter and spring-summer crops
Primary tillage	Reduction in weed emergence, Farm hygiene	Deep ploughing, alternation between ploughing and reduced tillage
Seedbed preparation	Reduction in weed emergence, Establishing good crop stand	False/stale seed-bed technique
Cultivation	Reduction in weed emergence	Post-emergence harrowing or hoeing, ridging
Cover crops	Reduction in weed emergence	Cover crops grown in-between two cash crops and used as green manure or dead mulch
Intercropping	Reduction in weed emergence, Improvement in crop competitive ability, Cover risk, Improve soil fertility	Cover crop used as living mulch, intercropped cash crops
Thermal treatment	Reduction in weed emergence, Pest control	Pre-emergence or localized post-emergence flam weeding
Mulching/soil solarization	Reduction in weed emergence	Use of black or transparent films (in glass house or field)
Crop genotype	Improvement in crop competitive ability	Use of cultivars characterized by quick emergence, high growth and soil cover rates in early stage
Sowing/planting	Improvement in crop competitive ability	Use of transplants, higher seeding rates, lower inter-row distance, anticipation of, or delay in sowing,/ transplant date
Fertilization	Reduction in weed emergence, improvement in crop competitive ability	Use of slow nutrient releasing organic fertilizers and amendments, fertilizer placement
Irrigation	Reduction in weed emergence, improvement in crop competitive ability	Irrigation placement (micro/trickle-irrigation)

STALE SEED-BED TECHNIQUE

The principle of flushing out germinabale weed seeds before cropping forms the basis of the stale or false seed-bed technique, in which soil cultivation may take place days or weeks before planting or transplanting a crop. This depletes the seed bank in the surface layer of the soil and reduces subsequent weed emergence. Where light rains occur for an extended period before the onset of the monsoon, or irrigation is available, it may be possible to kill several flushes of weed growth before planting. To ensure success, cropping should be delayed until the main flush of emergence has passed. The emerged weed seedlings may then be killed by flaming or light cultivation or application of contact non-residual herbicides like paraquat. It is vital not to cultivate below the top 1-2 cm soil otherwise a further flush of weeds may emerge.

TILLAGE

Tillage includes all operations and practices that are carried for the purpose of modifying the physical character of soil so as to provide favourable conditions. A major benefit of tillage is prevention and suppression of weeds. Tillage suppress weeds by breaking, cutting, or tearing from the soil, thus exposing the vegetation to desiccation, and by smothering them. Repeated tillage may deplete weeds from fields by diminishing seed or vegetative propagules in the soil and may also exhaust the food reserves of perennial weeds. Besides weed control, tillage is also important in crop production because of the following reasons,

- Seedbed preparation
- Burial of crop residue
- Control of insects, pathogens and rodents
- Improvement of soil physical conditions
- Incorporation of fertilizers, herbicides and soil amendments

Weed problem begins with weed seed in soil. It is the tillage, which directly affects the weed seed bank by physical mixing of soil and in turn reducing the weed number. Tillage influences the vertical distribution of weed seeds in soil layer. Weed seeds buried deep germinate but fail to emerge due to thick soil layer over it, resulting in death of weed seedling. One of the most important changes occurring in agriculture today is the increased awareness and acceptance by the agricultural community of conservation tillage practices. Conservation tillage, which advocates the maintenance of crop residue cover on 30% of the soil surface, is soundly based within the framework of sustainable agriculture. Johnson *et al.*, (1989) found that the presence of crop residues increased weed suppression and did not result in an increase in herbicide applications

NO TILLAGE AND NIGHT TILLAGE

A light requirement for seed to be released from dormancy is a wide spread mechanism found in many weed species. This requirement can be satisfied even after a brief exposure to light, such as that resulting from tillage. Thus, germination of weed species whose seed require light should be impeded if a no tillage system were implemented for crop production. Alternatively it is possible that night cultivation also could result in significant reduction in weed density (Scopel *et al.* 1994). The problem of *P. minor* is comparatively less, as the weed seed lying at lower depths does not come to the surface as in case of conventional tillage (Table 5). Infestation is also comparatively less due to non-availability of nutrients from well-placed fertilizer near to the crop seed. For many burried seeds, it is the exposure to light during soil cultivation that stimulates germination. The exclusion of light during seedbed preparation has been shown to reduce weed emergence. Covering an implement to prevent light reaching the soil at the point of cultivation may be sufficient to reduce weed emergence by up to 70% (Ascard, 1994). However, not all weed species have light-sensitive seeds. Welsh *et al.*, (1999) observed the reduced density of *Chenopoduim album* by cultivating in darkness.

Table 5: Effect of ZT and CT planting on weed population (no./m²) at 25 days after sowing (Yaduraju and Mishra, 2002)

Tillage	*P. minor*	A. ludoviciana	C. album	M. hispida	Total
Zero	9.82	4.00	12.73	4.42	17.67
Minimum	11.70	2.07	11.83	3.46	17.89
Conventional	15.16	2.11	13.52	2.60	20.80
LSD (P=0.05)	5.30	0.81	NS	NS	NS

Data subjected to square root transformation

FURROW IRRIGATED RAISED BED PLANTING SYSTEM (FIRBS)

The furrow irrigated raised bed planting system (FIRBS) is a method where crops are planted on raised beds. This system was introduced in India in 1995 on the pattern of wheat grown in the Yaqui valley of Mexico. Generally 2-3 rows of wheat are planted on the top of bed, 70 cm wide and irrigation is done through the furrows. The inter-row bed space is used to control weeds by mechanical weeding during the early vegetative growth of weeds. In the crop sequences where wheat follows soybean, maize or cotton, this system of reduced tillage can be followed by reshaping the same beds without opening up the soil for field preparation. In addition to improved weed management, the bed planting method was designed to improve input use efficiency. This technology is suitable for almost all type of soils except black cotton soils. Chauhan *et al.* (1998) obtained reasonably good control of *P. minor* in wheat but broad-leaved weeds in furrows were not controlled.

CROP ROTATION

Crop rotation is a fundamental tool of integrated weed management. Interest in using crop rotation to control weeds is gaining popularity especially among those focusing on sustainable agriculture Crop rotation introduces conditions and practices that are not favourable for a specific weed species and thus, thus growth and reproduction of that weed species are hampered.

Crop rotation under local conditions can be integrated by considering following points in view:

1. There is need to diversify not only cropping system but also farming system to introduce integrated weed management based on cropping system or farming system approach.
2. In the farming system approach, animal component may be integrated with crop component so that introduction of forage crops may help to diversify the weed management. Weed communities in the continuous cropping system tended to have greater total densities and were similar in composition than crop-fallow system (Derksen *et al.*, 1994).
3. Crop rotation allows use of different management options which help in diversity in weed flora.
4. In integrated weed management, the carry over effect of complex weed flora of weedy or weed free fallow should be defined in terms of holistic approach. The weed seed bank in the soil may be reduced or exhausted by crop rotation.

Liebman and Dyck (1993) reported that the success of rotation system for weed suppression appears to be based on the use of crop sequence that create varying patterns of resources competition, allelopathic interference, soil disturbance and mechanical damage to provide an unstable or inhospitable environment that prevents the proliferation of a particular weed species. The growth and reproduction of a troublesome weed species may be actively discouraged by introducing unfavorable practices in to a rotation. The inclusion of cover crops in the rotation at a time when land might otherwise lie uncropped will suppress weed development, yet maintain soil fertility and prevent erosion. In addition, the decomposing cover crop residues may release allelo-chemical that inhibits the germination and development of weed seeds.

In monocropping system, numerous weed species persist and expand rapidly but crop rotation helps in interrupting life cycle of weeds and prevents any weed species to become dominant e.g. *Sorghum halepense* become predominant weed in continuous maize but can be controlled by rotating with cotton (Dale and Chandler, 1979). Certain weed species often associated with particular corps (eg. Barnyard grass in rice, wild oat and *Phalaris minor* in wheat, dodder in alfalfa etc.). Therefore, population of such weeds usually will increase whenever the crop is grown on the same field continuously for several years. Growing in sequence crops may discourage weed associations with crops that have sharply contrasting growth and cultural requirements. Banga *et al.* (1997) reported higher incidence of *Phalaris minor* in rice-wheat rotation than in other rotations (Table 6). Parasitic weeds can also be managed through crop rotation by rotating host crop with trap crops, as they induce germination of weed seeds but they themselves are not parasitized. Rotation among crops having drastically dissimilar life cycles or requiring different management practices is useful in disrupting weed cycle. Shankaran and Chinnamuthu (1993) found that *Paspalum dilatum* was nearly eliminated after three crops of rice-maize-mung bean, whereas *Digitaria ciliaris* became dominant. In rice-rice-urdbean system, an increase of 70 per cent in *Cyperus rotundus* population has been reported (Laesino, 1980). *Cuscuta* is one of the important parasitic weeds causing lot of damage in rice-mungbean/urdbean cropping system in coastal Andhra Pradesh, Tamilnadu and parts of Madhya Pradesh. The reports of its infestation is also available in other pulses viz. lentil, chickpea and pigeonpea. The parasitic weeds can be effectively managed through crop rotation by rotating host crop with trap crops, as they induce germination of parasitic weed seeds but they themselves are not parasitised. Due to lack of suitable host, *Cuscuta* seedlings will emerge and die.

Table 6: Effect of crop diversification on population of *Phalaris minor* and wheat yield

Crop rotation	Population of *P. minor* (no./m²)	Wheat grain yield (kg/ha)
Rice-wheat(continuous for 10 years)*	2350	3000
Rice-berseem, rice-wheat	255	4200
Rice-wheat, rice-cotton, cotton-wheat*	2125	3200
Rice-berseem, sorghum-wheat	190	4500
Rice-potato, rice-wheat	255	4000
Cotton-wheat (for 4 years)*	38	4600
Rice-berseem, rice-berseem, rice-wheat	28	5000
Rice-winter maize, sorghum-raya, maize-wheat	18	5500
Maize-sunflower, rice-berseem, rice-wheat	35	5300
Rice-potato, chilly-potato, rice-wheat	24	5400
Rice-wheat (for 5 years), chilly-wheat*	2421	3200

*Isoproturon 1.0 kg/ha was also used in wheat at 30-35 days after sowing.

Crop Rotation and Seed Bank Dynamics

The composition and density of weed seed banks are frequently a reflection of long term crop rotations and management systems. Walker and Buchanan (1982) reported changes in the weed flora when mono-cropping has been replaced with crop rotation. However, in any given year the crop species, planting date, herbicide selection and weather patterns may alter species composition of the weed community. Therefore, knowledge of the role of crop rotation in seed bank dynamics would be useful in the refinement of an integrated weed management system.

INTERCROPPING

Intercrops increase the ecological diversity of farms and, by increasing the use of natural resources by the canopy compared with monocrops, often deprive weeds of the light, water and nutrients necessary foe their development. Intercropping suppresses weeds better than sole cropping and thus provides an opportunity to utilize crops themselves as tools of weed management. When two or more crops are grown together as intercrops, the total weed suppressing ability of a system will be higher than the sole cropping (Rao and Shetty, 1976). Finding an intercrop which will suppress the growth of the weeds but not the crop has been difficult. Two research approaches need to be considered;

i) Eco-physiological studies leading to the manipulation of intercropping systems to obtain better weed control, and
ii) Evaluation of various weed control methods including herbicides for important intercropping systems.

Many short duration pulses like, cowpea, greengram and soybean effectively smoother weeds without causing reduction in the yield of main crop. In wide spaced crops such as pigeonpea, maize, cotton, sorghum and sugarcane, intercropping is a common practice, which besides covering risk will reduce weed infestation. Patel *et al.* (1983) found that intercrops suppressed the weeds and increased the total productivity (Table 7). Intercropping of sorghum with cowpea and greengram smothered weeds and reduced hand weeding cost. Sugarcane is also a wide spaced crop. The row spacing ranges from 60-90 cm and the crop take nearly 3-4 weeks for germination depending upon temperature. Short duration, early canopy forming crops such as French beans, groundnut, soybean, cowpea and greengram can be raised as intercrops to suppress the weeds and to augment the total productivity.

Table 7: Effect of intercropping in pigeonpea on weed and yield of crop

Crops	Weed dry weight ($g\ m^{-2}$) at 60 DAS	Seed yield ($kg\ ha^{-1}$)	
		Pigeonpea	Intercrop
Pigeonpea sole	46	652	-
Pigeonpea + sorghum	18	593	514
Pigeonpea + maize	28	565	289
Pigeonpea + rice	32	597	125
Pigeonpea + blackgram	17	460	587
Pigeonpea + greengram	13	422	411
Pigeonpea + groundnut	29	592	592

MULCHING

Mulch is a protective covering of materials maintained or placed on soil surface. Materials used for mulch includes crop residues, straw, leaves, paper plastic films, gravel and dry soil. Mulching the soil surface can prevent weed seeds germination or physically suppress seedling emergence, but is not effective against established perennial weeds. Chemical effect of mulches on weed control includes allelopathy, toxic microbial products and pH changes in soil. Mulching with sugarcane trash at 2- 4-inch thickness proved to be very effective in controlling *Cyperus rotundus*, one of the most obnoxious weeds in sugarcane crop during the month of April-June. In maize (Table 8), addition of wheat straw mulch reduced the population and dry matter of *Echinichloa* spp., *Panicum capilare* and *Amaranthus albus*.

Table 8: Effect of wheat straw mulch on weed population and its dry weight in maize

Mulch level (t/ha)	Weed population ('000/ha)	Weed dry weight (kg/ha)
0	220	1500
1.7	160	1300
3.4	67	950
5.1	57	680
6.8	53	570

Living mulch consists of a dense stand of low growing-species, established before or after the crops. In cereals, an under story of clovers (*Trifolium* spp.) has been shown to improve soil fertility, and reduce pest and disease problems in addition to suppressing weeds (George, *et al.* 1997). A free-floating fern, *Azolla pinnata*, is used in low land and irrigated rice as a sort of companion crop. This fern has a symbiotic relationship with *Azolla anabenea,* a nitrogen fixing blue green algae. This symbiotic relationship not only contributes up to 100 kg N/ha but also reduces weed growth by forming an *Azolla* blanket over the water surface in a rice crop. The *Azolla* technique suppresses many annual weeds while leaving out certain annuals which have strong culms (*eg.* Barnyard grass) and perennial weeds.

COMPETITIVE CROPS AND CULTIVARS

Crops differ in relative growth rate, spreading habit, height, canopy structure and inherent competitive character and accordingly differ in their weed suppressing ability. A quick growing and early canopy-producing crop would be expected to be better competitor against weeds than crops lacking these characters. The competitive ability of wheat cultivars was found to be closely associated with their tallness and early canopy cover (Table 9). Tall cultivars apparently intercept more light and reduce weed growth particularly at later stages of crop growth. Seed size within a species also influences the competition through vigorous plants from larger seeds. Increased competitive ability of cultivars has been attributed to early emergence, seedling vigour, increased rate of leaf expansion, and rapid creation of dense canopy, increased plant height, early root growth and increased root size. Future breeding and variety testing programmes should take factors of crop competitive ability into consideration.

Among winter cereals, barley is reported to be more competitive than wheat or oat (Gupta, 1984). Rao and Shetty (1976) observed that maize had the highest weed suppressing ability (92%) followed by pearl millet (88%), Sorghum (81%) and Setaria (73%). Amongst non-cereals, cowpea (88%) was more competitive than groundnut (62%), castor (57%) and pigeonpea (54%). Similarly, *Cyperus rotundus* inflicted lower yield loss in maize and sorghum as compared to groundnut, soybean and grain legumes.

Table 9: Grain yield of wheat as influenced by cultivars and weed control practices

(Balyan and Malik, 1991).

Wheat cultivars	Wheat yield as affected by weed control practices		
	Wild oat infested	Isoproturon 0.75 kg/ha	Weed free
HD 2009	2794 (59.8)	4147 (40.3)	6948
HD 2285	4865 (27.0)	6692 (No loss)	6666
S 308	3916 (40.22)	5813 (11.3)	6551
WH 291	4653 (30.3)	6506 (2.5)	6676
WH 147	5044 (26.7)	6794 (1.3)	6884

Values in parentheses are per cent yield loss over weed free

SOWING TIME

Planting time considerably influences the occurrence and manifestation of weed species. In timely sown chickpea, the weed population by 30 days is generally quite high to attract manual weeding whereas in late sown chickpea the build up of adequate weed flora is only after 45 days. Delayed sowing of lentil and chickpea is also reported to reduce the infestation of *Orobanche* – a root parasite. Malik *et al.* (1988) reported that in chickpea, the maximum emergence of most competitive weed, *Chenopodium album* L. occurred when crop was sown on November 5 and declined gradually with delay in sowing. However, in most winter pulses this can not be a viable approach as delayed sowing invariably results in reduced yield. Sinha *et al.* (1988) reported that early sowing (10 August) and closer row spacing (30 cm) reduced the weed growth and increased the dry matter accumulation, LAI, NAR, CGR and grain yield of irrigated pigeonpea at Kalyani (West Bengal).

PLANT GEOMETRY AND PLANT DENSITY

Planting density and pattern modify the crop canopy structure and in turn influence weed-smothering ability. Narrow row spacing will bring variation in microclimate viz., light intensity, evaporation and temperature at soil surface. The establishment of a crop with a more uniform and dense plant distribution may result in better use of light, water and nutrients and lead to greater crop competitive ability. Crops grown in narrower rows start competing with weeds at an earlier stage than those in wide rows because of more rapid canopy closure and probably better root distribution. Increased shading at soil surface will smother weed growth. Narrow row spacing are known to suppress the weed growth (Sinha *et al.* 1988), however, degree of suppression varies greatly. In *Phaseolus vulgaris*, found that closer row spacing suppressed weeds effectively than wider row spacing.

Planting density and pattern modify crop canopy structure and in turn influence weed smothering ability. Narrow row spacing will bring variation in microclimate *viz*; light intensity, evaporation and temperature at soil surface. The increased shading of soil surface will smother weed growth. Increased population of maize (60.000 to 90,000 plants/ha) and decreased spacing were found to reduce the growth of weeds (Angiras and Singh, 1989). Bi-directional sowing and closer row spacing were quite effective in suppressing the growth of *Phalaris minor* in wheat (Table 10).

Table 10: Effect of cross-sowing and closer row spacing on weed growth and grain yield of wheat (Azad *et al.*, 1988).

Treatment	Weed dry weight (g/m^2)	Grain yield of wheat (kg/ha)
Unweeded	519	2050
Hand weeding	193	2950
Cross sowing	249	2900
Closer spacing (15 cm)	211	2900

NUTRIENT MANAGEMENT

Crop-weed interactions and weed community dynamics can be greatly influenced by crop fertilization strategy. Faster nutrient release is often advantageous to weeds, which are usually able to take up nutrients in earlier growth stages more quickly and efficiently than crops. Crops and weeds generally compete for the same nutrient pool. Increasing the levels of soil fertility can alter the competitive interactions between crops and weeds. Resource use by weeds often increases more rapidly with added nutrients, resulting in a greater ability of weeds to compete for other resources. Nitrogen, the major nutrient for which the plants compete could be banded close to the crops row, thus enhancing the crop accessibility to the nutrient. Increasing rates of fertilizer application encourage more weed growth than the crop growth if no weed control measure is followed. Under this situation, it is better to apply fertilizers at lower rates than needed to maximize yields. Inclusion of green manurs as an organic source of nutrient supply not only add nutrients and organic matter in the soil but also suppress the weed growth due to its dense foliage cover on the ground surface and incorporation of existing weeds in the soil. Higher incidence of parasitic weed *Striga* spp. was observed in impoverished soils. Application of nitrogen not only reduced the intensity of *Striga* but also increased crop yields (Table 11).

Table 11: Effect of nitrogen and compost on *Striga asiatica* population (at 90 DAS) and yield of sorghum (Guled, 1982).

Treatment	*Striga* population/m²	Grain yield (kg/ha)
Control	166	4300
Nitrogen 100 kg/ha	7	6300
Nitrogen 200 kg/ha	6	7000
Nitrogen 300 kg/ha	3	7300
Compost 7.5 t/ha	131	4400
Compost 15 t/ha	140	4700
Compost 7.5 t/ha + 100 kg N/ha	19	5700
Compost 15 t/ha + 100 kg N/ha	12	5800

WATER MANAGEMENT

Proper water management can reduce weeds in both wet and dry seasons. More weeds emerge earlier at high soil moisture than at low moisture level. Many weeds can not germinate under flooded conditions. This is used as weed control method especially in rice crop. Maintaining water submergence (10-20 cm) in rice field early in the season creates anaerobic condition, which will be unfavourable for weed seed germination (Table 12) however, it may increase problems of other weeds such as blue green algae. Method of irrigation also influences the weed density. In drip irrigation the weed density is greatly reduced due to the wetting of soil near the plant base only (Table 13). Rest of the area remains dry and free of weeds. In some situations, decreased water availability favours crop growth by reducing the impact of competing weeds (Patterson 1995). Thus, under these situations, manipulation of soil moisture may become a reasonable strategy for weed management.

Table12: Effect of submergence on weed emergence (No./m²) in low land rice (Hosmani, 1991)

Weed species	Depth of water submergence (cm)			
	0	2.5	12	17
Fimbristylis littoralis	794	390	170	0
Cyperus difformis	143	155	78	0
Cyperus iria	28	41	3	0
Scirpus latifolia	21	49	3	1
Monochoria vaginalis	55	42	38	0

Table 13: Effect of drip and furrow irrigation on weed intensity in cabbage (Kudarimani, 1977)

Days after transplanting of cabbage	Number of weeds/m²		Dry weight of weeds (g/m²)	
	Drip	Furrow	Drip	Furrow
30	1409	1844	128	157
60	948	1530	129	170
80	352	1165	44	61

SOIL SOLARIZATION

Solarization is a method of heating the surface by using plastic sheets placed on moist soil to trap the soil radiation. The process would raise the surface soil temperature by 8-10°C as compared to non-solarized soils. A duration of 4-6 weeks is sufficient to give satisfactory control of weeds. Many annuals, some perennials and parasitic weeds are sensitive to this treatment. Besides controlling soil borne pests, soil solarization is reported to enhance the availability of nutrients in soil and favour beneficial microflora, ultimately resulting in increased plant growth response in many crops.

The response to solarization on weeds varies with weed species. Winter weeds are generally more susceptible, whereas summer weeds especially *Cyperus* spp. and *Convolvulus arvenseis* are generally more resistant. In perennials with an established underground systems of deep roots, rhizomes or tubers, failures of solarization is probably due to the limited soil depth, <10 cm, where temperature reaches to lethal levels. Most perennials are capable of regenerating rapidly from partially damaged underground organs. While a Solarization for a mere 10 days gave complete control of *Phalaris minor* and *Avena ludoviciana* – the most dominant grass weeds in winter season crops in India, there was an increase in the population of *M. indica* even at 40-day solarization. The heat tolerance of *M. indica* has been attributed to its thick seed coat. Parasitic weed broomrape (*Orobanche* spp.) has been effectively controlled with solarization (Meti, 1993; Jacobsohn *et al.* 1980). Sauerborn *et al.* (1989) came out with promising data on *Orobanche* control in fababean and lentil with 20-50 days of solarization treatment (Table 14).

Table 14: Duration of solarization on Broomrape infestation in Fababean and Lentil

Days solarized	Fababean		Lentil	
	No./m²	Weight (g/m²)	No./m²	Weight (g/m²)
0	115	152	61	33
20	57	122	32	37
30	33	65	25	35
40	26	57	22	29
50	13	41	4	4

Soil solarization for 4-6 weeks in summer (Table 15) not only gave excellent weed control in soybean but also in the succeeding wheat crop (Yaduraju and Ahuja, 1996). The weed control in soybean was better in soil solarization alone than with manual weeding (1 or 2) and pre-emergence application of pendimethalin (0.75-1.5 kg/ha). Soybean yields were significantly higher with soil solarization as compared to pendimethalin. This was attributed to the poor control of *Commelina benghalensis* and *Digera arvensis* with the herbicide. Soil solarization, on the contrary gave excellent control of these weeds.

Table-15: Effect of soil solarization on yield (t/ha) of soybean and wheat

Treatment	Soybean (8)*	Wheat (5)*
Non-solarized	0.75	2.27
Hand weeding	1.47	3.77
Herbicide	1.29	3.33
Soil solarization	1.95	4.16

*Mean of number of experiments

MECHANICAL METHOD

Weed removal in India is largely restricted to mechanical and cultural methods. Mechanical weed control is a viable option only within a certain range of soil conditions, the amplitude of which varies upon soil types and implements used. Excessive soil moisture impedes field workability and may delay mechanical weed control until the crop is too high or weeds are too well developed. In certain cases high soil moisture also increases the risk of crop damage. Therefore, it is important not to rely on direct physical weed control only but to integrate it in an overall crop and weed management strategy arrange at the system scale. Integration of direct weed control with other cultural practices in a system context is the best approach for tackling weed problems anywhere. Mechanical weed control involves removal of weeds with various tools and implements including manual removal by hand weeding. Inter-cultural operations are performed primarily to destroy the weeds present in the field and create favourable soil conditions for crop growth. Hand hoeing and manual weedings are most common practices performed for weed control in almost all pulse crops. One or two hand weedings at critical crop-weed stage provide satisfactory control of weeds in most of the pulse. Extra weeding may be required if crops are infested with perennial weeds. The use of mechanical weed control demands experience and knowledge of how to use the soil as a weed control agent. Climate and soil type play an important role in the possibilities for mechanical weed control. Monitoring the early development of weeds is necessary for timing of weed harrowing at the optimum stage. Mechanical weeders include cultivating tools such as hoes, harrows, times brush weeders, and cutting tools like mowers. The success of mechanical weeding depends upon the stage of weeds, crop geometry and climatic conditions. Hand weeding may also be used after mechanical inter row weeding to deal with weeds left in crop rows.

FLAME CULTIVATION

It is another successful alternative method that is catching up in the developed countries in the recent past. Flame cultivation involves the use of liquefied petroleum gas, mostly propane for pre-emergence and post-emergence weed control. Pre-emergence flaming is relatively problem-free, although the timing is critical. Selective flaming in the growing crop requires precision and carries a great risk of damage to the crop. Flaming equipments to burn off weeds has been developed in several counters including Germany, Holland, Sweden and Denmark (Holmoy and Netland, 1994). Selective flaming is performed in a wide variety of crops such as cotton, corn, soybean, and grain sorghum. Flame is used to a limited extent in fruits such as blueberries, raspberries and grapes.

BIOLOGICAL CONTROL

Biological control of weeds is the deliberate use of natural enemies, primarily insects or fungi, to suppress growth or reduce the population of weed species. The utilization of biological weed control agents in weed management offers many advantages including,

(a) A high degree of specificity for the target weed.

(b) No effects on non target and beneficial plants or man.

(c) Absence of weed resistance development.

(d) Absence of residue build up in the environment.

The two most common approaches are; the classical or inoculative method and the bioherbicide or inaundative strategy. The classical strategy is an ecological approach, which relies on the ability of an organism to multiply and spread following small-scale release. The organism then remains in balance with the target weed, keeping the latter at an acceptably low level.

The bioherbicide strategy, the second major strategy of biological control of weeds, is to attack a whole weed population with a single, non-persistent application of inoculum. The inoculum inundates the whole of the area infested by a weed to increase the effectiveness of the agent. Generally, the pathogens considered for this strategy are fungi, hence the commonly used name-mycoherbicides. The major mycoherbicide products, which are in use, are given in Table 16.

Eradication of the target population is not possible with biological control. However, rapid and complete weed control is not required, as long as the target weeds are reduced in their vitality and their competitive ability and biological control is complemented by other measures within an IWM system. Consequently, biological control could constitute an important element in the suppression below the economic threshold level of a particular weed population escaping other IWM measures.

The control of pests through biological control agents has been considered as the ideal method for managing pests under organic farming. No doubt, it is environmentally benign and ecologically and socially acceptable. However, their efficacy and economics are questionable. In weed management, their application is limited as the target weeds are many and the bio control agents are not expected to be effective on all of them. Moreover, control of one weed will result in substitution with other weeds, some of which could be more dangerous than the one that has been controlled. The use of insect pests as bio-control agents has application in situations where only one weed is a problem such as in aquatic ecosystem, range lands, forestry or waste lands. The control of *Parthenium hysterophorous* with Mexican beetle *Zygogramma bicolorata* and the control of water hyacinth with weevil *Neochetina spp* are a couple of successful examples of use insect pests for weed control. The Mexican beetle is very effective on *Parthenium* and excellent control of the weed has been reported from several places in India. However, the insect is active only during the rainy season and it hibernates in soil during the dry and winter months. Several species of fungi have been developed as mycoherbicides (Table 16) in developed countries for the management of few problematic weeds. However, they have a number of limitations and their use is limited to a very small area.

Table 16: Mycoherbicide products in use

Products	Agents	Target weeds
DeVine	*Phytophthora. palmivora*	*Morrenia odorata* (Strangler vine or milkweed)
Collego	*Colletotrichum gloesporoides* f.sp. *aeschynomene*	*Aeschynomene virginica* (Northern jointvetch)

ALLELOPATHY

Allelopathy has potential in integrated weed management. It is a harmful, direct or indirect, effect of one plant on another through the production of chemicals that enter the environment. Crop plants have the capacity to produce and exude allelochemicals into their surroundings to suppress the growth of weeds in their vicinity. Olofsdotter and Navarez (1996) screened 111 rice cultivars for allelopathic potential against *Echinochloa crus-galli* (L.) Beauv. In the fields at IRRI and found that 11 cultivars in dry season and 21 in wet season had suppressed the weed growth by more than 50%. Similarly in wheat, Wu *et al.* (1998) reported that out of 38 wheat cultivars screened, some cultivars significantly inhibited both the germination and the radicle growth of *Lolium rigidum* Gaud. The use of allelopathic plants, or substances isolated from them and produced transgenetically, may become an important form of weed control in the future.

NATURALLY OCCURRING HERBICIDES

Many secondary plant metabolites and microbial toxins are reported to have good herbicidal properties. These include 'Bialophos' under the trade name 'Herbiaceae' isolated from the fermentation of broths of *Streptomyces hygrocopicus* and *S. viridochromogenes* and shows herbicidal activity against wide spectrum of grasses and broadleaved weeds following foliar application (Mase, 1984). Methoxyphenone, a synthetic derivative of microbial toxin anisomycin is marketed in Japan as a selective herbicide for control of barnyard grass (*Echinichloa crus-galli*) in rice and is easily degraded in soil (Munakata *et al.* 1973). The non-selective commercial herbicide 'Basta' contains ammonium salt of glufosinate, an analogue of 'Phosphinothricin', which is the active in gradient of the microbial herbicide bialophos (Herbiaceae) (Kochar, 1983).

USE OF ECO-FRIENDLY HERBICIDES

Herbicides are the most successful weed control technology ever developed. The eco-friendly herbicides may be defined as one which is less harmful to environment, including human health or has benign impact on the environment through the various stages of development, manufacture, packing, distribution, use, consumption, disposal and recycling. They are selective, cost effective, fairly easy to apply, have persistence that can be managed, and there are many formulations and kinds available. Herbicides are also eco-friendly to environment if applied at proper dose, method and time. With the normal use rate, the quantity of herbicide applied to soil at one time is too small in relation to total soil volume to have any detectable influence on a soil's physical or chemical state. Moreover, the herbicides are applied either before sowing or with in one month after sowing of the crop and degrade from environment in a few weeks. Results of the experiments conducted under All India Coordinated Research Project on Pesticide Residue and All India Coordinated Research Project on Weed Control indicated that herbicides when applied at recommended doses did not leave any toxic residues either in soil environment or in food chain. Jaykumar *et al.*(1986) found that butachlor at recommended rate of 1.5 kg/ha could degrade before harvest of rice. However, at levels of 4.0 kg/ha, it could persist beyond harvest. The residues in grain and straw were at detectable levels at higher rates of butachlor (2.0 and 4.0 kg/ha) but were below the prescribed maximum residue limit (MRL). He further observed that the butachlor content at immediately after application were 0.211 and 0.231 ppm, respectively for doses 1.0 and 2.0 kg/ha which became non detectable at harvest (Jaykumar *et al.*1998). Herbicides are registered for use after meeting all the requirements of their efficacy and safety to user and the environment. The safety of the herbicides is frequently reviewed and if need be banned from use if found to have any harmful effect on environment. Integration of herbicides with other methods such as manual, mechanical, cultural or biological will be more beneficial, economical and eco-friendly.

During the last seven years (1996 to 2002), global adoption rates for transgenic crops have been unprecedented and reflect grower satisfaction with the products that offer significant benefits ranging from more convenient and flexible crop management, higher productivity or net returns/hectare, and a safer environment through decreased use of conventional pesticides, which collectively contribute to a more sustainable agriculture. Glyphosate is toxicologically and environmentally benign with low toxicity to non-target organisms, low or no ground water movement and limited persistence (Reddy, 2001). Glyphosate resistant soybean and corn varieties have found success in the USA and some other countries of the world. Considering the several advantages of using the herbicide resistant crops, it is worthwhile exploring their possible use under Indian conditions.

UTILIZATION OF WEEDS

Weeds like Water hyacinth, Eupatorium, Lantana, Parthenium, Ipomoea, etc., are widely spreading in farm environment, ponds, forest plantation, grassland and roadside at the cost of many other useful vegetation. Proper utilization of such biomass through appropriate technology may help in supplementing chemical fertilizers and adding organic matter to the soil. The tender and succulent portion of many of the dominant weeds with wider C:N ratio may also be utilized for vermicompost preparation which has enough potentiality for use in the production of organic tea as well as for vegetable farming.

STRATEGY FOR INTEGRATED WEED MANAGEMENT SYSTEM

Integrated weed management (IWM) is a component of integrated crop management (ICM). IWM involves three elements:

(a) Multiple tactics (eg., competitive varieties, cultural practices and herbicide usage)

(b) Weed population maintained at levels below that cause economic damage, and

(c) Conservation of environmental quality.

All crop management practices starting from field preparation to harvesting have implications on weed competition. Judicious use of herbicides, therefore, is given top priority. Lower rates of herbicides integrated with other methods of weed management are invariably more efficient, economical and sustainable than use of higher dose of herbicides alone. The use of herbicides is being scrutinized by many public and private organizations. There is concern about food safety; ground water and atmospheric contamination, increased weed resistance to herbicides, destruction to beneficial organisms and concern about endangered species. IWM has been defined as the application of numerous alternative weed control measures, which include cultural, genetic, mechanical, biological and chemical means of weed control. None of the individual control measures on their own can be expected to provide acceptable levels of weed control. However if the various components of IWM are implemented in a systematic integrated manner, significant advances in weed control technology can be achieved.

Buchanan (1976) defined integrated weed management (IWM) as .... the application of many kinds of technologies in a mutually supportive manner. It involves the deliberate selection, integration and implementation of effective weed control measures with due consideration of economic, ecological and sociological consequences". The research approach to the development of an IWM system must take all aspects of the cropping system into consideration. Invariably, each cultural practice influences the competitive ability of both the crop and the weed community leading to a multitude of complex interactions. However, efforts must be made to work within the existing production practice to ensure a greater likelihood of acceptance by the farming community. Although every agronomic practice starting from planting to harvesting of crop has implications on weed competition, farmers are neither aware of it nor appreciate it. It is important to educate farmers about this. The effect may be small for individual practice but effect would be substantial if all such practices are combined effectively. IWM is a knowledge intensive exercise where scientists may also be discouraged to get involved. In India, at least we are yet to generate information on economic threshold limits (ETL) for different weeds, weed seed dynamics in soil, understanding of eco-physiology of crop-weed competition, weed flora shifts in crops and cropping systems etc. It is time we started working on these. Besides developing effective IWM packages for different crops and cropping systems, it is very important to demonstrate them to farmers on large scale.

Thus, it is important to change the existing system in a progressive manner. This progression must be reflected in the research strategy. This would allow for the transfer of specific components through education and extension, while research continues to refine and further develop the system. Such a strategy is outlined in Fig.2.

FUTURE THRUST AREAS OF RESEARCH

1. Studies on crop rotation and weed seed bank dynamics
2. Development of cover/smother crops that suppress weeds without reducing crop yield
3. Development of crop cultivars with enhanced crop interference potential
4. Modeling of crop-weed interference and development of economic threshold limits for different weeds
5. Biological weed management
6. Development of herbicide resistant crops
7. Education and extension

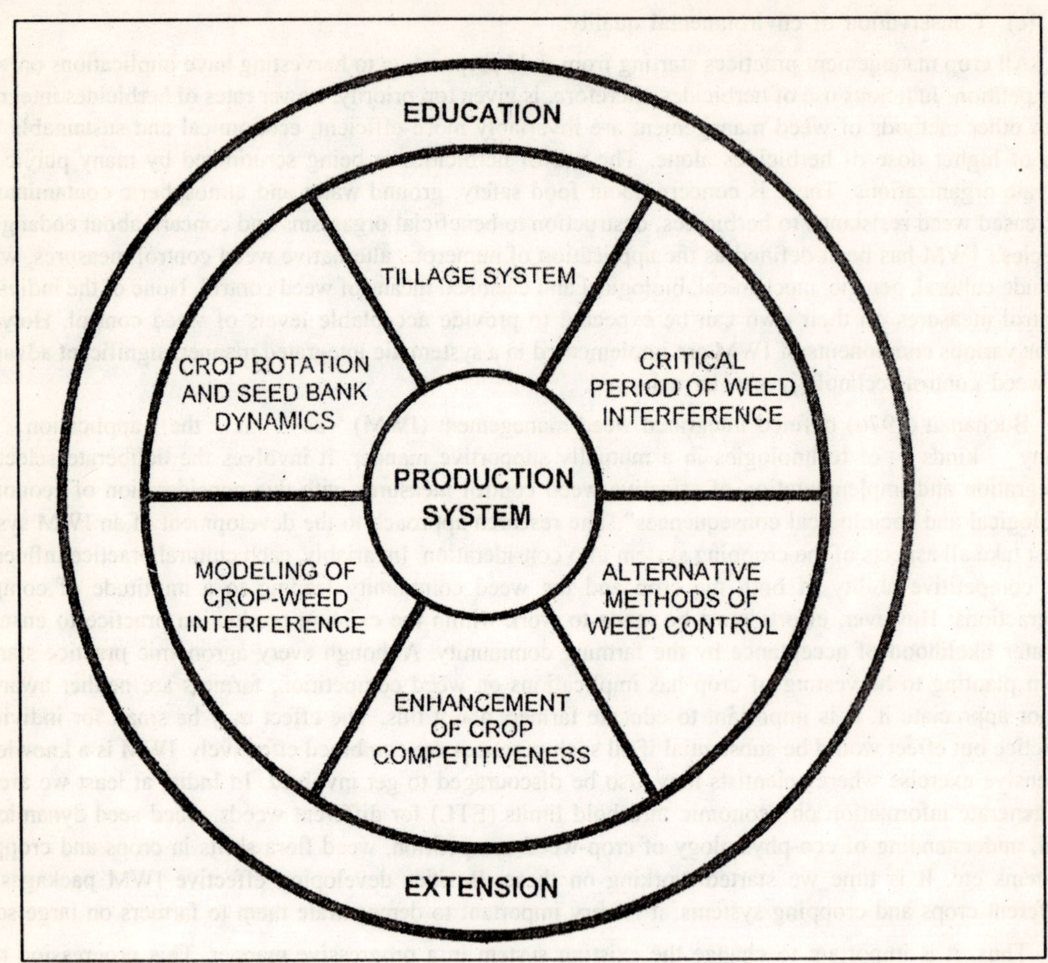

Fig 2: Research strategy for the development of an integrated weed management system

REFERENCES

Altieri, M.A. and M. Liebman.(1988). *Weed Management in Agro-ecosystems: Ecological Approaches* (Eds.). CRC Press, Inc., Boca Raton, FL

Angiras, N.N. and C.M. Singh (1989). *Indian J. Weed Sci.* 21:69-74

Ascard, J. (1994). Soil cultivation in darkness reduced weed emergence. *Acta Horticulturae* 372, Engineering for Reducing Pesticide Consumption & Operator Hazards, 167-177

Azad, B.S., H. Singh and A.N. Dubey (1988). *Pesticides* 22:41-42

Balyan, R.S. and R.K. Malik (1991). Competitive ability of wheat cultivars with wild oats. *Weed Sci.* 39: 154-158

Banga, R.S., A. Yadav and R.K. Malik (1997). Crop rotation-an effective mean to control resistant *Phalaris minor* in wheat. *Farmers and Parliament* 33 (3): 16-17

Barberi, P. (2001). Weed management in organic agriculture: are we addressing the right issue? *Weed Res.* 42, 177-193

Bhowmik, P.C, (1998). Herbicides in relation to food security and environment : a global perspective. In: *Proc. First*

Int. Agronomy Congress, Nov., 23-27, 1998, New Delhi, India. p : 244-258

Buchanan, G.A., (1976). Management of weed pests of cotton (*Gossypium hirsutum)*. In: *Proc. of US-USSR Symp.on the integrated control of the arthroped, disease and weed pests of cotton, grain, sorghum and deciduous fruits,* Lubbock T.X. pp. 168-84

Chauhan, D.S., R.K. Sharma, O.S. Verma, S. Nagarajan, Gyanendra Singh, and B.S. Tyagi (Eds.) (1998). Wheat: Research needs beyond 2000 AD. In: *Proc. of Int. Conf.*, 12-14 August 1997, Karnal, India, pp. 263-273

Derksen, D.A, A.G. Thomas, G.P. Layfond, H.A. Loeppky and C.J. Swanton (1994). Impact of agronomic practices on weed communities: Fallow within tillage system , *Weed Sci*, 42: 184-194

George, S.K., G. Donaldson and R.O. Clements (1997). Cover cereal bi-cropping with reduced agrochemical inputs; the yield and costs. *Aspects of Appl. Biol. 50, Optimising cereal inputs: its csientific basis*, 487-492

Guled, M.B. (1982). Effect of manuring and 2,4-D on *Striga* control in sorghum. *M.Sc.(Ag.) Thesis*, University of Agricultural Sciences, Bangalore

Gupta, O.P. (1984). *Scientific Weed Management.* Today and Tomorrow Printers and Publishers, New Delhi

Gupta, O.P. (1998). *Modern weed management*. Agro Botanica, Bikaner, India

Hall, M.R., C.J. Swanton and G.W. Anderson (1992). The critical period of weed control in grain corn (*Zea mays*). *Weed Sci.* 40:441-447

Holmoy, R. and J. Netland (1994). Band spraying, selective flame weeding and hoeing in late white cabbage , part 1. *Acta Horticulturae* 372, *Engineering for Reducing Pesticide Consumption & Operator Hazards*,223-234

Hosmani, M.M.(1991). *Cultural methods of Weed Management*. University of Agricultural Sciences, Dhrwad

Jacobsohn, R; A. Greenberger, J. Katan, M. Levi and H. Alon (1980). Control of Egyptian broomrape (*Orobanche aegyptiaca*) and other weeds by means of solar heating of the soil by polyethylene mulching. *Weed Sci.* 28: 312-316

Jaykumar, R., N. Kempuchetty, and S. Subramanian (1980). Quantitative determination of butachlor residues in rice and in subsequent greengram. In: *Abstract. Annual Conf.of ISWS*, Jaipur. p. 150

Jaykumar, R., S. Poongthai and A. Regupathy (1998). Studies on maximum residues of pesticides resulting from goog agricultural practices in rice. *Pestol.* 22: 48-51

Johnson, M.D., D.L. Wyse and W.E. Lueschen (1989). The influence of herbicide formulation on weed control in four tillage systems. *Weed Sci*. 37:239-249

Kocher, H. (1983). Influence of environmental factors on herbicide performance and crop yiel biology. (In) *Aspects of Appl Biol* 1. pp 227-234.

Kudarimani, H.B. (1977). A study on drip irrigation under Dharwad agro-climatic conditions. *M.Sc. (Ag.) Thesis*, University of Agricultural Sciences, Bangalore

Laesino, , R.Q. (1980). Ph.D., University of IOWA, Ames., pp . 170

Liebman, M. and E. Dyck (1993). Crop rotation and intercropping strategies for weed management. *Ecological Applications*. 3: 92-122

Malik, R.K., R.S. Balyan, and V.M. Bhan (1988). Effect of herbicides and planting dates on weeds in chickpea. *Indian J. Weed Sci.* 20:75-81

Mase, S. (1984). *Jpn. Pestic. Inf.* 45: 27-30

Meti, S.S. (1993). Control of *Orobanche cernua* is *bidi* tobacco by soil solarization. *M.Sc. Thesis*. UAS, Dharwad India

Mishra, J.S. (1997).Critical period of weed competition and losses due to weeds in major field crops. *Farmers and Parliament.* XXXIII (6):19-*20*

Mool Chand, R.L. Rajak, Radhey Shyam, O.R. Reddy, C.S.K. Vijaykumar and M. Indira (2000). Interception of weed seeds in wheat imported from Australia. *Indian J. Weed Sci.* 32 :210-212.

Munakata, K., O. Yamada, S. Ishida, F. Futasuya and H. Yamanota (1973). *Proc. Asian Pacific Weed Sci. Soc. Conf.* 4:215-219

Noda, K. (1977). In Fryer, J.D. and Matsunaka, S. (Eds.) *Integrated control of weeds*, University of Tokyo Press, Tokyo, pp. 17-46.

Oerke, E.C, H.W. Dehne, F. Schonbeck and A. Weber (1994). *Crop production and Crop protection: Estimated losses in Majoe food and Cash crops*, Elsevier 808 pp.

Olofsdotter, M. and D.C. Navarez, (1996). Allelopathic rice for *Echinochloa crus-galli* control. In: *Proc. 2nd Int. Weed Control Cong.*, Copenhagen, Denmark, 1175-1181

Patel, C.L., Z.G. Patel, and H.C. Patel (1983). Non-cash technology for minimizing weed growth in pigeonpea. *Indian J. Weed Sci.* 15:143-46

Patterson, D.T. (1995). Effect of environmental stress on weed/crop interactions. *Weed Sci.* 43: 483-490

Rao, A.N. and S.V.R. Shetty (1981). Investigations on weed suppressing ability of smother cropping systems in relation to canopy development and light interception. In: *Proc. 8th Asian Pacific Weed Sci. Soc. Conf.*, 357-364

Reddy, K.N. (2001). Glyphosate-resistant soybean as a weed management tool: Opportunities and Challenges. *Weed Biol. and Manage.* 1: 193-202

Sahoo, K.M. and V.N. Saraswat (1988). Magnitude of losses in the yields of major crops due to weed competition in India. *Pesticides Information*, 14(1): 2-9

Sankaran, S. and C.R. Chinnamuthu (1993). Integrated weed management in rice based cropping systems. In: *Proc. Int. Symp. Indian Soc. Weed Sci.*, Hisar, Vol I: 69-74

Sauerborn J., K.H. Linke, M.C. Saxena and W. Koch (1989). Solarization : a physical control method for weeds and parasitic plants (*Orobanche* spp.) in Mediterranean agriculture, *Weed Res.* 29: 391-397

Scopel, A.L., C.L. Ballare and S.R. Radosevich (1994). Photo-stimulation of seed germination during soil tillage. *New Phytol.*126: 145-152

Sinha, A.C., B.B. Mandal, and P.K. Jana (1988). Physiological analysis of yield variation in irrigated pigeonpea in relation to time of sowing, row spacing and weed control measures. *Indian Agriculturists*. 32:177-85

Walker, R.H. and G.A. Buchnan (1982). Crop manipulations in integrated weed management systems. *Weed Sci.* 30 (suppl.): 17-24

Welsh, J.P., L. Philipps, H.A.J. Bulson and M. Wolfe (1999). Weed control strategies for organic cereal crops. In: *Proc. Brighton Conf.- Weeds*, Brighton. U.K. 945-950

Wu, H., J. Partley, D. Lemerle, T. Haig and B. Verbeek (1998). Differential allelopathic potential in wheat accessions to annual ryegrass. In: *Proc. 9th Australlian Agronomy Conf.* Wagga, Australia, 567-571

Yaduraju, N.T. and J.S. Mishra (2002). Zero-tillage in rice –wheat cropping system on vertisols in Madhya Pradesh-Prospects and problems. *Paper for international workshop on herbicide resistance management & zero-tillage in rice-wheat cropping system.* March 4-6, 2002. CCSHAU, Hisar, India. pp 117-119

Yaduraju, N.T. and K.N. Ahuja, (1996). Effect of solarization with or without weed control on weeds and productivity in soybean-wheat system. In: *Proc. 2nd Int. Weed Control Cong.* Copenhagen. pp.721-727

Zimdahl, R.L. (1993). *Fundamentals of Weed Science*. Academic press, San Diego, CA,

16

Fungal Biocontrol of Plant-Parasitic Nematodes

D. PRASAD AND ARCHANA MITTAL

Plant-parasitic nematodes often cause plant diseases resulting in significant losses in crop yield and quality. However, uneven growth, stunting, chlorosis, poor quality and quantity of yield are general symptoms of severe infection and damage caused by plant-parasitic nematodes. Among the plant-parasitic nematodes, root-knot nematodes belonging to the genus *Meloidogyne* are regarded as serious pests constituting an important major limiting factor in the production of agricultural and horticultural crops. Though they attack almost every type of crop, causing considerable losses of yield and affecting the quality of produce, vegetable crops mostly suffer greatest damage. Bhatti and Jain (1977) found an estimated loss due to *M.incognita* throughout India in tomato is 46.2%.Thus, various measures have been in use for control of phytonematodes in the past are chemical, physical, regulatory, use of host resistance and biological methods. Fungi, bacteria and plant-parasitic nematodes constitute the significant component of soil ecosystem.

A number of chemicals both soil fumigants and systemic were successfully used for the management of plant-parasitic nematodes. However, most of the chemicals are banned and out of market due to their toxic effect on beneficial flora and fauna including man. They cause pollution hazards and their influence on non-target organisms have caused ecological imbalance which are sometimes difficult to be restored. Search for some alternative methods of nematode control thus received greater attention. In addition to the safety problems of pesticides, new crop protection technique, the IPM or Integrated Pest Management methods including organic amendments, fungal bioagents, botanical antagonists and reduced dose of nematicides has been proposed in order to reduce the effects of agrochemicals on the ecosystem. This is intended to maintain pest density at economic threshold injury level using a combination of various pest control methods.

Attempts are in progress in India to try fungal bioagents with very low dose of nematicides or organic amendments etc. as a part of integrated approach for the management of plant parasitic nematodes. The first report on fungal biocontrol agents was suggested by Lohde (1874) with observations on the endoparasite, *Harposporium anguillulae* in Germany. Kuhn (1877) reported *Catenaria auxillaris* in the females of beet cyst nematode, *Heterodera schachtii*. Contributions of Goswami and Rumpenhorst (1978), Kerry (1980), Khan (1990) and Leiz et.al, (1992) clearly showed that the biocontrol agents, particularly the nematode destroying fungi are common and abundant in both natural and agricultural soil as also in all kinds of decaying organic materials.

Fungal bioagents of phytonematodes are grouped into three categories—

A. Predaceous or trapping fungi

B. Endozoic or endoparasitic fungi

Division of Nematology, Indian Agricultural Research Institute, New Delhi 110012.

C. Egg parasitic or opportunisitic fungi

a.) Predaceous or nematode trapping ones mostly belong to the order Zoopagles (Zygomycetes) and from order Moniliales (Deuteromycetes) which capture nematodes by various devices as a) Sticky and b) mechanical traps. Sticky traps are of 3 types 1) sticky branches e.g., *Dactylella lobata* 2) sticky network e.g., *Arthrobotrys oligospora* and 3) sticky knobs e.g., *Dactylella ellipspora* while the mechanical traps are of 2 types a) non-constricting rings e.g., *Dactylaria candida* and b) constricting rings e.g., *Dactylella bembicodes*. Sucrose decomposition in the soil stimulates and increase in both the population of free living nematodes and the activity of indegeous nematodes trapping fungi and soil conditions viz., pH, temperature, moisture and availability of nutrients (Mankau,1968). The use of nematophagous fungi as bioagents of nematodes was shown by Upadhyay and Dwivedi (1988)

b.) Endozoic or endoparasitic fungi which are natural enemies of plant-parasitic nematodes produce either simple spores or flagellate spores as infectious agents and instead of producing hyphal development outside the body of the host, germination of spores are within the nematode body. Till they make contact with the host the spores remain viable but dormant. Simple spores are ingested by the nematodes, reach the oesophagus/ buccal cavity where they germinate, penetrate the oesophagus and colonise the body cavity e.g., *Harposporium anguillulae*. The sticky spores stick to the nematode at any point but mostly cause infection in the head region. They germinate, penetrate directly through the nematode cuticle and produce infective hyphae within the body cavity. E.g., *Meria coniospora, Nematoctonus bisporus*. The flagellate spores stick to the nematode and encyst before germination, penetration and colonisation of the host *Catenaria unguillulae, C. vermicola, Harposporium anguillulae* among the fungi was to be found effective in checking nematode population (Lodhe,1874). *Nematophthora gynophyla* in addition to *Globodera rostochiensis* and *Heterodera avenae* also parasitised females of *H. trifolii, H. schachtii,* etc. (Kerry and Crump,1977). In-vitro observations on the infection of *Meloidogyne incognita* eggs by *C. anguillulae* was reported by Wyss et. al. (1992) whose embryos were observed to be killed within a few minutes within mass aggregation and encystment of flagellate spores of fungus. However, the obligate nature of both predaceous and endozoic fungi difficulty in mass production and inability to establishment in soil are the characteristic that put a serious question mark for their candidacy as biocontrol agents.

c.) Fungal parasities or opportunistic fungi of eggs/cysts:

Several soil fungi found to be parasitic on eggs of plant parasitic nematodes colonise the reproductive structures of the nematodes particularly *Meloidogyne, Heterodera* and *Globodera* species whose sedentary stages are most vulnerable to fungal attack. These fungi occur in genera as *Paecilomyces, Verticillium, Fusarium, Phoma, Gliocladium* etc. These opportunistic fungi attack nematode eggs and/or cysts of the group Heteroderidae and those deposited in gelatinous matrix. Plant parasitic nematodes that become sedentary upon their maturity and have oviposition nature are considered to be the most vulnerable to attack to these fungi. As soon as these fungi contact egg masses or cysts, they rapidly grow and colonise these eggs that are in early embryonic developmental stages.

Arai *et al.* (1973) isolated peptide antibiotics lencinostatin, lilacinin from *Penicillium lilacinum. P.lilacinus* exhibits chitinase (Okafor, 1967; Gintis *et al.*, 1983) and proteolytic activity (Endreeva *et al.*, 1972). It also produces a peptidal antibiotic P168 which exhibited wide range of toxicity on fungi, yeast and gram-positive bacteria (Isogai *et al.*, 1980a, 1980b, 1981).The embryonated eggs treated with culture filtrates showed morphological changes in eggs and within 2-4 days, development of eggs stopped, embryo disintegrated and vacuole formed (Fitters *et al.*, 1993). The death of mature eggs, colonised internally was due to fungal metabolites (Carneiro and Gomes,1993). P.lilacinus can grow well in temperature ranges between 15 °C and 30 °C which is similar to its hosts. It can also survive well in wide range of soil pH, thus showing its competitive nature in most agricultural soils (Jatala,1986). This fungus is reported to be compatible with many fungicides and nematicides. (Davide and Batino, 1985). *P. lilacinus* showed growth at 3.0 to 11.0 while optimum mycelial growth occurred at pH 6.0 to 9.0 and sporulated at pH 4.0 to 6.0. The good sporulation was observed at 30 C and mycelial growth inhibited at 10 C and 40 C. The alternate light and darkness provided better fungal growth (Villanueva and Davide, 1984).

Cabanillas *et al.* (1989b) recorded maximum fungal growth of *P.lilacinus* between the temperature ranges of 24 C and 30 C and least was at 12 C and 36 C. The effect of soil temperature on fungal parasitism has not been studied so far. However, it has significant impact on the survival and biocontrol activity of *P.lilacinus* isolates (Cabanillas *et al.*, 1989b). Fioretto and Villacorta (1989) found that optimum temperature for *P.lilacinus* growth was 24 C to 25 C but biostatic at % C and 35 C. The lower threshold temperature was –10.9 C for development of the fungus. This attribute could be useful for storage of *P.lilacinus*. Liu *et al.*(1995) observed that among the 20 isolates of *P.lilacinus,* optimum temperature (25 C to 30 C) requirement for growth and sporulation was not different among the isolates. Certain chemicals viz., ethoprophos, fenamiphos, aldicarb, oxamyl and carbofuran were reported to be toxic to *P.lilacinus* in PDA-culture (Equiguren-Carrion, 1995).

Amoncho and Sasser (1995) found that benomyl stopped the *P.lilacinus* growth and chlorothalonil, capatan and PCNB suppressed the radial growth but fenamiphos slightly enhanced the fungal growth. The biocontrol potential of the fungus, *V.chlamydosporium* for root-knot nematodes has been reviewed by Leij and Kerry (1991). The fungus required some external nutrient for its establishment in soil. The colonisation of nematode egg masses depended on fungal inoculum and on galling caused by nematodes (Leij et.al, 1992). The efficiency of *P.lilacinus* in controlling *M.incognita* and *G.pallida* was demonstrated by Franco et.al, (1981). The eggs were deformed by the fungus with the help of toxic metabolites. Mature females of root-knot nematodes were usually penetrated by the fungus through the anus or vulva while cysts of *G.pallida* through the vulva and exposed or broken neck region (Jatala,1986).They reported that hyphal penetration of egg shells were brought about by mechanical pressure and/or enzymatic activity. Rate of fungal penetration into egg shell of *Meloidogyne* eggs is faster than *Globodera* where the egg shells show more complexity (Jatala, 1985). Work has been carried out for managing root-knot nematode infecting vegetables in India (Goswami and Mittal, 2000). Khan and Hussain (1986) reported that females, eggs and juveniles of *M.incognita* were infected by *Fusarium solani.*

There are various ways for soil-borne fungi to suppress nematode multiplication. A detailed review of fungi as biocontrol agents against plant-parasitic nematodes has been published (Kerry and Jaffee,1997). In summary, there are four mechanisms that fungi use to suppress nematodes. Some of these interactions are direct whereas others are indirect. The direct mechanism is performed by 1) fungi that feed on nematodes directly, known as nematophagous fungi: fungi interact with nematodes in an indirect manner by several mechanisms including: 2) fungi that kill nematodes by mycotoxin or 3) through the destruction of the feeding sites of sedentary nematodes in roots: 4) fungi that are nonpathogenic to plants, but compete with nematodes in roots and siginificantly reduce nematode multiplication. Many of these fungi are used as potential nematode biocontrol agents: 5) Mycorrhizal fungi improve the growth of nematode infected plants and may also affect nematode development.

This chapter also summarizes groups of nematophagous fungi according to their feeding habits:

A) Nematode-trapping fungi

Facultative fungi that form trapping structure to trap nematodes. There are 6 types of traps reported by Barron (1977).

Adhesive hyphae:

e.g. *Zygomycotina*

Stylopage

Cystopage

Adhesive traps:

e.g. Deuteromycota

Monacrosporium cionopagum (branches)

M. ellipsosporum (knobs)

Arthrobotrys oligospora (networks)

Non-adhesive traps

e.g. Deuteromycota

Arthrobotrys dactyloides (constricting ring)

(non-constricting ring)

B) Facultative parasitic fungi attacking sedentary stages of nematodes (Kerry and Jaffee, 1997)

These are facultative fungi that are commonly soil saprophytes and are opportunistic fungi isolated from the sedentary stages (female and egg stages) of sedentary nematodes such as *Heterodera*, *Globodera* and *Meloidogyne*. They do not form specialized infection structures except appressoria. They can survive and proliferate in soil in the absence of nematodes.

e.g. Hyphomycotina

Acremonium

Cylindrocarpon

Fusarium

Paecilomyces

Verticillium

C) Endoparasitic fungi (Kerry and Jaffee, 1997)

These are obligate parasitic fungi that have limited growth in soil outside the colonized nematode cadaver. They can infect vermiform nematodes by producing adhesive spores attached to cuticle of passing nematodes.

e.g. Hyphomycotina

Hirsutella rhossiliensis

Drechmeria coniospora

Verticillium spp.

Some can infect vermiform nematodes by producing conidia spores that can be ingested by nematodes.

Harposporium anguillulae

Some can infect vermiform nematodes by producing motile zoospores that encyst on the nematode's suface.

e.g. Oomcota

Myzocytium spp.

Lagenidium spp.

Chytridiomycota

Catenaria anguillulae

Some can infect sedentary nematodes when the nematodes were exposed on the root surface.

e.g. Oomycota

Nematophora gynophila

Methods for the isolation of Fungal bioagents

Stirling and White (1982) used baiting technique to isolate obligate parasites in which usually a large number of susceptible nematodes from laboratory culture are added to soil and then extracted after a period of incubation and assessed for infection. The egg-parasitic fungi which are consistently associated with the eggmasses of root-knot nematodes are isolated, subcultured and attempted for mass culture after testing their potentiality as biocontrol agent through invitro studies and pot expts. But, for the isolation of facultative parasites sprinkling techniques is suitable or in some cases selective media i.e, carrot extract, Potato Dextrose agar, etc. can be used.

Maintenance of stock culture

The culture of toxic and egg-parasitic fungi are maintained on Potato Dextrose Agar medium slants in test tubes and Potato dextrose broth in 250ml conical flasks after incubating in BOD at 25±2 °C for 10-15 days. Spores of source inoculum containing same medium serves as a seed culture for further mass multiplication.

Mass production of *Paecilomyces lilacinus*

Biocontrol programmes include enhancement of naturally occurring antagonists or augmentation of their activity by introducing other antagonist for suppression of target pathogen or pest of crops. Such augmentation could be achieved through release of mass produced parasites or predator. Sharma and Trivedi (1987) observed maximum growth of *P. lilacinus* on sesame cake followed by cotton, mustard and groundnut cake. Among the waste materials, mungbean husk exhibited maximum growth followed by cotton seed, guar powder, gram powder and rice husks. The oilcakes had additive effect because of their nematotoxic properties. Bansal *et al.* (1988) evaluated agro-industrial wastes for mass production of *P.lilacinus*. The rice husk produced of *P. lilacinus*. The rice husk produced least spores while the wheat bran provided maximum spores which was comparable with rice grain. The sporulation was found enhanced in rice husk and rice bran with addition of molasses amended with nitrogen and phosphorus.

Zaki and Bhatti (1988) recorded higher spore counts (55.5 × 10 /g) in neem leaves and concluded that incorporation of neem leaves infested with fungus may have synergistic or additive effect on nematode population. Mani and Anandam (1989) also evaluated plant leaves, oil cakes and agro-industrial wastes as substrate material for mass production of *P.lilacinus*. The leaves of *Leucaena leucocephala* and neem were found suitable and supported higher spore load. Mani *et al.* (1989) observed that wheat, bajra, jowar and rice as substrate media were suitable for mass production of *P. lilacinus. T.semipenetrans* population was markedly decreased with increasing the level of fungal inoculum and the spore suspension was found more effective than fungus infested wheat grains.

Vicente *et al.* (1989) compared among the available rice grains for mass production of *P.lilacinus* and found that pounded commercial rice and unpounded ground rice were most suitable on the basis of spore count while unpounded commercial rice had least spore load. Zaki and Bhatti (1991) found that *P.lilacinus* sporulated profusely on maize, gram, oats, rice and wheat grains. The maximum number of spores/g was observed on rice (52.8x10) followed by gram (24.5 x 10). The fungus grown on gram seed was the most effective against *M. javanica* infecting tomato plants. The reduction of gall index (78.6 %), population of *M.javanica* (94.2 %) and higher percentage of egg destruction were observed.

Patel *et al.* (1991) tested various organic amendments for their effects on growth and sporulation of *P. lilacinus* and found that karanj, neem and mahua cakes were most effective while pressmud and FYM were least effective. Bansal *et al.* (1992) observed that wood charcoal was a good carrier of *P. lilacinus* for applying in the field. The carrier material in low-density polyethylene pouch could support up to 10 spores/g for at least six months. The spore viability in charcoal packets was not affected by constant temperature 28 C and ambient temperature 14 C to 19 C or even under aerated / or unaerated condition.

Zaki and Maqbool (1991) obtained most effective control of *M. javanica* in chickpea with application of *P.lilacinus* (2g /pot) one week before nematode inoculation. Zaki (1994b) suggested that 4g of gram infested *P.lilacinus* per kg soil was optimum for significant reduction of gall index, and *M.javanica* population in tomato plants. The egg infection and destruction were 58 % and 66 % respectively. Jonathan *et al.* (1995) found greater reduction of *M.incognita* population on Piper betle with application of *P.lilacinus* infested rice substrate @ 8g/kg soil. Patel *et al.* (1995) observed that soil application of *P.lilacinus* at 3 % to 5 % (w/w) on neem cake was effective against *M.javanica* infecting groundnut whereas seed treatment was ineffective. Davide and Zorilla (1995a) found that tuber-dip and soil-mix method of application of *P.lilacinus* were equally effective against *G.rostochiensis* on potato and their combined application enhanced nematode control efficacy.

Work done by several researchers have proved that some microorganisms, especially fungi produced metabolites that are toxic to root-knot nematodes (Goswami and Mittal, 2000). Thus, Mankau in 1968 reported nematicidal activity of culture filtrates of *Aspergillus niger.* Singh and Sitaramaiah in the same year showed parasitisation of *M. javanica* by *Curvularia* species. Later, Singh *et al.* (1983) reported that culture filtrates of *Aspergillus niger, Curvularia lunata, Trichoderma viride* and *T. lignorum* obtained from rhizosphere of tomato plants proved to be nematoxic and inhibited hatching of *M. incognita* larvae. Tabreiz et. al., (1884) studied the effect of culture filtrates of 8 species of *Aspergillus* on hatching and mortality of *M. incognita* out of which *A. niger, A. terreus* and *A. fumigatus* were more toxic than the other species. Sharma and Saxena (1992) observed that culture filtrates of *Rhizoctonia solani* and *Trichoderma viride* adversely influenced hatching of *M. incognita* larvae. Both fungus showed relatively more toxicity. Khan *et al.* (1988) observed inhibition of hatching of *M. incognita* juveniles at different concentrations of fungal culture filtrates and complete inhibition was recorded at 50 % and 100% concentrations. The filtrates suppressed the root galling of tomato plants caused by *M. incognita.*

Chawla and Goswami (1998) in an vitro evaluation of undiluted *A.niger* culture filtrate, demonstrated the complete suppression of larval emergence for the first three days and the total death of juveniles within 24 hrs in case of *M. incognita.* Siddiqui and Mahmood (1995) while working on the management of root-rot disease of chickpea, demonstrated than filtrate of *A. niger* markedly reduced number of nematode in the soil which in turn was attributed to the concentration of oxailc acid produced by the fungus. Moreover, autoclaved culture filtrate also immobilized the nematodes, the heat stability of toxin principle.

Cayrol *et al.*, (1989) reported nematicidal properties of culture filtrates of *P.lilacinus.* The toxins production were different on different media and their effect differed on various nematode species. The mechanism of toxic activity was considered to be neurotropic. Acetic acid was identified from culture filtrates of *P.lilacinus* and it affected the movement of nematode juveniles (Djian *et al.*, 1991). Caroppo *et al.* (1990) found ovicidal activity of *P.lilacinus, P. maraquandi, P.variotii* and *P.coleus* against M.incognita. The culture broths of these fungi reduced 78 to 93 % egg hatching under exercised root culture of tomato. Khan and Khan (1992) observed that concentration of culture filtrates of *P. lilacinus* had direct relation to the inhibition of hatching and percentage mortality of *M. incognita.*

Siddiqui and Hussain (1991) tested that culture filtrate of *P.lilacinus* was quite effective against *M. incognita* and *Macrophomina phaseolina* while Zaki (1994a) reported that culture filtrate of *P.lilacinus* had immobilising effect on *M. javanica* and inhibited hatching of eggs. Oduor and Waudo (1995) also reported that *P.lilacinus, P.herbarum* and *Fusarium oxysporum* had suppressive effect on hatching of *Meloidogyne* spp. but hatched juveniles were not parasitized. A reduced hatching ranging between 21 % and 56 % was observed in *Heterodera glycines* by *P.lilacinus* (Chen-Sen Yu *et al.*, 1996).

Experimental results from the infested fields in different countries have indicated that *P.lilacinus* effectively controls *M.incognita* and even better than number of commonly used nematicides (Candanedo *et al.*, 1983; Jatala, 1983). The effect of single and multiple application of *P.lilacinus* against *M. incognita* under field trials revealed that only one inoculation was sufficient to establish it and to get substantial degree of nematode control (Jatala *et al.*, 1981). However, Cabanillas and Barker (1989) indicated that effective biocontrol for *M. incognita* required more than one application and at proper time.

Davide and Zorilla (1985) while investigating the biocontrol potential of *P. lilacinus* against *M. incognita* on okra found the fungus to be quite effective and economically better than nematicides. According to Shahzad and Ghaffar (1987) carbofuran @ 1kg a.i. / ha was less effective than *P.lilacinus* against *M.incognita.* This fungus also effectively controlled nematodes on okra and mung in subsequent seasons but nematicide was no longer effective. Similarly, in the fungus treated plots, higher yields were recorded than carbofuran @ 2.5 kg a.i./ha. The efficacy of *P. lilacinus* had been comparable to that of most commonly used nematicides (Jatala, 1985). The fungus, *P.lilacinus* reduced the soil population of *M. incognita* by 66 % to 77% where as isazophos reduced it by 89 % (Davide and Zorilla, 1986).

Dube and Smart (1987) obtained higher crop yields and reduced *M. incognita* population when *P. lilacinus* and *Pasteuria penetrans* were applied together in field microplots. Maheshwari and Mani (1988) also found that simultaneous inoculation of *P.lilacinus* and *Pasteuria penetrans* were more effective and capable of significant reduction of soil population of *M. javanica.* The introduction of *P.lilacinus* 10 days before at planting of tomato resulted in great protection against *M. incognita* and mid-season inoculation also provided higher percentage of egg mass infection (Cabanillas and Braker, 1989).

Khan and Esfahani (1990) found that *P. lilacinus* was more effective for controlling *M.javanica* infecting tomato when fungus was added prior to nematode or simultaneous inoculation. A great number of nematode eggs were infested, juvenile development inhibited, eggs devoid of juveniles and only filled with fungal mycelium. The developed juveniles were also found attacked and mycelial growth over their bodies were observed.

Walia *et al.* (1991) reported that application of *P.lilacinus* through wheat bran was better than carbofuran @ 1kg a.i./ ha in gall reduction on okra plants infected with *M. javanica.* Oduor *et al.* (1996) observed plant performance and low gall index with the application of *P. lilacinus* in combination with aldicarb or nematicidal plants.

The effect of *Calotropis procera* and *G. fasciculatum* for the management of *M.incognita* in tomato was studied by Rao *et al.* (1996). Significant reduction of galls, eggs/egg mass observed in interactive effect of these components. Higher colonisation of mycorrhiza in root of tomato inoculated with *G. fasciculatum* and *Calotropis* leaf was observed. Gaonkar and Sreenivasan (1994) observed a positive influence of locally available organic amendments on proliferation of *G. fasciculatum.* Devi and Goswami (1992) and later Bhagawati *et al.* (2000) demonstrated that both *G. fasciculatum* in case of cowpea while *G.etunicatum* in tomato respectively together with mustard cake helped in reducing the disease severity caused by *M.incognita* fungus complex in both the above hosts. They observed that the pre-establishment of VAM fungus checked the entry of *M.incognita* larvae as also colonisation of pathogenic fungi *Macrophomina phaseolina* and *F.oxysporum f.sp.lycopersici* respectively

Goswami and Mittal (2002) studied the effect of some fungal bioagents on root-knot nematode, *Meloidogyne incognita* infecting brinjal In vitro studies showed that *Aspergillus niger* was highly toxic followed by *Aspergillus fumigatus* exhibiting 85% mortality at 1: 10 dilution while *Paecilomyces lilacinus* and *Geotrichum candidum* showed least mortality. *Aspergillus niger* and *A.fumigatus* showed maximum inhibition of hatching and *G. candidum* the least. However, remarkably high egg parasitization was observed by *Paecilomyces lilacinus* (75%) and *Geotrichum candidum* (72%). In vivo studies showed that *Aspergillus niger* and *Paecilomyces lilacinus* along with mustard cake gave maximum reduction in root-knot nematode population both in soil and root.

Hemlata *et al.* (2001) studied the efficacy of biocontrol agents (*Trichoderma harzianum*, *Paecilomyces lilacinus* and *Aspergillus niger*) for the management of root-knot nematode on chickpea. Crop growth was best with *P.lilacinus* followed by *T. harzianum*, *A. niger* and a combination of all three biocontrol agents.

Maximum reduction of root-knot nematode was recorded with *P.lilacinus* followed by a combination of all biocontrol agents, *T. harzianum* and *A. niger*.

Ruqiya *et al.* (2000) studied the female parasitism of *Aspergillus* species and their effect on root penetration by root-knot nematode, *Meloidogyne javanica*. Maximum reduction in the root penetration of okra by *Meloidogyne javanica* was recorded with *A. niger*.

Rao *et al.* (1997) studied the management of root-knot nematode, *Meloidogyne incognita* on tomato by integration of *Trichoderma harzianum* with neem cake. Significant increase in plant growth and reduction in root galling and final population of *M. incognita* were observed in tomato seedlings transplanted in neem cake amended soil incorporated with *T. harzianum*.

Among the fifteen different fungi tested in vitro by Khan *et al.*, (2000) *Paecilomyces lilacinus* and *Trichoderma harzianum* exhibited the best performance followed by *Verticillium chlamydosporium*, *Aspergillus niger*, *Aspergillus flavus*, *Fusarium solani*, *Penicillium* spp, *Cephalosporium* spp, *Fusarium oxysporum*, *Harposporium helicoids*, *Rhizoctonia solani*, *Arachniotus* species and *Pythium ultimum*. *P. lilacinus* gave the maxiumum larval mortality of juveniles followed by *Trichoderma harzianum*.

Combination of fungal bioagents

In recent years with the development of a number of fungal bioagents isolated from egg masses of *M.incognita* infecting vegetables and categorization of two types i.e, a) toxic and b) egg parasitic ones .Attempts have been to test the performance of two types alone as well as together.The combined effect of a toxic bioagent *Aspergillus terreus* with an egg parasitic one, *Paecilomyces lilacinus* showed better performance in reducing nematode population with improved plant health of tomato than when either of the above fungi was used alone. (Goswami and Sharma,1999).

Commercialization of Nematophagous Fungi

In France two commercial agents, Royal 300 and Royal 350 were sold for control of *Ditylenchus myceliophagous* on mushrooms and *Meloidogyne* spp. on vegetables which being at a very high cost, have not been widely used. More recently, the facultative parasite, *P.lilacinus* has been produced in the Philippines as Bicon which can be applied for the control of several nematode species including root-knot nematodes. The American firm Mosanto is going to launch two commercial products one each of *P.lilacinus* and *V. chlamydosporium* respectively in the near future. *Trichoderma viride* has also been mass cultured and tried in farmer's (mainly in nursery) in India. Attempts for commercialisation of *T.viride* and *P.lilacinus* and other potential fungi are in progress.

FUTURE PROSPECTS

1. Integrating ecofriendly methods like organic amendments, fungal bioagents and botanical antagonists to phytonematodes would prove a very promising management package and also profitable to our farmers.
2. Since many of the biopesticide viz., *T. viride, P. lilacinus* and *A. niger* has already been commercialized both in India and abroad, it is advisable to search for an isolate which would prove effective against both important fungal diseases as well as nematode diseases in under field conditions.

REFERENCES

Amoncho, A and Sasser, J. N. (1995) Biological control of *Meloidogyne incognita* with *Paecilomyces lilacinus*. *Biocontrol*. 4: 51-61.

Arai, M., Mikami, Y., Fukushima, K., Utsumi, T and Yazawa, K. (1973). A new antibiotic, leucinostain, derived from *Penicillium lilacinum. J. Antibiot*. 26: 157-161.

Barron, G. L. (1977). The nematode – destroying fungi, Canadian Biological Publication Ltd.

Bansal, R. K., Walia, R. K and Bhatti, D. S. (1998). Evaluation of some agro-industrial wastes for mass production of nematode parasitic fungus, *Paecilomyces lilacinus. Nematol. Medit.* 16: 135-136.

Bansal, R. K., Walia, R. K. and Bhatti, D. S. (1992). Wood charcoal powder, a carrier of *Paecilomyces lilacinus* spores. *Nematol. Medit.* 20: 5-7.

Bhatti, D. S and Jain, R. K. (1997). Estimation of loss in okra, tomato and brinjal yield infested with *Meloidogyne incognita. Indian J. Nematol.* 1: 37-41.

Bhagwati, B ; B.K. Goswami and C. S. Singh (2000) Management of disease complex caused by root-knot nematode and wilt fungus through bioagents. *Indian J. Nematol.* (In press).

Cabanillas, E. and Barker, K. R. (1989) Impact of *Paecilomyces lilacinus* inoculum level and application time on control of *Meloidogyne incognita* on tomato. *J. Nematol.* 21: 115-120.

Cabanillas, E., Barker, K. R and Nelson, L.A. (1989b) Growth of isolates of *Paecilomyces lilacinus* and their efficacy in biocontrol of *Meloidogyne incognita* on tomato. *J. Nematol.* 21: 164-172.

Candanedo, E., Lara, J., Jatala, P and Gonzales, F. (1983). Control biologico del nematodo *Meloidogyne incognita* con el hongo *Paecilomyces lilacinus.* Abstr. XXIII Meeting, APS Caribbean Division.

Carneiro, R. M. D. G and Gomes, C. B. (1993) Methodology and pathogenicity tests effect of *Paecilomyces lilacinus* and *P. fumosoroseus* isolates to eggs of *Meloidogyne javanica. Nematologia Brasileira.* 17: 66-75.

Caroppo, S., Perito, B and Pelagatti, O. (1990) In vitro evaluation of nematicide activity by several fungi against *Meloidogyne incognita* eggs. *Redia.* 73: 451-462.

Cayrol, J. C., Djian, C and Pijarowski, L. (1989) Study of the nematicidal properties of the culture filtrate of the nematophagous fungus, *Paecilomyces lilacinus. Revue Nematol.* 12: 331-336.

Chawla, G and Goswami. B. K. (1998) In-vitro evaluation of culture filtrate of soil fungi on hatching of egg and mortality of *Meloidogyne incognita* and their in-vivo effect on root-knot nematode infecting mungbean. Proceedings of the " National symposium on rational approaches in nematode management for sustainable agriculture." November, 23-25, 1998. B. A. College of Agriculture, Anand Campus, Anand, India. pp17.

Chen-Sen Yu, Dickson, D. W., Mitchell, D. J and Chen, S. Y. (1996) Pathogenicity of fungi to eggs of *Heterodera glycines. J. Nematol.* 28: 148-158.

Davide, R. G and Batino, E. (1985) Biological control of root-knot nematodes on cotton through the use of fungi *Paecilomyces lilacinus* (Thom) Samson and *Gliocladium roseum* Bainler as seed treatment. *Phil. Agric.* 68: 159-167.

Davide, R. G. and Zorilla, R. A. (1985) Evaluation of a fungus *Paecilomyces lilacinus* for the biological control of root-knot nematodes *Meloidogyne incognita* on okra as compared with nematicide isazofos. *Phil. Agric.* 68: 493-500.

Davide, R. G and Zorilla, R. A. (1986) Evaluation of a fungus *Paecilomyces lilacinus* for the biological control of root-knot nematodes *Meloidogyne incognita* on okra compared with a nematicide isazofos. *Int. Nematol. Network Newsl.* 3: 32-33.

Davide, R. G and Zorilla, R. A. (1995a) Evaluation of *Paecilomyces lilacinus* and some nematicides for the control of potato cyst nematode, *Globodera rostochiesis. Biocontrol* 1(3): 69-76.

Devi, T. P and B. K. Goswami (1992) Effect of VA-mycorhiza on the disease incidence due to *Macrophomina phaseolina* and *Meloidogyne incognita* on cowpea. *Annals of Agric. Res.* 13: 253-265.

Djian, C., Pijarowski, L., Ponchet, M., Arpin, N and Favre-Bonvin, J. (1991) Acetic acid: a selective nematicidal metabolite from culture filtrates of *Paecilomyces lilacinus* (Thom) Samson and *Trichoderma longibrachiatum* Rifai. *Nematologica* 37: 101-112.

Dube, B and Smart Jr., G. C. (1987) Biological control of *Meloidogyne incognita* by *Paecilomyces lilacinus* and *Pasteuria penetrans. J. Nematol.* 19: 222-227.

Eguiguren-Carrion, R (1995) Control of *Meloidogyne incognita* and *Nacobbus* sp. with *Paecilomyces lilacinus* in the greenhouse and effect of nematicides on the fungus. *Biocontrol.* 4: 41-49.

Endreeva. N, A., Ushakova, V. I and Egorov, N. S (1972) Study of proteolytic enzmes of different strains of

*Penicillium lilacinum (*Thom) in connection with their fibrinolytic activity. *Microbiologiya* 41: 364-368.

Fioretto, A. M. C and Villacorta, A (1989) Thermal requirements for the development of the nematogenic fungi *Paecilomyces lilacinus. Pesquisa Agropecria Brasileira* 24: 975-978.

Fitter, P. F.L., Belder, E-den and Den, Belder, E (1993) A time-lapse technique to study the effect of fungal products on embryogenesis of nematode eggs. Mededelingen Van de Faculteii Landbouww-etenschappen, Universiteit-Gent. 58: 751-756.

Franco, J; Jatala, P and Bocangel, M (1981) Efficency of *Paecilomyces lilacinus* as a biocontrol agent of *Globodera pallida. Journal of Nematol*. 13 (4): 338-339.

Gintis, B. O., Morgan-Jones, G and Rodriguez-Kabana, R (1983) Fungi associated with several developmental stages of *Heterodera glycines* from an Alabama soybean field soil. *Nematropica* 13: 201-213.

Goswami, B. K and Mittal, A (2002) Effect of some fungal bioagents on root-knot nematode, *Meloidogyne incognita* infecting brinjal. *Pakistan J. Nematol.* 20: 1, 55-59.

Goswami, B. K and Rumpenhorst, H. J (1978) Association of an unknown fungus with potato cyst nematode, *Globodera rostochiensis* and *G. pallida. Nematologica*, 24: 251-256.

Goswami, B. K and Sharma, S. B (1999) Management of root-knot disease of tomato with integration of fungal bioagents (*Aspergillus terreus* and *Paecilomyces lilacinus)* Abstract. Nat. Symp. On Pl. Dis. Scenario under changing agro-ecosystems. Oct. 29-30, 1999. HPKV, Palampur, HP., 42pp.

Goswami, B. K and Mittal, A (2000) Effect of *Aspergillus fumigatus* and *Fusarium solani* alone and in combination along with neem oilseed cake against *Meloidogyne incognita* infecting tomato. *Annals of Agric. Res.* (in press).

Gaonkar, S. B. N and Sreenivasan, M. N (1994) Effect of inoculation with *Glomus fasciculatum* in conjunction with different organic amendments on growth and yield of wheat. *Microbiol. Res.* 4: 419-423.

Hemlata, P and Pandey, G (2001) Efficacy of bio-control agents for the management of root-knot nematode on chickpea. *Annals of Plant Protection Sciences*. 9: 1, 157-159.

Isogai, A., Suzuki, A., Higashikawa, S., Kuyama, S and Tamura, S (1980a) Constituents of a peptidal antibiotic P 168 produced by *Paecilomyces lilacinus* (Thom) Samson, Agr. Biol. Chem. 44: 3029-3031.

Isogai, A., Suzuki, A., Higashikawa, S and Tamura, S (1980b) Structure of peptidal antibiotic P 168 produced by *Paecilomyces lilacinus. Agr. Biol. Chem.* 44: 3033-3036.

Isogai, A., Suzuki, A., Higashikawa, S. Kuyama, S and Tamura, S (1981) Isolation and biological activity of a peptidal antibiotic P 168. *Agr, Biol. Chem.* 45; 1023-1024.

Jatala, P (1981) Jatala, P (1981) Biological control of *Meloidogyne* species: Methodology for preparation and establishment of *Paecilomyces lilacinus* for field inoculation. Int. *Meloidogyne* Proj. Proc. 3rd Res. Plann. Conf. On Root-knot nematodes, *Meloidogyne* spp. Jakarata Region VI, 228-231.

Jatala, P (1983) Potential and prospects of biological control of nematodes. Abstr. XXIII meeting APS Caribbean Division.

Jatala, P (1985) Biological control of nematodes. In: An Advanced Treatise on *Meloidogyne*. Vol. I. Biology and Control. (Eds. Sasser, J. N and Carter, C. C) N. C. S University Graphics, Raleigh, 303-308.

Jatala, P (1986) Biological control of plant parasitic nematodes. *Ann. Rev. Phytopathol*. 24: 453-489.

Jonathan, E. I., Padmanabham, D and Ayyamperumal, A (1995) Biological control of root-knot nematode on betelvine, Piper betle by *Paecilomyces lilacinus. Nematol. Medit.* 23: 191-193.

Kerry, B. R (1980) Biocontrol: fungal parasites of female cyst nematodes. *J. Nematol* 12: 253-259.

Kerry, B. R and Crump, D. H (1977) Observations on fungal parasites of females and eggs of the cereal cyst nematodes, *Heterodera avenae* and other cyst nematodes. *Nematologica.* 23: 193-201.

Kerry, B. R and B. A. Jaffee. (1997) Fungi as biological control agents for plant-parasitic nematodes. in The Mycota K. Esser and P. A. Lemke, eds., Vol. IV, pp. 204-218. Springer, verlag Berlin Heidelberg.

Khan, M. W., Asharf, S. S and Haider, S. R (1988) An analysis of biocontrol efficacy of *Paecilomyces lilacinus* in management of root-knot nematodes. Int. Conf. Res. Plant Sciences and its Relevance to the Future. Delhi, India, 67.

Khan, M. W and Esfahani, M. N (1990) Efficacy of *Paecilomyces lilacinus* for controlling *Meloidogyne incognita* on tomato in greenhouse in India. *Pakistan J. Nematol.* 8:95-100.

Khan, S.T and Khan, T. A (1992) Effect of culture filtrates of soil fungi on the hatching and mortality of root-knot nematode, *Meloidogyne incognita. Curr. Nematol.* 3:53-60.

Khan, M. W and Hussain, S. I (1986) Parasitism of *Meloidogyne incognita* by *Fusarium solani. Intern. Nematol. Netw. Newsl.* 3(2): 11-13.

Khan, H. U; Waqar, A; Ahmed, R and Khan, M. A (2000) Evaluation of culture filtrates of different fungi on the larval mortality of *Meloidogyne incognita. Pakistan J. Phytopathology.* 12:1, 46-49.

Kuhn, J (1877) Vorlaufiger Bericht uber die bisherigen ergebnisse der selt Jahre 1875 in auftrage des Vereins fur rubezucker Industrie Z. ver. Rubenzucker Indus. Disch. Reiches. 452-457.

Leiz, De. F. A. A. M and Kerry, B. R (1991) The nematophagous fungus, *Verticillum chlamydosporum* as a potential biocontrol agent for *Meloidogyne arenaria. Revue de Nematologie*, 14 (1): 157-164.

Leiz, De. F. A. A. M ., Kerry, B. R and Deunchy, J. A (1992) The effect of fungal application rate and nematode density on the effectiveness of *Verticillum chlamydosporum* as the biocontrol agent for *Meloidogyne incognita. Nematologica*, 38 (1): 112-122.

Liu, X. Z., Liu, W. M and Zhang, D. S (1995) Biological characteristics of *Paecilomyces lilacinus* associated with soybean cyst nematode. *Chinese J. Biological Control.* 11: 70-74.

Lodhe, G (1874) Einge heven parasitishehen pilze. Tageblatt der 47 Versamunhung deutscher haturforseher und Aerzste Berslan: 203-206.

Maheshwari, T. U and Mani, A (1988) Combined efficacy of *Pasteuria penetrans* and *Paecilomyces lilacinus* on the biocontrol of *Meloidogyne javanica* on tomato. *Int. Nematol. Network Newl.* 5: 10-11.

Mani, A., Murthy, I. R., Rao, P. K and Prasad, V. D (1989) Growth of *Paecilomyces lilacinus* on natural substrates and its efficacy against citrus nematode, *Tylenchulus semipentrans. J. Biological Control.* 3: 59-61.

Mankau, R (1968) Reduction of root-knot disease with organic amendments under semifield conditions. *Plant Dis. Reptr.* 52: 313-319.

Oduor, O. P., Sikora, R. A., Waudo, S. W and Schuster, R. P (1996) Effects of aldicarb and mixed cropping with *Datura stramonium, Ricinus communis* and *Tagetes minuta* on the control and integrated management of *Meloidogyne javanica. Nematologia* 42: 127-130.

Oduor, O. P and Waudo, S. W (1995) Effects of fungi on hatching and parasitism of eggs, juveniles and females of root-knot nematode. *Pakistan. J. Nematol* 13: 119-127.

Okafor, N (1967) Decomposition of chitin by microorganisms isolated from a temperate and tropical soil. *Nova Hedwigia* 13: 209-226.

Patel, R. R., Patel, D. J and Patel, B. A (1991) Influence of organic amendments on growth and sporulation of nematophagous fungus, *Paecilomyces lilacinus* (Thom) Samson. *Curr. Nematol.* 2: 39-40.

Patel, D. J., Vyas, R. V., Patel, B. A and Patel, R. S (1995) Bioefficacy of *Paecilomyces lilacinus* in controlling *Meloidogyne javanica* (Pathotype 2) on groundnut. *Int. Archais Newsl.* 15: 46-47.

Rao, M. S; Reddy, P.P and Das, S. M (1996) Effect of irrigation of *Calotropis procera* leaf and *Glomus fasciculatum* on the management of *Meloidogyne incognita* infesting tomato. *Nematol. Medit.* 74 (1): 59-61.

Ruqiya, A; Zareen, A; Zaki, M. J and Khan, S (2000) Female parasitism of *Aspergillus* species and their effect on root penetration by root-knot nematode *Meloidogyne javanica. Pakistan J. Phytopathology* 12: 2, 115-117.

Rao, M. S; Reddy, P. P and Nagesh, M (1997) Management of root-knot nematode, *Meloidogyne incognita* on tomato by integration of *Trichoderma harzianum* with neem cake. *Zeitschrift-fur-Pflanzenkrankheiten-und-Pflanzenschutz,* 104: 4, 423-425.

Shahzad, S and Ghaffar, A (1987) Field application of *Paecilomyces lilacinus* and furadan for the control of root-knot disease of okra and mung. *Int. Nematol. Ntework Newsl.* 4: 33-34.

Sharma, A and Trivedi, P.C (1987) Screening substrates suitable for the growth of *Paecilomyces lilacinus. Int. Netmatol. Network Newsl.* 4: 24-26.

Sharma, M and Saxena, S. K (1992) Effect of culture filtrates of *Rhizoctonia solani* and *Trichoderma viride* on hatching of larvae of root-knot nematode, *Meloidogyne incognita. Curr. Nematol.* 3: 61-64.

Siddiqui, Z. A and Hussain, S. I (1991) Control of *Meloidogyne incognita* and *Macrophomina phaseolina* on chickpea by fungal filtrates. *Pakistan J. Nematol.* 9: 131-137.

Siddiqui, Z. A and Mahmood, I (1995) Management of *Meloidogyne incognita* race 3 and *Macrophomina phaseolina* by fungus culture filtrates and *Bacillus subtilis* on chickpea. *Fundm. Appl, Nematol.*, 18: 71-76.

Singh, S. P; Veena, P; Khan, A. M and Saxena, S. K (1983) Inhibitory effect of culture filtrates of some rhizosphere fungi of tomato as influenced by oil-cakes on the mortality and larval hatch of *Meloidogyne incognita. Nematologia Medit.* 11(2): 119-123.

Stirling, G. R and White, A. M (1982) Distribution of a parasite of root-knot nematodes in South Australian vineyards. *Pl. Dis.* 66: 52-53.

Tabreiz, A; Khan, T. A; Hussain, S. I and Azam, M. F (1884) Effect of culture filtrates of eight species of *Aspergillus* on the hatching and mortality of *Meloidogyne incognita. Indian J. Nematol.* 14: 50-74.

Upadhaya, K. D and Dwivedi, K (1988) Role of fungi in biological control of plant-parasitic nematodes. *Brit. Crop. Prod. Conn.*, 3: 1205-1208.

Vicente, N. E., Acosta, N and Sanchez, L. A (1989) Rice (*Oryza sativa L.*) substrates and growth of the fungus *Paecilomyces lilacinus,* a nematode control agent. *J. Agric. Univ. Puerto Rico* 73: 79-82.

Villaneva, L. M and Davide, R. G (1984) Influence of pH, temperature, light and agar media on the growth and sporulation of a nematophagous fungus, *Paecilomyces lilacinus* (Thom) Samson. *Phil. Agric.* 67: 223-231.

Walia, R. K., Bansal. R. K and Bhatti, D. S (1991) Effect of *Paecilomyces lilacinus* application time and method in controlling *Meloidogyne javanica* on okra. *Nematol. Medit.* 19: 247-249.

Wyss, U; Vossand, B and Janson, H. B (1992) In-vitro observation on the infection of *Meloidogyne incognita* eggs by the zoospore fungus, *Catenaria anguillulae. Fundm. Appl. Nematol.* 15 (2): 113-139.

Zaki, F. A (1994a) Effect of culture filtrates of *Paecilomyces lilacinus* on *Meloidogyne javanica. Nematol. Medit.* 22: 41-43.

Zaki, F. A (1994b) Dose optimisation of *Paecilomyces lilacinus* for the control of *Meloidogyne javanica* on tomato. *Nematol. Medit.* 22: 45-47.

Zaki, F. A and Bhatti, D. S (1988) Economical method for mass culturing of *Paecilomyces lilacinus* (Thom) Samson. *Curr. Sci.* 57: 135.

Zaki, F. A and Bhatti, D. S (1991) Effect of culture media on sporulation *of Paecilomyces lilacinus* and its efficacy against *Meloidogyne javanica* in tomato. *Nematol. Medit.* 19: 211-212.

Zaki, M. J and Maqbool, M. A (1991) *Paecilomyces lilacinus* controls *Meloidogyne javanica* on chickpea. *Int. Chickpea Newsl.* 25: 22-23.

Management Strategies for Control of Plant-Parasitic Nematodes

ARCHANA MITTAL[1] AND D. PRASAD[1]

Plant-parasitic nematodes are microscopic invertebrate animals often called as threadworms, eelworms or roundworms. They are distributed all over the world in different kinds of habitats and is found in nearly every biological niche that supports life. They damage the crops not only by feeding on plants but also by interacting with various other organisms. They cause severe losses to economical important crops like vegetables, cereals, pulses, oilseeds, fruit crops, etc. Many estimates have been made on crop losses due to nematodes by research workers. (Table -1) Annual losses worth as high as Rs. 40,000,000/= in wheat and Rs. 30,000,000/= in barely due to cereal cyst nematode (*Heterodera avenae*) in Rajasthan alone have been reported (Seshadri and Dasgupta, 1980). Sasser and Freckman (1987) have indicated an annual crop losses due to phytonematodes on world wide basis to the tune of $100 billion. The yields of okra, tomato and brinjal 27.3 per cent losses respectively due to *Meloidogyne incognita* infestation @3-4 larvae g/soil under field conditions (Bhatti and Jain, 1977).

Palanisamy and Balasubramanian (1981) reported that reniform nematode, *Rotylenchulus reniformis*, cause 14.9, 6.0, 4.8, 13.2 and 8.7 per cent loss in yield of cotton, maize, fingermillet, cowpea and blackgram respectively. Mathur *et al.*, (1986) reported the losses in yield of wheat was upto 42.2% in sandy soil of Rajasthan due to *Heterodera avenae.* Van Berkum and Seshadri (1978) estimated a loss of 10 million dollars due to "earcockle" disease on wheat, 8 million dollars due to "Molya disease" in barley in Rajasthan alone and 3 million dollars to coffee due to lesion nematode, *Pratylenchus coffeae.* Ogunfowora (1987) reported a yield loss of 20-59% to cowpea by plant-parasitic nematodes.

A committee in 1971 on the estimation of crop losses constituted in the U.S.A estimated 6 per cent loss in field crops, 12 per cent loss in fruit and nut crops, 11 per cent loss in vegetable crops and 10 per cent loss in ornamental crops which gives an annual monetary loss of 1590.7 million dollars due to nematodes. Losses estimated due to nematode attack to different cultivated crops all over the world by F.A.O is around 400 million dollars (Anon, 1985). In India, root-knot nematodes (*Meloidogyne incognita* and *Meloidogyne javanica*) caused 51 per cent reduction in bidi tobacco transplant production (Shah *et al.*, 1981) and 50 per cent reduction in tobacco cured leaf (Patel *et al.*,1986).

The degree of damage caused by nematodes depends upon the population density of nematodes, susceptibility of the crop, environmental condition such as soil fertility, moisture and also the presence of other pathogenic micro-organisms which may interact with nematodes as they are known to cause complex plant disease in association with fungi, bacteria and viruses. Crop losses due to nematode appear in different forms. It may be the loss in yield, the deterioration in quality of the product and the increase in the cost of production, etc. Sometimes nematodes are involved in many disease complexes with fungi,

1. Division of Nematology, Indian Agricultural Research Institute, New Delhi-110 012, India.

bacteria and viruses, etc. It is now well established that the plant damage done by two or more pathogens is much more severe than what the pathogens cause independently. Continuous feeding of nematodes provides entry points to other pathogenic micro-organisms and breaks the physical as well as mechanical barriers of the host. In such a complex situation, nematodes may either act as wound causing agents, as resistance breaking agents or as modifiers of host tissues or as vectors. The losses in our agriculture production due to plant parasitic nematodes justify the need for adopting their different management practices to check their population and level of infestation.

Table 1: Estimated loss in yield of various agricultural crops due to phytonematodes.

Phytonematodes	Crop	Loss (%)	Reference
Aphelenchoides besseyi	Rice	12.22	Sivakumar, 1988
R. reniformis	Cotton	14.9	Palaniswamy and Balasubramanian, 1981
R. reniformis	Fingermillet	4.8	Palaniswamy and Balasubramanian, 1981
R. reniformis	Maize	6.0	Palaniswamy and Balasubramanian, 1981
M. javanica	Mung	42.1-93.4	Gupta and Verma, 1993
M. incognita	Okra	28.08	Parvatha Reddy and Singh, 1981
M. incognita	Okra	99.9	Bhatti and Jain, 1977
M. incognita	Cowpea	12.3	Patel *et al.*, 1993
M. incognita *M. javanica*	Greengram	12.7	Patel *et al.*, 1993
M. incognita	Brinjal	27.3	Bhatti and Jain, 1977
M. javanica	Pigeonpea	14.2	Patel and Patel, 1993
M. javanica	Urdbean	23.0	Ali, 1995

The various diseases caused by plant-parasitic nematodes are listed in Table-2.

Because of the serious losses caused by the plant-parasitic nematodes principal management strategies i.e, Physical, Chemical, Cultural, Biological, Regulatory and IPM are undertaken to minimize the crop losses caused by nematodes.

Nematode Management

A long/medium term strategies by combining low input components alone or together with other management practices to bring down the population level below threshold level should be the main aim for successful cultivation of crops. The management of phytonematodes can be achieved by the integration of different farming practices that fall into four broad strategies. (Bridges, 1996)

1. Using direct, non-chemical, cultural and physical methods particularly crop rotation and soil cultivation,
2. Encouraging naturally occurring biological control agents by an understanding of cultivation methods and appropriate use of soil amendments,
3. Preventing the introduction and spread of nematode-free planting material, and
4. Enhancing the biodiversity inherent in traditional farming systems that use multiple cropping and multiple cultivars to increase the available resistance or tolerance to nematodes.

Table 2: Diseases caused by Plant-parasitic nematodes in different agricultural crops

Causal organisms	Disease	Host plants
Anguina tritici (Wheat seed gall nematode)	Ear-cockle	Wheat
Ditylenchus angustus (Rice stem nematode)	Ufra	Rice
Ditylenchus dipsaci (Stem nematode)	Onion bloat	Onion
Tylenchulus semipenetrans (Citrus nematode)	Citrus die back	Citrus
Globodera spp. (Potato cyst nematode)	Potato cyst	Potato
Heterodera avenae (Cereal cyst nematode)	Molya	Barley
Rotylenchulus reniformis (Reniform nematode)	Reniform nematode	Pulse crops, Vegetables
Pratylenchus coffeae (Root lesion nematode)	Root lesion nematode	Coffee
Meloidogyne spp. (Root-knot nematode)	Root-knot nematode	Vegetables, Pulses, Ornamentals,
Hirschmanniella oryzae (Rice root nematode)	Mentek	Rice
Ditylenchus destructor (Stem nematode)	Red rot	Potato
Aphelenchoides besseyi (Foliar nematode)	White tip	Rice
Rhadinaphelenchus cocophilus (Coconut stem nematode)	Red ring	Coconut
Radopholus similis (Burrowing nematode)	Spreading decline	Banana

The following management strategies/practices bring down the nematode population significantly:

Cultural Methods

Fallowing

In many parts of India after harvest of rabi crops, the field is left fallow without host crops. This practice is widely used in Northern India which reduce the nematode population. If the fallow is coupled with summer ploughings, the decrease of endoparasite nematodes especially root-knot nematode was recorded to be as much as 40 per cent. The main principle behind this are starvation of nematodes due to non-availability of host crop and killed by solar heat and soil desiccation. Most of the phytonematodes do not survive in upper layers of soil. However, there are some problems with fallowing. It is a poor conservative practice and also cause monetary loss to the farmers.

Summer ploughing

This is a good practice for controlling plant-parasitic nematodes. At least two summer ploughings during May and June, under the north Indian conditions, will go a long way in reducing the soil population of nematodes as well as other soil borne organisms which will help the farmer in getting a profitable crop. Haque and Prasad (1982) reported drastic reduction of plant parasitic nematodes and significantly increased the yield with three deep ploughings.

Solarisation

The solar energy is preserved with the help of transparent polyethylene sheet to increase .soil temperature enough to kill the nematodes present in soil profiles. This is only applicable in limited areas. Prasad and Yaduraju (1993) however indicated that 30 days solarisation is enough to achieve 95% control of nematodes. Prasad *et al.* (1995) found about 90 per cent kill of root-knot and reniform nematodes when solarised for 30 day during May-June months. Solarisation coupled with triazophos application @ 1.5 l/ ha completely erradicated the plant parasitic nematodes under field conditions. Kumar *et al.* (1993) reported significant decrease of weed seed emergence by over 90% and reduced the population of plant parasitic nematodes by about 80%. The growth of soybean was improved and seed yield increased by 78% following solarisation. This method provided an economical alternative for multiple pest, disease and weed control in small areas like nursery beds, orchards, forestry plantation and high value crops in regions with hot summers. This practice is very effective in reducing plant-parasitic nematodes, fungus, weeds and soil insects population. Kumar *et al.*, (1993) reported that nematodes population could be reduced to 90% and weed population upto 75%.

Rotation of crops

Crop rotation can be one of the most effective means of controlling nematodes in low-input agricultural systems. Rotating crops in a sequential cropping system is the best known and most widely used cultural practice in traditional and subsistence agriculture. The basic principle of crop rotation for nematode management is to reduce initial populations of damaging nematode species to levels that allow the following crop(s) to become established and complete early growth before being heavily attacked. Mostly, rotation of plant crops (Table – 3) is widely practised but rotations can also involve the sequential planning of plant crops with others like green manure, cover and trap crops, antagonistic plants, grass and other fallows.

Table3: List of selected non-hosts plants useful in crop rotation.

Non-hosts or Poor hosts	Nematodes spp.
Wheat	*Tylenchorhynchus mirzai*
Lettuce, onion, radish	*Pratylenchus* spp.
Peanut	*Pratylenchus leiocephalus*
Potato, tomato	*Tylenchorhynchus brassicae*
Crotolaria spectabilis	*Dolichodorus heterocephalus*
Fescue, orchard grass	*Meloidogyne incognita*
Cabbage, chilli, eggplant	*Hoplolaimus indicus*
Corn, greenbeans, red clover	*Globodera rostochiensis*
Crotolaria spectabilis	*Radopholus similis*
Castor bean	*M. hapla* *M. thamesi*
Cotton, peanut, sorghum	*M. javanica*
Watermelon, tobacco	*Belonolaimus gracilis*
Corn, cowpea, potato, cotton	*Heterodera glycines*
Been, corn, onion, potato, alfalfa	*H. schachtii*
Corn, alfalfa	*Helicotylenchus dihystera*

Antagonistic plants and Trap crops

The most widely reported antagonistic plants are *Tagetes* spp., mustard, *Asparagus*, sesame, sunhemp and neem. Sukul (1992) reported that many plants belonging to 57 families possess nematicidal properties. (Table-4) Antagonistic plants also often act as trap crops, i.e, plants that reduce nematode populations by allowing invasion but only partial development in the roots. Mixed cropping either with non-host or antagonistic crops such as *Tagetes*, Sesbania, mustard etc. have also been reported to reduce root-knot and reniform nematodes. The use of trap crops as green manure, not only reduce the nematode population but improves soil fertility in many ways and promotes antagonistic microbial activities. For example, sunhemp (*Crotolaria spectabilis*) or Dhaincha (*Sesbania*) which allow invasion by *Meloidogyne* spp. but does not support development and can serve as good trap crop.

Naidu *et al.*, (2000) studied the management of Kalahasti malady (*Tylenchorhynchus brevilineatus*) of groundnut through crop rotation in Andhra Pradesh, India, in 1991-94. Rabi groundnut preceded by summer rice and kharif rice recorded the greatest reduction in the nematode population (59.7%) and disease severity (42.6%) and the greatest increase in pod yield (83.3%) compared to continous cropping with groundnut. A succulent herb common in lowland rice fields, *Spenoclea zeylanica* produces nematicidal exudates that can give a 99% kill of the rice root nematode, *Hirschmanniella* spp. in field soil.

Table-4 Nematicidal principles found in botanicals and effective against nematodes.

Name of the botanicals	Family	Nematode species	Active principles
Asparagus spp.	Liliaceae	*Trichodorus Christie*	Glucoside Asparagusic acid
Azadirachta indica	Meliaceae	*M. incognita A. Juss*	Nimbidine, Thionimone
Carthamus	Asteraceae	*Aphelenchoides tinctorius L.*	3-cis, 11-transtrideca -1,3, triene,5,7,9-triyne and 3 transis-
Cymbopogaon Flexuosus wats	Poaceae	*Meloidogyne* spp.	Citral, geraniol,
Nicotaiana tabacum L.	Solanaceae	*M. incognita*	Nicotine
Ocunium sanctum L.	Lamiaceae	*M. incognita*	Eugenol
Polygonum hydropiper L.	Polygonaceae	*M. incognita*	Formic acid, acetic acid, essential oil
Solanum	Solanaceae	*M. incognita*	Solamargine
Tagetes spp.	Asteraceae	*Meloidogyne* sp.	Polythienyes, X-terthienye
T. patula	Asteraceae	*Meloidogyne* sp.	Hydroxytrematone bithienye derivative

Resistant varieties

It is cheap, easy to recommend and most practical method of phytonemtaode control. During the last few years, nematologists and breeders followed the conventional approaches of plant breeding to transfer nematode resistance genes into agronomically superior cultivated species. Crop cultivars resistant to phytonematodes can be the most useful and cheapest means of nematode control for small-scale farmers. This involves a cross between a resistant source usually a wild type related species with a commercial type plant species. In subsequent steps, F_1 hybrids are back crossed with cultivated species to combine resistance factor however, it is now accepted that on long-term basis polygenic horizontal resistance to nematodes is more desirable than vertical resistance governed by mono gene. Secondly, the presence of biotypes of *Meloidogyne* spp. which break the resistance of cultivated crops. Ibrahim *et al.* (1991) found that Lupin cv. Giza 1 and Giza 2 were resistant to *M. javanica* and groundnut cv. Giza 4 was resistant to *M. incognita*.

Raut and Sethi (1980) reported that soybean cultivars like Forest, JS-2, Selection Ds74-20-1 were found resistant against *M. incognita*. Kaul (1977) indicated intervarietal differences in sunflower to the infection of *M. incognita* however, variety EC69874 and Sunrise selection exhibited lower damage.

Campos (1987) screened 25 sunflower cultivars against *M. incognita* or *M. javanica* in Brazil and found Conte GH 7833 and Uruguai Selecaco to be highly resistant, COEX 7811 and Contisol slightly resistant and the rest were susceptible to *M. javanica*. On the other hand, Pereolovile was highly resistant to *M. incognita* while IAC Experimental, Contisol, Coex7922 and ESAL was slightly resistant and the rest were susceptible. Dahiya and Gupta (1992) screened 24 cultivars/lines of sunflower and found no variety to be resistant. Three cultivars of *Helianthus annuus* i.e, Giza 1, Makhatt, Miak were found to be susceptible to *M. javanica*, *M. incognita* and *R.reniformis*. The lowest nematode multiplication occurring with *R.reniformis* on cv. Giza and Miak (Montasser *et al.*, 1985).

Among oilseed crops, Prasad (1989) screened 500 varieties/lines of groundnut crop against *M. arenaria* and exhibited two lines as resistant namely C-41(NRCG-31) and NCAC-2196(NRCG-1060). Sakhuja and Sethi (1985) also reported Local Ambali, G-201, Local-256, and Japtin 220-15 resistant against *M. javanica* under glasshouse conditions.

Soybean cultivars like Forest, JS-2, selection DS 74-20-2 were found resistant against *M. incognita* (Raut and Sethi, 1980). Tirupathi-1 and 2 of groundnut cultivar showed tolerant reaction against Kalahasti Malady disease of groundnut. Sakhuja and Sethi (1986) revealed that 5 groundnut cultivars viz., G-201, Japlin 220-15, Local 256, Local Ambali and 4018 exhibited resistance to *M. javanica*. Mehan *et al.* (1993) found that one of the resistant genotypes is a high yielding breeding line (TCG 1518) and this is being released for use in disease-affected areas of Andhra Pradesh to Kalahasti malady caused by stunt nematode, *Tylenchorhynchus brevilineatus*.

Table 5: List of nematode resistant varieties.

Crop	Nematode	Resistant varieties
Tomato	*Meloidogyne hapla*	PI270435 (L. peruvianum)
	M. arenaria	VEN-8, 569N-10, Nematex
	M. javanica	Nematex, SL-120 NTR-1, NTDR-1, VFN-8, Anahu, Atkinson
	M. incognita	Nematex, SL-120 Nemared, Canner, Big seven, Anahu, Anahu R, Merbein, Pelican, Red Glow (H) Karnataka Hybrid, VFN-8, VFN Bush, Sunburst, Rovita, Monarch, Bonus (H), Hawaii-55, Monte carlo NTDR-1
Potato	Cyst nematode	Kufri Swarna
Sweet Potato	*M. hapla*	Heartogold
	M. javanica	Heartogold, Maryland, Golden

Contd.

Crop	Nematode	Resistant varieties
	M. arenaria	Gold Rush, Heartogold, Porto Rico, Maryland Golden
	M. incognita	Apache, Red Jewel, Heartogold, Hopi, Jaspier, Jewel, White star
Pea	*M. incognita*	C-50, A-70, B-58
French bean	*M. incognita*	Springwater, Half runner, Coffee wonder, Blue lake, Brown Beauty, Cambridge, Kenya-3, Pinto WS-114, Seaferer, WS-114
Wingedbean	*M. incognita*	LBNC3
Cow pea	*M. incognita*	C-152, IHR 29-5, 82-1-B, Barsati mutant, Chinese Red, Floricream, Missippi Silver, New era, Red Rippe, Zipper cream, 32-1B, V-16
	Rotylenchulus reniformis	
	Heterodera cajani	Sel A, Sel 1-2, Sel 4-1, P 33-2-1
Soy bean	*M. incognita*	Forest, JS-2, Selection, DS-74-20-2, Dyer, Kaikoo, Jackson, Bragg, Hampton, Hill, Amredo, Dare, Laredo, Monroe
	M. javanica	Hampton
	M. hapla	Bragg, Ilbroy
	M. arenaria	Dyer, Bragg
Chick pea	*M. incognita*	Selection No-501
Groundnut	*M. javanica*	Ambali 4018
Pigeon pea	*M. incognita*	S-80, ICPL-8, ICPL-81, ICPL-94, Pant A-10,
	R. reniformis	ICPL-16, ICPL-81, Pant A-10
Barley	*H. avenae*	DL 349, DL375, DL376, DL 379, Rajkiran, BP627, BP639, BP644,
	BP630	
Citrus	*Tylenchulus semipentrans*	Swingle cirnmelo, Trifoliate orange
Tobacco	*M. javanica*	Nrepanda
	M. incognita	Florida 22, Coker 86, 254, 258 & 347, Virginia 080 and 770.

Adjusting Planting Time /Escape Cropping

This is a practice of reducing nematode damage by growing crops at times that are not favourable for the nematodes, such as when temperatures are too low or too high for infection or development or when populations of the nematodes have declined to low thresholds. Sethi and Gaur (1986) found that adjustment of dates or seasons of planting can help to avoid nematode damage to crops for example crops such as tomato, egg plant and other susceptible vegetables are planted in the cooler winter (rabi) months to avoid root-knot nematode damage.

Gaur *et al.*, (1979) reported that sowing of chickpea one month later than the optimal planting time when temperatures are low, minimized the damage by *Meloidogyne incognita* mainly because it lengthened the life cycle of the nematode and reduced second and third generation root invasion.

Organic amendment

The use of organic amendment is a well established and age old practice. The three basic principles involved in management of phytonematodes with organic amendments are – habitat management, host modification and direct toxicity of the allelochemicals which are released through decomposition of organic materials of the different modes of action of organic amendments, the nematodes may be affected by a single-factor, but in nature the factors usually run concurrently. Farmers use plant and animal wastes in the form of compost and farmyard manure, crop residues, oilcakes, municipal refuse and industrial wastes have been used by farmers as they provide better media for plants to grow, result in better soil texture, increase water holding capacity, supply the nutrients to deficient soil and stimulate microbial population of actinomycetes, bacteria, fungi and elements of which might be antagonistic to nematodes. The decomposition of residues results in the accumulation in the soils of specific compounds that may be nematicidal. But, the main drawback in using most soil amendments for nematode management is the large quantities required. The organic amendments have not only been found effective in normal cultivated crops but are also well suited in nurseries of vegetables fruits and ornamental plants.

Das (2000) compared different methods of using neem and neem products against nematodes infecting pigeonpea. He found that soil application of neem seed powder and different neem based formulations (achook, neemark, neemgold, nimbecidine and field marshal) were the best followed by seed soaking and seed coating. The neem seed powder was found to be highly effective against *Meloidogyne incognita*, *Heterodera cajani* and *Rotylenchulus reniformis.* It was followed by neemgold and neemmark which gave good result when used as seed coating and seed soaking. Mishra (1988) also reported that application of neem and karanj cakes at the rate of 100 kg N per ha were found very effective in reducing nematode population infecting chickpea and lentil.

Physical Control

In this method of control, plant-parasitic nematodes are basically eradicated by steam sterilization, hot water treatment, electrical soil heating, radiation treatments, seed cleaning, heat treatment, etc. But, soil solarization is the most recent practice for control of phytonematodes. The nematodes are effectively controlled but the method is restricted to summer seasons and that too only in tropical and sub-tropical regions of the world.

Biological Control

These method includes the use of predaceous and parasitic organisms such as fungi, bacteria, viruses, nematodes, tardigrades, mites, etc. The biological control of *M.arenaria* on groundnut under greenhouse conditions by using *Paecilomyces lilacinus* and *Verticillium chlamydosporium* has been reported by Godoy *et al.*, (1983). Coosemans (1976) reported synergistic effect of *P.penetrans* and *Verticillum* spp. on flax. Presence of *M. arenaria* in the pods of groundnut led to increased invasion of shells and kernels by *Aspergillus flavus*. However, the damage by root-knot nematodes did affect *A. flavus* infection. Wilting period of groundnut plants was reduced by 13 to 16 days on inoculation with *Fusarium solani* and *M. arenaria*. Also *Aspergillus flavus* along with *M.arenaria* showed synergistic affect on the growth of groundnut (Patel *et al.*, 1985).

Regulatory Control

Regulatory methods are used to prevent the spread of nematodes from one place to other place and from one area to another area through infested material. This strategy involves the strict imposition of plant quarantine laws. This method is useful for checking the introduction of pests and pathogen but it can not check the activity of nematode species already well established in an area of country. It is also ineffective when nematodes are spread by wind, water, insects, etc. *Globodera rostochiensis* was introduced into Europe from South America through infested tubers, soon it found its way to Asia, North America.

Table 6: Neem pesticides tested against plant-parasitic nematodes

Plant-parasitic nematodes	Crops	Neem pesticides	Methods
Meloidogyne	Cowpea	Nimbicidine	Seed treatment
Heterodera cajani	Okra		
M. javanica	Tomato Brinjal, Okra, Cowpea	Neemark	Seed treatment Root dip, Nursery bed treatment
M. javanica *M. incognita*	Cowpea Okra, Brinjal, Tomato	Achook	Seed treatment, Root dip, Nursery bed treatment
M. javainca	Brinjal, Tomato	Neemgaurd	Seed treatment, Root dip, Nursery bed treatment
Rotylenchulus reniformis *Pratylenchus delattrei*	Tomato	Margocide	Seed treatment, Root dip, Nursery bed treatment
M. incognita	Okra	Nimbasol	Seed treatment
Pratylenchus spp.	Soybean	Repelin	Seed treatment Root dip treatment

Foliar sprays of systemic chemicals namely carbofuran, carbosulfan and UC 54229 were compared with soil treatment for the control of *M. arenaria* on groundnut (Prasad, 1993). Treated plants had better growth than untreated control. The foliar sprays enhanced the growth parameters significantly and number of galls significantly reduced in both the methods of application. However, carbosulfan and carbofuran spray treatments were observed to be effective against *M. arenaria* without adversely affecting the plant growth. Residues in all three nematicides gradually declined in root, shoot and soil with the passage of time and within 15 days more than 75% of nematicides got degraded.

Sakhuja (1984) reported that seed soaking in phenamiphos, oxamyl, aldicarb or carbofuran at 1.5 kg a.i./ha gave reasonably good control of *M. javanica* and increased the growth of groundnut plant significantly over inoculated check. Seed dressing with carbofuran and isophenphos @ 2 and 4% w/w exhibited little phytotoxicity. Both the chemical at lower dose provided sufficient initial protection.

Prasad (1990) studied the control by seed pelleted with carbofuran (40 flowable) and aldicarb sulfone (75WP) at the @ 3 and 6 % w/w. Results obtained revealed no deleterious effect on germination or on the growth of seedlings. The penetration of *M. arenaria* larvae was significantly low with the duration of time of observation and increased dosages. Carbofuran seed treatment together with soil treatment of carbofuran, phorate or ethoprophos were evaluated for the control of *M. arenaria* on groundnut in field trials in Uttar Pradesh. Carbofuran (25 ST w/w) as seed treatment combined with soil treatment of phorate (10G) at @ 2kg/ha gave best control of *M. javanica* and increase in yield.

Sharma *et al.* (2001) worked on the management of *Meloidogyne incognita* on groundnut through nematicides like carbofuran, phorate and sebufos at 2.0 kg/ha in pots. Results indicated that all treatments were significantly superior to the untreated control with respect to plant growth. A similar trend was recorded with regard to nematode reproduction.

Prasad *et al.* (1992) studied the management of root-knot nematode, *Meloidogyne arenaria* infected groundnut by seed treatment with oxime carbamates namely Marshal 25 STD and Oncol 50 SP @ 1, 2, 3 and 4 per cent a. i on seed weight basis under pot culture conditions. Plant growth characters recorded after 45 days of inoculation indicated better growth in lower doses (1 and 2%) than higher ones (3 and 4%) of both the nematicidal seed treatment. As against this, gall index was low in highest dose. With respect to the plant growth and nematode control, 2 per cent is quite effective dose in groundnut. Pre-sowing application of benfuracarb and basudin on soybean @ 1, 2, 3, 4 and 5 kg a.i/ha resulted better growth in lower doses of both the chemicals. Out of these two chemicals, basudin was found to be quite effective at 2 kg which was at par with 3 kg of benfuracarb application (Prasad and Chawla, 1992). Amarnatha and Krishnappa (1992) used carbofuran, ethoprop or phorate @ 2 to 6 kg a.i/ha applied 7 days before nematode inoculation gave greatest reduction in nematode development and reproduction. Growth of sunflower plants was highest with phorate @ 6kg a.i/ha.

Prasad (1998) undertaken a study to evaluate benfuracarb and triazophos at 100, 200 and 400 ppm as seed soak on the penetration by juveniles of *M. incognita* on soybean. In general both the nematicides reduced the penetration of root-knot nematode, the effect increased with time. Out of two nematicides used as seed soak, triazophos was slightly better than benfuracarb in reducing the penetration of *M. incognita* juveniles into soybean roots.

Mittal and Prasad (1999) studied the effect of chemicals used as seed soaking on the development of *Meloidogyne incognita* and growth of soybean cv. Pusa 16. Three chemicals viz; Triazophos (40EC), carbosulfan (25 EC) and monocrotophos (36% SL) @ 400, 200 and 100ppm were tested as seed soaking against *M. incognita* infecting soybean. In general, higher concentrations of all the chemicals (i.e, 400ppm) was found most effective in reducing nematode population both in root and soil.

Dwivedi and Upadhaya (1992) evaluated carbofuran seed treatment together with soil treatment of carbofuran, phorate or ethoprophos for the control of *M. javanica* on groundnut in field trials in Uttar Pradesh. Carbofuran (25 ST w/w) as seed treatment combined with soil treatment of phorate at 2 kg/ha gave best control of *M. javanica* and increase in yield. The effect of soil drench with carbosulfan 25 STD and triazophos on penetration and development of *Meloidogyne incognita*. Race –1 on sunflower (cv. EC 68414) was studied in pots by Prasad and Narayana (2000). Development of root-knot nematode at lower concentration of nematicides was found to be quicker than at higher concentrations, 2nd stage changed to 3rd and 4th stages. On 7th day of observation both the chemicals were found to be quite effective in reducing the total penetration and subsequent development. This, in turn also affected the fecundity of the nematodes.

Sharma and Prasad (1995) under taken a study to evaluate the efficacy of deoiled cakes of cotton, sunflower and soybean (1, 5 and 10% w/w) and leaves of *datura* and *eucalyptus* (5 and 10% w/w) for the management of *Meloidogyne incognita* soybean cv. MACS –13 in plots. The results clearly showed reduction in penetration of 2nd stage juveniles and further development of galls and egg-mass with the increase in dose of botanicals. However, penetration and galling was least in cotton cake (10% w/w). Penetration, galling and egg-mass were found to be reduced by 94, 84 and 90% respectively with cotton cake (10%). The cotton cake application was most effective followed by eucalyptus, sunflower and datura.

Bhat and Krishnappa (1990) conducted a pot experiment by using carbofuran, sebufos, ethoprophos, carbosulfan or phorate to groundnut at sowing, at inoculation or at 14 days after inoculation for control of *M. javanica*. Most treatments reduced nematode populations and increased plant growth compared to control with sebufos at 4 kg a.i/ha applied at inoculation being the most effective. Ethoprophos at 2 kg a.i/ha 14 days after inoculation was the least effective.

Zhang and Yi (1992) conducted a trial on the control of *Meloidogyne arenaria* on groundnut in Shandong, China. They found that 5% aldicarb granules applied in a furrow at 3 kg/mu resulted in a groundnut yield increase of 17.6% compared with that of control. Joshi and Patel (1997) found that seed treatment with carbosulfan and quinalphos both at 1.5% (w/w) reduced *Meloidogyne javanica* penetration in roots and significantly improved plant growth of groundnut. Among the soil applications, phorate at 3 kg/ha was found most effective followed by carbofuran at 3 kg/ha in improving plant growth while fenamiphos at 1 kg/ha was found most effective in reducing host infestation. These chemicals which are registered and are available could be incorporated in INM schedule.

INTEGRATED NEMATODE MANAGEMENT

Integrated nematode management deals with combining of two or more strategies to manage one or more nematode species. The main aim of the integrated approach is :

1. To maintain the nematode population below the threshold level.
2. Maximising natural environmental resistance against plant-parasitic nematodes.
3. Providing maximum profits to growers with location and resource specific recommendation.
4. Utilizing several compatible and useful control techniques in combination which might have economical and social level.
5. Applying specific and drastic control measures only as and when necessary.

Integrated management strategies can be applied sequentially or simultaneously. The first approach includes the season to season or year to year integration of strategies and is particularly relevant to annual cropping cycles. The second major approach to INM involves simultaneous application of two or more strategies. It is one of the most feasible and reliable approach for the control of plant-parasitic nematodes.

Prasad and Khan (1995) used cotton seed cake, neem cake @ 0.5% applied one week before sowing with triazophos 1000 ug/ml sprayed on the foliage of groundnut. Results thus obtained indicate an increase of plant growth parameters and maximum was recorded in treatment where all three combinations were applied together with lowest gall index. Further, Prasad and Dube (1996) reported a significant reduction of *M. arenaria* and *Tylenchorhynchus vulgaris* populations under field conditions. The combination of more than one treatment were found to be efficient in managing *M. arenaria, T. vulgaris* and *Rotylenchulus reniformis* thus resulting in higher plant growth of groundnut. Prasad and Khan (1990) found that integration of aldicarb, neem cake and soobabul leaves under field conditions resulted in better growth and higher yield of soybean. All three combination resulted in maximum yield of soybean besides maintaining low equilibrum nematode population and maximum reduction being at 15 days of sampling.

Prasad (1999) conducted a field trial at State Govt. farm, Bilwa, Bareilly (U. P) during season of 1994 for management of plant-parasitic nematodes in groundnut. Various treatments significantly reduced population of *M. arenaria, T. vulgaris* and *Hoploliamus indicus* sampled just after 20 days of sowing. There was a drastic reduction of *M. arenaria* from 113 J2 to 7 in 20 days and 6 in 60 days of sampling in plots receiving summer ploughings + mustard cake 10g/ha + triazophos @ 2 litres a. i/ha. Treatments resulted in significant increase in yield, however, maximum yield was observed in mustard cake which account about 78.2 per cent increase over control.

Prasad *et al.*, (1997) studied an integrated approach for the management of plant parasitic nematodes in groundnut. He found that all treatments were effective in reducing root-knot and reniform nematodes significantly and increased the yield of groundnut. However, maximum increase was found in chopped madar leaves applied 15 days before sowing over untreated control. Oil cakes of mustard, sesame and cotton each at 20 and 30 q/ha and aldicarb, carbofuran and phorate at 1.5 and 3.0 kg/ha were applied before sowing groundnut cv. M-13 in a field heavily infested with *M. arenaria* by Grewal *et al.* (1985). Aldicarb, carbofuran and mustard oil cake were most effective in reducing the nematode population and root gall index and increasing the pod yields. Nematicides were generally more effective than oil cakes. A replicated field trial was carried out to manage *Meloidogyne javanica* race 2 on groundnut cv. GC-2' during Kharif seasons 1994 – 1997 in Gujarat, India. The lowest RKI (1.33) was recorded in *Paecilomyces lilacinus* @ 50 kg/ha treated soils. A significantly higher dry pod yield (up 25.56%) and fodder yield (up 21.61%) over the control was noted. Carbofuran (1 kg a. i/ha) and *P. lilacinus* at the lower dose (25 kg/ha) remained the second best effective treatment.

Future line of work

1. Intensive and extensive surveys on the presence of parasitic nematodes infesting agricultural crops should be conducted.
2. Use of cultural practices should be emphasized depending on area/locality.
3. Resistant lines/sources should be identified.
4. Emphasis should be given on the integration of nematode management practices by adoption of cultural practices, resistant varieties, nematicides, organic amendments, biological control agents, etc.
5. Role of secondary organisms in causing diseases along with nematodes should be studied and nematode management practices be evolved.

REFERENCES

Ali, S. S 1995. Estimation of yield losses in urdbean crop due to *Meloidogyne javanica* (Abstr.). Nat. Sym. On Nematode problems of India – An appraisal of the nematode management with eco-friendly approaches and bio-components. IARI, New Delhi, 20.

Amarnatha, B. S and Krishnappa K. 1992. Evaluation of certain pesticides for the management of *Meloidogyne incognita* infesting sunflower. *Indian J. Nematol.*, 22: 125-133.

Anonymous 1985. Annual Report, AICRP on tobacco. ICAR, New Delhi, 90-91

Bhatti, D. S and Jain, R. K. 1977. Estimation of loss in okra, tomato and brinjal yield due to *Meloidogyne incognita*. *Indian. J. Nematol.* 7(1): 37-47.

Bhat, A.I and Krishnappa, K. 1990 Chemical control of *Meloidogyne javanica* infecting groundnut., *Indian J. Nematol.* 20 (1) 95 –100.

Bridge, J 1996 Nematode management in sustainable and subsistence agriculture. *Ann. Rev. Phytopathol.* 34: 201-205.

Campos, V.P 1987. Effect of initial population of *Meloidogyne incognita* and *Meloidogyne javanica* on sunflower planted in microplots delimited by fibre glass in the field. *Nematologia Brasileria.*, 11: 204-214.

Coosemans, J., 1976. 2nd Intn. Verticillum Symp., University of California, Berkeley, 14pp.

Dahiya, R.S and Gupta, D. C., 1992. Reaction of some sunflower cultivars against root-knot nematode, *Meloidogyne javanica*. *Indian J. Nematol.*, 22: (1) 70.

Das, D., 2000. Interaction studies of pigeonpea cyst nematode (*Heterodera cajani*) reniform nematode (*Rotylenchulus reniformis*) and root-knot nematode (*Meloidogyne incognita*) infecting pigeon pea and management through neem formulations. Ph. D. Thesis, I. A. R. I., New Delhi, 12.

Dwivedi, K and Upadhyay, K. D., 1992. Management of root-knot nematode, *Meloiodgyne javanica* infesting groundnut with some promising nematicides. *Indian J. Nematol.*, 22 (1) 9-10.

Gaur, H. S., Mishra, S. D., and Sud, U. C., 1979. Effect of date of sowing on the relation between the population density of the root-knot nematode, *Meloidogyne incognita* and the growth of three varieties of chickpea, *Cicer arientinum. Ind. J. Nematol.* 9: 152-159.

Godoy,G., Rodriguez-Kabana, R., and Morgen-Jones, G., 1983. Fungal parasites of *Meloidogyne arenaria* eggs in an Albama soil. A Mycological survey and greenhouse studies. *Nematropica.,* 13: 201-213.

Grewal, P.S ; Chhabra, H. K. and Kaul, V. K. 1985. Control of *Meloidogyne arenaria* on groundnut with nematicides and oilcakes. *Journal of oilseed Research.,* 2: 22-28.

Gupta, D. C and Verma, K. K., 1993. Studies on avoidable losses in mungbean (*Vigna radiata*) due to root-knot nematode, *Meloidogyne javanica* and its control under field conditions. *Indian J. Nematol.,* 20: 148-151.

Haque, M. M and Prasad, D., 1982. Effect of agronomical practices and DD application to nematodes and crop yield. *Indian J. Nematol.* 13 : 226-229.

Ibrahim, I. K; Rezk, M. A. and Ibrahim, A. A., 1991. Reactions of some graminaceous and leguminous plant cultivars to *Meloidogyne incognita* and *M. javanica. Nematologia mediterranea.,*19 (2) 331-333.

Joshi, P. R and Patel, H. R., 1997. Nematicides in management of *Meloidogyne javanica* on groundnut. *Indian Journal of Nematol.,* 26 (1) 82-86.

Kaul, V. K 1997. Pathogenicity and control of root-knot nematode on sunflower by some systemic nematicides. M. Sc. Thesis, IARI, New-Delhi.

Kumar, B., Yaduraju, N. T., Ahuja, K. N. and Prasad, D.,1993. Effect of soil solarization on weeds and nematodes under tropical Indian conditions. *Weed Res,.* 33, 423-429.

Mathur, B. N., Handa, D. K., Swarup, G. , Sethi, C. L., Sharma, G. L. and Yadav, B. D 1986. On the loss estimation and chemical control of ' Molya' disease of wheat caused by *Heterodera avenae* in India. *Indian J. Nematol.,* 16: 152-159.

Mehan, V. K., Reddy, D. D. and Mc Donald, D.,1993. Resistance in groundnut genotypes to kalahasti malady caused by the stunt nematode, *Tylenchorhynchus brevilineatus. International Journal of Pest management.,* 39 (2) 201-203.

Mishra, S. D 1988. Management of plant-parasitic nematodes infesting chickpea and lentil by neem and karanj seed cakes. *Int. Soil. Zool. Colloquium.* 88-89.

Mittal, A and Prasad, D., 1999. Effect of chemicals as seed soaking on root-knot nematode, *Meloidogyne incognita* infecting soybean. *J. Oilseeds Res.,* 16(1), 157-158.

Montasser, S. A., Al-Sayed, A. A. and Kheir, A. M., 1985. Response of some sunflower cultivars to parasitism by three nematode species. *Pakistan Journal of Nematology,,* 3 (2) 91-94.

Naidu, P. H., Moses, G. J. and Sitaramaiah, K. 2000. Management of kalahasti malady (*Tylenchorhynchus brevilineatus*) of groundnut through crop rotation. *Indian J. Nematol.,* 30: (1) 58-61.

Ogunfowora, A. O 1976. Research on *Meloidogyne* at the institute of agricultural research and training. Univ. Ife. Moor Plantation Ibadan. In Proc. 1st IMP Res. Plant Conf. root-knot nematode, *Meloidogyne* spp. Region IV. 7-11 June. Int. Institute. Trop. Agr., 9-14.

Palaniswamy, S and Balasubramanian, P., 1981. Assessment of avoidable yield loss in cotton variety Suvin (*Gossypium barbadense L.*) by fumigation with metham sodium (Abstr.). Nematol. Soc. India Symp. Coimbatore, 52.

Parvatha Reddy, P and Singh, D. B. ,1981. Assessment of avoidable yield loss in okra, brinjal, frenchbean and cowpea due to root-knot nematodes (Abstr.) *IIIrd Int. Symp. Pl. Path.,* New Delhi, 93-94.

Patel, H. R; Vashnav, M. U. and Dhruj, I. U.,1985. Interaction of *Meloidogyne arenaria* and *Fusarium solani* on groundnut. *Indian J. Nematol.,* 15 (1) 98-99.

Patel, S. K., Patel, D. J. and Patel, H. V. 1986. Estimation of losses due to root-knot nematodes in Bidi tobacco field. Nat. Conf. on Plant Parasitic Nematodes of India: Problems and Progress. 12 (Abstr.)

Patel, D. J., Patel, H. V. and Patel, S. K. 1993. Estimation of avoidable yield losses in greengram and cowpea due to root-knot nematodes. *Indian J. Nematol.,* 23: 213-215.

Patel, G. A and Patel, D. J., 1993. Avoidable yield losses in pigeonpea cv. *Pusa ageti* due to *Meloidogyne javanica. Int. Pigeonpea Newslet.,* 17: 26-27.

Prasad, D 1989. Pathogenicity and screening of groundnut cultivars against *Meloidogyne arenaria. Pak. J. Nematol.* 7: 97-101.

Prasad. D 1990. Control of root-knot nematode, *Meloidogyne arenaria* on groundnut by chemical seed pelleting. *Curr Nematol*. 1: 31-34.

Prasad, D and Yaduraju, N. T., 1993. Solarisation of field soil for effective management of nematodes and weeds. Nat. Conf. Ecofriendly approaches Mgmt. Pests, dis, Ind. Effluents, Dec. 20-22., 67. (Abstr.)

Prasad, D., Dubey, K. N. and Mittal, A., 1997. An integrated approach for management of plant-parasitic nematodes in groundnut. (*Arachis hypogea L.*) *Ann. Pl. Protec. Sci.* 5(2): 217-219.

Prasad, D and Khan, E., 1990. Studies on integrated control of phytonematodes infesting soybean. (Abstr.) Nat. Sym.Problem and prospects of Bot. Pesticides in IPM, 21-22. Jan, C.T.R.I, Rajamundry, A.P., 44-45.

Prasad, D and Chawla, M. L. 1992. Pre-sowing nematicidal treatment for the management of root-knot nematode on soybean. *Curr. Nematol.* 3 (2) 201-202.

Prasad, D 1993. Efficacy of soil and foliar application of some carbamates against *Meloidogyne arenaria* on groundnut and their persistence. *Ann. Pl. Protec. Sci.* 1: 107-111.

Prasad , D 1998. Seed soaking in benfuracarb and triazophos – a prophylatic treatment against *Meloidogyne incognita* on soybean. *Ann. Pl. Protec. Sci.*, 6: 196-197.

Prasad, D 1999. Integrated management of plant-parasitic nematodes in groundnut. *Nature and Biosphere*, 4 (1 & 2): 16-19.

Prasad, D., Chawla, M. L. and Khan, E., 1992. Management of root-knot nematode, *Meloidogyne arenaria* infected groundnut by seed treatment. *Curr Nematology.*, 3 (2): 139-142.

Prasad, D and Narayana, R. 2000. Effect of carbosulfan and triazophos on the penetration and development of *Meloidogyne incognita* on sunflower. *Ann. Pl. Protec. Sci.*, 8 (1): 58-61.

Prasad, D., Mittal, A. and Yaduraju, N. T. 1995. (Abstr.) Nat. Symp. Perspectives in Eco-friendly approaches to plant protection. Soc. of Pl. Protect Sci. 7-8, Sept. IARI, New Delhi. pp. 75.

Prasad, D and Khan, E. 1995. (Abstr.) Global Conference on Advance in Research and Plant Diseases and their management. 12-17, Feb, 1995. at RAU, Rajasthan, Udaipur.

Prasad, D and Dube, K. N. 1996. (Abstr.) DAE-BRNS-Symposium on Nuclear Technology in increasing Crop and Animal Productivity Oct, 7-9, 1996 at BARC, Bombay.

Raut, S.P and Sethi, C. L. 1980. Plant-parasitic nematodes associated with soybean. *Indian J. Nematol.*, 16 (1): 126-128.

Sakhuja, P. K 1984. Studies on the inter relationship between root-knot nematode and root-rot fungi on groundnut. Ph.D. Thesis, IARI, New-Delhi.

Sakhuja, P. K and Sethi, C. L.,1986. Multiplication of *Meloidogyne javanica* as affected by *Fusarium solani* and *Rhizoctonia bataticola* on groundnut. *Indian J. Nematol.*, 16: 1-3.

Sakhuja, P. K and Sethi, C. L.,1985. Frequency of occurrence of various plant-parasitic nematodes and root-rot fungi on groundnut in Punjab. 15 (2): 191-194.

Sasser, J. N and Freckman, D. W., 1987. A world perspective on Nematology: The role of society. Vistas of Nematology, 7-14.

Seshadri, A. R and Dasgupta, D. R., 1980. Handbook of Agriculture, New Delhi: ICAR, 551-557.

Sethi, C. L and Gaur, H. S., 1986. Nematode management: An overview. In: Plant Parasitic Nematodes of India—Problems and Progress, 424-445.

Shah, H. M., Patel , D. J.and Valend, G. B., 1981. Assessment of loss due to root-knot disease in bidi tobacco industry. *Indian J. Nematol.* 11: 126-127.

Sharma, H. K and Prasad, D.,1995. Potentialities of a few botanicals in management of root-knot nematode, *Meloidogyne incognita* on soybean. *Ann. agric. Res., 16* (4): 473-476.

Sharma, S., Siddiqui, A. U. and Parihar, A., 2001. Management of *Meloidogyne incognita* on groundnut through nematicides. *Indian J. Nematol.*, 31 (1) 79-80.

Sivakumar, C. V 1988. Avoidable yield loss in rice due to *Aphelenchoides besseyi* in Kanyakumari Dist. Tamil Nadu. *Indian J. Nematol.*, 18: 123-124.

Zhang, S. A and Mq, Yi, 1992. Control of nematode diseases of groundnut with 5% aldicarb. *Plant Protection.*, 18 (2) 49.

18

Principles and Practices of Rodent Pest Management

R. S. TRIPATHI[1] AND V. CHAUDHARY[1]

Rodent pests are a serious impediment not only in crop protection and food storage but also responsible for spreading several dreadful diseases to man and his livestock. As long as the history goes back, rodents have had a bad reputation as man's enemy. Old Indian scriptures, dating back to 5 millennia, have mention of rodents. One of the shloka in Atharveda (60.50.1, 3000 BC) reads as, " O Ashwin, kill the burrowing rodents which devastate our food grains. Cut their heads, break their necks, plug their mouth so that they can never destroy our food. Rid mankind of them" (Jain and Tripathi, 2000). Such a long association of man with rodents has made these tiny mammals a serious problem in our food production and storage. They cause direct damage to crops/commodities by gnawing and feeding and also indirect damage by spoilage, contamination and hoarding during pre and post harvest stages.

Rodents form a very diverse group of mammals viz., porcupines, squirrels, rats, gerbils, mice, mole rats etc. Out of over 2000 species only a limited number of them are regarded as pests in agriculture. For example, in Australia and Western Europe only 4 out of 67 species and 5 out of 61 species have been reported to cause damage to agriculture respectively (Leirs, 2003). India, being a highly diverse country, this number is more as about a dozen species are regarded as pests of agriculture (Jain and Tripathi, 1988 and Tripathi *et al.*, 1992). Most of the pest species belong to family muridae, but even in this group only a few species fall in the category of 'pests'. Estimates of rodent damage to crops vary immensely depending upon the species, crop stage, agro-ecological region, availability of food and physical environment.

Management of rodent pests is a very intricate as well as a ticklish problem. Unlike insects rodent are very secretive and not easily observed because many of them are nocturnal. They infest an area throughout the year as compared to insects which may appear sporadically only for short periods in certain seasons. Often, the farmers must rely on various signs like damaged plants, tooth marks, burrows and traps to identify them. Moreover the rodents are highly mobile and a single individual may typically cause damage to several plants in one night by residing at some other place. Despite knowledge about their biology and control techniques, the fact however remains that the rodent continues to damage standing crops and stored food grains.

CROP FIELDS, AN IDEAL RODENT HABITAT

Pest status of rodent species is a combination of cropping pattern of the region vis-à-vis eco-biology of rodent species. Generally crop fields are homogeneous and man made ecosystem with a plant community dominated by one or two crop species. Depending upon the availability of water, climate suitability and the soil fertility, different crops are grown in an area, which regularly supply energy rich food to rodents in a limited area. The availability of food and cover to native rodents are continuous with different crops grown in different seasons. The rodent species that can take advantage of the temporarily abundant food

1 AICRP on Rodent Control, Central Arid Zone Research Institute, Jodhpur-342 003.

and at the same time somehow overcome the periods when the conditions are poor, make rodents' a good candidate as agricultural pests. (Leirs, 2003). Thus agricultural fields serve as a highly productive rodent habitat. Among various crops, sugarcane, rice, wheat, groundnut and fodder crop fields serve as an ideal habitat for rodent pests in India. Similarly threshing yards located near crop fields too act as an excellent abode for food and shelter of rodents. Pearl millet crop suffers higher damage by *Meriones hurrianae* when the harvested crop is heaped in the threshing yards than at crop standing stage. As many as 40 merion gerbils have been noticed in one such heap in a 15x40 m area (Prakash, 1976).

RODENT PEST SPECIES

As discussed above about a dozen species are regarded as pest in agriculture, however a pest complex of 2-4 rodents occurs in any particular agro-climatic region. *Bandicota bengalensis* is the most predominant rodent pest species and is well distributed in crop fields and residential areas all over the country, except the extreme hot arid regions (Jain *et al.*, 1993 and Chakraborty, 1992). During the last few years the bandicoots have knocked the doors of Indian desert by establishing their population in urban locales of Bikaner (Prakash and Mathur, 1987) and Jodhpur (Idris and Rana, 2001). Other species of national importance in the fields are *Millardia (=Rattus) meltada, Tatera indica* and *Mus booduga.* In arid areas of western Rajasthan Indian desert gerbil, *M. hurrianae* and *Gerbillus gleadowi* attain pest status in rain-fed crops. Similarly in plantation crops viz., coconut, coca, arecanut, *Rattus rattus wroughtoni* is a major problem (Bhat and Sujata, 1986). Among squirrels, *Funambulus pennanti* infest fruit orchard in North India (Prakash and Mathur, 1987) and *F. tristriatus* is a pest of fruits and plantation crops and spices in South India (Bhat, 1982). Indian crested porcupine, *Hystrix indica* is regarded as pest of tuber crops (Agrawal and Chakraborty, 1992). Another species, *Nesokia indica* has also been listed as pest of irrigated crops, fruit orchards and forestry plantation (Tripathi and Jain,1990; Jain *et al.*, 1995 and Ramesh, 1992). The two most common commensal rodents, viz., *Rattus rattus* and *Mus musculus* have shown atavism in their habitat by making their presence in crop fields also in different regions (Mathur and Prakash, 1987; Patel *et al.*, 1992 and Dubey *et al.*, 1992). *Rattus nitidus* is considered as a crop pest in higher reaches of the Northeastern hill region (Singh *et al.*, 1995). A brief account of predominant rodent pest complex in major cropping system is detailed in Table 1.

Table 1. Distribution of rodent pests in different cropping systems

(a) Dryland/rainfed crops		
(i) Arid zone	:	*M. hurrianae-G. gleadow- T. indica.*
(ii) Semi arid tracts (North Zone)	:	*T. indica-M .meltada.*
(iii) Semi arid tracts (South Zone)	:	*T. indica-M. meltada-B. bengalensis*
(b) Irrigated crops		
(i) Arid and Semi arid tracts (North Zone)	:	*T. indica-M. melatada-M. hurrianae.*
(ii) Semi arid tracts (South Zone)	:	*B. bengalensis-M .meltada.*
(iii) Deep water rice	:	*B. bengalensis-Mus spp.*
(c) Plantation crops		
(i) Coconut-cocoa-arecanut	:	*R.r.wroughtoni-F.palmarum-F.tristriatus- vandeleuria oleracea*
(ii) Sugarcane	:	*B. bengalensis-M. meltada-Mus spp.*
(iii) Spices	:	*F. palmarum-F .tristriatus.*
(d) Jhum crops and Pineapple (NEH Region)	:	*R.natidus-B.bengalensis-R.r.bunneusculus*
(e) Stores/godowns	:	*R.nitidus-M.musculus-B.bengalensis*

RODENT PROBLEM

Being herbivore, choice of food for rodents is unlimited. No crop is spared by these tiny vertebrates at any stage of its growth. Extent of rodent damage to crops largely depends upon the (i) species involved (ii) location and stage of crop (iii) population of the pest (iv) availability of crop and (v) physical environment. Their devastations have been well documented in food crops, pulses, oilseed, plantation crop, fruit orchards, grasslands and forestry plantations (Chopra *et al.*, 1996; Parshad, 1999; Jain and Tripathi, 1992; Srivastava, 1992; Awasthi and Agrawal, 1991; Vyas *et al.*, 2000; Advani, 1985; Bhat, 1992; Malhi and Parshad, 1987; Patel *et al.*, 1995; Rao, 2003). Perusal of data on the extent of damage as listed in Table 2 indicates that a plant, a seed, a fruit or its produce in any form is always exposed to the vagaries of rodent depredations. These are just pre harvest losses, which do not account for the damage inflicted by rodents between sowing and germination of seeds, when the rodents pinpointed feed on the sown seeds, because of their seedvorous behaviour. Thus one can broadly attribute 5-10% pre-harvest losses due to rodents in India. In Asian countries rice suffers a rodent damage 2-5% (in Malaysia); 10-20% (in Indonesia); upto 10% (in Vietnam); 6-7% (in Thailand) and 5-10% (in Bangladesh) (Leirs, 2003).

Table 2. Magnitude of losses caused by rodents

Commodities	Losses
Paddy	4.6-54.0%
Jhum Paddy	70.0-90.0%
Paddy hoarding in burrows	5.7 kg/ha
Wheat	11.1-11.9%
Wheat hoarding in burrows	1 q/ha
Sorghum	5.85-6.28%
Pearl millet	107.7 kg/ha
Groundnut	4.1-25.8%
Bengal gram	2.5%
Sugarcane	2.24%
Coconut	5.0-11.0%
Cacao	8.0-51.3%
Pomegranate (ripe fruits)	17.8 - 22.4%
Date palm	18-20%
Vegetables	8.7-10.0%
Desert forestry plantations	4.4 - 10%
Fodder grasses	1040 kg/ha

Population explosion of rodents leading to outbreaks has also been reported by Singh *et al.*, (1995), Jain and Tripathi, (2000) and Rao (2003). Rodent outbreaks in Notrheastern hill regions (Mizoram, Arunanchal Pradesh, Manipur etc.) leading to famine have been experienced during 1880-83, 1910-12, 1928-29, 1958-59 and 1976-77 (Singh *et al*, 1995). Rodent damage to rice, maize and other crops reached to the tune of 75-90% in these years. This outbreak is attributed to the flowering of certain bamboo species (*Bambusa tulda* and *Melocanna bambusoides*) in the region. It is locally believed that feeding of bamboo seeds enhanced the reproductive ability of native rodents causing severe pest outbreaks, however the relationship between rodent upsurge and bamboo flowering is still debatable. Similar situation is expected to arise in coming years again (2005-07). Large-scale devastation of crops has also been reported from other regions. For example rodent upsurge in Gujarat and Rajasthan during 1901-02; 1913-14; 1970-71; 1975-76 and 1997-98 have been documented by Mittal and Vyas, (1992); Prakash and Mathur, (1987); and Shah, (1979). In South India too rodent outbreak, as evident from upsurge of plague was recorded in1826, 1878-79, 1901-02 and 1913-14 (Kinnear 1919). Rao (2003) has listed eleven such events during the last ten years. Of these five outbreaks between 1999-2000 have occurred in Manipur, Arunanchal Pradesh, Mizoram and Nagaland and other outbreaks took place in Southern India viz., Cauvery Delta (1994-2001), Godawari Delta (1997-98 and 2001) and Pondicherry (1994). Saurastra region of Gujarat too suffered heavy devastation of groundnut, wheat, gram and cotton crop due to rodents' outbreak in 1990-91. According to Rao (2003) sudden upsurge of rodent population leading to outbreak may be due to (i) prolonged dry spell followed by heavy rains in Saurastra region (1990-91) (ii) failure of monsoon in preceding years in Cauvery delta (1994and2001) (iii) flash floods in Godawari delta in 1997-98 and 2001 and/or (iv) flowering of bamboo in NEH region.

Besides the crop losses, the rodents inflict severe damage to rangelands, afforestation plantation and cause soil erosion. In arid regions, where vast stretch of sandy waste/pasture lands are available the desert gerbil *M. hurrianae* cause havoc to various production systems. This diurnal rodent digs 450-15000 burrows/ha (Prakash, 1981) and can loosen 61,500-1,61,000 kg soil/day/km, which can be easily blown away by high velocity winds causing problem of desertification and soil erosion (Sharma and Joshi, 1975). In pasturelands also Prakash and Mathur (1987) reported a density of 477 gerbils/ha, which required 1040 kg/ha feed, whereas the annual productivity of the land was 1210 Kg/ha only, thereby leaving almost nothing for desert livestock. During summers, the rodents debark the stem and root of trees and feed on them to derive moisture. The short tailed mole rat *N. indica* was reported to slice the root of arid zone plantation to the tune of 4.4-10 per cent (Tripathi and Jain, 1990).

RODENT CONTROL TECHNIQUES

(i) Cultural control

This technique is primarily based on ecological concepts wherein the rodents' habitat is manipulated in such a way that creates stress among native rodent pests. The methods are low cost treatments and involve little modification in crop husbandry practices, like deep ploughing, removal of wild vegetation and refuge of previous crops (Pashehan and Sablok, 1987 and Jain and Tripathi, 1992), reduction in bund size (Sharma and Rao, 1989) etc. It helps in migration of pest rodents from crop fields. Moreover, the habitat stress created by these practices enhances the chances of rodents falling prey to predators (Green and Taylor, 1975 and Hansson, 1975).

(ii) Employment of rat catchers

Certain tribal communities in India viz., Mushars and Nats (Bihar); Yenadis (Andhra Pradesh), Irulas and Kuruwas (Tamilnadu) and many communities in Northeastern hill states consume rodents. Farmers of Andhra Pradesh and Tamilnadu employ such tribals for physical elimination of field rodents (Whitaker, 1987 and Jain and Tripathi, 2000). This practice is often done at crop maturity stage when the rodent's destruction is at its peak, however this eco-friendly method may prove effective if performed in a planned schedule.

(iii) Mechanical control

Mechanical removal of rodent population from any habitat is mainly done by use of different types of traps. This method is in vogue all over the world, but its success in field rodent control is doubtful. Trapping can no doubt provide, the information on species composition and density of pest population, hence can be successfully used for survey and monitoring purpose (Jain *et al.*, 1993). In indoor habitats, this method is quite effective (Table 3), however for fields this method may be used as a follow up measure after bringing down the rodent pest population by poison baiting. The traps may be of single (Sherman/ Snap traps) or multiple catch (Wonder traps) type. In snap traps the rodents are killed whereas others catch live rodents. Introduction of natural glues or sticky traps have opened new avenues in the field (Tripathi *et al.,* 1994). Rodents use visual and olfactory cues towards traps, therefore camouflaging the traps with bushes, grasses etc. in fields enhances the trap success considerably (Jain and Tripathi, 2000). Indigenous Tanjor kitty traps in South India and bamboo snap traps in Jhoom fields of NEH region have proved quite effective (Rao, 2003).

Table 3. Comparative efficacy of different traps

Habitat	Traps	No. of traps used	No. of rodents trapped on		Trap index
			Ist day	*In 5 days*	
Glue trap vis-à-vis Sherman trap					
Houses and godowns	Glue trap	30	76	180	118.8
	Sherman trap	30	8	20	13.3
Poultry farms (feed stores)	Glue trap	15	52	80	106.6
	Sherman trap	30	10	25	16.5
Glue trap vis-à-vis wonder trap					
Grain godowns	Glue trap	9	5	6	11.6
	Wonder trap	9	4	4	11.0
Poultry feed and egg stores	Glue trap	14	8	25	44.6
	Wonder trap	14	8	25	44.6

(iv) Biological control

(a) Disease

Use of disease agents for control of animals was successful through introduction of myxomytosis in rabbits in Australia, however after 10 years or so, the rabbits developed genetic resistance to disease. In case of rodents *Salmonella typhimurium* and *S. enteritidis* have been successfully used against rodents in forgien countries but in India, the pathogen was found ineffective against *R. rattus* and *B. bengalensis* yielding their mortalities to the extent of 16 and 18% respectively (Deoras, 1964). Bindra and Mann (1975) also expressed doubts about success of microbial control because the authors could not record any mortality of *R. meltada* and *B. bengalensis* whereas, in case of *M. musculus* and *T. indica* the mortality was over 20 and 40%, respectively.

Recently a protozoan parasite, *Trypanosoma evansi* causal organism of Trypanosomiasis in cattle, has been tried against *R. rattus* and *B. bengalensis*. The pathogen recorded two peaks of parasitismia in both the rodents, with cent per cent mortality after second peak (Singla *et al.*, 2003). Due to presence of pathogen in cauda epididymal fluid and spermatozoa, the authors indicated the possibility of sexual transmission of *T. evansi*.

Thus use of microbes has very little scope for rodent control particularly due to inherent possibility of spread of the diseases to man and livestock. WHO committee on Zoonosis has also doubted the practical application of microbial rodenticides due to possible public health hazards (Prakash and Mathur, 1987).

(b) Predators

Several vertebrates, mainly birds and mammals are listed as natural predators of rodents in literature (Deoras, 1964; Prakash and Mathur, 1987, Tripathi *et al.*, 1992; Whitaker and Dattari 1986; Neelanarayanan,

1997; Chaudhary *et al.*, 2001). However cats in domestic situation and snakes, owls, mongoose and varanids are predominant vertebrates predators of rodents (Rao, 2003). Information on stomach contents and feacal matters of predators indicated that rodents constitute over 75% diet of snakes viz., Cobra and Russel viper (Whitaker and Dattari, 1986) and 61% of spotted owlet (Kumar, 1985). Introduction and rehabilitation of barn owls, *Tyto alba* in oil palm plantation of Malaysia was reported to reduce the rodent damage to oil palm from 19.4% to 1.4% with in two years (Duckett and Karuppiah, 1991). In the Cauvery Delta of Tamilnadu, studies on barn owl as potential bio control agent of bandicoots revealed a predation rate of 1-6 (Av 1.58) rodents/night. *B. bengalensis* (40%) and *M. musculus* (33%) constituted the major prey items of barn owls (Neelanarayanan, 1997). Success rate of biological control of rodents through other predators is not very encouraging because rodents do not constitute the sole diet of most predators and exert relatively lower predatory pressure as compared to faster turnover of rodents. For example snakes in captivity require only one rodent in three days (Whitaker and Dattari, 1986) indicating a very poor potential.

Some examples of introduction of predators viz., monitor lizards in Pacific island to control rodents (Uchida, 1966); Indian mongoose to control *R. norvegicus* and tame ferret (*Putorius putorius*) for rabbit control in Newzealand (Gibb, 1981) etc. have not yielded a very encouraging success.

(v) Chemical Control

(a) Antifertility agents:

A host of chemicals with antifertility effect on pest rodents have been identified world over during sixties and seventies. Some of these are aniestrogen V-II, diphenylndane derivatives, metepa, tepa, tetradifon, furadentin, colchicine etc. (Wheatherbee, 1964; Gains and Kinborough, 1964; Kalra and Prasad, 1966 and Srivastava and Nigam, 1975). Erickson (1982), Andrews and Belknap (1983), Kassa and Jackson (1984) and Saini and Parshad (1988) have studied the effectiveness of alpha-chlorohydrins, as a chemosterilants on *R. norvegicus* and *B. bengalensis*. At lower dosages, it produces male sterility, but at higher dose it causes toxicity also. Despite sincere efforts worldwide, the chemical is yet to be practically used in fields. Recently the work on epi-chlorohydrin has been initiated under AICRP on Rodent Control. Since these chemicals have totally new approaches in integrated management of rodent pests, they are required to be field evaluated on a larger scale. Such chemicals seem to possess a bright future because of their compatibility with toxic rodenticides and are also less controversial from human/animal rights viewpoint.

(b) Rodenticides:

Use of rodenticides is most common, expedient and humane method to control pest rodents. They have greater scope in large-scale control operations, since mixed population of several species are encountered in fields. Zinc phosphide is most widely used rodenticide in India. It is recommended at 2% concentration in cereal baits. It yields around 60% control success (Rao *et al.*, 1998). It is highly toxic to a variety of rodent species. Major limitation in its frequent use is its high toxicity to non-target species and development of bait shyness/poison aversion in the target species after sub-lethal consumption (Prakash and Mathur, 1987). Bait shyness may persist for more than 2-3 months in different rodent species (Table 4). Its bait can be prepared by mixing the poison (2%) in oil smeared cereals viz., wheat, jowar, pearlmillet etc. Because of high toxicity, zinc phosphide is recommended for field rodents only, where the rodent infestation is very high. For controlling the residual rodents (surviving after zinc phosphide treatment), another rodenticide viz., aluminium phosphide (a fumigant) or second-generation anticoagulant rodenticide has been advocated (Rana and Tripathi, 1999).

Among second generation anticoagulants several products viz., bromadiolone (Anonymous, 1986), brodifacoum (Jain and Tripathi, 1988), flocoumafen (Jain *et al.*, 1992) and difethialone (Sridhara *et al.*, 2000, Chaudhary *et al.*, 2002 and Chaudhary and Tripathi, 2003) have shown their potency against Indian rodent species (Table 5), however only bromadiolone (0.005%) is registered with Central Insecticide Board since 1988 for rodent control in fields and also in commensal situations. Coumatetralyl, also an anticoagulant is used in cereal mixed baits at 0.0375% for indoor habitats. It has also been recommended for field use against *B. bengalensis* in Punjab (Rana and Tripathi, 1999). The second-generation anticoagulants are effective as single dose rodenticide and have very potent antidote in vitamin K_1. Moreover they are highly effective at a very small concentration in baits (0.0025-0.005%), thus are considered relatively safer to non-targets. They can be used for rodent management not only in fields but also in domestic situations, stores, godowns, poultry farm etc. (Anonymous, 1986 and Parshad, 1999).

Table 4. Persistence of zinc phosphide-induced bait shyness among Indian rodents.

Rodent species	Persistence of bait shyness (days)
F. pennanti	30
G. gleadowi	10-15
T. indica	115
M. hurrianae	35
R. rattus	75
R. meltada	135
Mus musculus	20
Mus booduga	95
B. bengalensis	30
B. indica	105

Table 5. Single dose toxicity of grain baits of second generation anticoagulant rodenticides (No choice)

Rodent species	Per cent mortality and (mean days to death) with			
	Bromadiolone (0.005%)	*Brodifacoum (0.005%)*	*Flocoumafen (0.0025%)*	*Difethialone (0.005%)*
B. bengalensis	100	100 (6.6)	100	100 (5.8)
T. indica	100 (9.5)	90 (6.9)	90 (8.2)	100 (7.9)
R. meltada	100 (6.4)	80-100 (6.6-7.1)	100 (5.7)	100
M. hurrinae	88	83-100 (4.7-8.0)	100 (7.0)	100 (5.9)
R. rattus	100 (7.5)	83-100 (8.0-10.8)	100 (8.6))	100 (5.9)
M. musculus	91 (8.1)	60-100 (4.6-8.4)	100 (6.8)	-
M. booduga	83	83 (9.4)		-
F. pennanti	-	66 (7.0)	100 (8.1)	100 (6.1)

(Source: Anonymous 1986; Parshad and Chopra, 1986; Jain and Tripathi, 1988; Parshad, 1988; Sridhra, 2000; and Chaudhary, 2002; Chaudhary *et al*, 2002)

COMPONENTS OF INTEGRATED RODENT PEST MANAGEMENT (IRPM)

Simply knowing the enemy and the control methods does not qualify for effective management of rodents. Though several methods of rodent control has been discussed in preceding pages, but no single method can prove its worth if applied in isolation. Since rodents are highly evolved mammals and its long association with the wisest mammal i.e. man, has made them wiser than man by learning various tactics to avoid many harmful activities performed by man against them. Trap avoidance, bait shyness/aversion to acute poisons, resistance/cross resistance to anticoagulants are only few examples of rodents behaviour, which very often questions the man's endeavors' for their control. Therefore an IPM approach based on sound eco-biology and ethology of pest species vis-à-vis population reduction technologies having economic viability and sociological acceptance needs to be evolved. Thus IPM is a bio-ecological system and is therefore dynamic one. Following components of population regulation require special mention for devising an IPRM system in different agro-ecosystems.

(a) Population ecology

Rodents exhibit a very reproductive potential. They are prolific breeders with a litter size of up to 20 young ones per female (Table 6). Such a high fecundity of rodents is countered by several biotic and abiotic factors operating in nature. Rodents sometimes regulate their own population by feeding on their own young-ones (cannibalism) during any type of stress depending upon the carrying capacity of their habitat. In drought years the rodent population maintains a low profile due to scarcity of natural food, however the population explodes in a good rainfall year succeeding any drought year.

Table 6. Peak breeding season and litter size of rodent pests

Rodent species	Peak breeding season	Litter size
H. indica	Monsoon and December	1-8
F. pennanti	March-April and July-September	1-5
G. gleadowi	May, June and October to January	2-5 (summer) 5-6 (winter)
T. indica	Monsoon (in arid regions) Oct-Nov (in Karnataka)	1-9 1-12
M. hurrianae	(i) February to March (ii) July and September to November	1-9 2-7
G. ellioti	March-August	5-10
R. rattus	April-September	1-9
R. meltada	Spring and Monsoon (in Rajasthan) March-May and July-Oct. (in Punjab)	3-9 -
M. musculus	All the year round	1-8
M. booduga	(i) September-October (ii)February and June	- 6-13
N. indica	January-March and August-October	2-5
B. bengalensis	All the year round	4-12

Most of the rodents species exhibit relatively lower population levels during winters and summers (Prakash and Kametkar, 1969; Jain, 1970; Prakash, 1981; Chopra 1980; Rana and Prakash, 1984 and Jain and Tripathi, 2000). Maximum pest population is encountered in monsoon and post monsoon seasons. However, *B. bengalensis* was reported to show very fast population growth rate from 108/ha in November to 446/ha in December during winters in West Bengal (Chakraborty, 1975). Such population upsurges are results of favourable climatic condition along with food availability leading to intensive breeding and immigration. A perusal of data on peak breeding season of Indian rodents (Table 6) indicates a bimodal breeding pattern, which coincides with the maturity/harvesting stage of field crops. Seasonal movements of *B. bengalensis* have also been reported from Uttar pradesh where the bandicoot migrates from rice to sugarcane to wheat fields after their harvest period (Rao, 1992). Bunds or fields with more weeds are an ideal place for burrowing rodents. Similarly denser fields with more crop density afford both cover and energy required for reproductive activity of field rodents. Mobility on the pest forms a limited social structure based on hierarchy. The members live in small territories called "home range" which depends on position of food reserves and cover condition (Frantz, 1984) and presence of other species. The home range of Indian rodents vary from 675 sqm for *M. musculus* to 1912 sqm for *T. indica* (Prakash and Mathur, 1985).

(b) Pest behaviour

Most of the field rodents are fossorial and nocturnal, which protect them from direct attack of predators and climatic vagaries. The burrows could be very simple (*T. indica, M. booduga* and *M. meltada*) or complicated (*B. bengalensis* and *M. hurrianae*). The mole rats (*B. bengalensis* and *N. indica*) keep their burrow openings tightly plugged with soil for safety against natural enemies and flooding. Food and feeding behaviour of rodents indicate their preference for various food items, which helps in selecting an effective carrier for poison bait preparation. Studies on bait preferences have revealed that almost all the 12 pest rodent species of rodents prefer wheat, rice and pearlmillet with slight variation (Rana *et al.*, 1992). Most preferred baits with their caloric values are detailed in Table 7. Similarly preference for oils, sugar, salts etc. have helped in selection of bait additives. Based on the preference of rodents, poison bait recommended by AICRP on Rodent Control is pearlmillet/wheat with groundnut oil in Rajasthan and Gujarat, wheat+sugar+mustard oil in Northern India, rice/ragi with coconut oil in Southern India (Rana and Tripathi, 1999). Even the effect of bait shyness induced by sub-lethal intake of acute poisons can be mitigated to some extent by changing the base and oil component of the baits.

Table 7. Most preferred baits and calorific requirements of Indian rodents (Rana *et al*, 1992)

Rodent species	Preferred bait	Calorific requirement (KJ/100g/day)
F. tristriatus	*Whole paddy or sorghum*	-
F. pennanti	Whole wheat grain or cracked pearl milt	104.97 (67.63-146.21)
Tatera indica	Millet and sorghum grains	(247.45-253.31)
M. hurrianae	Wheat floor or cracked pearl millet	150.48 (117.04-183.92)
Rattus rattus	Maize flour+10% arachis oil +1% sugar	121.21 (63.32-250.38)
Rattus meltada	Pearl millet flour+10% sesame oil; sorghum, maize and wheat grains	(58.52-83.6)
Mus musculus	Rice flour or pearl millet +1% arachis oil —	52.17 (145.08-372.18)
M. booduga	Rice flour+ 2% arachis oil 2% sugar; or pearl millet flour + arachis oil+ 2% sugar.	-
B. bengalensis	Wheat and rice (Punjab) Cracked rice +10% groundnut oil +2% salt (Karnataka)	-
B. indica	Maize flour or gram flour	43.47 (21.23-80.25)

(c) Damage threshold levels

Under IPM, pest population is to be manipulated to maintain them below economic threshold levels. Limited studies have been made in India in this respect. Buckle and Rowe (1981) reported that in rice 15% of hills or 2% tillers with rodent damage could be treated as threshold level. For working out these levels, rodent pest population needs to be regularly monitored in order to decide any management action. This may be done either on burrow count basis or actual damage done by pests. Prakash and Mathur (1987) opined that live burrow counts is a useful index for rodent population in crop fields. Rao (1992) has prepared an index which states that per hectare burrow density upto 25 can be regarded as low, 25-50 as medium and more than 50 as severe pest intensity for deciding management plans for pest rodents. Estimates of damage in wheat fields in Punjab revealed that a loss of Rs. 2.90 could be expected only with 2 bandicoot burrow thus rodent control is recommended if there are bandicoot burrows at 2-3 sites per acre of wheat fields (Anonymous, 1990). For damage appraisal diagonal transact method which require sampling of 25 rice hills and examining healthy and cut tillers was reported to be a feasible method by Singh and Rao (1982).

Rodent control operations at crop growth stage yield less success because the poison baits face stiff competition with standing crops especially in arid and semi arid regions. That is why poison baiting in crop fields are generally recommended before sowing/transplanting of crops. In rice and wheat crops major rodent depredation occurred at booting stage of the crop, thus poison baiting is effective before this stage.

(d) Ecological take over

The crop fields are mostly inhabited by a complex of 2-4 rodent species, however any one of them may be predominant. It has been generally observed that when any control operation is undertaken the predominant species experiences the first knockdown. Later in the absence of the dominant species, the co-dominants or subdominants take over the dominant status and multiplies un-checked in the absence of any competition. Secondly, the immigration of new rodents from neighborhood is also very fast. For example sustained trapping or poison baiting which may bring down the pest population to the extent of 80-90%, often fails to maintain this success any longer due to large scale immigration from surrounding untreated areas. Our experience in the arid regions indicated that after discontinuation of control operation after 3 crop seasons, the rodent pest population is restored to its original level with in few days (Jain, 1988).

(e) Impact of human activities on rodent fauna

Human disturbance of natural ecosystem have provided new opportunities for some rodent species to become pest and also for the current pest species to extend their range. Occurrence of a sylvan rodent, *T. indica* in the streets of Bikaner city in Rajasthan and introduction of more aggressive mesic species by replacing the xeric fauna in Indra Gandhi Canal Command areas in western Rajasthan are two most glaring examples of such an ecological transformation (Jain *et al.*, 1996 and Idris *et al.*,2001). Our studies in arid zones suggested that with change in land use pattern and urbanization the truly xeric species viz., *Gerbillus gleadowi, G. dasyrus* and *Meriones hurrianae* are driven away from their natural habitats (CAZRI, 2001). Similarly, the house mouse, *M. musculus,* which is considered to be a commensal species, has shown a reversal in its habitat by inhabiting crop fields in Punjab and Rajasthan.

(f) Planning for chemical control

Rodent pest management is mainly depends on the use of rodenticides. Timing of control and placement of poison baits plays significant role in yielding desired results. Jain *et al.*, (1993) advocated large scale rodenticidal baiting during summer months when most of the rodent species maintains lower population and the acceptance of the technology by farming community is also quite encouraging as they are relatively free during this period and can freely do the job on cooperative basis. Similarly second operation should be undertaken before sowing of rabi crops. Social engineering activity on rodent control initiated the AICRP on Rodent Control have recorded over 70% success with such schedule. Two methods of bait placement is recommended for fields i.e. burrow baiting and use of bait stations. The former method provides better opportunity to access fresh baits by target rodents and protects the non-targets from its exposure. Similarly, bait station made of locally available materials viz., bamboo or PVC pipes have also proved highly effective. The bait stations should be placed on the runways aligning vertical surfaces like bunds for better success.

Considering the havoc caused by rodents to agriculture, various option are available to Indian farmers to contain them, it is advisable to integrate these methods in proper scheme of things. Cultural control methods coupled with poison baiting and trapping at proper time and location is being recommended for long lasting solutions to rodent problem. The system evolved for arid and semi-arid regions is prescribed in Fig 1 (Jain and Tripathi, 2000).

FIG 1.INTEGRATED MANAGEMENT SCHEDULE FOR PEST RODENTS

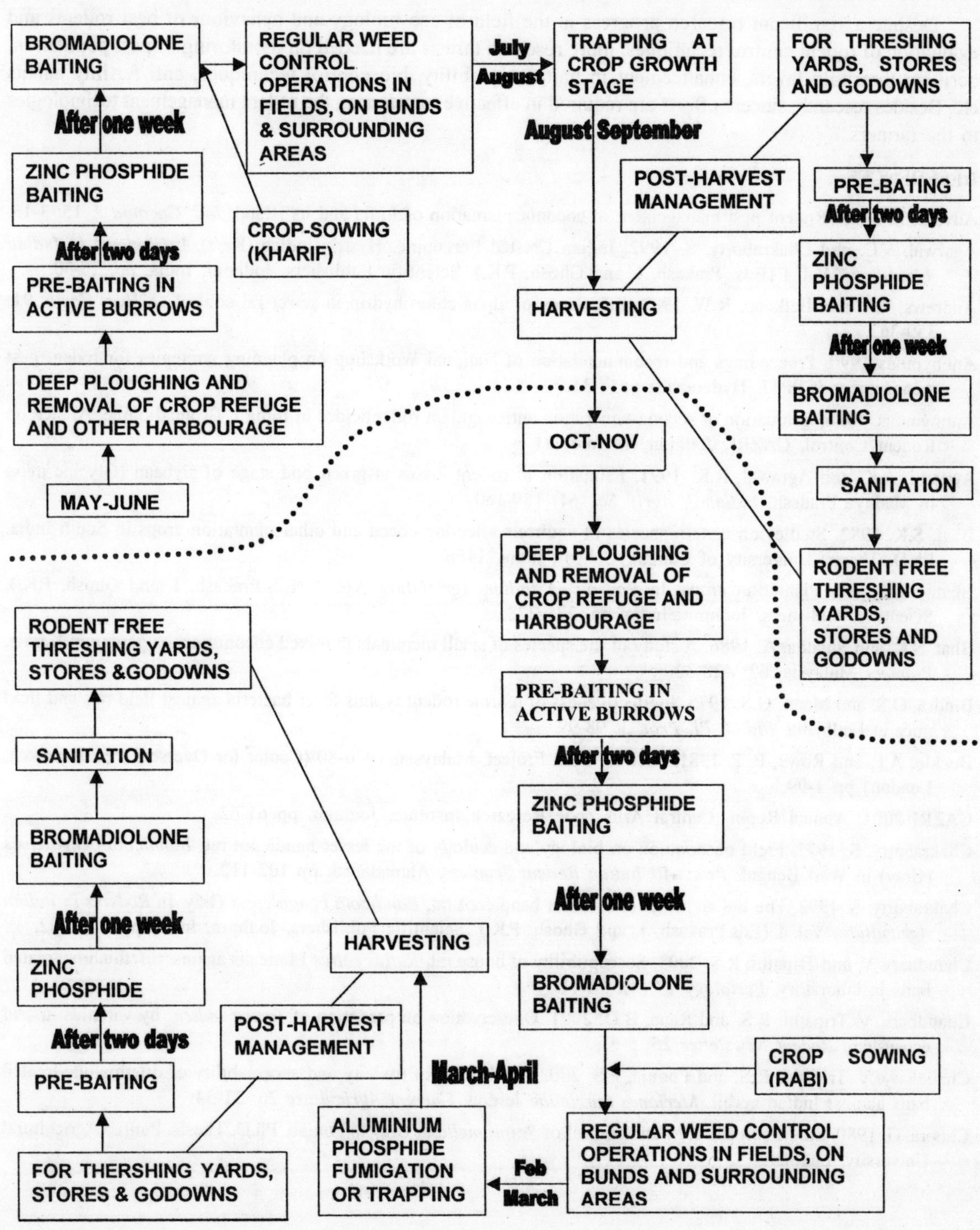

Needless to mention, the common propriety resources like Government lands viz., rail/road side land, panchayat land, forests etc. are the real breeding shelters and act as focal centers for their further advances towards crop fields. Since these do not belong to any individuals, the rodent control work, especially during summer months must be taken up by farmers in such areas also. The rodent damage remains at minimum levels when the rodent management is undertaken on whole village basis.

Despite a significant research progress in the field of eco-biology and behaviour of pest rodents and evaluation of rodent control techniques, more research thrusts are needed on monitoring of pest population, deriving threshold levels, enhancement in bait acceptability, bio control techniques, anti fertility agents etc. Besides research, sincere efforts are required in effective transferring the rodent management technologies to the farmers.

REFERENCES

Advani, R. 1985. Rodent pest management in coconut plantation of India and its island. *Ind. Coconut J.* 15: 3-15.

Agarwal, V.C. and Chakraborty, S. 1992. Indian Crested Porcupine, Hystrix indica (Kerr). In *Rodents in Indian Agriculture*, Vol. I (Eds. Prakash, I. and Ghosh, P.K.). Scientific Publishers, Jodhpur, India. pp. 25-30.

Andrews, R.V. and Belknap, R.W. 1983. Effeciency of alpha-chlorohydrin in sewer rat control. *J. Hyg. Camb.* 91: 359-362.

Anonymous 1990. Proceedings and recommendation of National Workshop on planning strategies for rodent pest management (CPPTI, Hyderabad) pp1- 31.

Anonymous, 1986. Evaluation of second-generation anticoagulant rodenticides in India I. Bromadiolone. AICRP on Rodent Control, CAZRI, Jodhpur (Memio) 21 p.

Awasthi, A.K. and Agrawal, R.K. 1991. Esimation of rodent losses at green pod stage of siybean (Glycine max) in Madhya Pradesh. *Indian J. Agril. Sci.* 61: 159-160.

Bhat, S.K. 1982. Studies on certain species of squirrels affecting cocoa and other plantation crops in South India. Ph.D. Thesis, University of Calicut, Calicut, India. 146p.

Bhat, S.K. 1992. Plantation crops. In *Rodents in Indian Agriculture*, Vol. I (Eds.Prakash, I. and Ghosh, P.K.). Scientific Publishers, Jodhpur, India. pp. 271-278.

Bhat, S.K. and Sujatha, A. 1986. A study of the species of small mammals in mixed coconut-cocoa garden in Kerala. *Planters* Malaysia. 62: 428-32.

Bindra, O.S. and Mann, G.S. 1975. Ineffectiveness of murine rodent typhus fever bacteria against field rats and field mice at Ludhiana. *Ind. J. Pl. Prot.* 3: 98-99.

Buckle, A.P. and Rowe, F. P. 1981. Rice field rat Project, Malaysia, 1970-80 (Center for Overseas Pest Research, London) pp 1-99.

CAZRI 2001. Annual Repot ,Central Arid Zone Research Institute, Jodhpur. pp 61-62.

Chakraborty, S. 1975. Field observation on biology and ecology of the lesser bandicoot rat, *Bandicota bengalensis* (Gray) in West Bengal. *Proc. All Indian Rodent Seminar*, Ahmadabad. pp 102-112.

Chakraborty, S. 1992. The Indian mole rat or lesser bandicoot rat, *Bandicota bengalensis* Gray. In *Rodents in Indian Agriculture*, Vol. I (Eds.Prakash, I. and Ghosh, P.K.). Scientific Publishers, Jodhpur, India. pp. 173-192.

Chaudhary, V. and Tripathi, R.S. 2003. Susceptibility of house rat, *Rattus rattus* Linneaus against difethialone treated baits in laboratory. *Pestology* 27 (4): 38-41.

Chaudhary, V. Tripathi, R.S. and Rana, B.D. 2001. Obeservation on predation of *Tatera indica* by varanids in arid ecosystem. *Rodent Newsletter* 25: 5-6.

Chaudhary. V, Tripathi, R.S. and Poonia, F.S. 2002. Evaluation of toxicity and acceptability of difethialone treated baits against Indian gerbil, *Meriones hurrianae* Jerdon. *Current Agriculture* 26: 31-34.

Chopra, G. 1980. Studies on ethology and control of *Rattus meltada* Gray in Punjab. Ph.D. Thesis. Punjab Agricultural University, Ludhiana.

Chopra, G., Kaur, P. and Guraya, S.S. 1996. Rodents: ecology, biology and control. R. Chand & Co., New Delhi. pp 1-96.

Deoras, P.J. 1964. Rats and their control. A review of an account of work done on rats in India. *Ind. J. Ent.* 26: 407-408.

Dubey, O.P. Awasthi, A.K. and Patel, R.K. 1992. Bengal Gram. In *Rodents in Indian Agriculture*, Vol. I (Eds.Prakash, I. and Ghosh, P.K.). Scientific Publishers, Jodhpur, India. pp. 269-270.

Duckett, J.E. and Karuppiah, S. 1991. Progress at PORIM/Australian Enterprise Berhad Barn Owl Project at Kok Foh Estate up to February 1990. Annual Report of the Negeri Sembilan Planters' Association. pp 62-70.

Ericsson, R.J. 1982. Alpha-chlorohydrin (Epibloc): a toxicant - sterilant as an alternative in rodent control. *Proc. 10th Vert. Pest Conf.* (Ed. Marsh, R.E.). Univ. of California, Davis, USA, pp. 6-9.

Gains, T.B. and Kinborough, R.D. 1964. Toxicity of metepa to rats. *Bull. WHO* 31: 737.

Gibb, J.A. 1981. What determines the number of small herbivorous mammals. *New Zealand J. Eco.* 4: 73.

Green, M.G. and Taylor, K.D. 1975. Preliminary experiments on habitat alteration as a means of controlling field rodents in Kenya. *Ecol. Bull.* Stockholm. 19: 175.

Hansson, L. 1975. Effect of habitat manipulation on small rodent populations. *Ecol. Bull.* Stockhlom. 19: 163.

Idris, I. S Tripathi, R.S. Chaudhary, V. and Rana, B.D. 2001. Incidence of Indian Field mouse, *Mus booduga* Gray in Thar desert. Proceed. National Symp. on Impact of human activities on Thar desert environment. Jodhpur (Feb. 15-17, 2001) pp 108.

Idris, M. and Rana, B.D. 2001. Occurrence of lesser bandicoot rat, *Bandicota bengalensis* in Jodhpur town. *Rodent Newsletter* 25(3-4): 9.

Jain, A.P. 1988. Social engineering activity Control- A case study. AICRP on rodent Control, CAZRI, Jodhpur (Memio).

Jain, A.P. and Tri[pathi, R.S. 2000. An integrated approach to rodent pest management. In *IPM systems in Agriculture, Vol. 7. Animal Key Pests* (Eds R.K. Upadhyaya, K.G. Mukherjee and O.P. Dubey). Aditya Book Pvt. Ltd. New Delhi. pp 275-307.

Jain, A.P. and Tripathi, R.S. 1988. Rodent Pests of Agriculture- An Illustrated Guide. AICRP on Rodent Control, CAZRI, Jodhpur (Memio) pp 1-12.

Jain, A.P. and Tripathi, R.S. 1992. Management of rodent pests in oilseed crops. In *Advances in oil seed Research.* (Eds. D. Kumar and M. Rai) Scientific Publisher, Jodhpur. Pp 360-367.

Jain, A.P., Mathur, M. and Tripathi, R.S. 1992. Bio-efficacy of flocoumafen against major desert rodent pests. *Indian J. Pl. Prot.*, 20: 81-85.

Jain, A.P., Rana, B.D. and Tripathi, R.S. 1993. Rodent Management- The state of Art. AICRP on Rodent Control, Central Arid Zone Research Institute, Jodhpur. 38p.

Jain, A.P., Tripathi, R.S. and Patel, N. 1995. Influence of aridity on burrowing and other behavioural traits of *Nesokia indica* Gray. *Ann Arid Zone*, 34: 49-51

Jain, A.P., Tripathi, R.S. and Rana, B.D. 1986. Impact of ecological transformation on the rodent fauna of the Thar desert. In *Faunal diversity in the Thar Desert: Gaps in Research* (Eds A.K. Ghosh, QH Baqri and I. Prakash) Scientific Publisher Jodhpur pp 389-402.

Jan, A.P. 1970. Body weight, sex ratio, age structure and some aspect of reproduction in the Indian gerbil, *Tatera indica indica* Hardwike in the Rajasthan desert, India. *Mammalia* 34:415-432.

Kalra, S.P. and Prasad, M.R.N. 1966. Effect of clomiphene on fertility in female rats. *Ind. J. Expt. Bio.* 5: 1-8.

Kassa, H. and Jackson, W.B. 1984. Bait acceptance and chemosterilant efficacy of alpha-chlorohydrin in Norway rats, *Rattus norvegicus*. *Int. Pest Control* 26: 7.

Kinnear 1919. Expected plague of field rats in 1920. *J. Bombay Nat. Hist. Soc.* 26: 663-666.

Kumar, T.S. 1985. A life history of spotted owlet (*Athene brama brama* Temminick) in Andhra Pradesh. Raptor Research Centre, 241p.

Leirs, H. 2003. Management of rodents in crops: the pied piper and his orchestra. In *Rats Mice and People: Rodent Biology and Management.* (Eds G.R. Singleton, L.A. Hinds, C.J. Krebs and D.M. Spratt) Australian Centre for International Agricultural Research, Canberra. pp 183-189.

Malhi, C.S. and Parshad, V.R. 1987. Pest damage to harvested wheat and paddy. *Indian Farming* 37: 29-30.

Mittal V.P. and Vyas, H.J. 1992. Groundnut. In *Rodents in Indian Agriculture*, Vol. I (Eds.Prakash, I. and Ghosh, P.K.). Scientific Publishers, Jodhpur, India. pp. 249-264.

Neelanarayanan, P. 1997. Predatory pressure of barn owl (*Tyto alba stertens* Hartert, 1928) on rodent pest- a field evaluation. Ph. D. Thesis, Mayiladuthurai, Bharthidasan University, Thiruchirapalli 92p.

Parshad, V.R. 1988. Laboratory and field evaluation of flocoumafen for rodent control. *International Biodeterioration* 24: 475-480.

Parshad, V.R. 1999. Rodent control in India. *Integrated Pest Management Reviews* 4: 97-126.

Parshad, V.R. and Chopra, G. 1986. The susceptibility of *Rattus rattus* and *Bandicota bengalensis* to a new anticoagulant rodenticide, flocoumafen. *J. Hyg.* Camb. 96: 475-478.

Pashahan, S.C. and Sablok , V.P. 1987. Effect of removal of wild vegetation on migration of field rodents. *HAU J. Res.* 17: 12-13.

Patel, N., Jain A.P., Tripathi, R.S., Praveen, F and Mathur, M. 1995. Losses caused to pomegranate fruits by rodents in arid horticulture. *Geobios New Report.* 14: 37-38.

Patel, R.K. Dubey, O.P. and Awasthi, A.K. 1992. Soybean. In *Rodents in Indian Agriculture*, Vol. I (Eds.Prakash, I. and Ghosh, P.K.). Scientific Publishers, Jodhpur, India. pp. 265-268.

Prakash, I. 1976. Rodent pest management- Principal and Practices. Central Arid Zone Research Institute Monograph Series No. 4, Jodhpur. 28p.

Prakash, I. 1981. Ecology of the Indian desret gerbil, *Meriones hurrianae*. Central Arid Zone Research Institue, Monograph Series No. 10. Jodhpur 88 p.

Prakash, I. and Kamethar, L.R. 1969. Body weight, sex and age factors in population of the northern palm squirrel, *Funambulus pennanti* Wroughton. *J. Bombay Nat. Hist. Soc.* 66: 99-115.

Prakash, I. and Mathur, R.P. 1987. Management of Rodent Pests, Indian Council of Agricultural Research, New Delhi. pp. 133.

Rana, B.D. and Prakash, I. 1984. Reproductive biology of *Rattus meltada pallidior* (Ryle, E. 1914) in the Rajasthan desert. *J. Bombay Nat. Hist. Soc.* 18: 59-70.

Rana, B.D. and Tripathi R.S. 1999. Recent advances in coordinated research on rodent control. AICRP on Rodent Control, CAZRI, Jodhpur, pp. 48.

Rana, B.D. Jain, A.P. and Soni, B.K. 1982. Bait preferences. In *Rodents in Indian Agriculture*, Vol. I (Eds.Prakash, I. and Ghosh, P.K.). Scientific Publishers, Jodhpur, India. pp. 419-426.

Rao A.M.K.M. 1992. Integrated rodent management. In *Rodents in Indian Agriculture*, Vol. I (Eds.Prakash, I. and Ghosh, P.K.). Scientific Publishers, Jodhpur, India. pp. 651-668.

Rao, A.M.K.M. 2003. Rodent problem in India and Stratigies for their management. In *Rats Mice and People: Rodent Biology and Management.* (Eds G.R. Singleton, L.A. Hinds, C.J. Krebs and D.M. Spratt) Australian Centre for International Agricultural Research, Canberra. pp 203-212.

Saini, M.S. and Parshad, V.R. 1988. Laboratory evaluation of a toxicant cum sterilant, alpha-chlorohydrin for the control of Indian mole rat. *Ann. Appl. Biol.* 113: 307-312.

Shah, M.P. 1979. Population explosion of rodents in groundnut field in Gujarat. *Rodent Newsletter* (ICAR) 3: 19-20.

Sharma, P.K. and Rao, A.M.K.M. 1989. Effect of bund dimension in rodent infestation in rice. *International Rice Research Institute (IRRI) Newsletter* 14:40.

Sharma, V.N. and Joshi, M.C. 1975. Soil excavation by desert gerbils, *Meriones hurrinae* (Jerdon) in Shekhawati region of Rajasthan desert. *Ann. Arid Zone* 14: 268-273.

Singh, C.D. and Rao, A.M.K.M. 1982. Composition of three methods for rodent damage survey in rice fields. *Ind. J. Pl. Prot.* 10: 98-99.

Singh, Y.P., Gangawar, S.K., Kumar, D., Kushwaha, R. B., Azad Thakur, N.S. and Kumar, M. 1995. Rodent Pest and their management in North eastern Hill Region. Monograph No. 37. ICAR Research Complex for NEH Region, Barapani. pp. 35.

Singla, N. Parshad, V.R. and Singla, L.D. 2003. Potential of *Trypanosama evansi* as a biocide of rodent pests. In *Rats Mice and People: Rodent Biology and Management.* (Eds G.R. Singleton, L.A. Hinds, C.J. Krebs and D.M. Spratt) Australian Centre for International Agricultural Research, Canberra. pp 43-46.

Sridhra,S., Raveendra Babu, T. and Ajay, P. 2000. Laboratory and field evaluation of difethialone. AICRP on Rodent Control, UAS, Bangalore (Memio) pp. 25.

Srivastava, A.S. and Nigam, P.M. 1975. Mechanism of action of chemosterilants in field rats. I. Role of furadentin and colchicines in causing sterility in lesser bandicoot, *Bandicota bengalensis* Gray. *Proceed. All India Rodent Seminar*, Ahmedabad. pp 239-246.

Srivastava, D.C. 1992. Sugarcane. In *Rodents in Indian Agriculture*, Vol. I (Eds.Prakash, I. and Ghosh, P.K.). Scientific Publishers, Jodhpur, India. pp. 231-248.

Tripathi R.S. and Jain A.P. 1990. Rodent damage to desert afforestation plantations. *Rodent Newsletter* 14: 7-8.

Tripathi, R.S. Jain, A.P. Kashyap, N., Rana, B.D. and Prakash, I. 1992. North Western Desert, Special Problem Areas. In *Rodents in Indian Agriculture*, Vol. I (Eds.Prakash, I. and Ghosh, P.K.). Scientific Publishers, Jodhpur, India. pp. 357-396.

Tripathi, R.S., Mathur, M., Jain, A.P. and Patel, N. 1994. Relative efficacy of glue and other traps for commrnsal rodent management. *Ann. Arid Zone*. 33: 143-145.

Uchida, T.A. 1966. Observation on monitor lizard, *Varanus indicus* (Daudin) as a rat control agent on Ifaluk, Western Caroline Islands. *Bull. WHO*. 5: 976.

Vyas, H.J., Kotadia, V.S. and Butani, P.G. 2000. Glimpses of research on rodent mangement in Gujarat. AICRP on Rodent Control, GAU, Junagadh pp 1-74

Wheatherbee, D.K. 1964. Vertebrate pest control by biological means. *Proceed. AAAS Symposium on Pest Control* Montreal, Canada pp 19.

Whitaker, R. 1987. Rodent control by Irula tribals. *Rodent Newsletter* 11: 1-2.

Whitaker, R. and Dattari, S. 1986. The role of reptiles in controlling food pests. *Hamadryad*, 11: 23-24.

Management Strategies for White Grubs in Agri-Horticultural Crops

C.P.S. YADAVA[1]

The term 'white grub' or 'root grub' is applied to the immature stage of scarab beetles popularly known as cock chafers, leaf chafers, chafer beetle, Lamelicorn beetle, May beetles or June beetles. They belong to family Scarabaeidae of order Coleoptera. It is the group which is causing concern to agriculturists throughout the world, as their grubs are root feeders which mainly belong to subfamily Melolonthinae and rutelinae; some grubs of dynastinae and cetoniinae are also root feeders.

White grubs are cosmopolitan in occurrence and significant pests in many agri-horticultural cropping systems throughout the semi-arid tropics and sub-tropics in the both, developing and developed countries. Some of the white grub species of international importance are:

Melolontha melolontha Linn., Europe; *Amphimallon majalis* (Razoumowsky) and *Popillia japonica* Newman, U.S.A.; *Heteronyx piceus* Blanch., Australia; *Costelytra zealandica* White and *Adoryphorus couloni* (Burm.) New Zealand; *Holotrichia parallels* (Mots.), Japan and China; *Leucopholis irrorata* Chevrolat., Philippines; *Phyllophaga crinita* (Buryn.), Mexico; and *Holotrichia consanguinea* Blanch. and *Holotrichia serrata* in India.

In India, white grub is one of the five pests of national importance and adversely affect the quality of life of millions of people in the country. As per very conservative estimates various crops over one million hectares are yearly infested by white grubs in the country and causing total loss of about Rs. 2000 million. Being polyphagous, it has become a serious constraint for production of almost all agri-horticultural crops grown during the rainy season *(kharif) viz.,* groundnut, sugarcane, pearl-millet, sorghum, cow-pea, pigeonpea, green grain, clusterbean, chilli, potato, brinjal, tomato, maize, ginger, upland paddy etc. The plantation crops like tea and coffee suffer similar damage in their seedling and early growth stages. They are distributed from high altitude of Himalayas to low altitude of coastal Kerala, from and semi-arid tracts of Rajasthan to high rainfall, humid regions of north-east hill region and west coastal peninsula.

NATURE AND EXTENT OF DAMAGE

The grubs are subterranean root feeders and beetles feed on foliage of trees, shrubs and bushes. Both of these stages are polyphagous. The grub stage has three instar. All the three instars feed on the roots of crops and grasses during rainy season. The feeding by first instar goes unnoticed; when the grubs reach to second instar stage the damage caused to some crops like groundnut becomes noticeable. However, maximum feeding is done by the third instar which is very voracious. The tap rooted crops suffer more damage as compared to the fibrous rooted crops. Besides root the grubs may feed on underground fleshy

---

1 Vice-Chancellor, Rajasthan Agricultural University, Bikaner.

plant parts as well. The symptoms of injury is root pruning by the grubs. Such plants depending on the type of the root systems show varying degree of yellowing, browning, wilting and eventually death. In the field dead patches appear in the early parts of the season which gradually extend outwards, and coalesce with other dead areas resulting formation of big patches of dead plants. In leguminous crops the grubs eat away nodules as well as fine rootlets and later cut and eat the main root. Because of this the tap rooted plants become highly susceptible to grubs whereas plants with adventitious root system like bajra, sorghum and maize can withstand relatively higher grub population. In groundnut, chillies, pulses and other such crops the plant damage may range from 60 to 100 per cent. In other crops the damage may vary from 10 to 50 per cent depending upon the grub population. If the grubs attack lawns, pastures and forage crops the injury occurs near the surface. Destruction of turf leads to subsequent invasion by weeds and damage to turf at hill sides is subjected to erosion.

The beetles, like the grubs, are also polyphagous and feed on about 250 different types of host trees. They are also very destructive particularly when they defoliate the orchards and other trees of economic importance, thereby adversely affect the fruit and fodder yield. Adults are leaf chafers, some are also feed on ripe fruits and developing grains. *Oryctes rhinoceros* L. and *Xylotrupes gideon* L. are the pests of coconut palm. Beetles of *Clinteria* sp., damages the red apple fruits. *H. consanguinea* defoliate the her and guava trees.

BIOLOGY AND BIONOMICS OF *HOLOTRICHIA CONSANGUINEA*

In Rajasthan, the most important scarab species is *Holotrichia consanguinea* Blanch. which is very serious pest in the semi-arid region of the State. It attacks all the crops grown in *kharif* season, however groundnut suffers the maximum damage (20-100%) by the pest. The beetles emerge from the soil during dusk after good rain in mid-May or later. Once active, the beetles follow the daily rhythm of emergence and congregation on host trees at dusk and return to soil at dawn. The beetles are polyphagous, and may feed on the foliage of a wide variety of host trees and bushes found in the close vicinity. However, they do exercise some preference for certain hosts. In some situations large number of beetles settle on a tree, which is not a host, only for mating and just after mating, shift for feeding to nearby preferred hosts. The host preference depends on the combination of hosts available in a particular locality. The beetles, usually after emergence, do not fly long distances in the evenings but try to avail of short twilight hour for mating and selecting suitable host as they are unable to see clearly in dark. It is the female, which first comes out of soil, settles on a host and secretes a pheromone which attracts the male for mating. If a female fails to mate in first 15 to 20 minutes of settling on the host trees, she tries mating next evening. The process of congregation of beetles on host continues every evening for a period of about four weeks to ensure mating of all females. The mated female returns to soil in early morning and starts egg-laying. Newly hatched grubs may feed on organic matter for some time till they come in contact with living roots.

For feeding, the grub makes a chamber by compressing the surrounding soil particles and then, eats the rootlets exposed into the chamber, thereafter it moves vertically a little, to eat more of the same root. After this, the grub moves horizontally making chambers and feeding on the exposed roots. The grubs continue active feeding from July to mid-October, some other species may continue up to April, and thereafter enter into pupation. Before pupation the grub ceases to feed and descends down in the soil to the depth of 40-70 cm or more in search of suitable moisture zone. A short prepupal period is spent before entering into pupation during which the grub prepares an earthen chamber for pupation. The beetles remain in the soil in inactive state up to the middle of May at a depth of about one meter. They emerge from their hiding abode after receiving first good rain. Thus, there is only one generation in a year.

OTHER MAJOR SCARAB PESTS OF INDIA

The other key pest species of scarabs attacking different crops in the country are as follows :

List of key pest species of scarabs attacking different crops in the country

Name	Distribution	Host range
Holotrichia serrata (F.)	Karnataka, Andhra Pradesh, Maharashtra, Tamil Nadu, Kerala, Tarai of U.P., Southern Rajasthan, Southern Bihar	Vegetables, pulses, millets, tobacco, sugarcane and sorghum
Holotrichia reynaudi Blanch.	Central peninsular region in the states of Andhra Pradesh, Karnataka, Tamil Nadu	Groundnut, pulses, oilseeds, vegetables
Holotrichia longipennis Blanch.	Uttaranchal	Small millets, upland paddy, maize, soy bean, potato, chilli
Leucopholis lepidophora Blanch.	Maharashtra (Kolhapur region), Coastal Karnataka	Sugarcane, arecanut, coconut
Holotrichia seticollis Moser	Uttaranchal and Sikkim	All rainy season crops, ginger, maize
Holotriehia coriacea Hope	Hill tracts of Himachal Pradesh, U.P.	Seed potato, cedar nursery
Anomala dimidiata Hope	Uttaranchal, H.P., and Jammu and Kashmir	Grubs - upland rice, barnyard millet, finger millet, chilli, wheat; adults- apple, walnut, plum, apricot, peach
Leucopholis coneophora Burm.	Heavy rainfall areas of coastal Karnataka and Kerala	Arecanut, coconut, cassava, sweet potato, yam, colocasia
Leucopholis burmeisteri Brenske.	South Kanara, Chikmagular and Shimoga (Western Ghats of Karnataka)	Arecanut, sugarcane
Holotrichia nilgiria Arrow	Chikmanglur, Coorg, Hassan, and western ghats in Karnataka	Young coffee plants.
Maladera insanabilis Brenske.	Western Rajasthan	Groundnut, chilli, okra, brinjal, pearl millet lucerne, onion, cucurbits

MANAGEMENT STRATEGIES

For managing the population of whitegrubs efforts can be directed against beetles or grubs. The *kharif* crops are cultivated in rainfed as well as irrigated conditions. In rainfed condition, the crops are sown with the onset of monsoon in June-July and this coincides with the emergence of beetles, but where irrigation is available, the farmers usually go for early sowing of the crop. Therefore, some of the practices which are applicable in rainfed condition may not be suitable for advance planted crops, as this does not coincide with the emergence of beetles. Looking to the variations in the conditions and suiting to these requirements, the management strategies have been devised separately for rainfed and irrigated conditions. Nevertheless, the beetle control strategy is suitable for both the conditions.

1. Management through beetle control (Suitable for advance as well as monsoon sown; rainfed/irrigated conditions)

Most of the scarab beetles which attack *kharif* crops emerge from soil with first good shower (pre-monsoon or monsoon) and congregate on some preferred hosts and can be killed conveniently by loading of pheromone and spraying of insecticides on the host trees. This technology is also successful where the number of host trees are more. Select one host tree from a group of host trees within the radius of 20 in and spray 0.05 per cent monocrotophos during day-time; pheromone dispensers be placed (2-3/tree) on this host tree in the evening after beetle emergence. No time gap should be allowed between onset of rain and spraying along with pheromone loading. This should be done for 3-4 days of first emergence of beetles. If there is emergence of beetles in response to pre-monsoon shower, then two sprayings will be required, first in response to pre-monsoon and second in response to regular monsoon. Where *H. consanguinea* is the pest species, no spraying is needed if there is rain prior to mid-May. For spraying, a foot-sprayer or rocking sprayer can be effectively used. To reach the height of trees, the nozzle of the sprayer may be fastened at one end of a bamboo pole and trigger valve socket be left open. This is quite useful in reaching heights of average-sized trees of *neem*, *ber*, *khejri* and *gular*.

The beetles which are attracted by pheromone, settle and feed on the sprayed host trees and get killed soon by contact as well as stomach action of the insecticide and are thus not able to return to soil next morning for egg-laying. This way number of trees to be sprayed is drastically reduced which make the beetle control operation more effective, eco-friendly, economical and less time consuming. If there are no host trees in the vicinity of an infested field, the twigs of easily available preferred hosts, loaded with pheromone and sprayed with the insecticides, should be planted in such fields to attract and kill the beetles. For beetle control, pheromone dispenser, insecticide, sprayer, bamboo pole etc. should be kept ready with the farmer prior to the onset of monsoon. This will enable the farmer to start the spraying operation immediately after the first shower, without waiting for outside agencies and thereby not loosing any time, as the first 3-4 days are crucial for the operation. The beetle control can be done without pheromone also, but in this case number of trees to be treated will be more. Some trees can be lopped completely to minimize the spraying cost and to cover more area. The lopping should be done before onset of monsoon. While lopping, some host trees should be left unlopped, which should be sprayed; the unlopped trees should be uniformly distributed in the area.

II. Management through grub control

(i) Monsoon sown crops (rainfed/irrigated):

Seed Treatment (only for groundnut): To protect groundnut crop from grubs, seed treatment with chlorpyriphos 20 EC or quinalphos 25 EC at 25 ml/kg seed is quite effective. For this, a known quantity of seed should be placed in a seed treatment drum or polyethylene bag ; the required quantity of insecticide be added to it and the drum/bag should be swirled gently to provide uniform coat of the insecticide on the seed. This seed should be sown in furrows. The interval between coating and sowing should be short (4-6 hours). Premonsoon or early sown crops may not get effective protection by the seed treatment.

Soil Treatment: Pre-sowing soil treatment with phorate 10 G at 25 kg/ha or quinalphos 5 G at 30 kg/ha effectively protects the crop. The insecticide be applied in furrows and seeds be placed in the same furrow. Since seed treatment is equally effective and cheaper, soil treatment may not be done in groundnut.

(ii) Advance sown crop: Standing crop treatment:

For advance sown crops, seed treatment or soil treatment may not be done as this may become ineffective by the time beetles emerge, lay eggs and grubs develop. For such crop situations, application of quinalphos 25 EC or chlorpyriphos 20 EC at 4 litre/ha with irrigation water be done three weeks after the first good monsoon shower. Insecticidal application in standing crop could be done in the following ways:

Required amount of insecticide be applied at the point from which the water enters into the plot. This can be accomplished by putting the insecticide in an earthen pitcher having a hole in the bottom. A cotton wick should be placed in the hole for controlling the drops. The drops should be so calibrated by the movement of the cotton wick that the required amount of insecticide is applied uniformly with irrigation. This pitcher can be placed on a tripod.

Soil collected from the field (approx. 80-100 kg/ha) be treated with required amount of insecticide (4 l/ha) and the insecticide mixed soil be uniformly applied in standing crop, near root zone. The application should be followed by light irrigation so that the insecticide percolates downwards. Insecticide mixed soil can also be used in rainfed crops provided it rains soon after the application.

III. MANAGEMENT IN OTHER CROPS

Pearl-millet

1. Phorate 10 G at 12 kg/ha to be mixed with required amount of seed and then the insecticide-seed mixture be sown.
2. Presowing soil application of phorate 10 G at 25 kg/ha or quinalphos 5 G at 30 kg/ha be done in furrows.
3. In standing crop, chlorpyriphos 20 EC or quinalphos 25 EC at 41/ha be applied with irrigation, 3 weeks after the first monsoon shower.

Kharif Pulses

1. For monsoon sown, soil application of phorate 10 G at 25 kg/ha be done.
2. In standing crop, chlorpyriphos 20 EC or quinalphos 25 EC be applied at 41/ha with irrigation, 3 weeks after first monsoon shower.

Sugarcane

1. Farmers be encouraged to give up sugarcane ratoon and in its place sunflower and /or vegetable crops be taken from January to March. It is more remunerative than sugarcane ratoon.
2. Soil application of phorate 10 G at 25 kg/ha be done before planting in November in furrows.
3. Severely damaged fields of sugarcane be ploughed in the months of May-June to expose the grubs to predatory birds and puddled paddy be taken in such fields in rotation.
4. Conserve and augment parasitic wasp, *Campsomeriella collaris,* which is effective in reducing white grub population. No insecticidal application be done in areas where the wasp has established.
5. Date of planting be shifted from December-January to June. Wherever possible ratoon be avoided. Rotation of puddled paddy with sugarcane be adopted (for Karnataka).
6. Soil application of phorate 10 G at 25 kg/ha or quinalphos 5 G at 30 kg/ha be done within one month from the first emergence of the beetles following rains in April-May in Karnataka. In Maharashtra, the insecticides be incorporated in the soil before sowing after receiving rain (June-July).

Plantation Crops

1. Hand picking and destruction of adults be done at beetle emergence (June-August).
2. Tapioca (cassava) be planted as a trap crop in coconut gardens in February-March. Third instar

grubs feeding on tapioca sets can then be exposed and killed.

3. Phorate 10 G at 20 g/tree be applied twice in a year to plants up to three years of age. The insecticide be incorporated in soil around the root zone, followed by light irrigation.

FUTURE THRUST AREAS

The nematode, *Heterorhabditis bacteriophora,* have been found pathogenic to whitegrubs, causing quick mortality (within 48 hours). After conducting tests through microplots and field experiments for appropriate doses, method of application etc., attempts should be made to develop nematode formulations with increased storage life and prolonged survival of nematodes in soil. Local strains of entomogenous nematodes be isolated from different climatic zones and their pathogenicity to key white grub species be established.

Future research work on search of better strains/isolates of *Metarhizium and Beauveria* should be done. Mass conidial production of promising isolates of these pathogens should be undertaken for their wide use under field conditions. Apart from this, attempt should be made to develop suitable formulations of virulent strains possessing prolonged storage and field stability. Also studies on application of bioagents formulation in relation to larval behaviour of white grubs in soil need to be intensified.

Behaviour regulators like pheromones and kairomones can be utilized very effectively for beetle control by using them along with insecticides. Presently a sex pheromone of *H. consanguinea* is available, however a dispenser which can release the pheromone, after one time loading, for 3/4 days, needs to be developed. Also efforts should be made to isolate kairomones from the host trees. Apart from all this, soil health condition should be constantly monitored. Till soil pest management programme is not juxtaposed with soil health management, sustainable agriculture would be a dream eluding the agricultural scientists.

20

Eco-freindly Management Strategies for Bacterial Diseases of Vegetable Crops

K. K. PANDEY[1] AND P.K. PANDEY[1]

Vegetable crops are more susceptible to disease as compared to cereals. Various biotic and ablotic factors are responsible for diseases in vegetables. Introduction of hybrid varieties in vegetable crops created several new disease problems and increased disease intensity of previously known minor diseases (Pandey and Pandey 2002). Intensive and monoculturing of vegetables also increased the disease pressure. The off-season practices of vegetable growing promoted survival period of the pathogen in the field. Seed home diseases and their inoculum in vegetables are increasing throughout the country Pandey *(et al.* 2002). Disease pressure from seedling stage to harvest and spoilage caused by microorganisms during transit, storage and marketing are the main constraints in total vegetable production.

Bacterial diseases of vegetables are common problem of all agroclimatic zones. It is worldwide problem. Phytopathogenic bacteria cause wilting of the plants, spotting of the leaves, cankers on the stem, leaf blights, scab on fruits, soft and root rotting. About 100 species of bacteria cause diseases in plants. Most plant pathogenic bacteria are facultative saprophytes and gram negative. They are successfully cultured on artificial nutrient media. Bacterial pathogens are having very fast multiplication in geometrical rate. It rapidly moves throughout the plant system and quickly destroys the plant. Phytopathogenic bacteria mainly confined to the vascular system of the plant and ultimately majority of them become seed home.

The understanding of epidemiology particularly disease initiation, factors influencing the disease development and its progress is pre-requisite for development of an effective, economic and viable management strategies of the disease. Several pre disposing factors are responsible for disease initiation which later on results in epiphytotic forms. Wind may influence the development of bacterial diseases primarily by increasing the spread of plant pathogenic bacteria. Wind blown soil, especially sandy soil can cause considerable injury of leaves. This injury can facilitate the entry of bacterial pathogen and leads to primary infection. Wind-blown rains splash are the prominent source for spread of bacterial disease. Sometimes wind may prevent the infection by accelerating the drying process of the wet plant surface on which bacterial pathogens may have landed. It is also helping the reduction of spread of secondary inoculum as bacterial ooze may dry on host itself. The influence of light on the development of bacterial plant disease under natural conditions is of less important than other environmental factors like temperature and moisture. Light intensity and duration are positively correlated with the intensity of the susceptible reaction. Nutrition is one of the most important components determining overall health of vegetable crops to defend themselves against attack of phytopathogenic bacteria. Application of unbalanced nutrients makes the plants susceptible for a longer period of time than the plants receiving balance nutrients. Excessive nitrogen makes the plants more susceptible to bacterial plant pathogens. Human being is also

1. Indian Institute of Vegetable Research, Varanasi-221 005, U.P.

responsible for secondary spread of the bacterial disease. During intercultural operations and frequent movement in the field particularly in the morning time when bacterial ooze is present along with dew deposition on leaves is main source of secondary spread in black rot of cole crops, bacterial bligbt and bacterial speck of tomato, bacterial blights of bean and pea (Pandey *et al.2003).*

In the recent past, the term management has replaced the terminology of control. Control means absolute check of a disease, which is neither true nor possible. No pathogen is completely eradicated from a natural ecosystem but only its population can be reduced. Bacterial diseases cannot be managed by a single method and always it requires holistic approach. Integrated management practices planned in such a way that it should be adjusted in the normal cultural practices and effective against more than one disease. Integrated disease management must be adopted for pathogen, host, time and environment. Pathogen management involves reduction in inoculum, eradication and prevention of inoculum while management of host involves improving plant vigour, inducing resistance through nutrition and genetic manipulation and protection by chemical means. The time factor could be managed by adjustment of sowing and planting date so that favorable time of disease should not coincide without any adverse effect on yield. Management of environments particularly microclimates involves water, soil and crop management (Pnadey 2001). Success of integrated disease management approach in vegetable crops is only possible when we apply all the above practices every year on community basis in large area by all vegetable growers (Pandey 2001).

Major bacterial diseases of vegetable crops having economic importance are described along with their effective and eco friendly management strategies.

BACTERIAL WILT OF SOLANACEOUS VEGETABLES

Symptoms

Bacterial wilt of solanaceous vegetables is the most destructive and widespread disease in tropical, subtropical regions of the world. In India, the disease is prevalent in Karnataka, Kerala, Maharashtra, Orissa, Bihar, Jharkband, West Bengal and States of North Eastern bills region. Bacterial wilt is very destructive disease for tomato, potato, bn'njal, chillies and capsicum. The disease has wide host range and recorded on several hundred species belonging to more than fifty plant families. Considering the economic importance, wide host range and worldwide distribution, bacterial wilt is ranked as one of the most important disease to be caused by phytopathogenic bacteria. The losses to potato crop alone have been reported to range from 30-70 per cent (Sinha, 1986) whereas in brinjal and tomato the losses are as high as 80 per cent and 90 per cent respectively under Indian conditions. (Rao, 1976) The symptoms characterized by the sudden wilting and death of infected plants. The petiole of older leaves droop down and the leaves showed pale yellowing symptoms accompanied by stunting of whole plant. Excessive development of adventitious roots is also take place in tomato. There is typical browning of vascular tissues of roots, stems or tubers. Whitish bacterial exudates come out from cross sections of infected plants.

Etiology

The bacterium was first described as *Bacillus solanacearum* (Smith 1896), later changed to *Pseudomons solanacearum* (Smith, 1914). Few years back the name was changed to *Burkholderia solanacearum* by Yabuuchi *et al.*, 1992 and later on further changed to *Ralstonia solanacearum* (Yabuuchi *et al.*, 1995). The species is divided into 5 races, 5 biovars and 2 RFLP groups (Cbakrabarti *et al.*, 2001). The bacterium is rod shaped, flagellate having 1-4 polar flagella and gram negative. The size is variable being influence by the growth conditions. The length is usually 1.5 to 3 times greater than the width. Extremely short rods frequently occurring in pairs and measuring 0.5 to 0.6 x 0.8 tol.2 @m occur in the infected tissues. Three distinct pathogenic races have been recognized. Race I affects solanaceous plants such as tomato, eggplant and potato and many non-solanaceous plants also. Race 3 is highly pathogenic on potato and weakly pathogenic on tobacco. Both the races are prevalent in India on solanaceous vegetables.

Epidemiology

The *R. solanacearum* bacterium is mainly soil home. It is also seed home in nature (Shakya, 1993). The pathogen survives in infected plant debris in wild host plants and weeds. Tuber is the source of primary infection in potato. Bacteria enter in plants through wounds occurring in roots due to cultural practices, nematodes, wind rain splashes, hailston-ns and insects. At high inoculum levels, infection may also take place in unwounded roots (Kelman and Sequelra, 1965). The bacteria reach in the xylem vessels and spread into the whole plant. These also enter in the intercellular spaces of the parenchyma cells and dissolve the cell walls. Various virulence factors of the pathogen include production of enzymes. The slimy masses of bacterial cells along with their polysaccharides cause occlusion and clogging of xylem vessels. Phenol oxidases cause oxidation of phenolic compounds into quinones and polymerise to form melanins resulting vascular browning. Relatively high soil moisture and soil temperature (25-35°C) favour the disease. Acidic soil has more severe problem of bacterial wilt. Spontaneous reversion of an avirulent phenotype of *R. solanacearum* to virulent wild types has been reported in Indian isolates (Shekhawat *et al.,* 1992).

Management

The bacterial wilt is very difficult disease to manage due to soil home nature. Most of the commercial cultivars are highly susceptible to the disease. There is no effective and feasible chemical control method. Reduction on soil inoculum is possible with chemicals, soil amendments, crop rotation, summer fallowing, escaping the disease, growing the crop in new fields, resistant cultivars and blocontrol. Chemical control has been tried by antibiotics, chloropicrin and bleaching powder but neither economical nor feasible. *Pseudomonasfluorescens and Bacillus polymyxa* have been found to delay the development and reduce the incidence of bacterial wilt (Shekbawat *et al.,* 1984). Reduced nitrogen application may also increase susceptibility to solanaceous plants to wilt incited by *R. solanacearum* (Agrios, 1997).

Resistance to bacterial wilt in tomato is available but the quality and quantity attributes is not satisfactory. However, genes governing resistance to bacterial wilt are known to linked with horticulturally undesirable traits. Moreover it breaks down in areas where virulent strains of the pathogen and root knot nematodes are present. Some of the tomato lines like, Hawa" 7996, Hawaii 7997, Hawaii 7998, BF-Okistu, R-3034, TML 114 and TML 46 to show 90% survival over different locations Wang *et al.* (I 998). They have also reported resistance sources to bacterial wilt in eggplant and found two varieties, Slim Jim and M 701 F, as resistant. Resistance sources to bacterial wilt in brinjal have also been identified in India. Eight accessions among the 95 namely Arka Nidhi, Arka Kashav, Arka Neelkantha, BB I, BB44, BB49, EP49 and Surya show high level of resistance (Sadashiva *et al.,* 1993). Crop rotation with non-solanaceous hosts is effective against this disease. Soil amendment with S-H mixture has been successfully used for the control of bacterial wilt of tomato (Hsu and Chang, 1989) in Taiwan. Soil amendment with CaO plus urea has been shown to be promising result by Elphinstone and Aley (1993). Akiew *et al.* (1996) have suggested that green manuring with *Brassica* species reduce the level of bacterial wilt inoculum in 11 and subsequent incidence of the disease on tomato, tobacco and other susceptible crops. The bacterial wilt of acidic soil can be managed effectively by increasing the soil pH up to neutral through soil application of lime (Pandey 2001).

BACTERIAL WILT OF CUCURBITACEOUS VEGETABLES

Symptoms and Etiology

Bacterial wilt of cucurbits occurs throughout the United States of America, Cananda, Europe, South Africa and Japan. The disease is sporadic in India and affects many cultivated and wild species of the family cucurbitaceae. Cucumber is the most susceptible host followed by muskmelon, squash and pumpkin while waten-nelon is least susceptible. Bacterial wilt of cucurbits caused by *Erwinia tracheiphila* is a common and destructive disease of cucurbits. This is a capsulated bacterium having four to eight peritrichous flagella. The first signs of wilt appeared as drooping of leaves, which become flaccid in sunny weather. Soon after dropping all leaves and entire plant is wilted. Affected stem becomes soft and gets mascrateted but later on plant dry up permanently. Wilted stems releases viscid, sticky, bacterial matrix exudates from the vascular bundles after cutting. Droplets of white bacterial ooze appear on the cut surface of the infected stem. This bacterial ooze can form delicate threads that may be extended for several centimeters. This feature is used as a means of diagnosis.

Epidemiology

This is a vascular pathogen and differs from other *Erwinia* spp. because it is unable to degrade middle lamella. The bacteria over winters in the bodies of adult cucumber beetles particularly red striped and spotted beetle. Primary infection is initiated when beetles feed on the young leaves or cotyledons of cucurbits plants causing deep wounds. The bacteria are deposited in such wounds with the faeces of the insects. This is the only means of natural infection known. This pathogen is one of the unusual ones in which the bacterium is completely dependent upon the insect for its survival. After infection, the bacteria swim through the droplets of sap present in the wound. It enters in the xylem vessels and move rapidly to all parts of the plant. Penetration of the host through stomata does not take place. The bacteria present in the vessels of infected plants die within I or 2 months after the dead plants dry up. The survival period of these bacteria is very short in infected crop debris.

Management

Polughing of soil is to be carried in summer to expose all the stages of beetles. For management of cucumber beetles from the soil at initial infestation stage, soil application of neem cake should be carried out. Control of cucumber beetles at high infestation level may be carried out by applying granular insecticides like carbofuran in surrounding soil of the plants or in the pits followed by dusting and follar spray of carbaryl. Tolerant cucurbit varieties with restricted use of exotic cucumber lines should be grown. Most of the exotic hybrids of cucumber are susceptible to the disease.

ANGULAR LEAF SPOT OF CUCURBITS

Symptoms and Etiology

Angular leaf spot occurs all over the world on cucumber, gherkin, muskmelon, pumpkin, squash, watermelon and related species. In India, it is a serious disease in the cooler areas (Jindal, 1994). The disease manifests as small water soaked spots on leaves, stem, blossoms and fruits. The numerous spots on leaves are angular due to delimitation by veins, become tan coloured and may enlarge to irregular shapes giving the leaves as blighted or scorched appearance. On squash and watermelon these are brown and black, respectively with a yellow halo. The bacterial ooze forms a crust on them and dead tissues may tear-off leaving shot holes. The spots on fruits are small, Circular, brown to black and superficial. The causal agent *Pseudomonas syringae pv. lachrymans* (Smith & Bryan) Young et a/.1996, is a gram-negative, aerobic, non-sporing, straight rot-shaped bacterium, motile with I to 5 polar flagella.

Epidemiology

Infested soil and contaminated seeds transmit the disease. The pathogen can survive for two years in the soil or debris from diseased plants. The bacteria can spread through irrigation water, rain splashes and can colonize the buds on plants. It can remain there for 2 to 3 weeks. A temperature range of 20 to 240C is optimum for disease development, but it may continue upto 37'C. Moisture is the most important factor governing disease incidence and development.

Management

Field sanitation and crop rotation reduces the disease incidence. To eliminate the risk of seed-bome infection, the seed should be obtained from disease free dry areas. Hot water treatment of seed is effective to kill the pathogen, but it may affect germination. Spraying of a fungicide containing cupric hydroxide plus zinc (0.5 - 1.0 %) gave good control of *Pseudomonas syringae pv. lachrymans* on cucumber in Romania. Copper fungicides are useful but care should be taken not to spray at high temperature or to young plants.

BACTERIAL BLIGHTS OF LEGUMNOUS VEGETABLES

Symptoms and Etiology

The bacterial pathogens associated with common blight is *Xanthomonas axonopodis pv. phaseoli* and with halo blight *Pseudomonas savastanoi pv. phaseolicola (Young et.al* 1996). Other bacterial disease of beans are fuscous blight *(X. phaseoli var.fuscans)* and brown spot blight *(P. syringae pv. syringae).* Common blight is the major disease of bean particularly in cowpea. Disease appear in the month of July to September in rainy season. Symptoms of common blight are first seen as small translucent, water soaked spots on leaves. Symptoms of common blight, fuscous blight and brown blight are so similar that it is not possible to differentiate on the basis of symptoms except, isolation of casual organism. However, the intervening tissue may turn yellow and die to form lesions of dead tissue of various sizes and shapes. Bigger blotch symptoms appear on the leaves in severe case of common blight particularly in rainy season. Small spots appear on pods of different ages. Vascular discoloration may become evident in the severe incidence. Halo blight has most of the symptoms of common blight. One striking point of difference is that around individual spots on the leaf a chlorotic halo appear. After one week of infection, bacteria ooze from the stomatal cavities and cause secondary spread of the disease. Water soaked reddish or brownish lesions appear on pods. The areas surrounding lesion remain green even on pod maturity. Kishun (1988) reported that on cultivars Contender and Arka Komal of french bean, the disease intensity varied from 4 to 71 per cent and yield loss of green pods estimated upto 84 per cent.

Bacterial blight of pea is widely distributed throughout the world. In India, the disease was observed for the first time in severe form during 1969-70 in Udaipur, Rajasthan (Chakravarti and Hegde, 1972) and later on in Kullu valley with 80 to 100 per cent incidence (Sood, 1989). Disease affects all the aerial plant parts. Symptoms on leaflets and stipules appear as small water soaked lesions usually developing into large irregular areas. Lesions are initially translucent, light green but later on become dark brown at the edge and a lighter brown shade in the centre. Sepals are readily infected which causes drooping of blossom and 'uvenile pods. It also penetarete in growing pods and become seed home. *Pseudomonas syringae pv. syringae and P. syringae pv. pisi* (Sackett) Young *et. al* 1996. The bacterium is gram-negative rod with rounded ends, 1.5 to 1.75 x 0.5 to 0.75 @m in´size and motile with 1-4 polar flagella. Bacterial blight of guar (cluster bean) caused by *X axonopodis pv. cyamopsidis* is a widespread in guar growing regions of India and is severe in rainy season crop. The disease appears as both leaf spot and blight. The leaf spots are inter-venial, round, water-soaked in appearance and later coalesce resulting in blight phase. Leaf infection advances systemically through the petiole to stem, producing black, longitudinal streaks leading to cracking and breaking of the stem.

Epidemiology

The bacterium over winters in the infected seeds and crop debris but seed is the main source of primary inoculum and an effective means of dissemination. The bacterium is a wan-n temperature pathogen and causes greatest damage to the plants at 28-32'C. High humidity and high rainfall increases the disease spread. Secondary spread of disease occurs by wind blown rain splashes, contact between leaves, irrigation water and insects. The infected seeds transmit bacteria] blight of pea externally and internally and it is the main source of the infection in the field. Cool, moist weather favours the disease while warm and dry weather retards the disease. The extent of disease spread depends upon the frequency of rainy periods and positively correlated. It enough rains occur, within the span of a week, the primary infection from a tiny amount of infected seed can spread over large areas. Most of the bacterial diseases in leguminous vegetables are seed home. Andrus (1948) reported that bean plants previously attacked by *Rhizoctonia and Fusarium become* more susceptible to common and halo blight caused by *X a. pv. phaseoli and P. s. pv. phaseolicola.* Reynolds and Laurence (1990) reported that the apparent infection rate of bush bean with *X a.* pv. *phaseoli* increased with higher hydrogen fluoride concentrations. Gandhi and Chand (1989) have reported positive correlations between disease index and relative humidity (RH), rainfall and frequency of rains for bacterial blight of guar. Continuous spell of rain coupled with high RH (80%) caused heavy disease incidence during monsoon season with 300C temperature. Higher doses of N from 20 to 40 kg/ba also increased the disease incidence, whereas phosphorus application has no any impact on disease severity. Young plants of 15-30 days old were more susceptible then 60-70 days old plant where the infection was confined to foliage only (Gandhi and Chand, 1987).

Management

Crop rotation of about 2-3 years is essential to dispose of over wintering bacteria and infected crop debris. Field sanitation in standing crop as well as after harvesting by removal and burning of all the infected leaves and crop debris to reduce the inoculum load in soil. It is very essential because bacteria survive for longer period in stubbles left over in field. Use of clean and disease free seed is the most effective control measure of bacterial blights in bean and peas. Use of tolerant varieties for each of the blight is desirable. Seed home inoculum is effectively checked by streptocycline seed treatment 250 ppm for 2 hours. Proper cultural practices like wide spacing between plants, removal of weeds and drainage should be maintained in the field to avoid water stagnation. Sowing time should be adjusted in such a way that fruiting stage should not coincide with heavy rainy period.

BLACK ROT OF COLE CROPS

Symptoms and Etiology

Black rot of cole crops is prevalent in all agro climatic regions of India. It occurs worldwide and attacks all cultivated brassicas and numerous cruciferous weeds. The disease is caused by the bacterium *Xanthomonas campestris pv. campestris.* First sign of the disease often appears along the margins of leaves as chlorotic lesions and chlorosis progresses in the direction of midrib. "V"shaped yellow area on leaves is a characteristic symptom of the disease at initial stage of infection in cauliflower. However, such symptoms is not necessary on cabbage because the highly susceptible varieties show irregular, marginal yellowing and burning of the leaves. Symptom may appear from any side and center of the leave. Severe infection of black rot is resulting complete burning of crop in cauliflower before the curd formation. Irregular round yellowing formed after hailstorm and insect feeding in the center of leaves. Some of the veins and veinlets within chlorotic area turn black in severe case. Black vascular scar observed on any detached infected leaves, midribs and veins. Yellowing of leaves was observed from lower portion of mid vein in severe infection.

Epidemiology

The bacterium is transmitted primarily through seed. The pathogen is also survived for not more than a year in diseased plant left in the field. Disease severity increases rapidly whenever hailstorm is taken place in cropping season. Cruciferous weeds are also source for perpetuation of bacteria in field. The bacteria usually enter the cotyledons through stomata and pass to the young leaves and progress systematically throughout the plant system. The foliage infection and secondary transmission is through water pores, insect injury, infested soil, hail stonn, cultural practices, seedlings and movement of farmers in the field particularly during morning time. Black rot development is more rapid between 20-300C temperatures with high humidity and sufficient dew. The incubation period vary between 7-20 days depending upon temperature. Clayton (1929) observed that stump-rot *(P. c. subsp. carotovorum)* disease followed severe black rot (X. *c.* pv. *campestris)* incidence in cauliflower plants.

Management

Always collect seeds from disease free plants for use in next year. Seed treatment with hot water at 500C for 30 minutes or seed dipping in 100 ppm streptocyclinesolution for 30 minutes is effective in eliminating the seed bome bacterial inoculum. Cotyledonary leaves of cabbage become purplish colour soon after emergence and cauliflower may show yellowing or whitening due to antibiotic seed treatment, which later on recovered. Crop rotation should be carried with non-cruciferous crops. Use intercrop of creeping type plant like green gram and black gram as mulches to reduce rain splash. Mulches of grass and straw can also be used for this purpose. Detach the lower infected leaves in afternoon when dew and bacterial ooze dried up from the leaves and then bum it. Nursery site should be changed frequently to avoid seedling infection. Use antagonistic bacteria in the soil. Spraying of antibiotic like streptocycline @ 150-200 ppm or kasugamycin @ 0.2% along with sticker @O. I % at 10- 1 5 days interval gives good control but spray must be given after detachment of leaves. Sprays of copper oxycholoride @ 0.3% at 10 days intervals reduce the spread of disease. Mixture of steptrocycline 100 ppm and copper oxycboloride @ 0.3% with sticker @ 0.1% gives good control of the disease. MR-1 variety of cabbage is highly tolerant to black rot. The cauliflower variety Pusa Snowball K-1, Super Snowball and cabbage variety Pusa Mukta has been reported to out yield Golden Acre by 75.7%, 39.4% and 28.8% at Nilgiris, Ranchi and Rahuri, respectively (Kalda *et al.,* 1991). Hsb-19 strain of *Bacillus* was reported an effective and safe means to suppress this important bacterial disease in *Brassica* and got 83.15% disease control over control Jalali and Parashar (1995). They reported higher protection than that obtained with the application of streptocycline when the antagonist was applied 24 to 74 hours before inoculation with *Xcc on Brassica.*

BACTERIAL SOFT ROT OF VEGETABLES

Symptoms and Etiology

Bacterial soft rot occurs on all vegetables and occurs through out world. It causes serious losses in standing crop, transportation and storage. This is the most important disease after post harvest and causes maximum losses in vegetables. Vegetable having more water content gets easily infected. The first symptom on affected tissue is small, water soaked lesion, which enlarge very fast engulfing entire plant part. The affected area becomes soft, mucilaginous, water soaked, translucent and depressed. Leafy vegetables are more susceptible to this disease. The cauliflowers curds are severely infected during ball storming, snow and rain fall in temperate climate as well as plains. Cabbage heads starts rotting from stump region and engulfs entire head. Infected fruits, tubers, leave, roots and other parts give repulsive bad odors. Infected part completely collapsed followed by growth of secondary bacteria on the decomposing tissues and produce foul smell. Lower part of the stem, roots and tubers in root crops are infected in the field as watery and shriveled. Such plants later on remain stunted and finally wilted. The causal bacteria is *Erwinia carotovora pv. carotovora.* The bacterium is straight rod shaped, measuring 0.5 - 1.0 x 1.0 - 3.0 @m, facultative anaerobe, capsulated and motile by several peritrichous flagella.

Epidemiolgly

The pathogen enters in the host through different types of injury caused by insects, snow, bail, wind, sun burning and other extreme environmental conditions, mechanical, physical, diseased lesions and natural openings. The pathogen over winters in infected plant parts, soil, stored tubers and planting materials. The bacterium is a typical week facultative anaerobe. Bacteria multiply on the exudates released by the wounded cells. The bacteria have strong potential to produce pectolytic and cellulolytic enzymes that results in degradation of middle lamella, plasmolysis of cells and finally collapsing of entire infected plant part. Infection is favored by high humidity and optimum temperature for disease development is between 25 - 30° C. Some previously occurring diseases like black rot, *Rizoctonia and Phythium* infection predisposes for further attack of this disease

Management

This disease can be only managed by checking all kinds of injury. Spraying of plantomycin or streptocycline @ 500 ppm in combination with copper oxychloride @ 0.2% at fortnightly interval is required soon after the disease initiation in standing crops. However, it is not possible in harvested products so the best method is to avoid injury, proper packaging, free air movement, low temperature transportation and delicately handling. Use of resistant varieties against soft rot bacteria may serve up to certain extent. Harvested vegetables can be dipped in a solution of sodium hypochlon'de containing 50 ppm of available chlorine is recommended for certain vegetables. The fields to be planted must be free from the previous years plant debris. As far as possible wounding stored planting tubers should be avoided. Plants subjected to curd rot infection should be kept free of surface moisture by planting in well drained areas, by providing sufficient plant to plant distance for adequate air circulation and by avoiding overhead irrigation. Chemical sprays for curd rot is effective at initial stage but later on in sever infection it is not effective. Chemical control in cauliflower and cabbage should be only recommended in seed crops. Spraying of insecticides to control the insects that spread the disease has been found effective in reducing infection.

BACTERIAL LEAF SPOT/BLIGHT OF TOMATO

Symptoms and Etiology

Bacterial spot of tomato is caused by *Xanthomonas campestris pv. vesicatoria.* The disease is prevalent throughout the state. Small, dark spots appear on leaflets and stems of seedlings transplanted crop. Numerous spots coalesced and cause severe burning of the leaves. Kharif crop of tomato is severely affected by the bacteria with an average incidence of 35-40% every year. Yield loss is recorded more than 40% in kharif tomato. Disease continues from July to October. Prominent yellow halo is observed around the spot. The most conspicuous and damaging phase is infection of immature fruits. Small water soaked spots appear on exposed surfaces of the green and turning red fruits. These are slightly raised and corky in appearance.

Epidemiology

The bacteria survive mainly in seeds, crop debris and several weeds hosts. Disease appeared in rainy season and continued up to fruit initiation stage. The disease is prevalent in wan-n humid weather when RH is more than 90% and temperature is between 25 and 35°C. Temperature also affects the host susceptibility in terms of the size of stomatal apertures. A low temperature less than 17°C limited the entrance and multiplication of *X. vesicatoria* by reducing the stomatal and water in the intercellular space of pepper leaves (Matthee, 1967). Vakili (1967) has shown that injury created by high-pressure fungicidal sprays enhances the infection of *X vesicatoria* in tomato. In general, injuries caused by other animate agencies like nematodes, insects, mites etc. increase the susceptibility of vegetable crops by providing wounds for entry of bacteria. It has been reported that limited supply of oxygen to pepper may result in decreased susceptibility to *X. vesicatoria* (Matthee, 1967).

Management

Seed dipping in streptocycline solution @ 100 ppm is required for elimination of seed bome inoculum. Summer ploughing to desiccate the bacteria and host. Soil solarization of nursery bed reduces seedling infection significantly. Change the site of nursery bed and main field every year. Crop rotation with non-host crop where tomato and chilli must be avoided. Seed should be collected from disease free plants.

Adopt inter cropping of creeping type variety of black gram and green gram as mulches to prevent rain splash. One spray of streptocycline @ 150-200 ppm followed by one spray of kasugamycin @ 0.2% and spray must be given in afternoon so that dew period of morning and evening could be avoided. Use of highly tolerant line like LE-415 and West Virginia is advisable against this disease. One spray of copper oxychloride @ 0.3% after fifteen days of antibotics application is essential. Remove infected leaves, fruits and crop debris in afternoon and anti bum them.

Table 1. Occurrence of bacterial diseases in vegetable crops in Utter Pradesh.

Crop	Disease	Pathogen	Severity (%)	Average incidence (%)
Brinjal	wilt	*Ralstonia solanacearum*	1-3	1
Tomato	Leaf spot	*Xanthomonas campestris* pv. *vesicatoria*	25-60	35
	Speck	*Pseudomonas syringae pv syringae*	5-80	1.0
	wilt	*R. solanacearum*	0-1	<1
	Fruit spot	*X campestris* pv. *vesicatoria*	1-5	1
	Fruit rot	*Erwinia carotovora* pv. *carotovora*	2-8	5
Chilli/Capsicum	Leafspot	*X campestris* pv. *vesicatoria*	5-38	22.8
	Soft rot	*E. carotovora* pv. *carotovora*	2-5	2
	wilt	*R. solanacearum*	1-5	3
Cowpea	Common blight	*X campestris* pv. *campestris*		
	Halo blight	*P. syrl,ngae* pv. *phaseolicola*		
Cole crops	Black rot	*X campestris* pv. *campestris*	5-30	10
	Soft rot	*E. carotovora* pv. *carotovora*	2-10	5
Cucurbits	Wilt	*E. tracheiphila*	0-2	<1
Root crops	Soft rot	*E. carotovora* pv. *carotovora*	5-10	5
Leafy vegetables	Soft rot	*E. carotovora* pv. *carotovora*	2-5	2

BACTERIAL SPECK OF TOMATO

Symptoms and Etiology

The disease is almost similar to bacterial spot but caused by *Pseudomonas syringae pv. tomato.* It is characterized by absence of chlorotic halo around spot, crusty and comparatively larger spots. Some time yellow halo is observed in this *Pseudomonas* species. Disease is occurring 10-15 % In this area every year. The spots are necrotic, circular to roughly circular. As the fruit approaches the pink stage, the tissue around the spot retains its green colour longer than non-nal. Small black spots appear on leaf, petiole, pedicel, and peduncle while irregular, elongated lesions on stem.

Epidemiology

Bacterial speck is prevalent from November to January and favoured by cool, moist, weather and temperature range betweenl5 and 25°C. Generally, disease is observed in winter season when fruit starts ripening.

Management

The recommendations for the management of the preceding diseases are also applicable for the management of bacterial aspect.

REFERENCES

Akiew, S., Trevorrow, P. R. and Kirkegaard, J. (1996). Mustard green manure reduces bacterial wilt. Bacterial Wilt Newsletter, ACIAR No. 13: 5-6.

Chakrabarti S.K., Gadewar A.V. and Shekhawat G.S. (2001). Regulation of virulence in *Ralstonia solanacearum* and its probable exploitation in manipulating host resistance. In Role of Resistance in Intensive Agriculture, ed. Nagaraj'an S, Singh DP, 37-52. Kalyani Publishers, Ludhiana.

Chakravar-ti, B.P. and Hegde, S.V. 1972. Outbreak and new records. FAO Plant Prot. Bull. 20: 21-22.

Sood, A.K. 1989. Outbreak of bacterial blight of peas in Kullu Valley. Indian Phytopathol. 42: 322.

Eilphinstone, J.G. and Aley, P. (1993). Integrated control of bacterial wilt of potato in the wan-n tropics of Peru pp. 276-283. In Hartman, G.L. and Hayward, A.C. (eds.). Bacterial Wilt. Proceedings of an International Conference held at Kaohsiung, Taiwan, 28-31 October, 1992. ACIAR Proceedings, No. 45, 381 p.

Gupta D.K. (1991). Studies on the black rot of cabbage in Manipur. Indian J. Mycol. PI. Pathol. 21: 203-204.

Hsu, S.T. and Chang, M.L. (1989). Effect of soil amendments on survival of *Pseudomonas solanacearm.* Plant Prot. Bull. (Taiwan) 31: 21-33.

Jalall, I..and Parashar R.D. (1995). Biocontrol of *Xanthomonas campestris pv. campestris in Brassica uncea* with phylloplane antagonist. Plant Dis. Res. 10: 145-147.

Kalda T.S., Tewari R.N., Singh R., Singh N. and Ven-na T.S. (1991). Evolution of Pusa Mukta a resistant variety to black rot disease in cabbage. Indian J. of Hort. 48: 247-250.

Kishun, R. 1988. Losses in yield of french beans affected with common blight. In: Gnanamanickam, S.S. and Mabadevan, A.J. (eds), Advances in Research on Plant Pathogenic Bacteria, Proceedings of the National Symposium on Phytobacteriology held at the University of Madras, Madras, India, march 14-15, 1986, 29-31 p.

Pandey, K.K. (2001). Important diseases of vegetable crops and their integrated management. National training course on vegetable production technology, *Directorate Of Extension, Ministry ofagriculture,* New Delhi, pp. 177-186.

Pandey, PK. and Pandey, K.K. (2002). Emerging Strategies to Control Fungal Diseases in Vegetables, In Applied Mycology & Biotechnology, Vol.2 (G.G. Khachatourians and Dilip K. Arora eds.) Published by Elsevier Science International B.V. pp. 197-217.

Pandey, K.K., Pandey, P.K., and S. Satpathy (2002). Integrated management of disease and insects of tomato, chilli and cole crops. Techinical Bulletin. No.9. Published by Indian Institute of Vegetable Research, Varanasi in collaboration of DBT New Delhi. pp.22

Pandey, K.K., Pandey, P.K., and S. Chandra. (2003). Sabllyon men Ekikrit Naseejev Prabandhan. Techinical Bulletin. No.11. Published by Indian Institute of Vegetable Research, Varanasi in collaboration of UPCAR, Lucknow. pp.29.

Rao, M.V.B. (1976). Bacterial wilt of tomato and eggplant in India. In: Sequelra, L. and Kelman, A. (eds.). Proceedings of the first international planning conference and workshop on the ecology and control of bacterial wilt caused by *Pseudomonas solanacearum, Raleigh,* North Carolina, 18-24, July, 1976, 92-94.

Sadashiva, A.T., Deshpande, A.A., Reddy, K.M. and Singh, R. (1993). New sources of resistance to bacterial wilt in eggplant. Capsicum and Eggplant Newsletter. 12: 94-96.

Shekhawat G.S., Gadewar, A.V. and Chakrabarti S.K. (1992). Spontaneous phenotypic reversion from affuldal to fluidal state in strain of *Pseudomonas solanacearum. ACIAR* Bacterial Wilt Newsl. 8: 5-9.

Shakya, D.D. 1993. Occurrence of *Pseudomonas solanacearum* is tomato seeds imported into Nepal. In: Hartman, G.L. and Hayward, A.C. (eds.), Bacterial Wilt. Proceedings of International Conference held at Kaohsiung, Taiwan, 28-31 October, 1992, ACIAR Proceedings No. 45: 371-372.

Shekhawat G.S., Kishore V, Garg I.D., Singh R. and Shan-na'A.K. (1983). Studies on brown rot, soft rot and common scab. Annu. Scl. Rept., 95-102. Central Potato Research Institute, Shimla.

Sinha, S.K. (1986). Bacterial wilt in India. In: Persley, G.J. (eds.), Bacterial wilt disease in Asia and South Pacific. ACIAR Proceedings No. 13: 28-29.

Sood, A.K. and Singh, B.M. (I 993). Prevalence of bacteria] wilt of solanaceous vegetables in mid hill subhum'd zone of Himachal Pradesh (India). In Hartman, G.L. and Haywards, A.C. (Eds.), Bacterial wilt 358-361. Proceedings of International Conference held at Kaoshing, Taiwan, Oct., 1992, 358-361 pp.

Wang, J.F., Chen, N.C. and Li, H.M. (1998). Resistance sources to bacterial wilt in eggplant *(Solanum melongena)*, 284-289. In: Prior, Ph., Allen, C. and Elphinstone, J. (Eds.). Bacterial wilt Disease: Molecular and Ecological Aspects, FNRA, Springer, 447 p.

Yabuuchi, E. Kosako, Y., Yano, I., Hotta, H. and Nishiuchi (1995). Transfer of two *Burkholderia* and an *Alcoligenes* species to *Ralstonia.* Gen. Nov: Proposal of *Ralstonia pickettii* (Ralston, Palleroni and Dounoroff, 1973) comb. Nov. and *Ralstonia eutrapha* (Davis, 1969) Comb. Nov. Microbiol. Immunol. 39: 897-907.

Young, WJ and Thomas HR (1957). A Monographic Study of Bean Disease and Methods for their Control. U.S. Dep. Agric. Tech. Bull. 868. 255 p.

21

Biological Control as a Component in IPM of Fruit Crop Pests in India

V.M. PAWAR[1], M.B. SARKATE[1] AND P.S. BORIKAR[1]

Insect pest complex constitutes the major limiting factor in the successful production of horticultural crops. Estimates of yield loss caused by pest attack varies from 25 to 40 per cent in various fruit crops during different seasons of the year. Efforts are being made to reduce the losses with adoption of cultural, physical/mechanical, biological, chemical and regulatory measures. With the advent of modern pesticides, the natural biological control based methods were neglected to least priority. The indiscriminate and excessive use of synthetic pesticides resulted in high pesticide reduces in food stuffs, development of pesticide resistance in major crop pets, pest resurgence of minor pests, ill effects on human health and non-target organisms such as parasites, predators, honey bees, pollinators and food chain and toxicity to natural enemies and wild life. The country has experienced from time to time the out break of citruc blackfly, *Aleurocanthus woglumi*; stripped mealy bug *Ferrisia virgata*; Grape mealy bugs, *Maconellicoccus hirsutus* and *Helicoverpa armigera*. This has created a new enthusiasm and increasing awareness about the incorporation of bioagents in an integrated pest management popularly referred as bio-intensive integrated pest management (BIPM) approach for pest management in which biological control has the major role.

Biological control is the core component of pest management. It includes natural and augmentative bio-control. Natural biological control is responsible for 40-60 per cent reduction in pest population. The optimization of established bio-control agents through conservation and periodic enhancement by augmentation offers the most immediate source of crop protection free from pesticide residues. It constitutes the main aspect of integrated control since 99 per cent of all potential pests are already under biological control (Debach, 1974).

Preliminary work on biological control of horticultural pests like wooly apple aphid, *Eriosoma lanigerum* and *Icerya purchasi* was started by importing *Aphelinus mali* and *Rodolia cardinais* Mulsant from England and USA from 1920-1930 in India (Sanskaran, 1974, Debach and Rosen, 1991).

To overcome the problems of side effects of pesticides and to provide a stable ecosystem, the Indian Institute of Horticuitural Research (IIHR) has been conducting research for the development of Integrated Pest Management of horticultural crops since two decades.

ORGANIZATIONS INVOLVED IN BIOLOGICAL CONTROL

'Augmentation' in biological control refers to increasing or strengthening of the biocontrol agents in the field in order to protect the crop from the pest to a desired level. Augmentation may involve inoculative

1. Marathwada Agricultural University, Parbhani.

releases and inundative releases of natural enemies for keeping down the pest population below the tolerance level. Augmentation can be achieved by either mass production in the laboratories and releasing them in the field or colonization and re-distribution of bio-control agents.

In India, the importance of production of bio-control agents has been realised for many year. The Common Wealth Institute of Biological Control (CIBC) established Indian Station in 1957 at Bangalore with the aid of Canadian-Colombo plan funds. The advent of CIBC marked beginning of organized and systematic biological control research in the country.

Six decades ago Indian Council of Agricultural Research (ICAR) started few coordinating centres for the production of *Trichogramma* (Hym. Trichogrammatidae) to the control of sugarcane borers. The only station which still continues the programme is located in Mandya district of Karnataka State.

The Department of Agriculture and Horticulture of several State governments and also few other organisations have started parasite breeding laboratories for the production of parasitoids of cotton bollworms, sugarcane borers, coconut caterpillars and few other parasites which are mass-multiplied and distributed. A few of these are still functioning.

Now most of the bio-control research is being carried out under All India Coordinated Research Project on Biological Control of Crop Pests and Weds (AICRP-BCCPW) which started functioning since 1977 under the control of Indian Council of Agricultural Research (ICAR), New Delhi. The status of Project is now elevated to an independent Project Director of Biological Control (PDBC) with its headquarters in Bangalore. PDBC, Bangalore, established in 1993 is presently acting as a nodal institute for the organization of biological control research on agricultural pests at the National level. It has 16 centres all over the country for conducting field studies on biological control of pests of crops such as cotton, sugarcane, rice, tobacco, pulses, oilseeds, coconut, horticultural crops (fruits and vegetables) and weeds. The Directorate carried out basic research on biosystematics of important natural enemies, evolving mass-multiplication techniques and artificial diets for natural enemies, genetic improvement of bioagents, use of semiochemicals for improving the efficacy of bioagents, use of entomopathogens and entomophilic nematodes for insect pests and fungal and bacterial antagonists for plant diseases and phytonematodes.

During the last decade, there has been an enthusiastic acclaim in the production and utilization of biocontrol agents in India. Pest Control (India) Ltd. made a pioneering venture by establishing a Biocontrol Laboratory (BCRL) in Bangalore in 1981. This is the only one first commercial biocontrol insectary established in India. Promising parasitoides and predators of major pests of sugarcane borers, cotton bollworms, coconut black-headed caterpillar, mealy bugs and scale insects are being produced at BCRL and supplied to farmers in different parts of the country along with professional advice.

Government of Maharashtra established first biocontrol laboratory at the Department of Entomology, College of Agriculture, Pune in 1965 for mass-production and utilization of *Trichgramma* sp. against cotton bollworms and sugarcane stem borers. Subsequently Maharashtra Government sanctioned a scheme for research on Insect parasitology at the Department of Entomology, College of Agriculture, Parbhani in 1974 for undertaking the research work on parasitoids of cotton of cotton bollworms, sugarcane borers and white grubs.

The Tamil Nadu Cooperative Sugar Federation established a laboratory at Chengalpattu in 1982 for the production and supply of *Trichogramma* to their sugar factory members/cane growers. In 1989, Simbhaoli sugar mills (UP) also started a *Trichogramma* production unit for its own use. Vasant Dada Sugar Institute (VSI), Pune established a biocontrol laboratory for pyrilla control in 1982 in Maharashtra State and achieve a significant reduction of pyrilla with *Epiricania melanoleuca*. Besides, biocontrol research is being carried out by other ICAR Institutes, State Agricultural Universities (SAUs), traditional universities, Govt. organizations such as Bhabha Atomic Research Centre, Mumbai (BARC), with the financial assistance from Indian Council of Agriculture (ICAR), Department of Biotechnology (DBT), National Agricultural Technology Projects (NATP), University Grants Commission (UGC), Department of Science and Technology (DST), Department of Environment and Forests (DE&F), etc.

INDIGENOUS NATURAL ENEMY COMPLEX

Citrus

Citrus is an important fruit crop being grown in all parts of the country where waster requirements are met and temperature does not fall below freezing. The perennial nature of the trees and the highly complex agro-ecosystem in which citrus is grown favour the occurrence of numerous species of insect and mites. More than 250 species of arthropods have been reported in infest the citrus in India (Pruthi and Mani, 1945; Wadhi and Batra, 1964, Nayer *et. al*; 1975). Of these leafminer, psylla, whiteflies, scales, mealy bugs have been observed causing serious losses. Besides these, bark eating Caterpillars, fruit sucking moths, fruit flies and butterflies occur occasionally causing substantial losses. Losses to the extent of 20 to 25 per cent in fruit production due to ravages of insect pests have been reported (Doharey and Butani, 1984). Radke and Ghuguskar (1982) estimated that citrus pyilla alone can cause a loss of Rs. 3 crores and fruit sucking moth Rs. 1 crore annually in Maharashtra State.

The first cottony cushion scale, *Icerya purchasi* which appeared in epidemic form in Nilgiri hills in 1927 infesting many crops including citrus, was brought under control by introduced lady bird beetle, *Rodolia cardianalis* (Muls) (Rao and Kamath, 1996). The fascinating success of this control has established biological control as valid method of pest control. About 40 species of parasites and predators have been recorded from the major pests in the citrus growing states in India (Rama subbaih, 1983, Kameswara Rao, 1983, Sharnagat *et al.*, 1983; 1982, 1983; Gupta, 1954; Sadana and Kanta, 1971; Anonymous, 1979, Pruthi and Mani, 1945; Tandon, 1992, Radke and Ghuguska,r 1984, Mani, 1992, Singh, 1992) (Table 1a to 1c).

Gauva

About 36 insect pests have been recorded on guava, out of which 4-5 species cause serious damage viz. fruit fly *Dacus dorsalis*, scale insects, *Pulvinaria psidi*, mealy bugs *Ferrisia virgata*, bark and short borer *Indarbella quadrinotoata* and *Aphis gossypii.*

As many as 20 natural enemies have been reported on *A. gossypii* infesting different crops in India. If they are saved from insecticidal drifts, they help in controlling the aphid population. Five cocoinellid predators viz. *Menochilus sexmaculata, Scymnus sp, Pseudaspidimerus circumflexa, Chilocoris nigritus, Cryptolaemus montrouzieri* and two syrphids viz., *Idvhifon scutellaris Paragus serratus* were observed feeding on *A. gossypii* (Mani and Krishnamoorthy, 1989). Eighteen species of parasitoids and predators were recorded from the different insect pests of Guava at Indian Institute of Horticultural Research, Hissarghatta from during the year 1989-91 (Anonymous, 1992).

As many as 26 parasitoids and predators have been recorded from the major pests i.e. *A. gossypii, Chloropulvinari, apsidii, Ferrsia virgata, Planococcus citri, P. lilacinus* and *Drephanococcus chiton* (Anonymous, 1991, Mani 1993). The details of various natural enemies recorded on the major pests of Guava are incorporated in (Table 2). Recently a conccinellid, *Harmonia octomaculata* has reported to feed successfully in *F. virgata* (Anonymous, 2001).

Mango

Mango, an excellent fruit since time immemorial grown in India is attacked by a large number of insects, mites and other vertebrate pests (Tondon and Abraham, 1985). The pest problems increased because of monoculture, high density planting, application of high dosages on nitrogenous fertilizers, frequent irrigation and use of broad spectrum insecticides. Major pest attacking mango are mango hopper, *Ameritodus atkinsoni*, mango fruit flies. *Dacus dorsalis*, stone weevil, *Sternochetus magiferae* and mealy bugs, *Drosicha mangiferae.*

Several species of spiders and coccinellids have been observed feeding on nymphs of mango hoppers, *A. atkinsoni*, besides fungal paathogens (Tandon, 1992). Eighteen species of parasitoids have been reported to parasitise fruit flies, Dacus dorsalis in India (Narayanam and Batra, 1960).

Table 1a: Natural enemies of insect pests of citrus.

Sr. No.	Parasite/Predator	Host stage Attacked	Family	Order
	Leaf miner, *Phyllocnistis citrella*:			
1.	*Sympiesis sp.*	Larvae	Eulophidae	Hymenoptera
2.	*Cirrospiloides phyllocnistoides*	Larvae	Eulophidae	Hymenoptera
3.	*Scolonix quadristriata*	Larvae	Eulophidae	Hymenoptera
4.	*Tetrastichus phyllocnistoides*	Larvae	Eulophidae	Hymenoptera
5.	*Bracon phyllocnistoides*	Larvae	Braconidae	Hymenoptera
6.	*Tetrastichus sp.*	Larvae	Eulophidae	Hymenoptera
7.	*Sympiesis purpurea*	Larvae	Eulophidae	Hymenoptera
	Citrus psylla, *Diaphorina citri*			
	Predators:			
1.	*Chilocoris nigritus*		Coccinellidae	Coleoptera
2.	*Coccinella septumpunctaata*		Coccinellidae	Coleoptera
3.	*C. rependa*		Coccinellidae	Coleoptera
4.	*Chilomenes sexmaculata*		Coccinellidae	Coleoptera
5.	*Brumus suturalis*		Coccinellidae	Coleoptera
6.	*Chrysoperla* sp.		Chrysopidae	Neuroptera
	Parasites:			
1.	*Tetrastichus radiatus*		Eulophidae	Hymenoptera
2.	*T. phyllocnistoides*		Eulophidae	Hymenoptera
3.	*Diaphorencyrtus aligharensis*	Larvae	Braconidae	Hymenoptera
	Citrus blackly, *Aleurocanthus woglumi*			
	Predator:			
1.	*Catana parcestosa*		Coccinellidae	Coleoptera
	Fruit sncking moth, *Ophiders materna* and *O. fulonica*			
1.	*Telenomus lucallus* Nixon	egg	scelionidae	Hymenoptera
2.	*Trichogramma chilonis* Ishii	egg	Trichogrammatidae	Hymenoptera
3.	*Euplectrus* sp.	larvae	Eulophidae	-do-
4.	*Snellenisus gelleus*	pupal		
	Leat eating caterpillar, *Papillio demoleus.*			
1.	*Melalophacharos tamanukii*	larvae		
2.	*Apanteles paplionis* Vierech	larvae	Braconidae	-do-
3.	*Pediobus imbreus* Walker	larvae		
4.	*Eurytoma* sp.	larvae	Eurytomidae	-do-
5.	*Euycia nymphalidophaga*	pupal		
6.	*Pteromalus puparum*	pupal	Pteromalidae	-do-
	Scale insects, *Coccus virdis.*			
1.	*Chilocorus nigritus*		Coccinellidae	Coleoptera
	Mealybugs, *Drosicha mangiferae*			
1.	*Rodolia cardinalis*	Nymph and adults	Coccinellidae	Coleoptera
2.	*Chrysops*	-do-	Chrysopidae	Neuroptera
	Leaf rolling caterpillar, *Psorostica ziziphi*			
1.	*Apanteles* sp.	larvae	Braconidae	Hymenoptera

Table 1a: Contd. . . .

Sr. No.	Parasite/Predator	Host stage Attacked	Family	Order
2.	*Pristomerus* sp.	larvae	—	—
3.	*Brachymeria euploea*	larvae	Chalcididae	-do-
Mites, *Eutetranychus orientalis*				
1.	*Scymnus gracilis*		Coecinellidae	Coleoptera
2.	*Scolothrips indicus*		Thripidae	Thysanoptera
3.	*Amblyseius cucumeris*		Phytoseiidae	Acarina
4.	*A. hibisci*		-do-	-do-

(Radke and Ghuguskar, 1984).

Table 1(b)

Pest	Parasite/predator
Cirtus black fly, *Aleurocanthus spiniferous*	*Amithus hesperidum,*
	Encarsia smighi
A. Woglumi	*Amitus hesperidum*
	Encarsia clypealis
	E. citrina
	E. divergens
	E. merceti
	E. opulenta
	Eretmocerus serius
California red scale, *Aonidiella aurantii*	*Aphytis melinus*
	Compriella bifasciata
Yellow scale, *A. citrina*	*C. bifasciata*
	Encarsia perniciosi
Red wax scale, *Ceropolaster rubens*	*Anicetus beneficus*
Florida red scale, *Chrysomphalus aonidum*	*Aphytis holoxanthus*
Fruit fly, *Dacus dorsalis*	*Opius incisi*
	O. longicaudàtus
	O. oophilus
White fly, *Dialeurodes citri*	*Encarsia lahorensis*
Citrus pryilla, *Diaphorina citri*	*Tamarixia radiata*
Mite, *Eutetranychus orientalis*	*Amblyseius rickeri*
Cottony cushion scale, *Icerya purchasi*	*Rodolia cardinalis*
Lepidosaphes beckii	*Aphytis lepidosaphes*
Lemon butterfly, *Papilio demoleus*	*Apanteles papilionis*
	Melalophachyarops sp.
Mealybug, *Planococcus citri*	*Cryptolaemus montrouzieri*
	Leptomastix dactylopii
	Lęptomastidea abnormis
Scale, *Pseuducoccus citriculus*	*Clausenia purpurea*
Mealybug, *Chloropulvinaria psidi*	*Cryptolaemus montrouzieri*
Tetranychic mite	*Phytoseiulus persimilis*

(Singh, 1992).

Table 1(c)

Pest	Parasite/predator	Family	Order
Aonidiella aurantii	*Aphytis sp.*	Aphelinidae	Hymenoptera
	Chilocoris nigritus	Coccinellidae	Coleoptera
Planococous citri	*Leptamastix dactylopii*	Encyrtidae	Hymenoptera
	Coccixenoides perigrinus	Encyrtidae	Hymenoptera
Nipaecoccus viridis	*Anagyrus dactylopii*	Encyrtidae	Hymenoptera
	Pseudaphycus sp.	Encyrtidae	Hymenoptera
	Alamella flava	Encyrtidae	Hymenoptera
	Aproscetus (Tetrastichus) purpureus	Eulophidae	Hymenoptera
Coccus viridis	*Encyrtus lecanarium*	Encyrtidae	Hymenoptera
Planococcus lilacinus	*Tetracnemoidea indica*	—	—
Papilio demoleus	*Trichogramma chilonis*	Trichogrammatidae	Hymenoptera
	Apanteles papilionis	Braconidae	Hymenoptera
	Telenomus incommodus	Scelionidae	Hymenoptera

Mani (1992).

Table 2: Natural enemies of major pests of Guava.

Sl. No.	Insect pest	Natural enemy	Family	Order
I)	**Aphid, Aphis Gossypii *Glover***			
1.		*Menochilus sexmaculata*	Coccinellidae	Coleoptera
2.		*Scymnus* sp.	Coccinellidae	Coleoptera
3.		*Chilocoris nigritus*	Coccinellidae	Coleoptera
4.		*Pseudospidimerus circumflexa*	Coccinellidae	Coleoptera
5.		*Cryptolaemus montrouzieri*	Coccinellidae	Coleoptera
6.		*Ischiodon scutellaris* (F)	Syrphidae	Diptera
7.		*Paragas serratus* (F)	Syrphidae	Diptera
II)	**Green shield scale, *Chloropulvinaris psidii* (Mask)**			
1.		*Coccophagus cowperii*	—	—
2.		*C. bogoriensi*	—	—
3.		*Aneristis* sp.	—	
4.		*Scymnus coccivora*	Coccinellidae	Coleoptera
5.		*Cryptolaemus montrouzieri*	Coccinellidae	Coleoptera
6.		*Ectopsocus* sp.	Ectopsocidae	Psocoptera
III)	**Hard scale *Drephanococcus chiton***			
1.		*Anicetus ceylonicus*	Encyrtidae	Hymenoptera
2.		*Diversinervus elegans*	Encyrtidae	Hymenoptera
3.		*Cephaleta brunniventris*	Pteromalidae	Hymenoptera
4.		*Chilocoris nigritus*	Coccinellidae	Coleoptera
IV)	**Striped mealybug, *Ferrisia virgata***			
1.		*Anagyrus quadrii*	Encyrtidae	Hymenoptera
2.		*A. indicus*	Encyrtidae	Hymenoptera
3.		*Aenasius advena*	Encyrtidae	Hymenoptera
4.		*Blepyrus insularis*	Encyrtidae	Hymenoptera
5.		*Chamaemyia* sp.	Chamaemyiidae	Diptera

Contd. . .

Table 2 Contd. . .

Sl. No.	Insect pest	Natural enemy	Family	Order
6.		*Scymnus coccivora*	Coccinellidae	Coleoptera
7.		*Mallada boninesis*	Chrysopidae	Neuroptera
8.		*Brumus suturalis*	Coccinellidae	Coleoptera
9.		*Spalgis epius*	Coccinellidae	Coleoptera
		Cryptolaemus montrouzieri	Coccinellidae	Coleoptera
b)	***Planococcus citri***	***Coccixenoides peregrina***		
			Encyrtidae	Hymenoptera
1.		*Allotropa citri*	Platygasteridae	Hymenoptera
2.		*Leptomastix dactylopii*	Encyrtidae	Hymenoptera
c)	***Planococcus lilacinus:***			
1.		*Tetracnemoidea indica*	Encyrtidae	Hymenoptera
2.		*Coccixenoides peregrinus*	Encyrtidae	Hymenoptera
3.		*Leptomastix dactylopii*	Encyrtidae	Hymenoptera

Shukla *et al.* (1984) reported that there are no effective natural enemies associated with stone weevil except a baculovirus on grubs and some fungi. Five species of conccinellids viz. *Aulis vestitus, Coccinella spetumpunctata, C. transversalis, Menochilus sexmaculatus, Rodolia cardinalis* and seven species of spiders namely *Phiodipus* sp, *Marpissa* sp., *Rhene indicus, Linphia* sp., *Oxyopes shweta Crytophora cicatrosa* and *Stegidypes sarasinorum* have been observed to feed on mealy bug (Tondon and Lal, 1978). *Spalgis epius* is commonly found feeding on *Rastrococcus iceryoides* while *mallaca astur* prey on *Chloropulvinaria polygonata.*

Various macro and micro organisms were documented as bioagents in control of pests of mango (Singh, 2000). These include 3 species or araneids (spiders), 8 species of coccinellids, 4 species of formicids (ants) and one species of mantid. The other natural syrphids, phytoseiids (predatory mitres), epipyropids eulophids and braconids (Venugopal Rao *et al.* 2000). Dhooria (2001) recorded predatory mite, *Typhlodromus caudiglans* during February-March actively freeing on mango gall mites, *Aceria mangiferae* from Punjab.

Grape

Grape cultivation or viticulture flourished in India during the moghul regime. Insect pests of grapes were not of much economic significance till sixties. By eighties, insecticides formed a major input in viticulture indicating that insect pests need to be controlled. Roving survey undertaken during 1985 and 1990 in Karanatak, Maharashtra, Tamilnadu and Andhra Pradesh served a basis for categorising seriousness, distribution and efficacy of control measures undertaken by the growers. The major insect pest infesting the crop are flea beetle, *Sceledonta strigicollis* thrips, *Rphipiphorothrips cruentatus,* mealy bugs, *Macanellicoccus hirsutus.*

Mealy bug, *M. hirsutus* has become a serious pest of grapevine in recent years. It poses serious damage in summer months and also in fruiting phase. According to Azam (1983) grape mealy bug alone caused a yield loss of 50 to 100 per cent in the field. A total of seven parasitoids and seven predators have been reported to attack the pest. Among parasitoids, *Anogvrus dactylopii* is the major parasite recorded on mealy bug. *M. hirsutus. Scymnus coccivora* Ayyar was found to be a predominant predator of the mealy bug (Mani and Thontadarya, 1988). A chrysopid larvae of *mallada bonienesis* voraciously feed on the mealy bug (Anonymous, 1992). Fourteen natural enemies have been reported from grape mealy bug *M. hirsutus* out of which *A. dactylopii* and *S. coccivora* are of considerable importance (Singh, 1992).

Pomegranate

Pomegranate aplid, *Aphis punicae* a suctorial pest of leaves, flowers and fruits is known to be attacked by 4 predators viz. *Coccinellids, Scymnus castaneus S. latemaculatus, Menochilus sexmaculata* and *P. serratus* which are responsible for destruction of aphid population within two months (Mani, 1992). Four natural enemies were reported on whitefly, *Siphoninus phillyreae* out of whcih an aphelinid, *Encarsia inaron* parasitisea the white fly from 80 to 90 per cent in the nature.

Several natural enemies were reported to feed on mealy bugs *Planoccocous lilacinus* and *P. astur* of which *S. epius* and *C. montrouzieri* were efficient in controlling the mealy bugs. (Mani and Krishnamoorthy, 1990).

Anar butterfly, *Deudrix epijarabs* Moore a serious pest in HP and J&K is known to be parasitiods and two larval parasitoids (Mani, 1992). The record of *Aphidius* sp. *Aphelinus* sp. And *Signiphora* sp. Appeared to be new on pomegranate aphid at Bangalore (Anonymous, 1994). Gupta and Sood (2001) carried out a survey at Solon for recording the natural enemies of *D. epijarabs* and recorded three egg parasitoids i.e. *Tehenomous cyrus Anastatus* sp. nr. *Kashmirensis* and *Trichorgramma chilonis* with egg parasitization of of 93.7, 2.7 and 6.3 per cent, respectively. Recently, an entomogenous fungus, *Entomophthora aphidis* has been found to attack the *A. punicae* on pomegranata in an epizootic form at the Indian Institute of Horticulture Research Farm, Hessarghatta, Bangalore (Gopalkrishnan and Mani, 2001).

Ber

Nearly 80 species of insect pests have been recorded to infest the crop in the country out of which 30 occur in the Maharashtra State. Important species in testing ber are fruit borer, fruitfly, tingid bug and hairy caterpillars. Scale insect *Dreponococcus chiton* is known to be a serious pest on shoots and leaves. Severe infestation may cause drying of the affected leaves. Two predator *Cryptolaemus montrouzieri* and *Scymnus* sp are known to feed on Crawlers of scale insects. Two key parasitoids viz. *Caphateta* sp. and *A. ceylonicus* occur in large numbers causing heavy reduction of scales in every generation.

Three species of mealy bugs are known to attack ber crop out of which *Nipacoccus viridis* and *Planoccous lilacirnus* cause severe economic losses when they infest the fruits. Many predators are known to feed on mealy bugs out of which *S. epius* is predominant. The local parasitoids of these mealy bugs play an important role in reducing the pest.

Apple

Apple an important temperature fruit crop is being grown in Jammu & Kashmir, Himachal Pradesh and North eastern of Uttar Pradesh. Amongst the various insect pests that attack the crop regularly are San Jose scale, *Quadriaspidiotus perniciosus;* wooly apple aphid, *Erisomma lanigeru,* codling moth *Cydia pomonella* and Gypsy moth *Lymantria obfuscata* in apple growing temperate areas of our country.

A number of coccinellid beetles, *Chilocoris horubidus, Cheilomenes bijugus* Mulsant, *Pharoscymnus flexibilis* Mulsant, *Sticholotis marginalis* Kapur are known to feed to San Jose scale, *Q. pernociosus*.

Wooly apple aphid, *E. lanigerum* an introduced pest has spread to all the apple growing regions in India. Various local predators like coccinellids, syrphids and chrysopids have been found to attack this pest. Codling moth, *C. pomonella* is a serious pest of apple and other fruit trees in Ladash district of J & K. About 30 to 70 per cent of apple fruit are damaged by this pest. Bird predators like *Parus caeruleus, P. major* and *Passer domesticus* feed on the over wintering larvae and number of natural enemies including granulosis virus cause considerable mortality.

Gypsy moth, *L. obfuscata* a common pest of apple is widely distributed in apple growing districts of Kashmir. Number of indigenous parasitoids viz. *Anastatus kashmirenis* Mathur from eggs, *Exorista russica*

Fall from larvae and *Pimpla* sp., *Theronia* sp. *Tetrastichus* sp. and *Brachymeria lasus* from the pupae have been reared (Singh, 1986).

Until 1989, phytophagous spider and mites mainly *Pannonychus ulmi* (Koch) and *Tetranychus utricae* (Koch) a pest of minor importance in Himachal Pradesh, have been appearing in apidemic form all over the State. Various type of predators have been associated with this pest. Six conccinellid species viz. black coccinellid beetle, *Stethorus punctum* (Leconte), mirid, *Chrysopa* sp, anthocorid, *Orius insidiosus* Say; black hunter thrips, *Leptothrsps mali* (Fitch), spiders and predatory mites, *Amblyseium fallacis* AH., *Zetzellia mali* Ewing *Typhlodromus* spp have been recorded by Khajuria and Sharma (2001).

INTRODUCTION AND ESTABLISHING OF EXOTIC NATURAL ENEMIES

Citrus

The first fascinating successful control of cottony cushion scale *Icerya purchasi* by introducing lady bird beetle *Rodolia cardinalis* in Nilgiri hills in 1927 laid down the foundation stone of biological control of insects in India (Rao and Kamath, 1966). *Leptomastix dactylopii* an exotic solitary endoparasitoid obtained from West Indies against citrus mealy bugs, *Planacoccus citri* was established released sites at Chethalli and Bangalore (Karnataka) and Pollachi (Tamilnadu). *Cryptolaemus montrouzieri*, an Australian lady bird beetle was also tried for the biological control of *P. citr* which gave good results in mix planted citrus orchards (citrus and coffee). *Encarsia clypealis* and *Amitus hesperidum* potential parasitoids of citrus black fly, *Aleurocanthus woglumi* Ashby sent from India in other countries proved highly successful in controlling the blackfly (Mani, 1992).

Guava

Cryptolaemus montrouzieri was highly effective against guava scale *Chloropulvinaria* which effectively controlled the guava mealy bug *Ferrisia virgata* (Mani *et.al*, 1990). The aphelinids, *Encarsia haitiensis* and *E. Guadeloupae* an introduced parasitoids of spiralling whitefly, *Aleurodicus dispersus* Russell were recorded in Kerala, Karnataka, Maharashtra and Tamil Nadu. *E. guadeloupae* adults parasitised white flies on guava near Bangalore (Anyonymous, 2001).

Mango

The coccinellid predator, *C. montrouzieri* consumed about 360 eggs or 500 nymphs of mango mealy bug. *Rastrococous iceryoides* in the laboratory and in the field. The predators were able to bring down the mealy bug infestation. *Leptomastix dactyopii* an exotic nymphal parasitoid was tried successfully for the biological control of mango meal bug. *R. iceryoides* at Indian Institute of Hessarghtta, Bangalore. *C. montrouzieri* predatory coccinellid beetles was found effective in suppressing the green shield scale, *C. polygonata* (Mani, 1992).

Grapes

Although a large numbers of indigenous parasitoids and predators were recorded on mealy bugs, they have developed density dependent relationship with the host. An exotic mealy bug predator, *Cryptolaemus montrouzieri* Mulsant has been extensively tested against *Maconellicocus hirsutus*. The performance of *C. montrouzieri* has been evaluated in Andhra Pradesh, Karnataka, Tamilnadu and Maharashtra (Mani, 1992). *Leptomastix dactylopiii* an exotic parasitoid (from West Indies) was recovered from *Planococcus citri* infesting citrus, grapes, guava, pomegranate and croton (Anonymous, 1989).

Pomegranate

A large number of parasites and predators are known to feed on impartant pests of pomegranate i.e. *Aphis punicae*, butterfly, *Deudrix epijarbs*. Two exoitic natural enemies mealy bugs on pomegranate were recovered (Anonymous, 1989, Mani and Krishnamoorthy, 1990).

Ber

An exotic coccinellid predator *Cryptolaemus montrouzieri* is known to feed on crawlers of scale insect *Drepanococcus chiton* an Indian Institute of Horticultural Research, Hissarghatta (Mani, 1992).

Apple

A number of coccinellid predatory beetles were recorded to feed an san jose scales, *Quadriaspidiotus perniciosus* Comstock, but were unable to control the pest to a desired level. Therefore, attempts were made in 1953 to introduce *Prospaltella (Encarsia) perniciosi* Tower from China and *Aphytis diaspidis* from California for the control of san jose scale. These parasites were established in Himachal Pradesh, Kashmir and Kuman hills of Uttar Pradesh, Predator, Chilocoris kuwanae from Japan was also evaluated in Jammu and Kashmir, Himachal Pradesh and Uttar Pradesh. Exotic eulophid *Aphelinus mali* from North America was introduced for the control of *Eriosoma lanigerum* in Saharanpur and Kuman region of Uttar Pradesh, Kullu valley and Simla of Himachal Pradesh, Nilgiri hills of Tamilnadu in India. This parasite was established in the released area by 1963 and fairly established by 1994. Two exotic egg parasitoids viz. *Trichogramma embryophagum* Htg and *T. cocociae palidum* Mayer were evaluated against codling moth, *Cydia pomenella.* These parasitoids when released during the egg laying period of the pest proved useful (Singh, 1986).

A number of parasites have been reared from Gypsy moth, *Lymantria obfuscata.* These are *Anastatus kashmirensis* Mathus from the eggs, *Exorista russica* Fall from larvae and *Pimpla* sp., *Theronia* sp. and *Brachymeria lasus* (Walkar) from the pupae (Singh, 1986). Biotic agents namely *E. Perniciosi, A. diaspidis, C. bijugus,* against san jose scale and *A mali, C. septumpuncta, Syrphus contrator* and C. *carnea* against wooly aphid were reported to be effective (Singh, 1993).

FIELD EFFICACY OF NATURAL ENEMIES AND THEIR USE IN IPM

The role of biocontrol agents such as parasitoids, predators, insect pathogens, antagonist organisms and weed killers is significant. The IPM strategy also involves the need based use of pesticides only when the pest reaches the economic threshold levels and this would promote the conservation of many biocontrol agents in the crop ecosystem. There are many successful cases of pest management through the use of biological control. About 60 per cent of natural control of insects in any crop is due to the biological control agents which need to be conserved by minimizing the use of pesticides.

Most of the fruit crops provide a stable environment for the survival and effective performance of biocontrol agents in the pest management programme. Pests of fruit crops are more amenable for biological control because many of key pests are sessile, soft bodied, with high biotic potential and are easily available throughout the year for attack by natural enemies. Successful examples of biological control of pests of fruit crops have created awareness in the filed of biological control. Biocontrol agents especially parasitoids and predators were used as a sole method of pest control or formed a core component in an integrated pest management (IPM) as per the new paradigm of pest management.

Citrus

About seven coccinellid predators and three prasitoids were recorded on citrus psylla (Kameswara Rao 1983). *Tetrastichus radiatus* were reported to be effective against citrus psylla in Punjab. Besides these, *T. phyllocnistoides* and *Diaphorencyrtus aligharensis* have been reported from Vidarbha (Anonymous, 1983). *Tamarixia radiata* caused up to 95 per cent parasitism of citrus psylla in nature (Mani, 1992). As reviewed by Ramasubbaiah (1983) few parasites viz., *Sympiesis* sp. *Cirrospiloideus phyllocnistoides* and *Scolonic quadristriata* have been reported to parasitise leaf minere larvae. Some mora parasites recorded from Vidharbha (MS) include *T. phylocnistiodes Cirrospilus quadristriate, Bracon phyllocnistidis, Tetrasticus* sp. and *Sympiesis purpurea.* A Maximum parasitisation to 43.94 per cent larvae and 27.95 per cent pupae was observed at Akola (Anonymous, 1983).

Predator *Catena pracesetosa* was found to feed on nymphs and pupae while its larvae were predators on eggs of black fly (Sharnagat *et al.* 1983). Egg parasitoids *Telenomus lucullus* Nixon and *Trichogramma chilonis* Ishii parasitized the eggs of fruit sucking moth to the extent of 46.4 and 64.4 per cent, respectively during 1980-81 and 1981-82 at Akola. *Euplectrs* sp. a larval parasite parasitized the host larvae from 5.6 to 7.25 per cent in both the years (Anonymous, 1982). At Katol, leaf eating caterpillar, *Papilio domoleus* was parasitized by a larval parasite *Melaophacharops tamanukii* Uchida to the extent of 10.1 and 9.1 per cent during 1975-76 and 1977-78, respectively. Besides other larval parasitoids viz. *Apanteles papilionis, Pediobus imbrues, Eurytoma* sp. and pupal parasitoids viz. *Euycia nymphalidophaga* and *Pteramolus puparum* were also need (Anonymous, 1983). Dipel 0.05 per cent (*Bacillus thuringensis*) was found to be highly effective against first, second and third nastar larvae of *P. domoleus* at Chethalli (Anonymous, 1982). *Chilocorus nigritus* suppressed the natural population of Florida red scale *Chrysomphalus aonidium* L. and purple scale *Lepidosaphes beckii* Newm at Chethalli (Anonymous, 1982).

Different combinations of entomogenous fungus *Verticillum lecanii* (Zimn) in comparison with recommended scalicides as well as their sub-ethal dosages on coorg mandar in indicated that even 1/10 of normal effective dose of Quinalphos + fungus (16 × 10^4 spores/ml) + teepol 0.05 per cent when sprayed just before the onset of monsoon controlled the pest effectively (Anonymous, 1982).

For the biological control of citrus mealy bug. *P. citri* releases of 10 beetles of *Cryptolaemus montrouzieri* were effective in mixed planted citrus orchards (Citrus and Coffee) (Anonymous, 1989). *Leptomastix dactylopii* (from West Indies) got established on *P. citri* at Chethalli, Bangalore (Karnataka) and Pollachi (Tamilnadu) and was recovered from *P. citri* infesting citrus, grapes, guava and pomegranate. In mixed planted orchards (citrus and coffee), the orchards with shade and less light interception (900 - 14100 lux) sprays of entomopathogenic fungus *Verticillium lecanii* at 10 × 10^6 spores/ml + 0.05 per cent Quinalphos + 0.05 per cent teepol were promising. In field trials Dipel a commercial formulation of *Bacillus thuingiensis* at 0.05 pr cent gave good control of citrus butterfly caterpillars. The citrus blackfly had effective natural enemies in less distributed ecosystem, i.e. Kodagu District (Karnataka) (Anonymous, 1989). *Coccidoxenoides peregrinus* (Timberlake) an important parasite of *P. citri* parasitizes to the extent of 10 to 30 per cent in field (Mani and Krishnamoorthy, 1989) *C. nigritus* could multiplied and released at 10-15/plant for the control of citrus scale (Krishnamoorthy, 1993). Plant products like Pongamia oil, Neem oil, neem seed extract neemark and Mahua oil were safe to *C. nigritus*.

An exotic encyrtid, parasitoid *Leptomastix dactylopii* How could be multiplied on nymphal and adult stages of mealy bus (Krishnamoorthy, 1988). This parasite released in Karnataka and Tamil Nadu on sweet orange, and lime and lemon resulted in a complete control of *P. citri* within four months. This parasitoid got permanently established in released orchards and migrated to near by orchards. A release rate of 5000 parasitoids per hectare was recommended to suppress the mealy bug population (Krishnamoorth and Singh, 1987). The mealy bug, *Nipaecoccus viridis* population declined on acid lime from 221.30 on 16th March to 1.40 in June to the activity of biotic agents *Anagyrus* sp and *C. montrouzieri*. A maximum population of 8 and 10 of *Anagyrus* sp. and *C. montrouzieri* was observed in the 2nd week of March (Mani and Krishnamoorthy, 2001). Rao *et al.* (2001) recorded *Cirrospilus quadristriatus* Subba Rao and *Tetrastichus phyllocnistoides* Narayanan as predominant parasitoids of citrus leaf miner from citrus growing tracts of Vidarbha. Activity of *C. phyllocnistoides* was significantly higher in Octo-November conceding with haste bahar flush.

Three egg parasitoids viz. *Trichogramma chilonis Telenomus incommadus* and *Ooenocyrtus papilionis* and a larval parasitoid *Apanteles papilionis* accounted for 88 per cent parasitism in nature (Krishnamoorthy and Singh, 1988; Jalali and Singh, 1990). *A. papilionis* is a common and gregarious parasitoid of citrus butterflies (Mani, 1992).

Guava

As many as 20 natural enemies have been reported on *A sossypii* infesting different crops in India. Five coccinellid predators and aphids effectively within 45 days of its appearance (Mani and Krishnamoorthy,

1989). An exotic predator *Cryptolaemus nontrouzieri* when released @ 20 beetles per tree was found to be effective against *C. psidii* (Mani and Krishnamoorthy 1990a). A total of 12 natural enemies recorded at Bangalore from the stripped mealy bug *Ferrisia virgata* and an encyrtid *A enasius advena* Comp. Were found parasitize the hosts to the extent of 50 per cent. Besides these, several local coccinellid predators like *S. cocivora, M. boninensis* (Okamoto), *Brumus suturalis* and *Spalgis epius* Westwool also played a significant role in controlling the *F. virgata* population (Mani and Krishnamoorthy, 1990 b). *Plananococcus citri* infesting guava could be controlled permanently by *Leptomastix dactylapi* along with the local parasitoids like *C. perigrinus, Allotrops citri* and predators like *M. boninensis* and *S. epius* (Mani, 1992). *E. guadeloupae* adults when released in February on guava in an orchard near Bangalore parasitized 14.67 per cent whitefly nymphs in March and parasitism gradually increased to 62.7 per cent in June.

Apple

Encarsia perniciosi was established on san jose scale of apple. Besides, *Aphytis proclia, Chilocorus bijugus, Pharoscymnus flexibilis* and *Cybocephalus* sp. played important role in the population suppression of this pest. Predators were early arrivers in the orchard.

Wooly apple aphid failed to multiply during spring and autumn due to heave predation by *Coccinella septumpunctata* and *Adonia variegata.* An anthocorid bug took heavy toll of apple blossom thrip. *Exorista rossica* was the most common parasitoid of *Lymantria obfuscata* and *Anastatus kashmirensis,* a promising egg parasitoid (Anonymous, 1989). The releases of *Aphytis proclia* and *E. perniciosi* in Kashmir an apple @ 2000 each per tree effectively reduced the infestination on san jose scale. The predator, *C. bijugus* @ 20/ tree reduced the scale density in Kashmir, while a single release of *C. bijugus* @ 50 neonate grubs per tree at Solan brought about a reduction in mean population of the scales. *Aphelinus mali* was introduced at Nauni in 1991 and has since then uniformly spread in apple orchardss and is providing satisfactory control of wooly aphid, as compared with the pest situation before the release of the parasitoid (Anonymous, 1994).

Encarsia perniciosi along with *A. proclia* was found to be very effective against san jose scale in north-west India (Singh, 2001). *Aphelinus mali* got established on *Eriosoma lanigerum. Trichogramma embryophagum* or *T. cacaeciaa* with pheromone baited traps proved effective against *Cydia pomonella* in Ladakh (Singh, 2001b). Wooly apple aphid population and its natural enemies monitored throughout the year at weekly intervals at Solan revealed that *A. mali* was consistently present while Chrysopids and coccinellids were observed during April-June, and syrphids during June, November and February. Similar surveys in Srinagar, Baramulla, Pulwama, Budgam and also the predators *C. infernalis, C. septumpumctata, P. batteatue* and *C. confrator.* Observations on the impact of the release of parasitoids of San Jose Scale showed a parasitisation of 20 to 54 per cent in above districts by both *Aphytis* sp. and *Encarsia* sp. (Anonymous, 2001).

Khajuria and Sharma (2001) reported six different species of predators. Of these predatory mites, *Amblyseius fallacis, Zetzellia mali* and *Typhlodromus* spp were most abundant and constituted 47.8 per cent of the total natural enemy population. Studies conducted a evaluate the performance of various pest management strategies against pest of economic importance in Himachal Pradesh indicated that an infestation of 2 per cent fruit damage of san jose scale was effective for initiating sprays of miscible oil during late dormancy provided good control of the pest without damaging natural enemies i.e. *C. bijugus* syrphid flies and *A. mali.* An apple orchard planted on a clonal root stock was found to be relatively free from the attack of wooly apply aphid in mashobra area (Kakar, 2001).

Pomegranate

Leptomastix dactylopii an exotic parasitoid from West Indies was established and recovered from *Planococcus citri* infesting pomegranate (Anonymous, 1989). Release of *Cryptolaemus montrouzieri,*

Coccidoxenoides pergrinus and *Leptomastix dactylopii* suppressed the mealy bug populations on pomergranate (Anonymous, 1994). The exotic parasitoid, *L. dactylopii* and the predator, *C. montrouzieri* were effective in suppressing the mealy bug *P. citri* and *Maconellicoccus hirsutus* on several fruits crops like citrus, grapes, guava and pomegranate (Singh, 2001a). Releases of *T. chilotraeae* reduced the fruit damage of pomegranate fruit borer *Deudrix isocrates* at Bangalore with just 14.15 per cent as compared to 24.24 per cent in control trees. The eggs of fruit borer at Solan were parasitised to the extent of 69 per cent by major parasitoids *Telenomus cyrus* and *Anastatus* sp. nr *kashmirensis* and *T. chilonis* (Anonymous, 2001).

Gupta and Sood (2001) carried out surveys during last 3 years on the incidence of pomegranate fruit borer and extent of egg parasitization by its natural enemies at Solan which revealed that attack continued from July to September to the extent of 45 per cent. Among 40 eggs sampled on fruits during different weeks revealed that 63 per cent were parazitized by three parasitoides viz. *Telenomus cyrus* Nixon (Scelionidae), *Anastatus* sp. nr. *Kashmirensis* Mathur (Eupelmidae) and *Trichogramma chilonis Ishii* to the extent of 93.7, 2.7, and 6.3 per cent, respectively.

Gopalkrishnana and Mani (2001) reported an epizootics of Entomogenous fungus *Entomophothora aphidis* Hoffman (Zygomycetes: Entomophthorales). On pomegranate aphid *Aphis punicae* (Homoptera: Aphididae) at IIHR, Bangalore, during the year 1998-99 and 1999-2000. Observations revealed that *E. aphidis* infected all four nymphal stages and late forms of *A. punicae* to the extent of 68 and 67 per cent in both years, respectively thus forming main biotic mortality factor of *A. punicae.*

Mango

Several parasitoids recorded on mealy bugs, *Rastrococcus iceryoids* caused 42 per cent parasitism in nature (Tandon and Lal, 1978). An exotic coccinellid predatory beetle, *Cryptolaemus montrouzieri* was found to be very effective in controlling the mealy bugs (Mani, 1992). *D. stebbingi* was predated by some coccinellids and syrphids and parasitized by *Phyadeuon* sp. But none of these were effective against the mealy bugs. Parasitoids like *Aneristus ceroplastae, Coccophagus* spp. and *Metaphycus* spp attacked green shield scale *C. polygonata.* However exotic predatory *C. montrouzieri* were effective in suppressing the scale population (Mani, 1992).

Five species of coccinellids and seven species of spiders were observed to feed on mealy bug crawlers (Tondon, 1992). *Spalgis epius* was commonly observed to feed on *R. iceryoides* and *M. astur* preyed on *C. polygonata* (Tandon and Lal, 1978).

Singh (2000) documented various macro and micro organisms associated with insect pests of mango, including 3 species of spiders, 8 species of coccinellids, 4 species of ants and one species of mantid. The other natural enemies of mango insect pests including tettigonids, chrysopids, vespids, carabids, syrphids, phytoseiids, epipyropofd, eulophids and braconids (Venugopal Rao *et al.* 2001). Dhooria (2001) recorded predatory mite, *Typhlodromus caudiglans* actively feeding on mango gall mites during February-March in Punjab.

FUTURE THRUST AND PROSPECTS

In India, most of the major pests of crops like citrus leaf miner, blackfly, scale insects, mealy bugs, mango hoppers and mealy bugs, grape mealy bugs, pomegranate fruit borer and mealy bugs, san jose scale and wooly aphids of apple are native. Unfortunately no suitable indigenous natural enemies or effective chemical pesticides are available for their control. Although exotic natural enemies viz. *Leptomastic dactylopii* and *Cryptolaemus montrouzieri* have been found to be effective in controlling the mealy bugs and scale insects of citrus, mango, guava, grape and pomegrante, augmentation and conservation of native natural enemies would go a long way in suppressing their population. There are several natural potential natural enemies which are not fully utilised in augmentation programmes for the want of techniques of their mass production.

Singh (2001) tested various biogents in controlling the insect pests of fruit crops. The details of the field releases are menthioned in Table 3. Puri *et al*,. (1997) published a resource inventory for the availability and supplies of bioagents of the important crop pests including pests of fruit crops. The details are provided in the following Table 4.

Table 3: Field release of parasites/predators for biological suppression of fruit crop pests.

Fruit crop	Insect pest	Parasite/predator	Host stage attacked	Rate of release/hect.
	Different pests of fruits	*Trichgramma*	egg	50,000-1.5 lac adults
Citrus	*Planocodccus lilacinus*	*Tetracnemoidea indica*	nymph	2000/adult parasitoids/tree
	Planococcus citri	*Leptomastix dactylopii*	nymph	2000/adult parasitoids/tree
	Aleurocanthus woglumi	*Encarsia clypealis*	nymph	2000/adult parasitoids/tree
	Aleurocanthus woglumi	*Eretmocerus serius*	nymph	2000/adult parasitoids/tree
	Aleurocanthus woglumi	*Amitus hesperidum*	nymph	2000/adult parasitoids/tree
	P. citri or *P. lilacinus*	*Coccixenoides peregrinus*	nymph	2000/adult parasitoids/tree
	Aonidiella aurantii	*Aphytis melinus*	nymph & adult	2000/adult parasitoids/tree
Grape	*Maconellicoccus hirsutus*	*Anagyrus dactylopii*	nymph & adult	60,000 adult parasitoids/ha
	Maconellicoccus hirsutus	*Allotropa japonica*	nymph & adult	60,000 adult parasitoids/ha
		Chrysoperla carnea	egg	10,000 larvae/ha
		Mallada boninensis	egg	10,000 larvae/ha
	Aphis craccivora	*Ischiodon scuttellaris*	nymph & adult	10,000 larvae/ha
Guav	*Ferrisia virgata*	*Scymnus coccivora*	nymph & adult	600-2500 adults/ha
	Ferrisia virgata	*Harmonia octomaculata*	nymph & adult	600-2500 adults/ha
Subabul	*Heteropsylla cubana*	*Curinus coeruleus*	nymph & adult	20 adults/tree
Grape, Citrus, Guava	*M. hirsutus, P. citri F. virgata*	*Cryptolaemus montrouzieri*	Nymph & adult	10 beetles/tree
-do-	*F. virgata*	*Coccinella septumpuncata*	nymph & adult	500 beetles/ha
-do-	*F. virgata*	*Chilomenus sexmaculata*	nymph & adult	500 beetles/ha
	Melanapsis glomerata	*Chilocoris nigratus*	nymph & adult	1500 beetles/ha
	F. virgata	*Brumoides suturalis*	nymph & adult	

(Singh, 2001b).

CONCLUSION

Biological control is the core component of an integrated pest management programme. Successful examples of biological control of pests of fruit crops have created awareness of biological control in the field. The IPM strategy involves not only the need based application of pesticides but also the conservation of the native biotic agents in the crop ecosystem. Fruit crops provide a stable environment for the survival and effective functioning of biocontrol agents. A brief resume of the research on use of natural enemies of fruit crop pests in an integrated pest management has been given in this paper.

This fascinating control of *Icerya purchasi* by introducing *Radolia cardinalis* in Nilgiri hills on citrus marked the real beginning of biological control in the country. Subsequently augmentative and inoculative release of two exotic parasitoids, *Encarsia perniciosi* and *Aphytis proclia* gave promising results in control of san jose scale on apples.

Table 4: Availability and supplies of bioagents for fruit crops in India.

Crop	Insect pest	Bioagent	Stage of host attacked	Sources
PARASITES				
Coconut	Black headed caterpillar	*Bracon brevicornis*	Larva	IARI, New Delhi, PCI, Bangalore, CIPMC, Baroda
Coconut	-do-	*Brachymeria nephantidis*	Pupa	APAU, CPCRI and CIPMCs
Coconut	-do-	*Bracon kirkpatricki*	Larva	CPCRI, PDBC, TN Coop. S.F. Chengalpattu
Cotonut	-do-	*Elasmus nephantidis*	Larva	APAU, CPCRI, IIHR, PDBC, and CIPMC
Coffee	Mealy bugs	*Elasmus nephantidis*	Nymph	BCL Simbhaoli (UP)
Coconut	Black header caterpillar	*Eulophid* sp	Pupa	PCIBCR Lab. Bangalore
Coconut	-do-	*Gonizus nephantidis*	Larva	PCI BCRL, Bangalore
Coffee, Custard, apple, Mango, Guava	Mealybugs	*Leptomastix dactylopii*	Nymph PDBC	IIHR,
Coconut	Black headed caterpillar	*Tetrastichus israeli*	Pupa	CIPMC, Bangalore TN Paraside Breeding Center, Coimbatore
Pomegranate	Annar butter-fly	*Trichogramma chilonis*	egg	PDBC, IIHR, MPAU, MAU, CIPMC, Nagpur
Apple	Coling moth	*T. embryophagum*	egg	CIPMC, Solan, Srinagar
Apple	Coling moth	*T. evanescens*	egg	-do-
Peach	Coling moth	*T. pretiosum*	-do-	-do-
Coconut	Black headed caterpillar	*Trichospilus pupivora*	Pupa	CPCRI, CIPMC, Bangalore
Coconut	Black headed caterpillar	*Xanthopimpla punctata*	-do-	-do-
Citrus, Pomegranate	Butterfly	*Telenomus remus*	egg	CIPMC, Indore, Lucknow, Nagpur
PREDATORS				
Apple	San Jose scale	*Chilocorus bijugus*	Nymph	SKUAS & T. YSPUH & F.
Vegetables	Sucking pests	*Chrysoperla carnea*	-do-	GAU, MPAU
Grape, Gava, Apple, Mango Jack fruit, Tamarind	Mealy bugs and scales	*Cryptolaemus montrouzieri*	Nymph and adult	APAU, CIPMC Indore, Lucknow, IARI, IIHR and PDBC
Vegetable and Fruit crops	Aphids and mealy bugs	*Coccinella septumpunctata*	Nymph and adults	Jeypee Biotech. Virudhunagar (TN)
Coffee, grapes and Citrus	Scales mealy bugs	*Chilocoris nigritus*	Nymph and adult	CIPMC, Burdwan, Baroda, Goakhpur and Jallandhar
Apple	Scales	*Encarsia perniciosi*	Adult	SKUAS & T. YSPUH & F. CIPMC, Bangalore and Srinagar

Contd. . . .

Table 4 Contd. . .

Crop	Insect pest	Bioagent	Stage of host attacked	Sources
Pineapple Tamarind	Mealy bug	*Mallada boninensis*	Nymph and adult	GAU, PDBC
Coconut	Scale	*Parena nigrolineata*	-do-	CPCRI
Coconut	Scale	*Rhinocoris fuscipes*	-do-	CIPMC, Burdwan, Bangalore, GAU.
Apple, Guava, Pear, Peach, Pomegranate	Aphils and mealybugs	*Scymnus sp*	-do-	CIPMC, Lucknow, Solan, Gorakhpur
Coconut	Scales and mealybugs	*Stethoconus praefectus*	Nymph	CPCRI, CIPME, Bangalore and Bhuvaneshwar
ENTOMOPATHOGENS				
Coconut	Red palm weevil, white grub	*Heterorhabditia sp. (Soil commandos)*	Grub	Ecomax. Agro Systems, Mumbai
Coffee	Berry borer	*Beauveria bassiana*	Larva	Biocontrol Lab. Simbholi (UP)
Coffee	Berry borer	*Dispel sp.*	Larva	Pesticides India Ltd. Udaipur
Coconut	Rhinoceros beetle	*B. bassiana*	Adults and grubs.	Jeypee Bio-tech. Virudhunagar (TN)

(Puri *et. al*, 1997).

Several parasitoids and predators have been recorded on key insect pests of citrus. *Tetrastichus radiatus* and *T. phyllocnistoides* have been reported to be effective against citrus psylla. *Tamarixia radidta* alone caused 95 per cent parasitism. *Catana paracesetosa* heavily predated citrus blackfly. *Telenomus lucullus* and *Trichogramma chilonis* parasitized eggs of fruit sucking moth to the extend of 46 and 64 per cent, respectively. *Chilocoris nigritus* fed voraciously on red and pruple scale, *Cryptolaemus montrouzieri* and controlled the mealybugs. *Leptomastix dactylopii* got established on *P. citri* in Karnataka and Tamilnadu and controlled the pest within four months permanently. An encyrtid, *Aenasius advena* parasitized striped mealy bug, *F. virgata* to the extent of 50 per cent. Besides these, several local *Coccinellids* also played a significant role in controlling the guava mealy bug. This pest could be controlled permanently by released of exotic *L. dactylopii* along with local parasitoids and predators. Demonstration trials conducted on farmers' field at Bangalore on guava and grape mealy bug to stand predated on *C. montrouzieri* proved successful. Exotic parasitoid, *Encarsia guadeloupae* parasitised nymphs of spiralling whitefly, *Aleurodicus dispersus* on guava to the extent of 62 per cent.

Although several parasitoids were recorded on mango mealy bugs causing 42 per cent parasitism in nature, *C. montrouzierii* an exotic predacious *coccinellied* was effective in controlling mealy bugs and scale insects. *L. dactylopii* got established and recovered from scales on pomegranate. *L. dactylopii* and *C. montrouzieri* effectively suppressed the scales and mealy bugs *P. citri* and *M. hirsutus* on several fruit crops like citrus, grapes, guava and pomegranate. *Trichogramma chilotraeae* parasitised pomegranate fruit borer eggs and reduced the fruit damage by 14 per cent. *Telenomus cyrus* parasitised 93 per cent of eggs of fruit borer. Epizootics of entomogenous fungus, *Entomophthora aphidis* was recorded on pomegranate aphid, *A. punicae* to the extent of 68 per cent in Bangalore.

Encarsia perniciosi and *Aphtis proclia* were effective against san jose scale insects in north-west India. *Aphelinus mali* was established on wooly apple aphid, *Eriosoma lanigerum* where as *Trichogramma embryophagum* with pheromone baited traps proved effective against *Cydia pomonella* in Ladakh.

REFERENCES

Anonymous 1979. *Annual Report of RFRS Katol*, submitted to Agresco. Horticultural sub-committee, Dr PDKV, Akola.

Anonymous 1982. *Annual Report of AICFIP on Citrus Dieback*, Akola sub-committee to "Fruits" sub-committee, PDKV, Akola.

Anonymous 1983. *Annual Report of AICFIP on Citrus Dieback*, Akola, sub-committee to "Fruit" sub-committee, PDKV, Akola.

Anonymous 1989. *Report on biological control of fruit and vegetable crop pests for 1989-91.*

Anonymous 1989. Achievements of AICRP on biological control. *Tech. Bull. No. 2*; p. 12.

Anonymous 1991. *Annual Report of AICRP on biological control of crop pests and weeds, 1990-91.*

Anonymous 1992. *Annual Report of AICRP on biological control of crop pests and weeds for 1991-92*, Biocontrol Centre, Bangalore.

Anonymous1994. Research Highlights of 1993-94 PDBC. Bangalore, *Tech. Bull.* No. 12, pp. 8-9.

Anonymous 2001. *Research Highlight of 2000-2001*, PDBC, Bangalore, pp. 10-11.

Azam, K.M. 1983. Losses due to pests in grapes. *Indian J. Ent* (special issue). 2:387-389.

Debach, P. 1974. Biological control by natural enemies. Maximizing biological control through research. pp. 227.

De Banch P. and Rosen D. 1991. *Biological control by natural enemies*. Cambridge University Press, Cambridge, UK.

Dhooria, M.S. 2001. Predacious mites on phytophagous mites. *Proc. of symposium on biocontrol based pest management*, Ludhiana, pp.74.

Doharey, K.L. and Butani, D.K. 1984. Newer approaches in the management of citrus pests. *Proc. National Seminar on pest management in citrus, cotton, sugarcane and sorghum* at PDKV Akola, pp. 47.

Gopalkrishnan, C. and Mani, M. 2001. Epizootics of *Entomophthora aphidis* Hoffman (Zygomycetes: Entomophthorales) on pomegranate aphid *Aphis punicae*, *Proc. of symposium on biocontrol contributed papers*, Ludhiana, pp. 140.

Gupta R.L. 1954. Life history of *Tonica zizphi* Stn. the citrus leaf rollere. *Indian J. Ent.* 16: 20-23.

Gupta, R.R. and Sood, A. 2001. Role of egg-parasitoids for the control of pomegranate fruit borer in mid-hills of Himachal Pradesh, Proc. of symp. on biocontrol contributed papers, Ludhiana, pp. 57.

Jalali and Singh, S.P. 1990. A new record of *Ooenocyrtus papilionis* (Hymenoptera: Encyrtidae) on the eggs of *Papilio demoleus* (Linn.) from India.

Kakar, K.L. 2001. Feasibility of using various pest management strategies on Apple in Himachal Pradesh. *Proc. of symp. on biocontrol contributed papers*, Ludhiana pp. 205.

Kameswara, Rao 1983. Management of citrus psylla. *Proc. of National Seminar on pest management in citrus, cotton, sugarcane and sorghum* at PDKV Akola, pp. 49.

Khajuria, D.R. and Sharma, J.P. 2001. Natural enemies of phytophagous mites and their role in integrated mite management in apple ecosystem. *Proc. of symp. on biocontrol, contributed papers*, Ludhiana, pp. 182.

Krishnamoorthy, A. 1988. Host range, development and sex-ratio of *Leptomastix dactylopii* on different stages of citrus mealy bug, *Planococcus citri J. Biocontrol*, 2: 8-11.

Krishnamoorthy, A. and Singh, S.P. 1987. Biological control of citrus mealy bugs with an introduced parasite *Leptomastix dactylopii* in India. *Entomophaga*, 32: 143-148/

Krishnamoorthy, A. and Singh, S.P. 1988. Observational studies on the occurrence of parasitoids of *Papilio* spp. in citrus. *Indian J. Pl. Prot.* 16: 79-81.

Mani, M. 1990. Collection of natural enemies of fruit crop pests. Rept. on biocontrol of fruit and vegetables for the year 1989-91, IIHR, Bangalore, pp. 2.

Mani, M. 1993. Studies on mealy bugs and their natural enemies on her orchards. *J. Biol. Contr.* 7: 75-80.

Mani, M. 1992. Parasitoids and predators as components in IPM of fruit crops. Compendium on summer. Inst. On IPM in hort. Crops. pp. 59-72.

Mani, M. and Krishnamoorthy, A. 1989a. Impact of insect predators in the control of *Aphis gossypii* Glov. On guava *J. Biol. Control*, 3: 128-129.

Mani, M. and Krishnamoorthy, A. 1989b. Occurrence of mealy bugs and their natural enemies on custard apple around Bangalore, S. India. *J. Biol. Control*, 3: 77.

Mani, M. and Krishnamoorthy, A. 1990a. Evaluation of exotic predator *Cryptolaemus montrouzieri* Muls. in the suppression of green shield scale *Chlroopulvinaria psidii*, *Entomon*. 15: 47-48.

Mani, M. and Krishnamoorthy, A. 1990b. Natural suppression of mealy bugs in guava orchards. *Entomon*, 15:245-247.

Mani, M. and Krishnamoorthy, A. 2001. Biological suppression of spherical mealy bug *Nipaecoccus viridis* Newstead. on acid lime in India. *Proc. symp. of biocontrol. contributed papers*, Ludhiana.

Mani, M. and Thontadarya, T.S. 1988. Development of the encyrtid parasitoid, *Anagyrus dactylopii* on grape mealy bug. *Maconellicoccus hirsutus*. Entomon, 14: 49-51.

Narayanan, E.S. and Batra, H.N. 1960. *Fruit flies and their control*, ICAR, publication, p. 68.

Pruthi, M.S. and Mani, M.S. 1945. Our knowledge of insect and mite pests of citrus in India and their control. Sci. Mongr. No. 16, 42 pp. ICAR, New Delhi.

Radke, S.G. and Ghuguskar, H.T. 1982. Losses due to insect pests in citrus crop. A Paper presented in National seminar on crop losses held at Hyderabad.

Radke, S.G. and Ghuguskar, 1984. Pest management in citrus. A Paper presented in national seminar on pest management in citrus, Nagpur 1984. pp. 1-43.

Rama Subbaiah, K. 1984. Management of citrus leaf miner, *Proc. of National Seminar on pest management on pest management in citrus cotton, sugarcane and sorghum at Akola*, pp. 51.

Rao, V.P. and Kamath, S.M. 1966. Record of cottony cushion scale *Icerya purchasi* Maskell in the cannore district of Kerala and its control. *Indian J. Ent.* 27: 552.

Rao, C.N., Shivankar, V.S. and Singh, S. 2001. Extent of parasitism on citrus leaft miner. *Phyllocnistis citrella* in Vidarbha (MS). *Proc. of symp. on biocontrol. contributed papers*, Ludhiana, pp. 56.

Sadana, G.K. and Kanta, V. 1971. Predators of citrus mite, *Eutetranychus orientalis* in India. *Sci. and cult.* 37 (11): 530.

Sankaran, T. 1974. Natural enemies introduced in recent years for biological control of agril. pests in India. Indian *J. agric. Sci.* 44 (7): 425-433.

Sharnagar, B.K. Kuchanwar, D.B. and Hardas, M.G. 1983. Management of citrus blackfly on nagpur mandarin in Vidarbha region. *Proc. National seminar on pest management in citrus, cotton, sugarcane and sorghum at PDKV Akola,* pp. 57.

Singh, S.P. 1986. Biological control of fruit crop pests. Compendium on Training programme on Biol. Control of crop pests and weeds, pp. 39-45.

Singh, S. P. 1992. Role of Biocontrol agents in IPM of horticultural crops. Compendium of summer Inst. on IPM in horticultural crops. 131-147.

Singh, S.P. 1993. Biological control of insect pests. In: K.L. Chadha and O.P. Parek (eds), *Advances in horticulture fruit crops*. 3(3): 1592-1615.

Singh, S.P. 2000. Bio-intensive approach helpful. Survey of Indian agriculture, 2000. *The Hindu* Chennai - 600018, pp. 216.

Singh, S.P. 2001a. *Advances in biological control in India*. Sourvenir, Advances in biocontrol. Publication of biocontrol based pest management.

Singh, S.P. 2001b. Innovations in the mass rearing technologies for mass releases, transportation and storage of natural enemies. *Biol. Control Symposium on biocontrol*, pp. 1-33.

Tandon, P.L. 1992a. *Integrated pest management in mango*. Compendium of lectures in Summer Inst. on IPM of horticultural crops, pp. 9-18.

Tandon, P.L. 1992b. *Integrated pest management in citrus*. Compendium of lectures on summer Inst. on IPM of hort. crops., pp. 19-26.

Tandan, P.L. and Verghese, A. 1985. *World list of insect, mite and other pests of mango*. IIHR. Tech. Doc. 5, 11-22 pp.

Tandon, P.L. and Lal, B.C. 1978. The mango Coccid *Rastrococcus iceryodies* Green. (Homoptera: Coccidore) and its natural enemies. *Curr. Sci.*, 47: 46-48.

Thakare, K.R. Radke, S.G., Borle, M.N. and Ghughuskar, H.T. 1984. Two decades of research on citrus pest management in Maharashtra, Brochure, DR, PDKV, Ankola.

Rao, N.V., Kuman, M.A., Babu, R.A. and Umamaheswari, T. 2001. Naturally Occurring bioagents in mango agri-ecosystems in Andhra Pradesh. *Proc. symp. on biocontrol, contributed papers*. Ludhiana, pp. 55.

Wadhi, S.R. and Batra, H.N. 1964. Pests of tropical and sub-tropical fruit trees. In *Entomology in India* (Ed. N.C. Pant's, Silver Jubilee No. Entomological Society, India, pp. 227-260.

Pest Management of Horticultural Crops under Arid and Semi-Arid Regions

B.G. BAGLE[1]

The production-protection axis constitutes the back-bone of nation's efforts to ensure proper food supply for it's millions. The land is limited and the population is alarming rising. Hence, increasingly intensive cultivation alone can provide a rational solution to our problems of food self-sufficiency. The job of the production specialists is the proper management of the inlet pipes while those on protection side have to keep the exit holes as narrow in size and as few in number as possible. Among the various factors that limit productive potential of arid horticultural crops synthesized by breeders or of intensive fertilizer programme evolved by soil scientists, entomologists all over the world are formulating their strategies and tactics of pest control. The strategy of pest control has to be thoroughly recast and changed from the existing pest wise approach to integrated crop wise approach. Thus to formulate adequate and rational policy regarding strategy and tactics of pest control; the most important pre-requisite is proper appreciation of nature and the problems involved.

The arid are considered as "reserve lands" in which the future agricultural production is based. This typical agro-climate has involved natural fruit and vegetables, which have the potential to survive harsh climatic conditions and still producing good quality produce. Information on the fruits and vegetable with known commercial potential and possibility for cultivation under rained conditions with suitable water harvesting techniques is now available. However, the productivity and production potential of these crops can be sustained only when appropriate pest management tactics are adopted at right time. The indigenous drought hardy type fruit-crops are now available for commercial exploitation. These are Ker (*Capparis decidua*), pilu (*Salvadora oleoides*), lasora or gonda (*Cordia gharaf*), ber (*Zizyphus* spp.), Karonda, (*Carrisa carandas*), pomgranate (*Punica granatum*) custard apple (*Annona squamosa*) and aonla (*Emblica officinalis*). Among the vegetable crops, most of the cucurbits, gourds, and melons; okra, clusterbean, cowpea and vegetables *viz*. chilli and brinjal have been found to be the best suitable under rainfed conditions in areas where annual precipitation is more than 300 mm and the monsoon starts in the first week of July.

With the limited attention paid to the improvement of these hardy fruit and vegetable crops under arid regions, no clear cut trend has so far been developed on insect pest control of these crops. Even, economic status of a number of insects recorded on the crops are not clearly known. Fruits crops because of perennial nature harbour and encourage the population through out the year. Thus pest carry over is relatively more in horticultural crops. The rapid stride of horticulture all over the country are bond to bring about a corresponding increase in pest complex. Keeping in view the fast peace of development of horticulture and rapid growth of orchards it will be wise policy to fully visualize the magnitude of pest problems and strengthen the pest control research to match the expected pest problems and not to allow

1. C.H.E.S., Vejalpur (Godhra), Gujarat.

them to become the limiting factor in out production programme. Forewarned is fore armed is an apt edge for our struggle against the insect-pests.

Scanning of literature clearly indicates that not much work has been done on the management of insect-pest. On these hardy fruit and vegetable crop under arid condition barring some examples of ber and pomegranate. A lot more is to be done on development of IPM strategies on these crops keeping in view the vast diversity of land, soil and agro climatic zones of this region. Efforts were however made in various parts of the country to control the insect-pest mostly by chemical means. The important insect-pest and available recommended control measures on the crops viz, pomegranate, ber, custard apple, aonla, etc. are given in the following paragraph which seems to be suitable, effective, economical and eco friendly IPM strategies for boosting up the production potential of these crops.

IPM IN ARID HORTICULTURE

Ber

Fruit fly - *Carpomyia vesyviana* Costa: The female flies puncture the fruits and the development around the puncture is arrested and this causes the deformation of fruits. The maggots feed on fleshy and juicy pulp. The infested fruits turn dark brown and rot and drop down.

Fruit borer - *Meridarchis scyrodes* Meyrick: On hatching, the caterpillars bore in to the fruits and feed on pulp filling the cavities with fecal matter.

Hairy caterpillar - *Thiacidas postica* Walkar: The young caterpillars feed gregariously but later they disperse and feed voraciously defoliating the entire twing.

Ber butter fly - *Taruchus theophrastus* Fabr: The caterpillar feed on leaves and flower buds and the damage is recognised by characteristic - white membranous streak on dorsal side of leaf surface.

Chrysomelid leaf miner - *Platypriya andrewese*: The grubs mine into the leaves eating out small cavities or pockets. Leaf defoliating beetle - *Holotrichia insularis Adoretius nibidus A. virsutius*: The leaf lamina are perforated by numerous small irregular holes which are bitten out by the strongly toothed maxillae,

Tortoise beetle - *Occasida cruentata* Febr.: Grubs and beetles scrape and bite holes in leaf lamina.

Bark eating Caterpillar - *Inderbela quadrinatata* Walker: The caterpillar feeds on trunks and main branches of ber trees, presence of silken webs comprising of excreta of larvae indicated the damage by the pest.

Leaf rollers - *Ancyles spp. Sunclera univocalis*: The caterpillars roll up the leaves and feed on epidermis of dorsal surface of rolled leaves.

Leaf Webber – *Psorosticha zizyphi*: On hatching the caterpillar mine the leaves, later they roll up longitudinally and web together several apical leaves the feed within first on epidermis and later on leaf lamina.

***Strategy of pest management*:**

1. Raking the soil around the trunk of expose maggots and pupae of fruit flies.
2. Removal and destruction of fallen fruits.
3. Insert in hole a cotton wool soaked in CS_2, Petrol, or kerosene dip oil and plug and hole with mud.
4. Foliar sprays with Monocrotophos or Dimethoate (0.05 per cent) followed by fenvalerate (0.05 per cent) or endosulfan (0.07 per cent) controls the pest effectively.

Pomegranate

Thrips - *Scirtothrips dorsalis*: Nymphs and adults lacerate the leaves, flower stalks petals and sepals and rasp the sap that oozes out resulting into leaf trip curl and dry away while flowers are shed.

Aphids - *Aphis punicae*: Both nymphs and adults suck the cell sap from tender leaves, and twigs. The affected parts get discolored and disfigured. They secrete honey dew on which sooty mould develops.

Mealy bugs - *Planococcus lilacinus*: Both nymphs and adults suck the cell sap and devitalize the plant resulting into shedding of buds and flowers and smaller size fruits.

Leaf eating - cater pillar - *Achaea janata*: Freshly hatched larvae congregate on larvaes later they segregate and feed voraciously devouring the entire leaf lamina. Adults are destructive, suck the juice of fruits.

Fruit sucking moths - *Othreis spp.*: The months are nocturnal in habit and are found during rainy season. The adults pierce the ripening fruits and the damaged fruits starts rotting as the punctured region are easily infected with bacteria and fungi, ultimately the fruits are dropped.

Anar Butterfly - *Virachola isocrates* Fabr. *Deodorix epijarbas*: On hatching the caterpillars bore inside the developing fruits subsequently the infested fruits are attacked by bacteria causing fruit to rot. The conspicuous symptoms of damage are offensive smell and excreta coming out of entry hole

Bark eating caterpillar - *Inderbela quadrinotata*: On hatching the larvae usually bore at joints between leaf stalk and twig and main stem and tunnel straight down ward. The larvae also cut a circular holes at various places through which they eject frass.

Shoot hole borer - *Xyleborus fornicatus Termites-Odontotermes obesus Rambur*: A stem boring ambrosia beetle causes damage to main trunk and branches. Maximum damage to the stem upto 30 cm height from ground level.

Strategy of pest management

1. Removal of wild creepers of Menospermaceae, reduced the incidence of fruit sucking moths.
2. Avoid fruiting in rainy season and destroy the fallen fruits.
3. Bagging of fruits before maturity prevents the damage by moths and borers.
4. Injection of petrol, CS_2, Diclorvos in holes and plugging the holes with mud controls the damage by bark eating caterpillar.
5. Inudndative release of *Trichogramma chilonis* is quite effective against anar butter fly.
6. Sprays with Monocrotophos, Dimethoate, (0.05 per cent) followed by Endosulphan (0.07 per cent) or Fenvalerate (0.05 per cent) effective against major pests of pomegranate.

Aonla

Aphids - *Cerciaphis emblica*: Both nymphs and adults suck the cell sap from tender shoots, leaves, and devitalize the plants. They exude honey dew on which sooty mould develops.

Mealy bugs - *Nipaecoccus vastator*: Both nymphs and adults suck the cell sap from tender shoots, leaves, and devitalize the plants.

Leaf twister - *Caloptilia acidula*: The caterpillar on hatching twist the leaf and feed within. In case of severe attack the entire plant show sickly appearance.

Hairy Caterpillar - *Euproctis flava*: The caterpillar on hatching feed voraciously and gregariously on tender leaves and defoliate the plants. The larvae later on migrates on entire plant and feed on leaves leading to defoliation.

Shoot gall maker - *Betousa stylophora*: The caterpillar on hatching enter the shoots and feed within causing a gall on the tender shoots.

Fruit borer - *Virachola isocrates-Meridarchis scyrodes*: The caterpillar bore into the fruits and feed within causing premature drop of fruits during monsoon season. Maximum damage is during July and August.

Bark eating Caterpillar - *Inderbela tetraonis* Moore: The larvae bore usually at the joints between twig and main stem and tunnel straight downward. Presence of silken Webs comprising of excreta of larvae indicated the damage.

Strategy of pest management

1. Inject petrol, Dichlorvos or CS_2, and plug the hole with mud.
2. Orchard sanitation is effective.
3. Foliar sprays with Dimethoate (0.05 per cent) or Monocrotophos followed by endosulfan (0.07 per cent) at tri weekly interval control the pest effectively.

Custard Apple - Mealy bug *Ferrisia virgata* Ckll.

Soft bodied insects covered with mealy wax secreted by the insect themselves. They infest leaves, shoots, bugs and even the fruits and suck the sap there from the affected plant parts get deformed and reduce in size besides they secrete honeydew on which sooty mould develops.

Fruit borer - *Abonaepestis bangalella* Regonot: On hatching, the caterpillars bore inside the fruits, making irregular tunnels and causing damage to melocarp. As a result the development of the infested fruits arrested and ultimately these fruits fall down.

Fruit fly - *Daecus zonatus* Saunders: The characteristic symptom of attack is the oozing out of fluid from the small fine punctures made by the flies. The infested fruits rot, show brown patches and ultimately fall down.

Thrips - *Rhipiphorothrips cruentatus* Hard: Nymphs and adults lacerate the leaf tissues and suck the exuding sap, severe infestation often results in leaves getting almost bleached.

Strategy of pest management

1. Ploughing around the tree during Jab-Feb kills the hibernating pupae of fruit flies.
2. Collect and burn the attacked fruits.
3. Spraying with 0.05 per cent Monocrotophos, Dimethoate or Dichlorvos is effective.

Jamun

Bark eating caterpillar - *Inderbela tetraonis* Moore: The caterpillar nibble the wood below the bark marking irregular galleries, then bore inside the stem trunk presence of large silken Webs comprising of chewed wood particles and excreta of larvae specially near the forking of trunk or the main branches.

Psyllids - *Trioza jambolanae* Crawford: If forms unilocular galls on dorsal surface of leaves. The gall may be separate or into groups of 2-8 galls. The mature galls loose their green color harden and split into several lobes. The pest sometimes causes malformation on shoots and foliage.

White flies - *Dialeurodes eugeniae* Maskell: Both nymphs and adults suck the cell sap thereby devitalize the plants.

Thrips - *Leeuwenia ramakrishnac*: The long tailed thrips, the nymphs of which feed on ventral surface of leaves motionlessly. Affected leaves show large yellowish patches which turn reddish brown and become brittle.

Leaf Roller - *Polychrosis cellifera*: The caterpillar fold or roll the tender leaves and Web the same with silken strands and feed within eating away patches of leaf lamina up to midrib.

Leaf Webber - *Argyriploce mormopa*: The caterpillar Web together leaves at short tips and feed within.

Leaf miner - *Acerocercops syngramma* Meyrick: The caterpillar mine the green leaves and cause blister like swelling subsequently dry and drop off leaving big holes in the leaves.

Fruit fly - *Dachas diversus Coquilett*: The pest migrates to jamun trees during monsoon and causes damages to fruits.

Fruit borer - *Meridarchis reprobata Meyrick*

Strategy of pest management

The caterpillars bore into the fruits and feed within.

1. Removal of silken Webs material and insert into the hole cotton wool soaked in CS_2, chloroform petrol or kerosene oil and plug the hole with mud.
2. Collection and destruction of fallen fruits prevents the build up of fruit flies and fruit borer in the orchard.
3. Spray 0.3 per cent dimethoate or oxydemetonmethyl (0.03 per cent) or phosphamidon (0.05 per cent) against sucking pests and foliage feeders.

Karonda

Mealy bugs Scales Defoliators: Both nymphs and adults suck the cell sap and devitalize the plant and also secrete honeydew on which sooty mould develops.

The larvae feed voraciously on foliage and defoliate the plants.

Strategy of pest management

1. Raking the soil around tree trunk to expose mealy bugs eggs.
2. Spray the crop with 0.05 per cent dimethoate or monocrotophos.

Passion Fruit

Mite: The mite generally feed on the underside of leaves and also on tender fruits. As a result a characteristic pale yellow depression in seen on attacked fruits which turn brownish and fetch lower price.

Strategy of pest management

1. Spray kelthene 0.03 per cent for the control of mite. Alternately spray of dimethoate of monocrotophos 0.05 per cent can be given.

Bael

Scales and Mealy bugs: Both nymphs and adults suck the cell sap and devitalize the plant.

Beside they secrete honeydew on which shooty mould develops.

Strategy of pest management

1. Spray monocrotophos 0.05 per cent or phosphomidon 0.05 per cent.

Date Palm

Rhinocerus beetle - *Oryctes rhinoceros* Lynn.: Beetles are destructive stage. At night these beetles feed on crowns of palms by boring into the unopened tender fronds and tunnel down wards feeding on soft tissues of growing plants. The growth of tree is arrested.

Red palm weevil

Rhynchophorus ferruginous Olivier: It is difficult to detect infestation in the initial stage but when noticed in advanced stage, the loss is irreparable, the grubs are voracious feeder once inside the trunk they seldom come out. They burrow in hardy woody portion. In case of severe infestation the trunk is completely hollowed and filled with decayed rubbish. In older trees the top portion of the trunk bend and ultimately breaks at bend. A thick reddish brown viscous fluid as also chewed up and discarded fiber pieces are thrown out from the holes.

Scale insects

Parlatoria blanchardi: The pest infests shaded pinnules and main veins of the fronds. In severe cases even fruits are attacked. The fruits become dry deformed and do not mature.

Black headed caterpillar

Nephantis serinopa Meyrick: The caterpillar feed gregariously on surface tissue of leaflets scraped out from lower surface and construct silken galleries and feed within these. Attacked leaflets turn brown and dry up in severe attack the leaflets are reduced to papery tissues.

Termites *Odontotermes obesus*: They attack root zone and gradually tunnel upwards making galleries inside the main trunk.

Strategy of pest management

1. Keep the orchard clean.
2. All potential breeding sites compost heap dung hills, refuse dumps must be destroyed.
3. Breeding substrate must be dusted with 5 per cent BHC or carbaryl.
4. Prompt destruction of dead and badly infested palms.
5. Filling up the leaf axis with 5 per cent BHC or lindane + saw dust or sand (1:1) three times a year.
6. Removal and burn all infested leaflets in the initial stage of attack.
7. In case of older trees, removal all the fronds and pour hot salty water over the crown.
8. Prompt cutting and burning of infested fronds.
9. Soil application of chlorpyriphos @ 5 ml/liter before planting the seed or transplanting the seedlings is effective.
10. Release of parasites, *Trichospilus pupivora,Bracon brevicornis* and *Perisierola nephantidis.*

Falsa

Mealy bugs - *Drosicha mangifera* Green: Both nymphs and adults suck the cell sap from shoot, twigs and tender leaves and devitalize the plant. The growth and fruiting capacity is adversely affected. The bugs also exude honey dew on which sooty mould develops.

Leaf eating caterpillar - *Euproctis fraterna* Moore: On hatching the caterpillars feed gregariously on leaf lamina, skeletonising the same and completely later, they segregate and gnaw the leaves. The entire plant may be defoliated.

Falsa caterpillar - *Giaura sceptica* Swinhoe: The caterpillar usually roll the leaves specially the lower ones and feed within. Some times a bunch of leaves may be Webbed together.

Leaf eating beetle - *Adoretus* Spp. *Anomala bengalensis* Blanchard: The grubs feed on root while the adults fee on fliage during night and hide in the soil or below the big clods during day.

Falsa beetle

Oxycetonia versicolor Fabr.: The adults feed on leaf lamina and make round holes.

Falsa bug (Gargara mixta)

Buckton: Nymphs move in to leaf axils and such the cell sap. the bugs also excrete honeydew which attracts the black ant and also favors shooty mould.

Membracid bug

Lepocentrus Taurus Fabr.: White fly - *Bemisia tabaci Gennadius*: Both nymphs and adults suck the cell sap from leaves causing discoloration and curling of leaves.

Strategy of pest management

1. Raking of the under the trees during winter helps in destroying the eggs.
2. Soil application of chloropyriphos helps in controlling the pest.
3. Spray 0.04 per cent monocrotophos, dimethoate of 0.07 per cent endosulfan for foliage feeders.

Orchards are contiguous both in space as well as in time-as these are grown over extensive contiguous and continue to grow for several decades, ecologically such conditions are very conductive for rapid multiplication of insect-pest. Hence utmost care, vigilance, surveillance and cooperative endeavor are necessary to keep the orchard free of pests. Following approaches can be adopted for successful management of different insect-pest or orchard.

(1) Orchard Sanitation: Phytosanitary methods keep in preventing the carryover of good many pests Eg. Bark cater pillar, stem borers etc.

(2) Frequent inter culturing is effective as it exposes the various immature stages of pests resting in the soil of Sun's heat and other natural enemies Eg. Eggs of mealy bugs, grubs of chafer beetle and pupae of fruit flies.

(3) Banding tree trunks with alkathene sheets (400 guage) in preventing the nymphs of mealy bugs for ascending during October-November.

(4) Cleaning the affected parts by bark and stem borers and insert in holes cotton wick soaked in a good fumigant like CS_2, chloroform or even petrol and seal the hole with mud.

(5) Use of pheromenes in controlling fruit sucking moths and anar buttery fly.

(6) Male sterile technique can be utilized against fruit flies.

(7) Use of organic manure like neem cake can be used for the control of white grubs.

(8) Use of bio-agent, microbials, (*Bt* K) and botanicals may find its place in an integrated Pest Management of any crop in question.

(9) Use of chemicals at appropriate time.

(10) Rationalize the use of pesticides - Avoid prophylactic sprays, use of selective and non-persistent pesticides.

Once the various integrated control programmes are developed, it is necessary that they should be out to practice. Implementation of these programmes will require scientists/trained technicians and skills which are needed to work with the farmers. As it is multidisciplinary approach, professional scientists are required to collaborate in its planning and execution.

CONCLUSION

The concept of Integrated Crop Management (ICM) should be promoted keeping in view the Integrated Nutrient management (INM) and Integrated Pest Management (IPM) essential ingredients under scientific crop production to ensure healthy crop and their products. Thus, there is a need to integrate the technology transfer from various disciplines, adopting a multidisciplinary approach to tackle to the problems of production and in this context, IPM strategy plays an important role in avoiding the losses caused by pests and diseases. Commodity research should be given prime importance, Nevertheless, research on basic aspects should not be altogether neglected while planning IPM.

REFERENCES

Bagle B. G. 1992. Incidence and control of fruit fly, *Carpomyia vesuviana* Costa of ber, *Zisyphus mauritiana* Lamk. *Indian. J. Pl. Prot.* 20: 205-207.

Bagle B.G. 1992. Studies on crop stage and number of application of monocrotophos in relation in thrips damage in chilli. *Indian., J. Ent.* 55(3): 241-244.

Bagle B.G. 1993. Seasonal Incidence and control of thrips *Scirtithrips dorsalis* Hood in Pomegranate. *Indian J. Ent.* 55(2): 148-153.

Bagle B.G. and Shashi Verma (1992. Persistence of phorate and monocrotophos in okra var. Pusa Sawani. *Advanced in Hort. & Forestry* 2: 144-157.

Bagle B.G. 1993. Effect of date of plating on the incidence of leaf curl caused by thrips *Scirtothrips dorsalis* Hood. *Indian. J. Pl. Prot* 21(2): 133-34.

Bagle B.G. 1993. Efficacy of fenvalerate on avoidable losses in brinjal as affected by shoot and fruit borer, *Leucinodes orbonalis* Guen. Scientific Hort. 3: 125-128.

Bagle B.G. 1995. Studies on assessment of avoidable losses in chilli due to thrips, *Scirtothrips dorsalis* Hood. *Scienfitic Hort.* 4: 87-91.

Bagle B.G. 1997. Evaluation of insecticides again thrips., *Scirt thrips dorsalis* Hood and fruit borer, *Irachola isocrates* Fabr. (Abst) National Seminar on orchards Management held at Patna, 22.

Bagle B.G. 1999. Seasonality of major pests of Aonla (*Emblica officinalis* Geartn) in semiarid region of Gujarat. (Abst). National Seminar on Sustainable Horticultural Production in tribal region held at Ranchi 30.

Hosagoudar, I.M., Nandihalli B.S., Lingappa S. and D.R. Patil 2000. Development of spray schedule in the management of fruit borer, *Meridarchis scyrodes* Meyrick and fruit fly, *Carpomyia vesuviana Costa. Pestology* 24(7): 71-74.

Kakar, K.L., Dogra G.S. and Amit Nath 1987. Incidence and control of pomegranate fruit borers, *Virachola isocrates* (Fabr.) and *Deudorix epijarbas* (Moore). *Indian J. agric. Sc.* 57(10): 749-52.

Lakra R.K. and Zile Singh 1984. Calender of losses due to her fruity fly, *Carpomuia vesuviana* Costa (Diptera: Tephritidae) in different zizyphus spp. in Haryana. *Indian. J. Ent.* 46(3): 261-269.

Mote, U.N., Tambe A.B. and C.S. Patil 1992. Management of fruit sucking moths in Pomegranate garden, *Maharashtra J. Horti.* 6(1): 220-24.

Mote U.N. and A.B.T. Tambe (1991. Observations on extent of damage caused by shot-hole borer of pomegranate and castor. J. Maharashtra agric. Univ. 16(3): 439-440.

Naitam N.R. and R.S. Sharma 2000 Evaluation of biorational insecticides in comparison with conventionals against the pest of Ber. *Pestology* 20(7): 66-68.

Patel, I.S., Upadhyaya, V.R. Vyas, H.N. and A.T. Patel 1990. Field Bioefficacy of Different Insecticides for the control of ber fruit fly. *Carpomyia vesuviana,* Costa. (Tephrutudae: Diptera). *GAU Res. J.* 16(1): 57-59.

Patel, B.H., Upadhyaya U.R., Murlidharan C.M. and G.S. Judal (1989. Comparative efficacy of different insecticides against ber fruit fly, *Carpomyia vesuviana* Costa. *Indian. J. Pl. Prot.* 17: 39-45.

Patil, P. 1994. Bioecology and management of ber fruit borer, Meridarchis scyrodes Meyrick (Lepidoptera: Corposinidae) M.Sc. (Agr.) Thesis: U.A.S. Dharwad.

Patil P. and B.V. Patil 1997. Management of ber fruit borer Meridarchis scyrodes Myrick *Karnataka J. Agric. Sci.* 10(1): 52-55.

Regumoorti, K.N. and Arumugam. R. 1992. Chemical control of ber fruit fly *Carpomyia vesuviana* Costa. *Indian. J. Pl. Prot* 20(1) 32-36.

Shewale, B.S. and D.N. Padule (1992. Comparative efficacy of various insecticides against fruit borer Meridarchis scyrodes Meyrick *Maharashtra J. Hort* 6(1): 17-19.

Yadava. C.P.S. 1969. Combating fruit sucking moths. Ind. Hort 13(3): 32-34.

Yelshetty, S. Baljkai. R.A. Biradar, B.M., and A.N. Mokashi 1996. Management of fruit borer, *Meridarchis scyrodes* Meyrick. *Karnataka J. Agric. Sci.* 9(3): 529-532.

23

Present Status and Management of Pests and Diseases of Temperate Fruits in Himalayas

S.S. SINGH[1], U.S. SINGH[2] AND N.C. NAINWAL[3]

The Himalayan region comprises of more than one eighth of the total land area of India which makes up the entire northern boundary running from Jammu and Kashmir in West to the Arunachal Pradesh in the east. In India, temperate fruits like apple, pear, peach, plum, apricot, almond, walnut and cherry are grown commercially in the North Himalayan Hill Regions of the country comprising the states of Jammu and Kashmir, Uttaranchal, Himachal Pradesh, Arunachal Pradesh and to some extent other North Eastern states such as Sikkim, Meghalaya, Mizoram, Nagaland and Manipur (Solanki, 1999; Pratap, 2003). The temperate fruits are the major crops in the hill areas of India and they have played a significant role in boosting the economy of the growers. The total area and production under temperate fruits has increased considerably, but productivity is low in comparison to other countries like USA, New Zealand and China. The productivity in other countries may be high due to deeper and more fertile soils and favourable climatic conditions, but even so our productivity is rather low.

Apart from several other factors for low productivity, the incidence of pests and diseases are extremely important. The major insect-pests infesting temperate fruits include San Jose Scale, *Quadraspidiotus perniciosus* Comstock, woolly apple aphid, *Erisoma lanigerum* Hausmann, European red mite, *Panonychus ulmi* Koch., peach leaf curl aphid, *Brachycaudus helichrysi* Kaltenbach, peach fruit fly, *Bactrocera zonata* Saunders, blossom thrips, *Thrips flavus* Schrank, codling moth, *Cydia pomonella Linnaeus*, apple root borer, *Dorysthenes hugelii* Redtenbacher and apple Stem borer, *Apriona cinearea* Cheverlot (Bhalla and Gupta, 1993; Singh and Lal, 2002). Among economically important diseases studies so far, premature leaf fall, *Marssonina coronaria*, white root rot, *Dematophora necatrix*, apple scab, *Venturia inaequalis* Cke., Collar rot, *Phytophthora cactorum*, peach leaf curl, *Taphrina deformans*, canker diseases, fruit spots and bacterial gummosis, *Pseudomonas syringae* cause huge losses and result in low yields (Gupta and Jindal, 1993; Verma and Bhardwaj, 2003). Besides these, injudicious use of pesticides in the farming system and pesticide resistance in potential pests and pathogens have caused heavy crop losses.

In view of the rising export potential of residue free fruits and the crisis situation created by indiscriminate use of pesticides, temperate fruits need much more attention in terms of IPM input than what it is getting now. It is therefore, highly imperative to develop and implement a rational approach with greater reliance on IPM to promote sustainability and to reduce the number of application of hazardous pesticides in temperate fruits.

---

1 Horticultural Experiment and Training Centre, Chaubattia-263 651, Ranikhet, Almora.
2 Deptt. of Plant Pathology, G.B. Pant Univ. of Agri & Tech., Pantnagar-263 145.
3 Agri-Export Zone, Diversified Agriculture Support Project, Haldwani-263 156, Uttaranchal.

PRESENT STATUS

The temperate fruits dominated farming system indicates the changing trends in agriculture and is gradually becoming the most powerful force shaping the environment in the Himalayan mountains these days. The importance of temperate fruits development has been recognized by many countries of the region notably India, Nepal, Pakistan, Bhutan and China. Mountain environment provides suitable niches for special cultivations. Harnessing these niches with appropriate location specific farming activities has comparative advantage in relation to the plains. The diverse agro-ecological conditions prevailing in the mountains is suitable for fruit crop farming especially temperate fruits. Ningnan in China, once considered below extreme poverty line is now an example of economic improvement within a very short time span due to large scale adoption of temperate fruits.

The temperate fruit crop acreage in Indian Himalayas is rapidly increasing and is developing into perennial stretches of monocropping systems. Parallely this has led to increase pests and diseases incidence and use of chemical pesticides, especially synthetic pyrethroids. As a consequence, pest resurgence, residues in fruits, deleterious effects on biological control agents and resistance buildup in insects and pathogens have all complicated pest management (Smith, 1970; Atwal and Dhaliwal, 1997). All these have reduced the quality of fruits affecting domestic and export market. Introduction of modern fruit crop production practices and increased use of pesticides have no doubt increased the fruit production many fold but has also resulted in changes in pest and disease scenario over space and time and many new pest and disease problems have emerged (Gupta, 1999). The premature leaf fall in apple was first detected in 1995 in some orchards, has now spread to all apple orchards in Himachal Pradesh and can cause up to ninety per cent of defoliation in unprotected orchards. All commercial Delicious cultivars are susceptible to this disease. Premature leaf fall in apple has also been reported recently from Kashmir, the neighbouring Uttaranchal and Bhutan (Sharma, 2003). Recently, two species of the Red Spider mites, European Red mite (*Panonychus ulmi*) and Two-Spotted mite (*Tetranychus urticae*) have emerged as new serious pest problem in apple growing areas of India. The survey revealed drastic increase in its population from barely negligible in 1992 to 46 per cent in 1996 and to 72 per cent in 1999. The pest has now appeared in an epidemic form in all apple-growing states of India. At present, the European Red mite and Two-Spotted mite occur in 70:30 ratio in the orchard ecosystem (Kakar, 2000). Intensity of white root rot, collar rot, canker diseases, powdery mildew and peach leaf curl has also increased. Similarly codling moth in apple and apricot known to be minor pest has become a regular and severe pest of Leh and Kargil districts of Ladakh region.

The incidence of pests is highest in the mid hills and valley areas and lowest in dry temperate regions. In the mid hills, San Jose scale is the most damaging pest followed by European Red Mite, Woolly Apple Aphid and Blossom thrips. In the high hills, Woolly Apple Aphid and European Red Mite are the most damaging followed by San Jose scale, European Red mite and Blossom thrips. In the mid hills and valley areas, high temperature is responsible for high multiplication of San Jose scale and Red Spider Mite owing to more number of generations per year, thus, causing extensive damage to the crop. However, under moisture stress conditions, woolly apple aphid, San Jose scale and Blossom thrips are more prominent (Thakur, 1999). According to a study by the Mountain Farming Systems Division at ICIMOD, Nepal reveals that main reasons of pests and diseases spread on temperate fruits have been identified as excessive and indiscriminate use of pesticides resulting in killing of predators, use of most toxic chemicals to achieve quick mortality, poor monitoring and inadequate knowledge about pests and diseases, etc. Atwal and Dhaliwal, 1997; Thakur, 1999). They reported that ignorance of improved cultural practices and faulty training and pruning systems lead to favourable conditions for development of diseases and pests and interference in proper penetration of spray material.

Pest and Disease Management Strategies need attention during establishment of nursery or orchard:

- Proper soil selection be made for raising orchards of respective fruit crops.
- Proper climatic factors/topography be considered for planting of the respective fruit trees.

- Plant material of suitable variety for laying quality fruit orchard should be obtained from registered nursery.
- Avoid planting of samplings infested with scales, borers and diseases.
- Make use of recommended quantity of FYM and chemical fertilizers in the soil for laying of nursery and orchard.
- Before laying or raising of plant nursery make use of *Trichoderma viride* and *T. harzianum* to control root rot diseases and at later stages for control of collar rot also.
- Ensure the use of neem cakes while raising plant nurseries to ward off any soil pest and pathogen.
- Solarisation of nursery beds be made to destroy soil pests and pathogens.
- Digging of pit at recommended size and at proper distance should be done atleast two months before plantation and left open to control soil borne pathogens and pests.
- Make use of disease free bud wood for raising healthy nursery plants.
- Do not grow the nursery at the same site every year.
- Destroy the affected seedlings.

PEST AND DISEASE MANAGEMENT STRATEGIES IN ESTABLISHED TEMPERATE FRUITS ORCHARD

Cultural Practices

- Orchard sanitation should be given top priority. Sanitation is a simple, but important preventive measure of pest control. Even simple removal of alternate hosts, weed hosts and crop residues greatly reduce the infesting potential.
- Cutting and burning of severely infested and dried branches.
- Digging the basins of the tree with spades during winter (December-January) to expose the diapausing pupae and spores to adverse climatic conditions.
- Fallen fruits should be collected throughout the season and either buried deep in the soil or destroyed otherwise.
- Overwintering caterpillars should be collected from the loose bark, other debris, bands used for trapping and destroyed.
- Collection and destruction of fallen leaves from the orchard floor during winter are found effective.
- Correct pruning during winter allows adequate air circulation round the tree canopy, thereby modifying the microclimate and reducing disease development.
- Plant early maturing varieties of peach and almond in case of peach fruit fly and gummosis endemic areas.
- Make use of balanced dose of chemical fertilizers.
- Mulching with dry grass or hay (10-15 cm in thickness) in the tree basins during March should be dome to conserve the moisture during fruit setting and fruit development period.
- Avoid excessive use of nitrogen and application of nitrogen should be made in split doses only.
- Apply white wash impregnated with glue regularly on the tree trunks to avoid sun burn and winter injury.
- If plants/trees exhibit the deficiency of micro nutrients go for the application of the same on the basis of soil and leaf analysis.

- Growing of flowering plants especially marigold and maize on the peripheries will help in conservation of both predators and parasites.
- Two to three Bee colonies per ha should be provided in each orchard at the time of bloom to increase pollination.
- Proper irrigation schedule should be adopted.
- Trees should neither be forced to drought nor water logging conditions.
- Rogue out infected plants.
- Maintain vigour of the tree to keep away shot hole/pin hole borers, bark beetles and other pest infestation.
- Grow marigold, berseem and mustard in between lines of Plantation to check the nematode population.
- Do not delay harvesting of fruits.
- Keep the trees as free as possible from mechanical wounds, winter injury, crotch separation and cankers.
- Pruning cuts should be made close to the branches leaving no stubs.
- Cut wounds should be covered with superior white lead paint or any other paint.
- Remove and destroy dead and drying fruit trees to ward off borer infestation.
- Clean the stem borer holes with flexible wire and apply the recommended chemical.
- Cover the exposed part of the stem to sun with dry grass or gunny bags soaked with methyl parathion 0.1 per cent once a month in March till October.
- Install a light trap near the orchard to collect and kill the beetles in kerosenized water.
- Shake non bearing trees over a cloth sheet at dusk, collect and destroy the collected beetles (defoliating and fruit eating beetles).
- Clip off terminal shoots with unshed cluster of dry leaves in winter for the destruction of shoot borers.
- Remove the dead bark and frass and apply water proof paint on hard wood to avoid borer attack.
- Add well rotten manure in case of termite infestation.
- Destroy termitaria in the vicinity of the orchards.
- Apply Chaubattia paste after cleaning the weeping wounds at the time of dormancy break for the control of bacterial gummosis.
- Cover fresh wounds with Chaubattia paste or copper oxychloride paint.
- Remove the infected roots and apply Chaubattia paste on the cut ends.
- Collect and destroy the grubs of root borers while preparing basins.
- Ensure complete collection and destruction of foliage and prunned wood in the orchards itself after leaf fall.
- Clip off mildewed twigs and destroy them.
- Remove all the mummified fruits, dead fruits and pruning from the orchards.
- Pruning of suckers and water sprouts be made regularly.

Biological Practices

Conservation

- Conserve the predators like Lace wings, Lady bird beetles, Carabids, Syrphids, Anthocorid bugs, Mirid bugs, Nobid bugs, Capsid bugs, Spiders, Predatory ants, Phytoseiid mites, Parasitoids like *Encarsia, Aphytis, Triohogramma, Telenomus* etc. in orchard.
- Growing of flowering plants, especially Marigold and Maize on the peripheries and legumes as inter-cropping help in conservation of both predators and parasitoids, especially *Chrysoperlla* and *Anthocorids* and also help in nematode management.
- Collection of egg masses of hairy caterpillars and putting them in a fine meshed cage for emergence of egg parasitoids, protect the parasitoids from orchard spray and predation. The larvae hatched may be destroyed.
- Collect the different larval instars of hairy caterpillars (suspected parasitized) from orchards and place them in wire meshed cages for emergence of larval parasitoids like *Exorista* sp., *Diriona* sp., *Sarcophage* sp., *Pollenia* sp., *Helina* sp., *Anilastus* sp., *Anthomyia* sp., *Euplectrus* sp., to protect them from scheduled spray effects and bird predation (Gupta, 1999).
- Collect the pupae of hairy caterpillars from the orchards and place them in wire mesh cages for emergence of pupal parasitoids like *Brachymeria* sp., *Monodontomerus* sp., *Hyposoter* sp., *Exorista* sp. and *Pimpla* sp. to protect the parasitoids from scheduled spray effects and predation (Gupta, 1999).

Augmentation

Monitor the incidence of pests like San Jose scale, woolly aphid, thrips, peach leaf curl aphid, mites, codling moth etc. and make releases of the following.

(i) Predators: Lady bird beetles – *Chilochorus bijugus, Pharoscymmus horni, Coccinella septumpunctata, Chilochorus tristis, Adalia bipunctala, Synharmonia* sp., *Exochomus quadripustulantus, Hippodamia convergens, stethorus* sp. @ 30-50 adults/infested tree (Sing, 2002).

– Green lace wings:*Chrysoperla* sp. and *syrphus* sp. 10-20 1st instar larvae/tree.

(ii) Parasitoids:

– *Encarsia perniciosi* and *Aphytis* spp. @ 2000 per infested tree at least 15 days of insecticidal sprays and 10 days after fungicidal sprays against scale insect. *Trichogramma embryophagum* (egg parasitoid) @ 2000 per infested tree at weekly interval. *Aphelinus mali* @ 1000 adults per infested tree. The dosage may vary depending upon the intensity and extent of damage (Singh and Mishra, 2003).

Chemical Control Measures

(i) Need based, judicious and safe application of pesticides are the most vital segments of chemical control measures under the IPM programme. It involves developing IPM skills to play safe with environment by proper crop health monitoring. Observing ETL and conserving natural biocontrol potential before deciding in favour of use of chemical pesticides as last resort.

(ii) Following suggestions have important bearing for success of control measures in the context of IPM strategy.

- The number of insecticidal/fungicidal sprays recommended can be minimized as per need after proper surveillance and pest intensity considering both biotic and abiotic factors.

- Avoid mixing of two or more pesticides/tank mixing as it may alter effectiveness of pesticides.
- Repeated application of same pesticides should be avoided as it may develop resistance in the target pest.
- Avoid use of synthetic pyrethroids which may cause resurgence of sucking pests.
- Use safer pesticides during early fruiting phase of crop growth.
- Encourage the use of neem based formulations and biopesticides which are comparatively environmental safer.
- Proper spray equipments should be used to achieve good result.

PEST MONITORING AND FORECASTING SYSTEMS

Necessity, economy and efficacy of control measures can be determined if there is a system of prognosis or forecasting the extent and time of occurrence of plant diseases and pests. Forecasting involves all the activities in ascertaining and notifying the farmers in an area/region that, i) conditions are sufficiently favourable for certain diseases and pests causing economic losses; that, ii) application of control measures most acceptable under integrated diseases/pests management will result in economic gain, and iii) the amount of diseases/pests expected is unlikely to be enough to justify the expenditure of time, energy and money to control. This concept of forecasting requires complete knowledge of biology and epidemiology on insect-pest and pathogen i.e. development of disease/pest in plant population under the influence of the factors associated with the host, the pathogen/pest and weather (Rajak, 1999; Thakur, 1999).

On date, apple scab, powdery mildew and premature leaf fall are the three principal diseases on apple crop and to control them as many as eight sprays of fungicides in a season involving an expenditure of Rupees 12,000-15,000.00 per ha or Rupees 70-80.00 per tree is required. The accumulated losses through these diseases are more than 40 per cent annually in all the apple growing States of India (Thakur, 1999). Likewise, insect-pests account for nearly 35 per cent of the total damage at pre and post harvest stages. Though the magnitude of pests incidence varies from region to region and orchard to orchard but in all, 24 insect-pests have been rated economically important of which about 12 account for the major share of the damage throughout the states. Among the insect-pests usually Woolly aphid and Red Spider Mite are the most dominant and principal insect-pests, affecting 82 per cent orchards each and followed by San Jose Scale (71%), Blossom thrips (70%) and Apple root borer (43%). These insect-pests are being managed through high potency insecticides every year to maintain the economic threshold level (Thakur, 1999).

Now, the question is that how long this practice will go on and what would be the consequences of these chemicals on human and plant health, soil fertility and crop productivity? Are these chemicals being made responsible for the current decline in temperate fruit crops productivity and quality? Yes, they are. They affect the soil microbial population, which transfer and make the nutrients available to the plant system. However, answer to this problem is to adopt need-based use of pesticides aided by an effective forecasting system through monitoring weather parameters for predicting diseases and pests outbreaks as well as production. This system has taken over the chemical management practices to judicious use of chemicals on temperate fruit crops world over.

AGRO ECOSYSTEM ANALYSIS (AESA)

Agro Eco System Analysis is an approach, which can be gainfully employed by extension functionaries and farmers to analyse field situation with regard to pests, defenders, soil conditions, plant health, the influence of climatic factors and their interrelationship for growing healthy crop. Such a critical analysis of the field situations will help in taking appropriate decision on management practices. The basic components of AESA are:

1) Plant health at different stages.

2) Built-in-compensation abilities of the plants.
3) Pest and defender population dynamics.
4) Soil conditions.
5) Climatic factors.
6) Farmers past experience.

METHODOLOGY OF AGRO ECO SYSTEM ANALYSIS

Field observations:

(a) Enter the orchard atleast 15-20 feet away from road/bund. Select fruit tree of medium size randomly.

(b) Record the visual observations on the following parameters :

(i) Flying insects (both pests and defenders)

(ii) Close observation on pest and defenders which remain on the plant.

(iii) Observe the appearance of major pests/diseases and their intensities and defenders like Lady bird beetle, Green Lace wing, Anthocorid, Bugs, Spiders, Predatory Ants, Phystoseiid mites and syrphids.

(c) Frequently visit the orchard during peak period of pests/diseases incidence and also record their level of incidence to adopt timely management strategies.

(d) Record soil conditions *viz.* dry, wet or water logged.

(e) Observe Rodent live burrows.

(f) Repeat the steps from b to e on atleast 5 trees in a diagonal row or cris cross fashion across the orchard to have overall average assessment of the orchard.

(g) Record the climatic factors like sunny, partially cloudy, cloudy, rainy etc. for the proceeding week.

Drawing

First draw the plant with actual number of main limbs in the center of a chart. Then draw pests diseases on left side and defender on the right side. Indicate the soil condition, rodent damage, etc. Give natural colours to all the drawings, for instance draw healthy plant with green colour, diseased plant/leaves with yellow colour. While drawing the pests and the defenders on the chart care should be taken to draw them at appropriate part of the plant, where they are seen at the time of observation. The common name of the pests, diseases and defenders and their population count should also be given alongwith diagram. The weather factor should be reflected in the chart by drawing the diagram of sun just above the plant, if it is sunny. If cloudy, the clouds may be drawn in place of sun. In case of partially sunny, the diagram of sun may be half masked with clouds.

Group Discussion and Decision Making

The observations recorded in the previous and current charts should be discussed among the farmers and apprentice trainees by raising questions relating to change in pest and defender population in relation to crop stages, soil conditions, weather factors, such as rainy, cloudy or sunny etc. The group may evolve a strategy based upon week AESA, ETL and corresponding change in pests, diseases and defenders ratio and take judicious decision for specific pest and disease management practices.

AESA By Extension Functionaries

The extension functionaries during the regular visit to the village should mobilize the farmers, conduct AESA and critically analyse the various factors such as the pest population *viz-a-viz* defender population and their role in natural suppression of the pest, the influence of prevailing weather condition/soil conditions on the likely build-up of defender/pest population. They may also take the decision based on the AESA, with IPM components like release of defenders, application of neem based formulations/safe pesticides are to be used for specific pest situation. Such an exercise may be repeated by the extension functionaries during every visit to the village and motivate the farmers to adopt AESA in the fields.

AESA by Farmers

After a brief exposure during IPM demonstrations/field trainings, farmers can practice AESA in their own fields. Wherever trained farmers are available, their experiences could be utilized in training their fellow farmers in their own villages. Thus a large group of farmers could be made efficiently competent in undertaking weekly AESA, thereby empowering themselves in decision making on any specific pest/disease situation. Farmer-to-farmer training approach will go a long way in practising IPM on a large area on sustainable basis.

PESTS OF NATIONAL SIGNIFICANCE IN TEMPERATE FRUITS

San Jose Scale (*Quadraspidiotus perniciosus* Comstock)

San Jose Scale is a highly destructive and polyphagous pest and is widely distributed in all the apple growing regions of the world. The scale has unsystematic distribution on the tree and forms a black or brown covering on its body. Underneath, a lemon yellow insect is visible when the covering is lifted (Singh, 1964). The scales cause damage by sucking cell sap from twigs, branches, fruits and are more destructive in nurseries and to young trees which become weak and ultimately die away. The scale has been found to be more on orchards planted at lower elevations than those at higher altitude. The pest is active from March to December and passes the winter in the nymphal stage (Atwal and Dhaliwal, 1997).

Management

- As the main source of infestation in the orchard in infested plant material from nursery, soil should be thoroughly treated with carbofuron @ 8g/m^2 before raising the nursery.
- Select healthy plants from nursery and then before planting in the orchard, treat them with chlorpyriphos 0.05 per cent.
- Spray during winter at later dormancy stage, of superior oils like Hindustan Petroleum Spray Oil or Indian Oil Spray oil at 2 per cent gave better control than mid or early dormant applications. The foliar sprays of oil can be given between end of February to early April coinciding with green tip stage and half inch green stages of growth with 2 litre of Agro emulsion or 4 litre HP Spray Oil or Servo orchard Spray Oil or Savo Spray Oil OEH or Agro Spray Orchard Oil or Ankur Tree Spray Oil or Loveen 3.6 by dissolving in 100 litre of water. On average, 5-6 litres spray solution is needed to cover full grown tree. Spray of oil is generally given on a clear sunny day with atmospheric temperature above 5°C (Kakar, 2000).
- Severe scale infestation under neglected orchard management requires additional summer sprays with thiamethoxam 0.5 per cent or imidacloprid 0.007 per cent or methyl demeton 0.05 per cent or carbosulfan 0.05 per cent.
- Heavily infested twings/branches should be prunned and burnt/destroyed.
- Release of *Aphytis proclia* and *Encarsia perniciosi* @ 2000 adults per tree during the second week of June and third week of July cause sizable scale reduction. Release of an indigenous predator *Chilocorus bijgus* @ 200 adults or 50 grubs per tree also reduced infestation of scale to a minimum level within a year (Singh and Mishra, 2003).

Woolly Apple Aphid (*Eriosoma lanigerum* Hausman)

In India, woolly apple aphid is found exclusively on apple and crab apple infesting aerial and subterean parts. The aphid lives in colonies on roots and aerial parts of the plant. It is seen as a white woolly mass on the aerial parts. Damage is caused by sucking of sap from stems, twigs and roots resulting in gall formation. Affected plants remain stunted with greatly reduced fruit bearing capacity.

Management

- Since infested nursery plants are main source for spread of the woolly aphid, soil should be thoroughly treated with carbofuron @ 8g/m^2before raising the nursery. Healthy plants should be selected from nursery and is treated with chlorpyriphos 0.05 per cent before planting in the orchard.
- Rootstock of M-21, M-25 and MM series, *Malus buccata* L. var. Himaliaca and crab apple are found tolerant against woolly aphid. Therefore, suitable rootstock of these groups should be utilized for vegetative propagation (Thakur and Dogra, 1980; Rai and Tripathi, 1984).
- During winter (January) all the aphids migrate from aerial parts to roots. This is the best period to control. During this period, spraying the roots by exposing the tree basin and aerial parts with thiamethoxam 0.05 per cent or imidacloprid 0.007 per cent or methyl demeton 0.05 per cent or carbosulfan 0.05 per cent.
- Release parasitoid, *Aphelinus mali* @ 1000 adults or mummies/infested tree once, as soon as infestation is noticed. The parasitoid is effective against aerial population of woolly aphid, it is more effective in vallies than on slopes.
- Infested twigs/branches should be prunned and burnt/destroyed.
- In areas where natural population of predatory Coccinellids is high during summer (April-June), application of toxic chemicals should as far as possible be avoided. Cover the cracks, crevices, wounds and cutends with protective paste.

European Red Mite (*Panonychus ulmi* Koch)

Both European Red mite, *Panonychus ulmi* Koch and the spotted mite, *Tetranychus urticae* are reported on apple in Himachal Pradesh. They feed on green matter and suck cell sap causing loss to chlorophyll. As a result, bronzing, leaf drop and weakening of buds takes place. Indiscriminate use of pesticides has been found to be an important factor responsible for outbreak of mite since some areas of Mashobra and Theog in Himachal Pradesh are still free from the attack. The mite causes maximum damage between May to September. There is gradual decline in population as monsoon advances (Khajuria *et al.*, 1999).

Management

- Cutting and burning of severely infested and dried shoots.
- The application of dormant sprays using miscible oils coinciding with green tip to tight cluster stage of apple. Plant with oil spray @ 3 litre of Hindustan Petroleum Spray Oil, 4 litre of IPOL orchard Spray Oil, 4 litre of Servo Orchard Spray Oil in 200 litre of water is effective against overwintering eggs. Mixing of Kanodane @ 500 ml improves the efficiency (Khajuria and Sharma, 1996).
- For effective control against mite, foliar application of Kanodane 20 EC @ 500ml at petal fall and colonol S 18.5EC or Nomite 18.5EC at fruit development have been found economical.
- Release of predatory mite, *Amblyseius tetranychivorus* @ 200 per plant has been found effective.

Table 1. Major Pests of Temperate Fruits in India Himalayas

Common Name	Scientific Name	Order	Family	Plant part damaged	Period of activity	Distribution	Host Range
San Jose scale	*Quadraspidiotus perniciosus* Comstock	Hemiptera	Diaspididae	Twigs, Branches, Fruits	March to December	All temperate fruits growing	Apple, Plum, Pear, Peach
Woolly Apple Aphid	*Eriosoma lanigerum* Hausmann	Hemiptera	Aphididae	Twigs, Branches, Roots	Throughout the year	All apple growing states	Apple, Carb apple
European Red Mite	*Panonychus ulmi* Koch	Acari	Eriophyidae	Leaves	May to September	Himachal Pradesh	Apple
Peach Leaf Curl Aphid	*Brachycaudus helichrysi* Kaltenbach	Hemiptera	Aphididae	Leaves	February to April	All Peach growing states	Peach, Apricot, Almond, Plum
Peach Fruit Fly	*Bactrocera zonata* Saunders	Diptera	Tephritidae	Fruits	June to August	All peach growing states	Peach, Plum, Apricot, Pear, Apple
Blossom Thrips	*Thrips flavus* Schrank	Thysanoptera	Thripidae	Flowers	February to April	All temperate fruits growing states	Apple, Pear, Peach, Plum, Cherry, Apricot
Codling Moth	*Cydia pomnella* Linnaeus	Lepidoptera	Tortricidae	Fruits	April to September	Temperate areas of Ladakh	Apple, Apricot, Pear, Walnut
Apple Root Borer	*Dorysthenes hugelii* Redtenbacher	Coleoptera	Cerambycidae	Roots	Throughout the year	All temperate fruits growing states	Apple, Aricot, Peach, Pear, Cherry, Walnut
Apple Stem Borer	*Apriona cinearea* Cheverlot	Coleoptera	Cerambycidae	Stems	March to October	All temperate fruits growing states	Apple, Peach

Peach Leaf Curl Aphid (*Brachycaudus helichrysi* Kaltenbach)

The peach leaf curl aphid is a highly destructive and polyphagous pest of peach, apricot, almond and plum in order of preference (Pruthi and Batra, 1960). The aphids suck the sap from young leaves due to which leaves curl up, remain smaller and turn pale. Consequently, poor set, premature drop and subnormal formation of fruit occur. Numerous light green aphids and nymphs can be seen inside the curled leaves. The pest is active from February to April on temperate fruits.

Management

- For effective management of aphid, removal of weeds from the orchard, which act as secondary host is essential.
- One pre bloom spray of thiamethorxam 0.05 per cent or imidacloprid 0.007 per cent or methyl demeton 0.05 per cent at pink bud stage (7-10 days before flowering) provides cent per cent control of peach leaf curl aphid. Post bloom spray should not be given to encourage biotic agents like Coccinellids, Syrphids and Anthocorid bug Which are very active in the month of March-April.
- Release of Green lace wing, *Chrysoperla carnea* @ 20 1st instar larvae per tree has been found effective.

Peach Fruit Fly (*Bactrocera zonata* Saunders)

The peach fruit fly is widely distributed and feeds on a number of fruits, viz., peach, plum, apricot, pear and apple. The damage is caused by the larvae only and the attack is characterized by the dark punctures, oozing of a fluid and rotting or dropping of fruits. The maggots found inside the fruits are dirty white, headless and legless. The attacked fruits are malformed, mis-shaped, undersized and are rotten within. The late varieties (mature in July-August) suffer more from the attack of this pest.

Management

- Early maturing varieties of peach like Flordasum, Alton, Red Haven, sharbati and Shan-i-Punjab should be grown in heavily affected areas as these varieties are mature before the attack of fruit fly.
- Hoe the orchard in May-June to expose the pupae which are present mostly at 4-6 cm depth in the soil.
- Apply bait spray of yeast hydrolysate (250g), crude sugar (2.5 Kg) and malathion 50EC (250ml) in 250 litres of water per ha. Spraying should be started two-three weeks before harvesting.

Blossom Thrips (*Thrips flavus* Schrank)

Thrips have been recorded as destructive pest from all the apple growing regions of the country. The infestation of this pest is associated with blossom of apple crop and its population increases with the advancement of flowering in spring. Both nymphs and adults feed on floral parts and delicate leaves of vegetative buds by sucking the sap therefrom, causing reduction in fruit set and yield. The set fruits may drop. Nymphs are elongate, yellow or white in colour and usually 1-2 mm long.

Management

- Digging the basins of the tree with spades during winter (December-January) to expose the diapausing pupae to adverse climatic conditions.
- Spraying of thiamethoxam 0.05 per cent or imidacloprid 0.007 per cent or carbosulfan 0.05 per cent at pink bud stage, before the opening of flower buds is very effective to control the thrips infestation.

Codling Moth (*Cydia pomonella* Linnaeus)

Codling moth is the most serious pest of apple confined to Ladakh region (Leh and Kargil District) of Jammu and Kashmir. This pest also attacks other fruits like pear, apricot and walnut. Aple is the most preferred host and damage may range as high as 95 per cent. All the varieties and cultivars of apple grown in Ladakh are attacked by this pest. Damage is caused by the larvae which burrow into the fruit and feed on the pulp. The infested fruits lose their shape and fall off prematurely. These fruits can not be marketed for human consumption.

Management

- All loose bark of apple trees should be scraped off to remove overwintering sites for the caterpillars. Vicinity of trees should also be kept clean of packing cases and all other debris which are likely to shelter the overwintering caterpillars during August to mid October.
- Bands of sacking (gunny bags) or corrugated cardboard about 150 mm to 240 mm wide can be tied round the tree trunks by late July till end of October to provide alternative over wintering sites for the caterpillars. These bands should be removed during the first week of November and either burnt or immersed in a pail of boiling water or kerosenised water.
- Fallen fruits should be collected throughout the season and either buried deep in the soil or destroyed otherwise.
- As far as possible the overwintering caterpillars should be collected from the loose bark, other debris, bands used for tapping and destroyed.
- Release of egg parasitoid, *Trichogramma embryophagum* @ 400 adults per tree per week should be undertaken from first fortnight of June to end of August.
- Apply two sprays – one in the 2nd half of June coinciding with hatching of 1st generation larvae i.e. after petal fall depending on climatic conditions and another two to three weeks later where regular infestation by codling moth occur depending on altitude and first emergence of codling moth. The best date for spraying varies considerably within a locality. The insecticides found effective are chlorpyriphos 0.05 per cent or malathion 0.05 per cent or thiamethoxam 0.05 per cent or imidacloprid 0.007 per cent.

Table- 2. Effective Natural Enemies of Major Pests of Temperate Fruits in India

Pest	Natural Enemy	Parasitoid/ Predator	Method of application
San Jose Scale	*Encarsia perniciosi* Tower	Parasitoid	2000 adults per infested tree
(*Quadraspidiotus*	*Aphytis* sp.(Proclia group)	Parasitoid	50 adults/Grubs per infested tree
perniciosus Comstock)	*Chilocorus bijugus*	Predator	50 adults/ Grubs per infested tree
Woolly Apple Aphid	*Aphelinus mali*	Parasitoid	1000 adults per infested tree
Eriosoma lanigerum	*Chrysoperla carnea*	Predator	50 adults/Grubs per infested tree
Hausmann)	*Chrysopa scelests*	Predator	50 adults/Grubs per infested tree
	Coccinella septumpunctata	Predator	30-50 Grubs/adults per infested tree
Codling Moth	*Trichogramma embryophagum*	Parasitoid	2000 adults per infested tree
Cydia pomnella Linn.)			at the interval of one week
European Red Mite	*Chrysoperla carnea*	Predator	50 adults/Grubs per infested tree
(*Panonychus ulmi*)			
Thrips (*Thrips flavus*)	*Anthocoris mincae*	Predator	30-50 Grubs per infested tree

Apple Root Borer (*Dorysthenes hygelii* Redtenbacher)

It is a common pest of apple, pear, apricot, peach and walnut. It prefers loam and sandy loam soils. The damage is caused by immature stages of the insect known as grubs who feed upon roots. They bore or girdle the roots as a result, plant become weak, branches start withering and finally plant dies. The infested trees become shaky and get easily uprooted with strong wind.

Management

- Collect and destroy grubs during December-January coinciding with the preparation of tree basins.
- Adult beetles can be collected from the light source (light trap) at night and killed in kerosenized water.
- Avoid dry sandy soils for establishing new Apply orchards as it provides congenial conditions for survival and multiplication of the pest.
- Treat the tree basin around the trunk with chlorpyriphos 0.05 per cent or monocrotophos 0.04 per cent in infested tree basin in December-January.

Apple Stem Borer (*Apriona cinearea* Cheverlot)

It is a long horned beetle found attacking apple and peach. The grubs bore into the branches and tree trunk below the bark and sometimes feed upon heartwood. The larva cuts 8-9 circular holes of the bark and tunnels through the ejects frass. The trunk in hollowed inside at several places. The affected trees become fragile and often break with moderate wind.

Management

- Clear the holes with flexible wire and then insert 0.5 gm of para-dichlorobenzene (PDCB) in the hole and plug it with puddling material or insert cotton wick soaked with patrol 1 ml per litre of water or dichlorovos 1.5 ml per litre water and plug the hole.
- Place small amount of crystals of PDB on the surface of the soil around the trunks of the tree during the leaf fall (September-October). The amount of PDB to be applied will vary with size of the tree. Apple the crystals in a ring completely encircling the trunk, neither closer to the bark than 1" nor at greater distance than 3". Cover the crystals with the earth to confine the PDB gas. Do not apply treatment during summer and late fall. The application is similar to peach tree borer.
- In case of localized infestation, swab the infested part with methyl parathion 0.2 per cent or spray with fenitrothion 0.05 per cent.
- Remove webbing and treat the main limbs and trunk of the attacked tree in March and September/October with 0.1 per cent methyl parathion and inject into the holes excavated by the caterpillars.

DISEASES OF NATIONAL SIGNIFICANCE IN TEMPERATE FRUITS

Premature Leaf fall (*Marssonina coronaria* Syn.)

The disease, which was first detected in 1995 in some orchards of Himachal Pradesh, has now spread to all apple orchards in the state and can cause up to ninety per cent of defoliation in unprotected orchards. It has also been reported recently from Kashmir, the neighbouring Uttaranchal and Bhutan (Sharma, 2003). Disease symptoms first appear as dark green circular patches on the upper surface of the mature whorl leaves, giving rise to 5-10 mm diameter brown leaf spots which turn dark brown in due course. When lesions are numerous they coalesce to form large dark brown blotches and the surrounding areas turn yellow. Severe leaf shedding follows these symptoms. (Verma and Bhardwaj, 2003). In affected orchards,

Table 3. Major Diseases of Temperate Fruits in Indian Himalayas

Common Name	Scientific Name	Plant part damaged	Period of activity	Distribution	Host range
Premature Leaf fall	*Marssonina coronaria* Syn.	Leaves	July to August	All apple growing states	Apple
White Root Rot	*Dematophora necatrix*	Roots	Round the year	All temperate fruits growing states	Apple, Pear
Collar Rot	*Phytophthora cactorum*	Collar region	Throughout of the tree	All temperate fruits the year	Apple, Pear, Peach, Apricot growing states
Apple Scab	*Venturia inaequalis* Cke.	Leaves & Fruits	April-August	All apple growing regions	Apple
Peach Leaf Curl	*Taphrina deformans*	Leaves, Flowers and Fruits	April-June	All temperate fruits growing states	Peach, Plum, Apricot, Almond, Cherry
Canker Diseases	*Botryosphaeria mali* *Coniothecium chomatosporum* *Corticium salmonicolor*	Trunk, Twigs, Branches	Round the year	All temperate fruits growing areas	Apple, Walnut, Almond, Peach, Apricot, Plum
Fruit Spots	*Gloedes pomigena* *Microthyriella rubi*	Fruits	June to August	All Apple growing states	Apple
Powdery Mildew	*Podosphaera leucotricha*	Shoots, Buds, Leaves	April to August	All Apple	Apple growing areas
Bacterial Gummosis	*Pseudomonas syringae*	Trunk, Branches, Fruits	June to August	All temperate fruits growing areas	Apple, Almond, Plum, Apricot, Cherry, Walnut

apples nearing maturity are commonly seen hanging from defoliated branches, and so the disease is according by called premature leaf-fall. Symptoms also appear on the fruit in the form of clear brown spots, which are initially circular (3 to 5 mm in diameter) and become oval, depressed and dark brown later on.

Management

The disease can be managed effectively by integrating different methods such as field sanitation, correct pruning and judicious use of fungicides.

- Collection and destruction of fallen leaves from the orchard floor during winter are found effective.
- Spraying with Urea 5 per cent on the leaf litter also helpful in reducing the primary inoculum by enhancing decomposition.
- Correct pruning allows adequate air circulation round the tree canopy, thereby modifyin the microclimate and reducing disease development.
- Spray schedule developed by Sharma, 2003 in Himachal Pradesh is :First spray with Dodine 0.1 per cent during fruit set (pea size), second with Mancozeb 0.2 per cent during fruit development (walnut size), third with Carbendazim 0.1 per cent, after 20 days of second spray and fourth spray with Mancozeb 0.2 per cent at 20-25 days before picking of the fruits. If level of disease incidence is low, number of spray can be reduced.

White Root Rot (*Dematophora necatrix*)

It is a widespread disease in the apple orchards and nurseries. The disease manifests itself underground and causes the roots to rot. The feeder roots are completely devoured and the infection spreads to bigger ones. In the advanced stages, bark as well as woody tissues decay. White mycelial growth of the pathogen is seen in soil attached to the roots during rainy season. Fifteen per cent disease incidence in orchards causing a loss of approximately Rs. 1.3 million (US $ 2,60,000) annually has been estimated. The disease is more prevalent in orchards located 1,800 to 2,400 m above mean sea level (Verma and Bhardwaj, 2003).

Management

- Before planting on infected sites, fumigation with 3 per cent formalin has been found effective.
- In the initial stage, chopping of roots followed by application of Chaubattia Paste or Bordeaux paint on roots is effective to check the disease incidence.
- The basin of the infected tree is brought to a moisture level of 30-40 per cent and 15-20 cm deep holes at a distance of 30 cm are dug with a crow bar all over the basin. These holes are closed after filling with 0.1 per cent suspension of carbendazim. The treatment must be repeated 3 to 4 times every year during the rainy season (Gupta and Jindal, 1990).
- In the initial stage of the disease, application of *Trichoderma viridae* @ 40 g per tree has shown good result (Gupta and Jindal, 1993).

Collar rot (*Phytophthora cactorum*)

Collar rot symptoms are often confused with white root rot because both the diseases occur in soil. In case of root rot, primary symptoms start from finer root and spread upwards up to collar region while in collar rot, primary infection is at collar region which spreads downwards to the roots. Infection shows in rough areas on the collar region of the tree and spreads mostly underground, and bark becomes slimy and cankerous. The wounds are irregular in outline but usually roughly oval, and extend rapidly, often girdling the tree. Susceptibility of the trees increases with age, and injury is a pre-requisite for the infection to occur. Fungi survive for a longer duration in orchard soils.

Management

- Approach grafting has been recommended for the control of collar rot. Raising the grafting point at least 20 cm above the ground level and improved drainage helps in reducing collar rot.
- Painting of wounds with copper paints and drenching the area around the trees with Mancozeb or Metalaxyl plus Mancozeb 0.3 per cent has proved quite effective (Rana and Gupta, 1981; Verma and Bhardwaj, 2003).
- Application of *Trichoderma viridae* @ 40 g per tree has been found to reduce growth of the disease at initial stage (Gupta and Garg, 1988)
- Apple rootstocks, M2, M4, MM105, M113 and MM114 have been found resistant under field conditions through production of disease in a natural environment (Gupta and Mir, 1983a; Gupta and Jindal, 1993).

Table 4. Biocontrol agents antagonistic to plant pathogens attacking temperate fruit crops in India

Crop/Disease	Antagonist	Plant parts where antagonist applied	Pathogen suppressed
Apple			
Grey mould	*T. harzianum*	Flowers	*Botryis cinerea*
Grey mould	*T. pseudokoningli*	Flowers	*Boiryis cinerea*
Crown gall	*Agrobacerium radiobacter* Strain K-84	Seed and root	*Agrobacterium tumefaciens*
Scab	*Athelia bonbacina*	Fallen leaves	*Venturia inaequalis*
Collar-rot	*Enterobacter aerogenes, Bacillus subtilis*	Root	*Phytophthora coctorum*
Collar-rot	*Trichoderma* spp. and *Gliocladium* spp.	Root	*Phytophthora coctorum*
Cankers	*Bacillus subtilis*	Leaf scar, wounds	*Nectria galligena*
Cankers	*Cladosporium cladosporoides*	Pruning cuts	*Nectria galligena*
White root-rot	*Enterobacter aerogenes*	Root	*Dematophora necatrix*
White root-rot	*T. viride and T. harzianum*	Root	*D. necatrix*
Silver leaf	*T. viride*	Tree wounds	*Chondrostereum purpureum*
Powdery mildew	*Ampelomyces quisqualis*	Leaves	*Podosphaera leucotricha*
White root-rot	*Glomus* spp.	Root	*D. necatrix*
Almond			
Fruit rot	*Sporobolomyces* spp.	Fruits	*Aspergillus flavus*
Apricot			
Cankers	*Fusarium lateritcum*	Pruning cuts	*Eutypa armeniacae*
Peach			
Brown rot	*Pseudomonas frequentans*	Twigs	*Monilinia laxa*
Cankers	*Epicoccum purpurascens*	Pruning cuts	*Necteria galligina*
Pear			
Fire blight	*Erwinia* spp.	Blossom	*Erwinia amylobora*
Fire blight	*Pseudomonas* spp.	Blossom	*Erwinia amylobora*
Strawberry			
Grey mould	*Cladosporium* spp.	Flowers	*Botrytis cinerea*
Grey mould	*Trichoderma* spp.	Flowers	*Botrytis cinerea*
Web blight	*T. harzianum*	Foliage	*Rhizoctonia solani*

Apple Scab (*Venturia inaequalis* Cke.)

The most striking symptoms of scab are commonly observed on leaves and fruits and rarely on 1-3 years old shoots. In severe cases, petioles, pedicels and blossoms also exhibit scab symptoms. Long wet periods (due to monsoon rains) nearing fruit harvest often help in the development of scab spots even on the mature apples. Fruits may show small, rough, black, circular lesions on their skin while on the tree or after keeping in cold storage. The lesions on the fruit epicarp enlarge and areas around them depress leading to fruit shrinkage. The affected fruits rot due to secondary infection of the lesions (Gupta, 1985; Gupta and Verma, 1985)

Management

- The removal of fallen leaves from the orchard floor helps to reduce the primary inoculum of apple scab.
- Potential of severe scab can be minimized by a spray of 5 per cent urea in autumn, shortly before leaf fall, accelerates the decomposition of fallen leaves. The sprays must be applied in all the adjoining orchards to minimize over wintering leaves and the inoculum.
- A protective programme of six sprays of nonsystemic (Mancozeb, Dodine, Dithianon and captan) and systemic fungicides (Carbendazim, Benomyl, Ergosterol biosynthesis inhibitors i.e., EBI) is recommended during the season from bud swell to pre-harvesting stage to control apple scab. Based on the prediction of infection periods certain fungicides, especially the ergosterol biosynthesis inhibitors (Bitertanol, Myclobutanil, hexaconazole, Penconazole and Fenarimol), are recommended which cure as well as eradicate the disease. All these prevent infection if sprayed promptly, do not allow scab lesions to increase if applied after infection, and also prevent sporulation in the developing lesions (Gupta, 1984; Gupta and Jindal, 1993; Verma and Bhardwaj, 2003).

Peach Leaf Curl (*Taphrina deformans*)

Symptoms of leaf curl are confined to current year's growth of leaves, twigs, blossoms and fruits. The leaf blade along the midrib becomes thickened and puckered. Infected twigs become swollen and are killed. Infected flowers and fruits drop rapidly. Peach is the main host but symptoms also appear on other fruits like plum, apricot, almond and cherry.

Management

- The removal of fallen leaves from the orchard floor helps to reduce the primary inoculum of the disease.
- Single spray of Carbendazim or Blitox-50 WP at 0.1 to 0.2 per cent before the break of dormancy (swelling of buds) is highly effective in controlling the disease (Sharma *et al.*, 1988a). The disease can not be controlled once the pathogen has entered the host tissue in the spring.

Canker Diseases

(i) Stem Brown Canker (*Botryosphaeria mali*)

(ii) Stem Black or Smoky blight Canker (*Coniothecium chomatosporum*)

(iii) Pink Canker (*Corticium salmonicolor*)

The cankers cause losses only in marginal apple belts where the chilling requirement is not fully met. Other factors affecting the prevalence of these fungi are tree age, duration of summer prevailing temperature, site and elevation of the orchard. (Sharma and Bhardwaj, 1999). The stem brown canker appears in two phases, i.e. stem canker and fruit rot. The former usually develops through sun burnt or wounded surfaces of the main trunk and branches as sunken dull black lesions and infection goes deep into the wood. Cankered bark turns black and brittle but remains attached to the wood. In stem black canker, reddish brown sunken lesions develop on trunk and branches which turn smoky with a series of alternate rings. The wood below is stained reddish brown. Spurs, branches and twigs above the canker are killed. Numerous pimple-like protuberanaces may appear over the bark of the blighted twigs or along the margins of the canker. Pink canker appears on trunk and twigs and causes canker, blight and die-back symptoms. Usually the lesions are sunken, dull brown in appearance with cob web like growth. The bark gets brown by numerous small protuberanaces consisting of white or buff coloured fungal structures which expand and form star shaped opening to expose the pastules (Gupta and Sharma, 1988; Gupta and Jindal, 1993).

Management

- Canker diseases can be reduced by destroying infected branches and removing them from the orchards where they would otherwise act as a source of inoculum.
- Correct training and pruning at the right time of year helps in reducing apple cankers considerably. In south or southwestern exposed sites, the trunk height must be kept short by allowing the first scaffold branches to grow at thirty cm above the ground, to protect the main trunk from direct sun. Heading back branches of more than 2.5cm diameter should be done as close to the main trunk as possible without creating a stub. The cut should be clean, slanting and sharp. All the pruning cuts should be covered with Chaubattia Paste or Bordeaux or Blitox-50 WP.
- Trees badly infected with canker should be sprayed after fruit harvest with 0.3 per cent copper oxychloride or 0.3 per cent captan. To treat existing canker wounds, first scarify, then wash with five per cent formaldehyde or rectified spirit and apply cow dung paint or Chaubattia or Bordeaux paint afterwards (Tamura, 1984; Verma and Bhardwaj, 2003).

Fruit Spots

(i) Sooty blotch (*Gloedes pomigena*)

(ii) Fly Speck (*Microthyriella rubi*)

Sooty blotch and fly speck appear with the onset of rains during July-August with more incidence in shady and foggy areas. Sooty blemishes and fly speck are circular, black dots. The presence of these spots impart dull colour to the fruits reducing their market value.

Management

- Since the spots are superficial, these can be removed by vigorous rubbing.
- Washing of the fruits with bleaching powder to remove the blemishes is also suggested (Verma *et.al.,* 1986).
- More effective control has become possible by using fungicides like Dithane – M-45 or captan or carbendazim 0.1 to 0.2 per cent. Normally the fungicides sprayed against apple scab take care of these diseases also but the last spray should be applied 15-20 days before the fruit harvest (Gupta and Jindal, 1993).

Powdery Mildew (*Podosphaera leucotricha*)

Powdery mildew is prevalent in all apple growing areas and causes the death of vegetative shoots and flower buds and die back. Small greyish or white patches of mycelium are seen on the underside of the leaves which soon become crinkled, curled, hard, brittle and are frequently killed.

Management

- Fertilizers, particularly nitrogen, affect the incidence of powdery mildew. Hence, balanced fertilizers should be applied on the basis of leaf and soil analysis (Kumar *et al.,*1987).

- Application of fungicides including karathane, carbendazim, Benomyl and Saprol at 0.1 to 0.2 per cent have been suggested for powdery mildew control by Gupta and Sharma, 1981b; Gupta and Jindal, 1993).
- In addition to the spray during the active season, dormant sprays of 0.1 per cent carbendazim alongwith a spray at bud swell stage have also been found quite effective in reducing the disease development (Gupta *et al.,* 1977; Gupta and Jindal, 1993).

Bacterial Gummosis (*Pseudomonas syringae*)

The disease is characterized by the presence of profuse oozing of gum on the surfaces of affected wood, back of the trunk and also on larger branches but more common on cracked branches and fruits. In severe cases, droplets of gum trickle down on stem and the bark turns dark brown with longitudinal cracks. Late varieties of peach and almond (mature in July-August) are affected severely with the incidence of this diseases. In case of severe infection inpeach and almond, almost all the fruits are get infected and lose their market value.

Management

- In Gummosis affected areas, early varieties of peach and almond should be planted.
- The diseased bark/portion should be removed, cleaned and covered with Chaubattia Paste or Bordeaux Paste or copper oxychloride paste.
- Application of copper oxychloride 0.3 per cent plus Borax 0.3 per cent should be done one month after fruit setting. Second spray can be done after 15-20 days of first spray.

CONCLUSION:

The temperate fruits are prone to attack by pests and diseases which are the major limiting factors in the successful production of temperate fruits. Estimates of yield losses caused by pests and diseases attack range from 10 to 30 per cent in various crops during different seasons of the year. Unlike agricultural crops, temperate fruit crops which are grown as monocultures, the pest and disease problems are entirely different and complex in nature. Such pest and disease situations have led to repeated and excessive use of chemical pesticides. This has resulted in development of resistance in the pest species, contamination of fruits, environmental pollution as well as rejection of export produce. The global concern regarding the over and misuse of chemical pesticides has resulted in adoption of Integrated Pest Management for containing the pest-disease problems. It is an ecofriendly approach encompassing cultural, mechanical, biological and need based use of chemical pesticides preferably in combination with botanicals and bio-pesticides. Agenda 21 of the United Nations Conference on Environment and Development (UNCED) at Riode Janerio in June, 1992 identified IPM in agriculture as one of the requirements for promoting sustainable agriculture/horticulture and rural development. Integrated Pest Management requires thorough preplanning even before establishing the orchard which includes selection of suitable site, soil type, improved and pest/disease free scion varieties, resistant rootstock compatible with scion, regular pest/disease and weather monitoring in the orchard, balanced used of fertilizers, water management, proper training and pruning, improved cultural practices, augmentation and conservation of parasitoids and predators, application of bio-pesticides and neem based formulation and need based use of selective pesticides. With respect to promote greater IPM implementation, special emphasis is necessary to generate increased awareness and transfer of available realistic IPM programmes. The strategic steps will include determination of farmers pest management knowledge, attitude and practices, IPM trial demonstrations and appropriate training of the extension personnel and farmers.

REFERENCES

Atwal, A.S. and Dhaliwal, G.S. 1997. Pest of Temperate Fruits. In: *Agricultural Pest of South Asia and their management.* Kalyani Publishers, New Delhi, pp. 274 286.

Bhalla, O.P and Gupta, P.R. 1993. Insect - Pests of temperate fruits. In: K.L. Chadha and O.P. Pareck (eds.), *Advances in Horticulture - Fruit Crops,* Vol.3 Malhotra Publishing House, New Delhi, PP. 1557-1589.

Gupta. G.K. 1984. Economics of apple scab control programme in Himachla Pradesh. *Pesticides,* 18 (6): 31-32.

Gupta, G.K. 1985 Recent trendes in forecasting and control of apple scab *(Venturia inaequalis) Pesticides,* 19 (1): 19-31.

Gupta, G.K. and Sharma, S.K. 1981a. Influence of preharvest fungicidal spray on sooty blotch, fly speck and storage rots of apple. *Pesticides,* 15 (2): 17-18.

Gupta, G.K. and Verma, K.D. 1985. Post symptom antisporulant activity of ftmgicides on apple scab foliage lesions. *Indian J Agric.,* 55: 3 81-3 82.

Gupta, P.R. 1999. Retrospect and Prospect of Biological Control of insect-Pests of Temperate Fruits in India. In Proc. National Seminar on Himalyan Horticulture. Jagmohan Singh, R.P. Sharma and V.K. Sharma (eds.) *Hortivision, PP.* 298-307. Gupta, V.K. and Garg, R.C. 1988. Effect of fungicides and antagonistic bacteria on *Phytophthora cactorum* and seedling rot of apple. *5h Int. Cong. Plant. PathoL,* Kyoto, August 1986 (Abster.) pp. 180.

Gupta, V.K. and Jindal, K.K. 1990. Management of white root rot of apple by biological and cultural methods. *23d Int Hortic. Cong., Florence,* Italy, 1990, Abstr. Pp. 3238.

Gupta, V.K. and Jindal, K.K. 1993. Fungal, Diseases of Temperate Fruits. In: K.L. Chadha and O.P. Pareek (eds.), *Advances in Horticulture - Fruit Crops,* Vol. 3. Malhotra Publishing House, New Delhi, pp. 1349-1372.

Gupta. V.K. and Mir, N.M. 1983a. Field testing of apple rootstocks against *Phytophthora cactorum* - A new technique. J *Trec. Sci.,* 2: 81-83.

Gupta, V.K. and Sharma, S.K. 1988. Diseases of apple and their management. In *Trec Protection (* V.K. Gupta and N.K. Sharma, Eds.). Indian Society of Tree Scientists, Solan, pp. 7-25.

Gupta, V.K., venna, K.D. and Munjal, R.L. 1977. Influence of fungicides on the development of powdery mildew in susceptible cultivars of *apple. Indian Phytopath.,* 30: 245-250.

Kakar, K.L. 2000. Pest status of Temperate Fruits and their Management. In: P.R. Gupta and K.K. Jindal (eds.), *Horticulture Technology:* Vision - 2000 and Beyond - 2 pp. 166-178.

Khajuria, D.R. and Sharma, J.P. 1996. Outbreak of Phytophagous mites on apple in Kullu Valley. Indian J. Plant Protec. 24 (1&2): 134-138.

Khajuria, D.R., Sharma, J.P. and Thakur, J.R. 1999. Integrated Mite Management in apple orchards: Present Scenario and Future strategy. *In Proc. National Se.,ninar on Himalayan Horticulture,* Jaginohan Singh, R.P. Sharma and V.K. Sharma (eds.), *Hortivision,* pp. 319 - 322.

Kumar, R., Srivastava, K.K. and Pandey, J.C. 1987. Economic spray schedule for apple powdery mildew control. *Int. J. Trop. Plant Dis.,* 5: 5 9 5.

Pratap, Tej. 2003. Horticulture in the Hills: Issues and Challenges. *In:* K.K. Jindal, Ranjan Bawa and Tej Pratap (eds.), *Apple Farming and Livelihoods in the Himalayas.* Bishen Singh Mahendra Pal Singh Press, Dehradun, pp. 17 - 28.

Pruthi., H.S. and Batra, H.N. 1996. *Important Fruit Pests of North-West India. Bull. 80,* ICAR, New Delhi, pp. 66.

Rai, K.M. and Tripathi, G.M. 1984. Resistance Of various rootstocks to the attack of apple woolly ah\phid, *Eriosoma lanigerum (Hausmam). Prog. Hort.,* 16: 359-62.

Rajak., R.L. 1999. Plant Protection Measures in Horticultural crops, Present Scenario and Future Needs. *In Proc. National Seminar on HimalaYan Horticulture.* Jagmohan Singh, R.P. Sharma and V.K. Sharma (eds.) *Hortivision,* pp. 285-292.

Rana, K.S. and Gupta, V.K. 1981. Cause and control of collar rot of apple in India. *Indian Phytopath.* 34: 17-19.

Sharma, I.M. and Bhardwaj, S.S. 1999. Canker Disease Management in apple: Present status and Future strategies. *In Proc. National Seminar on Himalayan Horticulture,* Jagmohan Singh, R.P. Shan-na and V.K. Sharma (eds.) *Hortivision,* pp. 323-329.

Sharma, J.N. 2003. Managing Premature Leaf fall in apple. In: K.K. Jindal, Rajan Bawa and Tej Pratap (eds.), *Apple Farming and Livelihoods in the Himalayas, Bishan* Singh Mahendra Pal Singh Press, Dehradun, pp. 289-292.

Sharma, R.C., Gupta, V.K. and Garg, R.C. 1988a. Dormant application of ftmgicides for the management of peach leaf curl. *Indian Phytopath,* 41:259-60.

Singh, C. 1964. Temperate Fruit Pests. In Entomology. In: E.S.1, New Delhi. pp. 266.

Singh, S.P. 2002. Role of Biocontrol Agents in Pest Management in the Millennium. In: D. Prasad and S.N. Puri (eds.), *Crop Pest and Disease Management-Challenges for the Millennium,* Jyoti Publishers, New Delhi, pp. 1-17.

Singh, S.S. and Lal, S.D. 2002. Integrated Pest Management in Temperate Fruits. In: D. Prasad and S.N. Puri (eds.), *Crop Pest and Disease Management-Challenges for the Millennium,* Jyoti Publishers, New Delhi, pp. 39-50.

Singh, S.S. and Mishra, P.N. 2003. Sitoshna Phalo me Ekikrit kit Prabandhan. *Phal Phool,* 26 (2): 17-19.

Smith, R.F. 1970. Pesticides: Their use and limitations in pest management. In: R.L. Rabb and F.E. Guthrie (eds.) *Concepts ofpest Management* North Carolina State University, Raleigh, North Carolina. pp. 103-113.

Solanki, A. 1999. Diversifying Mountain Horticulture. *Agriculture Today,* 2(5): 51-53.

Tamura, 0. 1984. Studies on the epidemiology and control of Japanese apple canker caused by *Valsa ceratosperma. Rep. Hokkaido Prefectural Agric. Exp. Sta: No.* 49. pp 81.

Thakur, J.R. and Dogra, G.S. 1980. Woolly apple aphid *Eriosoma lanigerum,* research in India. *Trop. Pest Management,* 26: 8 -12.

Thakur, V.S. 1999. Temperate Disease and Pest Management Through Weather Monitoring and Forecasting systems Horti - Pestvision 2020. *In: Proc. National Seminar on Himalayan Horticulture.* Jagmohan Singh, R.P. Sharma and V.K. Sharma (eds.) *Hortivision.* pp. 346-356.

Verina, K.D. and Bhardwaj, L.N. 2003. Apple Disease-Based Low Productivity in the Himalays. In: K.K. Jindal, Rajan Bawa and Tej Pratap (eds.) *Apple Farming and Livelihoods in the Himalayas.* Bishan Singh Mahendra Pal Singh Press, Dehradun, pp. 299-306.

Verma, K.D., Gupta, V.K. and Verma, H.S. 1986. Role of ftmgicides in the control of fruit spots and storage rots of apple. *In Advances in Research on Temperate Fruits,* (T.R. Chadha, V.P. Bhutani and J.L. Kaul, eds.). Dr. Y.S. Parmar University of Horticulture and Forestry, Solan. pp. 411-414.

Apple trunk infested with woolly aphid

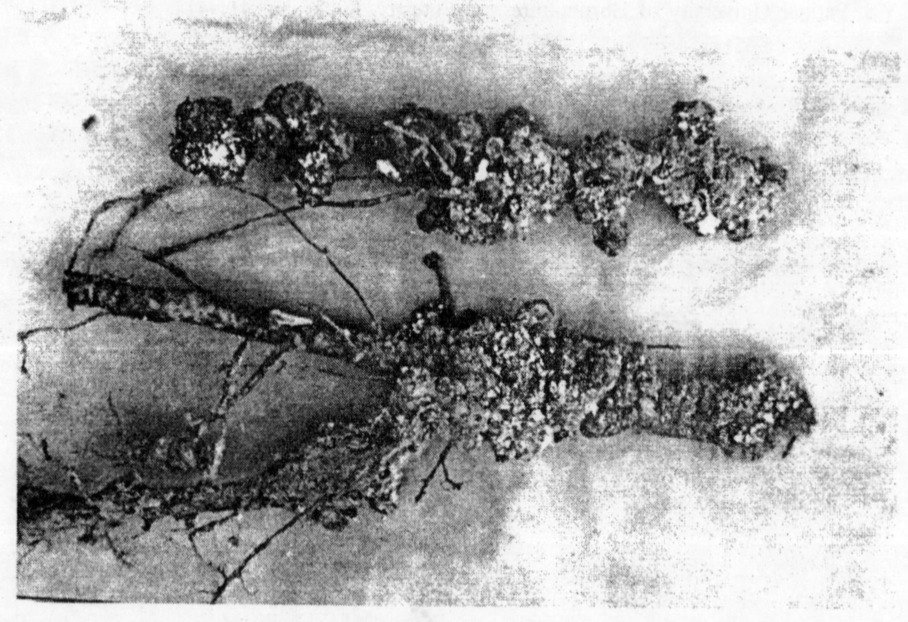

Incidence of woolly aphid on Apple roots

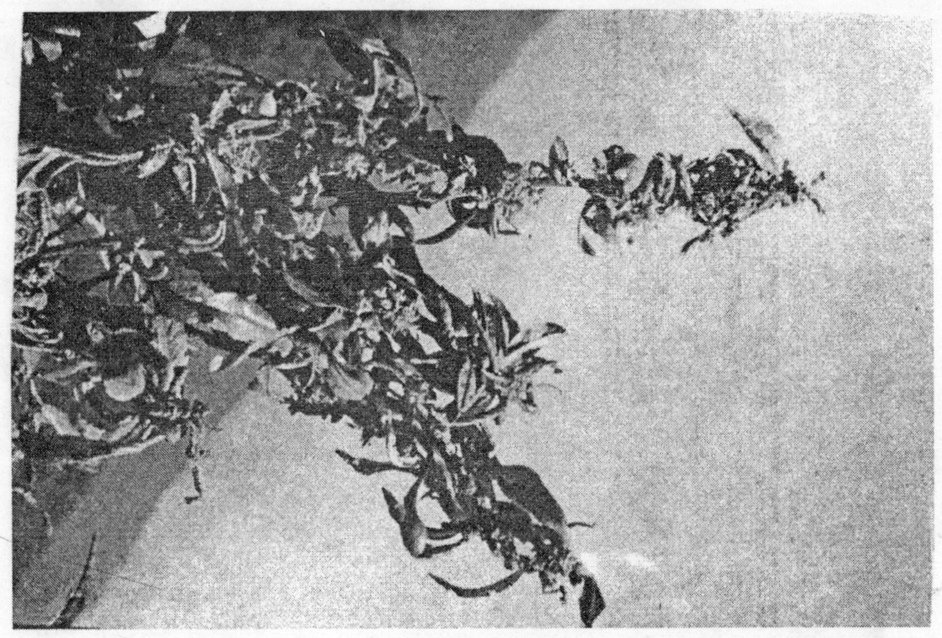

Infestation of leaf curl aphid on peach

Proper pruning at dormant stage (December-January) greatly helps in minimising the incidence of pests and diseases

Apple fruits infested with scab disease

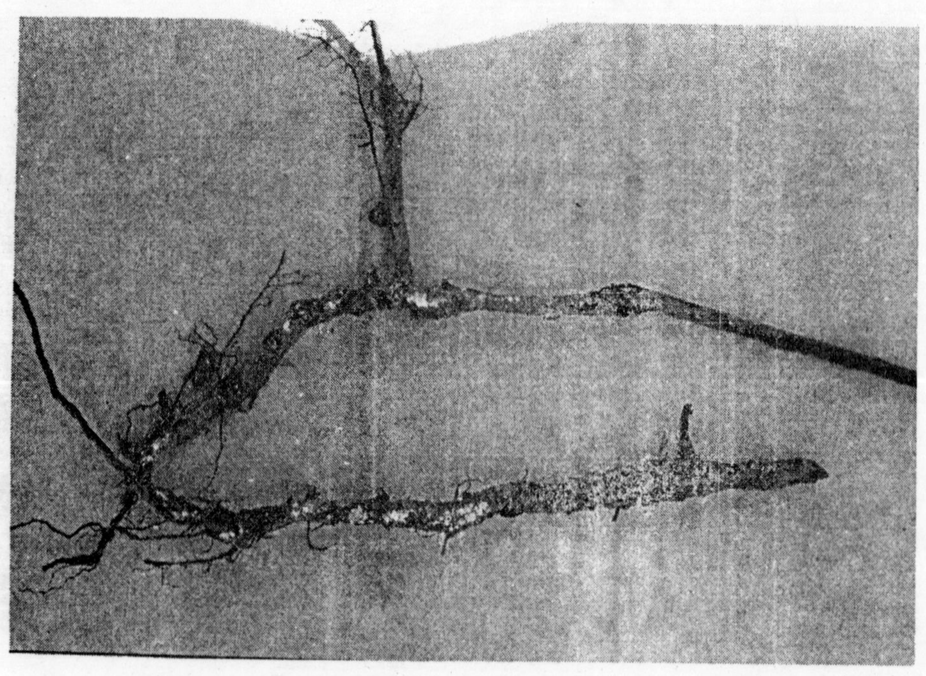

Infection of white root rot on apple

Apple trunk infected by black canrers disease

Infection of pink canker on apple

24

Integrated Management of Insect Pests, Fungal and Viral Diseases of Banana

B. PADMANABAN[1], R. SELVARAJAN[1], R. THANGAVELU[1] AND S. SATHIAMOORTHY[1]

India is the largest producer of banana and plantains in the world. Out of the 40 million tonnes of fruits produced in India, banana occupies number one position in India with an annual output of 14.0 m.t. from an area of 4,00,000 ha. Around 19 pests have been found associated with banana in India. Insect attack is noticed from planting to harvest. Among the insect pests, banana weevils, aphids and fruit scarring beetles are considered as of national importance and others of minor and regional significance. Among the fungal diseases, fusarium wilt , leaf spot and erwinia rot are important. Banana bract mosaic virus, banana streak virus and banana bunchy top virus are the important viral diseases causing considerable economic loss.

INSECT PESTS:

Banana corm weevil, *Cosmopolites sordidus* Germar

Banana corm weevil, *Cosmopolites sordidus* has been recorded from various parts of the country *viz.* Assam, Andhra Pradesh, Kerala, Tamil Nadu, Karnataka and Maharashtra. The grub burrow into the corm and corm gets riddled with holes. In the early stage of infestation one can notice yellow lines on the green leaf and in due course reduction in leaf size, where as in the advanced stage corm gets extensive patches of tunneling. Depending on the stage of crop and infestation, yield loss up to 85 per cent has been reported. Most of the commercial cultivars like Malbhog, Robusta, Kanthali, Thellachackrakeli and Nendran are susceptible to this pest. Eggs are laid singly in the crown region of the corm. Incubation period is 7 - 14 days. Larval stage lasts for two to six weeks. Pupation is completed in a week. The spread of weevil is manifested through suckers, hence planting material should be collected only from healthy garden.

Integrated Pest Management: Weed free cultivation is essential to avoid the spread of weevil. Remove the old and dried leaves. After harvest remove the pseudostem at ground level and treat it with carbaryl (2g/lit.), to avoid egg laying on the left over. Before planting treat the suckers with monocrotophos solution (14 ml in litre of water) for 30 minutes. Paring and pralinage to remove the adult weevils and grubs from the corm. Apply carbofuran 3G (40g/plant) before planting. Collect adult weevils by using banana pseudostem traps both (i) Longitudinal split trap (30cm length) (ii) Disc-on stump trap @100/acre. It is important to keep the split portion facing the ground, collect the trapped weevils once in a week and destroy. Install pheromone trap @ 25 traps/ha collect the trapped weevils and destroy. Otherwise install 4 traps/ha and each month move the tap 20 meters along the axis perpendicular to the trap line.

1. National Research Centre for Banana, Thayanur (Post), Tiruchirapalli - 620 102. T. N.

Small banana weevil, *Polytus mellerborgii* Boheman

A small weevil infests on rhizome as well as in the leaf sheath of leading commercial cultivars *viz.*, Nendran and Poovan Pseudostem trapping controls the pest.

Banana stem weevil, *Odoiporus longicollis* (Oliver)

Banana stem weevil, *Odoiporus longicollis* has been reported from Assam, West Bengal, Bihar, Eastern Uttar Pradesh, Kerala, Karnataka, Andhra Pradesh and North-Eastern hills region. Occurrence of banana weevil borers from non-traditional areas of Tamil Nadu has been reported by NRCB in 1998. The early stage symptoms include exudation of jelly like substance from pseudostem ,where as in the advanced stage of infestation plant show extensive patches of tunnelling, weak appearance, reduction in leaf and bunch size. The pest breeds throughout the year but it is more active during summer and monsoon season. Adult female lay eggs inside the leaf sheath. Incubation of larval, and pupal stage lasts about 6.7, 25.5, and 17.5 days respectively. Infestation occurs on five months pseudostem circumference of 25 – 50 cm are preferred by the weevil for egg laying. Susceptible cultivars include Nendran, Karpuravalli, Red banana, Rasthali and Robusta whereas Neypoovan and Poovan found less susceptible to weevil (Padmanaban and Sathiamoorthy, 2001, Padmanaban *et al.,* 2001).

Integrated pest management: Weed free cultivation helps in reducing the spread of infestation. Remove the old and dried leaves. After harvest, cut the plant from ground level and treat it with insecticide, Carbaryl (2g/l) to kill the egg laying weevils. Collect adult weevils by using banana pseudostem traps (i) Longitudinal split (30cm length), (ii) Disc on stump trap @40/acre. Keep the split portion facing the ground. Collect the trapped weevils and kill. Inject 4ml of solution (Monocrotophos 150ml in 350ml of water) using stem injector at 2 and 4 feet above the ground level at 30° angle. Injection can be given from fifth month to flowering stage, depending on the need.

Banana flea beetle, *Nodostoma subcostatum* Jac.

Banana flea beetle, *Nodostoma subcostatum* Jac. is a major insect pest infesting almost all banana cultivars grown in Bihar, West Bengal and North Eastern India viz. Jahaji, Borjahaji, Honda, Amritasagar, Manohar, Tulsi manohar and Dwarf Cavendish. Insect lives in the heart of the pseudostem and feeds on young leaves, and skin of young fruits. Severe scrapping of chlorophyll content of leaves and fruit peel leads to under developed fruit and inferior quality bunches, which loose the marketing quality, damage is noticed during may. Highest infestation was noticed in dwarf Cavendish and lowest in Bhimkol.

Management: Cover the bunch with polythene bag during shooting stage. Spray thiodan 35 EC (1.8 ml/lit) on the heart of the pseudostem within the roll of the central leaf as well as on the fingers.

Banana Leaf eating caterpillar, *Spodoptera litura* Fab.

The adult moth lay egg mass on the leaf and cover it with scales. Larvae after emergence congregate on the leaf and feed the leaf. When larvae grow, they move into different leaves and feeds on tender leaves. The larvae make holes whereas on half mature leaves, feed on surface of the leaf. Larvae make damage during dawn and dusk. All the commercial cultivars are found infested by this pest. Young plants of 3 – 6 months are highly affected by this pest.

Management: Collect egg mass, early larval stage and destroy. Collect and kill the male moth using pheromone trap @ 2 – 3 traps/acre at 3 mt. height. Spray thiodan 35EC (1.5ml/l of water) on the leaves. Bio control agents such as *Bacillus thuringiensis (Bt.)* (1 gm in 1 lit of water), Nuclear Polyhedrosis Virus (NPV) ($250x6^{10}$ ml) can be sprayed over the plant in the evening hours.

Banana aphid, *Pentalonia nigronervosa* Coq.

It is prevalent in Tamil Nadu, Karnataka, Kerala, Andhra Pradesh, Gujarat, Maharashtra and West Bengal. The following cultivars viz. Nendran, Poovan, Ney poovan and Virupakshi were found infested with the nymphs and adults. Thè nymphs and adults congregate under the outer base of the pseudostem. Brown shiny wingless aphids can be seen on the emerging out suckers. Aphids always accompanied by ants, which act as dispersing agents of nymph. The aphids suck the plants sap and reduces the growth and vigour. They also acts as a vector of Banana Bunchy Top Virus (BBTV).

Management: The virus affected plant should be rouged out. Ratoon and intercrops should not be taken up. Collect planting material from healthy plants. To control aphids spray Dimethoate 0.05% (1.8 ml/lit of water).

Banana finger thrips:

Rust thrips, *Chaetanaphothrips signipennis* Bagn.

Nymphs and adult feed on the fruit which develop as, rust coloured blemishes during fruit development. The rust blemish roughens and later cracks are developed. Infestation was very severe during the months of March to July. Cultivars such as Monthan, Saba, Rasthali and poovan are severely affected by this pest.

Management: To prevent thrips infestation cover the bunche with plolythene bags impregnated with chlorpyrifos (2.5ml/l) and soil application of phorate (30g per plant) is required to kill the thrips infesting on the leaf as well as the pupating thrips in the soil. The bagging should be done immediately after shooting of the bunch.

Flower thrips, *Thrips hawaiiensis* (Morgan)

Adults and nymphs infests the flower tips and fruits before the bud emerges from the plant and upto two weeks after emergence. Due to sucking by the larval and adults, the fruit develops elevated black spots. Poovan, Nendran and Karpooravalli are susceptible to this pest.

Management: Bagging of emerging bunch with polythene sleeves having 10 per cent ventilation prevents the bunch from infestation.

Banana silvering thrips, *Hercinothrips bicinctus* Bagn.

Nymphs and adult insects feed on the fruit peel and due to feeding and salivation on the fruit peel, gives silvery appearance. Thrips infestation is noticed on cultivars such as Robusta, Monthan and Pachanadan.

Management: Bunch covering can help to tackle the problem. Spraying Chlorpyriphos (2.5 ml/lit) over the bunch immediately after the emergence of all hands, controls the pest.

Leaf thrips, *Helionothrips kadaliphilu s* Ramk.and Marg.

The insect found infesting on the lower side of banana leaves. As a result of sucking by adult and nymphs, the lower side of leaf show white patches on the veins, continuous feeding leads to leaf drying.

Management: The severely infested leaves can be removed. Spray thiodan 35EC (1.5ml/lit.) towards the lower side of the leaves to control the thrips.

FUNGAL DISEASES :

Fusarium wilt / Panama disease (*Fusarium oxysproum* f. sp. *cubense*)

Fusarium wilt is the most destructive disease occurring in many banana growing regions of world. It was first recorded from southern Queensland of Australia in 1874. The fusarium wilt was recorded from India in 1911 and now the disease has been reported from Asia, Africa, South Pacific and Latin America. During 1940's and 50's the disease had been causing more damage in the western tropics particularly in central America and the Caribbean islands where it destroyed the Gros Michael banana plantations which was saved by growing Cavendish clone and later, the cavendish clone which was considered as resistant became susceptible because of the development of race 4 of the pathogen especially in the subtropics. However, this Cavendish group remains resistant to fusarium wilt (race 4) in the tropical regions. Fortunately in India so far the race 4 has not been reported.

Symptoms: The symptoms of the disease are expressed both internally and externally and the external symptoms normally can be seen only 4 -5 months after planting. However in the case of diseased suckers planted in the soil the symptoms are seen even two months after planting.

On leaves:

1) The symptoms on the leaves are characterized by appearance of yellowing at the leaf margins of lower leaves.
2) Then the yellowing progresses towards the mid rib and the whole leaf become yellowish and the yellowing spreads to upper leaves also.
3) The infected leaves gradually collapse at the petiole or towards the base of the midrib and hand down around the pseudostem and this gives 'Skirt" like appearance to the plant.
4) The top most leaves are the last to show the symptoms and often stand erect giving 'spiky' appearance to the plant.
5) The emerging leaves appear pale with reduced leaf lamina and finally the emergence of the leaf will be stopped.

On pseudostem:

1) Longitudinal splitting occurs in the pseudostem exposing the discolored vascular tissues (yellow or black or brown) in the sheath. The discolorations of vascular tissues can be seen even in the bunch stalk. In advanced stage the infected peudostem emit the smell of rotten fish.
2) Emergence of large number of side suckers also occurs before the death of infected plants.

On corm:

When the infected corm is cut crosswise, yellow, brown or black colored vascular discoloration can be seen. Generally no bunches are produced and if produced the fruits are small and only few fingers develop. Fruits ripen irregularly and flesh is pithy and acidic.

Pathogen

The causal organism produces micro and macro conidia and chlamydospores. Though this is primarily a pathogen of *Musa* sp. it can also cause wilt on *Heliconia.* The other alternatie host plants include *Paspalum*, *Panicum, Ixophorus* and *Commelina* but in these the symptom expression was not noticed.

Based on the reaction to different *Musa* sp. the pathogen has been grouped into four races.

Race 1: Occurs worldwide and infects Gros Michael (AAA), Rasthali (AAB), Pome (AAB), Karpuravalli (ABB) and Red banana (AAA).

Race 2: Infects only cooking type of bananas i.e., Bluggoe / Monthan (ABB) and it is widely distributed.

Race 3: It attacks the *Heliconia* sp. and presence is restricted to Central America.

Race 4: Mostly present in sub-tropics and susceptible cultivars are Cavendish and suscepts of race 1 and 2.

Susceptible varieties

Karpuravalli (Pisang Awak / Kanthali ABB), Rasthali (Malbhog / Silk AAB), Monthan (ABB), Suganthi (AAB), Pachanadan (Pome AAB), Ney Poovan (AA), and Virupakshi (Pome AAB) are susceptible to wilt disease in India.

Survival and spread

The pathogen survives in the soil upto 30 years as chlamydospores in plant debris or in the roots of alternate hosts. The spread of the pathogen is affected mainly by infected rhizomes or suckers, soil adhering the rhizome, implements and irrigation water. The spread of the pathogen is slower in fertile and permeable soils. Addition of crop residues may stimulate antagonistic microflora and reduce the survival of the pathogen. In light textured soils the pathogen survives for longer time than in heavy alkaline soils. Waterlogged condition for shorter period also increases the disease development. Burrowing nematode *Radopholus similis* and lesion nematode *Pratilenchus coffeae* are the predisposer of the disease.

Management strategies

No effective method is available to combat the disease, the following integrated management practices are recommended:

- Crop rotation with paddy or sugarcane once or twice followed by banana for 2-3 cycles
- Selection of resistant varieties such as Poovan, Robusta and Nendran for planting in endemic areas.
- Avoid susceptible varieties such as Rasthali, Monthan, Karpuravalli, Kadali, Rasakadali, Pachanadan etc.
- Extract healthy suckers for planting.
- Restricting the movement of infected suckers to non-infected areas by imposing quarantine regulations.
- Paring and pralinage (removal of roots and outer skin of rhizomes) and dipping of the suckers in a solution containing 0.2 per cent carbendazim + 0.5 per cent monocrotophos (i.e. 2g of carbendazim + 14 ml of Monocrotophos per litre of water).
- Removal of infected plants and application of lime 2-3 kg per pit.
- Provide good drainage.
- Drenching of carbendazim 0.2 per cent solution around the pseudostem at bimonthly intervals starting from five months after planting or injection of corm with 3 ml of 2.0 per cent carbendazim solution (20g / l water) or embedding 50.0 mg of carbendazim in a capsule in rhizome at fifth, seventh and ninth month after planting.
- Application of powder formulation of biocontrol agents such as *Trichoderma viride* and *Pseudomonas fluorescense* @ 15 grams along with 5 kg of well decomposed farm yard manure for 4 times once at a time of planting in the planting pit and remaining doses at 3rd, 5th and 7th months after planting around the plants.
- Application of cassava residue @ 10,000 kg/ha. + Rice bran with 21 days following before planting also found to check the occurrence.

Sigatoka leaf spot (*Mycospherella* spp)

Among the leaf spot diseases, Sigatoka is the very important, occurring in almost all the banana growing regions of the world. The pathogen causes three different leaf spot diseasesinIndia. Distribution This disease was reported first in Java in 1902 and subsequently in different countries such as Fiji,

Australia, Sri Lanka, Africa, Asia and America. Severe outbreaks were noticed in Uganda, Tanzania, Trinidad, Surinam, Guyana, Honduras, Venezuela, Colombia, Panama, Costa Rice, Guatemala, Mexico, Jamaica Haita, etc.

Symptoms

The first symptom is the appearance of pale yellow or greenish yellow streaks parallel to the veins on the upper surface of the leaves. Then these streaks darken and become more or less elliptical brown spots. Later on, the centre of these spots turns to light grey colour surrounded by yellow halo. Large number of these spots coalesces and thus kills the large area of midrib and petiole and the leaf hangs down. Experiment conducted in different parts of the country revealed that there should be at least 12 –15 functional leaves at the time of shooting to get normal, marketable bunch. But due to the severe infection of sigatoka leaf spot, there will be less than 5 leaves. A plant infected by the disease has scorched appearance due to necrotic spots on the leaves thereby reducing the phosynthetic area finally death of lamina and collapse of petiole, and the leaves hang from the pseudostem. All these results in the premature ripening, which became under sized and unsaleable fruits.

Susceptible varieties

All the commercial cultivars except Karpuravalli are susceptible to yellow sigatoka. Generally cultivars having more 'B' genome are tolerant to this disease.

Pathogen

The incitant of the disease is *Mycosphaerella musicola* (Imperfect stage – *Pseudoscercospora musae*). The pathogen produces both conidia and Ascospores.

Disease Management

(1) Removal and destruction of leaf spot infected leaves

(2) Intercultural operation such as timely removal of suckers, weed free cultivation, improving the drainage condition and adopting correct spacing are found to decrease the disease severity significantly.

(3) Spraying with systemic fungicides such as Propiconazole (1ml /litre) or Carbendazim (0.1%) + Calixin (0.1%) along with few drops of stickers like Sandovit or Teepol or Triton AE 1ml/litre of water was effective in containing the disease.

(4) Spraying of Paraffinic oil 5% + Propiconazole or Carbendazim @0.5 ml per lit of water at 25 days intervals for 4 times immediately after the notice of the symptoms controls the disease effectively.

FRUIT DISEASES

Cigar end rot – *Verticillium theobramae* (Turco)

The disease causes heavy losses in Cavendish and Nendran bananas mostly during rainy seasons in many of the banana growing regions in the country, particularly in Gujarat, Tamil Nadu, Maharashtra and Kerala.

Symptoms

The infection, which occurs in perianth, spreads to fingers causing blackening of the skin and shrinkage and folding of the tissues. The affected tissues are corrugated and covered with fungus conidiophores and powdery grey conidia resembling ash on a cigar end. Dry rot also occurs in the pulp and the affected tissues becomes dry and fibrous.

Favourable conditions

Warm, moist condition favours the disease occurrence and the disease spread is high in old and badly maintained plantations.

Pathogen

Verticillium theobromae and *Trachyshaera fructigena* are the two fungi involving in causing the disease.

Management practices

It includes removal of pistil and perianth by hand 10 days after finger emergence and the operation is called tipping dusting with copper based fungicides (0.3%).

Peduncle rot

The peduncle rot has become serious threat to banana production in Tamil Nadu as well as other banana growing regions of India. The problem is particularly pronounced in regions where the temperature is above 35 °C and leaf spot diseases are severe. The incidence of the disease in the banana gardens ranged from 5 to 25%. When the bunch emergence occur during summer months and the number of functional leaves reduced due to leaf spot diseases the upper portion of the peduncle is exposed to the hot sun. The peduncle become bleached and subsequently rot. The rotting continues, further its spread and results in reduced finger size and non-development or partial and filling up of fingers. When infection occurs at an early stage the disease can attain sufficient severity to spread into the fingers and result in the complete destruction of the bunch.

Pathogen

This disease caused by *Colletotrichum gloeosporioides* and *Botryodiplodia theobromae.*

Susceptible varieties

Karpuravalli, Rasthali, Pachanadan, Robusta and Dwarf cavendish.

Management Practices

The prophylactic spray of the peduncle with carbendazim at 0.1% immediately after shoot emergence and covering the peduncle with leaves to avoid sun scortch.

Crown rot

This is the most serious postharvest and complex disease occurring in all banana growing regions. If the dehanding is not done perfectly, the disease appears. The pathogens involved in causing the diseases are i) *Botryodiplodia theobromae* ii) *Ceratocystis paradoxa* iii) *Verticillium theobromae* iv) *Colletotrichum musae* v) *Fusarium* spp. (F. *moniliforme*, F. *pallidoroseum* and *F. m.* var. subglutinans) vi) *Nigrospora sphaeria.*

Symptoms

The spores of the above said wound parasites which may be present on the crowns, fingers and finger stalk at the time of harvest, infect the crown tissues after bananas are dehanded and causes blackening of the infected portions. Then the blackening spreads to pulp through the pedicel, which results in the rottening of fingers and / or separation of fingers from the hand causing heavy post harvest losses during transit and storage.

The pathogen is spread by air / rain , the disease is severe during hot and humid weather conditions.

Treating the harvested hands with Thiabendazole at 200-400 ppm is very effective in reducing the disease.

Anthracnose – *Colletotrichum musae*

The disease attacks the plant at all stages of its growth and found mostly on the leaves and on the maturing and ripened fruits.

At initial stage, develops as a small, circular, black spots. Then these spots enlarge in size, turn to brown colour and become sunkened and covered with orange or salmon coloured spore masses. On the ripened fruits, spots coalesce and make the fruit unsightly and unmarketable. When the spots occur on the green fruits, premature ripening results in. Occurrence of black lesions on the pedicel causes withering of the pedicel and dropping of the fingers from the hands.

Pathogen

The disease spreads by its air borne conidia. High atmospheric temperature and humidity favours the development of the disease.

Management practices

The disease can effectively be managed by avoiding injury to the peel, adopting good cultural practices including removal of dead trash leaves from the plants, cold storage of fruits at 13-14 °C, transporting the fruits in boxes with polyethylene lining and field spraying with 0.2% chlorothalonil or prochloroz at 0.15% or carbendazim at 0.1% four times at 15 days interval.

BACTERIAL DISEASE:

Rhizome rot or head rot or tip over – *Erwinia spp.*

The disease occurs in India, Central America, Jamaica, Israel etc. Generally the disease attack the young plants and causes germination failure in Cavendish (AAA) and diploid (AA) group of bananas. It occurs in severe form in high rainfall areas and in alluvial soils.

The symptoms can be seen both externally and internally. The external symptoms are I) failure of germination of planting materials when the rotting of rhizomes occurs extensively ii) Yellowing of leaves and stunting of the plants iii) The trunk base may become swollen or split and results in the breakage of rhizome at soil level causing toppling of over mature plants at the time of fruit maturing stage or during heavy wind periods. vi) Sometimes the base of the trunk becomes dark in colour with longitudinal splits and numerous transverse and other tissues. v) The diseased plants tend to produce have more number of suckers.

The internal symptoms are characterized by the presence of the pale grey water soaked area in the white cortical tissues at initial stage. These areas then spread as finger like process inwards towards the centre of the rhizome, upwards up to the growing point and outwards upto the daughter suckers. The affected areas become rottened and these areas are of irregular in outline, become yellowish brown and as the disease advances the colour of the rotten area turn to red. Brownish demarcation can also be seen at the junction of diseased and healthy tissues. The infected cortex may have a yellow or red marbled or freckled appearance.

Pathogen

Two species of Erwinia namely *E. chrysanthemi* and *E. carotovora* sp. *carotovora* are responsible for causing rhizome rot. The pathogen enters into the rhizome through the wounds or decaying pseudostem leaf sheaths surrounding the central heart bud.

Spread

Through infected suckers.

Management practices

1. Selection of 'B' genome clones for planting.
2. Planting of vigorous, disease free suckers.
3. Providing drainage facilities and improving aeration of the soil, application of more organic amendments and other cultural practices to improve the texture, aeration and steady supply of moisture to improve the growth and root development.
4. Removal of heart growth apex of corms suckers found to reduce the incidence.
5. Drenching the soil around the plant with either bleaching powder at the rate of 2g/lit or Emissan @ 1g/lit at an interval of 10-15 days found effective in controlling the disease.

VIRAL DISEASES:

Banana Bunchy Top Virus

Banana bunchy top disease appeared in epidemic form in Fiji in 1889 and it was the first report of the disease occurrence in the world. Subsequently the disease has been reported from many banana-growing countries viz., Taiwan, Australia and Sri Lanka etc. In India it was first reported occur in Kerala in 1943. Later it has been reported from many banana-growing states. Now the disease has been found even in the deep forest of North eastern hill regions. It has been reported that the disease was introduced to our country from Sri Lanka. It is also possible that the disease would have been present since long in many parts of the country unnoticed. Reports have not come from any western banana growing nations. Recently this disease has been recorded from Pakisthan. In Tamil Nadu and Kerala this disease has caused enormous loss in early periods. Even now the hill banana of lower palany hills wiped out due to this disease. One of the reasons for the severity in hill regions is the disease inoculum and the vector is present all round the year. The crop is grown as perennially for 10 to 15 years as large clumps.

Visual symptoms

Intermittent dark green dots and streaks of variable length is visible in leaf sheath, midrib, leaf veins and petioles. Sometimes the typical streaks may not appear in some cultivars. The leaves produced are progressively shorter, brittle in texture, narrow and give the appearance of bunchy ness. The affected plant fails to produce any flowers. If late infection occurs the plant can throw bunch but the fingers never develop to maturity. Marginal chlorosis in leaf lamina is also noticed. Vein flecking symptoms are noticed in some varieties. When infection occurs very late in the season the plant would show dark green streaks on the tip of the bracts. Sometimes the bracts tip of male bud is green and leafy. Symptom less infection or latency is reported to present for two seasons. Mild vein clearing and also blisters with dark green colour on mid rib and leaf sheath have also been recorded.

Virus

The BBTV virus belonging to nanoviridae family and babuvirs genus. The BBTV is the type species of the genus. The virus is an isometric particle measure 20nm diameter. The coat protein molecular weight found to be 20 K Dalton. The virus has multicomponant genome. There are six circular single stranded genomes have been known to be associated. The virus concentration is more present in phloem. The DNA component 3 has the coat protein gene.

Transmission

The virus is transmitted primarily by infected suckers and through tissue propagated material. Secondary transmission is by banana aphid *Pentalonia nigronervosa* in a semi persistent manner. As on date no mechanical transmission by sap has been reported. The efficiency of transmission is depends on stage of the insect and the other climatic factors.

Banana Bract Mosaic Virus

Samraj *et al* first noticed the banana bract mosaic virus in India in 1966. They have described the malady in the name of Kokkan, but failed to prove the association of a virus in Kokkan disease. It was Magnaye *et al* from Philippines to publish the occurrence of Bract mosaic in cultivar Cardaba in 1989. In Philippine also the disease was noticed in 1979 it self. Later In India NRCB in collaboration with INIBAP has found 'Kokkan' malady is nothing but the disease caused by BBMV. The EM and serology work done at NRCB in collaboration with IIHR, has clearly proved that this virus is the same BBMV reported in Philippines (Singh *et al.* 1996). The attempt to prove its etiology in past has taken a long time, which was sufficient enough for unexpected level of spread, not only in plantain Nendran, but also to many commercial cultivars such as Poovan, Robusta, Karpooravalli, Neypoovan and Monthan. Presently the virus is a major constraint in southern states of India, it requires quarantine restriction to avoid further spread to other major banana growing states. The disease is more prevalent in cooking and Cavendish bananas in Philippines where as in India it has been severe in plantain cultivar Nendran followed by Robusta, Poovan, Ney Poovan, and Red Banana. Cooking bananas and Pisang Awak are also affected but the yield loss is minimal. This viral disease has been recorded in neighboring countries like Bangladesh and Sri Lanka.

Visual symptoms

Banana bract mosaic disease is characterized by the presence of spindle shaped pinkish to reddish streaks on pseudostem, midrib and peduncle. Typical mosaic and spindle shaped mild mosaic streaks on bracts, peduncle and fingers also observed. In Nendran, the leaf orientation changes in such a way giving the appearance of 'Travelers palm' plant. The lower side of the leaf spindle shaped waxy coating, appears corresponding to these, on upper surface of the leaf spindle shaped mosaic pattern is seen. In Nendran, unusual long peduncle and very short peduncle, abort of bunches, raised corky growth on peduncle is also observed to be the symptoms of BBMV. Necrotic streaks on fingers, leaf, pseudostem and mid rib are found in some varieties. Extended female phase after a short male phase has also been observed in Nendran and Neypoovan cultivars. In Red Banana and Robusta the leaf stripping symptoms is common in some seasons in Tamil Nadu. In Robusta, fingers of infected plants stop to develop and give the appearance of 'pencil', explaining the local name 'pencil kai' (pencil-sized fruit). Though Thomas and Magnaye reported that this disease could only be diagnosed when symptom appears on bract, in Trivandrum district of Kerala farmers have named the disease Pola roga, means 'disease of pseudostem' in cultivar Nendran. Necrotic symptoms on pseudostem, midrib and petiole of infected plants of cv Neypoovan has been observed. When the pseudostem was cut horizontally, the necrosis was found to be deep seated.

Yield loss assessment

Around 40% reduction of bunch weight of kokkan - diseased plant over healthy was recorded by Prasannkumari *et al.* (1986). Yield losses up to 40% in cv Cardaba and cv Lakatan were recorded in the Philippines. In 1988 banana bract mosaic reached epidemic proportions around Santos city on Mindanao Island and 25,000 plants were destroyed (Jones, 1994). At NRCB, the yield loss was assessed due to BBMV. The yield loss of 68.34, 46.53 and 53.4% over control was also observed in Nendran, Robusta and Neypoovan respectively.

Virus

Banana bract mosaic virus is belongs to the family Potyviridae, genus Poty virus. It is distinct member of the genus. Cytopathological observations revealed the presence of pinwheel inclusions and flexuous rod shaped particles measuring 750x15nm in size, which are the typical feature of potyvirus genus. The genome of the virus is positive sense single stranded RNA having approximately 7.5 kb. The virus titre is more in bracts, mid rib than leaf sheath.

Transmission

This virus is transmitted through *Rhopalosihum maidi, Aphis gossypii* (Magnaye and Espino, 1990) and *Pentalonia nigronervosa* (Muñez, 1992) in a non-persistent manner. The vector requires brief probes to transmit the disease more efficiently. This virus is not mechanically transmitted by sap through tools used in cultivation. In India, recently Selvarajan *et al* have proved the transmission of BBMV by banana aphid (unpublished). Possibly other aphids may also transmit the virus in field. The virus primarily spread through infected suckers, tissue culture plants and ignorance of the farmers on the disease are the main cause of spread.

Banana Streak Disease: This disease has become a major threat to an important commercial group of banana, Mysore (AAB) cultivation. Though the disease was suspected to be present more than 40 years ago (Wardlaw, 1961), this had been unnoticed by *Musa* pathologist as this was considered as one of the genetic characters of Mysore group of banana. But recently the virus genome has been proved to serve as genetic marker for B genomic parents. The disease was first noticed on variety 'Poyo' under Cavendish sub group (AAA) of banana and was first reported by Lockhart in 1986 in Dwarf Cavendish and Giant Cavendish. In 1995, National Research Centre on Banana, Trichy has confirmed BSV infection in cv Poovan. The 'Mysore' group (Alpan, Champa, Chenichampa, Mysore Kadali, Poovan and Dasaman) express symptoms similar to BSV, which was reported by Lockhart. Symptom illustrations for CMV, given in the book *Banana* '99 show similarity to banana streak disease (Shanmughavelu *et al*, 1989). In 1996 we have observed both bacilliform particles of BSV and isometric particles of CMV under EM from the Poovan plants that had streak disease symptoms. The symptom less plants under stress like temperature, nutrients, etc exhibited severe symptoms.

Visual Symptoms

Initially small dots, with golden yellow colour develops later it extend to form long streaks. The chlorotic streaks become necrotic giving a blackish appearance on lamina. Necrotic streaks on midrib, pseudostem petiole is also observed. The symptoms are erratically distributed over individual leaves as well as in leaves of same plant. The midrib mosaic symptom, one of the symptoms of CMV, is not found with banana streak disease. Bunch choking, abortion of bunch and seediness in fingers are seen in infected plants. Symptoms appear on young emerging leaves, all leaves would not bear symptoms and also within the leaf intermitant symptoms are seen. Fruits get distorted, plant vigour/growth get reduced. Dieback and internal pseudostem necrosis have been associated with the disease. Plants/part of the leaves with pronounced symptom should be used for serological indexing. Virus particles are more in symptomatic portions than asymptomatic areas. Various changes in symptom have been noticed in different varieties. Leaf stripping symptoms are commonly noticed in cultivars Poovan, Grand Nain and Robusta. Typical symptoms of the virus can be noticed in cultivars Poovan, Cavendish, Nendran, and Red banana

Mild to severe mosaic pattern also been observed. The temperature plays a major role in symptom expression. At ambient conditions the expression of symptoms by the virus is very minimal, but Dahal *et al* has proved that the plants kept at 22° C express more typical symptoms and the virus titre also been increases considerably or sufficient enough to detect the virus by different serological assays. In tissue cultured Grand Nain, different types of symptoms have been noticed in Maharastra and Madhya Pradesh.

Yield Loss

It has been observed that the yield loss vary from variety to variety and also the climatic conditions prevail in the region. The plants derived from tissue culture exhibits severe symptom and consequently the yield loss also proved to be high. In India, the virus severely infects Mysore group of banana namely Poovan, Red banana, Robusta, Nendran, Cheni champa, Alpan, Chamba and also in Cavendish group (Basrai, Grand Nain) in Maharashtra region. The survey conducted past three years for BSV in Poovan revealed 80.28 per cent infection but the remaining healthy plants also found to carry without showing any symptoms.During 1996-1997 surveys of NRCB, based on severity, the BSV infection was categorised. Among 70 per cent of infected plants, 11.20 per cent found to have very severe symptoms leading to abortion of bunches, poorly filled fingers and also seediness in fingers. Seeded fruits are unfit for marketing. The yield loss due to infection of BSV is found to be 48.18 per cent over control (Selvarajan *et al.*, 1997 and Thangavelu *et al.*, 1996). The tissue-cultured plants supplied by some of the firms have infection with BSV.

Virus

Banana Streak Virus, a plant pararetrovirus, causes banana streak disease. Non-enveloped bacilliform particles, having the size of 130 x 30nm with double stranded DNA as the genome are the characteristic of the virus. Similar bacilliform badna viruses are reported to occur in Sugarcane, Citrus in India (Viswanathan and Selvarajan, 1996; Ahlawat et al 1996). In India presence of BSV in cultivar Poovan was proved by Selvarajan *et al* 1996. In interspecfic hybrids the infection occurs by activation of endogenous pararetroviral sequence integrated into Musa genome, but not from any external source of infection. One of the factors triggering episomal expression might be process of genetic hybridization especially from B genomic groups.

Transmission

BSV is transmitted in a semi-persistent manner by citrus mealy bug, *Planococcus citri* from banana to banana. Experimentally it has been proved that the sugarcane mealy bug, *Saccharicoccus sacchari* transmit ScBV from sugarcane to banana. This virus is not transmitted mechanically by sap or by tools used for inter-cultivation operations. But seed/vegetative transmission is the primary mode of transmission. Now it has been proved that the virus is transmitted to new hybrids through the activation of integrants present in parents having B genomes. Tissue culture processes induce the expression of BSV disease symptoms. Ndovora *et al* (1997) has proved that BSV infection in hybrids is not from external sources.

BANANA MOSAIC OR INFECTIOUS CHLOROSIS (CMV)

Banana mosaic or infectious cholorosis caused by cucumber mosaic virus, has been known to occur in early 1940's in Maharashtra states. This disease was reported from Poovan and Cavendish clones widely grown in the country. In Tamilnadu, an outbreak of this disease was reported by Mohan and Lakshmanan in Poovan and Robusta in 1988. Actually CMV is known to occur in wide range of hosts in crop plants, many weed hosts and cucurbitaceous crops grown as winter crop in banana are the actual source of the virus, and transmitted through aphid vectors.

Visual symptoms

CMV induces mild mosaic symptoms which appear in whole leaf or all the leaves of an infected plant, where as BSV induced symptoms are present only parts of leaf and irregularly distributed. CMV has also been related to heart rot symptoms in banana. Rotting of cigar leaf, internal necrosis of pseudostem and death of the whole plant are associated with CMV. Cholortic banding, rosette appearance of leaf arrangement are also the symptoms of CMV. In Poovan, it is often confused with BSV symptoms.

Management of viruses

- As viruses are difficult to control, eradication of infected materials is must.
- Reliable diagnostic kits are needed to detect the virus in mother plants used for mass multiplication.
- Indexing procedure should be adopted in quarantine labs and commercial firms involved in mass multiplication.
- Destruction of infected materials by incineration is still the better method. Once the inoculum is destroyed, even though vector insects are present transmission would not be there.
- Domestic legislation to control the spread to neighbouring states should be enacted for new virus diseases.
- In hill regions, replanting with virus free plants and destruction of infected plants and monitoring the vector population are needed to reduce the incidence.
- Banana growers should be made to know the symptoms of new viruses and while selecting suckers they should be advised to not to use infected materials.

ACKNOWLEDGEMENTS

Thanks are due to the Head (Crop Protection), National Research Centre For Banana ,Tiruchirapalli-620 102 for constant encouragement.

REFERENCES

Ahalawat Y.S., Pant, R.R., Lockhart, B.E.L., Srivastara M., Chakraborty, n. K., and Varma, A. 1996. Association of a badna virus with citrus mosaic disease in India. *Plant disease.* 80: 590-592

Jones, D.R. 2000. Diseases of Banana, Abaca and Enset. CABI publishing, Wallingford, Oxon OX108 DE , UK. pp. 544.

Moore, N.Y., Bentley, S., Pegg, K.G. and Jones, D.R. 1995. Fusarium wilt of banana. *Musa disease fact sheet. INBAP, Montpellier*, France, 5: 4

Padmanaban,B and S.Sathiamoorthy 2001 The Banana stem weevil, *Odoiporus longicollis Musa Pest fact sheet No.5* INIBAP,France.

Padmanaban,B, M.Kandasamy and S.Sathiamoorthy 2001 Banana stem weevil: Present scenario, National seminar on Emerging trends in pest and diseases and their management held at TNAU,Coimbatore, October 17-19, p. 21.

Selvarajan, R., S.J.Singh, H.P.Singh and R.Thangavelu 1997. Geographical distribution serodiagnosis and yield loss assessment of BSV in Inida. Poster paper presented at *Int. Conf. on Integrated disease management for sustainable agriculture* held at New Delhi on November 11-15.

Singh, S.J., R. Selvarajan and H.P. Singh 1996. Detection of Banana Bract Mosaic Virus (kokkan) by Electron Microscopy and Serology. In the proceedings on *"Challenges for the Production and Utilization of Banana in 21st Century"* held on 24-25 September, at Trichy. pp 381-383.

Thangavelu, R., Padmanaban, B and Sundararaju, P. 1999. Peduncle rot in India. *Infomusa.* 8: 35.

Thangavelu, R., R.Selvarajan and H.P.Singh 1996. Status of Banana Streak Virus and Banana Bract Mosaic Virus in India. In the proceedings on *"Challenges for the Production and Utilization of Banana in 21st Century"* held on 24-25 September, at Trichy. pp 364-376.

Viswanathan, R and R.Selvarajan 1996. A Badna Virus related to Banana Streak Virus in sugarcane. In the proceedings on *"Challenges for the Production and Utilization of Banana in 21st Century"* held on 24-25 September, 1996 at Trichy. pp. 397-402.

Advances in the Diagnosis and Management of Orchid Viruses

T. K. BAG

Orchids are one of the most fascinating creations of nature with over 800 genera and between 25000-3000 species (Chowdhery, 2001). The orchids represent the largest and highly evolved group of flowering plants. They are found in almost every type of habitat. Orchids have been attracting floriculturists since time immemorial due to their fads, fancies and fashions and this lead to "orchid mania" throughout the world. These are often considered to be the most luxurious flowers probably because many varieties of orchids are highly priced and difficult to cultivate commercially. The orchids are expensive and orchid industries in the world are well established and have further scope for enlargement.

Like other cultivated crops, orchids are also susceptible to infection by a variety of infectious pathogenic agents, which include fungi, bacteria, nematodes and viruses. Among the orchid diseases, viral diseases are very important as they are easily transmitted through planting material and get widely distributed detrimentally affecting their ornamental value. Till today approximately 31different viruses have been reported to infect orchids in different parts of the world (Table 1). Among orchid viruses Cymbidium mosaic virus (CyMV) is the oldest known virus and are reported to infect 26 species of orchids the world over (Francki, 1970). Orchid strain of Tobacco mosaic virus (TMV-O) is also distributed world wide and infect 16 species of orchids (Paul, 1975). Later it was proved that TMV-O and Odontoglossum ring spot virus (ORSV) are the same virus (Kado, *et al.*, 1968) and since then TMV-O is better known as Odontoglossum ring spot virus (ORSV). According to a recent report, ORSV infects about 20 genera of orchids the world over (Brunt *et al.*, 1996). Of the orchid viruses ORSV and CyMV are considered the most important ones due to their prevalence, world wide occurrence and the severe symptoms they can induce in several orchid genera, often reducing their vigor, lowering flower quality and leading to economic losses. Mixed infection are also quite common, because these viruses have similar biological and epidemiological characteristics and are frequently transmitted together (Freitas, 2003). However, the importance of other orchid viruses can not be over looked.

VIRUS SYMPTOMS IN ORCHIDS

The viruses induce several types of symptoms depending on the orchid genera/species, the plant parts infected and the environmental conditions. Inspite of this variability in expression of virus symptoms, a grower should be well aware of the types of symptoms most frequently associated with a particular virus in a particular genus. Cymbidium mosaic virus produces the typical mosaic mottle on young leaves of

Scientist Senior Scale (Plant Pathology), National Research Centre for Orchid (ICAR), Pakyong 737106, Sikkim, India

Cymbidium, which may later turn brown and necrotic, severely affected leaves may drop prematurely. In *Phalaenopsis* CyMV produces necrotic line patterns with sunken patches. In *Epidendrum* CyMV induction leads to stunting and development of brown or black oval spots which usually occur on the underside of the leaves. In *Cattleya* CyMV produces necrotic spots which appear as sunken spots on the underside of leaves. The spots later turn brown and coalesce turning the whole leaf black. Sometimes after the initial manifestation of the disease, the symptoms may disappear and the plant may remain normal to all outward appearances. The *Cattleya* flowers infected by CyMV develop necrotic spots and streaks (Faccioli and Marani, 1979). Infected flowers may appear healthy until several days after they open, when the symptoms begin to express. Necrotic spots and streaks appeared start to enlarge in number and size and eventually cover the whole flower (Inouye, 1972). Lee and La (1976) reported five kinds of symptoms evoked by CyMV on different orchid species. These are (I) mosaics, (ii) streaks with mosaic, (iii) ring necrosis, (iv) chlorotic rings and (v) necrotic spots. Oncidium hybrid: Golden Shower plant shows the infection as the sunken brown or black spots on the underside of the leaves. The spots sometimes may appear as concentric rings with breaking of few spots through the upper leaf surface (Gara *et al.*, 1996).

The first strain of Tobacco mosaic virus isolated and identified from orchids was from *Odontoglossum grande* in the USA by Jensen and Gold (1951). The virus induced a ring symptom on the leaves and was then named as TMV-O. Because this virus is a strain of TMV on orchid, it is referred to as an orchid strain or TMV-O. The TMV-O and Odontoglossum ring spot virus (ORSV) is the same virus (Kado *et al.,* 1968) and it is popularly known as ORSV. This virus induces a variety of symptoms on a range of orchid genera. Foliage symptoms include chlorotic and necrotic spots and streaks and rings. In lavender flowered *Cattleya* cultivars, the foliage patterns are often associated with distinct red or purple pigmentation. Leaf symptoms may appear as spots, chevron patterns or irregular line pattern of chlorotic or red pigmented tissues. Leaves of *Oncidium* spp may show a combination of sunken chlorotic and necrotic spots while *Epidendrum* shows only chlorotic spotting. ORSV may often occur in combination with CyMV. Sometimes double infection of these two viruses may induce other conditions known as blossom brown necrotic streak, blossom necrosis, *Cattleya* flower disease, colour break, flower break, flower necrosis and mild flower break. Major ORSV hosts are *Cattleya, Cymbidium, Epidendrum, Odontoglossum* and *Oncidium* (Lawson and Hsu, 1995). ORSV may also cause chlorotic streaks and/or mild mosaic on orchid species (Kondo *et al.*, 1996). In *Calanthe* the ORSV induce chlorotic mottle on leaves (Mitsumoto *et al.*, 1995). The growth of the plant is severely affected by CyMV than that of ORSV (Pearson and Cole, 1986). Orchid fleck virus (OFV) first reported in *Cymbidium* from Japan by Doi *et al.* (1969) induces chlorotic or necrotic flecks on leaves. In *Phalaenopsis* spp and hybrids it evokes chlorotic flecks. In *Renanthera* spp OFV induces chlorotic flecks, then sunken gray necrotic streaks with sharp margin and of variety size. Other major hosts of OFV are *Coelogyne, Dendrobium* species and hybrids, *Miltonia, Odontoglossum* species and *Oncidium* species, *Paphiopedilum* species and *Vanda* species. Hu *et al.* (1992) reported *Oncidium* orchid infected with TSWV showing chlorotic ring spots whereas according to Goff and Corbett 1977) TSWV exhibits faint chlorotic leaf streaks in *Cymbidium*. Impatiens necrotic spot virus (INSV) first reported on *Phalaenopsis* sp in California induces various types of yellow mottling or streaks, irregularly shaped necrotic (brown or black) spots and yellow to necrotic concentric ring spots on leaves of *Oncidium*. Infected *Oncidium* plants did not exhibit stunting or distortion of leaves and shoots. No flower symptoms is reported on *Oncidium* (Koike and Mayhew, 2001).

Although viruses induce definite symptoms on orchids, several orchid genera/species remain apparently healthy even after infection with virus and diagnosis based on the visual symptoms only may always not be reliable. Besides symptoms of virus infection may sometimes be confused with diseases caused by other types of pathogens or physiological disorders or micronutient deficiencies. Proper diagnostic methods must be used for the purpose of reliable detection of virus in orchids.

Table 1: Orchid viruses so far reported world over

Rod shaped viruses

Sl No.	Name of the virus	Virus group	Particle morphology	Dimensions (nm)	Geographical distribution	Mode of	Described by Transmission	References
1.	Bean yellow mosaic virus (BYMV)	Poty virus	Flexuous rods	750 × 13	Japan, USA, Germany	Sap and aphids	Inouye	Inouye, 1971 and 1973; Lesemann, 1985; Hammond and Lawson, 1988
2.	Calanthe mild mosaic virus	Poty virus	Flexuous rods	-	Japan	Sap and aphids	Gara, Kondo, Maeda, Inouye and Tamada	Gara, Kondo, Maeda, Inouye and Tamada, 1998
3.	Ceratobium mosaic virus	Poty virus	Rods like	-	Australia	-	Mackenzie, Nolan, Wei, Clements, Gowanlock, Wallace and Gibbs	Mackenzie, Nolan, Wei, Clements, Gowanlock, Wallace and Gibbs, 1998
4.	Clover yellow vein virus (CYVV-C)	Poty virus	Flexuous rods	750 × 12	Japan	Sap and aphids	Inouye,	Inouye,Maeda and Mitsuhata, 1988
5.	Cymbidium mosaic virus (CyMV)	Potex virus	Flexuous rods	415-475 × 13-18	World wide	Sap, nursery tools, pots and irrigation water	Jensen	Jensen, 1951 and 1970; Magee, 1943; Francki, 1970; Corbett, 1960, Kado and Jensen, 1964; Lawson, 1970
6.	Cypripedium filamentous virus (CFV)	-	Flexuous rods	772	Germany, Japan and French Polynesia	-	Lesemann and Koenig	Lesemann and Koenig,1978
7.	Dendrobium mosaic virus (DEMV)	Poty virus	Flexuous rods	750 x 13	Japan, USA	Sap and aphids	Inouye	Inouye, 1971; Hammond and Lawson, 1988
8.	Dendrobium vein necrosis virus (DVNV)	Closterovirus	Flexuous rods	1865	Germany, USA		Lesemann	Lesemann, 1977
9.	Habenaria mosaic virus	Poty virus	Flexuous rods	-	Japan	Sap and aphids	Inouye, Taeda, Mitsuhata, and Gara	Inouye, Taeda, Mitsuhata, and Gara, 1998
10.	Odontoglossum ring spot virus (ORSV) or Orchid strain of Tobacco mosaic virus (TMV-O)	Tobamovirus	Rigid rods	280-320 × 15-24	World wide	Sap, nursery tools, pots and irrigation water and plant debris	Jensen and Gold	Jensen and Gold, 1951; Jensen, 1970; Corbett, 1967, Lawson,1970

Sl No.	Name of the virus	Virus group	Particle morphology	Dimensions (nm)	Geographical distribution	Mode of Transmission	Described by	References
11.	Tobacco rattle virus (TRV)	Tobravirus	Rigid rods	65 and 193	Germany	sap	Lesemann and Vetten	Lesemann and Vetten, 1985
12.	Turnip mosaic virus	Poty virus	Flexuous rods	745	Germany	Sap and aphid	Lesemann and Vetten	Lesemann and Vetten, 1985
13.	Vanilla mosaic virus	Poty virus	Flexuous rods	767	French Polynesia	aphids	Wisler, Zettler and MU	Wisler, Zettler and Mu, 1987
Isometric viruses								
14.	Cucumber mosaic virus (CMV)	Cucumovirus	Isometric	30	Japan	Sap and aphid	Inouye	Inouye 1969
15.	Cymbidium chlorotic mosaic virus	Sabemovirus	Isometric	28	Japan	sap	Kondo, Maeda, Mitsuha and Inouye	Kondo, Maeda, Mitsuha and Inouye, 1996
16.	Cymbidium mild mosaic virus (Cy MMV)	-	Isometric	28	Korea	sap	Chang, Doi and Yora	Chang, Doi and Yora, 1978
17.	Cymbidium ring spot virus (CyRSV)	Tombusvirus	Isometric	30	England	Sap, soil borne	Hollings and Stone	Hollings and Stone, 1963; Hollings, 1972 and 1977
18.	Impatiens necrotic spot virus (INSV)	Tospovirus	Quasispherical	85	California (US)	Sap and thrips	Mayhew, Cooke and Raabe	Mayhew, Cooke and Raabe, 1992
19.	Masdevallia isometric virus	-		-	Colombia	-	Lesemann	Lesemann, 1981
20.	Tomato ring spot virus (TomRSV)	Nepovirus	Isometric	30	USA	sap	Goff and Corbett	Goff and Corbett, 1977
21.	Tomato spotted wilt virus (TSWV)	Tospovirus	Quasispherical	85	Hawaii (US)	Sap and thrips	Hu, wang, Ferreira and Ogata	Hu , Wang, Ferreira and Ogata, 1992
22.	Tricopilia isometric virus	-	Isometric	28	Germany	-	Lesemann	Lesemann (Unpublished)

(Contd...)

Table Cocld...

Sl No.	Name of the virus	Virus group	Particle morphology	Dimensions (nm)	Geographical distribution	Mode of	Described by Transmission	References
Bacilliform and Bullet shaped Viruses								
23.	Brazillian bacilliform rhabdovirus	Rhabdovirus	Rod shaped	100-120 x 40	Brazil	-	Kitajima, Blumenschein and Costa	Kitajima, Blumenschein and Costa., 1974
24.	Dendrobium virus (DV)	Rhabdovirus	Bullet shaped Bacilliform	120 x 50	New Guinea and Germany	-	Petzold; Duvel and Peters	Petzold, 1971; Duvel and Peters, 1971; Begtrup, 1972
25.	Grammatophyllum bacilliform virus (GBS)	Rhabdovirus	Bullet shaped Bacilliform	40-42 wide particle with variable length	Maryland	-	Corbett	Corbett, 1974
26.	Orchid Fleck virus (OFV)	Rhabdovirus	Bullet shaped Bacilliform	150 x 40	Japan, Germany, Brazil, Australia and USA	sap	Chang, Arai, Doi and Yora	Chang, Arai, Doi and Yora, 1976; Doi , Chang and Yora, 1977
27.	Short orchid rhabdo virus (SORV)	Rhabdovirus	Bullet shaped Bacilliform	105 x 47	Denmark, Germany	sap	Lesemann and Doraiswamy	Lesemann and Doraiswamy, 1975a
Large Rhabdoviruses								
28.	Dendrobium rhabdo virus (DRV)	Rhabdovirus	Bullet shaped Bacilliform	180, 320 x 85	Hawaii (US)	-	Ali, Lawson and Ishii	Ali, Lawson and Ishii, 1974
29.	Laelia red leafspot virus (LRLSV)	Rhabdovirus	Bullet shaped Bacilliform	190 –220 x 80		Germany	-	Peters Peters, 1977
30.	Long orchid rhabdo virus (LORV)	Rhabdovirus Bacilliform	Bullet shaped	176 X 83	Germany	- Doraiswamy	Lesemann and Doraiswamy, 1975	Lesemann and
Bacilliform virus								
31.	Phalaenopsis bacilliform virus (PhBV- unnamed)	-	Bacilliform	119 x 29	Germany	sap	Lesemann and Koenig	Lesemann and Koenig, 1980

DIFFERENT DIAGNOSTIC METHODS, THEIR APPLICATION AND SUITABILITY IN THE IDENTIFICATION OF ORCHID VIRUSES

Bioassay

Bioassay is an old and routine technique in the diagnosis of viral infection and quantification of virus in infected host plant. In this assay virus from infected orchid host to be tested is transferred to another species (known as local lesion host / bioassay host / indicator plant) that shows diagnostic symptoms. Extracts for inoculation of test plant are generally made from leaves but may also be prepared from roots and flowers of infected plants. Reaction indicating the presence of virus may appear as yellow or brown necrotic spots on the leaves of susceptible plant several days after the test extract is inoculated. CyMV and TMV-O are most frequently detected using this kind of test (Ganapathi *et al.,* 1985; Lee and La, 1976, Samad and Noraini, 1985). Bioassays are also widely used in certification programmes for diagnosis of short orchid rhabdovirus reported from Germany (Lesemann and Doraiswamy, 1975a) and the orchid fleck virus in Japan (Chang *et al*., 1976). In addition, Tomato ring spot, Cymbidium ring spot, Cucumber mosaic, Bean yellow mosaic, Dendrobium mosaic, Tomato spotted wilt and Impatiens necrotic spot viruses are detected in bioassay. A list of bioassay host/local lesion hosts of different orchid viruses is given in table 2 below.

The greatest advantage of the test is its low cost and no special skill is required to perform the test but need a little practice. However, this procedure has numerous limitations. It is reliable under proper conditions; it is time consuming, labour intensive and requires sufficient space in green house and proper selection of ideal hosts and environmental conditions (Ko *et al.,* 1985).

ELECTRON MICROSCOPY

Electron microscopy (EM) is one of the most important tools for virus detection. It is extremely useful, allowing visual observation of the shape and size of the virus particles. Therefore EM has been successfully used in detection of CyMV and ORSV and certain other orchid viruses.

EM can be used with leaf dip preparation, crude sap preparation and ultra thin section of virus infected plant tissues. In fact CyMV and TMV-O (ORSV) can be detected in dilution of plant sap that fail to reveal either virus in bioassay or serological tests (Lawson and Brannigan, 1986). Leaf dip is the most common method used to prepare samples for electron microscopy. In this method within a few minutes shape and size of the virus can be directly be observed and detected. Besides CyMV and ORSV, this techniques have been used to detect other rod shaped viruses including Bean yellow mosaic, Dendrobium mosaic, Dendrobium vein necrosis, Cipripedium filamentous virus, Turnip mosaic, Tobacco rattle and Vanilla mosaic viruses.

EM of thin section of pollinia of *Cymbidium* species and *Cattleya* species of infected plants with CyMV and ORSV was also tried to locate the virus on the pollen. 10% of the pollen grains in pollinia from ORSV infected orchids were infected with ORSV while < 1% of those from the CyMV infected orchids contained visible virions. Virions of both viruses were detected in leaf of orchids infected with these viruses but not in apparent virus free plants (Hamilton and Valentine, 1984).

This technique itself is a very rapid, taking only a few minutes per sample, but only one sample can be tested at a time. However, electron microscopy has also several limitations. Spherical or isometric viruses, and bacilliform and bullet shaped orchid virus particles are much more difficult to detect, as the particle shape of these viruses may not be distinguished with ease from normal host plant artifacts. The EM is an expensive instrument and has limited access for common people and only limited number of research institutions and universities have such facilities.

Table 2: List of orchid viruses and their known bioassay hosts

Sl No.	Name of the virus	Bioassay hosts	References
1.	Cymbidium mosaic virus (CyMV)	*Cassia sp, Chenopodium amaranticolor, C. quinoa, Datura stamonium* and *Tetragonium expansa*	Corbett ,1960; Francki, 1970; Isadpanah *et al.*, 1966; Hollings, 1966; Inouye, 1965 and 1968
2.	Tobacco mosaic virus strains including ORSV(TMV-O and ORSV)	*Chenopodium sp,* Gromphrena globosa, Tetragonium expansa and *Nicotiana spp*	Kado and Jensen, 1964; Inouye, 1965; Kado *et al*, 1968 and Thompson and Smirk, 1967
3.	Bean Yellow Mosaic Virus	*C. quinoa and Vicia faba*	Bos, 1970
4.	Dendrobium Mosaic Virus	*C. quinoa and Chenopodium amaranticolor*	Inouye, 1976
5.	Clover vein yellow virus	*Vicia faba and Pisum sativum*	Inouye et al., 1988
6.	Turnip mosaic virus	*Nicotiana clevelandii and C. quinoa*	Lesemann and Vetten, 1985
7.	Tobacco rattle virus	*Nicotiana tabaccum* cv. xanthi	Lesemann and Vetten, 1985
8.	Cymbidiun ring spot virus	*Chenopodium amaranticolor, Emilia sagittata, Nicotiana clevelandii and Phaseolus vulgaris* cv *the Prince*	Hollings and Stone, 1963; Hollings, 1966 and Hollings, 1977
9.	Cucumber mosaic virus	*Chenopodium amaranticolor, Citullus vulgaris, D. stramonium and Seamum*	Inouye, 1969
10.	Cymbidium mild mosaic virus	*Chenopodium amaranticolor*	Chang *et al.*, 1978
11.	Tomato ring spot virus	*Nicotiana tabaccum* cv. Kentuckey 35 and *Vigna unguiculata*	Goff and Corbett, 1977
12.	Short orchid rhabdovirus	*C. quinoa and Chenopodium amaranticolor and Nicotiana clevelandii*	Lesemann and Doraiswamy 1975a
13.	Orchid fleck virus	*C. quinoa , Chenopodium amaranticolor, Nicotiana tabaccum, N. glutinosa.*	Chang *et al.*, 1976; Doi *et a.l*, 1977 and
		Petunia hybrida, Tetragonium expansa and Vigna sinensis	Kondo *et al.*, 1995
14.	Tomato spotted wilt virus	*Petunia hybrida and Nicotiana clevelandii*	Hu *et al.*, 1992
15.	Impatiens necrotic ring spot virus	*Petunia hybrida and Nicotiana clevelandii*	Mayhew *et al.*, 1992.

With the advancement of serological techniques EM was used in combination with serology for detection of orchid viruses. Immuno Sorbent Electron Microscopy (ISEM) was found to be a rapid diagnostic method for CyMV and ORSV and was found to work by trapping low concentration of the viruses in crude sap preparations. The concentrations could be increased 160-200 folds by the homologous antibodies. This method also demonstrated that ORSV was more frequent than CyMV in orchid collections. No virus was detected on seedlings raised from seeds of infected plants as well as in *Cymbidium kanran* grown from meristem culture (Park *et al.*, 1990). In Japan, ISEM was used to detect CyMV, ORSV and OFV in different species of *Cymbidium*. Kondo *et al.* (1996) detected ORSV in 11 plants belonging to *Cymbidium ensifolium, C. forrestii, C. goeringii, C. kanran, C. koran, C. sinense* and other *Cymbidium* species showing chlorotic streaks and/or mild mosaic and CyMV in one plant of *Cymbidium* species showing mosaic and necrotic spots on leaves. Orchid fleck virus (OFV) was also detected on 4 plants of *Cymbidium formosanum, C. kanran, C sinense* and *Cymbidium* species. ISEM has also been routinely used by several other virologists to detect orchid viruses (Ali *et al*., 1974, Chang *et al*., 1976; Inouye, 1972, 1977; Goff and Corbett, 1977; Lesemann and Doraiswamy, 1975b; Samad and Noraini, 1985; Ryaboi and Polishchuk, 1990; Ryu *et al*., 1995)

A further improvement of EM using immunology and fluorescent dye was done for accurate detection of orchid viruses. Immuno fluorescence microscopy is fast and more accurate method for detection of plant viruses and has been successfully used to detect CyMV and ORSV infection in fresh orchid tissues. Another highly sensitive technique as Immuno gold labeling is also known today for precise identification of ORSV and CyMV (ko *et al*., 1988).

Techniques based on Serology

Serological techniques have been extensively tested for detection of low concentrations of virus in orchid bulbs. These techniques are especially useful for large scale testing because many samples can be tested at a time. Immuno diffusion technique has been developed and used successfully for detection of CyMV and ORSV in wide range of orchid genera. Ones the antiserum is prepared, Immuno diffusion technique required less labour and time than local lesion bioassays (Bodnaruk *et al*., 1978). Three orchid growers in the USA successfully used Immuno diffusion tests. Using this technique it was demonstrated that mature and cultivated orchids were relatively more infected with CyMV and ORSV (27% of 1328 samples in 17 genera). However no infection was detected in any of 438 samples collected from their native habitats (11 genera from 21 old and new world locations) or in 231 seedling orchids (Wisler *et al*., 1982). Micro agglutination test has also been used to diagnose CyMV and ORSV in 120 Cymbidiums and 46 *Cattleya* plants (of which 39 had CyMV and 46 ORSV and 9 had both virus infections) in Japan. Result obtained by biological tests and electron microscopy gave exact agreement and serological tests showed 99.4% agreements with the other tests (Inouye, 1977). The latex agglutination test (L) and a modification of this test, protein A –coated latex (PAL) were also standardized for detection of CyMV. The minimum detectable concentration of virus was similar for both L and PAL reaction at 0.104 mg/ml. The detectable limit in *Oncidium* sap was similar at a dilution of 1:2560. PAL appeared as more sensitive that L at a higher globulin dilution of 1:512 and in general, PAL demonstrated greater sensitive than L in the orchid species tested (Samad and Ari, 1993).

With the development of more sensitive serodiagnostic technique of Enzyme Linked Immuno Sorbent Assay (ELISA: Clark and Adams, 1977) virus testing became comparatively more easier and simpler not only in orchids but also in all other plants. ELISA can be used successfully by both amateur and grower with some of the readymade lest kits currently available in the market. Beetham and Moran (1988) developed and used an ELISA to indexed ORSV (using an antiserum of a Victorian isolate of ORSV) in *Cymbidium* genus from 10 commercial and home garden collections in Metropolitan Melbourne, Victoria, Australia. Urease was used for the conjugate and it was found to facilitate a highly sensitive colour change reaction. No ORSV was detected in 3 of the collections indicating that less than 1% of plants were infected. Estimated infection levels ranged from 3.45- 20.63% in the others.

ELISA can also be used for quick detection of ORSV released from the cut surfaces of leaf discs. Results of an experiment indicated that detection was possible but that multiple discs were prerequisite to obtain reliable detection at low virus concentration (Arunasalam and Pearson, 1989). Hu *et al.* (1993) used ELISA in large scale indexing of orchid viruses. He indexed approximately 3600 orchid plants representing 44 genera from 3 orchid collections, 22 commercial farms and 6 nurseries on the Islands of Oahu and Hawaii, USA for the presence of CyMV and ORSV, TSWV and Poty viruses. CyMV and ORSV were detected in 61% and 25% of 44 genera. Double infection with both viruses occurred in 20% of the genera. CyMV and ORSV were detected in 29 and 7 of 31 sites surveyed respectively. Besides, out of 330 clonal orchid samples he tested, CyMV was detected in 45% and ORSV in 9% of the clones. Most commercial

Dendrobium orchids grown in Hawaii are seed propagated hybrids produced by the Horticulture Department at the University of Hawaii (UH). Only 4% of *Dendrobium* hybrids (< 3 years old) were infected by CyMV and 94% 0f >3 years old UH Dendrobiums was infected with CyMV. ORSV was not detected from any commercial grown UH *Dendrobium* hybrids. TSWV infection was found in Oncidium in only 1 Hawaii nursery. Dendrobium mosaic Poty virus was detected in *Dendrobium superbum*.

ELISA can be operated using Monoclonal antibodies (McAbs) or polyclonal antibodies (PcAbs). It is recommended that a mixture of polyclonal and monoclonal antibodies in DAS –ELISA is mandatory for reliable detection of both ORSV and TMV-O in orchids (Dore *et al*., 1987). Vejaratpimol *et al*. (1998) tested indirect ELISA and DAS-ELISA using Polyclonal antibodies (PcAbs), egg yolk immunoglobulin G (IgY), rabbit antisera (IgG), or Monoclonal antibodies (McAbs) alone or any 2 in combination for their suitability to detect CyMV on orchids. It was found that indirect ELISA using IgG, IgY or PcAbs was able to detect 50-100ng of purified virus and detect virus in 2-8folds dilution of infected leaf extract. Detection of purified CyMv by DAS-ELISA using IgY as the detecting antobody gave 4-16 folds higher sensitivity that that using IgG on either McAbs or PcAbs coated microplates. The DAS-ELISA using McAbs as the coating antibody and IgY as the detecting antibody was shown to be the most sensitive, being able to detect 6.25 ng of purified virus and it could detect the virus in a 1024 folds dilution of an infected leaf extract.

Other sensitive and rapid immunological techniques used in the detection of orchid viruses are Dot Immuno Binding Assay (DIA) on nitrocellulose membranes and Rapid Immuno Filter Paper Assay (RIPA). These two techniques were evaluated to detect CyMV in several orchid genera including *Cymbidium, Cattleya, Dendrobium, Phalaenopsis and Vanda*. Minimum detection levels of CyMV using these methods were 100ng/ml in purified preparations, and CyMV was detected in a 10-4 dilution of extracts from infected orchid leaves. Though DIA took longer time to detect CyMV, it was reliable for diagnosis of a large number of samples. RIPA detected CyMV within a few minutes and sensitivity was similar to that of DIA. It is suggested that RIPA is suitable as a rapid tool for diagnosing virus diseases in orchids. RIPA could detect both CyMV and ORSV simultaneously in extracts from a plant infected by both viruses (Gara *et al*., 1997). RIPA was used in large scale detection of CyMV and ORSV in 43 nurseries in 8 provinces of Thailand (Tanaka *et al*., 1997).

La *et al*. (1999) applied Tissue Blot Immuno Assay (TBIA) to detect CyMV and ORSV from infected *Cymbidium* and *Cattleya* plant in Korea. CyMV and ORSV were readily detected in tissue blots made from leaves and roots of infected plants both direct and indirect TBIA. CyMV and ORSV were more efficiently detected in root than leaf samples by TBIA. Tissue blots made on polyvinylidene difluoride membrane gave much clearer specific colour reactions than those made on nitrocellulose or nylon membrane.

Nucleic acid hybridization based techniques using specific probes have been variously used for detection of orchid viruses. A Tissue Print hybridization Assay was standardized to detect and locate viruses in orchids. In this method orchid leaves are cut transversely and the cut surface are printed on to nitrocellulose paper. The nucleic acid on the nitrocellulose membrane is then hybridized to radio labeled cDNAs of CyMV and ORSV and the viruses are detected by autoradiography. The assay is able to trace as little as 20pg of purified virus that is 2-3 folds more sensitive than ELISA for CyMV. This method is also able to show the distribution of viruses within the plants (Chia *et al.,* 1992). Tissue print hybridization with cDNA probes was applied to detect viruses directly from infected leaves of *Phalaenopsis* species (Wang *et al.,* 2000).

Further improvement took place in the nucleic acid hybridization techniques and a Slot Blot Hybridization Assay was developed and standardized using either crude lead extracts or total RNA from infected leaves. Prior to that, DIG labeled sense and antisense cRNA probes were synthesized from cDNA clones of CyMV and ORSV for detection in infected plants. This assay detected 50pg and 250 pg of purified CyMV and ORSV RNA respectively. As little as 30mg of *Nicotiana benthamiana* infected leaves was sufficient to provide positive detection. CyMV and ORSV were detected at 3125 and 625 times dilution of leaf extracts respectively. The advantage of this method is that the DIG labeled cDNA probes is stable for >1 year. This method is sensitive and reliable and it is suitable for large-scale routine testing of plant viruses. By using the two DIG labeled cRNA probes *in situ*, CyMV and ORSV were located in systemically infected leaves and stems of *Nicotiana benthamiana* and orchids (Hu *et al.*, 1998).

When Dot Blot (DOT Immuno binding ELISA), DAS-ELISA and Indirect ELISA were compared with respect to sensitivity (Table 3), cost, time, requirements and potential for large scale application, it was found that all three were adequately sensitive but DOT Blot ELISA was faster and cheaper than others (Albouy et al., 1994). Detection of CyMV in crude sap preparation of infected leaves was 2 times more sensitive by ISEM than ELISA while DOT Blot Immuno Assay was 8 times more sensitive than ELISA. Bean yellow mosaic virus isolated from *Masdevallia* and *Calanthe* could not be detected by either DAS-ELISA or antigen coated indirect ELISA directly from the infected orchids using polyclonal antisera. But the detection was improved when McAbs were used in a DOT Blot ELISA or nitrocellulose membrane (Hammond and Lawson, 1988).

Table 3. Approximate sensitivity of various methods of virus detection in orchids

Sl No.	Method	Minimum detectable virus concentration
1.	Indirect ELISA	50-100 ng
2.	DAS - ELISA	6.25 ng
3.	Rapid Immuno Filter Paper Assay (RIPA)	100 ng
4.	DOT Immuno Binding Assay (DIA)	100 ng
5.	Slot Blot Hybridization Technique 250 ng (ORSV)	50n g (CyMV),
6.	Tissue print Hybridization Assay	20 pg
7.	PCR	Single copy of virus genome
8.	RT-PCR	10 fg
9.	Immuno Capture PCR	100 ng (CyMV) 10 ng (ORSV)
10.	Taq Man R real-time PCR	5 fg of each of CyMV and ORSV
11.	Molecular beacons	0.5 ng each of CyMV and ORSV

PCR BASED TECHNIQUES

PCR based techniques were recently introduced for the detection of orchid viruses. This technique is routinely used to detect orchid viruses due to its sensitivity and can detect even a single copy virus genome in the infected plant (Hull, 1993). Polymerase Chain Reaction was successfully used for rapid detection of CyMV using two ologonucleotide primers from regions homologous to potex viruses and CyMV in 30 cycles of PCR which resulted amplicons of 313 bp (Lim *et al.*, 1993a). PCR was also tested to detect CyMV and ORSV in plantlets obtained from thin section cultures of infected plants and PLBs with ribavirin treatment (Lim *et al.*, 1993b). A combined assay of Reverse Transcription and the Polymerase Chain Reaction (RT-PCR) was performed with 20 mer ORSV specific primers and crude nucleic acid extracts from virus infected orchids for rapid detection of the virus. The lowest concentration of template viral RNA required for detection was 10 fg. The RT-PCR is a 103 times more sensitive, reproducible and time saving method than ELISA. RT-PCR was also used to detect CyMV using 20 mer CyMV coat protein specific primers from crude nucleic acid extracts of virus infected orchid leaf tissue of *Cymbidium* spp., *Dendrobium* spp and *Oncidium* spp (Ryu *et al.*, 1995a, b). It has been established that RT-PCR leads to quick diagnosis of CyMV or ORSV (Jeong *et al.*, 2001).

Immuno Capture PCR (IC-PCR) assay, another improve version of PCR was developed and emerged as an efficient, reliable and rapid method to detect CyMV and ORSV and compared with direct binding PCR and ELISA. In this technique, Hawaiian strain of each of CyMV and ORSV was purified from *Dendrobium* and *Cattleya* orchid respectively to improve virus detection and to facilitate genetic engineering. The ORSV coat protein (CP) and 54k putative replicase component (54k) gene and CyMV CP and movement protein (MP) genes were cloned and sequenced. The sequenced of these genes were compared to those of CyMV and ORSV from Singapore, Japan, Korea and Germany. The high degree of sequence identity (91-99%) at the nucleotide level for all gene sequences analyzed, shows that CyMV and ORSV from different countries are closely related (Barry *et al.*, 1996). Barry and his associates observed that IC-PCR is 10-100 folds more sensitive than ELISA do and is able to detect CyMV from the equivalent of 100mg of *Dendrobium* leaf tissue. Therefore IC-PCR can be used to confirm samples which are inconclusive in ELISA.

Recently Cherng *et al.* (2000a) developed a highly sensitive novel detection technique called TaqMan R·real-time PCR, which combines RT-PCR with the power of fluorescent detection for simultaneous quantitation of two orchid viruses, CyMV and ORSV. Four TaqMan R probes were synthesized, targeting at the RNA dependant RNA Polymerase (RdRP) and CP genes of both the viruses. This technique could detect as little as 5 fg each of CyMV and ORSV simultaneously with either the RdRp or CP genes the target. Another identical technique is the development of molecular beacons. It is a new approach to plant virus detection. This approach has been successfully applied to detect as little as 0.5ng of viral RNA from *Oncidium* leaves simultaneously infected with CyMV and ORSV. *Cherng et al.* (2000b) claimed that this technique is applicable to orchid industry which routinely carries out virus indexing and screening for virus resistant cultivars.

Based on the molecular weights of coat protein of viruses, another rapid and simultaneous detection method of the two most prevalent orchid viruses, CyMV and ORSV, was developed using Liquid Chromatography/ Mass Spectrometry (LC/MS) and Matrix Assisted Laser Desorption-ionisation (MALDI) Mass Spectrometry. Using this method, molecular weights of coat proteins of CyMV and ORSV were accurately estimated and detected from 1g of virus infected flower of *Oncidium* (Tan *et al.*, 2000). Tan and his associates claimed that because of LC/MS and MALDI perform automated analyses of multiple samples with simple preparation steps, both the techniques are ideal for virus detection in large number of samples from infected plant extract.

Various methods are now available for virus detection in orchids with different levels of accuracy and sensitivity. One in a particular situation has to choose a viable technique cautiously. Today's multimillion-dollar orchid trade over the world is regulated with phytosanitary (biosafety) certificate and quarantine inspection of plant material as well as flower is mandatory before import or export. In that context, what is prime need is an easy to use, readily available and affordable diagnostic kit whether it based on coat protein or nucleic acid component for virus diagnosis. Suppose, a commercial orchid trader needs simply to know the presence or absence of any kind of virus in his import consignment, he certainly has to choose an imprint hybridization technique using a readily available DIG labeled broad spectrum probe which would help screening a large number of samples in a short time (Ramachandran and Varma, 2001)

MANAGEMENT STRATEGIES FOR ORCHID VIRUSES

Management through Sanitation

Strict sanitation programme is essential to prevent accidental transmission of orchid viruses. This includes using clean tools. Heat treatment is the safest and most effective method of sterilizing tools. Reusable razor blades or knives should be heated in an oven at 149^{0}C for an hour. The most stable orchid virus TMV-O can be inactivated by heating the blade of the cutting tools at 96-100 ^{0}C for few minutes. Propane torches may be used to treat cutting knives. The tools can also be chemicals sterilized by dipping them in alcohol before they are flamed. The cutting surface should be exposed to the flame for several seconds. Liquid virus inactivating solutions are available in the market and can also be used to disinfect tools. 2% Clorox can be used to inactivate CyMV in cutting tools but not TMV-O. A mixture of 2% sodium hydroxide and 2% formaldehyde has been recommended for treating cutting tools. 2% solution of Sodium hydroxide can effectively reduce TMV-0 infectivity by 86-96% Trisodium phosphate (5%) is another good virus inactivator because it produces a high pH in water solution. Disinfection of benches, on which diseased plants were maintaining previously, was recommended for reuse. Spraying of 10% solution of Clorox or Trisodium phosphate are advisable on the branch surfaces. TMV-O is stable and persists in plant debris for years, removal of all debris and roots from bench surface is essential prior to chemical treatment. Sanitation through proper sterilization of pots is also essential as orchids viruses are known to be transmitted from contaminated surface of pots. *Cattleya* seedlings planted in sphagnum from pots previously growing CyMV and TMV – O infected plants have become infected. Potting substrate should not be reused (Freitas, 2003).

Management through Vector Control

Although two most common and wide spread orchid viruses CyMV and ORSV are known to be transmitted mechanically through sap and till today there is no report of insect vector for these two viruses, several other orchid viruses are reported to be transmitted through their insect vectors. Therefore, management of insect vector may be effective in managing those orchid viruses. The green peach aphid (*Myzus persicae*) has been reported to transmit some strains of BYMV, Clover Yellow vein, Turnip mosaic, Vanilla potyvirus, Dendrobium mosaic, Cucumber mosaic virus and Dendrobium vein necrosis virus (Zettler *et al.,* 1987). Besides, Tomato spotted wilt virus and Impatiens necrotic spot virus are known to have thrips vector in orchids. Aphid and thrips control through insecticide sprays as soon as the insects are evident may be an important part of preventing the possible spread of several different viruses in orchids. Separation /isolation of affected plants to reduce the chance of aphids and thrips moving to other plants is always a good policy.

Management through Meristem culture

Truly speaking there was no definite control measures of orchid viruses except sanitation and vector management through chemicals until the early sixties of the nineteen-century. Indeed it was Morel (1960) who first developed revolutionary techniques of meristem culture to produce virus free Cymbidiums through meristem culture which established the techniques as a clonal propagation procedure (Morel, 1964). Morel standardized meristem tip culture to obtain virus free clonal material of *Cymbidium* (such as *C. talma, C. doris or C. candeur* based on the concept that majority of plant viruses do not infect the meristematic dome despite severe systematic infection of the plants and therefore, this region can be excised and regenerated to plantlet. Since then meristem culture has been used routinely for a large number of orchids to obtain CyMV free plants. Ishii (1974) successfully eliminated CyMV from double infected mericlone tissue pieces grafted onto callus tissue but ORSV remained infections. However, true meristem culture, which uses the actual meristem less than 0.5 mm takes longer time to regenerate plants. Sometimes, little or no survival may result when tissue pieces are less than 0.5 mm in length. On the other hand tissue pieces of 0.5-1.0 mm showed an increased in survival but remained virus infected. Consequently the meristem tip culture which was once thought to be the best method to obtain CyMV and ORSV free orchids has no longer been accepted as a means of clonal propagation techniques for virus free plants.

Management through Meristem culture with Chemotherapy

Further attention was paid to develop an alternative of meristem trip culture to get large quantity of virus free orchid stocks. In the early nineties new concepts of meristem tip culture with chemotherapy come into the picture. In this improved method meristematic tip plant tissue of Cymbidium hybrid Ayako Tanaka "Honodezuru" is cultured *in vitro* in association with antiviral agent ribavirin (Virazole). Results revealed that all plants regenerated from ribavirin free media were still infected by ORSV where as all treatments with ribavirin (with the exception of one) provided at least one virus free plant. However their efficiency varied with the concentration of the ribavirin, the way it was added to the medium and the type of medium to which it was added, resulting in 8.3 - 100% virus free orchids. Thus association of meristem tips culture and chemotherapy *in vitro* was effective for the production of ORSV free *Cymbidium* (Freitas and Rezende, 1998). Albouy *et al* (1988) added ribavirin (Virazole) in solid medium culture of Cymbidium Protocorms. They found that 5 subcultures of 18 days interval in the presence of 25mg/ litre of virazole produced 95% virus free plantlets and enhanced the efficiency of eradication of both CyMV and ORSV. Loi *et al.* (1991) reported that virazole was found effective for obtaining virus free Dendrobium shoot tips and callus from parent plants infected by CyMV and ORSV and it also eradicated ORSV from *in vitro* cultures of Cymbidium (Toussaint *et al.*,1993).

Kuehnle (1996a) tested virucides at various concentrations and treatment times using protocorms like bodies (PLBs) of *Dendrobium* (Jaquelyn Thomas "Uniwai Mist") infected with CyMV. These chemicals were added during the liquid culture phase of PLBs and also during solid medium phase of plant formation from the PLBs. Among the virucides, ribavirin or dithiouracil eliminated virus infection. In liquid culture chemotherapy only, the best results obtained with 53% virus free tissue in 0.1mM ribavirin and 63% virus free tissue in 0.2mM dithiouracil. A five weeks treatment was sufficient. In addition rapidly growing cultures in liquid medium also increased recovery of virus free material with upto 41% of virused control. Most exciting results were obtained with a combination of antiviral compound in liquid and solid medium tissue culture. Mericloning in 0.2mM dithiouracil (liquid) followed by plant production on 0.1mM dithiouracil (solid), produced upto 100% virus free platelets. Plant remained free of virus 2.5 years after chemotherapy and flowered normally. Hence chemotherapy is assured to be highly useful for ridding *Dendrobium* of CyMV at clonal propagation stage.

Yap *et al.* (1999) also demonstrated the eradication of CyMV from tissue culture plantlets of *Oncidium* Goldiana (Clone 018) through chemotherapeutic treatment of ribavirin. The authors found that 10 mg/ L of ribavirin for 30 days was sufficient to eradicate CyMV from 43% of meristem plantlets compared with 31% eradication from plantlets derived from callus tissues treated with 20 mg / L for 60 days and concluded that meristem treatments with ribavirin gave better eradication than that of callus tissues.

Management through Thin Section Culture with Chemotherapy

Thin Section culture with Chemotherapy was established as an efficient method of eradication of orchid virus by various researchers. It was reported that thin section cultures of infected plantlets and protocorms like bodies with chemotherapeutic treatment of ribavirin appeared to be better than meristem culture with chemotherapy. Lim *et al.* (1993a) demonstrated that regenerates from then section cultures of infected monopodial orchid hybrid (Mokara Char Kuan Pink) with or without ribavirin treatment was also free from CyMV and ORSV. However, the percentage of regenerates free from CyMV and ORSV varied. The regenerates from thin section (TS) culture of shoot tips were practically all free from both CyMV and ORSV whereas regenerants from TS of PLBs were only about 60% free. Prakash *et al.* (1996) have also tested the suitability of TS cultures with chemotherapy of ribavirin (virazole) for the production of virus free plantlets in several other orchids including *Aranda, Cattleya, Dendrobium, Oncidium* and *Vanda* and suggested that TS culture could be used for mass propagation of orchids as well as for the elimination of viruses from the infected plants alone or in combination with virazole.

Eradication of virus from orchids through Androgenesis

Eradication of CyMV from *Dendrobium* through androgenesis coupled with chromosome doubling with colchicine treatment during PLB culture was reported in Hawaii (Kuehnle, 1996a). Androgenesis refers to the production of haploid plant with only the parental chromosome set. In *Dendrobium*, ovule androgenesis occurs after fertilization of the egg cell with the elimination of maternal nucleus. It was first observed by

Kamemoto and reported in details by Amore and Kamemoto (1991) that amphidiploid plants crossed with the triploid D.x Louis Bleriot as female gave rise to plants identical to the original amphidiploid male parent but with half the usual number of chromosomes (i.e. haploid). Since the haploid plants go through seed, they become free of virus. Chromosome doubling of the androgenetic diploid with colchicine (during PLB culture) then restores the tetraploid chromosome number. This in effect enables recovery of virus free amphidiploid practically, identical to the original parental plants. This natural phenomenon was applied to obtain virus free breeding plants (Amore *et al.*, 1996).

Management through Genetically Engineered Virus Resistance

Genetic engineering coupled with tissue culture techniques offers a useful way to introduce specific genes into plants. The development of a suitable gene transfer system for *Dendrobium* Orchids would provide the breeder with greater opportunity to produce commercially desirable hybrids. Several gene transfer methods were tested with *Dendrobium* (Table 4). Among these micro particle bombardment (Klein *et al.,* 1987) was identified by two separate research groups as highly suitable for transformation of *Dendrobium* based on high frequency recovery of transgenic plants. DNA sequences thought effective against CyMV and ORSV were introduced into commercial cut flower varieties of *Dendrobium* (Barry et al., 1995). Selected regenerated, engineered plants were mericloned and are being grown in the green house to permit meaningful replicated disease trials. *Dendrobium* breeders can now proceed with increased confidence to provide producers with virus free starting material and in near future, new varieties possessing genes for virus resistance. With disease oriented molecular breeding programmes already well underway in Singapore or being established in South Korea, New Zealand and elsewhere using a variety of orchid genera (Kuehnle 1996b) and the threat of disease from CyMV promises to be reduced on a global scale in the new millennium.

Table 4: Selected methods of gene transfer tested with *Dendrobium*

Method	Short description of Procedure	References
Particle bombardment (gene gun)	Gold of tungsten spheres of about 1mm diameter coated with DNA are accelerated into target tissues of Protocorm or PLBs. Tissues are cultured on pant development medium with selection of transformants by media amendments or by expression of reporter genes. High stable transformation efficiency upto 11.7%.	Kuehnle and Sugii, 1992; Nan and Kuehnle, 1995a.
Protocorm electro-injection	DNA in solution enters target cell under the influence of an electric pulse. Protocorms are screened on plant growth medium containing selection agent. Acceptable stable transformation efficiency of 1.4%.	Nan and Kuehnle, 1995b.
Seed imbibition	Dry seeds removed from capsules are imbibed in DNA solution, then screened on germination medium containing selection agent. Low frequency recovery of transgenic tissue; 0% recovery of plants after more that one year on selection medium.	Nan and Kuehnle, 1995b
Pollen tube mediated DNA transfer	*In planta*: stigmatic surface of a pollinated flower is sliced and a DNA solution applied. *In vitro*: capsules are harvested, sliced at the column end and exposed to DNA solution, and cultured on ovary maturation medium. Resulting seeds are screened on germination medium containing selection agent. Low frequency recovery of transgenic tissue; 0% recovery of plants after more that one year on selection medium.	Nan and Kuehnle, 1995b

REFERENCES

Albouy, J., Flouzat, C., Kusiak, C. and Tronchet, M. 1988. Eradication of orchid viruses by chemotherapy from in vitro cultures of *Cymbidium*. *Acta Hort*. 234: 413-420.

Albouy, J., Kusiak, C., Louanchi, M., Wang, W., Mokra, V., Brunt, A., Derks, T. and Zaayen, A. V. 1994. Evaluation of rapid ELISA procedures for detecting viruses in propagation stock of *Pelargonia, Dahlia* and orchids. *Acta Hort*., 337:189-195.

Ali, S., Lawson, R. H. and Ishii, M. 1974. A bacilliform virus in white streaked *Dendrobium phalaenopsis* flowers. *Amer. Orchid Soc. Bull*., 43: 529-33.

Amore, T. and Kamemoto, H. 1991. Andogenesis in *Dendrobium*. *Lindleyana*, 6: 117-120.

Amore, T., Kamemoto, H., Kuehnle, A. R. and Sugii, N. 1996. Rejuvenation of virused amphidiploid *Dendrobium* breeding plants. *Univ. of Hawaii Hort. Digest*, No. 110:3-4.

Arunasalam, G. and Pearson, M. N. 1989. ELISA detection of Odontoglossum ring spot virus in mature plants and protocorms of *Cymbidium* orchids: Potential solutions to problems of sample preparation time and low virus concentration. *J. Phytopathol.*, 126: 160-166.

Barry, K., Hu, J. S., Kuehnle, A. R. and Sugii, N. 1995. Nuclear cloning and sequences analysis of Cymbidium mosaic and Odontoglossum ring spot viruses. *Phytopath*., 85: 1184.

Barry, K., Hu, J. S., Kuehnle, A. R. and Sugii, N. 1996. Sequences analysis and detection using Immuno capture PCR of Cymbidium mosaic and Odontoglossum ring spot viruses in Hawaiian orchids, *J. Phytopath.*, 144:179-186.

Beeetham, P. R. and Morani, J. R. 1988. ELISA indexing of Odontoglossum ring spot virus using urease antibody conjugate and its use in field surveys. *Acta Hort*. No.234: 243-248.

Begtrup, J. 1972. Structure of bacilliform virus in *Dendrobium* as revealed by negative staining. *Phytopathol. Z.*, 75:268-273.

Bodnaruk, W. H., Hennen, G. R. and Zettler, F. W. 1978. Applicability of immunodiffusion techniques in detecting Cymbidium mosaic virus and Odontoglossum ring spot virus in orchids. *Amer. Orchid Soc. Bull*., 47: 1113-1116.

Bos, L. 1970. Bean Yellow mosaic virus. CMI /AAB. *Description of plant viruses*. No. 40. Common Wealth Agricultural Bureaux Slough, England.

Brunt, A. A., Crabtree, K. Dallwitz, M. J., Gibbs, A. J. Watson, L and Zurcher, E. J. (Eds.) 1996. *Plant Virus online*: Descriptions and Lists from the VIDE Database. Version: 20th August 1996. URL http://biology.anu.edu.au/groups/MES/vide/.

Chang, M. U., Arai, K., Doi, Y. and Yora, K. 1976. Morphology and intracellular appearance of Orchid fleck Virus. *Ann. Phytopath. Soc. Japan*., 42: 156-57.

Chang, M. U., Doi, Y. and Yora, K. 1978. Cymbidium mild mosaic virus, isolation Cymbidium mild mosaic virus. *Korean J. Plant Prot.* ,17: 131- 138.

Cherng, Eun-Jin; SekMan Wong; Eun, J. C. A. and Wong, S. M. 2000b. Molecular beacons: a new approach to plant virus detection. *Phytopath.*, 90: 269-275.

Cherng, Eun-Jin; Seoh, M., SekMan Wong; Eun, J. C. A., Seoh , M. L. and Wong, S. M. 2000a. Simultaneous quantitation of two orchid viruses by TagMan R real time RT-PCR. *J. Virol. Methods*, 87: 151-160.

Chia, T. F., Chan, Y. S. and Chua, N.H. 1992. Detection and localization of viruses in orchids by tissue print hybridization. *Plant Path*., 41: 355-61.

Chowdhery, H. J. 2001. Orchid Diversity in North East. *J. Orchid Soc. India*. 15: 1-17.

Clark, M. F. and Adams, A. M. 1977. Characteristics of microplates method of ELISA for the detection of plant viruses. *J. Gen. Virol*., 34: 475-83.

Corbett, M. K. 1960. Purification by density gradient centrifugation, electron microscopy and properties of Cymbidium mosaic virus. *Phytopath.*, 50: 346-251.

Corbett, M. K. 1967. Some distinguishing characteristics of the orchid strain of tobacco mosaic virus. *Phytopath*., 57:164-172.

Corbett, M.K. 1974. Intracellular bacilliform virus likes particles in *Grammatophyllum scriptum*. *Phytopathology* (Abst.) Potomac Div. Meeting of Amer. Phytopathol. Soc., March 28-29, 1974.

Doi, Y., Chang, M.U. and Yora, K. 1977. Orchid fleck virus. CMI/AAB. *Description of plant viruses*. No. 183.

Doi, Y., Toriyama, S., Yora, K. and Asuyama, H. 1969. Ann. Phytopath. Soc. Japan, 35:388

Dore, I., Dekker, E. L., Porta, C. and Reggenmortel, M. V. H. 1987. Detection by ELISA of two tobamo viruses in orchids using monoclonal antibodies. *J. Phytopath.*, 120: 317-32.

Duvel, D. and Peters, K.R. 1971. Virusahniche partikelin dendrobium antenatum Ldl. *Gartenwelt,* 71: 52-54.

Faccioli, G. and Marani, F. 1979. Cymbidium mosaic virus associated with flower necrosis in *Cattleya* orchids. *Phytopath.Med.*, 18: 21-25.

Francki, R. I. B. 1970. CMI/AAB Description of plant viruses. No. 27. Common wealth Agricultural Bureaux Slough, England.

Freitas, A. J. 2003. Orchid viruses: Prevention remains the best defense against these pathogens. *Orchids,* 27: 578-580.

Freitas, A. J. and Rezende, J. A. M. 1998. Odontoglossum ring spot virus free *Cymbidium* obtained through meristem tip chemotherapy. *Fitopat. Brasi.*, 23: 158-60.

Ganapathi, A., Chinnathambi, C and Chullakaset. 1985. Some virus diseases of orchids and an attempt at their control. In: *Proc. Reg. Conf. Pl. Quar. Supp. Agric. Devel.* pp.87-91. Kuala Lumpur. Malaysia.

Gara, I. W., Kondo, H. and Maeda, T.1997. Evaluation of Dot Immuno binding assay and Rapid Immuno filter paper assay for detection of Cymbidium Mosaic virus in orchids. *Bull. Res. Inst. Bioreso., Okayama Univ.*, 4: 163-174.

Gara, I. W., Kondo, H., Maeda, T., Inouye, N. and Tamada, T. 1998. *Calanthe* mild mosaic virus, a new poty virus causing mild mosaic disease of *Calanthe* orchid in Japan. *J. Phytopath.*, 146:357-363.

Gara, I. W., Kondo, H., Maeda, T., Mitsuhara, K. and Inouye, N. 1996. Further characterization of CyMV from Vanda orchid. *Bull. Res. Inst. Bioreso., Okayama Univ.*, 5: 39-46.

Goff, L. M. and Corbett, M. K. 1977. Association of Tomato spotted wilt virus with a chlorotic leaf streak of *Cymbidium* orchids. *Phytopath.*, 67: 1096-1100.

Hamilton, R. I. and Valentine, B. 1984. Infection of orchid pollen by Cymbidium mosaic and Odontoglossum ring spot viruses. *Canadian J. Plant. Pathol.*, 6: 185-190.

Hammond, J. and Lawson, R. H. 1988. A strain of bean yellow mosaic virus is aphid transmitted. *Acta hort.*, No 234: 365-370.

Hollings, M. 1966. Local lesion and other test plants for the identification and culture of viruses. pp 230-241 in A. B. R. Beemster and J. Dijksta (eds), Viruses of plant, North Holland Pub. Co., Amsterdam.

Hollings, M. 1972. Orchids-Cymbidium ring spot. *GlassHouse crops Res. Inst. Ann. Report.*, pp 104.

Hollings, M. 1977. Cymbidium ring spot virus. CMI/ AAB. *Description of plant viruses*. No. 178.

Hollings, M. and Stone, O. M.1963. Cymbidium ring spot (previously undescribed virus). *Glasshouse crops Res. Inst. Ann Report.*, pp 89.

Hu, J. S., Ferreira, S., Wang, M. and Xu, M. Q. 1993. Detection of Cymbidium mosaic, Odontoglossum ring spot, Tomato spotted wilt virus and Poty viruses infecting orchids in Hawaii. *Plant Dis.*, 77: 464-68.

Hu, J. S., Wang, M., Ferreira, S. and Ogata, D. 1992. Tomato spotted wilt virus on *Oncidium* orchids in Hawaii. *Plant Dis.*, 76: 426.

Hu, W. W. and Wong, S. M. 1998. The use of DIG labeled cRNA probes for the detection of Cymbidium mosaic Potex virus and Odontoglossum ring spot tobamo virus in orchids. *J. Virol. Methods*, 70:193-199.

Hull, R. 1993. Nucleic acid hybridization procedures. In; *Diagnosis of plant viruses* (ed. R. E. F. Mathews) PP 253-71, CRC Press, Boca Raton.

Inouye, N. , Maeda, T. , Mitsuhata, K. and Gara. I. W. 1998. *Habenaria* mosaic virus, a new member of potyvirus from *Habenaria radiata* orchid. *Bull. Res. Inst. Biores.*, 5: 155-168.

Inouye, N. 1965. Virus diseases of orchids. II. Symptoms and properties of viruses in *Cymbidium. Japan Orchid Soc. Bull.*, 11: 1-6.

Inouye, N. 1968. Some experiments on the transmission of Cymbidium mosaic virus and Odontoglossum ring spot virus. *Nogaku Kenkyu*, 52: 89-97.

Inouye, N. 1969. Cucumber mosaic virus isolated from *Dendrobium. Agric. Res. (Kurashiki).*, 53:49-60.

Inouye, N. 1971. Virus diseases of orchids. V. Symptoms and properties of viruses in *Dendrobium. Japan Orchid Soc. Bull.*, 17: 3-7.

Inouye, N. 1972. On the causal virus of flower necrosis on *Cattleya* Alliance orchids. *Noga Kenk.*, 54: 71-78.

Inouye, N. 1973. A new virus isolated from *Dendrobium. Ann. Phytopathol. Soc. Japan,* 39: 367-368.

Inouye, N. 1976. Dendrobium mosaic virus. *Ber. Ohara Inst. Landw. Biol. Okayama Univ.* 16: 165-174.

Inouye, N. 1977. Serological diagnosis methods for Cymbidium mosaic and Odontoglossum ring spot virus in orchids. *Noga Kenk.*, 56: 1-13.

Inouye, N., Maeda, T. and Mitsuhata, K. 1988. A strain of Clover yellow vien virus isolated from *Calanthe* sp. *Acta Hort.*, 234: 61068.

Isadpanah, K. M. , Thompson, R. and Thornberry, H.H. 1966. A simple method for detection of flexuous rod virus associated with infectious *Cattleya* blossom necrosis. *Plant. Dis. Report.*, 50: 779-781.

Ishii, M. 1974. Partial elimination of virus from doubly infected orchids by meristem explant culture. *Acta Hort.*, No. 36: 229-33.

Jensen, D. D. and Gold, A. H. 1951. A virus ring spot of *Odontoglossum* orchid: Symptom, transmission and electron microscopy. *Phytopath.*, 41: 495-497.

Jensen, D. D. 1951. Mosaic or black streak diseases of *Cymbidium* orchids. *Phytopath.*, 41: 401-414.

Jensen, D. D. 1970. Virus diseases of orchids in Netherlands. *Netherlands J. Pl. Pathol.*, 76:135-139.

Joeng, J. H., Yang, D. C. and Paek, K. Y. 2001. Comparison of ELISA and RT-PCR for efficient diagnosis of Cymbidium mosaic and Odontoglossum ring spot viruses in *Cymbidium* hoosai Makino. *J. Korean Soc. Hort. Sci.* 42: 336-340.

Kado, C. I. and Jensen, D. D.1964. Cymbidium mosaic virus in *Phalaenopsis. Phytopath., 54: 974-977.*

Kado, C. I., Van Regenmortel, M.H.V. and Knight, C.A. 1968. Studies on some strain of tobacco mosaic virus in orchids. I. Biological, chemical and serological studies. *Virology*, 34: 17-24.

Kitajima, E. W., Blumenschein, A. and Costa, A. S. 1974. Rodlike particles associated with ring spot symptoms in several orchid species in Brazil. *Phytopathol. Z.*, 81: 280-286.

Klein, T. M., Wolf, E.D., Wu, R. and Sandford, J. C. 1987. High velocity microprojectiles for delivering nucleic acids into living cells. *Nature,* 327: 70-73.

Ko, N. J., Zettler, F. W. and Edwardson, J. R. 1988. Immunocytochemical identification of Cymbidium mosaic and Odontoglossum ring spot viruses in orchid tissues. *Acta Hort.*, 234: 299-306.

Ko, N. J., Zettler, F. W. and Wisler, G. C.1985. A simplified bioassay technique for Cymbidium mosaic and Odontoglossum ring spot viruses. *Amer. Orchid Soc. Bull.*, 54: 1080-1082.

Koike, S.T. and Mayhew, D. E. 2001. Impatiens Necrotic spot virus found in *Oncidium*: reporting a new ailment for orchids. *Orchids*, 70:746-747.

Kondo, H. J., Matsumoto, J., Maeda, T. and Inouye, N. 1995. Host range and some properties of orchid fleck virus isolated from oriental *Cymbidium* in Japan. *Bull. Res. Inst. Biores., Okayama Univ.,* 3:151-161.

Kondo, H., Maeda, T., Mitsuhara, K. and Inouye, N. 1996. Detection of the viruses occurring in Oriental *Cymbidium* in Japan. *Bull. Res. Inst. Bioreso., Okayama Univ.*, 4: 149-162.

Kuehnle, A. R. 1996a. Toward control and eradication of virus in *Dendrobium* by hybridization, tissue culture and Genetic engineering. *J. Orchid. Soc. India*, 10:43-51.

Kuehnle, A. R. 996b. Molecular Biology of orchids: In: *Orchid Biology: Review and Perspectives*, VII (ed. J. Arditti) Kluwer Academic.

Kuehnle, A. R. and Sugii, N. 1992. Transformation of *Dendrobium* orchid using particle bombardment of protocorms. *Plant Cell Reports*, 11: 484-488.

La, Y. J., Han, J. H., Sim, G. B., Kim, B. D. and Ahn, K. K. 1999. Detection of Cymbidium mosaic virus and Odontoglossum ring spot virus in orchid plants by Tissue Blot Immuno assay. *J. Korean Soc. Hort, Sci.*, 40: 481-484.

Lawson, R. H. 1970. Virus induced colour breaking in *Cattleya* orchid flower. *Amer. Orchid Soc. Bull.*, 39: 395-400.

Lawson, R. H. and Brannigan, M. 1986. Virus diseases of orchids. pp 2-49 in *Hand Book on Orchid Pests and Diseases* (revised edition) Amer. Orchid Soc. Bill. West Palm Beach, Florida.

Lawson, R. H. and Hsu, H. T. 1995. Orchid. In: *Virus and virus like Diseases of Bulb and flower crops*. (eds. G. Loebenstein, R. H Lawson and A. A. Brunt), pp 409-20. John Wiley and Sons, New York.

Lee, H. S. and La, Y. J. 1976 Studies on virus diseases of orchids in Korea. *Korean J. Plant Prot.*, 15: 137-145.

Lesemann, D. E. 1977. Long filamentous virus like particles associated with vein necrosis of Dendrobium *phalaenopsis*. *Phytopathol. Z.*, 89: 330-339.

Lesemann, D. E. 1980. Virus diseases in orchids: diagnosis, distribution, etiology, and epidemiology. *Annual rep. Biologische Bundesanstalt fur Land und Forstwirtschaft in Berlin und Braunscheig*. No. 14: H61.

Lesemann, D. E. 1981. Virus diseases in orchids: diagnosis, distribution, etiology, and epidemiology. *Annual rep. Biologische Bundesanstalt fur Land und Forstwirtschaft in Berlin und Braunscheig*. No. 13: H58.

Lesemann, D. E. 1985. Identification of bean yellow mosaic virus in Masdevallia. *Acta Hort.*, 164: 347-354.

Lesemann, D. E. and Doraiswamy, S. 1975a. Bullet shaped virus like particles in chlorotic and necrotic leaf lesions of orchids. *Phytopathol. Z.*, 83: 27-39.

Lesemann, D. E. and Doraiswamy, S. 1975b. Detection and electron microscopy characterization of long bullet shaped virus like particles in cultivated orchids. *Phytopathol. Z.*, 84: 201-214.

Lesemann, D. E. and Koenig, R. 1978. Virus diseases in orchids: diagnosis, distribution, etiology, and epidemiology. *Annual rep. Biologische Bundesanstalt fur Land und Forstwirtschaft in Berlin und Braunscheig*. No. 24: H70.

Lesemann, D. E. and Vetten, H. J. 1985. The occurrence of tobacco rattle and turnip mosaic viruses in *Orchis* sp., and of an unidentified potyvirus in *Cipripedium calceolus*. *Acta Hort.*, 164: 45-54.

Lim, S.T., Wong, S. M., and Goh, C. J. 1993a. Elimination of Cymbidium mosaic virus and Odontoglossum ring spot virus from orchid by meristem culture and thin section culture with chemotherapy, *Ann. Appl. Biol.*, 122: 289-297.

Lim, S.T., Wong, S. M., Yeong, C. Y., Lee, S. C. and Goh, C. J. 1993b. Rapid detection of Cymbidium mosaic virus by Polymerase chain reaction (PCR). *J. Virol. Methods*, 41: 37-46.

Loi, J. S., Lee, S. M., Lam Can, L. T., Fan, S. and Yiang, W. H. 1991. Eradication of orchid viruses in *Dendrobium sonia* using virazole. *Singapore J. Primary Industries,* 19: 16-22.

Mackenzie, A. M., Nolan, M., Wei, K. J., Clements, M. A., Gowanlock, D., Wallace, B. J. and Gibbs, A.J. 1998. *Ceratobium* mosaic potyvirus: another virus from orchids. *Archives Virol.*, 143: 903-914.

Magee, C. J. 1943. Orchid mosaic. *Austral. Orchid Rev.* 8: 51-52.

Mayhew, D. E., Cooke, A. L. and Raabe, R. D. 1992. Special report. A new virus is reported from *Phalaenopsis*. *Amer. Orchid Soc. Bull.* , 61: 574-577.

Mitsumoto, J., Maeda, T. and Inouye, N. 1995. Odontoglossum ring spot virus isolated from *Calanthe* sp in Japan. *Bull. Res. Inst. Bioreso., Okayama Univ.*, 3: 163-174.

Morel, G. M. 1960. Producing virus free Cymbidiums. *Amer. Orchid Soc. Bull.*, 29: 495-97.

Morel, G. M. 1964. A new means of clonal propagation of orchids. *Amer. Orchid Soc. Bull.* , 31: 437-77.

Nan, G. L. and Kuehnle, A. R. 1995a. Factor affecting gene delivery by particle bombardment of Dendrobium orchids. *In vitro Cell Devel. Biol.*, 31: 131- 136.

Nan, G. L. and Kuehnle, A. R. 1995b. Genetic Transformation in *Dendrobium* orchid. In: *Biotechnology in Agriculture and Forestry,* Vol. 34, Plant Protoplast and Genetic Engineering VI (ed. Y. P. S. Bajaj). pp 149-160. Springer Verlag, New York.

Park, W. M., Kim, W. G., Ryu, K. H., Yoon, K. E., Kwack, B. H. and So, I. S. 1990. Detection of Odontoglossum ring spot and Cymbidium mosaic viruses by ISEM. *J. Korean Soc. Hort. Sci.,* 31: 417-22.

Paul, H. L. 1975. Odontoglossum ring spot virus. CMI /AAB. *Description of plant viruses*. No. 155. Common Wealth Agricultural Bureaux, England.

Pearson, M. N. and Cole, J. S. 1986. The effect of Cymbidium mosaic virus on the growth of *Cymbidium* orchids. *J. Phytopath.*, 117: 193-197.

Peters, K.R. 1977. Orchid viruses: a new rhabdovirus in *Laelia* red leaf spots. *J. Ultrastruct. Res.*, 58: 166-171.

Petzold, H. 1971. Der elektronenmikroskopische nachweis eines bazilliformen virus an blattfleck enkranken dendrobien. *Phytopathol. Z.*, 70: 43-52.

Prakash, L., Lee, C. L. and Goh, C. J. 1996. In vitro propagation of commercial orchids: an assessment of current methodologies and development of a novel approach- thin section culture. *J. Orchid Soc. India*, 10: 31-41.

Ramachandran, P. and Varma, A. 2001. Viral diseases of orchids and their management. In: *Orchid: Science and Commerce*. (eds. P. Pathak, R. N. Sehgal, N. Skhar, M. Sharma and A. Sood) Bishen Singh Mahendra Pal Singh, Dehra Dun, India, pp 520.

Ryaboi, S. N. and Polishchuk, V. P. 1990. Development of identification methods for viral diseases of orchid of *Cymbidium* Sw. genus. *Ukrain Bot. Z.*, 47: 94-96.

Ryu, K. H., Park, W. M., Chung, S. Y. and Yoon, K. E. 1995. Occurrence of Cymbidium mosaic virus and Odontoglossum ring spot virus in Korea. *Plant Dis.*, 79: 321.

Ryu, KiHyun, WonMoK Park; Ryu, K. H. and Park, W. M. 1995a. Rapid detection and amplification of Odontoglossum ring spot virus by Polymerase Chain Reaction amplification. *FEMS –Microbiology Lett.*, 133: 265-269.

Ryu, KiHyun, Yoon, K. E., Park WonMoK; Ryu, K. H., and Park, W. M. 1995b. Detection by PCR of Cymbidium mosaic virus in orchids. *J. Phytopath.*, 143: 643-646.

Samad, N. A. and Ari, Z. 1993. The use of antibody sensitized latex to detect Cymbidium mosaic virus in orchids. *Pertanika. J. Trop. Agric. Sci.*, 16: 157-160.

Samad, N. A. and Noraini, A. S. 1985. Methods of detection of orchid virus. In: *Proc. Reg. Conf. Pl. Quar. Supp. Agric. Devel.*, pp 95-97. Kuala Lumpur, Malaysia.

Tan, WeiLing, SekMan Wong, Kini, R. M., Tan, W. L. S. and Wong, S. M. 2000. Rapid simultaneous detection of two orchid viruses using LC-and/or MALDI-Mass Spectrometry. *J. Virol. Methods*, 85: 93-99.

Tanaka, S., Nishi, H., Ito, S., Iwaka, M. Kameya and Sommartya, P. 1997. Detection of Cymbidium mosaic Potex virus and Odontoglossum ring spot tobamovirus from Thai orchids by Rapid Immuno filter paper assay. *Plant Dis.* ,81: 167-70.

Thompson, A. D. and Smirk, B. A. 1967. An unusual strain of tobacco mosaic virus from orchids. *New Zealand J. Bot.*, 5: 197-202.

Toussaint, A., Kummert, J., Maroquin, C., Lebrun, A. and Roggemans, J. 1993. Use of virazole to eradicate ORSV from *in vitro* cultures of *Cymbidium* Sw. *Plant Cell Tiss. Org. Cult.*, 32: 303-309.

Vejaratpimol, R., Channuntapipat, C., Liewsaree, P., Pewnim, T., Ito, K., Izuka, M. and Minamiura, N. 1998. Evaluation of ELISA for the detection of Cymbidium mosaic virus in Orchids. *J. Ferment. Bioeng.*, 86: 65-71.

Wang, H. I., Wang, Y. T. and Liao, L. J. 2000. Detection of Cymbidium mosaic Potex viruses by hybridization with cDNA probes or RT-PCR. *Plant Protection Bull. Taipei,* 42: 25-42.

Wisler, G. C., Zettler, F. W. and Mu, L. 1987. Virus infection of *Vanilla* and other orchids in French Polynesia. *Plant Dis.*, 71: 1125-1129.

Wisler, G. C., Zettler, F. W. and Purcifull, D. E. 1982. A serodiagnostic technique for detecting Cymbidium mosaic virus and Odontoglossum ring spot viruses. *Phytopath.*, 72: 835-37.

Yap, M. L., Lam, L. T. C., Yik, C. P., Kueh, J., Lee, S. M. and Loh, H. M. 1999. Eradication of viruses from commercial orchids. *Singapore J. Primary Industries*, 27: 11-19.

Zettler, F. W., Wisler, G. C., Elliott, M. S. and Ko, N. J. 1987. Some new, potentially significant viruses of orchids and their probable means of transmission. *Amer. Orchid Soc. Bull.*, 56: 1044 -1051.

26

Management Strategies for Common Plant Diseases of Cold Arid Region of Laddakh

H.V. SINGH[1]

Coid arid climate is easily distinguished form rest all, by cool summer and extremely chilling winter. The climate is readily experienced in Ladakh region of J&K and Lahaul spiti, region of Hamnchal Pradesh i.e. cold arid region of India. The region lied in north west Himalyas of the country and in Jammu and Kashmir is confined to Ladakh province, geographically located between 32° 15' to 36° north latitude and 75° 15' 80° 15' east longitude. It covers approximately 82,665 sq km area (SKUAST, Leh, 1998). The area with India is 45,110sq km., representing 87.4 pre cent of total cold arid region, however available area is small and scattered in small pockets (Table-1). Whole of the cold arid region may be divided into seven different into seven different valleys namely, Leh, Nubra, Changthang, Zanskar, Suru, Lahaul and Spiti.

Table 1: Cultivated area in Ladakh region and its allocation to different field crops

Cultivated land (ha.)	Kargil	Leh	Total
Net area	9,962.00	11,593.00	21,555.00
Double cropped area	2,430.00	220.00	2,650.00
Total area	12,392.00	11,813.00	24,205.00
Crop wise distribution			
Wheat	1,847.00	2,600.00	4,447.00
Barley	3,593.00	4,750.00	8,343.00
Pulse	634.00	200.00	834.00
Oilseed	2,689.62	100.00	2,789.62
Millets	88.20	450.00	538.20
Vegetables	242.00	860.00	1,102.00
Horticultural crops	608.12	863.00	1,471.12
Fodder	2,689.12	1,990.00	4,679.12

Source: 1996 *Statistical handbook*, Planning and Development Deptt., Govt. of J&K, Leh.

The soils of this region have formed from weathering of rocks. They are immature, highly porous with high proportion of sands in them. Prevalence of deserted conditions further impeded this development; consequently even today it lies very difficult to manage to large part of lands for higher agricultural plant productivity. Soil problem range from high permeability, poor nutrient status, sloping gradients, high salt

1. Regional Agricultural Research Sub-Station, Skuast (K) Kargil 194 103 (J&K).

contents in some soils, brackish nature of irrigation water, wind and water erosion and soil freezing to poor biological properties. In term of NPK status, soil are poor in nitrogen, low to medium in phosphorus and rich in potash. Soil and water analysis by Dubey (1998) have confirmed saline nature of soil and water having high pH and sand content.

Temperature and moisture are the two parameters mainly responsible for spare or reduced vegetation cover (Kachroo *et al.*, 1977). This vegetation is characterised by highly stunted, growth, which infers that moisture is the biggest factor determining area, density and quality of plant communities in cold arid region. Variation has been noticed in floral pattern with altitude (Sharma and Mir, 1997; Kachroo, *et al.*, 1977). The alpine, desert and oastic elements are the essential feature of cold arid flora. The alpine elements are visible in the narrow belts at upper beds of the mountains. They receive water from snow melting. Desert flora is common at higher reaches and the valleys while Oasis components are in arable areas, consisted of cultivated plants comprising local cultivars, land races, newly evolved varieties and exotic introduced strains. The important crop plants of cold arid regions are listed in table-2.

Table-2. Important crop plants grown in cold arid region

Fruit crops	Vegetable crops	Field crops	Forest trees/Bushes
Apricot	Cauliflower	Wheat	Poplar
Apple	Cabbage	Barley	Willow
Peach	Knolkhol	Small millets	Junipers
Plum	Potato	Field pea	Wild rose
Walnut	Tomato	lentil	Sea buck thorn
Grape	Carrot	Buckwheat	
Pear	Radish	Mustard	
Cherry	Swede	French bean	
Strawberry	Turnip	Oats	
	Cucurbits	Alfalfa	
	Onion		
	Spinach		
	Lettuce		
	Kale		
	Capsicum		
	Vegetable pea		
	Beans		

Several fruits, vegetables and field crops grown successfully in cold arid region (Sharma and Mir, 2000) but these crops are succumbs to some major and some minor diseases which are prevalent in this region and causes heavy losses in sever infection. To get maximum yield, the crops must be kept free form these biotic stresses. Several efforts have been made to check the diseases, however, the information are scattered and mainly based on the research done on tropical crops. Therefore, an appropriate management strategy should be made for each disease, which will worth full for growers as well as field workers.

MANAGEMENT OF FRUIT DISEASES

Apricot

The apricot (Prunus armeniaca L) is an attractive, delicious and highly nutritious fruit. It is particularly very rich source of vitamin A and also contains more carbohydrates, protein, phosphorus and niacin than majority of other common fruits (Teskey and Shoemaker, 1972). It ranks second next to plum among the stone fruits in India in area, production and popularity. The apricot suffers from numbers of fungal and bacterial diseases which can cause loss with sever infection (Mitra, et a.l, 1991).

Gummosis

The disease has world widespread and prevalent in all the stone fruits and is more serious when the temperature fluctuating during winter. Tree losses form 10 to 75 per cent have been observed in young orchards.

Symptoms: The bark and outer sapwood showed symptoms of circular to elongated water soaked lesions. Bark turn brown, gummy and sour smelling. The most characterstic symptoms of the disease is the formation of cankers accompanied by gum exudation. Cankers usually develop at the base of infected trees. Cankers are first notice in late winter or early spring. In the spring, gum is produced in most cankers, break through the bark and run down on the surface of the limbs.

The Pathogen: The pathogen cause the gummosis in apricot is pseudomonas syringae pv syringae. Though, several reports about gummosis complex in stone fruits. Fungi, bacteria and viruses have all been shown to causing gummosis of stone fruits (Anderson, 1956; Brown and Briton, 1986). Three fungal species namely Botryospheria dothidae, B. obtusa and B. radina have been shown to cause gummosis in stone fruits. The destructive form of disorder occurs when trees are attacked by pseudomonas syringae pv syringae. Dispite numerous investigations the cause of gummosis remain obscure because gum formation is associated with numbers of factors such as disease, injury, climatic condition and ethylene production in plant tissues (Boothby, 1983).

Disease development: The bacteria over winter in active cankers, in infected buds and leaves, systematically in the xylem of some hosts, epiphytically on buds and limbs of infected trees. Infection of limbs usually takes place during fall and early winter. In advance stages of infection, the bacteria break down parenchyma cells and cavities full of bacteria develop. Canker develops rather rapidly in the fall, slowly in winter and most rapidly in the period between the end of cold weather and beginning of rapid tree growth.

Management: No complete control of bacterial canker and gummosis of fruit trees can be obtained as yet by any single method. Certain cultral practices and chemical control measures help keep down the number and severity if infections. Gummosis disease can be managed by following ways:-

By increasing plant vigour by reducing stress caused by nutrients and specially moisture deficits. Balanced fertilizers should be used in the orchards. Only healthy bud wood should be used for propagation. Orchards should not be located in areas where trees are subjected to freeze damage, water logged soil or prolong drought. Fungal gummosis is reduced by applying Benlate-Captan sprays immediately after pruning (Brown and Briton, 1986). Bacterial gummosis is managed by applying of a streptocycline-based paste called Moshorbra paste after cleaning of weeping wounds at the time of dormancy break (Agrawala and Sharma, 1971). Copper oxychloride may be sprayed after leaf fall. Cankers on trunks and large branches can be controlled by cauterization with the hand held propane burner.

Wilt

Symptoms: The disease occurs through out the temperate zone. Gupta and Saharan (1973) reported the die back of apricot due to wilt. The leaves drop down from the affected limbs in early summer. The whole tree or part of it may be affected. No external evidence of disease except general stunting is visible. The infected tree is not always killed but may remain stunted and unproductive for several years (Anderson, 1956).

The Pathogen: The diseased caused by fungus Verticillium also-atrum. This is soil inhabiting fungus and infects through the roots. The fungus survives in the soil on plant debris.

Management: Since it is a soil fungus, the control through protective fungicides is not possible. Before new planting the pit soil should be sterilised with 5 per cent formalin.

Coryneum blight

Symptoms: The blight attacks twinges, leaves, blossom, fruits as well as un opened buds. Small, circular deep purple spots appearing on the fruit, become raised and rough (Pathak, 1980). The raised spots on the twigs expand into necrotic canker. Abscission of the diseased area results in shot hole effect. Diseases bud become darker in colour than normal ones and often shows a glistening layer of gum.

The Pathogen: The disease caused by pathogen Stigmina carpophylla. Pathogen survives on the plant debris in the soil.

Management: Clean cultivation is pre requisite for destruction of primary source of inoculum. The disease can be manage by two sprays of captan at leaf fall and bud swell stage (Pathak, 1980 Eisensmith and Jones, 1981). Copper oxychloride at rate of 300g/100 litre water should be sprayed before leaf fall and at bud swell.

Brown Rot

The disease attacks apricots in all the countries wherever it is grown. It is positively related to the rain fall in the first seven days after flowering begins.

Symptom: Both blossoms and fruits are affected. In case of sever infection, blossoms turn brown and die. After infection tiny brown spot enlarges to a large soft brown area within a few days.

The Pathogen: Two species of Monilinia i.e. M. laxa and M. fructigena and Sclerotinia spp. Are associated with the disease. The pathogen over winters in fruit mummies, in twig cankers and in rotten fruits. The penetration takes place via cuticle and stomata.

Management: Application of Zineb (0.3 %) and benomy (0.1%) during anthesis for effective control of disease. Burning of all infected fruits lies in the orchards. Propagation of resistant varieties provides good control of disease.

Powdery Milkew

Symptoms: The fungus' infects young leaves and shoots while older ones are fairly resistant. The white powdery appearance is caused by great number of conidia produced in chain arising from hypahe on leaf surface. The infected young leaf gets curled and narrowed as they expand. The disease checked growth of affected plant parts and reduces photosynthesis. In sever infection plants produces small fruits.

The Pathogen: The fungus Spherotheca pannosa is responsible for the disease. The pathogen survives as mycelium on other host.

Management: Three sprays of 2% wettable sulphur at different growth stages i.e. before flowering, at patel fall and two weeks thereafter are necessary to control the disease.

Boron Deficiency

The boron deficiency in the fruits has described by Demetriades et al., (1971). Internal browning, formation of corky tissue, fruit cracking, external browning and fruit shrivelling have been observed in the orchards with boron deficient soils. For the management of disease application of 225g Borax per tree is required.

Apple

Apple grown all the continents of the world. In India its production is mainly in Kashmir valley and in some areas in hills of Himachal Pradesh, Uttranchal, Skkim and Arunachal Pradesh. The apple in Ladakh comprising cold arid region succumbs to many diseases, however, following diseases are have economic importance.

Apple Scab: Scab is one of the most serious diseases of apple throughout the world. In India disease was first reported in Kashmir valley in 1935 by P. Nath, an epidemic of the disease occurred in Kashmir in 1973 (Gupta, 1975) and remained confined to Kashmir Valley till 1977. In late 1977 it was seen in a number of apple orchards in Shimla, Kullu areas of H.P. and increased from 150 to 40,000 hectares (Agrawala, 1985).

Symptoms: Typical symptoms of disease appear on leaves and fruits and in sever attack on one to tree years old shoots. The symptoms on leaves and fruits are very distinct. Symptoms appeared in spring on young leaves and flower buds as light brown to olive green irregular spots on lower surface. Later more pronounced spots are seen on the upper surface. There is premature fall of leaves and floral buds.

On fruits the scab spots are usually well defined in shape and appearance against shining and coloured background of the fruit skin. The spots are intially dull in appearance, brownish black and become almost black with passage of time. Early scab infection result in spiliting of the skin of fruit in the area occupied by the spot. Later, this forms corky layers with deep creaks. The shoot infection of scab was first reported in India by Gupta and Lele (2976) from Kashmir Valley.

The Pathogen: Apple scab is caused by the fungus Venturia inaequalis, the condial stage is Spilocaea pomi. The perfect stage (Ventruria) is saprophytic and grows n dead leaves, while condial stage (Spilocaea) is parasitic on living host tissues. The conidia are generally one celled but often they become two celled through seption. They are ovate to lanceolate with a truncate base and somewhat pointed apex and smoky brown in colour is maturity. They measure 12-22 × 6-9 microns.

The ascospores are oval to boat shaped, hyaline, yellowish or olivaceous in colour and two celled. The upper cell is shorter and somewhat wider than the lower cell. The un equal size of the cells gives the species its name inaequalis. This spore measures 11-15 × 5-7 microns.

Disease cycle: The apple scab fungus has two distinct stages in its life cycle: the saprophytic psedothecial or over wintering stage on dead fallen leaves and the ocndial summer stagae which is parasitic on leaves, flowers, buds fruits and shoots on the tree during spring and summer. The saprophytic stage is considered the major source of survival of the pathogen during dormancy. This intiates the active parasitic stage by formation of lesions on leaves on which millions of conidia are produced.

The primary infection by ascospores is the most important step in the management of apple scab. Elimination or reduction of perithecial production by destruction of fallen leaves, bearing stroma for ascospore development. Application of 5 per cent urea as spray an infected tree after fruit harvest at the pre leaf fall stage is effective in 95 per cent reduction in ascospore discharge from fallen leaves in spring. Collection and burning of fallen leaves in winter, where feasible is also recommended for elimation of primary source of inoculum. Fungicidal sprays to prevent primary and secondary infection are also recommended. A large number of fungicides have been found effective against the apple scab pathogen (Singh 1998). Among protective fungicides (Dikar, Polyram, Cypress (Dodin), Triforin, Glyodin, Glyodex) had been extensively used for apple scab in many countries. Gupta (1975) suggested two sprays of Dithan M-45 (0.3%) at 10-15 days interval after petal fall is effective. The systemic fungicide Bavistin have been found an excellent fungicide for apple scab.

Application of protectant fungicidal sprays first at green tip stage followed by the systemic fungicide at petal stagea and again protectants at walnut size stage and about 15-20 days harvest showed better protection than the only systemic fungicides (Sharma, 1995).

Powdery Mildew

The disease is common in apple orchards of North India. It damages the nursery plant more readily older plants. In certain years it causes considerable damage to leaves and young shoots resulting in reduced yield of fruits.

Symptoms: the disease appears soon after the buds develop in to new leaves and shoots. The new leaves and shoots are heavily mildewed and weakened. Early symptoms on such leavaes consist of small patches of white or grey powdery masses on the under surface but at the disease progress both surface of leaves and twigs become covered with powdery mass. Affected leaves grow longer and narrower than normal. The margin is curled. A severe attack of the disease results in partial defoliation of the tree. Fruit buds suffer more damage than the vegetative buds. They may not bear any flower at all. When fruits are attacked they remain small and deformed and tend to develop a rough surface.

The Pathogen: Powdery mildew of apple is caused by Podosphaera leucotricha. The ectophytic mycelium sends saccate haustoria in to the epidermal cells. Aerial conidiphores arise form this mycelium on leaves and shoots. These conidiophore bears a chains of oval, hyaline conidia which are measures 25-30 × 10-12 micron.

Disease cycle: The fungus survives in the form of resting mycelium or encapsulated haustoria in the buds. When the buds open in the next season these resting structure produce abundant of conidia, which are wind born and serves as secondary mycelium.

Management: Several fungicidal sprays have been found effective in controlling the disease. Four sprays of Bavistin (0.05 %) starting at bud swell stage and repeated as 15 days interval give control of diseases (Sharma and Gupta, 1980). They found maximum reduction in germination of spore and germ tube growth by Tridforine, followed by superior to Bavistin Sharma, et al. (1992).

Disease of peach, plum and walnut

Though several diseases are reported occurs on theses fruits crop but only few are economic importance.

Bacterial spot or canker: The disease attack peach, plum, walnut and some extent cherry plants.

Symptoms: Bacterial spot attacks leaves, twigs and fruits. Small water soaked area developed on leaves which are become irregular, circular lesions and formed shot whole. The disease forms cankers on twigs, which are the source of initial inoculum.

The Pathogen: the disease caused by bacteria Xanthimonas pruni. The bacteria survive in cankers formed on twigs and over winter there. The inoculum is disseminated by wind and air.

Management: The vigrous plant shows resistance to disease. Application of balanced fertilizers to enhance health of plants provides control of disease. Propagation of healthy propagating materials along with clean cultivation restricts the disease development. Late pruning also helped in minimizing disease. Applications of fungicidal sprays i.e. copper oxychloride (0.3%) dodin and captan mixture at leaf fall stage are good control measures of the disease.

Brown Rot

Symptoms: The symptoms appeared on blossom, twigs and fruits. Masses of ash gray podery spores appear on the surface of infected parts. Infected blossom weather and die. Ritted fruits or mummies left on the trees or on ground become a source of infection subsequently.

The Pathogen: Several species of fungus Monilinia are responsible to cause the disease in the trees. Fungus survives on rotten fruits left in the soil.

Management: The fungus survives on rotten fruits so the elimination of these infected fruits from orchards should be done. Chemical fungicide applications of Mancozeb (0.3 %) or Benomyl (0.1 %) are effective in controlling the disease.

Bacterial Gummosis

Symptoms: symptoms are similar to appear on apricots trees. Circular to elongated water soaked gumming lesions arise on bark or outer sapwood and fruits.

The Pathogen: the bacteria Pseudomonas syringae is responsible to cause the disease.

Management: application of fungicidal sprays i.e. copper oxychloride (0.3%) after leaf fall eliminate the infected leaves and fruits, the source of inoculum. Painting of infected portion of shoots by Boardeaux past checks the gumming.

Viral Disease: some viral diseases are also reported from some areas. There appearance varied from year to year as insect vectors spread them. The populations of insects are proportionally related to the severity of disease. For the management of the disease control of insect vectors particularly aphids by application of insecticides viz. Monocrotophos (0.3%), Rogar (0.1%) and Chlorophyriphos (0.3%) are effective in the checking disease severity are spread.

Disease of Grape

Grape is cultivated some parts of the cold region of Ladakh. Grapes has good potential in yield but suffers mainly the diseases namely, downy mildew and powdery mildew. In sever infection plant produces small bunch of berries.

Downy Mildew

Symptoms: symptoms of disease appeared on all aerial and tender parts if the vines. They are more pronounced on leaves, young shoots and immature berries. Downy growth of the fungus is seen under surface of the leaves. Later due to necrosis dark brown spot formed on leaves. Diseased shoots remain stunted and may be swollen due to hypertrophy. Infected leaves, shoots, tendrils and some times immature fruits become covered with whitish growth of the fungus.

The Pathogen: downy mildew of grape vine caused by fungus Plasmopara viticola. The sporangia are thick walled; oval or lemon shaped and measures 15-30 × 11-18 microns. They produced zoospores of p ear shaped and 7-9 microns in size. The oospores produced mostly in tissue adjacent to midrib.

Diseased cycle: the main source of survival is oospores formed in the leaf and shoot tissues. First crop of sporangia produced these oospores and cause infection of lower leaves near the soil line. The sporangia are detached by movement of leaves and are wind borne to new sites to cause secondary spread.

Management: sanitation is very important in the management of downy mildew of grape vine. Fallen leaves and twigs should be collected and burnt. Cutting of resistant varieties like, Amber Queen, Red Sultan should be propagated (Datar, 1986). Fungicidal sprays of Blitox-50 (0.3%), Mancozeb (0.3%) at immediately after pruning, 3-4 weeks after pruning, before bud open, when berries have formed and during growth of shoots are control disease. Spraying of Ridomil MZ (0.3%) before and after infection give effective control of the disease.

Powdery Mildew

Powdery mildew some times becomes more serious than downy mildew when occur in epidemic form.

Symptoms: the disease attacks the vines at the any stage of their growth. The appearvce of white, powdery patches on affected parts is the characteristic symptoms. All the aerials parts affected. Malformation and discoloration of the affected leaves are common symptoms. Diseased vines have a wilted appearance and remain stunted in growth. The stem turns brown. If disease has appeared early, the fruits may not develop at all.

The Pathogen: the disease is caused by the fungus Uncimula nector. The mycelium is entirely superficial on the attacked parts. The conidiophores arising from mycelium are simple and erect. They bear the chain of oval shaped conidia measuring 25-30 × 15-17 microns.

Disease cycle: the fungus penetrates through hyphae inside dormant vegetative bud and through cleitothecia where formed. Rain disperse the cleistothecia to bark of the vines where they retained between leaf fall and bud break in the next seasons (cortesi et al., 1998). After infection enormous number of conidia are produced to carry on the conidia to conidia life cycle.

Management: clean cultivation by removal and destruction of disease plant parts during pruning after leaf shading and thinning and cutting bark of lataerals are helped in meaning the disease. Among fungicides, Baylatan and Karathane were found effective against the disease (Bhujbal *et* al., 1982). Partial control of the fungus by the mycoparasite *Ampelomyces qusqualis* is reported by (falk, *et* al., 1995).

MANAGEMENT OF VEGETABLES DISEASE

In Ladakh few vegetables were grown before chinies aggression of 1962. There cultivation was confined to home gardens and only directly sown vegetables were grown. Vegetables, which are generally transplanted, become popular after 1965 when bŏth foreigner and Indian brought seeds of a number of vegetable crops and grew them successfully. Presently, about 50 vegetables have been found quite successful either in open field or in protected conditions in different pockets of the region. Increased areas in the vegetables increase of different diseases simultaneously.

Diseases of Potato

Potato grown on large area and have good potential in cold arid region. Though this crop suffers from a numbers of disease, but only two, late blight and early blight are economically important in the region.

Late Blight

Symptoms: the leaf lets when infected shows a water soaked indefinite spots along the margin. The spots enlarge rapidly until whole leaf lets is included. As the lesion is enlarged recurring weft of white mold appear adjacent to the healthy tissues on the under side the leaf lets. This is an important diagnostic symptom in the field. When disease progress tuber may get infected.

The Pathogen: the fungus *Phytophora infestans* is responsible to cause the disease. The fungus survives in the form of mycelium in the infected tubers buried in the soil or during storage. This plays as primary source of inoculum.

Management: sowing of healthy seeds tuber reduce the disease. Applications of fungicidal sprays are more common in the disease management. Sprays of mancozeb (0.3%) at 5-6 leaf stage and 15 days intervals give good control, for curative treatment application of Ridomil MZ (0.3%) quite effective in disease management. The variety Kufai Sindhuri is reported to shows moderately resistant to late blight.

Early Blight

The disease is less important to the late blight.

Symptoms: small, isolated scattered pale brown spots appeared on leaf lets. These become greenish blue and concentric ridge developed in the necrotic area to produce a target board effect. In dry weather the spot turn hard and leaves curl.

The Pathogen: the early blight of potato caused by fungus *Alternaria solani*, survives in the plant debris left in the soil.

Management: since the pathogen is soil born, crop rotation and field sanitation is essential for effective control of the disease. Dead infected parts should be raked together and burnt immediately after crop harvest. Among fungicidal sprays, Mancozeb (0.3%) is quite effective..

Diseases of Tomato

Tomato is grown in large area for commercial production. Several diseases have been reported to occur on tomato, however few are have economic importance.

Early blight

Symptoms: leaves near the soil are attacked first and disease progress upward. The disease becomes visible as small isolated pale brown spot on leaflets. In the necrotic area of the spots concentric ridges develop.

The Pathogen: the disease caused by fungus *Alternaria solani*. The hyphae in the host are first interecellular, later penetrating the cells. Conidiphore emerege throughthe stomata form the dead centres of the spots. Conidia are 120-296 × 12-20 microns in size, beaked muriform and dark colour.

Disease cycle: Mycelium and conidia are survives in the soil or diseased plant debris to cause primary infection in the next season.

Management: disease plant materials should be collected and burnt immediately after harvest. Spraying of fungicides, Mancozeb (0.3%), Blitox-50 (0.3%) is quite effective in managing disease.

Late Blight

Symptoms: symptoms are appeared on all aerial plant parts particularly leaflet, twigs and floral buds. In advance of the disease all the affected plant parts look likes burn. No fruits develop from infected flowers.

The Pathogen: the fungus *Phytophthora infestans* is associated to the disease.

Management: the disease managed by protective fungicides as Mancozeb (0.3%), or Blitox-50 (0.3%). The curative treatment of Ridomil MZ (0.3%) is most effective and two to three sprays at 15 days interval are required.

Wilt and Damping off:

Symptoms: the appearance of disease notice under nursery conditions. Some time wilt may appear in the old plants. In wilt symptoms, leaves become yellowish and soon die. The roots of the infected plant showing as black rot symptoms. Under moist conditions the dead stem may becomes covered with the pale pink mycelium of the parasite. Where as in case of damping off the plant showed a typical symptom as the stem near the soil surface become water soaked or discoloured and exhibits constriction or soft and plant fall at the base.

The Pathogen: wilt caused by fungus *Fusarium lycopersicae* and damping off caused by fungus pythum debarianum, both the pathogens are soil inhabiting fungi, survive in the soil on plant debirs.

Management: avoid excess stress help in controlling of disease. Plant should be transplanted on ridges. Seed treatment with Thiram @ 3g/kg or Bavistin 1kg/1kg eliminate the disease. Soil drenching with 1% Vitavax also help in the controlling the disease in field of conditions. Crop should be followed.

Mosaic: tomato mosaic caused by many different viruses.

Symptoms: Mottling is the main symptoms resulting from irregular patches of light green or yellow coming with similar patches of normal green. The edges of leaves turn downward and are less flexible. All the stages of this distortion may be found in an infested field.

Management: use of virus free seedlings is the most important step for control of tomato mosaic. Control of insect vectors with insecticides (Monocrotophos @ 0.3%) reduces the spread of disease.

Buckeye Rot

Symptoms: the fungus affects only fruits of the mature tomato plants. The lesions often are characterized by distinct zonate marking, which resemble the out line of an eye. Ordinarily the rot beginsat the blossom end, become in the soil reaches the tomatoes that touch the ground (Charles, 1998).

The Pathogen: it is caused y the fungus *Phytophthora nicotianae* as soil fungus which thrives best in the presence of abundant moisture.

Management: tying the vines to stakes will keep most of tomatoes from touching the soil and consequently will decrease the amount of rot.

Sun Scald

Sun scale may affect leaves, stems and fruits of tomatoes, when certain combination of environmental influences is preset. When very tender large plants are transplanted in the fields, a brown longitudinal steak of dead tissue may be produced on the sun-exposed side. The spots on the fruits, which are caused by the hot rays of the sun and aggravated by the lack of soil moisture, are serve as an open door way for the entrance of several fungal species. Usually the injured surface is shrunken, the skin is dry and papery and doses not redden normally when the tomato ripens, but retain a yellowish colour.

Disease of crucifers

It includes diseases of cauliflower, cabbage, knol khole and radish.

Damping Off

The disease appears in seedling stage of all crucifer's vegetables.

Symptoms: similar symptoms appear as appeared damping off of tomato. The disease noticed under nursery conditions. The stem at near the soil surface become water soaked or discoloured and exhibits constrictions or soft rot. Ultimately plants fall at the base.

The Pathogen: the disease caused by fungus Pythium spp.

Management: seed treatment with Thiram 3g/kg or Bavitin 1g/kg control the disease.

Downy Mildew

Symptoms: white cottony growth of the fungus seen under surface of the leaves. The upper surface of the spot becomes yellowish and necrotic.

The Pathogen: the fungus *Peronospora parasitica* is responsible to cause the disease. The fungus survives in soils as oospores.

Management: sprays of protective fungicides as Mancozeb (0.3%) is quite effective. For curative the sprays of Ridomil MZ (0.3%) gave the best control.

Powdery Mildew

This is most destructive under cold arid region and cause sever losses if appeared early plant growth stage.

Symptoms: the symptoms as tiny, white to dirty gray spots on leaves appeared. Powdery mass of the fungal conidia appeared on whole aerial plant parts. The effect of sever infection may be a mature defoliation of the plant.

The Pathogen: the disease cased by fungus *Erysiphae polygoni*. The fungus survives in the form of perithecia or as mycelium on volunteer host plants.

Management: fungicides are the only commercially available control option for the management of the powdery mildew. A spray of Karathane (0.5%) has been found effective to control the disease.

Disease of cucurbits

It includes the disease of bottle guard, ash guard, cucumber and squash.

Pythium Damping Off

Symptoms: disease appeared at seedling stage in the nursery. The plant showed water soaked lesions on stem near the soil surface. After some time plant fall due to rotting of stem.

The Pathogen: *Pythium debarianum* and *Phytophthora* spp, are main causes of disease. The Pathogen are soil inhabiting fungi and survive for long period in the soil on plant debris.

Management: seed treatment with Tirm at rate of 3g/kg seed or Bavistin 1g/kh seed provide complete control of disease. Seedling grow on raised bed reduce disease incidence.

Powdery mildew

Powdery mildew is a common and some times very serious disease of cucurbits in this region.

Symptoms: the first symptoms are tiny white to dirty gray spots on leaves and stems. Most extensive development of mildew occurs on the aboxial surface of the leaf. As these spots enlarge surface become powdery. In the later stage black pin point bodies representing the ascigerous stage of fungus appeared.

The Pathogen: Spharotheca fulifginia is responsible to cause the disease. The fungus survives in the form of cleistothecia or in the form of conidia on other host.

Management: fungicides are the only commercially available control options for management of powdery mildew in cucurbits. Some systemic fungicides such as Benomyl and Triozole are extensively used but the development of resistance against systemic fungicides has been reported (Mc Grath, 1996). Karathan (0.05% to 0.2%) provide effective control.

Downy Mildew

Downey mildew is an important disease of cultivated cucurbits.

Symptoms: the pale green areas are separated by islands of darker green, appeared on leaves. Soon the spot become well defined. They are angular, yellow and often restricted by veins on the upper surface. On the lower side of these spots a purpleish downey growth appears in moist weather. The entire leaf quickly dies.

The Pathogen: Pseudpernospore cubensis cause the disease. The pathogen survives in the form of active mycelium or as oospores on the sheltered places during severe winter.

Management: centain cultural practices that help in reducing losses from disease are early sowing, destruction of weed, proper nutrient management and reducing moisture among vines. Among chemical fungicides, Mancozeb (0.3%) or Dithan Z-78 (0.3%) as protective spray provides better control. Matalaxyl (0.1%) is effective as curative fungicide.

MANAGEMENT OF FIELD CROPS DISEASES

Among field crops mainly wheat, barley, small millets, mustard, oat, lentil, pea and alfalfa as fodder crop are cultivated in the cold arid region.

Diseases of wheat and barley.

Smut

Loose smut appeared on wheat whereas cover smut appeared on barley.

Symptoms: the infected heads of plants converted into a black spore and no grains are formed. The growth of the plants and its general appearance is not affected in any way until ears have appeared.

The Pathogens: loose smut caused by *Ustilago tritici* and covered smut of barley caused by fungus Untilago hordi the dormant mycelium in the seed become active when the infected seed are sown in the next season.

Management: seed treatment with Bavistin 3g/kg seed give good control.

Rust

Among three, rusts only yellow or stripe rust appeared on wheat and barley crops.

Symptoms: the disease appeared on leaves, but in severe form it appear on sheath, stem, glum also. The green colours of leaves are fades in long streaks on which rows of small urdosori appear. Each row consists of a series of oval, lemon yellow pustules, arrange end to end. The telia appear late in season as dull black patches or spot chiefly on under surface of the leaf.

The Pathogen: the yellow rust caused by Puccinia strigormis. The fungus survives as in the form of uredospore on grass host present on high altitudes. The grass host Bronus japonicus is reported to serve as host.

Management: elimination of host grass. Mixed cropping is also reduced the disease spread. Chemical control is quite effective. Spraying of Baylatan 25wp (0.2%) when first pustules are seen give good control. Mancozeb (0.3%) also provide satisfactory control.

Leaf Blight

The disease appeared on wheat and barley crops.

Symptoms: the disease first appears as small, oval, discoloured lesions, which are irregularly scattered on the leaves. The spots turn brown to gray in colour. Several lesions coalesce o cover areas and death of the entire leaf.

The pathogen: the leaf blight of wheat caused by *Helminthosporium sorokiniana* whereas some workers have reported that the disease caused by *alternaria triticina* (Prasada and Prabhu, 1962) and *A. triticola* (Vasant Rao, 1964). The pathogen survives in soil on plant debris. It may be seed born nature.

Management: cultivation of resistant varieties are good means of control. Varieties, UP 262, PBW 343, Sinchen and Mansrower are showed resistant to the disease. Chemical control by spraying of Mancozeb (0.3%) is quite effective.

Disease of Pea

Powdery Mildew

Symptoms: the powdery mildew first appears on leaves and then on other green parts of the plants. It is characterized by the formation of white powdery patches on both side of the leaf as well as on tendrils, pods and stems. The powdery mass consists of mycelium and spores of the fungus causing disease. The loss is in proportional to the disease severity and varies considerably depending on the stage at which disease occur.

The Pathogen: the powdery mildew of pea caused by *Erysihae polygoni*. The fungus survives in the form of cleistothecia in soil or on other host. The conidiophore arise vertically from the supeficial hyphae on the host surface.

Management: best protection against powdery mildew is a fungicidal cover of the foliage. Sulphur fungicides are most commonly recommended. According to Mathur, *et.al*. (1971) Elasal (0.5%), karathane WD (0.2%) and Morocide (0.1%) effective in controlling the disease. Several fungi such as *Ampelomyces*,

Cladosporium, *Telleetiopsis* and *Verticellium* have been found to parasitized the powdery mildew fungus on host surface.

Rust

Symptoms: the earliest symptoms are the development of aecia in round or elongated clusters. All the spore stages develop on every green parts of the host including pods. The formation of aecial stage is preceded by a slight yellowing, which gradually turns brown. The uredial pustules develop on both surfaces of the leaves as well as on other parts. They present a powdery, light brown appearance. The telia occur in the same sourus as he uredia and develop from same mycelium. They formed on leaves but most commonly on stem and petioles. They are dark brown or almost black in colour.

The Pathogen: the rust of pea is caused by fungus *Uromyces fabae*. The aecia are measure 14-22 microns in diameter. The uredospores are round to ovate, light, spiny with 3-4 germ pores and measure 20-30 × 18-26 microns. The pathogen may prennetes in forms of telial stage on diesease plants left after harvest with seeds.

Management: destruction of diseased plant debris after harvest, following suitable crop rotation and treating the seeds with suitable fungicide such as haxaconozole is recommended practiced to control the disease.

Disease of Alfalfa

Powdery Mildew

Symptoms: the powdery mildew first appears on leaves on second cutting. It is characterized by the formation of white powdery patches on both side of the leaf as well as stems. The powdery mass consists of mycelium and spores of the fungus causing disease. The loss is in proportional to the disease severity and varies considerably depending on the stage at which disease occur.

Pathogen: the powdery mildew of pea caused by *Erysiphae polygoni.*

Management: Sulphur fungicides are most commonly recommended. According to Mathur, *et al.* (1971) Elasal (0.5%), Karathane WD (0.2%) and Morocide (0.1%) effective in controlling the disease. Several fungi such as *Ampelomyces*, *Cladosporium*, *Telleetiopsis* and *Verticellium* have been found to parasitized the powdery mildew fungus on host surface.

MANAGEMENT OF FOREST PLANT DISEASES

The forest plants are not much succumbs to diseases, only powdery mildew and few leaf spots are seen in nursery as well as on old plant when rain fall occur. The disease is not much cause damage to old plants. The fungicidal applications to control this disease of forest trees are not economical.

REFERENCES

Agarwala, R.K. 1985. Apple scab. Directorate of Publicaion, Haryana Aggriculture University, Hisar, India. 18p.

Agarwala, R.K. and Sharma, V.C.1971. Further studies on mashorba paste for the control bacterial canker of stone fruit trees. Indian Phytopath. 94:131-133.

Anderson, H.W.1956. Disease of fruit crops. McGraw Hill Book Company Inc. New York, 501 p.

Becker, C.M., Burr, T.J. and Smith, C.A. 1992. Over wintering of conodia of *Venturia inaeualis* in apple buds in New York orchards. *Plant Disease,* 76:121-126.

Bhujbal, B.G., Patil, B.P. and Metane, N.S. 1982. Comparative efficiency of some fungicides against powdery mildew of grapes. *Pesticides* 16: 17-19.

Boothby, D.1983. Gumosis of stone fruit trees and their fruits. *J. Sci. food Agri.*, 34:1-7.

Brawn, E.A. and Britton, K.O.1986. Botryosphaeria disease of apple and peach in south eastern United States. *Plant Disease*, 70:480-484.

Chrles, C.1198 *Manual of Vegetable Plant Diseases*. Discovery Publishing House, New Delhi. Pp1-630.

Cortesi, P., Gadooury, D.M., Seem, R.C. and Pearson, R.C. 1995. Distribution and retention of cleisothesia of Uncinula nector on the bark of grapevine. *Plant Disease*, 79:15-19.

Dalar, V.V. 1986. Sources of resistance in grapevine against downey mildew caused by *Plasmopora viticola, Indian Phytopath*. 39: 102-121.

Demetriades, S.D., Holvas, C.D. and Dimpovlov-Lekha, 1971. *Annales del I. Institute Pytopathogique benaki*, 10: 226-228.

Debey, S.K.1998. Summary of the results and analysed soil and water samples of the Leh district collected on 06-09-1997. *CSCRI*, Karna, 3p.

Eisensmith, S.P. and Jones A.L. 1981. Infection model for liming fungicides application to control cherry leaf spot. *Plant disease*, 65: 955-958.

Falk, S.P., Gadoury, D.M., Pearson, R.C. and seem, R.C. 1995. Partial control of grape powdery mildew by the *mycoparasite Ampelomyces quisqualis. Plant Disease*, 9: 483-490.

Gupta, G.K.1975. epidemiology, forecasting and possible control of apple scab (*Venturia inaequalis*). *Pesticides*, 9: 31-34.

Gupta, G.K and Lele, V.C. 1976. The scab fungus (*Venturia inaequalis*) on grape twigs in Kashmir Valley. *Curr. Sci*., 45: 565.

Gupta, V.K. and Sahran, G.S. 1973. *FAO plant Protector Bulletin*, 21: 140-141.

Hill, S.A. 1975. The importance of wood scab caused by *Venturia inaequalis* as a source of infection of apple leaves in the spring. *Phytopathol*., 82: 216-223.

Kachroo, P.; Sapru, B.L. and Dhar, U. 1977. *Flora of Ladakh and ecological and taxonomica Appraisal*. International Book Distributors. Dehra Dun. 100p.

Matur, R.L.; Singh, G. and Gupta, R.B. 1971. Field evaluation of fungicides for the control of powdery mildew of pea. *Indian J. Mycol. Plant Pathol*., 1:95-98.

Mitra, S.K. Ratore, D.S and Boos, T.K. 1991. *Temperate fruits*. Horiculture and Allied Publishers, Calcutta. 744p.

Pathak, V.N. 1980. *Disease of fruit crops*. Oxford and IBH Publishing Co., New Delhi. pp 240-241.

Prasada, R. and Prabhu, A.S. 1962. Leaf blight of wheat caused by a new species of *Alternaria. Indian Phytopatho*., 15: 292-293.

Sharma, J.N. 1995. Efficacy of fungicidal spray schedule for the control of apple scab. *Indian J. Mycol. Plant Pathol*., 25: 250-253.

Sharma, J.P. and Mir, A.A. 1997. *Crop improvement perspective in cold arid zone-Ladakh*. SKUAST, RARS, Leh. 79p.

Sharma, J.P. and Mir, A.A.2000, *Dynamics of Cold Arid Agriculture*. Kalyani Pulishers, Ludiana. pp. 388.

Sharma, K.K. and Gupta, V.K. 1994. Influence of fungicides on spore germination, sporulaation and control of apple powdery mildew of apple. *Indian J. Mycol. Plant Pathol*, 24: 24-28.

Sharma, S.K. Gupta, S.K. and Kanth, D.S. 1992. Efficacy of fungicides in relation to number of sprays against powdery mildew of apple. *Indian J. Plant Pathol*., 22: 267-269.

Sharma, S.K. and Gupta, V.K. 1980. Efficacy of fungicides in relation to time of application against powdery mildew of apple. *Indian J. Plant Pathol*., 10: 173-174.

Singh, R.S. 1998. *Plant disease*. (7th Edition). Oxford and IBH publishing Co. Pvt.Ltd., New Delhi. 677 p.

Skuast, Leh. 1998. KVK, *Leh Annual Progress Report 1997-98 and Action 1998-99*, Directorate of Extension Education, SKUAST, RARE, Leh.32p.

Vasant, Rao 1964. Undescribed species of Alternaria on wheat. *Mycopath Mycol. Applied*, 23: 311-313.

Integrated Pest Management in Cotton

B.B. BHOSLE[1], MUKESH SEGHGAL[2], N.R. PATANGE[1] AND K. S. RATHOD[1]

Cotton, popularly known as "White gold" of India, occupies an enviable place amongst commercial crops of our country. India is a unique among the cotton growing countries of the world where four species of *Gossypium viz., G. hirsutum, G. barbadense, G.. arboreum* and *G. herbaceum* are cultivated commercially under diversified ecosystems. India accounts over 27 % of world's cotton area and ranks first, however is only 3rd with 14 % of world's cotton production. The world's cotton scenario is as below

Table 1. Global status of cotton

Country	Area (000 Hectares)	Production (000 M.T.)	Yield (Lint in Kg/ha)
USA	5433	3694	680
China (Main Land)	3750	3830	1021
Uzbekistan	1500	1150	767
India	8883	2650	298
Pakistan	2800	1800	643
Turkey	719	795	1105
Australia	440	733	1665
Brazil	806	648	804
Israel	15	25	1667
Egypt	274	229	836
Total	32489	18887	581

Source: Cotton:World statistics, October, 2000.

It is cultivated over 8.88 lakh million hectares with 70 % area under rainfed conditions. The production is about 13 million bales. However, the average productivity of lint 298 kg/ha is still low as compared to global average of 581 kg/ha (Mayee *et al.,* 2001). The cotton production remained stagnant over the last five years and productivity has also been very low, one of the major causes for low productivity is loss due to insect pests.

Impact of Insect Pests

The intensified production strategies have however aggravated the pest problems. Cotton crop is damaged by 1326 species of insects from sowing to maturity in different cotton growing areas of the world. Of which 166 species are associated with cotton in India. Amongst these 15 species are of major importance resulting in an annual loss, of 50-60 per cent of total production (Anonymous, 1995). The crop is infested

1. Cotton Research Station, Marathwada Agricultural University, Nanded.

2 National Centre for Integrated Pest Management, L.B.S. Building, I.A.R.I., New Delhi.

by various insect pests in its different growth phases, of which bollworms viz., Spotted bollworm, *Earias vittella* Fabricius, *E. insulana* Boisduval, Pink bollworm *Pectinophora gossypiella* Saunders, American bollworm *Helicoverapa armigera* (Hubner) and also jassid *Amrasca biguttula biguttula* Ishida, aphid, *Aphis gossypii* Glover, thrip *Thrips tabaci* Lind and whitefly *Bemisia tabaci* Gennadius are the major pests (Table 2).

Table 2: Major insect pests of cotton and season prevalence

Sl. No.	Insect Pest	Scientific name	Seasonal occurrence	Crop age	Growth phase
Sucking Pests					
1	Jassid	*Amarasca bigutulla bigutulla*	June – October	1-50	Vegetative phase
2	Aphid	*Aphis gossypii*	July-September	1-50	Vegetative phase
3	Thrip	*Thrips tobaci*	Sept-Oct	1-30	Square formation phase
4	White fly	*Bemisia tabaci*	July-Sep/Oct	35-110	Square formation phase
Boll feeders					
5	Red cotton bug	*Dysdercus cingulatus*	Sept- December	50-110	Boll bursting and maturity phase
6	Dusky cotton bug	*Oxycarenus laetus*	September-December	45-100	Boll bursting and maturity phase
7	Spotted boll worm	*Earias vitella* *E. insulana*	July-September	35-110	Square and boll formation phase
8	Pink bollworm	*Pectinophora gossypiella*	Aug-Dec/Feb	65-100	Flowering and boll formation phase
9	American bollworm	*Helicoverpa armigera*	Aug-February	65-110	Square and boll formation phase

The avoidable yield losses due to sucking pests and bollworms in cotton were estimated by several workers. They were found to be 52.9 per cent due to both types of pests, more or less equal to the losses caused by the bollworms only (Satpute *et al.,* 1988). The yield loss caused by the bollworms was worked out to be 60.42 to 71.43 per cent (Panchbhavi *et al.*, 1990). The total loss due to cotton pests (Jassid and bollworm complex) in North, central and south zone was estimated as 52.80, 46.53 and 47.56 per cent, respectively (Sundramurthy and Basu, 1990). The loss caused by cotton pests was estimated to be Rs.1200 crores during 1997 in Punjab alone, and due to *Helicoverpa* species only it was Rs.410 crores in Karnataka (Rao; 1999).

Production of cotton has been drastically reduced by insect pests so much so that about 54 per cent of pesticides used in the country are being used on cotton to reduce the pest populations and plant damage (Singh *et al.*; 2002).

Difficulty in controlling cotton bollworms with insecticides and problems associated with them has evoked enthusiastic acclaim in the use of bio-control agents. Although studies on many indigenous parasitoids have been reported from bollworms earlier (Desphande and Nadkarny, 1936, Dorge *et al.*; 1965, Raodeo 1971, Naik *et al.*; 1976, Manjunath, 1974, Varma *et al.*; 1984) no systematic studies were made to utilize these bio-agents against the cotton bollworms until late seventies. The excessive and indiscriminate use of chemicals on cotton have led to development of resistance in pests, resurgence of secondary and minor pests, adverse effects on natural enemies like parasitoids, predators besides pollinators and environmental pollution etc. In cotton, use of synthetic pyrethorids to control bollworms enhanced the population of whiteflies (Lal; 1996). Armes *et al.*(1994) and Kranthi *et al.* (1997) reported pyrethroid resistance in the population of *H. armigera*. Hence, better approach for pest control is needed in an integrated manner for obtaining economic and eco-friendly pest management as well as sustainable cotton production. Due to sole reliance on and continued use of pesticides, not only control measures have lost efficiency but also have become economically non-acceptable. Development and implementation of integrated pest management (IPM) technology hold promise to achieve sustainability as well as to maintain economic yield levels.

Integrated Pest Management

An integrated pest management is a system approach wherein all the control measures like cultural, mechanical, physical, biological, legal as well as chemical measures are utilized in an integrated manner in order to maintain the pest population below the economic injury level. It aims to minimize the use of insecticides keeping in view the agro-ecosystem. Actually, the system does not target hundred per cent kill of insect pest, but the pest is allowed to remain and feed on crop up to a level which can be tolerated i.e. economic threshold level. This also provides an opportunity for parasitoids and predators to survive by feeding upon the surviving insect pests keeping them at an equilibrium level. Thus, the integrated pest management concept aims at managing the pest through all the available tactics with favorable economic, social and ecological consequences.

KEY PESTS OF COTTON

Sucking Pests

Jassid: *Amrasca bigutulla bigutulla* Ishida (Cicadellidae :Homoptera): The pest attacks the crop during 1-50 days age. Adults are about 3 mm long and greenish yellow during summer whereas they develop reddish tinge during winter. The hind portion of the forewings has two black spots on the vertex. Nymphs are greenish yellow and wedge shaped. Both nymphs and adults walk diagonally to the body. The eggs are laid into the parenchymatous tissue of the leaves. Nymphal and adult stages last for 7-21 days and 35-50 days, respectively. There are a total of 7-8 generations in a year. Both adults and nymphs suck sap from the underside of leaves, later on which turn pale, red rust, curls downwards and dry up when infestation is severe.

Aphid: *Aphis gossypii* Glover (Aphidiae:Homoptera): The pest is active from June-october and February-April. Yellowish brown to black adults are 2-3 mm long with black cornicles and yellowish green abdominal tip. Nymphs are small and blackish. Aphids live in colonies and reproduce parthenogenetically. Adults live for 12-20 days and nymphal period lasts for 7-9 days. In all, the pest has 12-14 generations per year.

Both adults and nymphs suck sap from the tender leaves, twigs and buds, and weaken the plants. The leaves curl up. Each aphid makes several punctures and excretes honeydew which encourages development of sooty mold on the twigs and leaves, and this gives a blackened look of the plants. Honeydew attracts ants, and sooty mold helps in the development of pathogenic bacteria.

Thrip :*Thrips tabaci* Lind (Thripidae:Thysanoptera): The pest is generally active from June-September. The adults are slender, yellowish brown, 1 mm long. The tip of the abdomen is curved and abdominal segments are transversely banded with dark brown lines. Males are wingless and females have a long narrow strap like wings, fringed with hairs. Nymphs are smaller in size and blackish in colour.

The eggs are laid in slits in leaf tissue; eggs hatch in 5 days time, nymphal and pupal period lasts for 5 and 4-6 days, respectively. The adults survive for 2-4 weeks.

The nymphs and adults suck sap from the lower surface of the leaves lacerating the leaf tissues. Upper side of the older leaves turn brown and the lower side becomes silvery white. Leaves become curled wrinkled and finally get dried.

Whitefly: *Bemisia tabaci* Gennadius (Aleyrodidae:Homoptera): The pest occurs throughout the year. Nymphs and adults are clustered together on the under surface of the leaves. Nymphs are pale yellow and adults are yellowish with white waxy coating on the body. The hind wings are prominently long. Eggs are laid singly on the under surface of the leaves. Development period is greatly prolonged in winter and nymphal period lasts from 7-81 days. The pest has 12-15 generations per year. The nymphs(s) and adult(s) feed on the cell sap, reduce the vitality of the plant interfering with normal photosynthesis due to the excretion of honeydew and formation of sooty mold all over the surface of the leaf and lint of opened bolls resulting in process of blackening. Chlorotic spots develop on leaves and in severe cases the vein becomes translucent, thickened and in many cases, it drops off prematurely. Sooty mold contaminates the lint. The insect helps in transmitting and spreading of leaf curl virus (CLCV) disease.

Foliage and shoot feeder

Cotton leaf roller: *Sylepta derogata* Fabricius (Pyraustidae:Lepidoptera): It is an occasional pest and is active from August to November. The moths are nocturnal and lay eggs singly on the under surface of leaves. They are yellowish white in colour with black and brown spots on the head and thorax and a series of dark brown wavy lines on the wings. Wings span measure 28-40 mm the caterpillars are greenish grey or pink, have a semi translucent body with dark brown head, 25-30 mm when fully grown. Eggs are laid on the undersurface of leaf. Pupation takes place in soil or in rolled leaves. The life cycle is completed in 23-53 days. Damage is done by the caterpillars, which feed on the lower surface of leaves and as they grow, the edges of leaves roll inwards upto the midrib (trumpet shape) and feed on leaf tissues. In severe infestation, plants are completely defoliated.

Bihar hairy caterpillar: *Spilosoma obliqua* Walker (Arctiidae: Lepidoptera): The hairy caterpillar is a sporadic pest and is prevalent from July-November. The adult moths are dull yellow and are profusely covered with hairs, pale buff crimson abdomen with black spots, the caterpillars are gregarious in nature during the early instars brownish orange to yellow, 40-45mm long and with hairy head. Eggs are laid in clusters on the under surface of leaves and pupation takes place in soil or in plant debris in cocoons. The life cycle is completed in 6-12 weeks. Caterpillars feed on the leaves voraciously and defoliate the plants, leaving only the mid-ribs and veins in severe cases.

Semi looper: *Tarachae notabilis* Walker (Noctuidae: Lepidoptera): The insect is found through out the plains and is active during August-October. The adults are stoutly built white coloured with prominent grey and brown spots on the wings. Wing span measures 30-32 mm. They are nocturnal and lay eggs singly on the underside of leaves. The caterpillars are about 40 mm long, dark green with six pairs of black and bright yellow spots on the back. The life cycle is completed in 16-34 days. The caterpillars are most destructive for the crop and feed on the foliage voraciously and finally skeletonize them. In severe cases, the plants are completely defoliated.

Green semi looper: *Anomis flava* Fabricius (Noctuidae Lepidoptera): It is a sporadic pest and sometimes causes serious damage to the crop. The adult is reddish brown with forewings traversed by two dark zig zag bands, while the hind wings are pale brown. Eggs are laid singly on upper surface of leaf. The larva is semi looper, 25-30 mm long, pale yellowish green in colour. Pupation takes place in plant debris or in the soil. The life cycle is completed within 28-42 days. The young larvae congregate in groups and more actively, feed on the leaf lamina making small punctures. The grown up larvae feed voraciously leaving only the midrib and veins. They feed by chewing the leaves from margins towards the leaf veins, sometimes the caterpillars even feed on tender shoots, buds and bolls.

Stem borer: *Sphenoptera gossypii* Cotes (Buprestidae: Coleoptera): The pest is serious occasionally. The adult is a shiny coppery beetle, 8 mm long and the grubs are 2.5 cms, creamy white, curled with a swollen thorax. Grubs burrow into the stem feeding on internal tissues. The stem gets swollen at the base and dries up. The pest hibernates in cotton sticks or in the lower portion of stalk.

Stem weevil: *Pempheres affinis* (Curculionidae: Coleoptera): Grayish black weevils lay eggs in the hypocotyle region of the plant. Creamy white grubs tunnel the stems and feed on the soft tissues. Pupation takes place inside the stem. Grub aestivates in the cotton sticks. Development is completed in 35-57 days.

Boll Feeders

Red Cotton Bug: *Dysdercus cingulatus* Fabricius (Pyrrhocoridae: Heteroptera): Red cotton bugs are active throughout the year and comparatively more active in the months of September and October. They appear only when the crop is 50-90 days old. The bugs are elongate, slender, crimson red with white bands across the abdomen and are 1.5 -2.0 mm long, females being larger. The membranous portion of the forewings, antennae and scutellum are black with forewings having spots. Eggs are laid in moist soil or in crevices in the ground. Hatch in 7-8 days and the reddish orange nymphs emerge, nymphal period lasts for 5-7 days. The development is completed in 50-90 days. The adults and nymphs suck sap from the leaves and green bolls, stain the lint with their excreta or body juices that gives way for the growth of bacteria and fungi. They are called as "stink" bugs as they emit a foul smell when handled.

Dusky cotton bug: *Oxycarenus laetus* Kirby (Lygaeidae: Heteroptera): Dusky cotton bugs are often associated with red cotton bugs which are active from November to February. They vary in size and colour. Adults are 4-5 mm long, dusky brown with dirty white transparent wings and black spots on fore wings and having deep red legs. Nymphs have round abdomen and resemble adults except in size. Eggs are laid in moist soil or soil crevices, egg and nymphal period last for 7 and 26 days, respectively. Development is completed in 40-50 days. The nymphs and adults suck sap from immature seeds which do not ripe, remain light weight. Adults found in lint get crushed during ginning and stain the lint emitting bad odour.

Boll worms

Spotted Bollworm: *Earias insulana* Boisd *E.Vitella* Fabricius (Noctuidae: Lepidoptera): Generally, *E. vitella* is abundant in high rainfall areas and *E. insulana* in areas of scanty rainfall. The pest attacks the crop at 35-110 days of age. The moths of both the species have wings of about 25 mm. The forewings are grassy green in *E. insulana* and pea green with a wedge shaped white band running from base to outer margin in *E. vitella.* Larva is about 20 mm long, spiny, brownish with white streaks dorsally and pale yellow ventrally in *E. vitella* and greenish white with black marks and orange spots on prothorax in *E. insulana.* Moths lay bluish eggs singly on flower buds, branches and twigs, pupation takes place inside flimsy cocoon in fallen buds, plant debris or soil. The development is completed in 17-29 days in summer and greatly prolonged in winter (42-84 days). The caterpillars cause damage by boring into the growing shoots, buds, flowers and bolls. The attacked shoots wither, drop and ultimately die, and flowers and buds drop off. Infested bolls do not shed, open prematurely and the quality of the lint is spoiled. Sometimes pupation takes place in the bolls itself impairing the development of bolls.

Pink Bollworm: *Pectinophora gossypiella* Saunders (Gelechidae:Lepidoptera): The pink bollworm is one of the most destructive pests of cotton. The pest is active from August-October/November. The adult moths are dark with blackish spots on forewings. The margins of hindwings are deeply fringed. Wing span is 8-9 mm the caterpillars are creamy yellow when young and turn pink when grown, 8-10 mm long with distinct brown head. Eggs are laid on the underside of tender parts of the plant (shoots, flower buds, leaves and green bolls). Pupation takes place generally in half opened bolls. The egg, larval and pupal periods last for 4-15, 8-42, 8-12 days, respectively. The life cycle is completed in 3-6 weeks. The damage is caused by the caterpillars by feeding on the flower buds and bolls. The holes of entry close down by excreta of larvae which are feeding inside the seed kernels. They cut window holes in the two adjoining seeds thereby forming "double seeds" and finally damage them. The attacked buds and immature bolls drop off. Lint is destroyed, ginning percentage and oil content are impaired. The pest hibernates in "double seeds" and can be located in the cavities (hibernacula) impairing the development of the bolls

American Bollworm: *Helicoverpa armigera* Hubner (Noctuidae: Lepidoptera): The pest is polyphagous, most severe in attack, and is active from July – October and February-April. The adult moth is stout, yellowish brown with a dark speck and area on the forewings, which have grayish wavy lines and a black kidney shaped mark whereas the hind wings are whitish with blackish patch along the outer margin. The larva is about 35 mm long, greenish brown with dark grey yellow stripes along the sides of the body. Pupation occurs in the soil or in plant debris. Creamy white eggs are deposited singly on tender parts of plant, egg period lasts in 3-5 days, larval and pupal periods last for 17-35 and 17-20 days, respectively. The life cycle is completed in 25-60 days. The larvae feed on the leaves initially and then bore in to the squares, buds, flowers, bolls and seeds with its head thrust into the boll, leaving the rest of the body outside. A single larva can damage 30-40 bolls. The entry holes are large and circular at the base of the boll.

Tobacco caterpillar: *Spodoptera litura* Fabricius (Noctuidae: Lepidoptera): The pest is polyphagous and is occasionally serious on cotton. It is found throughout the year. The adult moth is stout with brownish forewings and whitish hindwings. The forewings are speckled with dark and wavy dots. The caterpillars are pale green with dark markings initially which later turn dark brown with numerous transverse and longitudinal bands, 25-35 mm long. The eggs are laid on the undersurface of the leaves in mass near the mid-rib, hatch in 5-8 days time. Larval period is of 20-30 days. Pupation occurs in soil or in plant debris. Pupal period ranges from 7 to15 days. The life cycle is completed in 50-60 days. The larvae feed gregariously on the undersurface of the leaves and skeletonize them leaving only the mid-rib and veins in severe cases. They also attack flowers, buds and squares causing considerable loss. Seasonal Prevalence and Succession of Cotton Pests. The cotton pest occurs in succession of various growth stages of the crop. The sucking pests viz., aphids, jassids and thrips are first to infest the crop followed by whiteflies. Successively the bollworms and bugs attack the crop after flowering and boll initiation starts (Table 2).

INTEGRATED PEST MANAGEMENT IN COTTON

Integrated pest management is a broad ecological approach of managing pests employing all methods and techniques viz., cultural, mechanical, biological and chemical in compatible manner to keep pest population below the economic injury levels. The important components of IPM are use of resistant varieties, conservation and augmentation of biological control agents, regular pest surveillance and use of sex pheromones for monitoring the pests built up. Host plant resistance has been used to suppress the population as cultivation of resistant or tolerant varieties is most economical with least disturbance to ecosystem. Adequate levels of resistance/tolerance are present against few pests viz., jassids and boll worms. However, only a few pest species can be controlled by the use of resistant varieties. Some of the varieties have subsequently failed to withstand the insect attack due to evolution of bio-types of pest species.

CULTURAL PRACTICES.

- Deep ploughing in summer to expose diapausing insect stages such as pupae of *H. armigera.*
- Avoid delayed sowings, sow crop between 3rd week of May to 1st week of June (Simwat, 1994) or as early as possible after onset of monsoon rains in rainfed cotton.
- Use certified seeds and acid delint the seeds and dress them with fungicides, captan or thiram @2-3 g/kg seed and imidacloprid 10 g/kg seed (Mote *et al.*; 1995) or thiamethoxam 70 WS @ 4.5 g/kg of seed; to safeguard against the seed borne fungi and sucking pests, respectively.
- Avoid monocropping and destroy alternate weed hosts like *Abutilons*, *Convolvulus*, *Sunchus sp*, *Lagasca mollis*, etc.
- Use trap crops and intercrops.
- Avoid excessive irrigation and nitrogen application.

– Terminate the crop as early as possible and allow grazing by animals to prevent the carry over population and dispose off the cotton stalks before April.

HOST PLANT RESISTANCE (HPR)

Host plant resistance helps in suppressing the pest population with least disturbance to ecosystem. Varieties with less trichomes and yellowish green canopy received less oviposition by insect pests (Patel *et al.*;1994). The population is negatively correlated with total sugars, gossypol, diphydroxy phenols and ferrous contents (Simwat, 1994). Several varieties are reported to be resistant to pests (Table 3). However, very few programmes were launched in the past for the development of resistant varieties. There is a strong need to initiate independent breeding programmes for insect pests resistance.

Table 3: Resistant /tolerant cotton varieties to insect pests

Pest	Variety/Hybrids	Area to which recommended	Reference
Jassids	Bikaneri Nerma,	North	Basu and Narayanan 1992
	Digvijay,	Central zone	Annoynous 1987
	Ganganagar agethi,	Central and south zone.	Basu and Narayanan 1992
	Jayalakshmi,	Central zone.	
	Krishna (AC 122)	A.P. Rice fallows	
	Mahalakshmi (1301 DD)	Rainfed areas of A.P.	Rao *et al.*, 1989
	Mahanandi (NA 297)	Rayalaseema region of A.P.	SundaramurthyandBasu 1990
	Priya	Rayalaseema region of A.P.	Rao *et al.*, 1990
	Savitha	Andhra Pradesh Tamil Nadu	Basu and Narayanan 1992
	Suguna	A.P.- Godavari tract	Rao *et al.*, 1989
	B 1007, DHY 286		Rao *et.al.* 1989
	ABH-466, H-777		Mayee *et al.* 2001
	G.cot-12, G.cot-10, G.Cot	Central zone	Puri *et al.*, 1998
	RS 875, RST-9, RG-8	Rajasthan	Mayee *et al.*, 2001
	Fateh Krishna, Mahanandi		Puri *et al.*, 1998.
	(355-E6), Priya (NA 920)		Mayee *et al.*, 2001
	Amaravathi (AV 1661)	A.P.	Mayee *et al.*, 2001
	Mahalakshmi 1301 DD		Puri *et al.*, 1998
	Lam 389, CAHH 468, H4,		Mayee *et al.*, 2001
	NHH 44,	Central and south zone	Puri *et al.*, 1998.
	H6, AHH 468	Central zone	Puri *et al.*, 1998.
	Savita,	A.P.	Puri *et al.*, 1998.
	NHB 12, Eknath (PA-32)	M.S.	Puri *et al.*, 1998.
	Sanjay		Puri *et al.*, 1998.
	Sujatha, DDH 2		Puri *et al.*, 1998.
	CICRHH 1(KIRTI)		Puri *et al.*, 1998.
	Jayalakshmi (DCH 32)		Puri *et al.*, 1998.
	Khandwa 2,	M.P.	Puri *et al.*, 1998.
	F-505,	Punjab	Puri *et al.*, 1998.
Thrips	Savitha	A.P.	Puri *et al.*, 1998.
	Sujatha,		Puri *et al.*, 1998.
	RG-8	Rajasthan	Puri *et al.*, 1998.
White fly	Kanchana (LPS 141)	South zone	Jayaswal and Sundermurthy; 1992
	Suguna	Central and south zone	Basu *et al.* 1992
	Kanchana, Supriya	A.P. Godavari tract	Rao *et al.* 1989
	LHH 144	Punjab	Mayee *et al.*; 2001
	Kanchana		Puri *et al.*, 1998
	Amravathi (LK 861), L 389,	A.P.	Puri *et al.*, 1998
	JKHY-1	M.P.	Puri *et al.*, 1998
	RS 875	Rajasthan	Puri *et al.*, 1998
	Vara Lakshmi, Lrk 516	Central zone	Puri *et al.*, 1998
	LRA 5166	M.S.	Puri *et al.*, 1998

Contd.

Pest	Variety/Hybrids	Area to which recommended	Reference
Bollworms	Abhaditha	Central and North zone	Anonymous 1987
	Mahalakshmi (1301 DD)	Rainfed areas of A.P.	Sundaramurthy and Basu 1990
	L 604, L 603,		Puri *et al.*, 1998
	LH-922, F-414	Punjab	Puri *et al.*, 1998
	Bikaneri Narma	Rajasthan	Puri *et al.*, 1998
	Digvijay	Central zone	Puri *et al.*, 1998
	CAHH 468		Puri *et al.*, 1998
	JKHY 1,	M.P.	Puri *et al.*, 1998
	AHH 468	Central zone	Puri *et al.*, 1998
	CICRHH 1 (KIRTI)		Mayee *et al.*, 2001
	NHH 1 (Godavari)	M.S.	Puri *et al.*, 1998
	Soubhagya (DS 59)		Mayee *et al.*, 2001

Biological Control

Several species of natural enemies have been recorded in cotton ecosystem (Table 4). However the potential of these has not been exploited to the core. High cost and non availability in sufficient quantities are apparently the major constraints.

Table 4: Natural enemies of cotton insect pests

Sl. No	Natural enemies	Pest	Stage of the pest	Reference
I. Parasitoids				
1	*Agathis* sp.	Boll worms	Larva	Bilapate 1985)
2	*Aleiodes aligharensis*	Boll worms	Larva	Bilapate 1985)
3	*Apanteles angaleti*	Boll worms	Larva	Annonymous 1987
4	*Apanteles pectinophorae*	Pink bollworm	Larva	Annonymous 1987
5	*Bracon brevicornis*	Boll Worms, Stem borer,	Larva	Singh 1985
6	*Bracon greeni*	Boll worms	Larva	Annonymous 1987
7	*Bracon kirkpatriciki*	Boll worms	Larva	Singh 1985
8	*Brachymeria sp.*	American and pink boll worms	Pupa	Singh 1985
9	*Campoletis chloridae*	Boll woms	Larva	Annonymous 1987
10	*Carcelia illota*	American Boll worm	Larva	Sundarmurthy 1990
11	*Chelonus blackburnii*	Bollworm	Larva	Raodeo *et al.*, 1983
12	*Cotesia kazak*	Bollworm	Larva	
13	*Encarsia formosa*	Whitefly	Nymph	Venugopala Rao and Reddy 1992
14	*Encarsia lutea*	Whitefly	Nymph	Venugopala Rao and Reddy 1992
15	*Encarsia shafeei*	Whitefly	Nymph	Kapadia and Puri 1990), Shanty and Natarajan 1991
16	*Encarsia subluneata*	Whitefly	Nymph	Venugopal Rao and Reddy 1992
17	*Encarsia transvena*	Whitefly	Nymph	Kapadia and Puri 1989
18	*Eretomocerus haldemani*	Whitefly	Pupae	Venugopal Rao and Reddy 1992
19	*Erectomocerus diversiliatus*	Whitefly	Nymph	Varma *et al.*, 1989
20	*Eretomocerus masti*	Whitefly	Nymph	Varma *et al.*, 1989
21	*Eretomocerus mundus*	Whitefly	Nymph	Kapadia and Puri 1993
22	*Eucelatoria bryani*	American bollworm	Larva	Shekarappa *et al.*, 1990

Contd. . . .

Sl. No	Natural enemies	Pest	Stage of the pest	Reference
23	*Goniophthalamus halli*	American bollworm	Pupa	Anonymous 1987
24	*Goniozus sp.*	bollworm	Larva	Anonymous 1987
25	*Hyposoter didymotor*	American bollworm	Larva	Anonymous 1987
26	*Prospaltella flava*	Whitefly	Nymph	Lal, 1980, Varma *et al.*1989
27	*Pyemotes ventricosus (mite)*	Pink bollworm	Larva	Sundarmurthy and Chitra1992
28	*Trichogramma achaeae*	Bollworms, Hairy caterpillars	Nymph	Agarwal and Gupta 1986
29	*T.brasiliensis*	Bollworms, Hairy caterpillars	Larva	Singh 1985
30	*T.chilonis*	Bollworms	Egg	Anonymous 1987
31	*T.pretiosum*	American bollworm	Egg	Anonymous 1987
32	*Telenomus heliothidae*	American boll worm	Egg	Anonymous 1987
33	*Trialeurodes vaporariorum*	Whitefly	Nymph	Anonymous 1987
34	*Trioxys indica*	Aphids	Nymph	Sundaramurthy and Chitra 1992
35	*Trichospilus pupivora*	Leafroller	Pupa	Sundaramurthy and Chitra 1992
II Predators				
1	*Amblysius* sp. (9.3 – 35.5 %)	Whitefly	Nymph	Sundaramurthy and Chitra 1992
2	Birds	bollworm	Larva	Anonymous 1987)
3	*Brinkochrysa scelestis* (as effective as insecticide)	Sucking pests and bollworm	Egg, Nymph, Adult	Dhandapani *et al.*, 1992
4	*Brumoides suturalis*	Whitefly	Nymph,	Varma *et al.*, 1989)
5	*Coccinella septumpunctata*	Sucking pests and bollworm	Egg, Nymph, Adult	Varma *et al.*, 1989
6	*Chrysoperla carnea*	Sucking pests and bollworm	Egg, Nymph, Adult	Kapadia and Puri 1989,1991a. Yadav and Patel 1987 Sundarmurthy and Chitra1992
7	*Chrysopa scelestis*	Sucking pests	Nymph, Adult	Varma *et al.*,1989)
8	*Euseius hibisci*	Whitefly	Nymph	Venugopal Rao and Reddy 1992
9	*Euseius soutalis*	Whitefly	Nymph, Adult	Venugopal Rao and Reddy 1992
10	*Geocoris bicolor*	Whitefly	Nymph, Adult	Varma *et al.*, 1989
11	*Mallada boninensis*	Whitefly	Nymph, Adult	Joshi and Yadav 1990
12	*Menochilus sexmaculatus*	Sucking pests and bollworm	Egg, Nymph, Adult	Sundarmurthy and Chitra1992
13	*Neosiama* sp (spider)	bollworms	Egg, Larva	Anonymous 1987
14	*Scymnus seriacus*	Aphids, Whitefly	Nymph, Adult	Kapadia and Puri 1991
15	*Seranginium cinctum*	Whitefly	Nymph, Adult	Venugopal Rao and Reddy 1992)
16	*Seranginium parcesetosum*	Whitefly Adult	Nymph,	Kapadia and Puri 1991)
17	*Syrphus* sp.	Aphids	Nymph, Adult	Anonymous 1987
18	*Sturmia macrphallus*	Semilooper	Larva	Anonymous 1987

Contd. . .

Sl. No	Natural enemies	Pest	Stage of the pest	Reference
19	Thomisus (Spider)	Bollworms	Egg	Anonymous 1987
20	*Verania vincta*	Whitefly	Nymph	Venugopal Rao and Reddy 1992
III Pathogens				
1	*Nosema* sp.	Spodoptera	Larva	Anonymous 1987
2	*Vairimorpha* sp.	American boll worm	Larva	Anonymous 1989
3	*Bacillus thuringiensis*	Bollworm	Larvae	Puri, *et al.* 1998
4	*Beauveria bassiana*	American bollworm	Larvae	Puri, *et al.* 1998
5	*Entomophthora aphidis*	Aphids	All stages	Puri, *et al.* 1998
6	Nuclear Poly hedrosis Virus	American bollworm Spodoptera	Larvae	Puri, *et al.*1998
7	*Nomouraea rileyi*	Spodoptera	Larvae	Puri, *et al.* 1998

Insecticidal Control

Conventional insecticides

Pesticides will remain as main tool to manage the pest problem in cotton as well as in an emergent situation. However, it should be used only when necessary in recommended dose and at right time. The broad-spectrum pesticides and its blanket application at specific interval should be avoided. The frequency of spray should be reduced down to protect parasitoids, predators, pollinators etc. as well as repetition of same insecticide should be avoided. Only one or two sprays of synthetic pyrethroids with a window concept when the cotton crop is about 75-90 days old should be applied and need not be repeated.

Chemical insecticides (Table 5) are required in situations of high levels of infestation. However a careful selection and restricted need based use would help to solve many of the adverse effects.

Table 5: Recommended insecticides against cotton insect pests

Pest	Insecticide	Formulation	Qty. of insecticide/ha
Jassids, Aphids, Thrips	Methy demeton	25 EC	500-750 ml
	Dimethoate	30 EC	500-750 ml
	Phosphamidon	100 EC	100-250 ml
Whiteflies	Methyl demeton	25 EC	500-750 ml
	Neem oil + Teepol		3.0-3.51+500 ml
	Fish Oil resin soap		14-15 kg
	Phosalone	35 EC	2.5-3.0 litres
Spotted, Pink and American bollworm	Endosulfan	35 EC	2.5-3.0 litres
	Chlorpyriphos	20 EC	2.5-3.0 litres
	Quinalphos	25 EC	2.5-3.0 litres
	Monocrotophos	36 WSC	2.5-3.0 litres
	Carbaryl	50 WP	1.5-2.5 kg
	Fenvalerate	20 EC	400-500 ml
	Cypermethrin	10 EC	800-1000 ml
	Decamethrin	2.8 EC	600-700 ml
Spodoptera	Chlorphyriphos	20 EC	1.5-2.0 litres

Contd. . . .

Pest	Insecticide	Formulation	Qty. of insecticide/ha
	Fenvalerate	20 EC	400-500 ml
	Cypermethrin	10 EC	800-1000 ml
	Decamethrin	2.8 EC	600-700 ml
Ash weevil	Aldicarb	10 G	10 kg/ha
	Cabofurna	3 G	30 kg/ha
Stem weevil	Stem application on 20th and 35th day with		
	Monocrotophos	36 WSC	1.5 ml/litre of water
	Phòsalone	35 EC	2.0 ml/litre of water
Mite	Dicofol	25 EC	1.5-2.0 litres

Botanicals

The eco-friendly nature of natural products and their ready availability at low cost coupled with effectiveness as pesticides has made them more reliable. Several natural products have been reported to be effective against pests of cotton. Neem based insecticides have attracted the attention of all. More than 63 neem products are in market, however, there is no consistency in their concentration and dosages (Table 6). The question of their biodegradation is also important. Another important point associated with yield of seed cotton treated with neem products are low compared to that of insecticide treated plants. Therefore, improvement in toxicity is important.

Table 6: Recommendations of some neem based insecticides on Cotton

Sl. No.	Insecticide	Manufacturer	Concentration of Azadirachtin (%)	Dosage/ha (formulation)	Target Pest
1	Bioneem	Ajay Bio-Tech. Pune	0.03	2.5 litres	American bollworm
2	Godrej Achook	Godrej Agrovet Ltd. Mumbai	0.03	2.5 litres	American bollworm
3	Margocide OK 80 % EC	Monofix Agro Products, Hubli	0.03	5.0 litres	American bollworm
4	Nimbo BAS 300	Basarass Bio Control Labs, Pennadanam (T.N.)	0.03	3.0 litres	Bollworms
5	Nimbicidine	T. Stanes and Co., Coimbatore and whitefly	0.03	500 ml	American bollworm
6	Neemark	West Coast Herbochem, Bangalore	0.03	2.5 litres	American bollworm and whitefly
7	Repelin RD-9 93 % EC	M/s ITC Limited, Hyderabad	0.03	5.0 litres	American bollworm and whitefly
8	Margocide CK 20 % EC	Monofix Agro Products, Hubli	0.15	2.5 litres	Bollworms
9	Neemgold	SPIC, Madras	0.15	1.25 litres	American bollworm and whitefly
10	Nimbasol	Nimba Foods and Chemicals, New Delhi	0.15	1.5 litres	American bollworm and whitefly
11	Rakshak	Murkumbi Mfg., Belgaum	0.15	5.0 litres	American bollworm and whitefly
12	Neemazal F	EID Parry Ltd., Madras	5.00	1.0 litre	American bollworm and whitefly

Bio-pesticides

The production of bio-pesticides and bio-agents and the area covered is very low (below 2%) as compared to the chemical insecticides. Nevertheless, the problems in production and supply could be overcome and a large area could be brought under these eco-friendly pesticides by establishing smaller units in selected clusters of villages with the help of technical know how from the Research Institutes/ S.A.Us

The use of *Bt.*, pheromones and viral insecticides have gained importance for the control of pests and avoiding the developing resistánce. Their compatibility with conventional insecticides and safety have made them reliable. Many *Bt.* preparations and NPV offer vast scope in IPM packages. The biological pesticides commercially available in the market and their relative prices (Brithal, P. S.;2003) are given in Table 7.

Table 7: Biological pesticides and their prices

Bio-pesticide	Unit	Market price	Govt. price	Average price	Share of public sector in total supply (%)
Trichogramma	Rs/cc	17	7	10	75
NPV	Rs/100 LE	159	130	142	58
Neem pesticides	Rs/litre	280	200	268	17
Bt	Rs/kg	797	610	775	12

Role of Bt *cotton in IPM*

The *Bt* is a short form of soil bacterium, *Bacillus thuringiensis*. This is a gram positive and spore forming bacterium that forms parasporal crystals during stationary phase of its growth cycle. The bacterium produces endotoxins, which are highly toxic to insect-pests. They kill insects by acting on epithelial tissues of mid-gut of caterpillars. These proteins are characterized by their insecticidal activity

1. Lepidoptera specific Cry-1.
2. Lepiodoptera and Diptera specific- Cry-2
3. Coleptera specific Cry-3
4. Diptera specific Cry-4

Bt cotton a genotype or an individual, which is developed by the techniques of genetic engineering, is referred, as transgenic. A transgenic may be a plant, an animal or a microbe. Transgenic plants contain foreign gene or genetically modified gene of the same species. The foreign gene may be distantly related, closely related or unrelated species or from microorganisms such as fungi, bacteria and viruses.

Bt cotton refers to transgenic cotton, which contains endo-toxin protein inducing gene from soil bacterium, *Bacillus thuringiensis*.

Attributes of *Bt* Cotton

Bt cotton has several unique features like

- *Bt* plant has in built control mechanism independent of weather factors
- Protein is expressed in whole plant day and night throughout the plant life
- Neonates cease feeding within hour of ingestion the plant tissues and die within 2 to 3 days. Since the pest is killed in its very early stage any potential serious damage to the plant is prevented
- Larval populations developing resistance to chemical insecticide are also killed.
- Increase in yield owing effective bollworm control

- Safe to biological control agents, non-target organisms, plants and environment
- The technology is available in the seed itself. No special efforts are required to grow *Bt* plants as the seeds are to be sown like any other normal seeds and hence illiterate farmers easily adopt the technique.
- Healthier and safer produce can be achieved
- Genetic stability
- Reduced exposure of farmers, farm laborers and non target organisms to pesticides and less pesticide residues in cotton
- Increase natural enemies activity due to reduction in pesticides.
- Induced earliness in cotton eventually result in raising another crop in rabi season.
- Reduction in use of chemical insecticides (Table 8).

CONSTRAINTS OF *Bt* COTTON

1. Possibility of development of resistance to *Bt* protein in insects.

The resistance can be managed by:

- High dose of *Bt*
- Growing refuge crops (non-*Bt* cotton)
- Pyramiding (stacking) of genes i.e. introducing second *Bt* proteins gene into Cry-1 Ac cotton to ensure greater lethal effect for bollworms.
- IPM:*Bt* cotton is not a stand-alone technology to manage cotton pests. It may be one of the most dependable components in the IPM for bollworms management.
- By rotating *Bt* plants with non-*Bt* plants
- Tissues specific expression of *Bt* proteins:Using tissue specific promoters or only in response to insect feeding or application of chemical trigger.
- Alternate host.
- Destruction of carry over populations.
- Use of economical thresholds
- Molecular genetic approach
- Establishment of base line susceptibility to ICP

Table 8: Reduction in number of insecticide sprays for *Bt* cotton in selected countries per season

Country	Reduction in no. of sprays
China	Upto 14
Indonesia	Upto 8
South Africa	Upto 7
Australia	Upto 7
INDIA	Upto 5
Mexico	Upto 3
Argentina	Upto 3
USA	Upto 2

Source: Compiled by Clive James, 2002 various sources and country case studies.

2. Secondary pests will no longer be controlled, there is need to control secondary pests through chemical sprays, which may kill natural enemies and thus offsets one of the major advantage of transgenic.
3. Likely hood of monopoly of seed producers
4. The cost of producing and deploying transgenic is very high
5. Proximity of transgenic crops sprayed fields may reduce effectiveness of transgenic due to insect migration.
6. Possibility of evolution of new insect biotypes
7. Escape of resistant genes into wild relatives.

FUTURE THRUST AREAS

1. Isolation of insect resistance genes
2. Transgenic cotton for herbicide resistance
3. Transgenic cotton for disease resistance
4. Transgenic cotton for modifying cotton fibre
5. Hybrid cotton
6. Identification of multiple sources of resistance
7. Availability of cost effective *Bt* cotton seeds
8. Development of *Bt* cotton Resistance to drought and salinity condition
9. Development of *Bt* cotton in straight varieties.
10. Development of *Bt* varieties / hybrids of diploid cotton.

Conclusion

- *Bt* cotton is extremely valuable pest management tool
- It can be integrated with other pest control techniques and thus is an essential component of integrated pest management.
- *Bt* cotton is not a panacea for all cotton pests and therefore insect management with other pest control approaches is necessary.
- *Bt* transgenic is an additional tool in combating lepidopterous pest problems in cotton.

SUSTAINABLE COTTON IPM

Conservation of natural enemies and their augmentation to avoid the use of insecticides, residual toxicity, resurgence of pest is of prime importance. Hence, to maintain ecological balance and manage the pest problems more efficiently in an eco-friendly manner, the use of bio-agents and bio-pesticides/ botanicals have to be emphasized. The application of chemicals should be need based on the basis of ETL (Table 9).

Considering the meager availability of biopesticides and botanical pesticides, it is advisable to adopt IPM module using botanicals, biopesticides and need based application of synthetic insecticides against cotton pests to achieve maximum C: B ratio through environment friendly IPM package (Fig. 1).

Keeping in view the pest scenario, the cost economics and environmental safety; the following package comprising the use of different IPM tools is suggested for sustaining the productivity of seed cotton.

Table 9: Economic threshold level (ETL)for major insect pests of cotton

Insect		Age of crop	Threshold level
Ahpid	(*Aphis gossypii Glover*)	1-50	15-20% infested plants
Jassid	(*Amrasca bigutulla bigutulla* Ishida)	1-50	50-100 nymphs/50 random upper, middle and lower leaves
Thrips	(*Thrips tabaci* Lindeman)	1-30	15-20 heavily infected plants
Whitefly	(*Bemisia tabaci* Gennadius)	35-110	8-10 adults or 20 nymphs/leaf
Spotted bollworm	(*Earias vittella* Fabricius) (***E.insulana*** Boisd.)	35-110	10% of attacked shoots or reproductive parts
American bollworm	(*Helicoverpa armigera* (Hubner))	65-110	10% of attacked reproductive parts
Pink bollworm	(*Pectinophora gossypiella* Saunder)	65-110	10% of attacked bolls or flowers

Intercropping

Intercropping with mung bean, soybean, groundnut, setaria, maize, cowpea and onion (Sundaramurthy and Basu, 1990). Two rows for every ten rows of cotton reduced infestation of boll worms and helped to colonise the bio-control fauna.

Trap cropping

Trap cropping with okra (1:10), castor (Thondadarya and Jai Rao, 1980), marigold, *Nicotiana rustica* (Sundaramurthy and Chitra, 1992), wild brinjal (Puri *et al.*;1994) have offered protection to the main crop. Insects feeding on the trap crops should be periodically removed.

Border crop

Planting marigold or cowpea around cotton crop ensures reduced incidence of bollworm and whitefly populations. Flowering of the border crops should synchronise with initiation of square formation in Cotton.

Bird perches

Predatory birds have been reported as effective agents of controlling the population of the bollworms. The birds feed on the larvae and reduce the population considerably. About 84% predation of American bollworm has been reported (Anonymous, 1987). Hence, provision of bamboo pegs as perches @5/ha at random in different places in the field would help predation and conserve natural enemy populations. Intercropping of setaria serves as live bird perch in cotton.

Intensive use of bio-pesticides and bio-agents

The following schedule of application of bio-agents, bio-pesticides and botanicals would offer sustenance without the use of chemical insecticides.

a. Release of egg parasitoids *Trichogramma* sp. @ 1.5 lakh/ha (35-65 days) to control the population of spotted bollworms. Subsequent releases at weekly intervals.

b. Release of *Chrysoperla carnea* predator @ 50,000 eggs / ha to control the nymph and adult population of spotted bollworms. Subsequent releases at weekly intervals.

c. Spraying of NSKE 5 % to combat the on slaught of bollworms (50-75 days)

d. Spraying of HaNPV on marigold sown as border crop to curtail the migratory population of American bollworm (75-100 days).

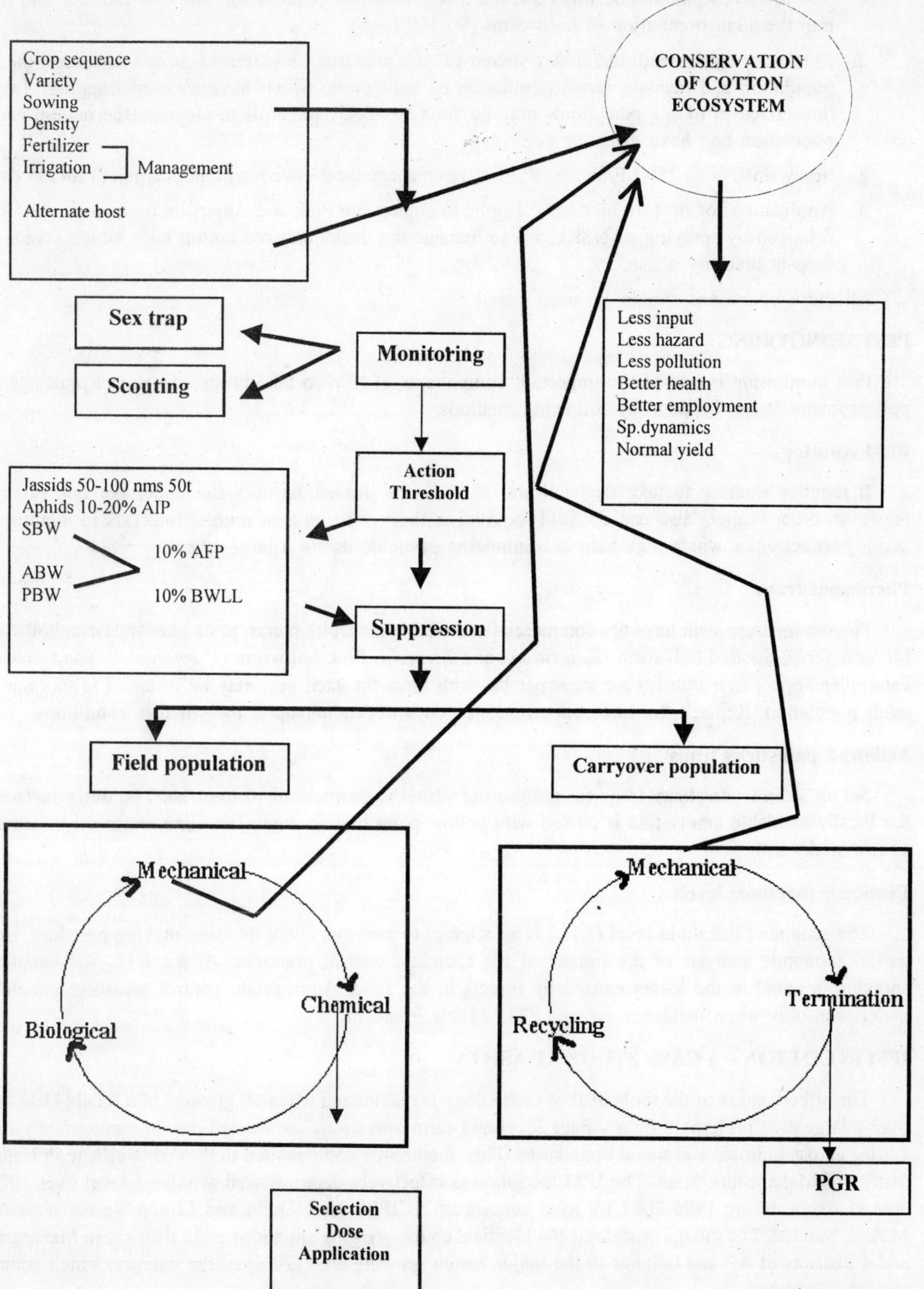
Crop sequence
Variety
Sowing
Density
Fertilizer
Irrigation
Management
Alternate host
CONSERVATION OF COTTON ECOSYSTEM
Less input
Less hazard
Less pollution
Better health
Better employment
Sp.dynamics
Normal yield
Sex trap
Scouting
Monitoring
Action Threshold
Jassids 50-100 nms 50t
Aphids 10-20% AIP
SBW
ABW
PBW
10% AFP
10% BWLL
Suppression
Field population
Carryover population
Mechanical
Biological
Chemical
Mechanical
Recycling
Termination
PGR
Selection
Dose
Application

Fig.1

e. Use of baited pheromone traps 5-8 /ha along with live perches for birds, to monitor and mass trap the adult population of bollworms (90-110 days).

f. Release of egg parasitoid *Trichogramma* sp. and predator *Chrysoperla carnea* to control the egg population and neonate larval population of bollworms, which have escaped trapping. Two to three releases of egg parasitoids may be done at weekly intervals to augment the natural enemy population and have effective control.

g. Spray HaNPV @ 250 LE/ha (2x10^9 PIB's) to manage the *Helivcoverpa* population (120-135 days).

h. Applications of *Bt.* formulation @ 1kg/ha to control the Pink and American bollworms (130-140) followed by spraying of NSKE 5% to manage the dusky and red cotton bugs which invade the crop at maturity phase.

All these operations should be need based.

PEST MONITORING

Pest monitoring is one of the important components of IPM to take proper decision to manage any pest problem. It can be done by following methods.

Field scouting

It requires training to take decision and so only the trained farmers can undertake this exercise. However, other farmers also can do field scouting in their own fields at regular intervals to monitor the major pest activities which may help in minimizing pesticide use to a large extent.

Pheromone traps

Pheromone traps with lures are commercially available for lepidopteran pests like American bollworm, (*H. armigera),* Spotted bollworm *(E. insulana* and *E.vittela*) Pink bollworm (*P. gossypiella*) and Tobacco caterpillar *Spodoptera litura.* Five traps per ha. with lures for each pest may be installed to monitor the adult population. Replace the lures once in 15-25 days depending upon the whether conditions.

Yellowish pan/sticky traps

Set up yellow pan/sticky traps for monitoring whitefly, thrips etc@10 traps/ha. The outer surface of the locally available empty tins is coated with yellow paint having grease/Vaseline/castor oil be used as yellow sticky trap.

Economic threshold levels

The economic threshold level (ETL) is an attempt to improve quick decision making practices, using partial economic analysis of the impact of the chemical control practices. At the ETL, the benefit of spraying is equal to the losses caused by insects in the field. Appropriate control measures should be undertaken only when incidence crosses ETL. (Table 9 and Fig 2).

IPM IN COTTON – A CASE STUDY AT ASHTA

The effectiveness of the tools of IPM technology is maximized when all growers of a locality use them over a large area preferably on a village level and farm operations are carried out on a community basis cutting across political and social boundaries. This theme was implemented in the Ashta village in Nanded district of Maharashtra State. The IPM module was effectively demonstrated at village level over 100 ha area at Ashta during 1998-2001 by joint venture of NCIPM, New Delhi and Cotton Research Station, M.A.U, Nanded. The village represents the identical cotton growing situations of 11 districts in Maharashtra and 4 districts of AP and belongs to the major cotton growing area (70%) of the country, which is under rainfed conditions.

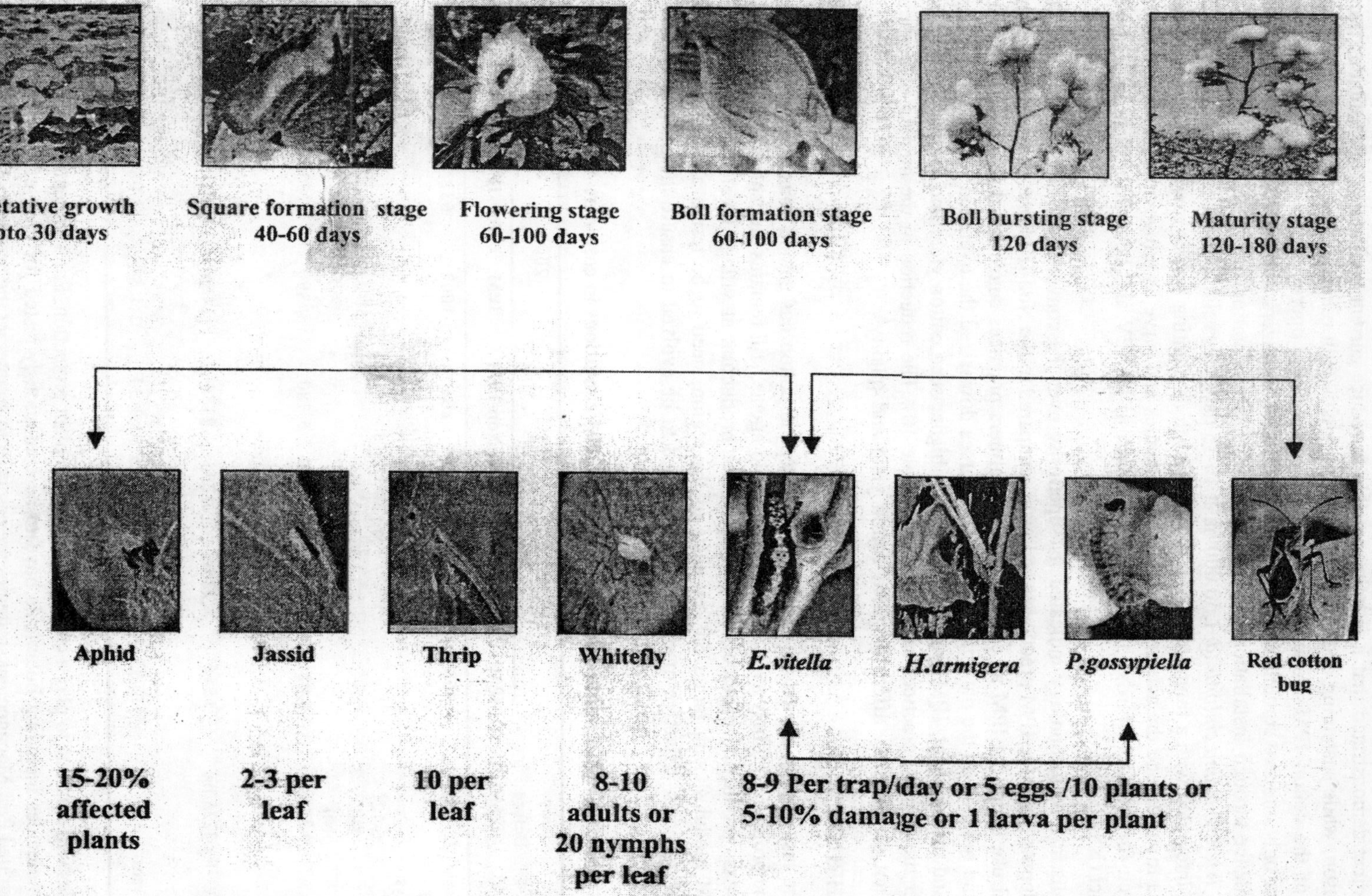

Fig. 2: Growth Phases of Cotton in Relation to Pests

The project demonstrated that it is quite possible to reduce current application of pesticides from 6 to 2 per season, while sustaining and or even increasing yields with reliance on eco-friendly tactics. The key status of the cotton IPM module lies in completing sowing operations in shortest possible time, utilization of only one-two suitable cultivars, balanced use of fertilizers, regular scouting and monitoring of pest incidence through installation of pheromone traps, augmentation of natural enemies (use of egg parasitoids, *Trichogramma chilonis* @ 1,50,000/ha at 10-15 days intervals) to reduce reliance on chemicals. These were integrated with a range of cultural methods such as planting border row of maize interspersed with cowpea to encourage the activity of the natural enemies and serve as refuge and sataria as intercrop (as live perch for predatory birds), use of microbial such as HaNPV and botanicals such as neem seed kernel extract (NSKE 5%).

The holistic IPM technology indicated its superiority over the conventional farming practices. The cost of the inputs under the IPM practices was low as compared to the Non-IPM practices. The low capital on the input use under the IPM as compared to the farmers practices suggested that judicious and need based use and reliance on the naturally available resources downsized the overhead expenditure on crop protection and management. IPM practices resulted in higher seed cotton yield and the cost benefit ratio of IPM over the farmers practices also improved (Table 10). The reduction in the number of pesticidal sprays did not significantly affect the seed cotton yield and quality.

FUTURE THRUSTS

Over the last decade, bio control based IPM concept in cotton has gained considerable momentum. However, the problem of timely availability of quality bio-agents, still remains unresolved. Low cost, semi-automated mass production protocols for bio-agents for production at village level is needed. Location specific IPM modules, with target on key pests of the region, need to be tested and passed on to the development agencies. Effective monitoring system need to be evolved to ensure the quality of the bio-agents.

Table 10: Comparison of IPM and Non-IPM conditions in cotton at Ashta

Particulars	1997 Before IPM	1998		1999		2000		2001	
		IPM	**Non-IPM**	**IPM**	**Non-IPM**	**IPM**	**Non-IPM**	**IPM**	**Non-IPM**
Seed cotton Yield (Kg/ha)	25-125	963	593	1075	806	1002	632	1032	564
No. of sprays									
insecticides	6-12		Nil		1		Nil		1
HaNPV	Nil		1		1		1		1
NSKE	Nil		2-3		2		1		1
Quantity of Pesticides used (actual) (kg/ha)	9.285	0.021	5.70	0.720	5.780	0.390	7.750	0.247	4.945
Cost of plant Protection Inputs (Rs.)	Rs.3225/ha	1545	2280	1255	2156	1465	1908	928	1217
Cost-benefit ratio		1:2.03	1:1.05	1:2.05	1:1.34	1:1.84	1:1.08	1:1.91	1:1.01

Pesticides will continue to play a crucial role in cotton plant protection. Environmentally and ecologically acceptable molecules including botanicals have to be continuously tested for their efficacy in the cotton system. Integrated Pest Management strategies are essential to counteract the problem of pesticide resistance among cotton pests.

CONCLUSION

IPM is a need of time, specially in cotton for sustainable production. The biotechnological products like transgenic cotton, NPV, pheromones, bio-agents like *Trichogramma* and *Chrysoperla*, plant products-neem and its formulations along with other eco-friendly measures have been found very effective in IPM approach for cotton pests and their use need to be encouraged in order to reduce down the reliance on chemical pesticides for sustainable agriculture as well as increasing the cotton quality and productivity.

REFERENCES

Agarwal, R.A. and Gupta, G.P. 1986. *Recent Advances in Cotton Pest Management.*

Plant protection Bulletin, 36 (1-4):51-54.

Amerika Singh, Sharma, O.P., Lavekar, R.C., Bambawale, O. M., Murthy, K.S. and Dhandapani, A 2002. *IPM technology for rained cotton.* 1 p.

Anonymous, 1987. *Annual Report.* AICRP on Biological Control of Pests and weeds. Indian Institute of Horticultural Research, Bangalore.236 pp.

Anonymous 1989 *Technology for increasing cotton production in India.* AICCIP, CICR Regional Station., Coimbatore. 63 pp

Anonymous 1995. *Highlights of Cotton Research*, UAS, Dharwad-580005 31 pp.

Armes, N.J., Banerjee, S.K., Desouza, K.R., Jadhav, D.R,; King A.B.S., Kranti, K.R., Regupathy, A., Surlivelu, T. and Venugopal Rao, N. 1994. Insecticide resistance in Helicoverpa armigera in India: Recent development *Proc. Brighton crop Prot. Conf-Pest and Diseases* 4: 437-442.

Basu A.K. and Narayanan 1992. Achievements of All India Co-ordinated Cotton Impovement project during 1967-92. silver jubilee celebrations, CICR, Nagpur. 176 pp

Basu A.K. and Narayanan S. S., Kapadia S.N. and Charyulu N.R. 1992. Breeding achievements in south zone. In: *All India cotton improvement project, Achievements*, silver jubilee 1967-92, Nagpur. Pp.39-52.

Bilapate, G.G. 1985. Studies on overlapping generation of Earias vitella in pesticide free cotton field. *Journal of Maharashtra Agricultural Universities*, 10:147-150.

Birthal, P. S.2003. *Economic Potential of Biological Substitutes for Agrochemicals.National Centre for Agricultural Economics and Policy Research* (ICAR) New Delhi, India. 132 pp.

Despande, B.P. and Nadkarny N.T 1936. *The spotted bollworms of cotton, Earias fabia stoll and E. insulana Boisd in south Gujarat*, Bombay Presidency, ICAR, Sci.Monograph 10,208.

Dhandapani, N., Kalyanasundaram, M., Swamiappan, M., Sundarbabu, P.C. and Jayaraj, S.1992. Experiments on management of major pests of cotton with bio control agents in India, *Journal of applied Entomology*, 114:52-56.

Dorge S.K. Dalaya, V.P and Khandge, V.S. 1965 *Annonated list of some parasities, predators and other natural enemies recorded on crop pests in Central Maharashtra Agril. College Mag*, Pune 57(1-2&3): 77-84.

Jayaswal A. P. and Sundarmurthy, V.T. 1992. Achievements in Insect Management in Cotton. All India Cotton Improvement Project. Silver jubilee celebrations (1967-1992), September 17-19 Nagpur. pp. 117-151

Joshi, B.C. and Yadav, D.N. 1990. Biology and feeding potentiai of Malada boninensis, a chrysopid predator of white fly *Bemisia tabaci* Genn. *Journal of Biological Control*, 4 (1):18-21.

Kapadia, M. N. and Puri, S.N.1989. Seasonal incidence of natural enemies of *Bemisia tabaci. Indian Journal of Ecology*, 16: 164-168.

Kapadia, M.N. and Puri, S.N. 1990. Development of relatice proportions and emergence of Encarsia transvena and Erectmocerus mundus important prasitoids of *Bemeisa tabaci* Genn. *Entomon*, 15 (3-4): 240.

Kapadia, M.N. and Puri, S.N. 1991. Biology and comparative predation effiency of three hymenpteran species recorded as predators of *Bemisia tabaci* in Maharashtra. *Entomophaga*, 36(4): 555-559.

Kapadia, M.N. and Puri, S.N. 1991 a. Parasitization of Chrysoperla Carnea (Stephens) in cotton field under Parbhani conditions. *Journal of Mahrashtra Agricultural University*, 16(3): 453-454.

Kapadia, M.N. and Puri, S.N. 1993. Parasitization of whitefly in relation to variety, plant protection and sowing dates of cotton. *Journal of Maharashtra Agricultural Univertsity*, 18(2): 313.

Kranti, K.R., Armes, N.J., Rao, N.G.V. Sheoraj and Sundarmoorthy, V.T. 1997: Seasonal dynamics of metabolic mechanisms mediating pyrethroid. Resistance in Helicoverpa armigera. *Pesitic. Sci*. 50: 91-98.

Lal.S.S 1980. A Note on Prospaltella flava Shafee an aphelinied parasitoid of *Bemisia tabaci* Genn. *Madras Agricultural Journal*. 67: 557-558.

Lal, O.P. 1996: *Integrated pest management for sustained crop production.* APC Publications Pvt. Ltd., New Delhi, 1996 pp. 29-44.

Manjunath, T.M. 1974. Evaluation of the effectiveness of two insecticidal schedules on important pest of Hy.4 cotton and their natural enemies. M.Sc(Agri).Thesis, G.A.U. Anand, Gujrat, India (In workshop on biological control of Heliothis, 1985, pp-225).

Mayee, C.D, Rao, M.R.K. and Yadav, M.S., 2001. *Cotton-March Towards New Millenium* C.I.C.R, Nagpur (M.S.) India. pp. 19-24.

Mote, U.N., Dankhar, R.V. and Lolage, G.R. 1995. Efficacy of Imidocloprid as seed treatment against initial sucking pests of cotton. *Pestology*, 19 (1): 5-8.

Naik, L.M.; Parlekar, G.V. ; Khandg. S.V. and Dhumal, S.V. 1976. Further records of predators and parasites of crop pests in central Maharashtra. *J. Maha. Agric. Univ.* 1(adi): 301-304.

Panchabhavi. K.S. 1990. *Indian Journal. Agric Sci*, 60.

Patel, M.S. Yadav, D.N., and Rai, A.B. 1994. Neemark (Azadiarchtin) as ovipositional deterrent against cotton pests. *Pestology* 18(8): 17-19.

Puri, S.N., Bhosle, B.B., Fartade, M.K., Kolhal, R.N., Butler, G.D. and Henneberry, T.J. 1994. Wild Brinjal (*Solannum khassainum*) as a potential trap crop for the Integrated Management of Bemisia tabacion cotton. *Phytoparasitica*, 22(4): 358-359.

Puri, S.N., Sharma, O. P., Murthy, K. S. and SheoRaj. 1998. *Hand Book on Diagnosis and Integrated Management of Cotton Pests*. NCIPM, New Delhi. India. 100-104 pp.

Raodeo, A.K. 1971. Preliminary note on the parasites reared on cotton bollworms in Maharashtra State. Agric. *Coll.Magz. Parbhani*, 12-54-55.

Raodeo, A.K., Sarkate, M.B., Kausale, P.K., Bilapate, G.G. and Shinde, K.S. 1983. Population dyanamics of cotton bollworms in cotton in Marathwada. *Cotton Development*, 13 (2): 31-34.

Rao, N.V., Reddy, A.S., and Rao, K.T. 1989. Natural enemies of cotton whitefly, *Bemisia tabaci* Genn., in relation to pest population and whether factors. *Journal of Biological Control*, 3(1): 10-12.

Rao, N.V. Reddy, A.S., Ankaiah, R., Rao, Y.N. and Khasim, S.M., 1990 . Incidence of whitefly *Bemisia tabaci* in relation to leaf characters of upland cotton *gossypium hirsutum*. *Indian Journal of Agricultural Sciences*, 60(9): 619-624.

Rao, M.V. 1999. In Proc. National worshop Coimbatore 1992.Satpute, U. S. Sarnaik, D.N. and P.D. Bhalerao 1988. Assessment of avoidable field losses in yield of cotton due to sucking pests and bollworms. *Indian J. Plant Prot*. 16(1) : 37-39.

Shanti, P. and Natarajan, K. 1991. Population dynamics, behviour and development of *Encarsia Shafeei*, a potential parasitoid of *Bemisia tabaci*. In: *Emerging trends in Biological Control of Phytophagous Insects*. (Ed) Ananthakrishnan, T.N. National Symposium, Madras, Oxford and IBH publishing company, New Delhi. pp: 55-58.

Shekarappa, K., Jai Rao, K. and Lingappa, S. 1990. *Proceedings of National Workshop of Heliothis Management*. TNAU, Coimbatore. pp. 212-215.

Singh. S.P. 1985. *Proceedings of National Seminar on Integrated Pest and Disease Management* (Ed). Jayaraj.S. September, 1984. TNAU, Coimbatore. pp. 309-318.

Simwat. G.S 1994. Modern concepts in Insect Pest Management. In.*Trends in Agricultural Insect Pest Management.* Pp: 186-237.(Ed) Dhaliwal, G.S and Ramesh Aurora. Commonwealth Publishers, New Delhi-India.547 pp.

Sundarmurthy, V.T. 1990. Proceedings of First National Workshop on Helicoverpa management. *Current Status and Future Strategies*. (Ed.) Sachan, J.N., 30-31th August. Indian Institute of Pulses Research, Kanpur.pp: 84-91.

Sundarmurthy, V.T. and Basu, A.K. 1990. *Proceedings of National Workshop on Heliothis Management*. TNAU. Coimbatore.PP: 290-296.

Sundarmurthy, V.T. and Chitra, K. 1992. Integrated Pest Management in Cotton. *Indian Journal of Plant Protection*, 20(1): 1-17.

Thodhadarya. T.S. and Jai Rao, K. 1980. Bhendi as trap crop in the Integrated Pest management. *Andhra Agricultural Journal*, 27(3&4): 100-101.

Venugopala Rao, N. and Reddy, A.S. 1992. The natural enemies of cotton whitefly – *Bemisia tabaci* Gen - A Review, *Agricultural Reviews*, 13 (1): 12-20.

Verma, G.C.; Sekhon, B.S. and Shenhnar, M 1984. Role of parasitoids in the control of cotton bollworms Pectinophora gossipilla (Saunders) and Earis spp in the Punjab. *Cott. Dev.* 18-21.

Verma, R.S, Das, S.B., Shaw, S.S., Mandloi, K.C. and Badaya, A.K. 1989. Chemical control of white fly, *Bemisia tabaci* on cotton. *Journal of Cotton Research and Development*, 3(1): 49-51.

Yadav, A. N. and Patel, R.S 1987. *Proceedings of Seminar cum workshop on Biological control of crop pests and weeds*. Gujarat Agricultural University. June 29-2nd July, 1987., 163 pp

KEY PESTS OF COTTON

Jassid with infested leaf

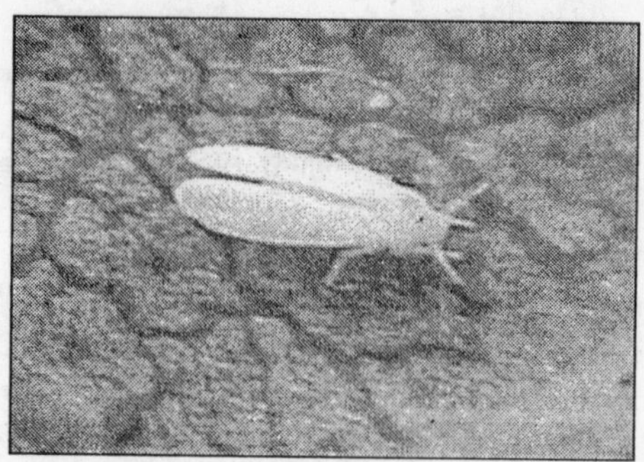
Whitefly

Thrip with infested leaf

Aphid with infested plant

H. armigera in square

E.vitella in square

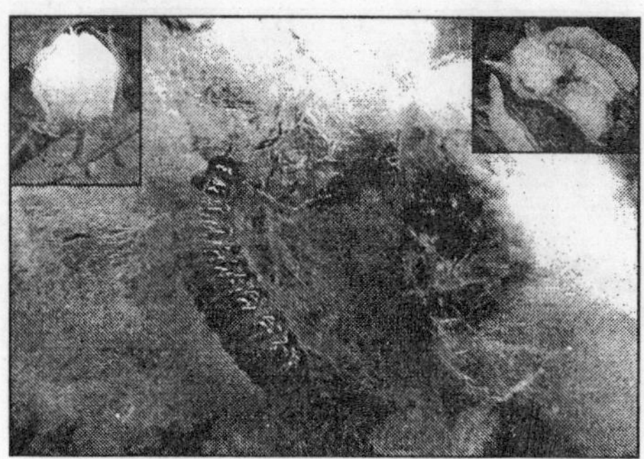

.gossypiella in festing flower, locule and seed

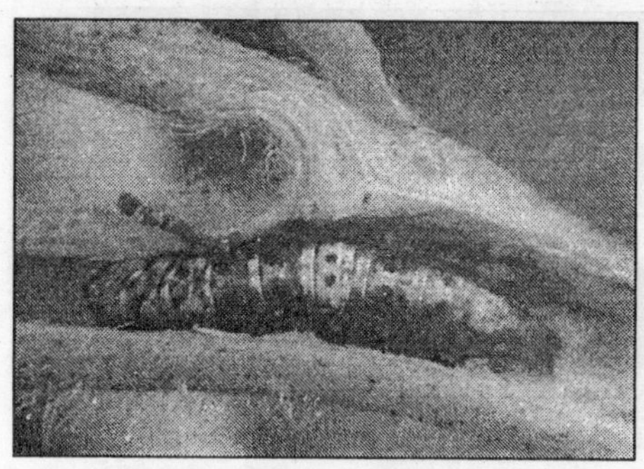

E.vitella in growing shoot

Infested plant by Semilooper

Infestation by Spodoptera

Boll infestation by *E. vitella*

Boll infestation by *H. armgeria*

Dusky cotton bug

Red cotton bug

Grey weevil

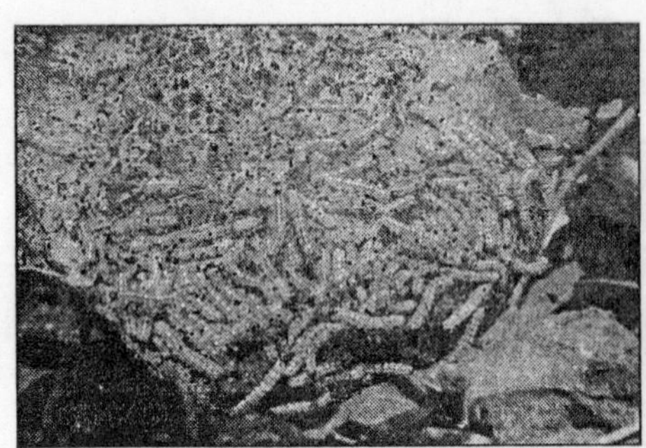

Bihar hairy caterpillar

NATURAL ENEMIES OF COTTON PESTS

Chrysoperla carnea

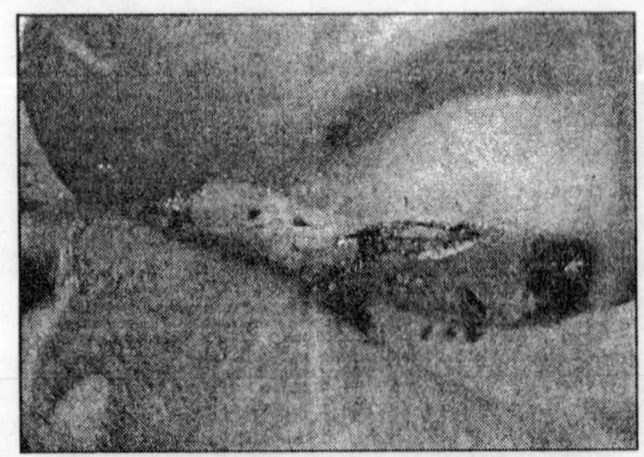

Pupa of *Campoletis cloridae*

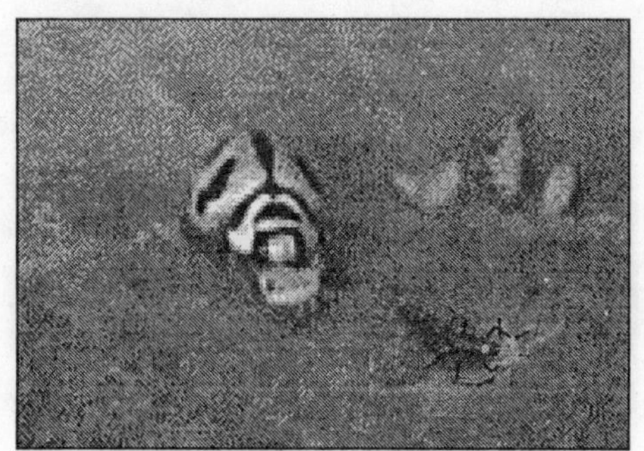

Lady bird beetle

Pupae of *Apenteles*.

Pupa of *Rogus*

Reduviid bug

Nest of predatory bird in cotton

Bird's nest in IPM field

COMPONENTS OF IPM

Clean up campaign

Seed treatment

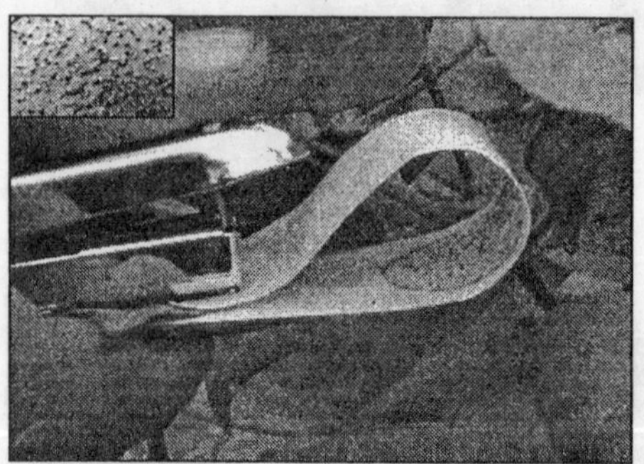

Trichocards

Use of Sataria, Maize and Cowpea

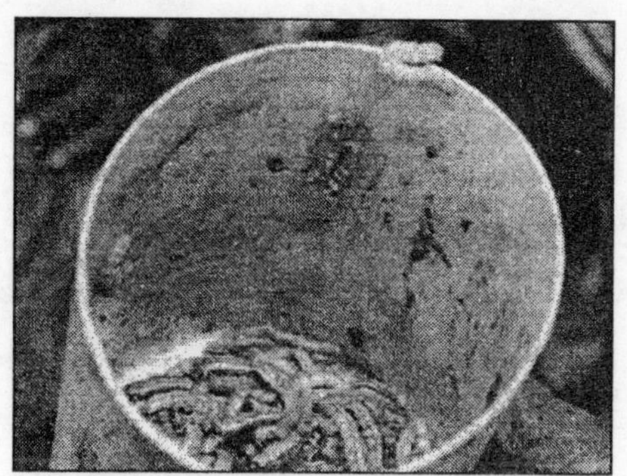
Mechanical collection

Preparation of NSKE

HaNPV affected larva

Pheromone trap

28

IPM in Rice Ecosystem

D.K. GARG[1], M.D. JESWANI[1] AND MAHESH PATHAK[1]

Rice crop is adaptable to a wide range of environmental conditions because it has an efficient system of air passage from shoot to root. This system allows rice to grow in water logged soils. Because of the air passage systems roots respire aerobically and utilize carbohydrates to produce energy even when they are growing in an anaerobic environment. It is an annual grass (belonging to family Gramineae of plant kingdom), cultivated from 53° N to 40° S latitude. Annually rice is grown in about 42 million hectares in four major ecosystems; irrigated (19 m ha), rainfed lowlands (14 m ha), flood prone (3 m ha) and rainfed upland (6 m ha) ecosystems (Prasad *et al.*, 2001). The total production of paddy is 120 million tonnes, thus India's share in world's paddy production is 21.4 per cent and it ranks 2nd in the world. No other country in the world has such diversity in rice ecosystems. More than half of the net rice area is rainfed, making production vulnerable to unpredictable rainfalls. The demand for rice in India is projected at 128 million tonnes for the year 2012 (Tiwari, 2002). For increasing rice production, there is negligible scope by increase in rice area, hence, productivity of rice is of great concern. Two important strategies, location specific high-yielding varieties and improved crop management practices, are essential for achieving increased rice production. A number of varieties suitable for various ecosystems are now available to cater specific needs of farmers. Inadequate crop stand/establishment, poor water management, weed competition, insect and disease problems and low fertilizer use efficiency are basic problems in realizing higher productivity levels. If these yield-guiding constraints are addressed in the holistic manner through integrated approach it can lead to sustainable rice production.

Rice crop is mostly grown under warm and humid climate, which is also very conducive to many pests. Pests are organisms that attack the rice crop or compete with it for nutrients, thereby causing yield reductions. The common organisms that are pests of rice include insects, weeds, mollusks, mammals, birds, fungal and bacterial pathogens and viruses (Heong *et al.*, 1995). Infact, among food crops, rice is the only one, which receives maximum pesticides. Pests are important constraints to rice production and the increasing yield loss has resulted in quest of developing synthetic pesticides. Moreover, Basmati rice, which has high export potential and low degree of tolerance to pests, has resulted in high level of pesticide use. However, pesticides are not a perfect answer for managing pest problems. Over reliance on highly toxic, hazardous pesticides has created higher magnitude of environmental pollution leading to imbalance in natural ecosystem. Problems of pest resurgence and development of resistance to insects, disease causing organisms, weed flora, adverse effects on many beneficial organisms in the rice ecosystem and creation of the problem of minor pest becoming major due to indiscriminate use of pesticide have been well documented.

[1] National Centre for Integrated Pest Management, L.B.S. Building, I.A.R.I., New Delhi - 110 012

PEST SCENARIO/PEST STATUS IN DIFFERENT RICE ZONES

The rice plant passes through various stages during its growth cycle so is the pest development. The different stages of the plant are vulnerable to one or more pests or with varying magnitude of infestation (Tables 1 and 2). The knowledge about insect pest scenario, disease distribution in different eco zones is of great help in formulating effective IPM module.

Table 1: Insect pest scenario in different states of India

Insect pest	Intensity	States
Major pests		
Yellow stem borer *Scirpophaga incertulas*	Severe	Andhra Pradesh, Kerala, Maharastra, Orissa, Punjab
	Moderate to severe	Haryana, Madhya Pradesh, Uttar Pradesh
	Low to moderate	Andaman and Nicobar Islands, Assam, Bihar, Gujarat, Himachal Pradesh, Jammu and Kashmir, Karnataka, Meghalaya, Manipur, Pondicherry, Tamil Nadu, West Bengal
Leaf folder *Cnaphalocrosis medinalis*	Moderate– severe	Andhra Pradesh, Haryana, Kerala, Maharastra, Pondicherry, Punjab, Tamil Nadu
	Light-moderate	Assam, Bihar, Goa, Gujarat, Himachal Pradesh, Jammu and Kashmir, Karnataka, Kerala, Madya Pradesh, Meghalaya, Orrisa, Uttar Pradesh, West Bengal
Brown planthopper *Nilaparvata lugens*	Severe	Andhra Pradesh, Kerala
	Moderate-severe	Karnataka, Madya Pradesh, Meghalaya, Orissa, Pondicherry, Tamil Nadu
	Low-moderate	Assam, Bihar, Goa, Gujarat, Maharastra, Punjab, Rajasthan, Uttar Pradesh, West Bengal
White backed planthopper *Sogatella furcifera*	Moderate-severe	Haryana, Madhya Pradesh, Punjab, Uttar Pradesh
	Light-moderate	Andhra Pradesh, Assam, Bihar, Gujarat, Himachal Pradesh, Jammu and Kashmir, Maharashtra, Orissa, Pondicherry, Tamil Nadu West Bengal
Green leaf hopper *Nephotettix virescens*	Moderate-severe	Andhra Pradesh, Madhya Pradesh, Pondicherry, Tamil Nadu
	Low-moderate	Assam, Bihar, Goa, Gujarat, Haryana, Himachal Pradesh, Kerala, Jammu and Kashmir, Maharastra, Orissa, Tamil Nadu, Uttar Pradesh, West Bengal
Gall midge *Orseolia oryzae*	*Severe*	*Andhra Pradesh, Madhya Pradesh, Manipur, Orissa*
	Moderate-severe	Assam, Bihar, Goa, Karnatka, Kerala, Meghalaya
	Low-moderate	Goa, Maharashtra, Tamil Nadu, Uttar Pradesh, West Bengal
Minor pests		
Rice hispa *Dicladispa armigera*	Moderate-severe	Assam, Gujarat, Himachal Pradesh, Punjab, West Bengal
	Low-moderate	Andhra Pradesh, Bihar, Haryana, Madhya Pradesh, Meghalaya, Orissa, Uttar Pradesh, West Bengal

Insect pest	Intensity	States
Gundhi bug *Leptocorisa acuta*	Moderate-severe	Andhra Pradesh (Telangana region), Uttar Pradesh
	Low-moderate	Assam, Bihar, Gujarat, Haryana, Karnataka, Madhya Pradesh, Orissa, Pondicherry, Tamil Nadu, West Bengal
Case worm *Nymphula depunctalis*	Low-moderate	Andhra Pradesh, Goa, Karnataka, Kerala, Madhya Pradesh, Maharastra, Orissa, Tamil Nadu, Uttar Pradesh, West Bengal
Whorl maggot *Hydrellia philippina*	Low-moderate	Andhra Pradesh, Assam, Bihar, Haryana, Karnataka, Goa, Madhya Pradesh, Maharastra, Orissa, Tamil Nadu, Uttar Pradesh

Source: Reports of Production Oriented Survey, Directorate of Rice Research, Hyderabad (Several years)

Table 2: Disease scenario in different rice ecologies

Disease	**Ecology**	**Intensity**
Blast *Pyricularia oryzae*	Rainfed uplands, Irrigated and favourable lowlands	Severe - moderate
Brown spot *Helminthosporium oryzae*	Rainfed uplands, Irrigated and favourable lowlands	Severe – light
Sheath blight *Rhizoctonia solani*	Irrigated and favourable lowlands, uplands	Moderate – severe
Sheath rot *Sarocladium oryza*	Irrigated and favourable lowlands	Moderate
False smut *Ustilaginoidea virens*	Irrigated and favourable lowlands	Light – Moderate
Bacterial blight *Xanthomonas campestris pv oryzae*	Irrigated and favourable and unfavorable lowlands, uplands	Severe – Moderate
Bacterial leaf streak *Xanthomonas* sp.	Irrigated and favourable lowlands	Moderate
Rice tungro virus	Irrigated and favourable lowlands	Moderate – severe

(Source: Mathur *et al.*, 1999).

Some common nematode diseases in rice ecosystem are: Stem nematode/ Ufra disease (*Ditylenchus angustatus*), white tip disease (*Aphelencoides besseyi*), Rice root nematode (*Hirschmainella* spp.), Root knot nematode (*Meloidogyne* spp.).

LOSSES DUE TO PESTS

Several reports are available on losses due to insect pests and diseases damage in traditional rice cultivars. On an average yield losses due to insects, rodents and nematodes is estimated to be around 20 per cent. The losses due to the attack of various diseases put together is estimated to be around 15-20 per cent. It has been reported that losses were in the range of 5-10% due to blast (Padmanabhan, 1965), about 80% due to brown spot during great Bengal Famine (Padmanabhan, 1973), and 6-60% due to Bacterial blight (Srivastava, 1967). An analysis of 135 multi locational trials conducted in rice during 1969-1977 under the All India Coordinated Rice Improvement Project (AICRP) revealed that the avoidable losses in rice have been estimated to vary between 21 and 51 per cent (Pathak and Dhaliwal, 1981; Singh and Dhaliwal, 1994). In recent past Oerke *et al.* (1994) reported higher losses of 51.4% of the attainable yields of which 15.1% was due to pathogens, 20.7% due to animal pests including insect pests and 15.6% due to weeds. Savary *et al.*, 1997, reported of yield losses in rice caused by various agents like insect pests, diseases and weeds, 0.46 tonnes/ ha is contributed by stem borer alone. Puri (2000) reported that brown plant hopper, gall midge and yellow stem borer were the key pests in rice causing 10-70%, 15-60% and 25-30% loss, respectively. Under puddled and direct seeded/ transplanted conditions, the yield losses due to weeds are 10-40 per cent.

Atwal and Dhaliwal, (2002) reported on an average loss of 100-500 kg of paddy per hectare in Andhra Pradesh due to yellow stem borer. In Orissa the cumulative loss has been estimated at 44 per cent with an increase of one per cent incidence of the pest at the earing stage, the yield is reduced by 0.604 per cent. The pest has been reported to cause as high as 51 per cent loss in yield (Patel *et al.,* 1991). White backed plant hopper has been recorded to inflict 11-39 per cent loss in crop yield (Khatri *et al.*, 1983). Brown plant hopper may cause 10-70 per cent yield losses under severe outbreak. Rice root weevil is reported to cause upto 10 per cent loss in yield. Gundhi bug during severe attack may cause upto 53 per cent yield decline in the crop (Rai *et al.*, 1991). However, high yield losses due to insect pests is now considered an artifact of over estimation based on worst case scenarios, short term small plot trials at single sites frequently unrepresentative of farmers field conditions (Matteson, 2000).

COMPONENTS OF IPM

In view of the fact that farmers have been mostly relying on single tactic of chemical control for managing pest problems, it has become imperative to develop a holistic system of tackling pests to make it more environment friendly, economically viable and socially acceptable for the farmers. Rice IPM in broad sense involves two parts:

a) Pest surveillance and

b) Pest management.

Pest Surveillance

Pest surveillance is one of the most important and integral parts of IPM technology in rice. It is mainly ecologically based and plays important role in forewarning system for pest outbreaks and subsequently timing the needed action. It involves quantifying the insect pest as well as natural enemies. Mathur *et. al.*, (1999) reported that depending on the pre-monsoon rainfall in March-May, the activities of rice gall midge in any year can be predicted. Similar forecast of cutworm incidence is possible from the study of occurrence and pattern of flooding and its subsequent recession. Rainfall in Aug.-Nov. and rainy days in Oct.-Nov. are negatively correlated with the occurrence of YSB population. Knowledge of ecological understanding of overwintering larvae in the rice stubble helps predict the emergence of rice stem borer in winter. Apart from weather parameter light trap and field data aid in detecting the occurrence and subsequent prevalence of stem borer, leaf folder, gall midge, leaf and plant hoppers. The pests monitoring using pheromone traps are also gaining popularity, especially for YSB and leaf folder (Krishnaiah, 1995). Pheromone mediated mating disruption is also attempted. This technique at certain locations has been found promising in preventing the build up of YSB population. These form the base for making decisions for proper interventions by taking economic thresholds as guidelines. Pest surveillance also helps in estimating the pest damage.

Pest Management

Integrated Pest Management, a sound alternative to the conventional use of chemicals, is an attempt to promote natural, economical and social farming techniques through effective blending of appropriate tactics like growing of pest resistance cultivars, suitable cultural practices, use of eco-friendly pesticides, conservation of *in situ* natural biological control and other novel pest control techniques like use of pheromones etc.

Host Plant Resistance

Host plant resistance is a vital component of integrated pest management in rice. Resistant varieties being ecofriendly and with no additional cost burden on production input are usually compatible with other IPM components. Now various resistant cultivars for one or more pests (multiple resistance) with acceptable traits suitable to various agro climatic conditions are available (Table 3).

Table 3: Some multiple resistant varieties to rice pests (Insect pests and diseases)

Variety	State in which Released	Resistant Against
Bhuban	Orissa	GM, BLB
Daya	Orissa	GM, BPH, GLH, BLB
Kunti	West Bengal	GM, BL
Lalat	Orissa	GM, BPH, GLH, BL
Rasmi	Kerala	GM, BPH, BL
Samalei	Orissa, Madhya Pradesh	GM, BPH, GLH, BL
Shaktiman	Orissa, West Bengal	GM, BPH, WBPH, BL
Suraksha	Andhra Pradesh, Orissa, West Bengal	GM, BPH, WBPH, BL
Vikramarya	Andhra Pradesh	GM, GLH, RTD

(GM= Gall midge, BPH= Brown planthopper, WBPH= White backed planthopper, GLH= Green leafhopper, BL= Blast, RTD= Rice tungro disease, BLB= Bacterial leaf blight)
(Source: Reports of Directorate of Rice Research, Hyderabad, Several Years).

Cultural Practices:

Rice varieties can provide an inherent resistance to insect-pests. Another way of reducing pesticides is through cultural practices. These are the tactics adopted by the farmers to harvest the potential production of the crop. This includes all those techniques that will make the environment less favouable for pest to develop or multiply, but which still favours rice production. The manipulation of cultural practices with appropriate knowledge of pest response at times offer direct benefit to farmers in pest suppression at no additional cost. Cultural practices such as rotations, cultivar mixtures, crop mosaics, planting times, flooding, stubble burning, fallow, intercropping, terracing and plant spacing used on the basis of host pest interaction knowledge, can help in pest management through pest avoidance or minimizing the pest build up. Host genetic diversity when used with knowledge of environmental diversity, can greatly enhance the durability of disease resistance and reduce the use for chemical use. Some of the common cultural practices like early and synchronous planting (for managing stem borer, gall midge, BPH, WBPH and GLH), judicious use of fertilizer; cropping pattern (properly planned crop sequences are help in breaking disease cycle or pest build up), alley ways (responsive in managing BPH/WBPH); field sanitation (for managing carry over of pests viz. stem borer and gall midge) and water management (for the pests where micro climate plays important role like case worm, termites, plant hopper, sheath blight, bacterial blight etc.) are positive in managing the pest problems. Removal of stubble, which harbour the overwintering larvae of stem borer, helps in bringing down their population. Split application of potash fertilizer improves the plant ability to resist the buildup of rice hoppers and stem borer. However, over dose of Nitrogen favours the buildup of many pests like BPH, GM, LF and stem borer (Mathur *et al.*, 1999).

Chemical control:

Among the curative method, pesticides are most effective in managing the pests. In case of outbreaks it is considered to be the sole alternative to save the crop from the pest menaces. The high inputs to the crop in the quest of harvesting high yields are also responsible for the crop susceptibility to various pests. The use of pesticide is considered as the easy way to minimize the pest menace. Insecticide use among the rice farmers seems to be based on perceived needs and perhaps fear, rather than real needs. This has resulted in unnoticed damage to non-target organisms and natural enemies. Pesticide application assisted by predictive system can be helpful/useful to check its indiscriminate use. Some of the effective wrecticides against various rice insect pests are listed in Table 4.

Table 4: Effective Insecticide against insect pests

Insecticide	Rate (g a.i./ ha)	Stem borer	Gall midge	Whorl maggot	Leaf folder	Hispa	Brown plant hopper	Green leaf hopper
I. Granular Formulations								
Carbofuran	750	***	**	***	-	**	***	**
Cartap	750	***	-	-	***	-	-	***
Fipronil	75	***	***	**	**	**	***	**
Phorate	1250	***	***	-	-	-	**	**
Quinalphos	1000	***	***	-	-	-	-	-

II. Spray Formulations

Insecticide	Rate (g a.i./ ha)	Stem borer	Leaf folder	Hispa	Brown plant hopper	White backed planthopper	Cut worms	Green leaf hopper
Quinalphos	500	**	***	**	-	**	-	-
Monocrotophos	400	***	***	**	***	***	**	***
Cholrpyriphos	500	***	***	**	-	**	**	-
Carbaryl	750	*	**	**	***	***	-	**
Phosphamidon	500	**	***	**	***	***	-	**
Ethofenprox	75	*	-	-	***	***	-	***
Cartap	300	***	***	-	-	-	**	**
Fipronil	50	***	***	**	***	***	**	***

* Moderately effective, ** Effective, *** Highly effective
(Source: Reports of Directorate of Rice Research, Hyderabad, Several Years)

Pesticide use concept:

A widely acceptable concept of economic decision making to avoid crop loss is that of the economic threshold (Stern *et al.*, 1959). They defined economic damage as the amount of injury that will justify the cost of control. In most cases, the threshold represents time for control (Norgaard, 1976), assuming that the pest population will increase sustainability, barring some exceptions. Early season insecticide use not only does not benefit rice production but it can be detrimental to the ecological balance, causing secondary brown planthopper problems. Thus an immediate strategy may be to reduce early spraying of insecticides.

In the 1960s and 1970s when there was an urgent need to increase the food production, agriculture research was strongly oriented to chemicals (Rossiter, 1975). Now, the time has come as the knowledge about pest build up and adversities of indiscriminate use of pesticides are well documented and even felt by the farmers, with the help of monitoring tools and some level of forewarnings available, the judicious and need based use of the pesticides should be part of our package and the farmers should be accordingly educated.

Table 5: Bio-efficacy of some commercial fungicides against major rice diseases

Fungicide	Blast	Sheath Blight	Brown Spot	Sheath Rot
i) Organophosphorous group				
Hinosan 50 EC	***			
Kitazin 48 EC	**			
Kitazin 17 G	***			
ii) Carbendazim group				
Bavistin 50 WP	***	***	**	**
Denosal 50 WP	***			
iii) Others				
Dithane M 45 (75 WP)	*	*	***	**
Pyroquilon (Fongorene 50 WP)	***			
Tricyclozole (Beam 75 WP)		*		
Contaf 5 EC		***		
Tilt 25 EC		***		

*** Highly effective, ** Moderately effective, * Effective.
(Source: Reports of Directorate of Rice Research, Hyderabad, Several Years)

Seed treatment with fungicides is used for reducing the incidence of seed borne diseases and has been found effective in checking initial inoculum. Effectove fungicides against major rice diseases are listed in Table 5.

Biological control:

Biointensive IPM package has lot to offer both in terms of pest management and environment safety. The affectivity of bioagents on stem borer and leaf folder is very well known and it also can have considerable effect on other important pests like gall midge and plant hoppers. Maximizing the impact of *in situ* natural enemies as an essential part of IPM programme needs emphasis. In India, the release of bioagents / natural enemies is restricted to egg parasitoids particularly *Trichogramma japonicum* and *T. chilonis*, mainly because suitable techniques are available for their mass multiplication in laboratory and subsequent release in the field. In rice, releases of *Trichogramma* are useful against the stem borer and leaf folder complex (Mathur, 1983; Gupta *et al.*, 1987; Bentur *et al.*, 1994). In order to overcome the deleterious effect of insecticides, insecticide resistant strain of *T. chilonis* with multiple tolerance to endosulfan and monocrotophos has been developed by PDBC (Anonymous, 1999). Some of other promising microbial includes fungal pathogens such as *Beauveria bassiana* against rice hispa (Hazarika and Puzari, 1990), *Pandona delphacis* against BPH (Narayansamy, 1995).

Conservation of Natural enemies:

The main steps to conserve the indigenous natural enemies and enhance the biological action in nature, the following strategies could yield significant effects (Beevi *et al.*, 2000; Katti *et al.*, 2001; Tanwar, 2003).

Altering pesticide use

The prophylactic application should be reconsidered in the light of the information available about the pest build up and as far as possible it should be only on the need basis than the blanket ones. The curative use of pesticide should be the last resort. Beside this, the selectivity of the pesticide keeping in view its effect on natural enemies, temporal and spatial separation of natural enemies and pesticides, use bio-pesticides wherever possible. Broad spectrum hazardous insecticides should be avoided.

Habitat Management

The ecological compensation in the form of natural habitats in the vicinity of crop lands can prove to be important in providing alternative food and refugia for natural enemy populations.

Habitat Manipulation

Intercropping, use of trap crops, weed conservation, crop sanitation, retention of crop stubble and providing supplementary diet helps in conservation of natural enemies. Artificial structures like rice straw bundles (loosely packed), bamboo cage cum perches, nesting boxes for wasps and predatory birds can be helpful in promoting natural enemy build up.

Cultural practices

In adopting the appropriate cultural practices the knowledge of these practices about their ill and beneficial effect on non target insects and natural enemy build up can be very useful. Some of the much talked about cultural practices are conservation tillage, avoidance of thrash burning, strip harvesting, use of organic manure should be practiced, which are known to have some kind of effect on natural enemies.

Use of bio-pesticides

Use of microbial pesticides like Bt formulation with endotoxins, which are specific to insect pests. Evaluation of some of these formulations has revealed that they are quite effective against leaf folder and moderately effective against stem borer. Some of the other promising microbial include fungal pathogens such as *Beauveria bassiana* against rice hispa and *Pandora delphacis* against BPH.

Utilization of semio chemicals (Insect sex pheromones)

Sex pheromones are volatile chemicals produced at very low concentrations. They are usually very specific to the species produce them. Pheromones are unaffected by the rainfall and have no adverse effect on the biota or the environment. Sex pheromones have been found promising for management of yellow stem borer, in monitoring as well as direct control either by mass trapping or disrupting mating communication.

Weed Management:

Weed problems in rice differ from region to region because of the variations in rice culture, agro-climatic conditions and management practices. Inspite of wide variation, the number of species that constitute the major portion of the weed flora causing economic concern are *Echinichloa crusgalli, E. colona, Cyperus rotundus, C. diformis, C. iria, Eleusine indica, Fimbristylis miliacea, Ischaemum rugosum, Monochoria vaginalis* and *Sphenoclea zeylanica*. Because of the close morphological similarity, *E. colona* and *E. crusgalli* are the most important weeds of rice worldwide. Grasses are usually the most dominant in competition during early season while sedges and broad-leaved weeds dominate later in the season. The critical density which reduces yields in rice ranges from only 1-3 plants/m in highly competitive species like red rice (*E. crusgalli*) to about 100 plants/m in less competitive species like *M. vaginalis.* The full impact of weed competition occurs early in the growing season.

The problem of changing weed flora need to be tackled in an integrated manner involving cultural practices, crop rotation and use of herbicide mixtures. Closer row spacing in combination with 1.0-1.5 kg/ ha butachlor has been found quite effective in controlling weeds and increasing the grain yield of rice (Table 6). Reduced doses of pre-emergence herbicides followed by hand weeding at 30 DAS are equally effective. Herbicide mixtures are effective tools to enhance herbicide efficiency, lower optimum doses of herbicides, reduce cost of weed management, arrest weed shifts, prevent herbicide resistant in weeds and facilitate improvement in overall weed management. Mixed stands of grasses, broad leaf and sedges in rice necessitates use of herbicide mixture.

Table 6: Promising herbicides for weed control in rice

Herbicides	Dose (ga.i./ha)	Time of application	Remarks
Butachlor	1000-1500	6-7 DAS/DAT	Controls annual grasses and some broad leaved weeds. Ensure sufficient moisture at the time of application.
Anilophos	400	3-5 DAS/DAT	Controls annual grasses and some broad-leaved weeds. Ensure sufficient moisture at the time of application.
Butachlor + propanil (Butanil)	1120 + 1120	10-12 DAS	Excellent control of annual grasses and some dicot weeds. Apply after light rains when weeds have just started emerging
Chlorimuron+metsulfuron (Almix)	4	15-20 DAS/DAT	Controls annual broad-leaved weeds and sedges
2,4-D	500-750	20-25 DAS/DAT	Controls annual broad-leaved weeds and sedges
Pendimethalin	1000-1500	6-7 DAS/DAT	Controls annual grasses and some broad-leaved weeds. Ensure sufficient moisture at the time of application.
Pretilachlor + safener	750	3-7 DAS/DAT	Controls annual grasses and some broad-leaved weeds
Pyrazosulfuron	150	20-25 DAS/DAT	Controls annual grasses and some broad-leaved weeds
Thiobencarb	1000-1500	6-7 DAS/DAT	Controls annual grasses and some broad-leaved weeds

(Source: Yaduraju and Mishra, 2003)

Economic threshold levels:

IPM in reality does not indent to eliminate the pest but to manage it, that too in such a way so as to reap the fruits of all inputs (material and labour) with high profitability and social acceptability by following environment friendly tactics. Besides increasing the farmer's awareness about the pest behaviour. The IPM methodology helps in increasing the risk bearing capacity of the farmers. The ETL is the scientific yardstick to measure the level at which the pest can result into economic loss to the crop. ETL is the critical level or rather the junction from where the proper IPM interventions are taken up to manage the pest and prevent its buildup. Monitoring is the key guiding factor to workout the ETL and that too at right time. The information regarding ETL of some key insect pests and diseases is presented in Table 7 and 8:

Table 7: Economic threshold levels of different insect pests

Insect	Economic Threshold
Brown planthopper/ Whitebacked planthopper	10 insects per hill at vegetative stage or 20 insects/ hill at later stages
Gall midge	5% Silver shoot (at active tillering stage)
Green leafhopper	2 insects/ hill in tungro endemic areas. 20-30 insects/ hill in other areas
Gundhi bug	1 nymph/ adult per hill
Leaf folder	3 damaged leaves/ hill (post active tillering stage)
Stem borer	10% 'Dead hearts' or 1 egg mass or 1 moth/ m^2

Table 8: Economic threshold levels of different diseases

Diseases	Threshold levels
Sheath blight	Pre tillering to mid tillering- 5% or more affected tillers/ m^2
	Panicle initiation and booting- 5% or more affected panicle/ m^2
Bacterial blight	Pre tillering to mid tillering- light 2-5% disease severity
	Panicle initiation and booting- light to moderate
Rice blast	Pre tillering to mid tillering- light 2-5% disease severity
	Panicle initiation and booting- 2-5% leaf area damaged
	Mid tillering to booting- 5-10% leaf area damaged
	Flowering and after- 5% leaf area damaged or 1-2% neck infection
Brown spot	Flowering and after- 2-5% affected tillers

IMPLEMENTATION OF IPM

Implementation of IPM on a large scale needs action of government agencies, NGOs, industry and farmers. As IPM requires community action, cluster approach of selecting villages and farmers in contiguous area needs to be followed.

Through Operational Research Project

ICAR initiated 6 operational research projects on IPM for rice long back in 1975 under the supervision of All India Coordinated Rice Improvement Project, Hyderabad, Kerala Agricultural University and Department of Agriculture, West Bengal. The components of IPM included monitoring of the pests, parasite and predator populations, minimal use of pesticides at selected times to encourage natural enemies buildup, ploughing of rice stubble and the use of early maturing, short duration resistant varieties of rice. Adoption of IPM practices resulted in increase in rice yield from 34.88 to 49.33 q/ha in A.P. during 1981-86 (Krishnaiah and Reddy, 1989), while in Kerala, the number of insecticide sprays on rice had been cut from 4-6 to average of two per crop (Sankaran, 1987). There are also a few other examples of successful implementation of IPM in rice in selected districts of different states like Andhra Pradesh, Haryana, Kerala, Madhya Pradesh, Tamil Nadu and Uttar Pradesh (Razak, 1986).

IPM Through Farmers Participatory Approach

Large scale implementation of IPM through farmers participatory approach has been reported recently by DRR, Hyderabad in rice bowl district of West Godavari (A.P). A total of 100 acre was covered. The trial involved two treatments by dividing each farmers field into two equal blocks (i) IPM block with need based application of insecticide based on periodical surveillance and tentative economic thresholds and

(ii) non-IPM block with conventional practices adopted by farmers. The results revealed that the IPM plots yielded 10 q/ha higher than non-IPM despite cut in insecticidal application by two to three times compared to farmers practices (Katti, 2000). Thus, the trial revealed the scope of convincing the farmers to reduce the number of application of insecticides, as it was beneficial to him in terms of cost, safety to human health as well as to natural enemies and other non-target organism. In the field trials conducted in the Haryana villages of Kurali Rembha, Darar, Araipura and Salaru during 1983-90, the effects of IPM were evaluated on pesticide expenditure and yields of rice with varieties PR-106, PR-107, PR-109, HKR-120, Jaya and Basmati-370. It was recorded that the cost of plant protection was 58.2 per cent higher in non-IPM fields than in IPM fields. Moreover, rice yields increased by 3.3 per cent in IPM over the non-IPM fields (Quadeer and Tomar, 1993).

IPM in Basmati Rice

Many pests attack the Basmati rice in the predominant basmati rice growing areas of Haryana, western U.P. and Punjab. For controlling these pests farmers follow mainly the chemical approach, which is quite expensive and many a time leads to residue problem. Being an export oriented commodity, residue problem may hamper its export potential also. To overcome this menace, IPM is the apparent option. On the basis of information available on rice pest management strategies, the location specific IPM modules were prepared and field evaluated by NCIPM in Haryana and Western U.P. The IPM treatment was compared with sole pesticidal treatment and untreated control. The results of Haryana revealed that IPM approach was almost at par with the insecticidal control during all the 3 years. The yield data showed that highest yield was secured in chemical control, which was 31.51, 34.33 and 28.55q/ha during 1994,1995 and 1996 respectively, while in IPM 29.89, 31.96 and 28.25q/ha yield was obtained. However, the economic analysis indicated that IPM was slightly superior to chemical control as the mean cost benefit ratio of IPM over untreated control was 1:5.7 as compared to 1: 5.03 in chemical control (Garg and Baranwal, 1999).

IPM IN PUSA BASMATI-1

Location specific IPM modules were synthesized by NCIPM for the Basmati rice growing areas of Uttar Pradesh and evaluated on a small scale during *Kharif* season of 1997 and 1998 followed by 40 hectare during 1999. Encouraged with the success of these IPM trials (Garg *et al* 2000), the large scale validation of IPM approach has been carried out during *Kharif* 2000 to 2002 in a village Shikohpur of Baghpat district in western Uttar Pradesh, where farmers solely depended on the use of intensive pesticidal applications. This farmers participatory approach was validated over an area of 120 and 160 hectare during the years 2000 and 2001 respectively. With the adoption of IPM approach, which mainly comprised of seed treatment, regular pest surveillance and monitoring, conservation and augmentation of natural enemies, need based application of pesticides and some improved crop management practices, there has been drastic reduction in pesticidal application and the yield levels were enhanced to a significant extent. The actual IPM interventions are presented in Table 9 alongwith the farmers practices (non-IPM).

In IPM fields, more abundance of natural enemies was recorded as compared to non-IPM due to least pesticidal interventions. The common predators like spiders, dragonflies and damselflies, grasshoppers, crickets, carabids and *Cyrtorhinous lividipennis,* were recorded in high numbers in IPM fields, while among parasitoids *Apanteles* spp., *Telenomus* spp., *Tetrastichus* spp., *Stenobracon* spp. and *Xanthopimpla* spp. were quite common. The population of all these natural enemies was almost negligible in non-IPM fields during the period of study (Garg *et al.,* 2002). The consistent higher yield levels of 58.04 q/ha (2000) and 57.40 q/ha (2001) in IPM, as compared to 48.21 and 45.60 q/ha for the same years in the farmers practices (non-IPM) reveals the success of this approach (Table 10). Besides this, input cost was significantly lowered in IPM with higher cost-benefit ratios. During *Kharif* 2002, IPM validation by the centre was not carried out directly but the records were taken on the farmers IPM practices, to evaluate the impact of earlier years IPM validation on farmers mind. It was found that after first year of withdrawl, majority of farmers continued to follow the IPM tenets, which led to enhanced yield and economic returns (Table 11). They even procured the critical IPM inputs like *Tricho* cards at their own and pesticides use was almost negligible.

Table 9: Main interventions in IPM and Farmers practice (non-IPM) fields at village Shikohpur

IPM	FARMERS PRACTICE
IPM inputs	
Seed treatment with carbendazim @ 2.0 gm/ kg seed	No seed treatment
Monitoring for insect pests and diseases	Systematic monitoring lacked
Release of parasitoid *Trichogramma japonicum* @ 1.5 lakh/ha against stem borer and leaf folder after initiation of infestation	No release of bioagents
Need based application of recommended insecticide and fungicide/ bactericide (Only spot applications)	Pesticides used indiscriminately in the entire field
Crop management inputs	
Green manuring with 'Dhaincha' (*Sesbania* sp.)	Green manuring was not taken
Balanced fertilization (N 100: P 60: K: 40), $ZnSo_4$: 25 kg/ ha) and judicious irrigation	Fertilizer(N 120: P 40: K 00), $ZnSo_4$ 25 kg/ ha)
Planting of 2-3 seedlings per hill	Planting of one seedlings per hill
Drain off excess standing water	No draining of excess water

Table 10: Yield and economics of Pusa Basmati –1 in IPM and Farmers practice (FP) or Non-IPM, *Kharif* 2000 and 2001.

	I P M Village Shikohpur		F P Village Sarurpur (2000) and Badka (2001)	
	2000	2001	2000	2001
Total Cost (all inputs)	17781.54	17733.00	20553.52	20980.00
Mean Yield (q/ha)	58.04	57.40	48.21	45.60
Total Returns	56589.00	55965.00	47005.00	44460.00
Net Returns	38807.46	38232.00	26451.48	23480.00
Cost-Benefit ratio	1:3.18	1:3.16	1:2.28	1:2.12

Table 11: Yield and economics of Pusa Basmati –1 in Farmers' own IPM and Non-IPM (FP) fields *Kharif*-2002.

	I P M (Shikohpur)	F P (Badka)
Total Cost (all inputs)	24469.00	27723.00
Mean Yield (q/ha)	51.60	43.50
Total Returns	54138.00	45675.00
Net Returns	29669.00	17952.00
Cost-Benefit ratio	1:2.21	1:1.64

* The paddy was sold @ Rs. 975 / q (2000), Rs. 975/q (2001) and Rs. 1050 / q (2002).

** Total cost included labour cost for land preparation, nursery sowing, puddling, transplanting, fertilizer application, hand weeding, pesticide application etc. and material cost like seed, fertilizer, pesticides, biocontrol agents, irrigation etc.

Economic impact of IPM adoption in Pusa Basmati-1

In IPM, farmers additional gain was Rs. 12356, 14752 and 11717/ ha during 2000, 2001 and 2002 respectively. Thus, by following IPM approach farmers secured additional income of Rs. 1482720, 2360320 and 1406040 from 120, 160 and 120 ha area respectively. Besides, cost of production also came down due to judicious use of fertilizers, pesticides and irrigation (Table 10 & 11). Labour absorption was also less in IPM.

IPM IN TARAORI BASMATI

After the success of IPM at Shikohpur in Pusa Basmati – 1, a new village Chhajpur Khurd, Panipat (Haryana) was selected for the validation of IPM in another Basmati rice variety 'Taraori local' which is being exported in large quantities. The yield potential of this variety is less in comparison to Pusa Basamti – 1, however, it fetches higher price in domestic and export market. Like Pusa Basmati – 1, this variety is also susceptible to many pests *viz.* stem borer, leaf folder, white backed plant hopper, blast (leaf and neck), BLB, sheath blight etc. Farmers were dependent on the advise of pesticides vendors and State Agriculture Deptt. for pest control. They used to apply pesticides 4-5 times to overcome the pest menace. Farmers used to get a harvest in the range of 20-25 q/ha of Taraori, they showed their unawareness about the IPM.

During *Kharif* 2002, a total of 70 acre area was under IPM involving 17 farmers. Simultaneously, 30 acres was marked under Farmer's practices (FP) in which farmers used control tactics at their own judgment.

The IPM farmers were advised to follow some improved crop management practices like transplanting 2-3 seedlings per hill and judicious use of nitrogenous fertilizer to avoid lodging as Taroari local has the tendency to lodge. Details of interventions in IPM and FP fields are given in Table 12.

Yield data showed that higher yields (28.35 q/ha) were secured from IPM fields as compared to 22.20 q/ha in FP. It seemed that in addition to less pest incidence in IPM, some of the proper crop management practices like judicious application of nitrogenous fertilizer saved IPM fields from lodging which was much higher in FP and responsible for reducing the yield to a certain extent. The economic analysis indicated that cost benefit ratio was much higher in IPM as compared to FP (Table 13).

Table 12: Interventions in IPM and Farmers Practices (FP) in Taraori Basmati

IPM	FP
♦ Seed treatment with carbendazim	♦ No seed treatment
♦ Planting of 2-3 seedlings/ hill (only few farmers)	♦ Planting of 1 seedling /hill
♦ Use of recommended fertilizer dose	♦ Fertilizer dose higher than recommended
♦ Release of *Trichogramma japonicum* and *T. chilonis* (2 times)	♦ No release of bio agents
♦ Need based pesticides applications based on monitoring, one spray of streptocycline for BLB (about 1.0 ha) and one spray of tricyclazole for blast (.05 ha)	♦ Use of pesticides (3-5 times) indiscriminately (Phorate, fipronil, dichlorovos, methyl parathion, dithane M-45, carbendazim etc.)

Table 13: Yield and Economics of Taraori Basmati Rice Production in IPM and FP

Expenditure/ Returns	I P M	FP
Total Cost** (all inputs)	16209	19607
Mean Yield (q/ha)	28.35	22.20
Total Returns	56700	44400
Net Returns	40491	24793
Cost Benefit ratio	1:3.50	1:2.26

Rate of paddy: Rs. 1800 / q .

**Total cost included labour cost for land preparation, nursery sowing, puddling, transplanting, fertilizer application, hand weeding, pesticide application etc. and material cost like seed, fertilizer, pesticides, biocontrol agents, irrigation etc.

Economic impact of IPM adoption in Taraori Basmati

In IPM, farmer's additional gain was Rs. 15,698/ha, thus, by following IPM approach, farmers secured an additional income of Rs. 4,39,544 from 28 ha area. Increase in income was 63 per cent over FP. Cost of production in IPM on account of judicious use of fertilizer, pesticides and irrigation has come down to 20 per cent. Moreover, in IPM labour absorption was also less (139.16 man days/ha) as compared to FP (145.66 man days/ha).

The outcome of IPM validation in Pusa Basmati-1 and Taraori Basmati emphasizes the need to further spread the message of IPM in Basmati rice growers by developing and implementing area specific, cost effective and environmental friendly IPM tactics.

The coverage under IPM in rice at present is estimated to be less than two per cent of the total 43 million hectares of cultivable area in the country. Very few efforts have been made to synthesize the location specific IPM modules to take care of insect pests, diseases and weeds, and other crop damaging organisms together (Mathur *et al.* 1999, Pathak *et al.* 1998). IPM in basmati rice is all the more crucial as on one hand it will save lot of input cost, on the other it will help to overcome the pesticide residue problem as it has been reported recently that large consignments of basmati rice for exports to other countries are routinely returned as they invariably contain high levels of fertilizer and pesticide residues

(*The Economic Times*, Feb. 15, 2001). The only solution to this problem is the development and implementation of area specific, cost effective and environmental friendly IPM strategies.

REFERENCES

Anonymous, 1999. *Annual Report*, 1998-99. Project Directorate of Biological control, Bangalore. p. 206.

Atwal, A.S. and G.S. Dhaliwal. 2002. *Agricultural Pests of South Asia and their Management*. Kalyani publishers, New Delhi, pp. 498.

Beevi, S.P., Lyla, K.R. and Prabhakaran, P.V. 2000 a. Quantification of pests and natural enemies of rice ecosystems. In: *Proc. of the Twelfth Kerala Science Congress*, January 2000, Kumily p. 617-621.

Bentur, J.S., Kalode, M.B., Rajendrann, B. and Patil, V.S. 1994. Field evaluation of the egg parasitoid *Trichogramma japonicum* Asm. (Hym: Trichogrammatidae) against the rice leaf roller, *Cnaphalocrosis medinalis* (Guen.) (Lep: Pyralidae) in India. *J. of Appl. Ent.*, 117: 257-261.

Garg, D.K. and V.K. Baranwal. 1999. An integrated approach for pest management in Basmati rice in Haryana. In: *Ecological Agriculture and Sustainable Development (Vol. 2). Proceedings of International Conference on Ecological Agriculture: Towards Sustainable Development*, Nov. 15-17, 1997 Chandigarh, India pp. 216-221.

Garg, D.K., M.D. Jeswani, P. Kumar and R.N. Singh. 2000. IPM approach enhances the yield of Basmati rice. *ICAR News*, 6 (2): 12-15.

Garg, D.K., Pradyumn Kumar, R.N. Singh and Mahesh Pathak. 2002. Role of parasitoid *Trichogramma japonicum* and other natural enemies in the management of yellow stem borer and leaf folder in Basmati Rice. *Indian J. Ent.*, 64 (2): 117-123.

Gupta, M., Rao, P.P. and Pawar, A.D. 1987. Role of parasites and predators in rice ecosystem. *Proceedings of the National Symposium on Integrated Pest Control Progress and Perspectives*, October 15-17, 1987 (Eds) Mohan Das, N and George Koshy., pp. 545.

Hazarika, L.K. and Puzari, K.C. 1990. *Beauveria bassiana* (Bals). Vuill for biological control of rice hispa (RII) in Assam, India. *Int. Rice Res. Newsl.*, 15:31.

Heong, K.L.; Teng, P.S. and Moody, K.1995. Managing Rice Pests with Less Chemicals. *GeoJournal*. 35.3. 337-349.

Katti, G. 2000. On farm pest management in rice with emphasis on natural biological control. In National Training Course on Integrated Pest Management, New Delhi. Training Manual: 78-79.

Katti, G., I.C. Pasalu, N.R.G. Verma and K. Krishnaiah. 2001. Integration of pheromone mass trapping and biological control as an alternate strategy for management of yellow stem borer and leaf folder in rice. *Indian J. of Ent.*, 63 (3): 325-328.

Khatri, A.K., G.A. Gangrade and V.S. Rathore. 1983. Studies on the crop losses in rice caused by the white-backed planthopper (*Sogatella furcifera*) and the economic injury in India. *Trop. Pest Manage*, 29: 220-223.

Krishnaian, K and Reddy, P.C. 1989. Operational Research Project on Integrated Pest Management in Rice Pests in Medchal area, Ranga Reddy district, A.P. Achievements and Constraints. Annual Rice Workshop, HAU, Hissar, April 19-22, 1989, pp. 9.

Mathur K.C., Reddy, P.R., Rajamani, S. and Moorthy, B.T.S. 1999. Integrated pest management in rice to improve productivity and sustainability. *Oryza* 36(3): 195-207.

Mathur, K.C. 1983. Biological Control in Pest Management. In: *Proc. Annual Rice Workshop* 1983, Hyderabad, India.

Matteson, P.C. 2000. Insect Pest Management in Tropical Asian Irrigated Rice. *Annu. Rev. Entomol.* 45: 549-574.

Narayanasamy, P. 1995. Mycoinsecticide: A novel biopesticide in Indian scenario. *Biotechnology and Development Review*, Jan.-June, 1995. pp. 15-20.

Norgaard, R.B. 1976. The economics of improving pesticide use. *Annu. Rev. Entomol.* 21:45-60.

Oerke, E.C., D.W. Dehne, F. Schonbeck, and A. Weber. 1994. Crop production and crop protection: Estimated losses in major food and cash crops. Amsterdam, Netherlands; Elsevier.

Padmanabhan, S.Y. 1965. Estimating losses from rice blast in India. *In*: *The Rice Blast Disease*. *Proc. Symp.* held at the International Rice Research Institute in July 1963. The Johns Hopkins Press, Baltimore, Maryland, USA, 203-221 pp.

Padmanabhan, S.Y. 1973. The Great Bengal Famine. *Annu Rev Phytopathol*. 11: 11-26.

Patel, K.G., C.B. Patel and S. Raman. 1991. Avoidable losses in rice due to stem borer. p. 157. *Proc. Intern. Symp. Rice Research. New Frontiers*. November 15-18, 1990. DRR, Hyderabad.

Pathak, M.D. and G.S. Dhaliwal. 1981. Trends and Strategies for rice insect problems in Tropical Asia. *IRRI Research Paper Series no. 64*, International Rice Research Institute, Los Banos, Philippines.

Pathak, M.D., Rao, Y.R.V.J., Row, K.V.K. and Mukhopadhyay, S.K. 1998. Integrated pest management in rice. In: *Proc. International Symposium on Rainfed Rice for Food Security*. Central Rice Research Institute, Cuttack: 259-300.

Prasad, G.S.V., Muralidharan, K. and Rao, C.S. 2001. Stability and yield performance of genotypes: A proposal for re-grouping world rice area into mega-environments. *Curr. Sci*. 81: 1337-1346.

Puri, S.N. 2000. India (1). pp. 83-96. In: *Farmer-Led Integrated Pest Management in India and the Pacific*. Asian Productivity Organization, Tokyo.

Quadeer, G.A. and Tomer, R.S. 1993. Rice IPM demonstration in Haryana- a success story. *Pl. Prot. Bull.*, Faridabad, 45 (10: 21-23.

Rai, A.B., J. Singh and Lallan Rai. 1991. Evaluation of gundhibug damage, Leptocorisa varicornis in rice. pp. 159-161. *Proc. Intern. Symp. Rice Research: New Frontiers*. November, 15-18, 1990. DRR, Hyderabad, India.

Razak, R.L. 1986. Integrated Pest Management in Rice crop. *Pl. Prot. Bull*. 38: 1-4.

Rossiter, M.W. 1975. The Emergence of Agricultural Science. Yale University Press, New Haven.

Sankaran, T. 1987. Biological control in Integrated Pest control-Progress and Perspective in India. *Proceedings of the National Symposium on Integrated Pest Control. Progress and Perspectives*, Oct. 15-17, 1987 (Eds) Mohan Das, N. and Ggeorge Koshy. pp. 545.

Savary, S., R.K. Shrivastava, H.M. Singh and F.A Elazegui. 1997. A characterization of rice pests and quantification of yield losses in the rice-wheat systems of India'. *Crop Prot*. 16: 387-398.

Singh, J. and G.S. Dhaliwal. 1994. Insect pest management in rice: A perspective. pp. 56-112. *In* Trends in Agricultural Insect Pest Management. Dhaliwal, G.S. and Arora, R. (Eds.), Commonwealth Publishers, New Delhi.

Srivastava, D.N. 1967. Epidemiology and control of bacterial blight of rice in India. In: *Proc. Symp. On Rice Dis. and their Control*: 11-18

Stern, V.M.; Smith, R.F.; van den Bosch, R.; Hagen, K.S. 1959. The integrated control concept. *Hilgardia*. 29: 81-101.

Tanwar, R.K. 2003. Conservation of natural enemies of rice, cotton and vegetables in different agro ecosystem. Lecture compendium. National Training on IPM in cotton, rice and vegetable crops. National Centre for Integrated Pest Management, New Delhi. pp. 190-197.

Tiwari, K.N. 2002. Rice production and nutrient management in India. *Better crop International*. 16: 18-22.

Way, M.J. and K.L Heong. 1994. The role of biodiversity in the dynamics and management of insect pests of tropical irrigated rice-a review. *Bull. Entomol. Res*. 84: 567-587.

Yaduraju, N.T. and J.S. Mishra. 2003. Integrated weed management in rice, cotton and vegetables. Lecture compendium. National Training on IPM in cotton, rice and vegetable crops. National Centre for Integrated Pest Management, New Delhi. pp. 190-197.

Eco-Friendly Management of Insect Pests of Pulse Crops in India

RAM UJAGIR[1]

Pulses play very important role in maintaining soil health and supplying protein to large masses of the people in the country. India is major country which grows pulses with a share of 30-35% and 27-28% of area and production, respectively in the world. It produces about 12-14 million tonnes of pulses from an area of about 22-24 million hectare. The commonly grown pulses in India include chickpea *(Cicer arietinum)*, pigeonpea *(Cajanus cajan)*, urdbean (*Vigna mungo*), mungbean (*Vigna radiata*), lentil (*Lens culinaris*),field peas (*Pisum sativum*), horse gram (*Macrotyloma biflorus)*, mothbean (*Vigna aconitifolia*), lathyrus (*Lathyrus sativus*), cowpea (*Vigna unguiculata*), and drybean (*Phaseolus vulgaris)*.These crops are highly susceptible to a large number of insect pests at various stages of their growth. On an average 2.5 to 3.0 million tonnes of pulses are lost every yea due to attack of insect pests. Pulses are grown under diverse agro climatic regions. Majority of pulses growing farmers are resource poor which grow these crops under rain fed, low input, mixed/ intercropping with a high risk of pest attack without adopting pest control measures. When crop is grown under progressive farming it is much more prone to pest attack due to high input, rainfed, monocropping and late planting of the crop. Most of the insect attacking to pulse crops are common except very few. However, varying number of insect pests have been reported to occur on pulses at various stages of crop growth consisting of inverteberate and verteberate pest species. They may attack root, stem, foliage, buds, flowers, pods of plants and stored grain.

Crop phenology

Pulses are grown in all the three season - *kharif, rabi* and spring/ summer. Pigeonpea is sown mostly during kharif season, June–July and harvested from November to April depending on growth habit. The insects pests attacking at seedling/vegetative stages of the crop do not usually cause serious loss to pigeonpea crop. Activity of insects pests particularly pod borer complex and pod sucking bugs at flowering and podding stage pose serious problems and crop is in economic injury phase. Mungbean and urdbean are grown both in *kharif* and spring/summer seasons. In *kharif* season these crops are planted from mid-July to mid-August and harvested from October to November whereas spring/summer season crops are sown during mid-February to mid April and harvested between mid-may to mid June before onset of monsoon. At initial stages of its growth, stemflies and galerucid beetle respectively may cause seedling mortality and defoliation of the plants in certain areas. At vegetative stage of the crop, Bihar hairy caterpillar, tobacco caterpillar and white fly are common in *kharif* season crop. At reproductive stage thrips may drop the flower which is more pronounced in spring/summer seasons followed by pod feeding insects particularly, sucking bugs and pod borer complex at podding of the crop. During *rabi* season chickpea, field pea and

1 Department of Entomology, G. B. Pant University of Agriculture and Technology Pantanagar-263145 (Uttaranchal)

lentil are grown during October to April. Chickpea has serious problems of gram pod borer, *Helicoverpa armigera* in all the areas. It can attack at both vegetative and podding stage of the crop. In southern and central India it attacks at both the stages of the crop. The damage caused at vegetative stage may be called non economic injury phase as it is insignificant to justify control. When it occurs at podding stage it is called economic injury phase where reduction in yield or quality at least equal to pest controls. At initial stage of pea crop growth, stem fly may cause seedling mortality in early sown crop. Next are defoliators, aphid and pea leaf miner. The heavy infestation by these insects may reduce the yield of the crop. Late planted crop generally suffer from pod borer complex comprising of *Eitella zinckenella, Lampides* sp. and *H. armigera*. Lentil receive attack of aphids at vegetative and pod borers, E. zinckenella and H. armigera at podding stage of the crop.

Insect pests complex

The important insect pests attacking pulse crops have been described below crop wise.

Pigeon pea

It is attacked by a large number of insect pests more than 300 species of insects (Lal and Sachan, 1997). The most economical pests are those who attack at flowering and podding stage of the crop including, legume pod borer (*Maruca vitrata*), gram pod borer (*Helicoverpa armigera*), pod fly (*Melanagromyza obtusa*) and pod sucking bugs (*Clavigralla* sp.. and *Nezara viridula*). The leaf tier, *Grapholita critica, M.Vitrata* and *H.armigera* are the most important pests of early maturity crops. The gram pod borer and pod sucking bugs are serious pests of medium and late maturity crop of pigeonpea.

Mung bean and urd bean

Mungbean and urdbean and other related *Vigna* spp. pulses are attacked by a diverse range of insects during various stages of their growth. The stemflies at seedling stage may cause killing of some plants and stunting others, causing slow growth and gaps in the crop. During vegetative stage a number of insects including, Bihar hairy caterpillar (*Spilosoma obliqua*), tobacco caterpillar (*Spodoptera litura*) and some other lepidopteran may defoliate the growing plants. White flies (*Bemisia tabaci*) are vector of mungbean yellow mosaic virus (MYMV). The thrips (Megalurothrips *distalis, Thrips* sp., *Caliothrips indicus, Scirtothrips* sp.) cause non setting of pods by flowers drop. At podding stage, the *M. Vitrata, Lampides boeticus* are important pests which bore into the pods. The bugs, *Clavigralla gibbosa, Nezara viridula, Piezodorus hybneri, P. rubrofasciatus* may cause shrivelling and drying of pods.

Chickpea

Chickpea has been reported to be attacked by more than 52 species of insects at various stages of its growth but gram pod borer (*Helicoverpa armigera*) is the regular and most dominant pests all over the India.

Field peas

The stems of plants at seeding stage are mainly attacked by stemfly, *Melanagromyza phaseoli*. The direct damage to foliage is mainly caused by leaf miner (*Chromatomyia horticola)*, pea semilooper (*Plusia orichalsia*), aphids (*Acyrthosiphon pisum* and *Aphis craccivora*), thrips (*Caliothrips indicus*). Pods are damaged by *E. zinckenella, H. armigera, L. boeticus* and *Euchrysops cnejus*.

Lentil

The important pests attacking foliage are leaf miner, *C. horticola* and aphid, *A. craccivora*. The developing pods are mainly attacked by pea pod borer, *E. zinckenella* and gram pod borer, *H. armigera*

Pest monitoring

The use of attractants, particularly pheromones in traps to monitor pest population specially Lepidoptera (*H. armigera*), so that pesticide application can be timed more precisely in relation to pest incidence is well established and form an integral part of pulses insect pest management scheme. The work on pheromone traps was initiated at ICRISAT then under All India Coordinated Project on Pulses. Monitoring through pheromone is helpful in determining the peak activity of the pest. With the help of pheromone the peak activity of *H. armigera* across the country has been established through net working of pheromone studies. Pheromone traps should be set up before moths started to enter the crop i.e. before flowering at above canopy height of the crop. In chickpea 3-4 and in pigeon pea 5-6 moths are recommended as action threshold. The traps may be inspected daily or every two days to know the number of moths. Crop monitored in this way only need to be sprayed when the threshold catch is exceeded that is when 5-6 moths or more are caught in trap during 2-3 consecutive days period. For *H.armigera*, the insecticide spray are applied one week after the catch reaches to threshold level to kill newly hatched larvae before they start to feed into the pods. Consequently the date the 'threshold' number of moths is achieved is considered synchronous with start of substantial oviposition in the crop.

Cultural control practices

Ecofriendly pest management have generated renewed interest in cultural control methods and these have been extended to include the analogous concept of ecological control (Beirne, 1967) and ecosystem management (Speight, 1983). These terms emphasize in particular, diversification of agricultural system by rotation, inter and intra crop diversity and the manipulation of other environmental factors that affect the pest species detrimentally either directly or indirectly by benefiting their natural enemies. To design and implement cultural control measures it becomes necessary to know weak links in the pests life cycle with the knowledge of the crop, phenology, pest biology and ecology.

Cromartie(1986) outlined the cultural practices like diversity can influence pest control. This also applies for control of pests of pulses. The crop or habitat becomes unacceptable to pest by interfering with pest activities like oviposition and host plant discrimination or location by both adult and larval stages. The crop may be made unavailable to pest in space and time by utilizing knowledge of its life history, especially its peak activity, migration and over wintering. The reduction in survival of pest on crop could be made enhancing the number and activity of natural enemies.

Some of examples from these have been used to illustrate the role of cultural control practices in pulses.

Crop rotation

Isolation of new crop from previous sources of infestation has not been very successful against major insect pests, largely because the female of like *H. armigera* are capable of dispersing to very far places. This means the offspring from previous infestation can influences population in the surrounding, since many pulse growers have relatively small farm, it is usually impossible to obtain desired level of isolation. In addition to cultivate crop, however, isolation also depend on the number and types of wild host plant s in the region. Several species of weed host plants like, *Chenopodium, Vicia, Melitotus* found in *rabi* season support population of *H.armigera*. It is difficult to generalize the contribution of wild hosts plants but such plants definitely act as reservoir for the pests and hence prevent isolation in the true sense. Obiviously, complete destruction of alternate/wild host plants from large areas is impossible.

Planting and harvesting time

Date of sowing could help in manipulation of pest population on particular crop in reducing the pest incidence without sacrificing the optimum yield of the crop. Pigeonpea crop sown between mid- May and mid- June received low pod borer damage as compared to July sown crop. Manipulation of date of sowing in blackgram has been reported in reducing the pest population of jassids and galerucid beetle (Saxena *et.,al* 1973). Early planting of chickpea in most areas except *Tarai* of northern India causes low borer incidence due to asynchrony with peak activity of gram pod borer population (Yadava *et al.,* 1983). Early sowing of chick pea synchronize the *Helicoverpa* larval population with its parasitoids, *Campoletis chloridae*. Early sowing of chickpea and pigeon pea flower and mature early which avoid pest attack to the crop. Similarly late planting (November) escape the attack of stem fly damage thus schedule could be improve by the use of thermal acceleration unit model which predicts when *Helicoverpa* and stem fly etc. optional mating has occurred and the probability of accumulating sufficient thermal unit to establish the crop. So rather than using calendar years which are too late in some years earlier late planting could be made without risk of attack

Early and timely harvest can often reduces losses from pest damage in some case. Thus, optimum planting time to avoid pest may also influence optimum harvest time. Timely harvest of pigeon pea may escape from the damage of pulse beetle *Callosobruchus* spp. which have attracted to ripening grain and seed legumes in the field.

Rotation

This determines the frequency of annual crops and, in general, the more frequent the rotation the more damaging endemic pest are likely to become. The practice of crop rotation involving host and non host crop of insect pest is particularly effective against species having a limited host range and dispersive ness: it is unsuitable for aphids, blossom beetle and *Helicoverpa* like Lepidoptera or species with a range of wild host. The intensive farming rotation a time of the past the intensive cropping system has triggered problem of *H. armigera* in many parts of the country.

Trap crop

Small and early planting than main planting may divert the insect attack away from the crop attack at risk by using more attractive food sources. Trap crop can either be destroyed or sprayed necessarily before the pest reproduce and the susceptible stage of the crop is available to them.The marigold, sunflower, sesame, carrot, and Solanum viarum has been found as trap crop for *H. armigera.*

Intercropping

Farmers have been using traditional system of intercropping / multicropping. That is growing more than one crop on the same pieces of land, for centuries and intercropping or polyculture is one type of cultural system that is a common feature in the Indian agriculture. Intercropping generally result in higher yields due to reduced weed flora, soil conservation, better use of incident radiation water and soil nutrients (Wiley,1979). Another advantage is that insect attack is often less than that on pure crops. There appear to be two reasons why intercropping can reduce phytophagous pest level (Root, 1973). Polyculture encourage a greater number of natural enemies and the more favour conditions of associated plants on ability of the insect herbivores to find and utilize its hosts plants.

Attempts to reduce pest attack by increasing diversity have been made by growing of linseed, barley, safflower, coriander, mustard with chickpea; urd, mung sesame, maize with pigeonpea.

Mungbean intercropped with sorghum and pearlmillet decreased incidence of jassids and other pests.Similarly, intercropping of urdbean with pigeonpea and cowpea delayed the appearance of jassid, greenbug and membracid (Singh and Singh, 1977). Similarly intercropping of moongbean with blackgram had reduced the incidence of MYMV (Dasgupta and Chaudhary, 1985). Although, overall effects have been encouraging, it has not always been easy to predict the outcome of any particular intercropping system. It has been also found that coriander and linseed with chickpea and sorghum with pigeon enhances the number of natural enemies, parasitoid and predatory arthropods. The differences may be also result from interference by intercropped plants during attraction and oviposition phase of the pest and also for the growth influence of the intercropping on microclimate of the plants. It s main drawback may be competition from the intercrop often, though always reduce the crop yield. Such competition may not be as important in future, however, if lower yield lower quality produce become accepted on the previous that it is entirely insecticide free. More recently intercropping experiment involving growing coriander with chickpea have produced some of the benefits of intercropping without severely reducing crop yield (AICPIC, 2000-02).

Tall crop like sorghum and sunflower serve as perching site for insectivorous birds as droryo which prey upon larvae of *H. armigera*.

Thus manipulation of dates of planting and harvest to escape peak pest populations; intercropping particularly with non-host crops of pests; hand picking of egg masses of hairy caterpillars and bugs; summer ploughing; destruction of crop residues; growing of trap crops and balanced use of fertilizer will help in bringing down the harmful pests population.

HOST PLANT RESISTANCE

Plant resistance has been used in integrated pest management system to replace broad spectrum insecticides and provide desired level of control for individual major pests of the crop. Development of resistant varieties that are resistant or tolerant to major insects pests to *H. armigera*, *M. obtusa* and stem flies has received major attention in recent past particularly for chickpea, pigeon pea and moongbean.

Varieties of chickpea, pigeonpea and moongbean showing varying level of resistance have been reported in India. Few have been adopted for cultivation by farmers in various part of the country (Lateef, 1985, Reed and Lateef,1990,and Sharma *et.al.*,1999). Shukla and Yadav (1998) have discussed the advances made in plant resistance to *H. armigera* in chickpea and pigeonpea. Chauturvedi *et.,al.* (1997) have summarized the progress made in breeding for resistance to *H. armigera*. Talekar (1990) has listed the resistant cultivars of mungbean against stemflies. Chhabra and Kooner (1998) and Chhabra *et al.* (1988) reported mungbean and urdbean resistance to white fly, jassids,thrips and lepidopteran pod borers.

Some of the information on identification and utilization of resistance to *H. armigera* to major pulses is summarized below.

Chick pea

Genotype ICC506, ICCV7, ICCV10, ICC6663, Dulia,ICC10667 and ICC5264(Lateef,1985) PUSA261(Reedy *et.al.* 1996); C235 (Ahmad and Kotwal,1996); JG74 (Das and Kataria,1999); BG256 (Kotikal *et.al.* 1996a,b) ; ICCV890109-20-BP, ICC506, ICCL86102, ICCV95992 and L Beg 3 (AICPIC,1998) have been reported to be less susceptible to *H. armigera* in India.

Pigeon pea

Several cultivars have been identified as least susceptible to *H. armigera* such as ICPL 332, PPE45-2, MA-2 and ICPL84060(Reed and Lateef,1990) ; PDA88-2E and ; PDA99-3E, PDA89-2E and SL 21-9-2 (Das,1998) ; GAUT85035 and BDN-2(Kalariya *et.al.*,1998); ICPL4 (Venketaswarlu and Singh,1999); Lal and Katti (1997) identified pod fly resistant doners such as PDA88-2E,PDA89-2E and PDA92-3E.

Moong bean

Screening of moongbean cultivars at Asian Vegetable Research and Development Centre (AVRDC), Taiwan has led to identification of 3 accessions-V 2396, V3495 and V4281 as resistant to stem flies (Talekar,1990). Balboa (1972) reported Mungong Taglog in Philippine against *O. phaseoli* and V1360 and V2816 as least susceptible in Indonesia (Tengkano and Soekarna, 1976)

In India several cultivars have been reported as resistant to MYMV complex, jassids and pod borers. They include such as ML1, ML2, ML5, ML170, ML194,ML195,ML235 and 15 127 (Chhabra and Kooner,1980;Chhabra *et. al.*1980). These genotypes ML337, ML423 and M428 proved resistant to whitefly, jassid, MYMV and pod borer complex (Chhabra *et.al.*, 1998).

Urd bean

The cultivars -5-2, LU-15,LU-178,LU-190, LU196, LU130,LU-397, LU426,LU436,UL29, UL257, UL119, UL187, UL29302, UL407, UL29 and Pant U30 were identified as least susceptible against whitefly- MYMV and jassids in the summer season at Ludhiana (Punjab) by Chhabra and Kooner (1981) and Chhabra *et.al* (1984). The entries LU335, LU274, LU332, LU470 were reported as moderately susceptible to aphid, *Aphis craccivora*.

Biological control

All the major pests on pulse crops are attacked by one or more species of natural enemies. In most of the pulse crops natural enemies have been limited to the identification of natural control agents including parasitoids, predators and pathogens. An attempt to use these bio-agents in biological control is limited.

Helicoverpa

A wide range of parasitoids, predators and pathogens have been identified as important bio- agents of *H. armigera*. The most of the parasitoids belong to Hymenoptera, Diptera. The microbes include, Nematodes (3spp.), Bacteria (9spp.), Viruses (4spp.), Fungi, (5spp.), Protozoa (4spp.), which attack larval stages of the pest. The predators of egg and larvae include Coleopterans, Dermaptera, Hemiprtera, Orthoptera, Hymenoptera, Neuroptera, Arachnida, Amphibians and Birds.

Among the several parasitoids, the egg parasitoids, *Telonomous, Trichogramma* and Trichogrammatodea are important parasitoids. Trichogrammatids rarely attack *H.armigera* on chickpea and pigeonpea. Information on natural enemies and biological control of *H. armigera* has been reviewed by Bhatnagar *et.al.* 1982; King *et.al.*1982; Van der Bera *et.al.* 1988; King and Coleman 1989; King and Jackson, 1989; King1994; Romies and Shanower1996).

The relatives importance of parasitoids and predators varies between location and crops. Larval parasitoids *Campoletis chloridae* causes substantial parasitism of young larvae at early reproductive stage of the crop when pod damage due to *H. armigera* has just started in Northern India. Very few larval paradsitoids occur at critical stages of the crop to suppress the larval population of *H. armigera.*

Predators include Anthocorids and chrysopids feeding on eggs and Hemiptera and Formicidae on eggs and larvae. Augmentative release of *Chrysopa carnea* has been attempted without success. The type of response by both specific and polyphagous predators to change of their prey provides a framework for our considerations of predators effectiveness.

Nematode parasites belonging to Mermithidae have been recorded from various parts of the country. Commercial formulations DD136 and Green Commends have been evaluated in chickpea and pigeon pea.

The use of pathogens, Viruses, bacteria and fungi has been effective against *H. armigera*. Bacteria and viruses used along with adjuvant and ultra-violet retardant have proved effective. The effectiveness of these two pathogens is dependent on prevailing climatic condition and pest pressure of the area. It has been observed that under pest pressure (74%damage) they may not able to give desired control of *H. armigera*. Number of formulations containing *B. thuringiensis* var Kurstaki @ 0.5-1kg/ha alone or combinations have been evaluated on chickpea and pigeonpea etc.

Among the viruses, Nuclear ployhedrosis viruses are most widely used viruses in the control of *H. armigera.* It has been used at varying level, 250-500LE/ha in chickpea and 500LE/ ha in pigeon pea in integrated manner with other control measures. Relatively high cost, rapid inactivation by ultra-violet light often lead to poor performance under field conditions further lack of consistency high level of purity and virulence have reduced the use of NPV. However, where *H. armigera* is a problem, early spraying with *B.t* and NPV may contribute to great extent in overall IPM programme.

Genetic Engineering

Transgenic plants are normal plants with addition of some genes conferring resistance to pests. Most of the attempts have been made on development of crop plants against *H. armigera*. Considerable progress in other crops like cotton, corn, tomato and groundnut etc. has been made. In pulses limited work on chickpea and pigeon pea has been done against *H. armigera*. Chickpea genotypes, ICCV1 and ICCV6 have been developed with incorporation of Cry1A gene and plasmid containing *npt*II gene (Kar *et.,al*, 1997). Larval feeding assay shown that gene expression was inhibitory to the development and feeding by *H. armigera*.

Chemical control

The adoption of chemical control measures was not a matter of chance but it provides users with an apparently highly successful means of achieving their objectives. A wide range of chemical insecticides has been reported to control insect pests of pulses. Nowadays chemical control is just one means of limiting pest population which must be combined constructively with all the other ecologically acceptable methods within the concept of integrated pest management. Among the chemical tested and recommended, the endosulfan, fenvalerate, phosalone, quinalphos, spinosad, lufenuron, methoxyfenozide, novaluron, diflubenzuron, teflubenzuron and proclaim are the safe to natural enemies. Under botanical insecticides,Neem seed kernel extract and commercial formulations-Achook, Nimbicidine and Rakshak gold have proved effective under the All India Coordinated Project on Chickpea (AICPIC,1998).

In peas and urdbean and moongbean the soil application of various granular and dust insecticides against stem flies can be replaced with lesser quantity of insecticides through seed treatment with newer insecticides like imidachloprid and thiamethoxam.

Progress towards eco-friendly pest management in pulses

A number of features of pulse growing act to retard the adoption of IPM. Chief of these low priorities of insect pest in relation to other husbandry consideration decision making though economic benefits are high. Most of the available methods of pest control in comparison to chemical control are not highly effective in suppressing the pests population. Cultural practices that are compatible for all pests in a given situation may be impossible. *Bt* formulations, Dipel8L and Delfin WG, and HaNPV have reported to be effective in controlling *H. armigera* on chickpea (Singh *et al.* 1999). The alternate use of quinalphos and HaNPV varied across the seasons and the genotypes (Cowgill and Bhagwat, 1996). Similarly the combination of HaNPV and endosulfan was found superior to endosulfan or HaNPV alone or NSKE or 2% neem oil (Sivaprakasam, 1998). Initial spray with either NPV or NSKE, followed by endosulfan proved effective in chickpea crop against *H. armigera* (Sanap and Pawar, 1998). Integration of cultural practices, microbial pesticides, NSKE and need based use of chemical insecticides have shown spectacular results in controlling pests of pulses. In recent years main focus has been to develop IPM options against major pests in important pulse crops. In chickpea integration of early planting, intercropping, NSKE, NPV and need based use of endosulfan gave cost effective and ecofriendly management of *H. armigera* population at most of the locations under the All India Coordinated Project on Chickpea (AICPIC, 2002). Similar options have been developed in other crops. Even with present level of knowledge IPM can significantly reduce uses of pesticides. Further if crop diversification and crop rotation may be adopted due to non profitability of cereal crops. This will result low incidence of pest population with high number of natural enemy population, even partial resistance may play very significant role. Decision making based on more accurate (field based forecast) will help to maintain yield and profit under such conditions.

REFERENCES

Ahmad, H., and Kotwal, D.R. 1996. Screening of chickpea varieties against gram pod borer, *Helicoverpa armigera* Hub. *Ann. Pl. Protec. Sci.* 4: 171-172.

AICPIC, 1998. All India Coordinated Project on Improvement of Chickpea, Rabi Pulses Group Meet. Consolidated Reports on Chickpea (Entomology) Indian Institute of Pulses Research, Kanpur (U.P.)

AICPIC, 2002. All India Coordinated Project on Improvement of Chickpea, Rabi Pulses Group Meet. *Consolidated Reports on Chickpea (Entomology) Indian Institute of Pulses Research*, Kanpur (U.P.)

Balboa, F. C. 1972. Comparative resistance of five varieties of mung, (*Phaseous aureus*) Roxb., to the attack of the bean fly, *Agromyza phaseolis* Coq *Araneta research journal* 19: 66-79.

Beirne, B. P. 1967. *Pest Management*, Leonard Hill, London

Bhatnagar, V.S.,Lateef,S.S., Sithananatham, S., Pawar, C.S., and Reed, W.1982. Research in *Heliothis* at ICRISAT. pp. 385-386. *In* proceedings of the international workshop on *Heliothis* management, 15—20, Nov1981, ICRISAT Center, Patencheru, AP, India (Reed,W., and Kumble, V. eds) patencheru, AP, India: International crops research institute for the semi arid tropics (ICRISAT).

Chabbra, K S. and Kooner, B. S.1980. Sources of white fly, *Bemisia tabaci* G. and yellow mosaic virus resistance *Vigina radiata. Tropical Grain Legume Bulletin No.*19:26-29.

Chabbra, K. S., Kooner, B. S. and Brar, J. S. 1980. Resistance behaviour of moongbean, *Vigna radiata* (L) W. cultivars against insect pests complex and yellow mosaic virus. *Indian J. Eco.* 7(2): 276-280.

Chabbra, K.S., Kooner, B. S. Saxena, A. K.and Sharma, A.K.1984. Influence of biochemical components on the incidence of insect pests and yellow mosaic virus in blackgram. *Indian J. Entomol.* 46(2): 148-156.

Chhabra, K. S. and Kooner, B. S. 1988. Varietal resistance in mungbean (*Vigna radiata* (L.) wilczek.) to thrips (*Megalurothrips distalis* (Karny). *TVIS NEWS* (The Tropical Vegetables Information Service), 3(2): 10-11.

Chhabra, K. S., Kooner, B. S., Sharma, A. K. and Saxena, A. K. 1988. Sources of resistance in moongbean (*Vigna radiata)* to insect pests and moongbean yellow mosaic virus. Pages 308-314 *In: Mungbean Proceedings of the second international symposium*, Bangkok. (S. Shanmugasundaram, ed.)

Chhabra, K.S. and Sehgal. V. K. 1994. Integrated pest management in *Vigna* spp. crops. *In* Proceeding of International Symposium on Pulses Research, April 2-6, 1994, New Delhi.

Chaturvedi, S. K., Dikshit, H. K., Gupta, S. R., Asthana, A. N., and Pankaj. 1997. Recent advances in chickpea breeding for biotic stress in India. *Crop Improvement* 24: 143-150.

Cowgill, S. E., and Bhagwat, V.R. 1996. Comparison of the efficacy of chemical control and *Helicoverpa* NPV for the management of *Helicoverpa armigera* (Hubner) on resistant and susceptible chickpea. *Crop Protection* 15: 241-246.

Cromartie, W.J. 1986. The environmental control of insects using crop diversity. In Handbook of Pest Management in Agriculture (Ed. D. Pimentel), 223-251. CRC Press, Florida.

Dasgupta, B. and Chaudhary, A. K. 1985. Use of inter- varietal intercropping to minimise yellow mosaic virus of urd and mungbeans. *Indian J. plant Proct.* 3 (1): 100-101.

Das, S.B. and Kataria,V.P. 1999 Relative susceptibility of chickpea genotypes against *Helicoverpa armigera* (Hub.). *Insect Environment* 5(2): 68-69.

Das, S.B.1998a. Impact of intercropping on *Helicoverpa armigera* (Hub.): Incidence and crop yield of chickpea in West Nimar Valley of Madhya Pradesh. *Insect Environment* 4: 84-85.

Das, S.B.1998b. Synergist action of neem seed kernel extract and insecticides against pests of piegeonpea. *Insect Environment* 4: 83-84.

Kalaryia, G.B., Judal, G.S. and Patel,G.M. 1998. Reaction of pigeonpea genotypes against important insect pests. *Gujrat Agric. Univ. Research J.* 23(2): 33-38.

Kar, S., Basu, D.S., Ramkrishnan, N.A., Mukherjee, P., Nayak, P. and Sen, S.K. 1997. Expression of cryIA gene of *Bacillus thuringiensis* in transgenic chickpea plants inhibits development of pod borer (*Heliothis armigera*) larvae. *Transgenic Research* 6: 177-185.

King, A. B. S.1994. *Heliothis/ Helicoverpa* (Lep.: Noctuidae) pages 39-106 in insect pests of cotton (Mathews, G.A. and Tunstall, J.P. eds.) Wallingford UK: CAB International.

King, E. G. and Coleman, R. J. 1984. Potential for biological control of, *Helicoverpa* spps. *Ann Rev.Entomol.*34: 53-75

King, E. G., Powell, J. E. and Smith, J.W.1982. Prospects for utilization of parasites and predators for the management of *Heliothis* spp Pages103-122 in proceedings for the international workshop on *Heliothis* management 15-20 Nov1981, ICRISAT Center, Patancheru, AP, India (Reed, W., and Kumble, V. eds) Patancheru, AP, India:

International Crops Research Institute for the Semi Arid Tropics (ICRISAT).

King, E. G.,and Jackson, R. D. (eds.) 1989. *In* Proceedings of workshop on biological control of *Heliothis*: Increasing the effectiveness of natural enemies, Nov1985, New Delhi, India (King, E. G., and Jackson, R. D. eds.), New Delhi, India: Far Eastern Regional Research Office, US Department of Agriculture.

Kotikal, Y. K., Shantappanvar, N. B.,, Lingappa, S., and Yelshetty,.1996. Screening chickpea for the resistance to pod borer in Karnataka, India. *Int. Chickpea and pigeonpea Newsl.*. 3: 41-43.

Lal, S. S. and Katti, G. 1997. Podfly, *Melanagromyza obtusa* Malloch- A Key Pest of pigeonpea. Indian Institute of Pulses Research, Kanpur, 26 pp.

Lateef, S. S. 1985. Gram pod borer (*Heliothis armigera* (Hub.) resistance in chickpea. *Agric. Ecosystem and Environment* 14: 95-102.

Lateef, S. S. and W. Reed. 1990. Insect pests of pigeonpea. *In* Insect pests of tropical legumes (Singh, S. R., ed.). Chichester, UK: John Wiley.

Reddy, P.V.R., Singh, Y., Singh, K.M. and Singh, S.P.1996. Chickpea varietal reponse to pod borer, *Helicoverpa armigera*. *Indian J Entomol.* 58: 60-65.

Reed, W. and lateef, S. S.1990. Pigeonpea: Pest management. pp. 349-374 *In: The pigeonpea* (Nene, Y.L., Hal,S.D. and Shiela, V.K., eds.), Wallingford, UK: CAB International.

Romeis, J. and Shanower, T. G.1996. Arthropod natural enemies of Helicoverpa *armigera* (Lep: Noctuidae) in India. *Biocontrol Science and Techn.* 6: 481-508.

Root, R. B. 1973. Organisation of a plant-arthropod association in simple and diverse habitats: the fauna of collards (*Brassica oleracea*). *Ecol. Monogr.*, 43: 94-125.

Sanap,M.M., and Pawar, V.M. 1998. Integrated management of *Helicoverpa armigera* on gram (*Cicer arietinum*). *Indian J. Agric. Scien.* 68: 162-164.

Saxena, H. P., Singh,Y., and Phokela A. 1973. Effect of date of sowing on pest incidence. p. 115 in Proceeding of IX Workshop (Kharif pulses), New Delhi.

Sharma, H.C., Sharma, B.U., Hariprasad, K.V., and Bramel-Cox, P. J. 1999. Host-plant resistance to insects in integrated pest management for a safer environment. *Proceedings, Academy of Environmental Biology* 8: 113-136.

Shukla, A. and Yadav, H.S. 1998. Resistance in crops against *Helicoverpa armigera* (Hubner) - a review. *Advances in Plant Sciences* 11: 265-270.

Singh, K. M. and R. N. Singh. 1977. Succession of insect- pests in moongbean and blackgram under dryland condition at Delhi. *Indian J. Entomol.* 39: 365-370.

Singh, P. P., Monobrullah, M., and Sigh, B. 1999. Field efficacy in some microbial pesticides against gram pod borer (*Helicoverpa armigera* Hubner) in chickpea. *Shashpa* 6: 63-66.

Sivaprakasam, N. 1998. Control of tomato fruit borer (*Helicoverpa armigera*) by nuclear polyhedrosis virus in comparison with other treatments. *Indian J. Agric. Sci.* 68: 801-802.

Speight, M. R. 1983. The potential of ecosystem management for pest control. *Agric. Ecosystem and Environment* 10: 183-199.

Talekar, N. S. 1990. *Agromyzid flies of food legumes in tropics*. Wiley Eastern Ltd., New Delhi, p. 297.

Tangkano, W. and Soekarne, D.1976. Screening of moongbean (*Phaseous aureus*) Roxb., varieties for tha resistance to the bean fly, *Agromyza phaseolis* Coq Pages 39-47. *In:* Annual report central research Institute for Agriculture, Ministry of Agri.Bogor, Indonesia.

Van den Berg H., Waage, J. K., and Cock, M. J. W. 1998. Natural enemies of *Helicoverpa armigera* in Africa – a review. Ascot. Berks. UK: CAB International Institute of Biological Control. p. 81

Venkateswarlu, O. and Singh, T. V. K. 1999. Response of pigeonpea genotypes against important insect pests. *Insect Environment* 5: 123-124.

Wiley, R. W. 1979. Intercropping – its importance and research needs. *Field Crop Abst.* 32 (1), 1-10.

Yadava, C. P., Lal, S. S., Dias, C. A. R. and Nigam Rajani. 1989. Host evasion a prospective approach for suppressing *Heliothis armigera*. *Int. Pigeonpea Newsl.* 2: 62-64.

Eco-friendly Insect Pest Management in Sugarcane

H.R. SARDANA

Sugarcane is an important cash crop in many tropical and sub-tropical countries and is one of the main sources of sugar production in the world. In India also sugarcane plays a pivotal role in country's agricultural and industrial economy. During the period of last five decades sugarcane production and productivity has shown remarkable resilience and in the year 1999-2000 there was a record cane production of 300 mt in an area of 4.22 mha. Today there is growing concern about sustainability of sugarcane production as the growth rate has not only stagnated and rather declining trend is seen in States like Maharashtra, Gujarat, Bihar and Orissa. These insect pests are the most important biotic constraints in sustaining sugarcane productivity particularly in sub-tropical agricultural ecosystem where because of a more conducive environment for pest build-up, losses incurred are substantially very high. About 212 insect pests and 76 non-insect pests are reported to attack this crop (David and Nandagopal, 1986). Out of these, about dozen pests are recoganised as major pests and economically important. These pests are moth borers like early shoot borer, top borer, stalk borer, root borer, sucking pests like black bug, pyrilla and scale insect and others like white grub and termites. On a conservative estimate, the losses due to insect pests is about 20 per cent in yield and about 15 per cent in sugar recovery. Long duration of sugarcane plant crop sometimes extending from 10-18 months followed by one or two ratoons which provide a sort of monoculture and continuous stable agro-ecosystem for multiplication of insect pests staggered planting of the sugarcane to meet the long duration crushing needs of sugarcane factories and extensive cultivation of only a few varieties which may sometimes be susceptible for some pests over large areas make it ideal for build up of pests. Moreover, in this crop the application of insecticides is very limited, i.e. less than 3.0 per cent of insecticides used in India, partly because of stability of pest population maintained through natural control by number of natural enemies present in the sugarcane fields and also difficulties in application of insecticides in a grown up crop.

CHANGING SCENARIO OF PESTS

Though sugarcane eco-system is stable one but periodically it is affected by biotic and abiotic changes in the environment. Besides the above mentioned factors, indiscriminate transport of seed material, growth of sugar mills, improved farm practices and agro technologies and occurrence of droughts and floods have played important role in changing insect pest situation in certain regions. Many minor pests have become major in some areas like root borer which was a minor pest has become a serious problem in Punjab, Haryana and Bihar because of introduction of high sugared early maturing varieties like Co 89003. Scale insect which was considered to be a minor pest in North has become a major problem in certain areas of Haryana like Palwal, Rohtak and Sonepat. Stalk borer was originally confined to water logged areas of Eastern UP has now spread to Punjab and Haryana. Gurdaspur borer was confined to Punjab has spread

1. National Centre for Integrated Pest Management, L.B.S. Building, I.A.R.I., New Delhi - 110012

to Haryana. Similarly Jassids, *Arboridia* sp and giant white grub, *Leucopholis lepidophora* are causing great concern to the sugarcane growers of the Kolhapur (Maharashtra) area. Recent outbreak of sugarcane wooly aphid in Maharashtra and to certain extent in UP is glaring example of changing pest situation in sugarcane. Gap between potential and actual yield can be partly narrowed down through effective management of the pests and this will ultimately increase sugarcane productivity in the country.

SUGARCANE ECO-SYSTEM

Sugarcane ecosystem has several unique features compared to other cash crops in India. The physical characteristics of the plant limit the use of chemical insecticides once the canopy has closed. The insecticide application in sugarcane is mostly limited in the early stages when the crop is young. Despite 20 per cent loss in sugarcane due to insect pests (Avasthy, 1983) the uncommon features of sugarcane and its pests and their interaction render plant protection impractical or uneconomical (Mukunthan, 1991). This of course makes sugarcane ecosystem ideal for functioning of natural enemies. Stability of pest population is maintained through natural control by number of natural enemies present in the field. However, in spite of several natural enemies operating at various stages of key pests of sugarcane (David *et al., 1986)* they are unable to contain the pest.

PEST COMPLEX, THEIR DISTRIBUTION AND LOSSES

As mentioned earlier there are about a dozen economically important pests and there is much variation in geographical distribution and period of activity. Pest problems are much more severe in sub-tropical than tropical India. Though sugarcane agro-ecosystem is stable but periodically it is affected by different biotic and abiotic changes in the environment. These changes individually or jointly influence the insect pest complex over a period of time. Pest damage occurs at all the stages of crop growth. The heavy losses caused by pests include total elimination of plant setts by termites, young shoots by shoot borers and older canes by top and stalk borers and also by white grubs, termites and rats. In grown up canes, the pests damage results in heavy losses in yield, sucrose content and sugar recovery and also the quantity of juice extracted. Major pests, their distribution and losses are given in Tables 1 and 2.

Table 1. Status of Sugarcane Insect Pests in Different Sugarcane Growing States of India

State	Major Pest	Pest assuming major Status	Minor Pest
Assam	Termite, shoot borer, Plassey borer, Stalk borer	White flies, wooly aphid	Thrips, mealy bugs
Bihar	Termite, shoot borer. Top borer, Stalk borer	Pyrilla, White flies, Beetle grubs	Army worm, mealy bugs, thrips, grass hoppers
Gujarat	Pyrilla, scale insect, shoot borer	Top borer, white flies, root borer	Mealy bugs, termites, top borer
Haryana	Termite, root borer, shoot borer, Top borer, Stalk borer, Gurdaspur borer	Grubs, pyrilla, root borer	Scale insect, mealy bugs, thrips, white flies, mites
Karnataka	Termite, shoot borer	White flies., mealy bugs	Grubs
Madhya Pradesh	Termite, shoot borer	Pyrilla, Top borer, Internode borer	Mealy bugs, mites, Army worms
Maharashtra	Shoot borer, pyrilla, scale insect	Internode borer, Jassids, *Arboridia* sp, white flies, internode borer	Mites, Mealy bugs, Army worms

State	Major Pest	Pest assuming major Status	Minor Pest
Nagaland	Shoot borer, Stalk borer	Wooly aphid, pyrilla, mealy bugs, weevil	Thrips, white flies, top borer
Orissa	Termites, shoot borer	Stalk borer, Top borer, pyrilla	Mealy bugs, white flies, Top borer, Army Worms
Punjab	Black bug, root borer, shoot borer, Top borer, Stalk borer, Gurdaspur borer	Pyrilla, White flies	Termites, mealy bugs, grubs
Rajasthan	Termite, root borer, shoot borer, Top borer	Pyrilla	mealy bugs, grubs, white flies, Gurdaspur borer
Tamil Nadu	Termite, shoot borer, Internode borer	Pyrilla, root borer	Mealy bugs, white flies, mites, defoliators
Uttar Pradesh	Black bug, shoot borer, Top borer, Stalk borer, Gurdaspur borer	Pyrilla, root borer, white flies, thrips, scale insect	Mealy bugs, weevil, Slug caterpillar, Army worm.
West Bengal	Termite, root borer, shoot borer	Pyrilla, white flies, thrips	Army worms, grass hoppers, Plassey borer

Table 2. Sugarcane pests and the losses caused by them

Pest	LOSS (%)	
	Cane yield	Recovery (Units)
Borers		
Early shoot borer	22.0 – 33.0	Up to 2.0
Top borer	20.0 – 35.0	0.2 – 4.0
Stalk borer	Up to 33.0	1.7 – 3.7
Root borer	Up to 10.0	0.2 – 1.9
Gurdaspur borer	5.0 – 15.0	0.1 – 0.8
Sucking pests		
Pyrilla	Up to 28.0	0.2 – 5.0
Black bug	Up to 35.0	0.1 – 2.8
White fly	Up to 23.4	0.1 – 2.9
Scale insect	5.0 – 55.0	1.7 – 9.1
Subterranean pests		
Termites	2.5 – 33.0	Up to 4.5
White grubs	80 – 100	-
Defoliators		
Grass hoppers	-	-
Weevils	Minor	-
Non-insect pests		
Mites	-	-
Rats	Up to 16.7	-
Jackals	-	-

FACTORS AFFECTING PEST BUILD UP

The important factors that influence the build up of pests and natural enemies in sugarcane are summarized below:

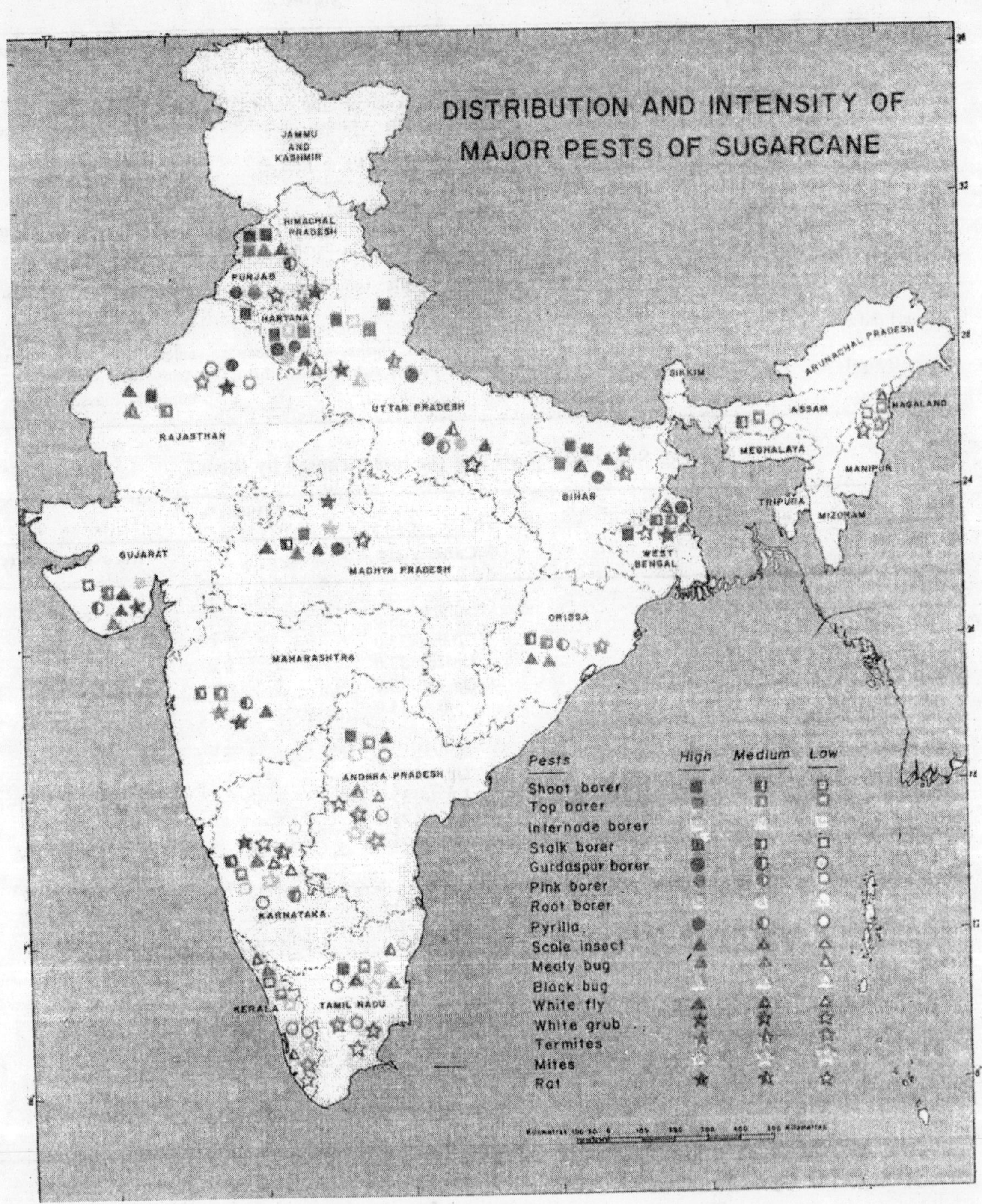
DISTRIBUTION AND INTENSITY OF
MAJOR PESTS OF SUGARCANE
JAMMU AND KASHMIR
HIMACHAL PRADESH
PUNJAB
HARYANA
RAJASTHAN
UTTAR PRADESH
SIKKIM
ARUNACHAL PRADESH
ASSAM
NAGALAND
MEGHALAYA
MANIPUR
BIHAR
TRIPURA
MIZORAM
WEST BENGAL
GUJARAT
MADHYA PRADESH
ORISSA
MAHARASHTRA
ANDHRA PRADESH
KARNATAKA
KERALA
TAMIL NADU
Pests
High
Medium
Low
Shoot borer
Top borer
Internode borer
Stalk borer
Gurdaspur borer
Pink borer
Root borer
Pyrilla
Scale insect
Mealy bug
Black bug
White fly
White grub
Termites
Mites
Rat

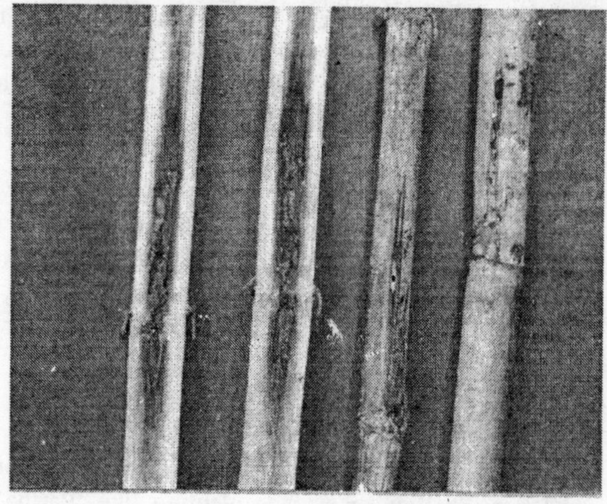

Climatic Factors

Thus rainfall and cloudy weather is not suitable for the build up of shoot borer while rainfall and subsequent waterlogging increase the population of stalk borer, top borer, root borer, internode borer and white flies. However, very high rainfall reduces the population of Gurdaspur borer, root borer, and sucking pests. Rainfall reduces mealy bug population by washing off the nymphs and also by promoting growth of the entomophagous fungus, *Aspergillus parasiticus*, which considerably reduced the bug population. Wind aids in the dispersal of neonate larvae of borers, crawlers of scale insect and mealy bugs etc. Hot and dry climate favours the build up of shoot borer, green borer, black bug and red mite i.e. the period from March to June is most favourable for population build up. Moderate temperature and high humidity are suitable for heavy build up of top borer, internode borer, stalk borer, pyrilla and yellow mite. That is the period from July to Sept. are most suitable for pest multiplication. Mealy bug, white fly and scale insect population increases during summer in hot and dry conditions as well as and during monsoon and post monsoon months under water logged conditions.

Micro-climate

Highly manured crop with luxurious growth are more attracted by pyrilla, stalk borer and internode borer. Conversely nitrogen deficiency favour multiplication of white fly population. Large scale introduction and cultivation of susceptible varieties like CoJ 64 and Co 8021 for top borer, Co 8021 in Tamil Nadu, Co 89003 and Co 7314 in Haryana for root borer, Co 527, Co 975, CoA 7602 and CoR 8001 for scale insect, CoC 671 for white fly, Co 7704 for mealy bugs and Co 1148 for stalk borer increases the pest incidence often leading to pest epidemics. Inter cropping with coriander reduced the incidence of top borer and root borer while inter-cropping with green gram reduces incidence of early shoot borer (Sardana, 2000). Similarly inter-cropping with wheat enhances pink borer problem in sugarcane, as this borer first builds up in wheat and then migrates to adjoining sugarcane shoots (Easwaramoorthy, 1986).

Edaphic factors

Sandy to sandy loam soils show more root borer, termites and white grubs activity while heavy soils favour Plassey borer, *C. tumidicostalis* build up. Infestation of shoot borer is greater under saline conditions. Alkaline soils are favourable for the build up of white flies. Soil pH may not affect root borer activity.

Cropping Pattern

Planting of cane at different times of the year is becoming popular with the sugar factories. Late planting is favourable for heavy multiplication of shoot borer while autumn planted crops are more attacked by stalk borer and special season planting by internode borer. Late planted crop is subjected to heavy infestation of top borer in Madhya Pradesh. Mid-January planting in Maharashtra is less infested with shoot borer than the one's planted in mid-February and Mid-March. Internode borer in Tamil Nadu affects more the autumn planted crop.

Natural enemies

A number of natural enemies play a key role in the regulation of key pests. However, their populations densities may also be influenced by climatic factors. Low humidity reduces the rate of parasitisation by the *Isotima javensis*. Humidity also affects the egg parasitoides of pyrilla which emerge more efficiently when the relative humidity is above 65 per cent. Hot weather is detrimental to the survival of *Trichogramma* sp.

Other factors

Trash mulching reduces incidence of shoot borer but increases the incidence of termites, armyworm and rats. Shallow or surface planting like 'Partha' method increases shoot borer infestation, and close spacing and high population of cane/unit area results in high population of shoot, internode and stalk borers. Repeated ratooning leads to severe damage by shoot, stalk and top borers, black bug, white fly, scale insect, mealy bugs and white grubs. In case of pyrilla summer fodder sorghum acts as real breeding ground during off-season. Neem acts as a food source for white grubs hence these trees bordering sugarcane fields promote heavy white grub damage.

NATURAL ENEMY COMPLEX

A large number of egg and larval-pupal parasitoides and predators have been reported to play an important role in natural regulation of sugarcane pests (Table 3).

Table 3. Extent of mortality caused by natural enemies

Pests	Natural enemies	Extent of kill (%)
Shoot borer	*Trichogramma* sp.	4 - 5
	Telenomus spp.	15 - 22
	Apanteles flavipes	4 - 14
	Sturmiopsis inferens	6.5 - 22.2
	Stenobracon deesae	5 - 7
	Egg predators	13.4
	Granulosis virus	2.6 - 14.6
Top borer	*Trichogramma* sp.	4 - 5
	Telenomus spp.	5 - 85.5
	Apanteles flavipes	4 - 14
	Elasmus zehntneri	5 - 15
	Rhaconotus scirpophagae	7 - 12.2
	Isotima javensis	5 - 40
Stalk borer	*Apanteles flavipes*	10 - 14
	Campyloneurus mutator	60 - 78
	Sturmiopsis inferens	20 - 50
Internode borer	*Trichogramma* sp.	2 - 78.5
	Telenomus spp.	15 - 90
	Apanteles flavipes	4 - 14
	Rhaconotussp nr.signipennis	25
	Egg predators	9.4 - 22.4
	Granulosis virus	3.8 - 31.5
Pyrilla	Egg parasites	2.7 - 97.6
	Epiricania melanoleuca	15 - 100

Some of these occur through out the year, while others occur during certain seasons and decimate pest populations. Over 185 spp of parasites, 51 predators and 31 entomopathogens have been reported on some of the major pests of sugarcane crop. The tachinid parasite, *Sturmiopsis inferens* Tns. is active throughout the year in nature in Coimbatore area. Together with granulosis virus (Easwaramoorthy, 1986), there is natural suppression of shoot borer. *Isotima javensis* Roh. is active throughout the year on top borer in Tamil Nadu and Karnataka. Natural suppression of white grubs is caused by the milk disease bacterium, *Bacillus popilliae* Dutky in Tamil Nadu and Gujarat and by *B. poplliae* and an entomopathogenic nematode, *Heterorhabditis indicus* in Karnataka (Karunakar, 1990). In the absence of these natural enemies the losses due to the pests would have been several fold. So, disruption of naturally occurring biological control agents by poor choice and/or application technique or insecticides and or employing harmful cultural practices should be avoided. Strategy for IPM in sugarcane is to conserve and maintain the naturally occurring biological control so that the insecticidal applications are kept at the lowest level. When these are properly done, the need for other control measures may be greatly reduced or even eliminated for some pests. List of natural enemies and their extent of kill/parasitisation recorded on major pests in sugarcane is enclosed (Table 3).

ECONOMIC THRESHOLD LEVEL

Precise estimation of losses due to individual pests in specific regions is essential to determine the economic threshold level and to take decisions for their control operations. ETL is the minimum level of pest in the crop, which can justify adoption of control operations so that the pest is prevented from reaching economic injury levels (EIL). ETL and EIL are highly location specific for each region, variety and plant/ratoon crop so these population estimates should be determined for each region locally for each variety. However, an average value càn be obtained which may help in decision making process for taking pest management operation. The hardy nature of sugarcane crop withstands sizeable population densities of different pests. Therefore, the values of economic threshold and economic injury levels of major pests are too high. This attribute of the crop provides enough room for the sustained action of natural enemies.

The ETL determined for major insect pests are given below:

Early shoot borer: = 16.8% for early variety – Co 89003 (Sardana, 1998)

= 15.0 – 22.8 % for late variety – Co 1148 (Sardana, 1997)

Internode borer = 28.5, 12.2 and 6.2 larvae

Or

28.39, 24.4 and 17.15 bored internodes per 6 meter row for varieties CoJ-64, Co 8606 and Co6304, respectively (Nandagopal, 1983).

Stalk borer = 12.20, 17.83 and 15.16 larvae

Or

= 17.08, 19.61 and 16.67 bored internodes of 6 m row for varieties CoJ 64, Co 7717 and Co 1148, respectively (Sardana, 1996).

Pyrilla = 3-5 individuals per leaf (Pawar, 1983).

DISTRIBUTION PATTERN AND SEQUENTIAL SAMPLING PLAN

The efficiency of control measures can be enhanced after determining the nature of spatial distribution of various borer pests and distribution studies are also helpful in establishing relationship between density and various factors affecting mortality and vitality. Early shoot borer and stalk borer in sub-tropical region in three cultivars viz; CoJ 64 (early), Co 7717 (mid-late) and Co 1148 (late) follow aggregated distribution pattern which results from environmental heterogeneity. A plant sample size of 126 of 6 tillers each has been found to sufficient for studying aggregated distribution (Sardana, 1997).

For the effective use of ETL, sequential sampling plan techniques which are ready recknor for sampling of population of early shoot borer and stalk borer have been developed which are as follows:

	'Co 7717'	'CoJ 64'	'Co 1148'
	$d_0 = 8.43n - 10.11$	$5.69n - 12.17$	$7.06n - 15.62$
	$d_1 = 8.43n + 10.11$	$5.69n + 12.17$	$7.06n + 15.62$

These equations can be used in the field for taking samples.

PEST MONITORING AND QUANTITATIVE ESTIMATION

Periodical monitoring and estimating the size of insect pest population are fundamental, very –2 essential and basic pre-requisites for deciding the need based control measures and to develop IPM strategies. This is the key for successful implementation of IPM. Such survey and surveillance should be regular so as to know the changing pest situations due to agronomic practices and climatic factors. Pest build up may be carefully and systematically monitored regionwise at periodic intervals, taking into consideration the principal seasonal influence and timely decisions made based on economic threshold levels of the pests and cost-benefit analysis. There are several ways for estimating pest population. Walking diagonally into the field examining about 50 shoots in 5 randomly selected locations (250) is a simple way of estimating the pest infestation. For borers, the infestation is expressed as percentage of tillers infested and canes bored and intensity is expressed as percanetage of internodes bored.

$$\text{Borer infection index is expressed as} = \frac{\text{Per cent incidence} \times \text{Per cent intensity}}{100}$$

For scale insects and mealy bugs the infestation is expressed as percentage of joints infested when the incidence is light. For *Pyrilla* 20 clumps are selected at random. The total number of adults, nymphs and egg masses, both healthy and parasitised, on the fifth leaf from spindle of 20 selected canes at one per clump is counted and expressed as mean per cane. For white fly, the total number of nymphs/puparia in the third leaf is counted and expressed as mean population per leaf or when the population is heavy it is counted in 1 cm^2 in the top, middle and bottom regions of the leaf and expressed as mean per cm^2. Sometimes visual grading system (very light, light, moderate, severe, and very severe) may be used for scales and mealy bugs based on the extent of infestation as under:

For scales:

i)	Few scales here and there without well established colony	Very light
ii)	Scale encrustation covering approximately ¼ internode	Light
iii)	Scale encrustation covering approximately 1/2 internode	Moderate
iv)	Scale encrustation covering approximately 3/4 internode	Severe
v)	Scale encrustation covering more than 3/4 internode	Very severe

For mealy bugs the rating is as under:

i)	Stray occurrence	Very light attack
ii)	Below 5 mealy bugs/node or internode	Light
iii)	3-10 mealy bugs/node or internode	Moderate
iv)	10 to 15 mealy bugs/node or internode	Severe
v)	Above 15 mealy bugs/node or internode	Very severe

POPULATION DYNAMICS AND PEST FORECASTING

Estimating the size of insect population is fundamental to a complete understanding of the pest insects. A thorough understanding of effect of various climatic factors on pest build up can also help in forecasting the outbreak of insect pests and appropriate time of action in the devising the effective integrated schedule. Information on forecasting on long term basis for any of the sugarcane pests is not available in India. However, short-term predictions are possible for pests like pyrilla, *P. perpusilla* and top borer, *S. excerptalis*: Two models have been developed for predicting pyrilla outbreaks. The model based on quantitative relationship between temperature and population flucutations is equated as ($Y = (3.4753 \times 10^{-28}) (33.4965) \times (0.9562^{x2})$). This model predicts a decline in pyrilla population when temperature reaches 39°C (Bains *et al.*, 1982). The second model is developed for the population of pyrilla in relation to population of its nymphal-adult parasitoid, *Epricania melanoleuca* (Brar *et al.,* 1978). Both these models hold good for forecasting pyrilla outbreaks in situations of favourable weather conditions. In sub-tropical India, a fall in temperature coupled with moderately high environment humidity during June can predict pyrilla and top borer populations to a fair accuracy. Since both temperature and humidity are critical to to top borer and pyrilla abundance during monsoon period (Kalra, 1964) these occupy prime place in prediction models. Similarly Sardana and Das (2001) gave a short term prediction model for stalk borer wherein temperature and rainfall prevailing during July-August were the main factors which influenced the pest build up however, once the population is sufficiently build up, these factors don't play any role.

INTEGRATED PEST MANAGEMENT

The requisites for the effective implementation of IPM in which biological control has a great role to play, are:

i) Knowledge of crop/enviroment/pest complex interaction,

ii) Monitoring of pest population

iii) Exploitations of ecological knowledge to control pest complexes.

Integrated pest management aims at sequential use of all methods in as compatible manner as possible to suppress the pest population levels below economically damaging levels at every stage of their life-cycle. IPM may be defined as Intelligent Pest Management that will ensure ecologically sound, economically viable and socially acceptable. However, in order to practice IPM, it is necessary to have an idea about the key pests in an area, their biology and sound knowledge of pest management techniques available and also the means to monitor the pest activity by ecological or management practices.

COMPONENTS OF IPM

Some of the components like host plant resistance and biological control simply optimize the naturally occurring phenomenon of insect control while other components like cultural, mechanical and chemical control are artificial. In integrated pest management all or most of the components are to be blended in a harmonious manner with minimum disruption to the eco-system.

Host Plant Resistance The development and use of resistant varieties which possess other desireable economic traits is an economically cheap and environmentally sound technique of pest management. In sugarcane crop, however varieties having good commercial potential combined with desirable level of resistance have not been identified so far. The major reasons for this are:

a) non-availability of pest resistant sources

b) pecularities of the sugarcane genetics and morphology

c) unique feature of sugarcane plant, its pests and their interaction

d) unique difficulties involved in host plant resistance programme in sugarcane.

Inspite of all these constraints, in case of top borer some progress has been made in developing varieties with acceptable level of resistance. The sources of resistance to top borer include Nargori, Pansahi and Saretha clones of *Saccharum barberi*. Varietal manipulation in sugarcane chiefly envisages the avoidance of lage scale cultivation of susceptible varieies in pest prone areas rather than using resistant ones. Accordingly, the following varieties idicated against the different pests have been found to be highly susceptible to them and hence may be avoided in areas wherever the incidence is heavy.

i) Top borer : CoJ 64, CoJ 85, Co 8021

ii) Stalk borer : Co 1148, Co 7717, Co 89003

iii) Root borer : Co 89003, 7314, CoC 671

iv) Scale insect : Co 740, Co 8021, Co 7290, Co 62175

Another aspect of this method of pest suppression is utilising, wherever possible, less susceptible varieties in areas of high pest severity. Some of which are cited below:

i) Top borer : CoJ 85, CoS 767, Co 89003

ii) Stalk borer : CoS 767, C0 7303

iii) Root borer : CoS 767, Co 1148

iv) Scale insect : Co 8014, CoC 671

Hybrids in general show better resistance to borers compared to noble canes, because of the inheritance from *S. spontaneum* clones. Varieties showing relative resistance to some of the major pests include Co 281, 285, 312, 421, 449, 453, 527, 617, 775, 853, 975, 1007, 6510, 6601 (shoot borer); Co 205, Co 285, 331, 421, 453, 7224, CoJ 67, C0 S767, E50, E128, E132 and E 276 (top borer); Co 395, 951, 1007, 7302, CoJ 46 and CP 43/47 (Stalk borer). Sugarcane lines segregate and probably never breed true because of the large irregular number of chromosomes in this heterozygous polyploid plant. Introduction of a parent with a high degree of stem hardness usually increases average hardness of variety. 245 accessions against top borer, 247 accessions against stalk borer and 203 against stalk borer comprising *Saccharum officinarum*, *S. barberi*, *S. spontaneum*, *S. robustum*, ISH clones of Indian and Foreign origin were evaluated for identifying newer sources of resistance under natural conditions for exploitataion in hybridization programme (Table 4).

Incorporation of foreign/native genes such as Bt toxins, protease or amylase and other insect toxins (Meeusen and Warren, 1989) in desired varieties to produce required titer of toxins at tissue specific location of plants to tackle target pests are the future line of research in sugarcane host plant research. Incorporation of resistance to multiple pests is desireable and it is more probable by this technology.

B) Cultural Control: Cultural control is one of the oldest methods of crop protection and these methods are compatible with most other control measures. Many of the cultural practices used earlier are not in use now but some of these methods still hold good in IPM system in sugarcane. Planting of insect free setts is one of the most important cultural practices. Various other cultural methods are useful as adjuncts to other management tactics.

a) Time of Planting

Early planting during February reduces shoot borer incidence drastically throughout the country. Autumn planting should be avoided where stalk borer and internode borer is intensity is high.

b) Irrigations

Frequent irrigations during March to June reduce the shoot borer and root borer incidence considerably.

Table 4. Natural Performance of Sugarcane World Accessions showing Resistance Behaviour against Shoot, Top and Stalk borer

Shoot Borer (*Chilo infuscatellus*)	Top Borer (*Scirpophaga ecxcerptalis*)	Stalk Borer (*Chilo auricilius*)
Bo 38, Bo 76, Bo 95, BO 111 Co 611, Co 846, Co 1157, Co 1187, 6520, Co 6607, Co 62399, CoS 703, CoS 767, CoS 7922, CoS 8009, CoS 8315, CoS 8408, CoS 8432, CoLk 7701, Nco 320, B 37172, CP 44/153, 58, H 54/775, CP 29/116, CP 44/155, Levoseold, Senebile, Bhurli, Fizi-40, Hemza, Seratha, Khakai, Pararia – 257	Buxeria, Pathari, Pararia-257, Putlikhajee, CoS 770, CoS 8315, B 49119, Bo 36 Nco 376, M64-68, H 53/363.	Co 213, Co 443, Co 842, Co 846, Co 854, Co 857, Co 859 Co 1007, Co 1158, 1223, Co 1349, Co 6415, Co 6611, Co 6613, Co 6614, Co 6616, Co 6802, Co 7708, Co 7915, C0 8008, Co 87262, Co 87267, Co 88039, CoS 370, CoS 564, CoS 570, CoS 630, CoS 733, CoS 786, CoS 788, CoS 796, CoS 832, CoS 853, CoS 856, CoS 7913, CoS 7922, CoS 8016, CoS 8103, CoS 8118, CoS 8119, CoS 8122 CoS 8123, CoS 8301, CoS 8315, CoS 8408, CoS 8701, CoS 85233, CoS 88236, Bo 38, Bo 43, Bo 45, Bo 59, Bo 61, Bo 70, Bo 74, Bo76, Bo 80, Bo 8, NG-77-38, Pararia Shaj, Kansar, Saretha, ISH- 62, ISH-111, ISH-118, ISH-129.

c) Earthing up

A light earthing up usually given at 45 days after planting serves to suppress shoot borer population.

d) Trash mulching

Spreading of trash after germination in between rows helps in suppression of shoot borer. However, this practice may increase termites, army worm, cut worms and rats damage in areas where these pests are prevalent.

e) Manuring

Blanket spray of 2 per cent urea spray during May reduces incidence of early shoot borer as it enhances the growth of crop which thus escapes attack. Spraying of endosulfan with 2 per cent urea against black bug provides greenery to damaged plants and also reduces the attack by this pest. Urea application during June end may reduce the top borer incidence. However, excessive dose of nitrogenous fertilizers increases the incidence of stalk borer internode borer and pyrilla. Nitrogen deficiency increases white fly attack.

Trash mulching

f) Water management

Heavy irrigations and water logging leads to lodging of canes which enhance stalk borer attack.

g) Detrashing

Removal of lower leaves during the 5th and 9th month stage of the crop is highly effective in reducing the population of stalk/internode borer, scale insects, mealy bugs, white fly and pyrilla.

C) Mechanical Control: Mechanical control is also compatible with other pest management practices. Despite scarcity of labour and increased cost, mechanical control is highly rewarding in the management of many pests of sugarcane.

i) Collection and destruction of infested tops of Gurdaspur borer and Plassey borer during July to October on campaign basis has been found to be very successful. This practice done on a compaign basis has eliminated the Gurdaspur borer from the Yamunanagar area of Haryana (Agarwal, 1980).

ii) Collection of egg masses of 3rd brood of top borer in mid-July in sub-tropical region helps in suppression of incidence of top borer.

iii) Spectacular suppression of white grub population may be obtained by mass collection of grub beetles from their host trees at night. The collected beetles may be killed in water mixed with kerosene.

iv) Installation of light traps in the field helps in trapping of large number of moths of root borer, top borer and Plassey borer. Both the sexes of these pests are attracted to light.

Adoption of mechanical control measures has to depend much on the timely availability of labour and cost benefit analysis.

D) Biological Control: The important components in biological control are i) Conservation and encouragement of natural enemies ii) Introduction and artificial release of parasites and predators and colonization and iii) Propagation and dissemination of specific viral, bacterial and fungal diseases. Biological control forms an effective base for successful pest management. Nearly 60% of the natural control of the pests is due to entomophagous occurring in nature. In sugarcane crop biological control is more successful in the light of the long duration of crop, which offers advantage like a less disturbed system, thick canopy of crop which offer conducive micro-climatic conditions for more effective population development and sustained activities of natural enemies and also there is less use of insecticides in the crop thus there is safety to parasites and predators. When the natural control fails to suppress the pest population below economic injury lével, which often happens in the case of key pests, the natural enemy population can be artificially increased to bring down the pest population below ETL. This is called 'Classical biological control'.

It is important to note that in integrated pest management systems, natural enemies need not provide complete control of the pest since the system integrates many other tactics to achieve the goal. So there is need to educate the farmers that they need not expect the natural enemies to cause cent percent mortality of the pests and in fact there should be some pest populations for the survival of the natural enemies.

In sugarcane, colonization of specific parasites has resulted in the effective suppression of some key pests. Owing to the cheapest and easiest mass breeding techniques available for *Trichogramma*, this parasite has received main emphasis in the biocontrol programme against the sugarcane borers in the country. In Harinagar, Bihar, releases of *T. chilonis* @ 4.0lac/ha/week in 8 hectares has resulted in minimising the borer population. Inundative release of *Trichogramma chilonis* Ishii for the control of internode borer in Tamil Nadu (Sithanathan *et al*; 1978) has been found to be very beneficial and is being utilised against this pest since 1967. In Punjab also inundative releases of *Trichogramma chilonis* Ishii into sugarcane fields resulted in significant reduction in the infestation of *Chilo auricilius* Dudgeon (Mann and Doomra, 1996). An outstanding success is that of *Isotima javensis* in Tamil Nadu and Karnataka for the suppression of top borer, which has virtually eliminated the use of insecticides for the past 3 decades in these States (David *et al.,* 1991). The efficiency of the parasite, under sub-tropical conditions is however, limited. Redistribution and mass release of *Epiricania melanoleuca* has provided effective suppression of pyrilla in several States like Haryana, Uttar Pradesh, Gujarat and Maharashtra. In Uttar Pradesh field releases of *Cotesia flavipes* @ 2000 mated females/ha in 4 doses at weekly intervals from July to November has resulted in a significant reduction in the incidence of *C. auricilius*. This parasitoid is an example of new association in the biological control success. Among the insect pathogens granulosis virus has given promising results against shoot borer in Tamil Nadu and the milky disease bacterium against white. Entomophagous fungi, *Metarrhizium anisopliae* and *Beauveria bassiana* are found to be promosing against pyrilla, *E. depressella* and *C. auricilius* but their field potential on large scale is yet to be proved.

Table 5. Doses and period of field releases of *Trichogramma* spp. against sugarcane borers in India.

Trichogramma sp	Host	Releases	No. and period of releases (adults/ha)	State	Reference
T. chilonis	*C. infuscatellus*	12,500	6 weekly release from 6 weeks after planting	Karnataka	
	A. stenielius	1,25000	At 10 day interval from July to October	Punjab	Varma *et al* 1979
	C .auricilius	50,000	Weekly releases from July to October	Punjab	Shenimar and Brar 1996
	C. sacchariphagus indicus	50,000	Weekly releases from 4th to 11th month of planting	Tamil Nadu	Sithanathan *et al* 1978
	C. sacchariphagus indicus	50,000	From July to October at 10 day interval	Uttar Pradesh	Tiwari *et al* 1997
	E. depressella	50,000	From July to October at 15 day interval	Haryana	
T. chilonis and ***T. japonicum*** *R. ablutellus* *S. excerptalis* *S. inferens*	*C. infuscatellus*	50,000	Weekly releases	Bihar	Misra and Pawar 1987

A number of exotic egg and larval parasitoids have also been evaluated against various insect pests specially borers in different parts of the country (David and Nandagopal, 1986). None of these parasites have established in the field to provide the economic suppression of the sugarcane pests. Thus the trials with parasitoids remained a matter of academic interest, however, use of *T. eldanae* is being continued against the sugarcane shoot borer in Maharashtra area with success. The various exotic parasitoid evaluated against sugarcane borers are given in Table 6.

The natural enemies recommended for the suppression of important pests are given below:

Shoot borer

i) Four applications of granulosis virus at 10^6-10^7 inclusion bodies/ml alongwith a surfactant like teepol or sandovit 0.05 percent on 35th day of planting. If necessary the application may be done at fortnightly intervals.

ii) Sequential release of 125 gravid females of the tachnid parasite, *S. inferens* per ha. from 30th to 50th day of planting.

iii) Weekly release of 0.5 Lakh adults of *Trichogramma chilonis* per ha. (when planting is done during autumn or special season)

However, these natural enemies against shoot borer are effective under tropical conditions only.

Table 6. Exotic parasitoids evaluated against sugarcane pests

Parasitoid	Origin	Current Status
Egg stage		
Trichgramma brasilensis Ashmed	South America	Tried and discontinued
T. chilonis Ishii	Taiwan	Tried and discontinued
T. fasciatum (Perkins)	Barbados	Tried and discontinued
T. japonicum Asgmed	Phillippines	Tried and discontinued
T. perkinsi Girault	Colombia	Tried and discontinued
Trichogrammatoidea eldanae Viggiani	Mexixo	Being used against shoot borer
Larval stage		
Allorhogas pyralophagus Marsh.	Mexico	Tried and discontinued
Cotesia flavipes Cameron	Indonesia	Being evaluated against borers
Diatraeophaga striatalis Tns.	Indonesia	Tried and discontinued
Lixophaga diatraea Tns.	Antigua, Cuba, Jamaica, Dominican Republic and Taiwan	Tried and discontinued
Metagonistylum minense Tns.	Brazil	Tried and discontinued
Paratheresia claripalpis Wulp.	Trinidad	Tried and discontinued
Sturmiopsis parasitica Curran	Ghana	Tried and discontinued

Top borer

Inoculative release of 125 females of *I. javensis* per ha in fields showing more than 10 per cent top borer infestation. The technique is useful only in tropical India.

Stalk borer and internode borer

Four to five releases of *T. chilonis* @ 1.0 lakh parasites/ha during July to October.

Pyrilla

Release of 8 to 10 thousand cocoons or 8 to 10 lakh eggs of *E. melanoleuca* per ha.

Scale insect

Release of at least 1500 beetles of *Pharoscymnus horni* or *Chilocorus nigritus* per ha at the first appearance of pest in tropical India as the pest is a problem in this region only.

Release Technique

Trichogramma spp are generally released inundatively with an objective that the parasite population outnumber the host eggs before the target pest could cause serious damage to the crop. The prasites can be conveniently released in the pupal stage. The 'Tricho cards' bearing the parasitised eggs are cut into small pieces and are stapled to the underside of leaves in the evening or early morning throughout the field. However, in this method the eggs are sometimes destroyed by the predators in the field. Alternatively, the egg cards are placed in closed cups fitted with wire mesh at bottom and are hung on to canes at different places. A recent technique is to mix the eggs parasitised on different days and prepare the cards which ensure sequential emergence of parasites in the field for several days.

The mated females of *S. inferens* and *I. Javensis* can be released by holding the container open and slowly moving in the field in all directions.

For effective control of pyrilla, leaf bits bearing 2 to 3 egg masses or 5 to 7 cocoons of *E. melanoleuca* can be stapled to the underside of sugarcane leaf at several places in the infested field.

The granulosis virus can be sprayed in the evening hours on the leaf whorls and stem preferably using high volume sprayer.

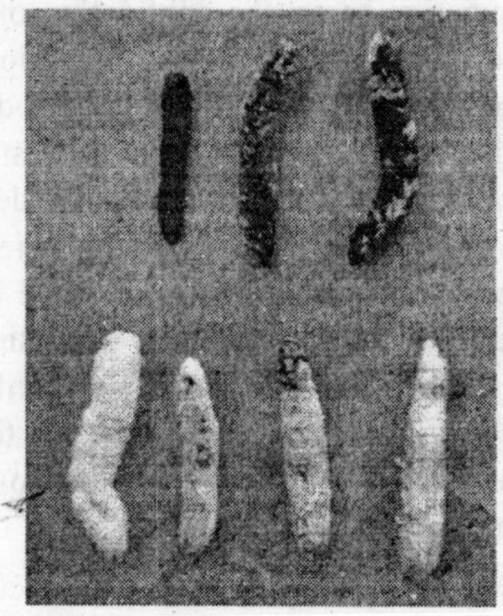

E) Chemical Control: Application of insecticides is essential in an overall pest management programme, especially in endemic areas. Insecticides are the most reliable tools and farmers get easily convinced. They give rapid control especially in endemic and emergency situations specially against sucking pests, which build up to injurious levels quickly. In fact the insecticides are the most effective component of IPM programme especially when the population has exceeded economic threshold levels. However, the farmers have to be educated that 100 per cent mortality is not required and he should avoid indiscriminate and unjudicious use of pesticides. Emphasis on need based use of only recommended insecticides with minimum effect on natural enemies should be given. Broad spectrum insecticides should be avoided. Method and timing of insecticide application should be properly selected.

Soil application of insecticides and sett treatment do not interfere with natural enemies. Whorl application is also less disruptive. Sardana and David (1992) recommended drenching of chlorpyriphos 20EC on the setts at the time of planting or application at 35 and 65 days of planting along the plants for reducing early shoot borer. Similarly soil and whorl application of regent (fipronil) insecticide is recommended for its suppression (Sardana, 2001). These insecticides because of method of application are selective and don't harm the activity of *Sturmiopsis inferens* Trans. parasite. Another example of use of selective insecticide is for top borer control. Effective control of top borer is achieved with carbofuran granules @ 1.0 kg a.i./ha or phorate granules @ 3.0 kg a,i. /ha during the first week of July. Insecticide application shpuld be followed by earthing up and light irrigation. Insecticide application does not effect its parasite, *Isotima jevensis.* This also reduces the margin of parasite:host ratio and promote the parasitic activity and efficacy in the subsequent broods as well. Similarly soil application of aldicarb, phorate, carbofuran and quinalphos don't effect the parasite of scale insect, *Adelencyrtus mayuri* Subba Rao. While foliar sprays of dimethoate, monocrotophos, quinalphos etc. reduce the level of parasitism under field conditions.

Some insecticides are less harmful to certain stages of parasites even as foliar sprays. For instance, foliar sprays of malathion (0.5 kg a.i./ha) and dimethoate (0.4 kg a.i./ha) are reported to be less harmful to for the cocoons of *Epiricania melanoleuca* (Verma and Bindra, 1980). Toxicity also depends upon the development stage of the parasite during which the insecticide is applied. Similar results were obtained in case of *Trichogramma spp.* (Sithanathan and Navarajan Paul, 1980). Chemical treatment sometimes becomes necessary during summer for shoot borer, during June-July for top borer (3rd brood) and during epidemic conditions for scale insect, white fly, Pyrilla and white grub and may be resorted to in such conditions only.

F) Behavioural Control: *Pheromones:* The sex pheromones produced by early shoot borer, stalk and internode borer have been identified and synthesised which can be exploited for monitoring their activity, mass trapping of male moths and mating disruption. (David *et al*, 1985, Mukunthan, 1991, Govindachari *et al.*, 1992). In China, based on the number of moths trapped, release of egg parasites, *T. chilonis* is regulated. In Mauritius mating disruption has been found feasible in case of internode borer. Mass trapping is very profitable against internode borer in Tamil Nadu.

Light traps: Fluorescent light traps are also useful in attracting the beetles of white grubs, *Holotrichia consanguines* Blanch. (Gupta, 1958). They can be used during the beetle emergence season in areas where pest is a serious problem. Black light traps are useful in attracting both the sexes of root borer moths and can be employed for monitoring the activity of borer and also as a tool in its suppression (Sardana, 2000).

IMPLEMENTATION OF IPM

Effective organization of the programme is necessary for timely and quick implementation under proper supervision on the application of different management practices. Field application of pest management technologies need active participation by cultivators, extension workers of sugarcane factories and staff of Department of Agriculture and trained scientists. Farmers and cane managers are trained to understand that in IPM, eradication of pest is neither necessary nor possible and the management programme is implemented taking into consideration the cost/benefit involved, compatibility of the treatments, their synergistic effects, multiple benefits and hazards accruing, and the effects likely to follow. The farmers are to be taught about the need based application of insecticides, more as a corrective measure than as a prophylactic operation. The ultimate success depends upon the proper execution of the whole programme. For instance, during the last pyrilla epidemic in UP and Haryana during 1999 based on the advice of the scientists, the factories refrained from going for aerial spraying of insecticides. This has paid rich dividends, as pyrilla could be controlled over a period of time by *E. melanoleuca* without the pollution of toxic molecules in the ecosystem.

Planning for mass production and easy availability of natural enemies on a large scale and also periodic monitoring of pest activity are needed.

FOLLOW UP

Care may be taken to assess adequately the over all efficacy and utility of methods implemented and the benefits accruing highlighted so that more area may be brought under such pest management and intensify the programme through proper motivation.

PEST MANAGEMENT STRATEGY AND RECOMMENDATIONS

The basic goal of commercial farmers is to maximize farm profits. The most productive strategy is to maximise natural control forces and supplement the same with suitable unified management practices for the locally prevalent key pests, taking into consideration the economics of control, the benefits accruing and the compatability of treatments, hazards and the adverse effects likely to follow. Some of the management strategies suggested for the suppression of major pests in sub-tropical and tropical India are given in Table 7 and Table 8. But it should be clearly borne in mind that IPM is only a concept, idea, philosophy and not a fixed schedule plan of control operations. The appropriate control strategy should be chosen based on crop pest situation in relation to local conditions.

Table 7. Schedule of pest management practices for Sub- tropical India

Stage of Crop	Target insect pests	Recommended cultivation practices
Pre-planting	Borers, mealy bugs, Scale insects White flies	Select setts from uninfested fields. Reject scale insect and borer infested cane pieces when setts are selected from infested fields. Avoid planting in low lying areas endemic to white flies .Avoid autumn planting to reduce stalk/Gurdaspur. Use less susceptible varieties.
	Scale insect	Immerse the setts in 0.1 per cent malathion or 0.08 per cent dimethoate solution for 15 minutes
	White grubs	Give two extra ploughings for exposing white grubs for predation.
At planting	Termites, shoot borer	i) Prefer early sowing for ESB ii) Grow quick growing varieties iii) Apply chlorpyriphos 20EC @ 0.1% on setts in furrows in 1500 litres of water/ha or
On 30th, 45th day	Shoot borer	i) Trash mulching
On 45th day	Shoot borer	i) Apply chlorpyriphos 20EC @ 0.1%. ii) Give frequent irrigations.
On 90th day	Borers, pyrilla, scale insect, mealy bugs	Detrash lower leaves, remove egg masses and infested canes.

Stage of Crop	Target insect pests	Recommended cultivation practices
April-May s	White grub	i) Use of aggregating pheromone after first heavy summer rain and collection and destruction of beetles. ii) Use of light traps and collection and destruction of beetles from host trees immediately after first summer showers.
June-July	Pyrilla	i) Release of 4,000-5,000 cocoons and 4-5 lakhs eggs of *Epiricania melanoleuca*/ha or ii) Dusting 5% folidol @ 20-30 kg/ha after removing lower leaves iii) Collect and destroy lower leaves with egg masses /nymphs.
	Scale insect	Spray 0.08 per cent dimetioate or 0.1 per cent malathion after detrashing lower leaves.
	Gurdaspur borer, Plassey borer	Collection and destruction of infested tops.
	Top borer	i) Collection and destruction of egg masses and mid-rib tunneled leaves. ii) Soil application of carbofuran @ l.0 kg a.i./ha or phorate @ 3 kg a.i./ha. during June end followed by light irrigation.
	Stalk borer	i) Apply recommended dose of nitrogeneous fertiliser. ii) Avoid water logging and cane lodging.
August – October	Stalk borer	i) Spray monocrotophos 0.75 kg a.i/ha after detrashing ii) Detrash the lower dried leaves once a month from September to October. iii) Remove late and water shoots. iv) Release *T. chilonis* at fornightly intervals from July till October.
	Root borer	i) Apply chlorpyriphos @0.1% or NSKE 5% along the roots followed by light irrigation ii) Install light traps from March till September. iii) Flood the field for 24 hours once a month.
	Pyrilla, scale insect	Spray malathion 0.1 per cent after detrashing with power sprayer.
December- January	Pyrilla, scale insect, Borers, mealy bugs	i) Harvest heavily infested crop early and burn crop residue. ii) Give early ploughing and burn stubbles iii)Before ratooning give stubble shaving and burn trash. iv) Detrash lower leaves

Table 8. Schedule of pest management practices for tropical India

Stage of cultivation	Target insect pests	Recommended management practices
Upto May	Same as under Table 7.	
May-June	Shoot borer	i) Spray granulosis virus of shoot borer @ 10^8 inclusion bodies/ lml in 500 litres/ha or ii) Release 125 gravid females of *S. inferens* around 45th day.
June-July (120th day)	Internode borer white flies	i) Spray 0.1 per cent monocrotophos or Spray 0.1 per cent endosulfan
July-August (On 135th day)	Internode borer	i) Release *Trichogramma chilonis* @ 1.5 lakh/ha at fortnight interval from 4th month until a month before harvest.
Sept.-Oct. (180th day)	Internode borer, scale insect, mealy bugs, white flies, pyrilla	i)Detrash the crop followed by spraying with monocrotophos @ 0.75 kg a.i/ha ii) Malathion 0.1% spraying may be given after detrashing for sucking pests.
On 270th day	– do –	Detrash the crop

REFERENCES

Agarwal, R.A. 1980. Eradication of major pests of sugarcane. *Int. Sug. J.,* 82: 177-179.

Avasthy, P.N. 1983. Insect pest management for sugarcane in India. In '*Sugarcane Pest management in India*" (eds. Balasubramaniam, M. and A.R. Solyappan.) Tamil nadu Cooperative Sugar Federation., Madras P.71-77.

Bains, S.S., Dilawari, Brar, R.S. and Dhaliwal, Z.S. 1982. A model for prediction of sugarcane pyrilla and its outbreak in relation to high temperature during pre-monsoon period in Punjab. *Indian. J. Ecol.* 4: 298-302.

Brar, R.S., Bains, S.S. and Vaishney, R.K. 1978. Mathematical models for prediction of pyrilla epidemic in relation to its parasite, *Epipyropes melanoleuca* Fletcher. *Proc. Natn.Systems Conf.,* Ludhiana 5: 6-9.

Chaudhary, J.P. and Sharma, S.K. 1988. Biological control of pyrilla using parasites and predators, pp. 186-206 in David, H. and Easwaramoorty, S. (eds.) *Biological Control Technology for Sugarcane Pest Management.* SBI, Coimbatore.

David, H. 1981. Recent advances in sugarcane entomology in India. In: *Recent advances in entomology in India* (ed. T.N. Ananthakrishanan) Loyola college, Madras. pp. 37-47.

David, H. 1995. Integrated Pest Management. In "*Sugarcane Production Manual*" Eds. Alexander, K.C. and Arulraj, S. Sugarcane Breeding Institute, Coimbatore pp. 102-111.

David, H. and Alexander, K.C. 1986. Integrated pests and diseases management in sugarcane. *Plant Prot. Bull.* 38: 37-39.

David, H. and Nandagopal, V. 1986. Defoliators. Published in Sugarcane Entomology in India. (Eds. David, H, Easwaramoorthy, S. and Jayanthi, R.) SBI, Coimbatore. pp: 321-342.

David, H., Easwaramoorthy, S and Jayanthi, R 1986. Sugarcane Entomology in India. SBI, Coimbatore, ICAR.

David, H., Easwaramoorthy, S and Jayanthi, R 1991. 'Integrated Pest Management in Sugarcane with special emphasis on Biological Control'. Sugarcane Breeding Institute, Coimbatore, ICAR.

David, H., Nesbitt, B.F., Easwaramoorthy, S. and Nandagopal, V. 1985. Application of sex pheromones in sugarcane pest management. *Proc. Indian Acad. Sci.* (B) 94: 333-339.

Easwaramoorthy, S. and Shantalalshmi, G. 1988. Efficacy of granulosis virus in the control of shoot borer, *Chilo infuscatellus* Snellen. *J. Biol. Cont.* 2(2): 106-108.

Easwaramoorthy, S., Kurup, N., Shanmugasundaram, M. and David, H. 1980. An exotic Tachinid parasite, *Diatraeaphaga striatalis* Trs. estabilishes on sugarcane internode borer. *Proc. A. Conv., Deccan Sug. Technol. Assoc.,* 70: 147-148.

Easwaramoorthy, S. 1986. New approaches in sugarcane pest control. In: *Sugarcane Entomology in India.* (eds. David, H., Easwaramoorthy, S and Jayanthi, R) Sugarcane Breeding Institute, Coimbatore, TN. pp. 438-57.

Easwaramoorthy, S. 1986. Ecology of sugarcane pests. In: *Sugarcane Entomology in India.* (eds. David, H., Easwaramoorthy, S and Jayanthi, R) Sugarcane Breeding Institute, Coimbatore, TN. pp. 31-67.

Gupta, B.D. 1958. A new fluorescent light trap. IISR, Newsletter. pp. 4: 17.

Govindacharichari, T.R. Narasimhan, S., Suresh, G. and Ganeshawar, P. K. 1992. Field capture of the males of the sugarcane shoot borer, *Chilo infuscatellus* Snell. (Crambidae: lepidoptera) using synthetic pheromone components. *Hexapoda,* 4(2): 113-118.

Kalra, A.N. 1964. Recent advances in studies on influence of temperature and humidity on incidence and population of some of the major sugarcane pests. *Ent. Res. Workers Conf.,* Delhi. (Unpubl.)

Karunakar, G. 1990. *Studies on entomophilic nematode for the control of two spp of sugarcane white grubs.* MSc (Ag.) Thesis, TNAU, Coimbatore.

Mann, A.P.S. and Doomra, S. 1996. Biological control and management of sugarcane borers in Punjab. *Coop. Sug.,* 28: 288-292.

Meeusen, R.L. and Warren, G.W. 1989. Insect control with genetically engineered crops. *Ann. Rev. Entomol.* 34: 323-381.

Misra, M.P. amd Pawar, A.D. 1987. Biological control of early moth borers of sugarcane by *Trichogramma* in North Bihar. *J.Biol. Cont.* 1: 70-71.

Mukunthan, N. 1991. Scope of pheromone technology in sugarcane pest management. In: *'Biocontrol technology for sugarcane pest management'*. (Eds. H. David and S. Easwaramoorty. Sugarcane Breeding Institute, ICAR. p. 337-346.

Nandagopal, V. 1983. Studies on yield loss assessment and determination of economic threshold for sugarcane internode borer, *Chilo sacchariphagus indicus* (K.. MSc. (Ag.) Thesis, TNAU, Coimbatore.

Pawar, A.D. 1983. National efforts in bio-control of sugarcane pests. In '*Sugarcane pest Management in India*' (Eds). Balasubramaniam, M.and A.R. Solyappan. T. N. Coop. Sugar Federation., Madras pp. 34-35.

Raja, Rao, S.A. 1964. Biological control of top borer, *Scirpophaga nivella* F., of sugarcane at Pugalur. (Madras State). *Proc. Bienn. Conf. Sug. Cane Res. Dev. Works*. 5: pp 588-592.

Sardana, H. R. 1996. Economic injury level and sequential sampling plan for sugarcane stalk borer, *Chilo auricilius* Walker. *Ann. agric. Res.* 17(2): 125- 129.

Sardana, H. R. 1997. Determination of economic injury level for early shoot borer, *Chilo infuscatellus* Snellen in late maturing variety of sugarcane. *Ann. Pl. Prot. Sci.*, 5 (2): 188-190.

Sardana, H. R. 1997. Distribution pattern of early shoot borer in sugarcane. *Insect Sci. and its Applic.* 17 (3): 368-371.

Sardana, H.R. 1998. Economic injury level and sequential sampling plan for sugarcane shoot borer. *J. Insec. Sci* 11(1): 30-33.

Sardana, H.R. 2000. Impact of companion cropping on the incidence of early shoot borer, *Chilo infuscatellus* Snellen of sugarcane. *Ann. Pl. Prot. Sci 8(2): 239-40.*

Sardana, H. R. 2000. Population dynamics of sugarcane root borer in Haryana. *Ann. Pl. Prot. Sci.* 9(2): 304-306.

Sardana, H. R. 2001. Evaluation of a new insecticide regent (Fipronil) against sugarcane shoot borer, *Chilo infuscatellus* Snellen and root borer, *Emmalocera depressella* Swinhoe. *Pestic. Res. J.* 13(1): 74-78.

Sardana, H.R. and. Das, D.K. 2001. Regression modelling to predict sugarcane stalk borer, *Chilo auricilius* Dudgeon, *Indian J. Ent.* 63(3): 345-349.

Sardana, H.R. and David, H. 1992. Efficacy of some insecticides for the control of shoot borer, *Chilo infuscatellus* Snellen (Crambidae: Lepidoptera) of sugarcane. *Indian J. Ent.* 54(2): 608-618.

Shenhmar, M. and Brar, K.S. 1966. Evaluation of *Trichogramma chilonis* Ishii for the control of *Chilo auricilius* on sugarcane. *Indian J. Pl. Prot.* 24(1): 47-49.

Shinde, V.K.R. and Sharma, S.K. 1971. *Bacillus popillae* Dutky pathogenic to *Lachnosterna consangunea* Blanch. *J.Eco. Ent.*, 64: 1301-1302.

Sithanatham, S., Muthuswamy, S. and Durai, R. 1978) Experiments on inundative release of *Trichogramma australicum* Gir., in biological control of sugarcane stem borer, *Chilo indicus* (Kapur). *Madras agric. J.*, 60: 467-471.

Sithanatham, S. and Navarajan, Paul, A.V.N. 1980. Toxicity of some insecticidal sprays to the developmental stageas of *Trichogramma confusum*. *Pestologg.*, 5: 21-24.

Tiwari, N.K., Tanwar, R.K. and Varma, A. 1997. Effectiveness of *Trichogramma chilonis* Ishii against sugarcane borers in Uttar Pradesh. *Indian J. Sug. Cane Technol.* 11(1): 42-44.

Varma, G.C. and Bindra, O.S. 1980. Studies on the chemical control of *pyrilla perpusilla* and effect of insecticides with natural enemies. *Pesticides*, 14: 9-10.

Varma, G.C., Bindra, O.S., Singh. S. and Singh, B. 1979. Comparative efficacy of biological, mechanical, integrated methods for controlling sugarcane borers in Punjab. *J. Res. Punjab Agric. Univ.* 16: 45-48.

Integrated Insect Pest Management in Major Crops of Andhra Pradesh

K. MANJULA

Andhra Pradesh endowed with diversified agro-ecosystems, produce many cereal, pulse, oilseed, fibre, vegetable and fruit crops besides tapioca, oil palm, grapes coffee, the selective ones. However, 60-65 per cent area of the 11 million hectares of cultivable land is occupied by crops like Rice *(Oryza saliva)*, groundnut *(Arachis hypogaea)*, cotton *(Gossypium* spp) pigeon-pea *(Cajanus cajan)*, mango *(Mangifera indica)* and tomato *(Lycopersicon esculentum)*. The expression and performance of these crops undergo cyclic fluctuations due to the influence of abiotic and biotic factors. Among the biotic factors, insects are predominant in influencing the yield of these crops. Hence, in many areas of the state, the production of these crops become risk-prone due to involvement of more inputs oriented towards plant protection. In that process only the state has a distinction of dumping more than 13,500 tonnes of pesticides annually. This ultimately led to serious repercussions in the shape of insecticide resistance in many problem insects and insect resurgence as well as deleterious residues of chemicals in the end products.

Keeping these things in view integrated insect pest management (IPM) that emphasizes the role of naturally occurring bio-agents as well as cultural practices in as compatible manner as possible with other location specific strategies was proposed. ANGR Agricultural University and Regional Research Centres of ICAR are in fore front in developing IPM modules in major crops and spread these technologies at field level as alternatives to the existing chemical oriented insect control. The IPM modules proposed by ANGRAU are presented herein.

RICE *(Oryza sativa)*

India is the second largest producer of rice in the world, covering nearly 42 million ha of area in diverse ecological situations and accounts for more than 42 per cent of food grain production. Rice is the staple food for 65 per cent of the population. In Andhra Pradesh, the crop is being grown in about 40 lakh ha. The crop is well adopted in almost all regions of the state *viz*., Coastal Telangana and Rayalaseema where the ecosystems are characteristic due to the sources of irrigation as well as other environmental factors of the area. Hence the pest scenario vary and key pests differ from region to region. However, the following are the commonly occurring insect pests on rice in the state (ANGRAU, 2002a).

Department of Entomology, S.V.Agricultural College, Tirupati - 517502, A.P.

Insects pests associated with rice crop in Andhra Pradesh

Nursery	Tillering	Panicle Emergence	Grain Maturity
Rice gall midge, *Orseolia oryzae*	Rice gall midge, *O. oryzae*; yellow stem borer, *Scripophaga incertulas*	Yellow stem borer, *S. incertuals* leaf folder, *Cnaphalocrosis medinalis medinalis*	Gunni bug, *Leptocorisa acuta*
Cutworm, *Spadoptera mauritia*	Thrips, *B. biformis*; Grass hoppers, *H. banian*	Brown plant hopper (BPH), *Nilaparvata lugens*	Yellowstemborer, *S. incertulas* BPH, *N. lugens;* WBPH, *S. furcifera*
Thrips, *Baliothrips bifromis*	Rice hispa, *Dicladispa armigera*	White backed plant hopper (WBPH), *Sagatella furcifera*	Climbiung cutworm, *Mythimna separata*
Grass hopper, *Heiroglyphus banian*	Horned caterpillar, *Melanitis leda ismene,* whorld maggot, *Hydrellia sesaki*	Panicle mite, *Steneotarsonemums spinki*	
	Case worm, *Nymphula depunctalis*	Leaf mite, *Oligonychus oryzae*	
	Root weevil, *Enchinocnemus oryzae*		

KEY PESTS IDENTIFIED IN THE STATE

Region

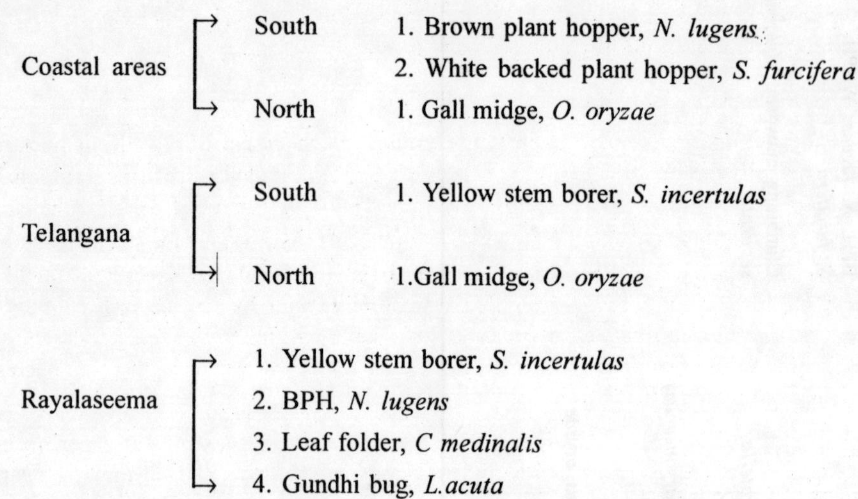

IPM strategies recommended keeping the key insect pests in view

- Growing moderately yellow stem borer resistant varieties like Sasyakee, Ratna, Kauveri; against gall midge, Pothana, Kavya, Kesava, Erramallelu, Vorugallu, Badrakali, Siva, Rudrama, Varsha; against BPH, Chaitanya, Krishnaveni, Chandana, Triguna, Deepti, Nandi, Vijetha.
- Nursery application of cartaphydrochloride or phorate granules @ 1.5 kg ai/ha five days before transplanting against stem borer and gall midge.
- Clipping the tops of leaf blade before transplanting since the egg masses of stem borer are laid on upper half of the leaves.
- Maintenance of alley ways of 30 cm width after every 2-3 meters crop to facilitate basal spraying against hoppers (BPH and WPBH) that infest the basal part of the plant at water level (ANGRAU, 2000a).
- Balanced use of nitrogenous fertilizers in 2-3 splits to reduce the build-up of pests like plant hoppers, leaf folder, leaf hopper.
- In transplanted crop, application of cartap hydrochloride or phorate granules @ I kg ai/ha (If the crop is in vegetative stage) or spraying with monocrotophos (0.07%), phosphamidon (0. I%) etc. against stem borer, plant hopper and leaf hoppers.
- Intermittent draining of the field for I or 2 days reduces the population of plant hoppers. In scarce rain fall zone, the practice of postponing the irrigation for I or 2 days can be followed for the purpose.
- Use of pheromone traps against yellow stem borer @ 10/ha to monitor the moth activity.
- Use of *Trichogramma japonica* eggs @ 1,25,000/ha on receipt of considerable moth catch suppresses stem borer population.
- Conservation of predators of plant hoppers viz., a mirid bug, *Cyrtorhinus lividipennis* and a spider, *Lycosa pseudoannulata.*

- Spraying chlorpyriphos (0.05%) or quinalphos (0.05%) against leaf folder and gundhi bug.
- At maturity stage, application of carbaryl (0.15%) (at least 20 days prior to harvesting) against plant hoppers.
- Harvesting crop close to the soil surface, ploughing or flooding the field after harvest kills the larvae of stem borer hibernating in the stubbles (ANGRAU, 2000a).

GROUNDNUT *(Arachis hypogaea)*

Groundnut is an important oilseed crop of India being grown to the extent of 9 million hectares. Andhra Pradesh is in fore-front to adopt groundnut crop even by sacrificing several of the traditionally grown crops like millets, oilseeds and pulses. The crop is grown in two million hectares of land in Anantapur, Chittoor, Warangal, Khammam, Cuddapah, Guntur, Nalgonda, Karimnagar, Visakhapatnam, Viziayanagaram, Srikakulam and Mahaboobnagar districts. Its virtues to withstand low moisture levels, replenishment of soil fertility, usefulness as cattle feed (both raw and processed), suitability for inter or mixed cropping of local systems, usefulness as raw and altered food products etc., have given scope for continuous and repeated cultivation of this crop in dry (rainfed) areas of Rayalaseema where more than 1.2 million hectares of cultivable land is covered by this crop.

However, the cyclic fluctuations of its productivity levels ranging from 680 to 1150 kg, per hectare during the past one and half decades disappoint both the producers and planners of agriculture sector. The situation has become much worse during the past three seasons particularly in Rayalaseema districts, where uneven distribution of precipitation as well as sudden flare-up of insect pests and microbial diseases became a common feature. The insect pests that commonly infest the groundnut crop in the state are enumerated below.

Insect pests that attack groundnut crop in Andhra Pradesh

EARLY GROWTH PERIOD	POD DEVELOPMENT STAGE	MATURITY PERIOD
Root grub (white grub), *Lachnosterna cansaguinea*	RHC, *A. albistriga*	*S. litura*
False wire worms, *Gonocephalum elongatum*	Leaf miner, *Aproaerenia motlicellti*	Pod sucking bug, *Eltisniolomus sortli(lus*
Red hairy caterpillar (RHC), *Amsacta albistriga*	Grm pod borer, *Hclicoverpa armigera* tobacco cut warm, *Spodoptera litura*	
Thrips, *Scirtotlirips dorsalis, Frankliniella schultzii, Caliothrips indicus* and *Thrips palmi*		Stern borer, *Sphenoptera parotetti*
Jassids, *Enipoasca kern* Aphids, *Aphis craccivora*	Mites, *Tetranychus* spp Asliy weevil, *Myllocerus* spp	Earwig, *Anisolabis statti* Ants, *Dorius orienttalis* Termites, *Odentotermes* sps, *Microtermes* sps

(ICRISAT, 1995). (ANGRAU, 2000b)

KEY PESTS IDENTIFIED

Region

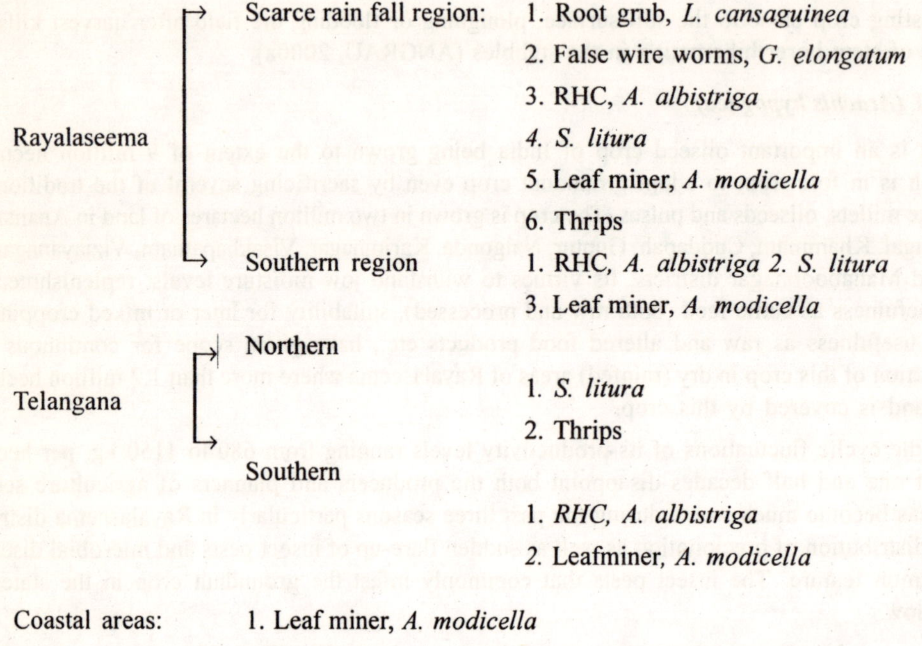

Coastal areas: 1. Leaf miner, *A. modicella*

2. *S. litura*

3. Jassids, *E. kerri*

The following are the IPM strategies for key pests of groundnut

- Deep ploughing in summer to expose soil borne pupae of RHC, root grub, Helicoverpa, Spodoptera, Leaf miner etc., to sun radiation and bird predators.
- Application of contact insecticidal dust like methyl parathion 2% under perennials and ploughing the soil up to 10-20 cm to destruct the hibernating adults of root grub before receiving monsoonic rains. Shaking of branches of perennials after getting rains and destruction of fallen adults of root grub.
- Resorting to bonfires to trap adult moths of RHC (1 fire/10 acres) during 7.00-9.00 P.M from 2nd day after receiving monsoonic rains (30 nm or above) during June-August. This practice should be continued for 2-3 days. All the moths emerge from soil pupae at a time (as a single brood) in response to the good rains. If the rainfall is scarce and intermittent, the moth emergence will be in spells (ARS, ATP, 2000).
- Installation of light traps @ 1 trap/20 ha when the facility of electricity is available.
- Collection and destruction of RHC egg masses from weeds, walls of farm houses, polls, walls at houses etc.
- Inter-cropping of groundnut with cow-pea or castor is useful as these are the most attractive crops to RHC larvae. When the larval population is moderate, the caterpillars can be collected and destroyed. The farmers in Anantapur and Chittoor districts are well adopted to inter-crop groundnut with red-gram and red-gram + cow-pea respectively. The defoliation of intercrops due to early instar larvae of RHC is also an indication of RHC brood activity (ARS, ATP, 2000).

- Border cropping with fast growing cereals like bajra gives response as they act as barriers to thrips as these carry over peanut stem necrosis disease (which occurred in epidemic form during 2000 *kharif in* Anantapur district, caused by tobacco streak virus, first record in groundnut in the world) from weeds (ARS, ATP, 200 1).
- Maintenance of optimum plant population by using appropriate seed rate and spacing as it increases the pest density by providing favourable micro-climate (ARS, ATP, 2000).
- Seed treatment with chlorpyriphos @ 6 ml/kg or soil application of phorate @ 15 kg/ha against root grub. This treatment also useful against false wire worm (ARS, ATP, 200 1).
- Clean cultivation practices like periodical removal of weeds in crop and on bunds as they act as alternate hosts to insect pests.
- Spraying the bunds and crop (if available) with quinalphos 0.05% at 10-15 days after the emergence of RHC moths helps in reduction of early instar larvae (ARS, ATP, 2001).
- Formation of trench (5-10 cm depth) with plough around the field and application of insecticidal dust in it to prevent migration of RHC larvae from one field to another (ARS, ATP, 2000).
- Hand collection and destruction of grownup larvae of RHC, *Helicoverpa*, egg masses and early larvae of Spodoptera.
- Use of pheromone traps @ 10/ha to monitor the moth activity of *Helicoverpa*, *Spodoptera* and leaf miner.
- Use of yellow sticky traps against adults of leaf miner and thrips.
- Use of *Trichogramma* eggs @ 1,25,000/ha against *Helicoverpa* and *Spodoptera.*
- Spray of NPV @500 LE/ha / 0.5% neem oil / 5% NSKE against early larvae of *Helicoverpa* and *Spodoptera.*
- Conservation of natural enemies: Several hymenopteran parasitoids of leaf miner like *Goniozus sps, Bracon* sps etc. checks the leaf miner larvae. If the parasitization is >50%, insecticides need not be applied.
- During September-October months (second generation/ 2nd and 3rd broods), if monsoonic rains are received, the maximum RHC larvae (grown up) will be naturally checked by a dipteran parasite, *Exorista* spp. The caterpillars will find to dead in the fields with detached hair, as cut pieces etc. (ARS, ATP, 2001).
- Poison baiting with chlorpyriphos against later instars of RHC and *Spodoptera*:

Ingredients	Quantity/acre
Rice bran	6 kg
Chlorpyriphos	400-500 ml
Jaggery	0.5 kg

Chemical has to be sprinkled and uniformly mixed with the bran on a tarpaulin. Jaggery has to be made in to syrup with required quantity of water and has to be added to the rice bran - chemical mixture. When mixed well, it forms in to loose balls, some flakes and also powder. The poison bait then should be applied (broadcasted) uniformly to one acre field (ARS, ATP, 2001).

- Spraying with quinalphos (0.07%) or thiodicarb (0.05%) against grown up larvae of RHC and Helicoverpa.
- Application of systemic insecticides like monocrotophos 0.06%, imidacloprid 0.002% when thrips population is high (ANGRAU, 2000b).

PIGEONPEA *(Cajanus cajan)*

Pigeonpea (Red-gram) is the most important pulse crop in India. Of about 5 lakh ha of area of Red-gram in A.P., nearly 115/h is in Prakasam and Guntur districts, as mixed crop of pulses and oil seeds. Most of the remaining red-gram production comes from rainfed lands of Rayalaseema and Telangana districts. The diseases like wilt, dry root rot and insects like ash weevil, pod borers are the constraints to red-gram output. The following are the insect pests that attack red-gram in the state (ANGRAU, 2000c).

Seedling Stage	Active Growth Stage	Flowering to Pod Formation	Pod Maturity
False wireworm, *Gonocephalum* sps Ash weevil (cuts the seedlings) *Alcidodes collaris* Red hairy caterpillar, *Amsacta albistriga* leaf miner cum folder, *Aproaerenta Modicelia*	Ash weevil, *A. collaris*; cow bugs, *Oxyrachis tarandus;* red mite, *Tetranychus* spp.; Eriophid mite, *Aceria cajani;* Leaf tier, *Eucosma critica;* Grass *hopper* Colemania *sphenarioides*	Spotted pod borer, *Maruca testulalis;* blister beetles, *Mylabris* spp; gram pod borer, *Helicoverpa armigera;* pod fly *Melanogramyza abtusa*	*H. armigera* green pod borer, *Etiella zinkinella;* Plume moth, *Exelastis atomosa;* Blue butter flies, *Echrysopes* spp.; Pod bug, *Clavigralla gibbosa*

Identified Key pests:
1 . Ash weevil, *A. collaris*
2. *H. armigera*
3. Spotted pod borer, *M. testulalis*
4. Pod fly, *M obtusa*
5. Pod bug, *C. gibbosa*

IPM MEASURES FOR KEY INSECT PESTS OF RED-GRAM

- Deep ploughing to expose the soil pupae of pod borers to sun radiation and bird predators.
- Growing marigold as trap crop of *Helicoverpa* in borders and in between the rows.
- Use of relatively higher seed rate to compensate the loss of seedlings due to ash weevil.
- Application of methyl parathion 5% dust in red-gram rows immediately after germination against ash weevil.
- Use of pheromone traps g 10/ha to monitor the activity of *Helicoverpa* and *Maruca* in adult stages: Scouting of eggs of pod borers.
- Use of *Trichogramma* eggs g 1,25,000/ha in synchrony with moth activity predicted through lure traps.
- Collection of blister beetles attacking flowers of red-gram with nets and destruction.
- Spray of dichlorvos 0.1% or carbaryl 0.15% against blister beetles on need basis.
- Spray of NPV g 500 LE/ha or neem oil 0.5% or NSKE 5% at flower initiation stage against early larvae of *Helicoverpa.*
- Shaking of plants by spreading a tarpaulin under the Red-gram plants and destruction of fallen caterpillars.
- Mechanical collection and destruction of grownup larvae of *Helicoverpa.*
- Spray of acephate (0. 15%) or thiodicarb (0.05%) against *Helicoverpa* in case of heavy population on need basis.

- Intervention by dichlorvos + chlorpyriphos sprays to reduce spotted pod borer and pod fly damage (ANGRAU, 2000c)

COTTON *(Gossypium* spp*)*

Cotton is an important commercial fibre crop grown in different states of India with an area of about 90 lakh ha. In A.P., as an important cash crop, cotton is being grown in 11 lakh ha of area in fertile black soils mainly in the districts Guntur, Prakasham, Krishna, Nalgonda, Warangal, Adilabad and Kurnool. Now-a-days this crop suffers very much from sucking insects pests in early stages and boll worms in later stages. Earlier, only boll worms were the major pests. Due to indiscriminate use of synthetic insecticides against them, the problem of insecticide resistance in Heliocoverpa (bollworm) and resurgence of whitefly *B. tabaci* has surfaced in many regions of the state. Subsequently a well defined module of IPM on cotton was developed by ANGRAU.

Cotton crop experiences the following insect pests in its different growth periods

Early Growth Phase	Squaring Phase	Boll Formation Phase	Boll Bursting Phase
Jassids, *Anirtisca biguttula bigtuttula*	Spotted boll worm, *Earias* spp: tobacco cut worm, *Spodoptera lititra*	*H. arniigera:* wiliteflies, *B. tabaci*	Pink bollworm, *P. gosvypiellti:* Red cotton bug, *Dysdercus singulatus*
Thrips, *Tlirips tabaci*	American boll warm, *Helicoverpa amiigera;* Wllteflies, *Bentisia tabaci*	Jassids, *A. biguttula biguttula*	Dusky cotton bug, *Oxycaranus hyalinipennis*
Surpentine leaf miner, *Lirioniyza trifolii*	Jassids *A.* biguttula *biguttula:* Pink bollworm, *Pectinophora gossypiella*		Red mite, *Tetranychus* spp
Apliids, *Aphis gossypi*	Leaf roller, *Sylepta derogota*		

Key pests: 1. Jassids, *A. biguttula biguttula*

2. White flies, *B. tabasi*

3. Boll worms: *H. armigera, Earias spp., P.gossypiella.*

IPM strategies for key pests of cotton

- Cultivation of whitefly resistant varieties like LPS-141 (Kanchana), LK 86 1, NA 1280 etc and jassid resistant varieties like L-604, LRA 5166, NA 1325 (Narasimha), MCU 5, L603 and hybrids NHH44, Savitha, H8 etc (ANGRAU, 2000d).
- Growing intercrops like groundnut, cowpea, greengram, soybean etc on which natural enemies of insect pests can develop well.
- Border cropping with the cereals like jowar or maize in 2 or 3 rows not only serves as barrier for migration of whiteflies and thrips but also provide pollen and nectar to *chrysoperla carnea* (a predator of sucking pests and eggs of boll worms).
- Growing *Tegetes* (African marigold) and castor plants @ 250/ha that become trap and guard crops for *Helicoverpa* and *Spodoptera* respectively. Okra can be sown as trap crop for spotted boll worm, *E.vitella* and jassids, *A. bigutula bigiittula.*

- Acid delinting of cotton seed to reduce the chances of continuity of pink bollworm like pests.
- Seed treatment with imidacloprid @ 5 g/kg or carbosulfan @ 25 g/kg.
- Stem application of imidacloprid or acetamiprid (1:20), monocrotophos (1:4) in water at 40 and 60 days using a small brush.
- Topping cotton plants after 80/90 DAS to reduce the egg load of Helicoverpa, that refers to lay eggs on terminal parts (young leaves and squares).
- Use of pheromone traps @ I 0/ha to monitor the activity of boll worm moths.
- Use of *Trichogramma* of egg parasitoid @ 1,25,000 /ha against boll worms.
- Collection and destruction of egg masses and early larvae of *Spodoptera* and grown up larvae of *Helicoverpa.*
- Use NPV (500 LE/ha) /Bt(0.1%)/neem oil 0.5%/NSKE 5% against early instar larvae of boll worms.
- Spray of quinalphos (0.07%) or thiodicarb 0.05% or indoxacarb (0.05%) in case of heavy damage by boll worms (ANGRAU, 2000d).

MANGO *(Mangifera indica)*

Mango is considered to be the king of fruits and it is believed that it originated in India. Mango occupies more than 8 lakh ha of area in India. It is the major fruit crop of Andhra Pradesh. In terms of area and production, the state is next to Uttar Pradesh. The major mango growing districts of Andhra Pradesh are

Coastal	Rayalaseema	Telangana
Krishna, East Godavari, West Godavari, Visakhapatnam, Vizianagaram, Srikakulam.	Chittoor, Cuddapah	Karimnagar, Adilabad, Khammam.

Insects form as major constraint in the production of mango under the hot humid climate of the state. The following are the predominant insect pests in different stages of the crop (ANGRAU, 2000e).

Vegetative stage	New flush stage	Flowering	Fruit setting-harvesting
Stem borer, *Batocera rufomaculata:* Bark eating caterpillar, *Indarbela tetraonis*: leaf webber, *Orthaga exguinacea*: Termites, *Microtermes* spp.: Snow scales, *Aspidiotes destructor*: gall insects, *Procontarinia matteriana:* Red ant, *Oceophylla smaragdina*	Mealy bugs, *Drosicha mangiferae*: Thrips *Caliothrips indicus*: *Rhipiphorothrips creuentatus Scirtothrips dorsalis*: Tea mosquito bug, *Helopeltis antonii*: Leaf miner, *Acrocercops syngramma*: Ash weevil, *Myllocerus* spp.: Slug, *Latoia lepida*	Leaf hopper, *Amritodes atkinsoni, Idioscopus clypealis I. niveosparsus* Thrips, *C. indicus R. cruentatus, S. dorsalis*	Leaf Hopper: stone weevil, *Sternochaetus mangiferae:* Immature fruite borer, *Deanolis albizonalis*: Fruit fly, *Bactrocera* spp

Key pests

1. Leaf hopper, *A. atkinsoni, I. clypealis, I. niviosparsus*
2. Stone weevil, *S. mangiferae*
3. leaf webber, *O. exguinacea*

→ In all the growing tracts

Lepidopteran fruit borer, *Deanolis albizonalis:* Coastal Andhra Pradesh

Integrated management strategies of mango insect pests

- Keeping the orchards clean and avoiding overcrowding and water logging to reduce the hopper infestation.
- Repeated digging or ploughing of orchards exposes the eggs of mealybugs, pupae of fruit fly, adults of stone weevil etc.
- Removal and destruction of leaf webs formed by leaf webber periodically.
- Cutting and destruction of stem borer affected branches with grubs and pupae.
- Collection and destruction of stone weevil infested fruits.
- Adult flies of fruitfly can be controlled by placing poison bait containing 500ml malathion 50 EC + 200 g molasses in 2 l of water. Bait spray may sprayed adding an additional 18 li of water to poison bait. Addition of 10ml methyl enginol in place of molasses is also useful.
- If holes of stem borer are located, insert cotton wool soaked in petrol, kerosene or dichlorvos and seal the holes with mud.
- Against leaf hoppers, 3-4 sprays may be necessary first at the time of bud swelling, 1-2 at flowering at 15-20 days interval and another after fruiting. When the hoppers move to the tree trucks, direct the spray fluid should direct on to trunks. Phosphamidon (0.1%), monocrotophas (0,06%) or carbaryl (0.15%) are the effective chemicals (ANGRAU, 2000e).
- Spray of carbaryl (0.15%) at fruit maturing stage reduces stone weevil damage. The performance of endosulfan (0.07%) and azadirachtin (0.0015%) also promising when repeated at every stage of flower-fruit development.
- On required basis, spraying of contact insecticides against tea mosquito bug, leaf webber and red tree ant and systemics like dimethoate, phosphamidon against mealy bugs, scales, thrips, leafininer and gall midge is to be practiced.
- Collection and destruction of dead wood after fruit harvest even though laborious, reduces infestation by lepidopteran fruit borer.

TOMATO *(Lycopersicon esculentum)*

Tomato is one of the three (Okra and Brinjal) commonly and popularly consumed vegetables grown throughout the India. This is being widely cultivated in cooler regions of Andhra Pradesh *viz.,* parts of Visakhapatnam, Chittoor etc. The state produces nearly 2.5 lakh tonnes of fruits annually. Apart from diseases, the major concern with respect to insect pests is Helicoverpa which act as fruit borer. The crop is attacked by other pests like *Spodoptera*, surpentine leaf miner, thrips, lady bird beetles etc.

Vegetative phase	Reproductive phase
Surpentine leaf miner, *Lyriomyza trifolii.* Thrips, *Frankliniella schulzii, Scirtotlirips dorsalis:* Lady bird beetles *Henosepilachna vigintioctopuctata*	*H. armigera: S. litura:* suspentine leaf miner, *L. trifolii*

Key pests

Helicoverpa armigera, surpentine leaf miner, *L. trifolii*

IPM strategies for key insect pests of tomato

- If infestation of *Liriomyza* is severe, spray of neem oil (0.5%) and triazophas 0.05% are useful.

For an effective management of *Helicoverpa*, an integrated package has been suggested by IIHR.

- Transplanting 2 rows of marigold for every 16 rows of tomato, as a trap crop as the females will be attracted to mary gold flowers and lay eggs. To facilitate synchronous flowering, planting 40 and 25 days old seedlings of marigold and tomato respectively at a time. Also monitoring the pest through pheromone traps and spray 5% NSKE to kill early stage larvae. Placement of 15-20 bird perches ('T' shaped) per hectare helps in inviting the insectivorous birds. Spray of NPV @ 500 LE/ha during evening hours. Jaggery (1kg), Teepol (100ml) are to be mixed with NPV. On need basis, spray of endosulfan 0.07% quinalphos 0.05% against larvae (ANGRAU, 2000f).
- The above practices like bird perches, NSKE and insecticidal application are also useful against *Spodoptera*. In addition, castor plants can be grown as trap crops for egg laying. Collection and destruction of egg masses and early larvae can be practiced. Pheromone traps and NPV of *Spodoptera* can also be utilized.

REFERENCES

ANGRAU, 2000a. Insect pests. In: *Rice - insect pests, diseases, nutritional disorders*, Acharya N.G.Ranga Agricultural University, Hyderabad, A.P.

ANGRAU, 2000b. Insect pests. In: *Groundnut - insect pests, diseases, nutritional disorders*, Acharya N.G.Ranga Agricultural University, Hvderabad, A.P.

ANGRAU, 2000c. Insect pests. In: *Cotton - insect pests, diseases, nutritional disorders*, Acharya N.G.Ranga Agricultural University, Hyderabad, A.P.

ANGRAU, 2000d. Insect pests. In: *Redgram, Bengalgram - insect pests, diseases, nutritional disorders*, Acharya N.G.Ranga Agricultural University, Hyderabad, A.P.

ANGRAU, 2000e. Insect pests. In: *Mango - insect pests, diseases, nutritional disorders*, Acharya N.G.Ranga Agricultural University, Hyderabad, A.P.

ANGRAU, 2000f. Insect pests. In: *Tomato - insect pests, diseases, nutritional disorders*, Acharya N.G.Ranga Agricultural University, Hyderabad, A.P.

ARS, APP, 2000. *Annual report*, Agricultural Research Station, Anantapur, A.P.

ARS, APP, 2001. *Annual report*, Agricultural Research Station, Anantapur, A.P.

ICRISAT, 1995. "Identification of insect pests in groundnut", *Information bulletin no.39*, ICRISAT, Patancheru, Hyderabad.

32

Alternatives to Methyl Bromide for Post-harvest Treatment of Fresh Agricultural Produce

R.D. GAUTAM[1]

Commonly known as PMB the pink mealy bug, *Maconellicoccus hirsutus* (Green) (Pseudococcidae: Homoptera) was rated as one of the most serious pests in the Caribbean since its first report from Grenada in 1994 (CARDI,1995-96). It has now spread through out the region and has attacked 290 plant species, mainly in Trinidad and Tobago, Grenada. St. Kitts and Nevis, Anguilla, British Virgin Islands, Montserrat, US Virgin Islands, Lsser Antilles and Guyana (Gautam *et al*, 2000). Several exporters involved in agri-horti sector especially within the region remained virtually with no trading while many traffickers' jobless or looking alternatives for their lively hood. The pest was subsequently contained with a great success after introduction of bio-intensive program using predatory ladybird beetles *Cryptolaemus montrouzieri* Mulsant and *Scymnus coccivora* (Ayyar) from India and the parasitic wasp, *Anagyrus kamali* Moursi from China in the region (Gautam,'Wayne and Maraj, 1996; Kairo and Peterkin, 1998). However, trade in agriculture which was banned due to the devastating *M. hirsutus* remained upset. It was estimated that about 30(exporters were involved in Trade before the PMB infestation in Grenada alone who are adversely affected due to trade restrictions. Crop losses resulting from reduced yields projected for 1995, 1996, 1997 in Grenada were US$ 1.77 million, US $ 1.80 million and US $1.83 million, respectively. However, potential losses are suggested to be over US $10 million annually. Similarly, Ministry of Agriculture Land and Marine Resources, Trinidad and Tobago estimated that potential loss to agriculture and forestry could exceed US $125 million annually (MALMR, 1996).

A Caribbean Regional meeting on the pink mealy bug and its impact on the Fresh Produce Trade was held in St. Lucia on 19-20 December, 1996 where in CARDI (Caribbean Agricultural Research and Development Institute) was identified to develop a Post-Harvest Treatment protocol including validation of methyl bromide fumigation (Anonymous, 1997).

METHYL BROMIDE DILEMMA

The widespread use of methyl bromide has always been considered a most effective weapon, providing quick and 100 per cent kill to a wide range of insect-pests. Further, methyl bromide is the preferred fumigant in plant quarantine work because of its excellent penetrative property, quick dispersal, general plant tolerance, high toxicity to insects and it's effectiveness over a wide range of temperature (FAO, 1984). By the way, methyl bromide is enlisted under **Montreal Protocol-1992** on substances that depletes the ozone layer and in the process of phasing out due to its serious ill effects on ozone layer depleting chemicals, contributing about 6 per cent as against rest of the enlisted chemicals (Taylor, 1996; Anonymous, 1998). Several countries including Caribbean are signatory to the agreement of Montreal Protocol. In order to safeguard and resume trade, CARICOM (Caribbean Communities) took a decision to investigate use of methyl bromide for

1. Plant Health Clinic, Division of Entomology, Indian Agricultural Research Institute, New Delhi-110 012.

quarantine treatment of fresh fruit and vegetables infested or suspected of being infested with a recently introduced pest of serious quarantine importance, *M. hirsutus*. Unfortunately, commodities treated with it were found to have phyto-toxicity except tubers probably due to presence of 2 per cent chloropicrin in it hence was rated for limited use. A point need to be noted that grace period of, up to 2005 for developing countries for it's specific quarantine use per capita is only 300g annually.

IMPORTANT OPTIONS FOR POST HARVEST TREATMENT

With a view to achieving quarantine security against insect pests, several methods namely, irradiation, hot-water treatment, coldwater treatment, pesticidal dip, fruit coating with carnuba wax emulsions, vapor heat treatment and fumigation have been advocated (Jeffery, 1994). Some of these options studied against PMB threatened fresh agricultural commodities in the Caribbean are as follows:

1. A Systems Approach
2. Fumigation
3. Hot Water Treatment
4. Pesticidal Dip

A systems approach

A systems approach include, pre-harvest crop management, post-harvest culling and final product inspection and certification (Moffitt, 1989). More confidently, certificate may be issued if the systems approach to quarantine security is followed as demonstrated by Hata *et al*;1992 for red ginger (*Alpinia purpurata)* infested with mealy bugs, aphids, ants, thrips and earwigs where in field infestation was reduced by spray of Chlorpyrifos at two weeks interval and pesticidal dip using fluvalinate in combination with insecticidal soap (potassium salts of fatty acids) with agitation for 5 minutes and culling of infested material in order to achieve zero level of pest, considered to be the ideal situation for certification.

Culling: The process of culling becomes an integral part of post-harvest treatment activities as it is experienced that in spite of precautions undertaken in harvesting the healthy produce along with certain fruits either undersized or infested are bound to come. Such fruits may be discarded for outside trades in order to maintain quality and pest free produce and thus reducing the pressure on pest risk during subsequent post-harvest handling. Notification or demarcation of Pest Free Areas (PFA) as a guideline in case of quarantine pest need to be followed, as suggested by FAO (1998) for harvesting.

Brushing: Certain commodities like avocado, dasheen, yam, eddoes, ginger, sweet potato, golden apple, tannia, plum, mango and citrus need to be thoroughly brushed (CARAPHIN NEWS, 1997) which eliminates 80 – 90 per cent of PMB population. Besides manual brushing, power hose may also be used wherever necessary to remove all traces of visible pest.

Washing: The washing of produce as mentioned above is recommended in combination with brushing. Washing water should contain soap detergent (5 ml in 10 gallons of water) plus bleach (10 ml of any commercial bleach containing sodium hypochlorite 5.2%) traded as clorox in order to kill microbes on surface of the fruits, change the water solution frequently. A good indication is when water becomes cloudy or with high level of floating of PMB.

Pilgrim (1997) and Gautam *et al.* (2000) elaborated reception area soaking produce (washing site), sorting and grading, treatment site, port of exit and entry being the critical inspection points for PMB. He successfully demonstrated the activity levels in the light of systems approach to achieve quarantine security from pink mealy bug targeted for revival of inter-regional trade. However, they cautioned that the systems approach will vary for different commodities and its effectiveness in reducing the infestation at each component level *viz* field cleaning site (90 % PMB mortality), washing with soap and bleach (95 - 100 % PMB mortality) and treatment with methyl bromide or hot water (100 % PMB mortality) is highly dependent on the vigilance of management.

Fumigation

Treatment with a chemical agent that reaches the commodity wholly or partially in a gaseous state. The prominent fumigants in use are aluminium phosphide, magnesium phosphide and methyl bromide with restrictions world over.

Fumigation with methyl bromide

Pink mealy bug and fresh agricultural produce trade

Gautam and Pilgrim (1998) revealed that methyl bromide @ l.5 and 0.75 lb per l000 cu ft (containing 2% chloropicrin) with 2 and 4 hr exposure period, respectively were most effective for obtaining l00 per cent kill of *M. hirsutus,* irrespective of stages. Results of residue analysis of fresh produce *i.e.*, de husked dry coconut, sweet potato, dasheen, tannia, yam, eddoe and ginger indicated that treatment with methyl bromide (l.5 lb/l000 cu ft, for 2 hr) was within the acceptable limit. The bromide residue was below the maximum residue limit and overall quality of the produce not affected. Thus, sapodilla, tubers and white potato could safely be fumigated. However, avocado, orange, plum, carambola, grapefruit, pineapple, breadfruit, sour sop, banana, golden apple, sugar apple, mango, plantain, bluggoe were not suitable for methyl bromide fumigation due to phyto-toxicity and quality deteriorations, although pesticide residue was with in the maximum residue limit. Similarly, shelf life and quality of *Heliconia* cut flowers was affected due to fumigation with methyl bromide.

Fumigation with Aluminium Phosphide

Gall midge complex and hot pepper trade

The gall midge complex, representing *Contarinia maculipennis* Felt and *Prodiplosis longifila* Gagne (Cecidomyiidae:Diptera) have become pests of great concern to hot pepper growers-exporters in Jamaica. The fly immatures can be seen between the fruit and calyx as well as in the pedicel either single or in groups of several congregating maggots. During the inspections of shipments in Miami, a substantial number of boxes containing hot pepper were rejected persistently by USDA quarantine staff due to presence of aforesaid pests. In the mean time, USDA/APHIS had proposed to provide fumigation facility of traded hot pepper on exorbitant cost basis, using methyl bromide at the rate of 2.5 lbs/1000 cu ft volume with possible risk of phyto-toxicity affecting consumer's acceptability. This posed a threat to hot pepper trading partners of Jamaica and USA. Besides, expected decline in production due to high level of fruit (West Indian Red) infestation ranging between 13.26 to 62.18 per cent is a perpetuating problem. The number of maggots per fruit was recorded up to 108, below or around the calyx. The pest is widely distributed in Jamaica on both West Indian Red and Scotch Bonnet Peppers of export market.

In response to this potential crisis, a national task force was formed with representatives from CARDI, Ministry of Agriculture Jamaica, agri-business council, Jamaica Export Association (JEA), APHIS and Rural Agricultural Development Authority (RADA). CARDI was mandated to develop IPM strategies including pre and post-harvest treatment protocols (Gautam, 1998) which comprised hot pepper harvesting and grading, selection of site, checking the barrels for air tightness, calculation of barrel volume and dosage and placement of fumigant. The standard dose of aluminium phosphide @ 1g/ M^3 was used as a base to calculate fumigant dose. The volume of barrel was calculated based on formula 22/7 x radius^2 x height. The cardboard boxes filled with 5 lbs of hot pepper (West Indian Red) were placed inside the barrels and fumigated with a pellet (0.6g) in individual barrels for 2 days, 3 days and 5 days.

Studies indicated that a dose of phostoxin pellet (0.2g gas) per barrel (0.22-0.26 M^3 volume) with 3 days exposure at room temperature was found to kill 100 per cent maggots and other stages. The fumigation of hot pepper (2 and 3 days) infested with the gall midges revealed that there were no adverse effect on shelf life of the produce as well as on phyto-toxicity except on fruit pedicels and thus almost comparable with control on the 5th day. Hence, fumigation was recommended during shipment.

Effect of ph Phostoxin (PH3) emission from the barrel at one M distance was below the detectable level (0.05 ppm). The fumigant also killed larvae of *Helicoverpa* and *Spodoptera* spp. found along with fumigated hot pepper fruits.

Fumigation with Magnesium Phosphide

Use of magnesium phosphide @ 5g per barrel (100 x 58 cm) with the exposure period of 10-24 hr was found effective in providing 100 per cent mortality to all the stages of *M. hirsutus* besides a dozen of other associated insect species with the fumigated commodities under prevailing conditions of 26-31ºC in Grenada Interestingly, no phyto-toxicity after fumigation with magnesium phosphide was observed on 100 fresh agricultural produce except ginger lily flowers. Shelf –life of these produce (cut flowers, fresh fruits, vegetables, salad and seasonings, planting materials etc.) and other quality parameters like taste, colouration and flavour of the fruits (Table-1) were also not affected (Gautam, 2003).

Table 1. Effect of magnesium phosphide on quality attributes of fresh fruits.

Freshly harvested Produce	Quality Parameters				Remarks
	Rind colour	Flesh colour	Smell	Taste	
Mango (Julie)	*	*	*	*	Plums about to mature got matured and suffered with dampness inside the barrel.
Mango (Graham)	*	*	*	*	
Banana	*	*	*	*	
Bluggoe	*	*	*	*	
Pineapple (Montserrat)	*	*	*	*	
Avocado	*	*	*	*	
Sapodilla	*	*	*	*	
Papaya	*	*	*	*	
Goldenapple	*	*	*	*	
Soursop	*	*	*	*	
Red plum	*	*	*	*	
Breadfruit	*	*	*	*	
Plantain	*	*	*	*	
Sweet orange	*	*	*	*	
Green coconut	*	*	*	*	
Pomegranate	*	*	*	*	
Sugar apple **	*	*	*	*	
Dry coconut	*	*	*	*	
Dunks	*	*	*	*	
Mature Tamarind	*	*	*	*	
Fig (*Ficus*)	*	*	*	*	

* Normal condition of fumigated fruits with regards to coloration, taste and smell
** Slight phyto-toxity noticed on rind and not hardening as in methyl bromide

Hot Water Treatment

The treatment of banana and plantain at 45°C for 5 minutes (Parker, 1997) while dipping crates containing produce *viz*.; banana, bluggoe, carambola, dasheen, eddoes, ginger, golden apple, mango, plantain, sapodilla, soursop, sugar apple, sweet potato, tannia, white (Irish) potato and yams at 48 - 50°C for 5 - 10 minutes provided 100 percent mortality to all the stages of PMB. The studies were conducted using hot-water pilot plant developed by CARDI and installed at Marketing Board, Grenada. However, for soursop 15 minutes dipping is suggested. (Pilgrim,1997). Place fruits on rack and allow to drop out water.

Pesticidal Dip

It is one of the most common post-harvest disinfestations treatments for tropical cut flowers. The insecticides effective on specific pest like PMB in the field should also be effective as dip. The fluvalinate is one of the synthetic pyrethroids registered in the US as a dip for flower and foliage cuttings. It is effective against white flies (*Trialeurodes vaporariorum*) two spotted spider mites (*Tetranychus urticae*), egg plant mealy bug (*Phenacoccus solani*), green peach aphids (*Myzus persicae*) providing at least 70 per cent kill of these pest species dipped for one minute and is non-phyto-toxic against most of the tropical foliage plants (Osborne, 1986), Fluvalinate in combination with insecticidal soap (potassium salts of fatty acids) for five minutes gives 95 – 100 per cent kill against mealy bugs in red ginger (Tenbrink *et al;* 1991). Hata *et al;*1992 showed that a post harvest insecticidal dip of fluvalinate and insecticidal soap is 100 per cent effective against various pests of red ginger as long as field populations are reduced with pre-harvest insecticide application. Synthetic pyrethroids also have repellent activity hence act as protectant. Some of the chemicals (Table 2) reported to be effective against P.M.B.may be validated for pesticidal dip of cut flowers and other planting materials.

Table-2: List of common Insecticides for pesticidal dip of cut flowers and cuttings infested with the pink mealy bug

Name of Insecticide	Insecticide Group	Mode of Action	Dose ml/litre
Chlorpyrifos 20 EC	organophosphate	contact, stomach poison	1.25 – 1.50
Salute – a combination of Dursban and Perfekthion	organophosphate	contact, stomach, systemic	2.00 –5.00
Actellic 50 EC (primiphos methyl)	organophosphate	contact, some fumigant action	2.00 – 3.00
Fenitrothion 50 EC		stomach, contact	2.00 – 3.00
Fluvalinate	synthetic pyrethroid	contact, stomach	2 .00 – 3.00
Cypermethrin 25 EC	synthetic pyrethroids	contact, stomach	2.50 – 3.00
Dichlorvos |76 WSC	organophosphate	contact, stomach, fumigant	2.00 - 2.50

Note:Phyto-toxicity to be validated for each flower and material

The dipping has following disadvantages:

1. Flowers need absolute drying after dipping to avoid fungal problems.
2. If insects are inside flower's tight folds they escape from insecticidal contact.
3. It is labour intensive.
4. Phyto-toxic to certain flowers.
5. Efficacy of insecticide depends on susceptibility of the insect species.
6. Insecticides used in field for same insect cannot be used due to risk of developing resistance.

The pink mealy bug and cut flowers trade

The cut flower export from Trinidad to Barbados was affected due to PMB in1995-96 years to the tune of 3 million dollar. Studies on post harvest treatment with Pesticidal dip of cut flowers namely, Anthurium, orchids, ginger lily and Heliconia indicated that five minutes dipping of heliconia, gingerlily and anthurium with salute (5ml in a litre water) did not adversely affect quality and shelf-life of these flowers and killed 100 per cent PMB while actellic and decis gave phyto-toxicity (Gautam and Cooper, 2003) .The insecticidal dip against high valued commercial cut flowers in Trinidad and Tobago were effective for providing the pest mortality indispensable for quarantine security, leading to the considerations for the resumption of agricultural trade.

MAGNESIUM PHOSPHIDE: A BETTER OPTION

Magnesium phosphide (Mp) when used reacts with OH radicals in the atmosphere and releases hydrogen phosphide commonly known as phosphine (PH_3), which is extremely poisonous gas for the insects. Benzing (1994) stated that decomposition of magnesium phosphide is nearly 100 per cent and more rapid than aluminum phosphide and almost no toxic residues are left in the fumigated food- stuffs. Half-life value of phosphine in the atmosphere is about 28 hrs. The tolerance limit for phosphine residues in fumigated produce recommended by FAO/WHO for cereals, vegetables, fruits etc varies from 0.01 - 0.1 ppm. Phosphine gets diluted at a distance of 10 - 40 meters from the emission source. Effect of magnesium phosphide on the mortality of *M. hirsutus*, and other insects associated with fumigated produce was 100 per cent while treated produce safe and marketable (Gautam *et al*, 2000; Gautam, 2003). Moreover, being a solid fumigant it is easier to handle as compared to liquid methyl bromide, which is almost phased out and also under restricted use like ethylene di bromide.

Thus, recommendation of magnesium phosphide as an effective fumigant mainly for post harvest treatment of fresh agricultural produce was considered for commercialization and quarantine security in the Caribbean in lieu of methyl bromide's phasing out status and phyto-toxic effect. Introduction of such a chemical in our country may be useful in solving certain pest problems especially for post harvest treatments in the light of methyl bromide's disappearance in near future.

FUTURE OUT LOOK

The stringent requirements for the registration of new pesticides makes it unlikely that many new fumigants are likely to be commercialized in the near future. Introduction of magnesium phosphide or other chemicals, which are being used for quarantine treatment in countries like Australia, US or by other trading partners for safeguarding the trade in our country may be useful in solving several pest problems especially for post harvest treatments in the light of methyl bromide's disappearance in the forthcoming period and liberalization on Global Trade.

Development of Pesticidal Dip Protocols for cut flowers and planting materials of export potential besides, treatment of cartons and other packaging material are to be emphasised. In general, there is a need to have more emphasis on researches in these directions as well as policy decisions at Government level in the light of foreign trade in agriculture. We may have to rely less on the true fumigants and more on alternatives.

CONCLUSION

Studies revealed that use of magnesium phosphide @ 5g per barrel (100 × 58 cm) with the exposure period of 10-24 hrs at 26-31°C and 60 ± 10 per cent R H was found providing 100 per cent mortality to all the stages of *M. hirsutus* and a dozen of other insect species associated with the experimental produce. Interestingly, no phyto-toxicity after fumigation with magnesium phosphide was observed on several fresh agricultural produce except ginger lily flowers. Shelf-life of these produce (cut flowers, fresh fruits, vegetables, salad and seasonings, planting materials etc.) and other quality parameters like taste, colouration and flavour of the fruits were also not affected. Fumigation of fresh produce with methyl bromide @ 1.5 and 0.75 lb per 1000 cu ft with 2 and 4 hrs exposure periods, respectively, no doubt provided 100 per cent kill to various stages of *M. hirsutus,* also resulted in severe phyto-toxicity except to tuber crops. However, results of residue analysis of de husked dry coconut, sweet potato, dasheen, tannia, yam, eddoes and ginger indicated that it was below the maximum residue limit and the overall quality of the produce was not affected. A dose of 0.2g gas (0.6g pellet) per barrel of aluminium phosphide was observed to kill 100 per cent maggots of hot pepper gall midge complex (*C. maculipennis* and *P. longifila*) in Jamaica while no phyto-toxicity and adverse effect on shelf-life of fumigated peppers. In other study, five minutes dipping of *Heliconia*, ginger lily and *Anthurium* with salute (a combination of Dursban and Perfekthion) @ 5ml/litre water did not adversely affect quality and shelf-life of these cut flowers and killed 100 per cent *M. hirsutus* while actellic and decis gave phyto-toxicity. Thus, aforesaid options and magnesium phosphide in particular were recommended for post harvest treatment of fresh agricultural produce as other methods were observed to provide limited relief for quarantine security. Introduction of such chemicals in our country may be useful in solving several pest problems especially for post harvest treatments and quarantine in the light of methyl bromide's disappearance in near future.

ACKNOWLEDGEMENTS

I would like to place on record my great debt of gratitude to Dr. Compton Paul, Director Technical Program (Acting.), Caribbean Agricultural Research and Development Institute, St. Augustine, Trinidad; Dr. G. V. Pollard, FAO Regional Plant Protection Officer and Secretary, Caribbean Plant Protection Commission, Barbados; CARDI Team Leaders of Jamaica and Grenada as well as concerned Governments and Ms. Cynthra Persad, Coordinator HMB National Program, Trinidad and Tobago for their interactions and constant encouragement. Fumigation assistance rendered by Mr Whiby Leroy and Ms Antoinette James, authorized fumigation operators (Field Officer) for Food Storage and Infestation, Ministry of Industry, Investment and Commerce, Jamaica is gratefully acknowledged. My special thanks to the Govt. of Grenada and Grenada Cooperative Nutmeg Association for encouraging the study and complementing the fumigant, Magnesium phosphide for post-harvest research with regards to resumption of agricultural trade, respectively. The facilities provided by Mr. Reginald Andall, CTL, Grenada and technical assistance rendered by Mr.Ronald Reuben are duly acknowledged. Finally, I express my sincere gratitude to Mr. Hayden Blades, then Executive Director, CARDI for his encouragement and my placement to head the program and making this task possible.

REFERENCES

Anonymous, 1997. Regional Meeting on pink mealy bug. *CARAPHIN News* 15 March,1997, pp.5-6.

Anonymous, 1998. Methyl bromide phase out accelerated by Protocol, *Fresh Produce Exporter* (Ad.. Andre Satney), March, 1998, Roseau, Dominica, p. 4.

Benzing, L. 1994. Advantages of Products based on Magnesium phosphide for the fumigation of food stuffs. *Tecnica molitoria*, 45:2, 139 - 143.

CARDI 1995 - 96 - Managing the pink Mealy bug - An example of Regional Collaboration in Agriculture R & D - CARDI Annual Report, pp 23 - 27.

CARAPHIN News, 1997. Regional Meeting on Pink Mealy bug, March 15, 1997, pp.5-6

FAO, 1984. Food and Agriculture Organisation of the United Nations. International plant quarantine Treatment Manual. FAO Plant Production and Protection Paper, 55, p. 220.

FAO, 1998. Update on New Pest Introductions. Circular letter No. 2/98,June 1, 1998 from Dr. G. V. Pollard, Regional Plant Protection Officer and Technical Secretary CPPC, FAO, Barbados, p.1.

Gautam R D, 1998. Mission Report to Jamaica for developing Post Harvest Treatment Protocol against hot pepper infested with gall midge complex.CARDI, Grenada, September 8-14, 1998, CARDI Annual Technical Report, Grenada Unit, West Indies, 4 p. (unpublished).

Gautam, R.D. 2003. Fumigation of fresh agricultural produce with magnesium phosphide for quarantine security. *Indian J Ent.*.65(2): 193-201.

Gautam, R.D. and Cooper Bertrand, 2003. Insecticidal dip of some tropical cut flowers for quarantine security against pink hibiscus mealy bug, *Maconellicoccus hirsutus*. *Indian J. Ent.*, 65(2):259-263.

Gautam R.D., De Chi Wayne and Maraj Champa. 1996.A successful Journey of ladybirds from India to Trinidad. *IICA Caraphin News,* June 14, 1996:1-2.

Gautam R.D. and Pilgrim, R.N. 1998. Effect of methyl bromide on mortality of the pink mealy bug, *Maconellicoccus hirsutus* (Green) and quality of the fresh agri-tropical produce. CARDI Annual Technical Report, Grenada Unit, West Indies.

Gautam, R. D., Pilgrim, R. N. and V. Stewart, 2000. *The pink Mealy bug:Protocols for the protection of Agricultural Production and Trade*:The Systems Approach (SA) The Production Manual .CARDI , St Augustine, Trinidad and Tobago, p. 53.

Hata, T.Y; Hara,A.H; Jang,E.B; Imaino,L.S; Hu,B.K.S. and Tenbrink, V.L.1992. Pest management before harvest and insecticidal dip after harvest as a systems approach to quarantine security for red ginger. *J. of E. Ent.*,85:2310-2316.

Jeffery, K.B.l994. Vegetable. In *Insect Pests and fresh Horticultural Products* (ed. Robert, E. Paul and John W. Armstrong). CAB International, Wallingford, UK. 360 p.

Kairo, M.T.K. and Peterkin, D. D. 1998. Dossier on *Anagyrus kamali* Moursi an exotic natural enemy for Biological Control of the Hibiscus Mealy bug in the Caribbean (Revised). International Institute of Biological Control, Curepe, Trinidad and Tobago. p. 24.

MALMR,1996. Press Clippings on HMB work pioneered by Dr. R. D. Gautam, Member-Consultant. Compiled by the Ministry of Agriculture, Land and Marine Resources, Research Division/Library Services Unit, Central Experimental Station, Centeno, P. O. Arima, Trinidad and Tobago, November 1996, 70 p.

Moffitt, H. R. 1989. A Systems Approach to Meéting Quarantine Requirements for Insects Pests of Deciduous Fruit. *Proceedings of the Washington State Horticultural Association* 85, 223-225.

Osborne, L.S. 1986. Dip Treatment of Tropical Ornamental Foliage Cuttings in Fluvalinate to Prevent Spread of Insect and Mite Infestations. *J. of Eco. Ent.* 79, 465-470.

Parker, L 1997. *CARDI Annual Technical Report*, Grenada Unit pp. 18-23.

Pilgrim, R. N 1997. A Systems Approach to Achieve Quarantine Security of the Pink Mealy bug for the Revival of Regional Trade. *CARDI Annual Technical Report*, p. 33 (unpublished).

Pollard, G.V. l997. Summary paper on methyl bromide and the Montreal Protocol with reference to Caribbean Countries (unpublished). Circulated in a PMC meet held in Holiday Inn, Trinidad on 2l-22 August, p. 6.

Taylor, B. l996. Trade and environment, protecting the environment from dangerous chemicals. Methyl bromide-The dilemma for developing countries. International trade Centre UNCTAD/WTO, Export Quality UK No. 5O, pp. l-4.

Tenbrink, V.L. Hansen, J. D. and Hara, A.H. 1991 Post-harvest control of mealy bugs using Dips, 1990. *Insecticide Acaricide Tests* 16, p 258.

Wang,C.L.and Lin.R.T.1984.Study on the quarantine treatments of insect-pests on Chrysanthemum cut flowers-fumigation and smoking methods. *J. Agri. Res.*, China 34, 88-93.

IPR Issues in Indian Agriculture

A.K. GANGULY[1], VIKAS KANWAR[2] AND D. PRASAD[1]

Recognition of agriculture as profit making institution, enterprise of investment and to realize the potential for knowledge as a creator of wealth and subsequently inclusion in the discussion of GATT in Uruguay round conference (1986-94) lead to the establishment of WTO in January 1995. The important function of WTO is to ensure that trade flows smoothly and freely. Decisions in the WTO are normally taken by consensus among all member countries and they are ratified by respective parliaments or senates of the member countries. There are set of rules for trade in WTO principle which are being updated every year at various meetings. These meetings have created new rules for dealing with trade in services, IPR, dispute settlement and trade policy reviews. The complete set of rules runs into some 30,000 pages consists of 60 agreements and separate commitments (called schedule) made by individual members in specific areas such as reducing custom duties and services, market openings (Roychoudhry 2003). Through these agreements each country receives guarantees that its exports will be treated fairly and consistently in the market of other countries. About half a dozen agreements are on agriculture which rest on three pillars like market access, domestic support and export subsidies and others like Application of Sanitary and phytosanitary measures (SPS) Trade related Intellectual property rights (TRIPS), Agreements on Agriculture, Technical barriers to trade, Antidumping safeguard subsidiaries, etc. The rules governing agriculture trade are strengthened which will lead to improved predictability and stability for importing and exporting countries alike In this context India will have to pay urgent attention to substantially enhance our levels of innovation and creativity since agriculture is important in India, not only in terms of business or commerce but agriculture for us is a matter of livelihood and food security. Nearly 650 million people (65% of total population) in this country are still dependant on agriculture. There is need to understand the provisions of internationally accepted legal instruments of TRIPS and others. Generation of intellectual property, its capture, documentation, protection, evaluation and its exploitation assumes a crucial importance for us.. Indeed, there will have to be a sea change in our ability to manage our intellectual property be it patents, copyrights, designs, and so on. The greatest challenge will be posed by patents in the sphere of agriculture research and development.

The Intellectual property is loosely defined as product of mind. It is similar to other properties wherein the owner or the proprietor may use this property at his wish (Erbisch and Carlos 1998). Nobody can lawfully use it without the permission of the owner. The WIPO (World Intellectual Property Organisation) a specialised body of United Nations established in 1967 provided that Intellectual Property Right (IPR) shall include the "rights relating to (1) Artistic, literary and Scientific work, (2) Performance of performing artist, phonogram, broadcast, (3) Inventions in all fields of human endeavour, (4) Scientific discoveries, (5) Industrial design, (6), Trade marks, Service marks and commercial names and designation, (7) Protection

Planning, Publication and Information Unit, IARI[1] NCIPM[2] and Division of Nematology[1], Industrial Agriculture Research Institute, New Delhi 110 012.

against unfair competition and all other rights resulting from IP activity in the industrial, scientific literary or artistic field".

Intellectual property rights provide an essential incentive for private investment in agricultural applied research and development, which can take a number of years and invoice a large sum of resources. Thus the success of our agricultural development in the current scenario depends on the institutions which will facilitate protection of IPR as well as strengthen the capacity for dealing with the adverse effect of globalisation negotiation in international forum such as WTO. The effects of trade liberalisation in agriculture could be positive in promising agricultural development, it can also have an adverse short run effects on specific groups of small and marginal farmers. Policies and institutions are needed for protecting these vulnerable groups. For example, social safety nets and investments in education and skills that equip them to diversify their income-generating activities. There is also a need to put into place an institutional framework that manages natural resources that have implications for agricultural development.

An understanding of implication and application of these agreements particularly TRIPS is an important in every stage of planning, research project formulation, upscaling and commercialisaiton of agricultural technology. In future when TRIPS will be in use in different areas of agriculture we will face problems, until efforts are made for awareness, IPR literacy, higher education capacity building process. The Trade Related Aspects of Intellectual Property Rights of World Trade Agreement has provisions on patents, plant varieties protection, trade secrets, trade marks and geographical indication which have relevance to agricultural research and development. TRIPs obliges signatory member countries for patenting of microorganisms and microbiological processes for production of plants and animals, though it allows the exclusion from patents of plants and animals and essentially biological processes for their production. It also obliges members to provide protection for plant varieties either through patents or through an effective sui generis law or any combination of these two. The Convention on Biological Diversity (CBD) allows sovereign rights on a nation's genetic resources; its stated objectives are: the conservation of biological diversity, the sustainable use of its components, and the fair and equitable sharing of benefits arising out of the utilization of genetic resources. The provision needs to be consistent, adequate and effective in Convention on Biological Diversity on compulsory access to genetic resources and transfer of technologies. The latter obviously means that any transfer of technology would not be on terms other than those set commercially in the market. Naturally, there is requirement of internationally accepted legal instruments to implement these provisions (Rai 2002).

Plant Breeders' Rights

The Plant Breeders' Rights (PBR) on a variety is given when the variety is new, distinct (D), uniform (U) and stable (S). In general, the purpose of a PBR certificate is entitled to prohibit others from selling or offering the variety for sale, importing or exporting the variety, sexually multiplying the variety for marketing, and using it for producing another variety. The scope of prohibition under PBRs to harvested products, leading to think among breeders to extend PBRs on crop varieties in countries where that particular crop is not grown. That is thought with a view to take action against imports derived from unlicensed propagation of the variety in countries where plant variety protection (or PBRs) is not available. There are, however, two exceptions : (i) researchers' exemption- allowing use of any protected variety for conducting experiments or research to develop (as opposed to produce) a variety that is distinct from the protected variety; (ii) farmers' exemption - to save, sow and re-sow seed on farmers' own holdings. The concept of essentially derived varieties (Edv) arrived to discourage plant breeding which involves only cosmetic changes in a protected variety. For this purpose there is need to take authorization of the original breeder for marketing the essentially derived varieties right. However, PBRs are considered to provide a minimum form of protection adapted to the needs of the traditional breeder as well as the farmers as it can allow scope for exemption for both breeders/researchers and farmers.

To encourage crop improvement programme researches, seed availability of varieties protected through PBRs as well as easy procedures of authorization for developing EDVs are thus necessary. The provision in the Indian legislation has taken care of the special situation of farmers in the country and their immense contributions, and allowing the flexibility to Indian farmers to even sell unbranded seed, as their right rather than privilege. The research exemption clause is obviously also kept to facilitate continuity in required crop improvement researches. The clause on 'extant' varieties and provisions for farmers' rights and benefit sharing are unique to Indian legislation. Indian legislation has also cared to guard the interests of breeders/breeding institutions by reducing financial harm to them to any substantial extent by providing the essentially derived varieties clause (Rai 2002). The clauses on compulsory license further safeguard the interest of farmers to ensure uninterrupted availability of quality seed. The protection of plant varieties and farmer's right Act 2001 provide for a liability claim which seeks to protect farmers from being deceived by producers/sellers of the seed. The act states that other conditions being stable if the seed sold to the farmer fails to realize its expected potential - in yield, quality pest resistance, etc - as had been disclosed by the breeder, the farmers will retain the right to claim compensation from the breeder of the variety.

Copyright

Copy right is awarded for original literary, dramatic, musical artistic works, cinematographic films and records. According to the Copyright Act 1999 agreement, the term of protection, available to the performers, shall last least until the end of a period of fifty years computed from the end of the calendar year in which the performance takes place (website : indialawinfor.com). The new amended Act also provides for power to the Government to extend the provisions of the Copyright Act to broadcasts and performances made in other countries, provided these countries extend similar protection to broadcasts and performances made in India. This has ushered in comprehensive changes and brought the copyright law in line with the developments in satellite broadcasting, computer software and digital technology. Several measures have been adopted to strengthen and streamline the enforcement of copyrights like the setting up of a Enforcement Advisory Council, training programs and setting up special policy cells to deal with cases relating to infringement of copyrights.

It will be worth mentioning that the rich heritage of our country can be protected in this form by documenting the original art like Kalamkari art, Tanchoi, Madhubani painting identifying creators at grassroots through extension network. Traditional motifs like Kalamkari art, Tanchoi, Madhubani paintings, are examples. In the age of Information Technology, this form of IPR is also relevant to agricultural researchers involved in development of software, web site development and all forms of electronic publishing. Institutions need to be stringent on use of legal versions and respect the rights of the owners as per the amended Act of Copyright 1999 (Sastry and Rao 2003).

Design

Design means features of shape, configuration, pattern, ornament or composition of lines and colours applied to article in two or three-dimensional form by any industrial process. The process can be manual, chemical, mechanical or combination of all. The grant is for 10 years initially and non-renewal leads to it becoming public property. The Indian Design Act, 2000 with its subsidiary legislation is in force in compliance with TRIPs provisions (IDA, 2000) The designs of textiles patterns, ornaments, embossments that are new, original, and industrially applicable. The traditional heritage in textiles, folklore, toys, village artifacts are all areas to which protection can be extended under this Act. Researchers can identify based on indigenous knowledge of communities of weaker section of people for grant of IP protection in this form.

Patent

Patent is a government granted and secured legal right to prevent others from practicing i.e making, using, hiring or selling the inventions covered by the patent. It is an exclusive right granted by a country to the owner of the invention provided the invention satisfies certain conditions stipulated in the law. This right is available only for limited period of time. The use or exploration of a patent may be affected by other laws of the country which has awarded the patent .These laws may relate to health, safety food security further existing patents in similar area may also come in the way. To be patentable, an invention must be new, industrially applicable, non-obvious, and the subject of complete disclosure, the last item requires that the invention be described so that a person skilled in the art to which the invention relates can reproduce it. This also enables researches to develop further useful products or service, The awareness of patenting will not only encourage the scientists to avoid duplication of routine type of work but also to do meaningful research. The term of protection for a patent in India is 20 years from the date of patent (i.c the date of filing the complete specifications) for all inventions. Indian patent law permits only process patent but drugs and medicines ,agrochemicals could go for product patent after the year 2005. Currently all the Applications for product patent are kept in mail box which will be open only after 2005. There is no exemption to farmers, as the rationable to consider patents on plant materials is to reduce the perceived discrimination between plant developers and industrial inventors. Even the breeders' privilege is theoretically to the extent of exemption for experimental use, permitting only purely academic or non-commercial research. These exemptions are narrow or less cleâr than its Plant Breeders' Rights analogue. With respect to plant breeding per se, patents on varieties or inbred parental lines and on genes particularly emerge as having substantial significance. The patent protection can obviously have greater effective force for the inventor than a PBR certificate, as it does not permit farmers' and breeders' privilege in the manner desired (Sastry and Rao 2003).

The gene-rich countries expressed their great concern to utility patents on plants granted in countries like the USA. Utility patents are granted to products consisting of plant material as well as to inventions of processes relating to or using of plant material, and entitles the patentee to exclude others from making, using or selling the invention. The scope of such protection on a plant material is broader, and covers a broadly described category of plants or parts of plants, of plant material, and one or more plant varieties. Such broad-based patents severely restrict the receipt and use of potential germplasm. The effort made by the Government of India to fight judicial cases of patents granted on neem, turmeric and basmati rice in the USA are well known facts to the Indian citizen. These cases of infringement on our traditional indigenous knowledge with little novelty and non-obviousness in them made researchers /NGO's aware at large about patent aspect.

Patents are probably the most important IPR today for agricultural goods and services as they provide the strongest protection on patentable plants and animals and biotechnological processes for their production. Patents universally give the patentee the right to prevent third parties for making, using or selling the patented product or process. The term of a patent is not the same for all inventions. For example process patents relating to drugs and food are allowed five years from the date of sealing of patent, and even years from the date of patent whichever is the shorter. For all other patents the term is fourteen years from the date of the patent. This is subjected to the payment of a renewal. The recent amendment on patent Act in 2002 (IPA, 2002) now provides a uniform term of 20 years to all to all inventions.

Some of the areas cannot be patented such as: An invention (1) contrary to well established natural laws; (2) either primary or intended use or commercial exploitation leads to public order or morality or to serious prejudice to human, animal or plant life or health or to the environment; (3) discovery of a scientific principle or the formulation of an abstract theory or discovery of any living thing or non-living substance occurring in nature; (4) The discovery of any new property or new use for a known substance or of the mere use of a known process, machine or apparatus unless such known process results in a new product or employs at least one new reactant; (5) A substance obtained by a mere admixture resulting only in the aggregation of the properties of the components thereof or a process of producing such substance; (6) The mere arrangement or re-arrangement or duplication of known device each functioning independently of one another in a known way; (7) A method of agriculture and horticulture; (8) Any process for a medicinal, surgical, curative, prophylactic, diagnostic, therapeutic or other treatment of human beings or (9) process for a similar treatment of animals to render them free of disease; (10) to increase their economic value or that of their products; (11) Plants and animals in whole or any part thereof other than microorganisms but including seeds, varieties and species and essentiality biological process for production of propagation of plants and animals; (modified by human interaction e.g. transgenic microorganism); (12) A mathematical or business method or a computer program per se or algorithms; (13) A literary, dramatic, musical or artistic work or any other aesthetic creation whatsoever including cinematographic works and televisions productions; (14) A mere scheme or rule or method of performing mental act or method of playing game; (15) A presentation of information; (16) Topography of integrated circuits; (17) An invention which, in effect, is traditional knowledge of which is an aggregation or duplication of known properties or traditionally known component or components (Sastry and Rao 2003).

Trade Secrets

Trade secret provides protection to persons/Institutions on information which is lawfully under their control from being disclosed to acquired to or restrict others to use without their consultation. When submitting undisclosed test or other data for the marketing of new pharmaceutical or of agrochemicals, it is expected to protect such information against unfair commercial use. There is no obligation to disclose the invention or creative ideas to the society.

In agricultural sector, Trade Secret Protection Centre need to protect hybrid/plant varieties. Trade secrets have enabled private plant breeding companies to maintain the exclusive use of certain genetic materials and information relating to such materials. This has resulted in certain varieties of plants, particularly F1 hybrids, only being available from a particular supplier. Component lines of synthetics and composites, and vegetatively propagated varieties based on a limited number of parent clones, provide other examples, where the exclusive possession of elite source materials by a breeder, together with information related to their combination, ensures that only the breeder can offer authentic seed of a particular variety in the market place. The time frame of protection under trade secrets is not limited. The moment it is being discovered by somebody, it looses its protection under trade secret.

In agricultural research, a material transfer agreement for exchange of plant/animal material become evident in the light of multi porty research programmes involving testing at multilocations and manpower. MTA's are agreement between collector and appropriate authorities provides confidence for the ownership or acquisition without affecting research for development.

Trademarks

Trademark is an invented word, a visual symbol, a word. a name, a picture , device, a label, or a colour scheme applied on goods or services to indicate the consumer that they belong to a particular manufacturer or provider of services, to distinguish them form similar goods manufactured or services provided by others.

Trademark helps to identify the product and its origin, guarantees its unchanged quality. It also confers on the proprietor a kind of monopoly right over the use of the mark, essential to protect it and the goodwill attached to it, prevents the use of fraudulent marks on merchandise. Such distinguishing marks constitute protectable subject matter under the provisions of the TRIPS Agreement. The Trade and Merchandise Marks Act, 1958 was made a Bill to repeal, the 1958 Act passed by Parliament and notified in the Gazette in 1999. This Act not only makes Trade Marks Law, compliant with TRIPS but also harmonizes it with international systems and practices. The Agreement provides that initial registration and each renewal of registration shall be for a term of not less than 7 years and the registration shall be renewable indefinitely.

The inclusion of service marks provision for registration for collective marks in TM Act 1999 for first time in India. (TM Act, 1999). The new act would help the service industry comprising information technology, banks, investment banks, chartered accountants, consultants, hotels, advertising agencies, airlines, hospitals and like to register their respective trade marks and service marks. The salient features of the amended act the jurisdiction of the high courts for appeals has been done away with and such things would now be entertained by an Intellectual Property Appellate Board. Marks used in commerce can be applied to both agricultural and industrial products and services. For instance, trademarks are used to market seeds or spraying services and need to be built for an advantage. The trademark may not be directly relevant as on date to research activities but it will become important with value addition of agricultural commodities and to make the rural sector more competitive in their product and services. Indigenous communities can sensitise to build for protection of local signs for local goods.

Items likely to cause confusion, marks, obscene, religious sentiments, and of society official seals, identical marks are not considered for registration under this protection.

Geographical Indications

Geographical indications identify the geographic origin of a product (Chaturvedi 2002). These are indications upon goods, of their origin from a country, region, or locality; where the quality, reputation or other characteristic of the goods are essentially attributable to their geographical origin. The geographical indications influence the consumer in his choice of products, to guide him for use of an authentic quality product. Any association of persons or producers or any organisation or authority established by or under any law for the time being in force, representing the interest of the producers of the concerned goods who are desirous of registering geographical indication in relation to such goods may apply for such registration in a prescribed procedure. The registration of a geographical indication shall be a period of 10 years but may be reviewed from time to time for an unlimited period by payment of renewal fees (Rashtriya 2003).

Appellation of origin is a geographical indication that declares the quality of goods derived essentially or exclusively form an area of production. All appellations are geographical indications but all geographical indications need not be appellations. Plant varieties developed with traditional knowledge and associated with a particular region can also be protected as geographical indications. The advantage in such protection is that there is no limitation of time but the benefits from this protection emanates over time and familiarity.

The Registrar of Geographical indications maintains the national register. A modern geographical indications registry (GIR) has been established at Chennai in July 2001 having a website (G.I. Act 1999).

A Geographical indication does not belong to a particular enterprise. It is used by several enterprises at the same time for the product, which originates in a particular geographical area. Geographical indication, which is a matter of public law and is non-assignable to any particular owner, but trademark fails in the realm of private law and is framed as per requirements of the owner.

The GI can be developed into a very powerful instrument especially in agriculture including traditional goods (Sahai 2000). With the national legislation in place, researchers would need to concentrate on building data and evidences for grant under this asset. The implication of this instrument is that it gives right to the communities/people who produce products to prevent others from using the geographical name in marketing produce not originating in defined areas (Gupta, 2003), especially in areas of traditional and ethnic knowledge. Short-listing items for their registration as GIs is an immediate concern and researchers would need to build strong portfolios based on scientific evidences.

Farmers' and Community Rights

The conservation of biodiversity is of critical importance, not only because the very diversity is under threat of extinction and erosion, but also because it can be of direct and indirect benefit to mankind. An estimated 1.7 million species have been described till date. It is estimated that total number of species existing on earth may, however, at present vary from 5 million to around 12.5 million. Genetic resources in terms of biodiversity are the most precious assets in natural resources are available for humankind. They consider to be the nature's tools for harvesting solar energy and processing mineral resources into food, feed fibre, fuel, fertilizers, timbers, medicines, industrial, chemicals, etc. They encompass all plants, animals and microbial diversity which is crucial for human survival in the physical environment. It has thus offered avenues for the protection of community rights (farmers' rights in agrobiodiversity) and national rights in biodiversity for the gene-rich countries. Obviously, a critical requirement is to simultaneously operationalize sovereign rights on biodiversity and Intellectual Property Rights in agriculture. The contemplated stipulations in Convention on Biological Diversity do require the evaluation of biodiversity at the national species and intraspecific level for which indigenous efforts are absolutely necessary. In the overall realm of biodiversity, far more emphasis is required to be placed on biodiversity of economic importance as this is the area from where maximum mileage is to be derived based on utility, value and worth of the materials. Thus, economically important biodiversity has to be specifically addressed to in research and development of agriculture and other biodiversity-based sectors (Gene Campaign 2003).

Biodiversity bill passed by Indian Parliament is to regulate access to genetic resources and associated knowledge to foreigners. The bill also ensures that the benefit for tapping such traditional resources are equitably shared with the country as well as local communities where knowledge is originated. Under the Bill, foreign users would need approval of national biodiversity authority for any form of intellectual property right on inventions based on biological resources for India's traditional knowledge. State level Biodiversity Board would also be set up to regulate access to genetic resources. Survey have revealed local communities were using more than 7500 plants for medicines, 3900 for edible purpose, 300 for gum and dyes, 100 for incense and perfumes out of 46000 plant species available in India (Mudur 2002). It is expected these will attract large number of multinational companies to exploit plant resources and traditional knowledge.

In implementing farmers' and community rights, plant and animal improvement programmes and institutions related to their genetic resources have to play a pro-active role in providing basic information on locations/regions providing useful genetic resources as initial raw material. Such passport data of raw genetic resources used can help in extending benefits to such regions for their vital contributions. However, understanding of precise contribution of a community or region is necessary before considering for rewarding them. The latter is stated as general contributions can only be recognized as general obligations; and all cannot be rewarded, as rewarding all would mean subsidizing all. In cases where origin of basic genetic resources used is not traceable, any portion of income generated from exploitation of the finished products developed can be considered only for general development requirements of different regions as general obligation towards all indigenous communities. Basically, commercialization of improved agricultural inputs or products, duly including a mechanism to assign a portion of received remuneration for honoring farmers'/ community rights has to be considered. Besides, direct collaborative industry or government sponsored programmes with indigenous communities are required for resource generation for their upliftment.

Time will tell how effective the new mechanism would be in curbing access to the country's biological resources but neither the Bill nor any other Act could prevent anyone from adding value to traditional knowledge based on biological resources and claiming genuine innovation. India's research community and industry are yet to add value to traditional knowledge. But it is such value addition that can ensure real economic gain. India's traditional plant based medicines promise quite a lot against various serious human diseases like asthma, diabetes even HIV. But we lag behind in staking claims to Intellectual property when it comes to herbal medicines. One survey has shown that China holds the highest share (45 per cent) of herbal medicinal products followed by Japan (22 per cent) and Russia (16 per cent) (Mudur, 2002).

CONCLUSION

The main objectives of spectacular advancement in Science and Technology is to provide maximum benefit to mankind relating the basic need of food, health, water, energy, employment, shelter and ecological security. There is also requirement to use Science and Technology to create wealth both by enterprise and as a profit making institution. The contribution of Indian scientist to the growth and development of Science and Technology are well known and highly appreciated. But our scientists are poor in economic sense. Our science-business link is very poor. There is a need to change mind set of the scientists move so as to create an awareness among the scientists about IPR. Some of the steps can ensure better understanding the IPR regime :

- Manpower planning for IPR protection will need emergency measures; IPR must be made a compulsory subject matter in the law courses including professional courses like Engineering and technology stream and establishment of patent training institutes will have to be set up; China has set up 5000 patent training institutes whereas we have none in India at the moment. It is a matter of deep concern that with a 100year old system on patents in India only 4000 patents were filed last year whereas with just a 10 year old system, China had 90,000 patents in 2002 (Mashelkar 2002);
- Awareness to protect our inventions and its implications. Launching of a patent literacy mission is need of hour. Development of skills in filing, reading and exploiting patents will be most crucial in the years to come;
- There is need to change the attitude and mindset of industrial community to upgrade the technologies, improve product design and finally, strategic partnerships not only within India but outside also to face the challenges of global competition in trade;
- Technology licensing agreement and its conditionalities in terms of technology markets, stake holding, product swap, technology being broken into different packages has made the situation complex and the steps are needed to put the house in order;
- We must develop strength in terms of technology, foresightedness capacity building to counter in the comity of nations.
- The growing movements of plant and genetic resources for improvements of crop and livestock under globalisation and libealisation of trade have given rise to probability of introduction of new key pests and diseases. However, quarantine restrictions on movement of plants and soils will likely to become more important in the fact of expanding trade and global climatic change. It is only recently the legal standards have been promulgated in the form of Sanitary and Phytosanitary measures for regulating international trade on agricultural products to avoid introduction of key pest and diseases in the importing countries for the protection of human, animal and plant life or health. Pest diagnostic, biodiversity inventories and tools of bioinformatics are essential for assessment and some important aspects related to pest organism like conservation management and utilisation of biodiversity in sustainable manner, SPS measures and PRA under IPCC. The high level of technical skill is required to support all these functions to ensure that the country is able not only to meet its international obligations under SPS but also to protect India from the inadvertent introduction of economically important exotic pests and to gain market access for agricultural commodities in an increasingly competitive global trade environment;
- Urgent need to document the traditional knowledge in the light of granting of patents elsewhere;
- Required to set up the fast track patent granting institution which can reduce the time taken in granting the patent as the time taken in India for grant of patent.

Dr. M.S. Swaminathan noted Agricultural Scientist in his Convocation address to University of Madras (Published in Hindu, October 2003), suggested to establish a patent bank for peace and happiness under the umbrella of University Grants Commission. Scientists and technologists from all our Universities and public research Institutions should be encouraged to assign their patents to such a bank so that fruits of scientific discoveries are available for public good. Such a patent for peace and happiness bank would stimulate scientists to consider themselves as trustees of their intellectual property sharing their invention for the benefit of poor.

Indian farmers also need to be more competitive and quality oriented to find a place in the global market. Support of farmer's friendly Govt. policies, good infrastructural facilities, technological innovations and their access to end user's and clean post harvest management practices, millions of small and marginal farmers will stand up to meet the demand of international market. Production by the farming community both men and women spirit of cooperation skill as well as knowledge based development and zest of quality standards should be the hallmark of Indian agriculture when it enters the global market.

REFERENCES

Chaturvedi, S. 2001. India, the European Union and Geographical Indications (GI) Convergence of Interests and Challenges Ahead Rls-Dp # 35-2002.

Erbsch, F.H. and Carlos, Velazquez 1998. Introduction to Intellectual Properties. In:Erbisch, F.H. and K.M. Maredia (Eds), *Intellectual Property Rights in Ag. Biotechnology*, CAB Institute, pp. 3-18.

G.I. Act 1999. http: www.patentoffice.nic.in/ipr/gi/geo-ind.htm

Gene Campaign 2003. India's Protection of Plant Varieties and Farmers' Right Act, Chapter in Section 111, An article by Ujjawal Kumar and Suman Sahai, pp. 57-58.

Gupta, A.K. 2003. Rewarding conservation of biological and genetic resources and associated traditional knowledge and contemporary grass roots creativity. IIMA W.P. No. 2003-01-06.

ID Act 2000 http://www.patentoffice.nic.in/ipr/design/designs.htm

IPA 2002. http://www.patentoffice.nic.in/ipr/patent/patents.htm.

Mahanti, Subodh 2001. Intellectual Property Rights. Employment News XXVI No.14, pp. 48.

Mashelkar, R.A. 2002. Intellectual Property Rights – Indian Agriculture. J. Ragotam Reddy Memorial Lecture Nov.25, 2002. pp. 26.

Mudur 2002. Unnecessary Anxiety. The Telegraph. 23rd December.

Rai, M. 2002. Intellectual property rights, farmers' rights and Agricultural Research & Development. Indian Farming 52: 10-17.

Rashtriya, T. 2003. Intellectual Property Rights Part 1. Employment News XXVIII No.6, pp. 40.

Raychoudhury, P. 2003. World Trade Organisation – Some facts. Employment News. Vol. XX VIII No.25, pp. 64.

Sastry, R. and Rao, N.H. 2003. Protection of Intellectual Property and Agricultural Research, NAARM, Hyderabad 14-17 October, Resource Material. Vol.1., pp. 48

Swaminathan, M.S. 2003. Patents for peace and happiness. 'Hindu'October.

website : http://www.indialawinfo.com

Index

N

O

P